武汉大学百年名典

自然科学类编审委员会

主任委员　　李晓红

副主任委员　卓仁禧　周创兵　蒋昌忠

委员　（以姓氏笔画为序）

文习山　宁津生　石　兢　刘经南
何克清　吴庆鸣　李文鑫　李平湘
李晓红　李德仁　陈　化　陈庆辉
卓仁禧　周云峰　周创兵　庞代文
易　帆　谈广鸣　舒红兵　蒋昌忠
樊明文

秘书长　　李平湘

社会科学类编审委员会

主任委员　　韩　进

副主任委员　冯天瑜　骆郁廷　谢红星

委员　（以姓氏笔画为序）

马费成　方　卿　邓大松　冯天瑜
石义彬　佘双好　汪信砚　沈壮海
肖永平　陈　伟　陈庆辉　周茂荣
於可训　罗国祥　胡德坤　骆郁廷
涂显峰　郭齐勇　黄　进　谢红星
韩　进　谭力文

秘书长　　沈壮海

杨弘远

（1933—2010）院士，1933年9月26日出生于湖北省武汉市，1950年至1954年就读于武汉大学生物系，毕业后留校任教，1982年任武汉大学生命科学学院教授，1984年任博士生导师，1991年当选为中国科学院院士。曾任国务院学位委员会学科评审组（生物学）召集人、全国博士后管理委员会专家组召集人、教育部基础研究规划组（生物学）副组长、教育部科学技术委员会委员、国家自然科学基金委员会植物学科评审组专家、中国细胞生物学学会常务理事、湖北省科协常委等。曾担任《中国科学》、《植物学报》、《细胞研究》、《细胞生物学》、《波兰植物学报》、《整合植物生物学杂志》等国内外权威学术杂志编委。

杨弘远院士是我国被子植物生殖生物学的开拓者和奠基人之一，也是推动我国实验胚胎学迈向实验生殖生物学的先驱。他将毕生的精力奉献给了我国的植物生殖生物学研究，为我国生物学相关专业的创建和发展作出了巨大贡献。20世纪90年代，他率先在我国倡导植物发育生物学研究，主持了我国第一个植物发育生物学重大课题，创建了我国第一个植物发育生物学教育部重点实验室和第一个发育生物学全国重点学科。他几十年如一日地坚持科研与教学，并取得了丰硕的成果：1985年获国家教委科技进步一等奖，1987年获国家教委科技进步二等奖，1991年获国家自然科学奖三等奖，2004年获国家自然科学奖二等奖。2003年获中国图书奖，2006年获浙江省树人出版奖特等奖。曾被评为国家级有突出贡献的中青年专家，全国优秀教师。著有《植物有性生殖实验研究四十年》、《水稻生殖生物学》、《植物生殖寻幽探秘》；主编《被子植物受精生物学》；翻译《双受精——有花植物的胚和胚乳发育》等多部专著，为我国生物学教育和科研事业作出了突出贡献。

杨弘远院士是一位具有远见卓识的教育家。执教50多年来，他孜孜以求，诲人不倦，提携后辈，甘为人梯。当他主持的重点项目在2001年结题后，他就坚决不再申请新的科研项目，而让年轻教师挑大梁，为年轻人创造脱颖而出的环境。杨院士曾说："培养学生不能像放羊。"几十年间，他"精耕细作"，培养了18位博士和12位硕士，很多学生现已成为知名的学者。他提出"要抢占学科制高点"，所在的重点学科和教育部重点实验室已有2名"长江学者"特聘教授，2名珞珈特聘教授，2名国家杰出青年基金获得者，4名教育部跨世纪人才或新世纪人才基金获得者。杨弘远院士非常注重学生的培养和教育，他常常告诫学生："要防止急功近利的思想，要从自己的专业中体会到学习的乐趣"。他还说："走自己的路，完全靠强迫是不行的，更重要的是培养对专业的兴趣。强迫只能奏效于一时，很难做到几十年如一日。"多年来，许多素昧平生的青年给他写信，或是登门拜访，他都会逐一回复或接待。

杨弘远院士一生淡泊名利、学风严谨、胸怀坦荡、潜心科研。他曾专门撰文抨击学风浮躁与学术腐败现象。他始终坚守科学信仰，保持清醒与宁静，将自己定位于一名普通的科学家。1980年，他与周嫦教授首次从水稻未受精子房中培养出单倍体植株。1990年，他作为被美国植物学会特邀的中国专家，在第41届生物科学联会上作专场报告。虽然他在植物发育生物学研究领域作出了突出贡献，但他并不喜欢别人给他附上"权威"、"泰斗"这样的头衔。他说："人的生命是有限的，学术生命更有限，能在科学的大山上添一粒土，就很知足了。"

杨弘远院士一直坚持开展对社会公众的科学普及工作，他经常应邀到兄弟院校和中小学作科学报告，讲述成才之道，普及科学知识。他从来都把自己当作一个普通人，为人低调，不事张扬。他公私分明，严于律己，把自己放在一个平凡人的位置上，从不搞特殊化。他高风亮节的师德风范，教育和影响了一代又一代年轻学子。

杨弘远院士出生在珞珈山下，在武汉大学度过了他的小学、中学、大学时代，一生与武汉大学相伴，对武汉大学充满了无限挚爱，奉献了毕生精力。

杨弘远院士的一生，是精心育才、无私奉献的一生；是锐意创新、硕果累累的一生；是辛勤耕耘、奋发有为的一生。他不仅为我国的教育科学事业作出了卓越贡献，而且给我们留下了宝贵的精神财富。杨弘远院士的逝世是武汉大学的巨大损失，更是我国教育界、科学界的重大损失。我们要深切缅怀杨弘远院士，永远铭记他的卓越贡献，更要化悲痛为力量，继承他"百年树人"的遗志，加快武汉大学的建设和发展，慰藉长眠的杨弘远院士。

武汉大学
百年名典

植物有性生殖实验研究四十年

杨弘远　周嫦　编著

武汉大学出版社
WUHAN UNIVERSITY PRESS

图书在版编目(CIP)数据

植物有性生殖实验研究四十年/杨弘远,周嫦编著.—武汉:武汉大学出版社,2013.10
武汉大学百年名典
　ISBN 978-7-307-11513-2

Ⅰ.植… Ⅱ.①杨… ②周… Ⅲ.植物—有性繁殖—实验胚胎学 Ⅳ.Q945.53

中国版本图书馆 CIP 数据核字(2013)第 210949 号

责任编辑:黄汉平　　　责任校对:刘　欣　　　版式设计:马　佳

出版发行:武汉大学出版社　　(430072　武昌　珞珈山)
　　　　　(电子邮件:cbs22@whu.edu.cn　网址:www.wdp.com.cn)
印刷:湖北恒泰印务有限公司
开本:720×1000　1/16　印张:62　字数:887 千字　插页:9
版次:2013 年 10 月第 1 版　　2013 年 10 月第 1 次印刷
ISBN 978-7-307-11513-2　　定价:124.00 元

版权所有,不得翻印;凡购我社的图书,如有质量问题,请与当地图书销售部门联系调换。

FORTY YEARS' EXPERIMENTAL RESEARCHES ON SEXUAL PLANT REPRODUCTION

YANG Hong-Yuan and ZHOU Chang

WUHAN UNIVERSITY PRESS

《武汉大学百年名典》出版前言

百年武汉大学，走过的是学术传承、学术发展和学术创新的辉煌路程；世纪珞珈山水，承沐的是学者大师们学术风范、学术精神和学术风格的润泽。在武汉大学发展的不同年代，一批批著名学者和学术大师在这里辛勤耕耘，教书育人，著书立说。他们在学术上精品、上品纷呈，有的在继承传统中开创新论，有的集众家之说而独成一派，也有的学贯中西而独领风骚，还有的因顺应时代发展潮流而开学术学科先河。所有这些，构成了武汉大学百年学府最深厚、最深刻的学术底蕴。

武汉大学历年累积的学术精品、上品，不仅凸现了武汉大学"自强、弘毅、求是、拓新"的学术风格和学术风范，而且也丰富了武汉大学"自强、弘毅、求是、拓新"的学术气派和学术精神；不仅深刻反映了武汉大学有过的人文社会科学和自然科学的辉煌的学术成就，而且也从多方面映现了20世纪中国人文社会科学和自然科学发展的最具代表性的学术成就。高等学府，自当以学者为敬，以学术为尊，以学风为重；自当在尊重不同学术成就中增进学术繁荣，在包容不同学术观点中提升学术品质。为此，我们纵览武汉大学百年学术源流，取其上品，掬其精华，结集出版，是为《武汉大学百年名典》。

"根深叶茂，实大声洪。山高水长，流风甚美。"这是董必武同志1963年11月为武汉大学校庆题写的诗句，长期以来为武汉大学师生传颂。我们以此诗句为《武汉大学百年名典》的封面题词，实是希望武汉大学留存的那些泽被当时、惠及后人的学术精品、上品，能在现时代得到更为广泛的发扬和传承；实是希望《武汉大学百年名典》这一恢宏的出版工程，能为中华优秀文化的积累和当代中国学术的繁荣有所建树。

《武汉大学百年名典》编审委员会

再 版 说 明

《植物有性生殖实验研究四十年》第一版于 2001 年由武汉大学出版社出版，该书被纳入"十一五"国家重点图书，由"华夏英才基金"资助出版，并于 2002 年荣获第十三届"中国图书奖"。适逢武汉大学 120 周年校庆，武汉大学出版社决定将该书纳入"百年名典"再次出版，以怀念著名学者——杨弘远院士，并向武汉大学校庆献礼。

此次再版基本保持第一版全部内容，对原书中少数差错进行了订正。为了与"百年名典"其他图书在版式上保持统一，将原书开本由正 16 开改成异 16 开。

序

好几年前就有不少朋友建议我们将多年研究的成果汇集成书出版,我们则一直忙于"向前看",把作出新成果、发表新论文当做主要任务,无暇去做"向后看"的整理总结工作,所以对于编书一事虽心有所动而踟蹰不决。现在,20世纪即将逝去,新千年的钟声即将敲响,而我们在科学道路上也已跋涉了40年有余了。回顾过去,展望未来,"温故而知新",已经到了一个恰当的时刻。正在这时,"华夏英才基金"给了我们一个出版学术著作的机会。经过思考,我们决定利用这一机会,认真梳理一下以往的工作,理出一条研究思路的脉络,并纳入国际研究总趋势之中,使读者于一斑中窥全豹。于是便拟定了书名和全书的结构。

早在20世纪50年代之初,植物实验胚胎学作为植物胚胎学中最活跃的研究方向刚露头角,就展现了强大的生命力。半个世纪的经历证明,这个学科生长点仍在不断发展,并且上升到新的高度。在本书的"前言"中对此有所阐述。需要提到的是,这个前言,是我们在10多年前所发表的一篇专题论文。我们感到其内容与观点今天仍未过时,因此借用作为本书的前言,原文基本不变,仅个别词句作了修改。我们回顾自己几十年的研究历程,发现自己很幸运地顺应了学科发展的世界趋势,作为中国科学工作者由青年而中年到老年,艰难而不无自豪地在我国这片相对贫瘠的学科园地上耕耘播种,结出了一批果实。由于发表的文章散见于中外刊物,许多年轻的同行很难全面了解,更难把握其整体脉络,所以我们力求加以精选、分类、系统化为一书。希望通过本书反映植物有性生殖实验研究从性器官操作到性细胞操作的演进轨迹,并提供我们自己的研究方法和结果,以飨读者,并望不吝批评指正。

本书分五章。前三章分别汇集植物有性生殖过程中三个环节（雄性细胞发育、雌性细胞发育、传粉受精与胚胎发育）的实验研究。每章均按由性器官操作到性细胞操作的趋势分节。例如第一章，大致按花药培养——花粉人工萌发——花粉原生质体操作——生殖细胞操作——精细胞操作的思路编节，等等。第四章关于植物有性生殖过程的超微结构与细胞化学研究，则是按由结构研究向结构与功能相结合的研究趋势编节。第五章"专题论述"精选若干代表性文章，着重从理论与方法上开阔思路。

为了便于读者把握各章、节的思路，在每一章、节均安排一段简短的"提要"，说明当时的研究背景、目的与特点。论文是以往发表的，"提要"是现在编写的，它们可以起"粘合剂"的作用，把原来零散的论文组合起来形成有机的体系。当然，所有的研究工作都是几十年间自然发展的产物，并非从一开始就按预定计划实现的，在发展中还受诸多主观与客观条件的制约，出现残缺不全是势所难免的，不可能是一部天衣无缝的体系。回顾过去，我们自己也有不少遗憾。

书中收集的论文，是从20世纪50年代开始长达数十年期间陆续发表的，每篇论文只能反映当时的研究状况，因而也只能用历史的眼光加以看待。之所以将一些早年的论文也收进本书，一则是为了反映研究由低至高、由浅入深的时代轨迹；二则也因为有些论文即使今天看来也不失可取之处，读者也许还可从中发掘某些可资利用的东西。科学研究的本质是探索未知的客观事物，只要是通过严谨的方法所揭示的新现象，只要它们还没有被彻底认识，哪怕埋没多年，也还值得加以发掘。举一个例子：我们在1959年发表的一篇"处女作"中，揭示了气态化学分子诱导稻花开颖的现象，至今仍为一个悬而未决的问题，读者如有兴趣可以查阅第三章第一节的"提要"和第一篇论文。

作为单篇论文，自有其严格的结构规范。但收入书中就不必保持原有形态了。我们在编纂中保留了原来论文中的英文摘要、引言、材料与方法、结果、讨论等主要部分，精选了一部分重要的表格与插图以及少数具有代表性意义的照相图版；删除了中文摘要、关键词、大部分图、表、照相图版和全部参考文献目录，以便全书总篇幅不至过分膨胀。为了弥补以上缺陷，采取了如下一些措施：(1) 在正文中凡引证的前人姓名和带方括号的文献序号一概保留，以便知其有所出处。(2) 正文中引证的图、表、照相图版与参考文献，仍保留其原有的序号，以便读者了解它们原来的存在。在每篇论文末尾注明该论文的作者，原载期刊的年份、卷、期、页，原论文包含的图、表与图版总数和被选用的号码，以便读者必要时查阅期刊中的原始论文。(3) 各个时期在不同刊物中发表的论文，格式差别很大，编书中尽可能按统一规范加以修正，但内容则保持原貌，以忠于历史事实。(4) 选编的125篇论文，大多全文收录；少部分论文以摘要形式录入；还有不少相关的论文未能选入，只在每章之末列出目录，计70篇。总之，尽管采取了以上各种措施，也难以消弭一切缺陷。由此给读者带来的诸多不便，编著者谨表歉意。

本书得以出版，有赖"华夏英才基金"的大力支持与各级统战部的帮助。武汉大学出版社为本书作了精心的编排出版工作。书中的研究成果，是在国家自然科学基金、国家教委博士点基金等的长期资助下完成的；省、校、院、系各届领导也给予了许多关怀。书中汇集的论文曾在国内、外各种期刊中发表，尤其是《植物学报》为我们提供了发表论文的主要阵地。我们的研究成果是一个研究集体、包括众多研究生多年劳动的结晶，部分研究成果是与国内、外同行合作完成的。在工作中，不少亲友给予了热情的鼓励。编著者谨在此一并致以诚挚的谢意。

编著者
2000年6月

曾参加该课题研究的人员（按姓氏笔画为序）：

王 劲　于凡立　田惠桥　卢 萍　刘中来　孙 芹　孙蒙祥
李仕琼　李国民　李昌功　李师弢　何才平　吴 燕　吴新莉

汪　泳　　陈以峰　　陈绍荣　　范六民　　董　健　　施华中　　赵　洁
张劲松　　梁　立　　梁世平　　莫永胜　　徐秉芳　　夏惠君　　黄群飞
阎　华　　傅　缨　　傅春梅　　蔡得田　　魏正元　　韩红梅

前　言

从1968年召开第一次"种子植物有性生殖的细胞生物学"国际讨论会算起,到植物生殖生物学被公认为一门新学科,经历了20多年的孕育才瓜熟蒂落。现在,在生殖生物学中又出现了一个活跃的研究领域,它有演进为一门新分支学科、并导向一个新技术领域的趋势,值得我们提出来讨论。这里所说的新分支学科,可以命名为"植物实验生殖生物学";新技术领域可以命名为"植物生殖细胞工程"或简称为"植物生殖工程"。

一

植物实验生殖生物学是实验胚胎学发展的更高阶段,是生殖生物学时代的实验性分支学科。从实验胚胎学到实验生殖生物学,有继承的一面,更有发展的一面,这里着重讨论后一方面。从近年研究的动态看来,有两个主要的特征说明其质上的变化。

第一个特征是实验操作技术由器官、组织水平提高到细胞、原生质体水平。实验胚胎学所依赖的操作技术,如花药培养、子房与胚珠培养、试管受精、胚胎培养等,基本上是器官与组织的操作,这从Johri(1982)主编的《维管植物的实验胚胎学》一书中的各章内容可以清楚地看出。但是所有这些器官、组织水平的操作都在以不同程度的势头向更加精密化的、细胞与原生质体水平的操作技术迈进。例如就雄性系统而言,20世纪60年代突破了花药培养;70年代发展到花粉培养;80年代又在花粉原生质体、精细胞和生殖细胞的操作方面取得长足的进展,成为研究的热点。这里仅举出一些突出苗头:通过花粉四分体原生质体和体细胞原生质体融合以进行"配子-体细胞杂交"已经再生杂种植株；大量分离和纯化花粉原生质体的技术已在几种植物中完全成功；幼嫩花粉原生质体培养已诱导了多细胞团与类似原胚的构造，突破了其孢子体发育的第一关；大量分离精子及其前身生殖细胞的技术已在多种植物中过关，等等。以上种种，在不久前的实验胚胎学著作中还没有反映或仅属推测，今天已经成为现实，足见其进展之迅猛。就雌性系统的操作而言，20世纪60年代成功地实现了传粉后的子房与胚珠培养；70年代后期又实现了未传粉的子房与胚珠培养；到了80年代，人们已经采取实际步骤向胚囊和卵细胞的操作前进，其第一步——生活胚囊、胚囊细胞原生质体的分离——已经突破。雌、雄两性细胞与原生质体操作技术都达到相当成熟的地步。以此为基础，开展雌雄细胞体外融合的实验，终于在90年代实现了以往所提出的"受精工程"的设想。过去名为"离体受精"实则离体子房或胚珠授粉的技术（器官水平的操作）现已演进为高等植物真正含义上的离体受精（原生质体水平的操作）。

实验生殖生物学的第二个主要特征是其多学科综合性研究的性质较以往更为显著。以往的实验胚胎学，主要是在胚胎学中融入了生理学和遗传育种的成分。Maheshwari 与 Rangaswamy（1965）所撰长篇综述"与生理学和遗传学相关联的胚胎学"中的内容即体现了这一特征。诚然，在某些（尤其是近年）实验胚胎学研究中也加强了细胞生物学与分子生物学的内容，但无论广度与深度都是很不够的。当代的实验生殖生物学研究中，细胞生物学、分子生物学等多学科的渗透上升到重要地位，各种新的研究方法不断被用来揭示生殖过程的奥秘。如应用超微结构和超微细胞化学技术、荧光显微术、活体观察与视频显微术、图像处理术、显微定量测定术、放射自显术、免疫学技术以及分子生物学技术，从不同角度对同一问题作了深入细致的探索，在有关研究报道中日有所见。这里仅以围绕精子分离所进行的多学科研究为例加以说明：分离精子的目的最初是为了观察它们在离体条件下的形态变化与运动性能。以后发展到观察在分离状态下的雄性生殖单位与精子二型性。最近，由于雌、雄识别的研究逐渐由花粉与雌蕊的识别向配子间的识别深入，加之关于"倾向受精"的认识表明，被子植物的一对精子可能分别具有与卵细胞或中央细胞识别的能力，目前已着手分离精子的免疫学鉴定，旨在探讨精子的识别物质。利用精、卵体外融合实验系统，已能构建受精前后卵细胞与合子的cDNA文库，筛选受精过程中优势表达的基因。应用共聚焦激光扫描显微术，已能测出体外受精过程中出现的钙波。由此可以看出，精子分离不只是一个操作技术的问题，而是涉及细胞生物学、生理学、生物化学、分子遗传学等多个相邻学科的范围，囊括了分离提纯、活体观察、超微结构、生活力测定、冷冻保存、单克隆抗体制备、免疫学测定、由微量细胞构建cDNA文库的分子生物学技术、蛋白质分子分离与鉴定、原生质体融合等多种研究技术方法。这个例子可以很好地反映当代实验生殖生物学多学科综合性研究的特色。当然，所谓多学科综合性研究是就整体而言的，并不意味着排斥了为特定目的而进行的单项研究。这正如上文所述，实验生殖生物学的操作水平从整体上已进入细胞与原生质体操作的阶段，并不意味着排斥器官与组织的操作，其道理是不言自明的。

二

现在再讨论植物实验生殖生物学向应用方面延伸的问题。实验生殖生物学的强大生命力不仅表现在探索未知方面，而且表现在通过它开拓生殖工程应用技术的前景方面。依我们看，"生殖工程"的含义包括整个有性生殖过程的控制与利用，比"受精工程"的范围更广。它标志着人类通过生殖过程改造植物遗传性的能动性达到一个新的高度。

首先让我们从宏观上考察一下近代植物改良基本途径的演变动态。近代植物改良的主要途径是通过有性杂交结合双亲的遗传物质，即通常所称的常规杂交育种。20世纪60年代以后，出现了两方面的重大突破：一是细胞与原生质体培养技术的建立，证实了植物细胞的全能性，初步实现了高等植物细胞的"微生物化"；二是在分子遗传学的基础上发展了基因工程技术。二者的结合导致了植物细胞工程技术的兴起。后者由于是以体细胞及其原生质体为主要实验体系的，实际上可说是"体细胞工程"。体细胞工程的成就表明人类可以在常规有性杂交之外增加一条通过体细胞培养改良植物的新途径。与此同时，实验胚胎学所创造的一系列方法也成功地用于单倍体育种和远缘杂交育种。在这样的背景下，人们对生殖系统的机能与利用潜力就投以新的眼光，试图通过实验生殖生物学的努力和借鉴体细胞工程的经验开辟生殖细胞工程或性细胞工程的新局面。可以预测，从现在开始的一二十年内将会有一个研究生殖工程的高潮。

既然有体细胞工程，何必还需要生殖工程呢？为了回答这个疑问，有必要分析生殖工程的特点和优点。我们认为至少以下三点是很明显的：第一，有性生殖毕竟是植物的自然生殖过程，是无性的或准性的生殖过程无论如何不能完全取代的。换言之，就多数植物来说，通过有性过程繁衍后代是一条顺乎自然的、因而也是阻力较小的途径。并且近年有些研究表明，通过这一过程施行基因工程也是可行的。现在认为：花粉管不仅是雄配子的运载工具，而且具有外源基因载体的功能；卵细胞由于部分地缺壁，类似天然原生质体状态，这不仅是对受精的适应，而且也具备了基因工程受体的特征；至于合子，作为天然的胚胎发生原始细胞，其优越性更是毫无疑义的。因此，借助有性生殖施行基因工程的设想在理论上是经得起论证的。现在的问题还是在实际操作技术的有些方面尚未过关。第二，生殖系统中拥有许多天然的单倍性细胞，由它们可以直接制取单倍原生质体。小孢子四分体、小孢子、花粉粒、花粉管、生殖细胞、精子、大孢子、胚囊、卵细胞与其它胚囊成员细胞，都是单倍性细胞，其中有些细胞（如精子）还是天然的原生质体。单倍原生质体在遗传工程中的价值勿庸赘言。但迄今获得单倍原生质体的方法是先通过花药培养或其它途径诱导单倍体植株，然后再由后者的细胞制取原生质体，这种间接的方法是相当麻烦的。如果生殖系统所提供的单倍原生质体能够培养和操作，将会在遗传工程中显示巨大的优越性。第三，生殖系统中蕴藏着丰富多样的细胞类型，尤其在细胞质方面差异很大。如卵细胞有丰富的细胞质;精子的细胞质则相对贫乏,犹如天然的核质体。在一粒花粉中的两个精子,其细胞器的种类与含量也可能有很大差别,如白花丹的二型精子,一个富含质体、另一个富含线粒体,等等。这将成为开展细胞器工程研究的好材料。因此,把生殖系统中可供利用的细胞开发出来,可以建立各种有特色的细胞工程实验系统,潜力是不小的。

最后作一个简略的归纳:植物生殖生物学是以认识植物生殖过程自然规律为内容的基础研究学科。植物生殖工程是通过生殖过程改良植物的应用技术领域。而实验生殖生物学则是把二者联结起来的桥梁,是基础研究与高技术的一个结合点,我们应当给予足够的重视。

(作者:杨弘远、周嫦。原载:植物学通报,1989,6(4):193~196。现略加文字修改,作为本书前言。)

目　录

第一章　雄性器官与细胞的实验研究 ………………………………… 1
提要 ……………………………………………………………………… 2
第一节　花药培养与雄核发育 ………………………………………… 3
  提要 ……………………………………………………………………… 3
  1　水稻花粉去分化新激素的探索 …………………………………… 5
  2　外源激素在水稻花药培养中的作用 ……………………………… 12
  3　水稻花粉两条发育途径的实验研究 ……………………………… 19
  4　水稻幼花直插培养诱导花粉孢子体发育的实验 ………………… 27
  5　大麦的花药培养与雄核发育 ……………………………………… 31
第二节　花粉萌发与花粉管生长 ……………………………………… 33
  提要 ……………………………………………………………………… 33
  1　Fluorescein diacetate used as a vital stain for labeling living pollen tubes …………………………………………………………… 34
  2　用荧光染色与冬青油透明技术显示花粉细胞核 ………………… 40
  3　玉帘离体萌发花粉管中生殖细胞核的有丝分裂和无丝分裂 …………………………………………………………………… 47
  4　pH 值与聚乙二醇对芸苔属花粉人工萌发的影响 ……………… 49
  5　外源 Ca^{2+} 对烟草花粉管生长和生殖核分裂的调节 …………… 50
  6　Ca^{2+} 载体 A23187 对烟草生殖核分裂的调节 ………………… 52
  7　Nifedipine 对烟草花粉萌发、花粉管生长及生殖核分裂的影响 …………………………………………………………………… 53
  8　外源钙调素和钙调素拮抗剂对烟草离体花粉管生长和生殖核分裂的调节 ………………………………………………… 55

第三节　花粉原生质体操作 …………………………………… 61
　　提要 …………………………………………………………… 61
　　1　三种植物花粉原生质体的大量分离与初步培养………… 64
　　2　Cell divisions in pollen protoplast culture of *Hemerocallis*
　　　　fulva L. ………………………………………………… 70
　　3　唐菖蒲花粉原生质体及其萌发花粉管的超微结构研究…… 78
　　4　萱草幼嫩花粉原生质体培养启动细胞分裂的超微结构
　　　　研究 …………………………………………………… 87
　　5　玉米花粉与唐菖蒲花粉原生质体中肌动蛋白微丝的荧光
　　　　显微观察 ……………………………………………… 96
　　6　甘蓝型油菜和紫菜苔花粉原生质体的大量分离 ………… 104
　　7　紫菜苔花粉超低温保存及其原生质体分离 …………… 109
　　8　烟草花粉原生质体的分离 ……………………………… 115
　　9　烟草幼嫩花粉原生质体分离与早期离体发育 ………… 118
　　10　烟草单个花粉原生质体透射电镜样品制备及其超微结构
　　　　观察 …………………………………………………… 124
　　11　紫菜苔花粉原生质体的电激转化及 Zm13-260-GUS-NOS
　　　　融合基因的时序表达 …………………………………… 129
　　12　芸苔属花粉-下胚轴原生质体融合再生杂种小植株 …… 133
　　13　Development and molecular identification of pollen-somatic
　　　　hybrid plants in *Brassica* spp. ………………………… 143
　　14　烟草属花粉-叶肉原生质体的融合及杂种植株再生 …… 147

第四节　脱外壁花粉操作 …………………………………… 152
　　提要 …………………………………………………………… 152
　　1　芸苔属脱外壁花粉的分离与人工萌发 ………………… 154
　　2　芸苔属脱外壁花粉作为研究花粉萌发的新实验系统 …… 162
　　3　Preparation of exine-detached pollen in *Nicotiana tabacum* … 171
　　4　烟草脱外壁花粉人工萌发与离体授粉实验系统的建立 … 174
　　5　烟草花粉发育过程及不同组织中的内源 GUS 活性 …… 184
　　6　烟草脱外壁花粉的电激基因转移 ……………………… 185

目 录

    7 β-glucuronidase gene and green fluorescent protein gene expression in de-exined pollen of *Nicotiana tabacum* by microprojectile bombardment ………… 187

第五节 生殖细胞操作…………………………………… 193
    提要………………………………………………………… 193
    1 Direct observations on generative cells isolated from pollen grains of *Haemanthus katherinae* Baker ………… 195
    2 Isolation and purification of generative cells from fresh pollen of *Vicia faba* L. ………………………………… 201
    3 花粉生殖细胞的大量分离与纯化………………………… 206
    4 多种被子植物花粉生殖细胞大量分离技术的比较研究 … 211
    5 分离的黄花菜花粉生殖细胞在培养条件下的核分裂…… 219
    6 几种植物花粉生殖细胞的融合实验………………………… 225
    7 Fluorescent vital staining of plant sexual cell nuclei with DNA-specific fluorochromes and its application in gametoplast fusion ……………………………………………… 226
    8 Isolated generative cells in some angiosperms: a further study ………………………………………… 234
    9 Microtubule organization of *in situ* and isolated generative cells in *Zephyranthes grandiflora* Lindl ……………… 242
    10 Microtubule changes during the development of generative cells in *Hippeastrum vittatum* pollen ……………… 250

第六节 精细胞操作……………………………………… 259
    提要………………………………………………………… 259
    1 Isolation of viable sperms from pollen of *Brassica napus*, *Zea mays* and *Secale cereale* ……………………… 261
    2 紫菜苔精细胞的大量分离和生活力保存………………… 266
    3 几种具二细胞型花粉植物精细胞的分离和融合………… 277

其它论文目录………………………………………………… 288

第二章　雌性器官与细胞的实验研究……291

提要……292

第一节　未传粉子房与胚珠培养和离体雌核发育……293

提要……293

1. 从水稻未授粉的幼嫩子房培养出单倍体小植株……296
2. Induction of haploid rice plantlets by ovary culture……300
3. 水稻未受精胚囊的离体胚胎发生……304
4. 水稻子房培养时助细胞的无配子生殖和卵细胞的异常分裂……310
5. 水稻子房培养中的胚状体与愈伤组织形态发生特点……317
6. 水稻离体无配子生殖的进一步胚胎学研究……318
7. 毒莠定作为外源激素促进水稻子房培养中胚状体的分化……325
8. 水稻胚囊植株染色体倍性及其它性状的研究……330
9. 水稻的未传粉子房培养……337
10. *In vitro* production of haploids in rice through ovary culture……348
11. 大麦未授粉子房培养的胚胎学观察……363
12. 由向日葵幼花或胚珠培养出单倍体小植株与胚状体……365
13. 向日葵未受精胚珠培养时胚状体发生的显微观察……368
14. 向日葵离体孤雌生殖过程的组织化学研究……374
15. 几种因素对向日葵离体孤雌生殖和体细胞增生的调节作用……376
16. 用整体染色与透明技术观察胚囊、胚、胚乳和胚状体……385
17. An electron microscope study on *in vitro* parthenogenesis in sunflower……392
18. *In vitro* production of haploids in *Helianthus*……407
19. 韭菜未传粉子房培养中单倍体的胚胎发生和植株再生……419
20. 韭菜孤雌生殖和反足细胞无配子生殖的超微结构观察……426

第二节　胚囊操作……437

提要 ··· 437
　　1　被子植物胚囊酶法分离的研究：固定材料的分离技术与
　　　　显微观察 ··· 439
　　2　用酶解技术观察泡桐与芝麻的大孢子发生和雌配子体
　　　　发育过程 ··· 446
　　3　金鱼草胚囊的人工分离 ································· 453
　　4　酶法分离胚囊的荧光显微观察 ························· 457
　　5　Observations on enzymatically isolated living and fixed embryo
　　　　sacs in several angiosperm species ···················· 462
　　6　A study of fertilization events in living embryo sacs isolated
　　　　from sunflower ovules ································· 473
　　7　金鱼草珠被绒毡层壁囊的分离与鉴定 ················· 480
　　8　分离烟草胚囊的新方法及诱导卵细胞与助细胞原生质体的
　　　　原位融合 ··· 491
　其它论文目录 ··· 501

第三章　授粉、受精与胚胎发育的实验研究 ··············· 503
　提要 ··· 504
　第一节　授粉与受精 ··· 505
　　提要 ··· 505
　　1　水稻去雄方法的初步研究 ····························· 507
　　2　小麦受精过程中若干问题的胚胎学研究 ············ 512
　　3　芝麻授粉方法和性因素年龄的初步研究 ············ 524
　　4　芝麻花粉在雌蕊上萌发与生长的研究 ··············· 526
　　5　花粉数量对芝麻受精结实、胚胎发育和后代的作用 ······ 533
　第二节　离体受精 ·· 550
　　提要 ··· 550
　　1　用聚乙二醇诱导选定的成对原生质体间的融合 ···· 552
　　2　烟草雌性细胞原生质体的融合实验 ·················· 559
　　3　烟草中央细胞离体受精过程中雌雄核融合生活动态的

5

记录 ··· 569
第三节　合子、中央细胞与幼胚的操作······················· 572
　　提要 ··· 572
　　1　烟草受精后胚囊和合子的分离及合子的离体分裂 ····· 574
　　2　烟草合子与二胞原胚在离体培养中的发育 ············· 581
　　3　烟草未受精中央细胞及其它胚囊细胞的离体分裂 ····· 589
　　4　植物幼小原胚的电激转化 ·· 596
　　5　Gene transfer into isolated and cultured tobacco zygotes by
　　　　a specially designed device for electroporation ········ 602
　　6　小麦分离合子与幼胚中膜钙和钙调素的分布 ············ 608
　　7　水稻卵细胞与合子的分离 ·· 609
　　8　In vitro development of early proembryos and plant
　　　　regeneration via microculture in *Oryza sativa* ········ 611
　　9　Isolation and *in vitro* culture of zygotes and central cells
　　　　of *Oryza sativa* L. ··· 623
　　10　水稻原胚和刚启动分化的幼胚 cDNA 文库的构建与
　　　　分析 ·· 633
　　11　一种适于植物幼胚 mRNA 整体原位杂交的方法 ······ 638
　　其它论文目录 ·· 642

第四章　有性生殖的超微结构与细胞化学研究··············· 645
　　提要 ··· 646
　　第一节　胚囊和珠孔的超微结构研究························· 647
　　　提要 ·· 647
　　　1　水稻胚囊超微结构的研究 ····································· 648
　　　2　向日葵胚囊的超微结构和"雌性生殖单位"问题 ····· 662
　　　3　Ultrastructure of the micropyle and its relationship to pollen
　　　　　tube growth and synergid degeneration in sunflower ····· 673
　　　4　陆地棉珠孔的结构及花粉管在其中的生长途径 ······ 687
　　　5　甘蓝型油菜珠孔与胚囊的超微结构研究 ················· 690

第二节　雌蕊中钙与钙调素的细胞化学定位 …………… 692
提要 ……………………………………………………………… 692
1　Ultracytochemical localization of calcium in the embryo sac of sunflower ……………………………………………………… 693
2　向日葵柱头、花柱和珠孔中钙分布的超微细胞化学定位 … 708
3　陆地棉雌蕊的花粉管生长途径中钙分布的超微细胞化学定位 ……………………………………………………………… 712
4　甘蓝型油菜授粉前后珠孔和胚囊中钙分布的超微细胞化学定位 ……………………………………………………………… 722
5　水稻雌蕊与胚囊中钙的超微细胞化学定位 ………………… 730
6　钙调素 mRNA 和蛋白在水稻花药和雌蕊发育过程中的原位定位 ……………………………………………………………… 737
7　烟草受精前后胚囊中钙调素的免疫细胞化学定位 ………… 748
8　钙调素 mRNA 在受精前后分离的烟草胚囊中的定位 …… 756

第三节　雌蕊中 ATP 酶与植物激素的细胞化学定位 ……… 762
提要 ……………………………………………………………… 762
1　金鱼草胚珠中 ATP 酶活性的超微细胞化学定位 ………… 763
2　向日葵胚珠中 ATP 酶活性的超微细胞化学定位 ………… 769
3　烟草雌蕊的花粉管生长途径中几种植物激素的免疫金电镜观察 ……………………………………………………………… 781
4　受精前后烟草卵细胞内玉米素和三类酸性植物激素分布的免疫金电镜观察 ……………………………………………… 787
5　烟草原胚中赤霉素 GA_7 与 GA_4 分布的免疫电镜观察 …… 799

其它论文目录 …………………………………………………………… 803

第五章　植物实验生殖生物学专题论述 …………………… 805
提要 ……………………………………………………………… 806
1　Experimental plant reproductive biology and reproductive cell manipulation in higher plants: now and the future ………… 808
2　*In vitro* induction of haploid plants from unpollinated ovaries

7

	and ovules ···	820
3	*In vitro* gynogenesis ···	832
4	花粉原生质体、精子与生殖细胞的实验操作 ················	847
5	植物性细胞原生质体的分离、培养和融合 ····················	857
6	花粉原生质体与配子原生质体操作的研究进展与前景 ···	870
7	被子植物离体受精与合子培养研究进展 ·······················	878
8	Some approaches to the experimental manipulation of reproductive cells and protoplasts in flowering plants ········	887
9	受精过程中助细胞退化机理的研究进展 ·······················	901
10	钙在有花植物受精过程中的作用 ···································	907
11	荧光显微术在当代植物细胞生物学研究中的应用 ········	919
12	荧光显微术 ···	932
13	植物胚胎学中的整体透明技术 ·····································	948

其它论文目录·· 953

彩色图版

CONTENTS

Chapter 1   Experimental researches on male organs and cells ······ 1
Note ······ 2
Section 1   Anther culture and androgenesis ······ 3
Note ······ 3
1. An investigation on new exogenous hormones for pollen dedifferentiation ······ 5
2. The role of exogenous hormones in rice anther culture ······ 12
3. Experimental researches on the two pathways of pollen development in *Oryza Sativa* L. ······ 19
4. Induction of sporophytic development in rice pollen by flower cutting ······ 27
5. Anther culture and androgenesis of *Hordeum vulgare* L. ······ 31

Section 2   Pollen germination and pollen tube growth ······ 33
Note ······ 33
1. Fluorescein diacetate used as a vital stain for labeling living pollen tubes ······ 34
2. A fluorescence staining-methyl salicylate clearing technique for demonstrating pollen nuclei ······ 40
3. Mitosis and amitosis of generative cell nuclei in artificially germinated pollen tubes in *Zephyranthes candida* (lindl.) Herb. ······ 47
4. Influence of medium pH and polyethylene glycol on the artificial pollen germination in *Brassica* ······ 49

5   Exogenous Ca^{2+} regulation of pollen tube growth and division of generative nucleus in *Nicotiana tabacum* ⋯⋯⋯⋯⋯⋯⋯⋯ 50

6   Regulation of generative nucleus division by calcium ionophore A23187 in *Nicotiana tabacum* pollen tubes ⋯⋯⋯⋯⋯⋯⋯ 52

7   Effects of nifedipine on pollen germination, pollen tube growth and division of generative nucleus in *Nicotiana tabacum* ⋯⋯ 53

8   Regulation of *in vitro* pollen tube growth and generative nucleus division by exogenous calmodulins and calmodulin antagonist in *Nicotiana tabacum* L. ⋯⋯⋯⋯⋯⋯⋯⋯⋯⋯⋯⋯⋯⋯⋯ 55

Section 3   Manipulation of pollen protoplasts ⋯⋯⋯⋯⋯⋯ 61

Note ⋯⋯⋯⋯⋯⋯⋯⋯⋯⋯⋯⋯⋯⋯⋯⋯⋯⋯⋯⋯⋯⋯⋯⋯⋯ 61

1   Mass isolation and culture of pollen protoplasts from three plant species ⋯⋯⋯⋯⋯⋯⋯⋯⋯⋯⋯⋯⋯⋯⋯⋯⋯⋯⋯⋯ 64

2   Cell divisions in pollen protoplast culture of *Hemerocallis fulva* L. ⋯⋯⋯⋯⋯⋯⋯⋯⋯⋯⋯⋯⋯⋯⋯⋯⋯⋯⋯⋯⋯⋯⋯ 70

3   An ultrastructural study on pollen protoplasts and pollen tubes germinated from them in *Gladiolus gandavensis* ⋯⋯⋯⋯⋯⋯ 78

4   An ultrastructural study on the triggering of cell division in young pollen protoplast culture of *Hemerocallis fulva* L. ⋯ 87

5   Fluorescence microscopic observations on actin filament distribution in corn pollen and gladiolus pollen protoplasts ⋯⋯⋯⋯⋯⋯ 96

6   Release of pollen protoplasts in large quantities in *Brassica napus* and *B. campestris* var. *purpurea* ⋯⋯⋯⋯⋯⋯⋯⋯ 104

7   Pollen cryopreservation and pollen protoplast isolation in *Brassica campestris* var. *purpurea* ⋯⋯⋯⋯⋯⋯⋯⋯⋯ 109

8   Isolation of pollen protoplasts in *Nicotiana tabacum* ⋯⋯⋯ 115

9   Isolation and early *in vitro* development of young pollen protoplasts in *Nicotiana tabacum* ⋯⋯⋯⋯⋯⋯⋯⋯⋯⋯⋯⋯⋯⋯⋯ 118

10   Specimen preparation of single pollen protoplasts for TEM and their ultrastructural observation in *Nicotiana tabacum* ⋯⋯ 124

CONTENTS

11 Transient transformation of pollen protoplasts via electroporation and temporal expression of Zm 13-260-GUS-NOS chimeric gene in *Brassica campestris* var. *purpurea* ················ 129

12 Regeneration of hybrid plantlets via pollen-hypocotyl protoplast fusion in *Brassica* spp. ················ 133

13 Development and molecular identification of pollen-somatic hybrid plants in *Brassica* spp. ················ 143

14 Pollen-mesophyll protoplast fusion and hybrid plant regeneration in *Nicotiana* ················ 147

Section 4 Manipulation of de-exined pollen ·············· 152

Note ················ 152

1 Isolation and artificial germination of de-exined pollen in *Brassica* ················ 154

2 *Brassica* de-exined pollen as a new experimental system for studying pollen germination ················ 162

3 Preparation of exine-detached pollen in *Nicotiana tabacum* ··· 171

4 Establishment of an experimental system for artificial germination and *in vitro* pollination with de-exined pollen in *Nicotiana tabacum* ················ 174

5 Intrinsic GUS activity in various tissues and during pollen development of *Nicotiana Tabacum* ················ 184

6 Exine-detached pollen of *Nicotiana tabacum* as an electroporation target for gene transfer ················ 185

7 β-glucuronidase gene and green fluorescent protein gene expression in de-exined pollen of *Nicotiana tabacum* by microprojectile bombardment ················ 187

Section 5 Manipulation of generative cells ·············· 193

Note ················ 193

1 Direct observations on generative cells isolated from pollen grains of *Haemanthus katherinae* Baker ················ 195

    2   Isolation and purification of generative cells from fresh pollen of *Vicia faba* L. ·········· 201

    3   Mass isolation and purification of generative cells from pollen grains ·········· 206

    4   A comparative study on methods for isolation of generative cells in various angiosperm species ·········· 211

    5   Nuclear divisions of isolated, *in vitro* cultured generative cells in *Hemerocallis minor* Mill. ·········· 219

    6   Fusion experiments of isolated generative cells in several angiosperm species ·········· 225

    7   Fluorescent vital staining of plant sexual cell nuclei with DNA-specific fluorochromes and its application in gametoplast fusion ·········· 226

    8   Isolated generative cells in some angiosperms; a further study ·········· 234

    9   Microtubule organization of *in situ* and isolated generative cells in *Zephyranthes grandiflora* Lindl ·········· 242

    10   Microtubule changes during the development of generative cells in *Hippeastrum vittatum* pollen ·········· 250

Section 6   Manipulation of sperm cells ·········· 259

    Note ·········· 259

    1   Isolation of viable sperms from pollen of *Brassica napus*, *Zea mays* and *Secale cereale* ·········· 261

    2   Mass isolation and preservation of viable sperm cells in *Brassica campestris* var. *purpurea* ·········· 266

    3   Isolation and fusion of sperm cells in several bicellular pollen species ·········· 277

List of other papers ·········· 288

Chapter 2   Experimental researches on female organs and cells ·········· 291

    Note ·········· 292

Section 1 Unpollinated ovary or ovule culture and *in vitro* gynogenesis 293

Note 293

1 *In vitro* induction of haploid plantlets from unpollinated young ovaries of *Oryza sativa* L. 296

2 Induction of haploid rice plantlets by ovary culture. 300

3 *In vitro* embryogenesis in unfertilized embryo sacs of *Oryza sativa* L. 304

4 Synergid apogamy and egg cell anomalous division in cultured ovaries of *Oryza sativa* L. 310

5 Morphogenetic aspects of gynogenetic embryoid and callus in ovary culture of *Oryza sativa* L. 317

6 Further embryological studies on the in vitro apogamy in *Oryza sativa* L. 318

7 Picloram as an exogenous hormone promotes embryoid differentiation in rice ovary culture 325

8 An investigation on ploidy and other characters of the gynogenic plants in *Oryza sativa* L. 330

9 Unpollinated ovary culture in rice 337

10 *In vitro* production of haploids in rice through ovary culture 348

11 Embryological observations on ovary culture of unpollenated young flowers in *Hordeum vulgare* L. 363

12 *In vitro* induction of haploid embryos and plantlets from unpollinated young florets and ovules of *Helianthus annuus* L. 365

13 Microscopical observations on the embryoid formation in cultured unfertilized ovules of *Helianthus annuus* L. 368

14 Histochemical studies on *in vitro* parthenogenesis in *Helianthus annuus* L. 374

15 Regulation of *in vitro* parthenogenesis and somatic proliferation in sunflower by several factors 376

16 The use of a whole stain-clearing technique for observations on embryo sac, embryo, endosperm and embryoid 385

17 An electron microscope study on *in vitro* parthenogenesis in sunflower 392

18 *In vitro* production of haploids in *Helianthus* 407

19 Haploid embryogeny and plant regeneration in unpollinated ovary culture of *Allium tuberosum* 419

20 Ultrastructural observations on parthenogenesis and antipodal apogamy of *Allium tuberosum* Roxb 426

Section 2 Manipulation of embryo sacs 437

Note 437

1 Enzymatic isolation of embryo sacs in angiosperms: isolation and microscopical observation on fixed materials 439

2 Observations on megasporogenesis and megagametophyte development in *Paulownia* sp. and *Sesamum indicum* by enzymatic maceration technique 446

3 The enzymatic isolation of embryo sacs from fixed and fresh ovules of *Antirrhinum majus* L. 453

4 Fluorescence microscopic observations on enzymatically isolated embryo sacs 457

5 Observations on enzymatically isolated living and fixed embryo sacs in several angiosperm species 462

6 A study of fertilization events in living embryo sacs isolated from sunflower ovules 473

7 Isolation and identification of integumentary tapetal wall sac in *Antirrhinum majus* L. 480

8 A new method for embryo sac isolation and *in situ* fusion of egg and synergid protoplasts in *Nicotiana tabacum* 491

List of other papers 501

Chapter 3 Experimental researches on pollination, fertilization and embryo development ······ 503

Note ······ 504

Section 1 Pollination and fertilization ······ 505

Note ······ 505

1 A preliminary study on the emasculation method in rice ······ 507
2 Embryological studies on some problems in wheat fertilization process ······ 512
3 A preliminary study on the pollination method and the viability of pistil and pollen in sesame ······ 524
4 A study on pollen germination and pollen tube growth in the pistil of sesame ······ 526
5 Effect of pollen-grain number on fertilization, embryo development and progeny's characteristics in *Sesamum indicum* L. ······ 533

Section 2 *In vitro* fertilization ······ 550

Note ······ 550

1 Polyethylene glycol-induced fusion of selected pairs of single protoplasts ······ 552
2 Single-pair fusion of various combinations between female gametoplasts and other protoplasts in *Nicotiana tabacum* ······ 559
3 The first record of dynamics of male-female nuclear fusion in viable tobacco central cell during *in vitro* fertilization ······ 569

Section 3 Manipulation of zygotes, central cells and young embryos ······ 572

Note ······ 572

1 Isolation of fertilized embryo sacs and zygotes and triggering of zygote division in vitro in *Nicotiana tabacum* ······ 574
2 In vitro development of zygotes and two-celled proembryos in *Nicotiana* ······ 581
3 In vitro divisions of unfertilized central cells and other embryo

sac cells in *Nicotiana tabacum* var. *Macrophylla* ············ 589
4 Transformation of tobacco young proembryos by
    electroporation ·· 596
5 Gene transfer into isolated and cultured tobacco zygotes
    by a specially designed device for electroporation ············ 602
6 Distribution of membrane-bound calcium and activated
    calmodulin in isolated zygotes and young embryos
    of *Triticum aestivum* ··· 608
7 Isolation of egg cells and zygotes in *Oryza sativa* ············ 609
8 In vitro development of early proembryos and plant regeneration
    via microculture in *Oryza sativa* ······································· 611
9 Isolation and *in vitro* culture of zygotes and central cells
    of *Oryza sativa* L. ··· 623
10 Construction and analysis of cDNA libraries from proembryos
    and just differentiating young embryos in rice ············ 633
11 An approach to mRNA whole-mount *in situ* hybridization for
    plant young embryos ·· 638
List of other papers ··· 642

Chapter 4 Ultrastructural and cytochemical studies on sexual reproduction ·· 645

Note ·· 646

Section 1 Ultrastructural studies on embryo sac and micropyle ·· 647

Note ·· 647
1 An ultrastructural study of embryo sac in *Oryza sativa* L. ··· 648
2 Ultrastructure of sunflower embryo sac in respect to the concept
    of female germ unit ·· 662
3 Ultrastructure of the micropyle and its relationship to pollen tube
    growth and synergid degeneration in sunflower ············ 673

    4   Structure of micropyle in *Gossypium hirsutum* and the pathway of pollen tube growth ········ 687

    5   Ultrastructural studies on the micropyle and embryo sac in *Brassica napus* L. ········ 690

Section 2   Cytochemical localization on calcium and calmodulin in the pistils ········ 692

    Note ········ 692

    1   Ultracytochemical localization of calcium in the embryo sac of sunflower ········ 693

    2   Ultracytochemical localization of calcium in the stigma, style and micropyle of sunflower ········ 708

    3   Ultracytochemical localization of calcium in the pollen tube track of cotton gynoecium ········ 712

    4   Ultracytochemical localization of calcium in micropyle and embryo sac of *Brassica napus* before and after pollination ········ 722

    5   Ultracytochemical localization of calcium in the gynoecium and embryo sac of rice ········ 730

    6   *In situ* localization of calmodulin mRNA and protein in the developing anthers and pistils in rice ········ 737

    7   Immuno-localization of calmodulin in unfertilized and fertilized embryo sacs in *Nicotiana tabacum* var. macrophylla ········ 748

    8   Using isolated embryo sacs and early proembryos for localization of calmodulin mRNA before and after fertilization in *Nicotiana* ········ 756

Section 3   Cytochemical localization of ATPase and phytohormones in the pistils ········ 762

    Note ········ 762

    1   The ultracytochemical localization of ATPase activity in the ovules of *Antirrhinum majus* L. ········ 763

2 The ultracytochemical localization of ATPase activity in the ovules of sunflower ······ 769
3 Immunogold electron microscopic observations on distribution of several phytohormones in pollen tube growth track through pistil in tobacco ······ 781
4 Distribution of trans-zeatin, $GA_{7/4}(+)$ ABA and IAA in tobacco egg cells before and after fertilization: immunogold electron microscopic observations ······ 787
5 Distribution of gibberellins A_7 and A_4 in tobacco Proembryos using immunoelectron microscopy ······ 799

List of other papers ······ 803

Chapter 5 Review articles on experimental reproductive biology in plants ······ 805

Note ······ 806

1 Experimental plant reproductive biology and reproductive cell manipulation in higher plants: now and the future ······ 808
2 *In vitro* induction of haploid plants from unpollinated ovaries and ovules ······ 820
3 *In vitro* gynogenesis ······ 832
4 Experimental manipulation of pollen protoplasts, sperms and generative cells ······ 847
5 Isolation, culture and fusion of sexual plant protoplasts ······ 857
6 Advances and perspectives of the manipulation of pollen protoplasts and gametoplasts ······ 870
7 Recent advances in *in vitro* fertilization and zygote culture of angiosperms ······ 878
8 Some approaches to the experimental manipulation of reproductive cells and protoplasts in flowering plants ······ 887
9 Recent advances in research on the mechanism of synergid

	degeneration during fertilization process	901
10	The role of calcium in the fertilization process in flowering plants	907
11	Application of fluorescence microscopy in contemporary studies of plant cell biology	919
12	Fluorescence microscopy	932
13	Whole clearing techniques in plant embryology	948
List of other papers		953

Color photographs

第一章 雄性器官与细胞的实验研究

提要

本章介绍从20世纪70年代到20世纪末的20多年期间,我们在植物雄性系统方面的实验研究由器官操作向细胞与原生质体操作的发展。

早期的花药培养,是我们应用离体培养技术进行实验研究的开端。在20世纪70年代全国性的花药培养与单倍体育种研究热潮中,我们作为参加者曾做了一定的工作。我们的特色不在于大规模地培养和单倍体选育,而是以胚胎学的观点与方法研究花粉两条发育途径的控制条件和雄核发育过程不同阶段对培养条件的需求。换言之,属于实验胚胎学研究的范畴。

花粉人工萌发是极好的实验系统,是国际上长盛不衰的研究热点。我们在20世纪80年代曾发表一项用荧光染色标记生活花粉管的研究结果;提出了一项用荧光染色结合透明技术显示花粉细胞核的新方法。20世纪90年代,对花粉管中生殖核的分裂行为、特别是对钙及其有关因素在花粉管生长与生殖核分裂中的作用进行了研究。还研究了pH值与PEG对芸苔属花粉人工萌发的影响,提出了芸苔属植物的高效花粉培养基配方。

从20世纪80年代中、后期开始,我们在雄性系统方面的实验研究重点转入花粉原生质体和配子原生质体操作。

早在20世纪70年代,国际上即有人尝试分离花粉原生质体,但由于未能找到克服花粉外壁障碍的有效途径,一直没有解决大量分离的方法。到了80年代后期,日本研究者在百合、我们在其它三种植物中各自独立地闯过了这个难关,将这一研究方向推进到新的层次。在此基础上,我们相继开展了花粉原生质体的培养、融合、转化及细胞生物学等一系列有特色的研究,特别是首创了幼嫩花粉(小孢子)原生质体的分离、诱导孢子体发育与"花粉-体细胞杂交"实验系统。

在进行花粉原生质体研究的过程中,衍生出一个新的研究方向,即脱外壁花粉的研究。后者是介于完整花粉粒与花粉原生质体之间的结构单位,它脱去了外壁而保留了内壁。国际上迄今尚无人开展这方面的研究。我们之所以将它列入计划,是考虑到它有两方面的意义:一是作为花粉转化的新系统;二是在花粉生物学研究中具有其独特的价值。为此,我们相继建立了芸苔属和烟草脱外壁花粉的分离、人工萌发、离体授粉和外源基因导入等实验系统,开展了有关细胞生物学研究。

雄配子原生质体是指包藏于花粉或花粉管中的精细胞及其前身生殖细胞的离体状态。20世纪80年代中期,美国研究者首次建立了大量分离精细胞的方法。同时,我们也首次分离生殖细胞成功。以后我们又在众多植物的生殖细胞分离、培养、融合、细胞骨架研究等方面进行了系统的研究,创立了多种操作技术。在三细胞型与二细胞型花粉中的精细胞分离技术也获成功。这些结果为深入认识生殖细胞与精细胞的生物学特点以及进一步开展离体受精奠定了基础。

第一节　花药培养与雄核发育

提要

20世纪60年代中期,印度研究者由曼陀罗花药培养出单倍体胚状体,是植物实验胚胎学史中一项重要突破。由此在70年代掀起一股花药培养与单倍体育种的世界性热潮。中国的"花培热"尤其突出,上至科研院所与高等学校,下至基层农业科研单位,全国几乎没有一处不被卷入这股热潮中。湖北省也成立了水稻单倍体育种协作组,武汉大学作为参加单位承担了为协作组培养花粉植株的部分任务。然而,借助花药培养实验系统进行实验胚胎学研究才是我们真正的兴趣所在。

外源激素是花粉去分化的关键因素,当时在水稻花药培养中常用的是2,4-D、NAA等。为了探索更多的花粉去分化新激素,我们设计了一种简易的"种胚预测法",以当时所收集到的分属7种类型的

11种除草剂,分别试验稻胚的反应,从中筛选出3种有促进愈伤组织形成的作用。以之试验花药培养,结果发现其中 MCPA 比 2,4-D 有更强的去分化效果,TCP(picloram)和 2,4,5-T 的效果也不低,从而扩大了花粉去分化激素的范围。MCPA 以后不仅在我们的花培试验中,而且在未传粉子房与胚珠培养研究中一直作为主要的外源激素使用;picloram 在子房培养中也被证明有独特的价值(参看第二章第一节)。

在花药培养的全过程中,不同阶段对外源激素的要求与反应如何? 为了探讨这个问题,设计了专门的花药转移实验,即在接种后不同阶段,分别将花药由有激素培养基转移到无激素培养基上,或者相反。通过比较各种转移实验处理对愈伤组织诱导和分化的影响,提出了外源激素在培养不同阶段有不同作用特点的观点。

在一般花药培养中,小孢子大部分死去,一部分被诱导启动孢子体发育,很少继续原有的配子体发育。为了探讨水稻花粉在离体培养条件下两条发育途径的控制因素,我们设计了幼花直插培养、平贴培养两种方式,而以一般花药培养作为对照。胚胎学观察表明:直插培养时多数花粉沿配子体途径发育成成熟的雄配子体,授粉后可萌发花粉管;少数花粉发生细胞分裂,显示向孢子体途径偏移的趋向。平贴培养时部分花粉走配子体途径;部分花粉走孢子体途径形成花粉愈伤组织。一般花药培养时配子体途径被阻断,孢子体发育成为唯一的途径。以上三种培养方式显示由配子体途径向孢子体途径偏移的三个不同级别。由此得出看法:花粉孢子体发育(雄核发育)依赖于摆脱整体制约与花药直接接触培养基两个因素。仅仅前一因素即可启动孢子体发育,而孢子体发育的继续进行则还需要后一因素。

进一步,设计了两个培养步骤的实验:第一步是直插培养;第二步是将直插培养的花药取下,继续进行花药培养。结果很有趣:直插培养时间愈长,花粉细胞团的频率愈高,并且在后续花药培养中形成花粉愈伤组织的时间愈短;反之亦然。这说明两步培养是一个既有不同特点而又前后连续的过程。

除水稻花药培养外,也研究过大麦花药培养与雄核发育。通过 50

多个不同类型品种的实验，比较了它们在花药培养中的反应，研究了影响培养效果的因素，观察了雄核发育的途径和细胞学特点，为前期诱导频率高而后期急剧降低的原因寻找细胞学依据。

为了便于雄核发育的细胞学观察，我们提出了爱氏苏木精整体染色后进行整体透明或埋蜡切片的技术（见本章其它论文目录）。爱氏苏木精在各种苏木精染剂配方中以其无需分色的优点最适于整体染色。我们利用这一优点于雄核发育观察上，取得良好效果，并在后来的雌核发育观察中长期应用，大大节省了人力与物力，使大规模的胚胎学观察成为可能（参看第二章第一节）。

1 水稻花粉去分化新激素的探索

An investigation on new exogenous hormones for pollen dedifferentiation

迄今花药培养基的研究主要包括基本培养基中无机盐类、有机成分、补加物质、简化培养基、碳源和激素浓度等方面。但是，目前应用的激素种类极为有限，理论研究也较缺乏。激素是诱导去分化形成愈伤组织的重要因素，这类激素我们暂称为去分化激素。广泛探索花药培养去分化新激素，不仅可能发现更有效地促进愈伤组织形成的物质，进一步提高诱导效率，而且通过对各类激素作用性质的比较研究，将更加深我们对花粉去分化机理的认识。

为了有效地开展这项工作，我们首先研究了筛选去分化激素的预测方法，试验表明，利用水稻成熟种子胚培养对供试药品的反应，可以预测后者对花粉去分化的作用。这一方法简称为种胚预测法。根据预测结果，选择其中部分药品进行花药培养试验。

材料与方法

种胚预测法的试验材料为6个粳稻品种。试验了多种除草剂、一些激素和杀菌剂共22种药品。

种胚预测法的做法是：将成熟谷粒去壳，切去胚乳，用95%乙醇（1～2min）和0.1%升汞（10～12min）先后对种胚消毒，无菌水洗4次。在小三角锥瓶中用无菌水将种胚浸泡1d，接种于含3%蔗糖中，并分别加入各种供试药品的 N_6 培养基上。对照Ⅰ加入2mg/L 2,4-D。对照Ⅱ不加任何激素。

花药培养试验材料主要是粳稻品种、少数粳稻品种间杂种、籼粳杂种与籼稻品种，共16个材料。根据种胚预测结果，从22种药品中选择部分药品做花药培养试验（表1）。第一培养基为含5%蔗糖的分别加入各种供试药品的 N_6 培养基。对照Ⅰ加入2mg/L 2,4-D。对照Ⅱ不加任何激素。第二培养基为含3%蔗糖的 N_6 培养基，不加任何激素。

结　果

一、种胚预测法的试验

种胚对供试药品的反应　种胚对供试药品的反应大致可归纳为以下两类。

(1)产生愈伤组织：在含2,4-D、MCPA、2,4,5-T和TCP的培养基上，胚的生长发生明显变化。接种后不久，芽鞘伸长，但根芽均受抑制。以后接近芽鞘基部处逐渐产生愈伤组织（图版Ⅰ,4、5、6、7）。

在含NAA的培养基上，芽的生长比较正常，也能生根，但根的伸长受阻，一般短而多，常有根毛。以后往往在接近根的基部处产生一些愈伤组织（图版Ⅰ,3）。

(2)不产生愈伤组织：在无激素的培养基（对照Ⅱ）上，种胚迅速生长，正常生根出苗（图版Ⅰ,1）。加入丁二酸、阿魏酸等，胚也能正常出苗。7431浓度较高时,根芽生长受到一定抑制(图版Ⅰ,2)。在含赤霉酸的培养基上,幼苗十分细长。有草枯醚时,苗虽生长,但变黄衰弱。而有镇草宁时,根芽均不生长,种胚长期停留在接种时的状况。药品浓度对胚的生长有明显的不同影响。总之,所有上述药品对种胚的作用无论如何不同,其共同特点均为不能产生愈伤组织。

种胚预测法的效果 根据上述试验结果,有意识地选用种胚反应不同的两类药品进行花药培养,以检验种胚预测法的效果。结果表明(表1),在种胚培养时,能够诱导种胚产生愈伤组织的药品(2,4-D、MCPA、2,4,5-T、TCP和NAA),在花药培养时,也都能够程度不同地诱导花粉形成愈伤组织。不能促使种胚产生愈伤组织的药品,一般没有高于无激素培养(对照Ⅱ)的花粉去分化诱导能力。7431情况比较特殊,详见下文。

表1　　　　　　　种胚培养与花药培养结果的比较

药品名称 (普通名称与化学名称)	种胚培养		花药培养	
	药品浓度 (mg/L)	愈伤组织 产生与否	药品浓度 (mg/L)	去分化 效果
2,4-D(2,4-二氯苯氧乙酸)	2	+	2	+
MCPA(2-甲基-4-氯苯氧乙酸)	2~50	+	1~8	+
2,4,5-T(2,4,5-三氯苯氧乙酸)	1~10	+	1~8	+
TCP(毒莠定,4-氨基-3,5,6-三氯-2-羧基吡啶)	10	+	5~100	+
NAA(萘乙酸)	4	+	4	+
7431(对氨基苯磺酰氨基甲酸甲酯)	0.01~100	−	0.01~0.1	?
N-异丙基-O-乙基-O(2-硝基-4-甲氧基)苯基硫代磷酰胺酯	0.2~100	−	0.05~1	−
对碘苯氧乙酸	2~50	−	50~200	−
非草隆	5~50	−	10~50	−
丁二酸	10~100	−	10~100	−
阿魏酸	1~100	−	1~5	−

种胚预测法的优点是:第一,反应迅速。大约接种 4~5d 即已显出初步趋势,10d 以后,结果明显。第二,反应整齐。每个处理 10~20 个种胚,就可以大致看出种胚对药品的反应。因而可以大大减轻工作量。第三,不受生育期限制,可以常年进行。总之,根据我们的初步试验,种胚培养作为预测花粉去分化新激素的简易方法是有价值的。

二、水稻花粉去分化新激素的试验

二甲四氯(MCPA)　根据四个材料的花药培养试验,发现 MCPA(钠盐粗制品,折算成纯品浓度)在 1~8mg/L 的浓度范围内,愈伤组织诱导率呈一曲线,一般以 2mg/L 效果最好(表2)。

MCPA 和 2,4-D 对比试验表明,MCPA 的愈伤组织诱导率一般超过 2,4-D,它所诱导的愈伤组织分化率和绿苗率与 2,4-D 大体相近(表2)。

表2　　　　　MCPA 的 诱 导 效 果

材　料	处理 (mg/L)	接种 花药数	愈伤组织 诱导率 (%)	转移愈伤 组织块数	愈伤组织 分化率 (%)	绿苗率 (%)
(鄂晚3号×TDK)PO$_2$ × 鄂晚3号 F$_1$	MCPA 1	760	37.8	103	32.0	18.2
	MCPA 2	720	55.8	129	57.4	27.2
	MCPA 4	720	53.9	129	42.6	30.9
	MCPA 8	720	45.3	100	32.0	25.0
	2,4-D 2	760	36.3	73	31.5	34.8
廉江密早	MCPA 1	480	24.8	32	65.5	19.0
	MCPA 2	510	32.9	38	42.1	37.5
	MCPA 4	510	36.3	55	25.5	28.6
	MCPA 8	510	29.4	37	18.9	14.3
	2,4-D 2	510	24.7	37	51.4	31.6

续表

材料	处理 (mg/L)	接种 花药数	愈伤组织 诱导率 (%)	转移愈伤 组织块数	愈伤组织 分化率 (%)	绿苗率 (%)
农林15	MCPA 1	320	10.9	17	58.8	60.0
	MCPA 2	360	15.8	24	45.8	45.5
	MCPA 4	360	12.5	23	26.1	33.3
	MCPA 8	520	11.7	25	52.0	38.5
	2,4-D 2	400	11.5	15	46.7	57.1
珍汕97	MCPA 4	520	1.3			
	MCPA 8	520	1.3			
	2,4-D 2	600	0			

MCPA与2,4-D相比,愈伤组织出现早而多,而且持续地保持这一优势(图1)。不仅如此,MCPA诱导的愈伤组织生长也较快,在同一时间内达到转移分化的愈伤组织数也较多。

2,4,5-T 水稻花药培养基中加入2,4,5-T,也能促使愈伤组织产生。根据4个材料的花药培养试验。2,4,5-T的愈伤组织诱导率、分化率以及绿苗率和2,4-D相近。2,4,5-T的浓度在1~8mg/L的范围内,以2~4mg/L的效果较好(表3)。

毒莠定(TCP) 我们用TCP(Picloram)作水稻花药培养试验,发现它有诱导水稻花粉形成愈伤组织的作用。在1~60mg/L的浓度(粗制品,未经折算)范围内,愈伤组织诱导率逐步上升。其最适浓度尚待进一步确定。根据7个材料,接种近万花药的试验结果,TCP与2,4-D相比,诱导效果大体相近,部分试验结果列入表3。

表3　　　　　　　　TCP 的诱导效果

材料	处理 (mg/L)	接种花药数	愈伤组织诱导率（%）	转移愈伤组织块数	愈伤组织分化率（%）	绿苗率（%）
农垦4号	TCP 1	280	1.1	3	0	
	TCP 5	360	4.2	14	35.7	60.0
	TCP 12	320	5.6	16	56.3	33.3
	TCP 30	360	5.0	16	87.5	57.1
	TCP 60	400	6.8	26	76.9	40.0
	2,4-D 2	360	5.0	16	62.5	40.0
早粳19	TCP 1	160	0			
	TCP 5	160	2.5	4	25.0	100.0
	TCP 12	240	10.8	26	15.4	50.0
	TCP 30	160	10.6	16	100.0	62.5
	TCP 60	160	20.0	22	59.1	7.7
	2,4-D 2	280	18.6	44	38.6	23.5

其它药品　除了上述药品外,我们还试验了6种药品,它们在种胚培养时均不能诱导愈伤组织,而在花药培养时,一般产生少数小型的愈伤组织。其中以含7431和非草隆的培养基上,愈伤组织稍多。

讨　　论

1. 如何从浩瀚的化学物质中探索花粉去分化新激素是一个值得探讨的问题。历史经验启示我们,从各类除草剂中发现新的植物激素可能是比较有希望的途径之一。20世纪20年代末期发现苯基氨基甲酸酯具有植物生长调节剂的性质。40年代以后,苯胺灵、燕麦灵等相继问世,导致苯基氨基甲酸酯大量用于除草[1]。Zimmerman 等(1942)合成了2,4-D等,指出它具有植物激素的性质[3]。两年后,Hammer等用它作除草剂获得重大成功,为化学除草剂的应用开辟了广阔的前景[8]。从而使除草成为植物激素应用的一个重要方面。Hameker 等

(1963)介绍了一种除草剂——Tordon(毒莠定 TCP),它对阔叶类植物比 2,4-D 和 2,4,5-T 有更大的毒性[7]。Diaz-colon 用大豆组织培养试验,比较 2,4,5-T、毒莠定和麦草畏三种除草剂的效果。他指出,它们之中 2,4,5-T 毒性最大。毒莠定和麦草畏在低浓度(10^{-6}M)的条件下有促进大豆组织生长的作用[6]。Чернова(1975)在几种豆科与禾本科植物组织培养中,将毒莠定用作去分化激素[12]。Hassan(1975)曾将多种均三氮苯类除草剂用于促进高粱愈伤组织生长,指出它们有类似细胞分裂素的性质[9]。此外,有人还发现 N-1-萘邻氨羰基苯酸(N-1)等除草剂具有激素性质[3]。总之,植物激素与除草剂研究的某些方面是互相渗透的。有些化合物往往兼具刺激与抑制植物生长两种作用(因施用浓度等条件而异)。因此,从激素中发现新除草剂或从除草剂中发现新激素,都是有可能的。当然,激素和除草剂的类型与作用机理都很繁复,兼性的药品只是其中一小部分。而去分化激素又是植物激素中的很少一部分。这就需要进行大量的筛选,才能从众多的除草剂中发现少数有价值的去分化激素。

2. 本试验从七类除草剂中选取 11 种药品参加种胚预测,从中挑选 7 种进行花药培养试验。初步确定其中 3 种除草剂(分属两种类型)有去分化的作用。在花药培养的文献资料中,我们还没有看到类似的报道。根据初步比较,这 3 种除草剂中 MCPA 比 2,4-D 诱导愈伤组织的效果更好。TCP 和 2,4,5-T 的诱导效果不弱于 2,4-D。在前人的其它组织培养研究工作中,TCP 比 2,4-D 有较好的诱导去分化与再分化的作用[12]。2,4,5-T 与 2,4-D 相比,有加速愈伤组织生长、防止愈伤组织变褐等优点[11]。本文在水稻花药培养方面,上述效果不甚明显,这可能与试验材料、药品纯度以及施用浓度等多方面因素有关,值得进一步研究。至于 MCPA 与 7431,我们还未看到把它们当做去分化激素的研究报告。但 MCPA 作为除草剂,曾有不少有关的资料。

本试验所选用的其它各种药品,有些是常用激素如增产灵、矮壮素、阿魏酸、赤霉酸、三碘苯甲酸和丁二酸等,它们没有诱导水稻花粉去分化的作用。其中增产灵(对碘苯氧乙酸)和 2,4-D、2,4,5-T、MCPA 同属苯氧乙酸类物质,但它并没有像后三者那样的去分化作用。巴比妥酸类物质曾被发现有诱导兰科植物产生愈伤组织的效果[5]。丁二酸处理玉米、小麦和黄瓜等种子，能够增产，被归为植物生长调节剂[2]。但在水稻花药培养中巴比妥酸和丁二酸均无去分化效果。试验的杀菌剂多菌灵虽不能诱导去分化，却有防止培养基霉染的作用（详见另文）。

3. 种胚预测法的基本成功，不仅在实践上可大大减轻筛选花粉去分化新激素的工作量，而且在理论上可以揭示花粉与种胚对外源激素反应的某种程度的一致性。正是基于这种相对的一致性，预测才有成效。一致性主要表现在性质方面，即能否诱导产生愈伤组织；至于在浓度方面，我们初步看到，诱导二者去分化的浓度范围比较接近（不是完全相同）。当然，种胚与花粉毕竟属于植株的不同器官，它们之间必然存在细胞生理的异质性，因而上述一致性显然是相对的，即存在不一致的方面。Inoue 与 Maeda（1976）曾指出，诱导水稻不同器官产生愈伤组织需要不同浓度的 2,4-D[10]。因而，种胚培养只可作为大致而非绝对准确的花粉去分化激素预测方法加以应用。大野（1975）为了探索愈伤组织形成的最适培养基，在进行花药培养之前作了适于诱导种子愈伤组织的培养基筛选。看来，这方面工作还有继续研究的必要。

（作者：周嫦、于登洲、张均燕、陈家治。原载：花药培养学术讨论会文集，科学出版社，1978，86～92页。图1幅，表5幅，参考文献12篇。仅选录表1、2。）

2　外源激素在水稻花药培养中的作用

The role of exogenous hormones in rice anther culture

外源激素在花药培养中的作用是一个重要问题，引起不少作者的注意与研究。近几年来我们在研究水稻花粉去分化激素的过程中感到这个问题还有不少未被认识的领域,值得进一步探索。作为第一步,需要了解外源激素在水稻花药培养中有哪些方面的作用?是否在培养的全过程中都有积极的作用?在培养的不同阶段激素作用的大小与特点如何?本文的目的就是试图对上述问题作初步的探讨。

材 料 与 方 法

试验材料主要是粳稻品种。花药培养采用常规固体培养技术。去分化培养基为 N_6,以 2 甲基-4-氯苯氧乙酸(MCPA,农药标准品,纯度 99% 以上,抚顺农药厂产品,2ppm)作为外源激素,以不加 MCPA 作为无激素培养对照。分化培养基采用 MS,不加激素。培养温度一般为 25℃ 左右(夏季温度 25~30℃ 之间)、弱光照。花药接种时期一律为单核靠边期,不经低温预处理。每处理接种花药数为 300~1700 粒,但一组实验中各处理接种花药数大致相近。

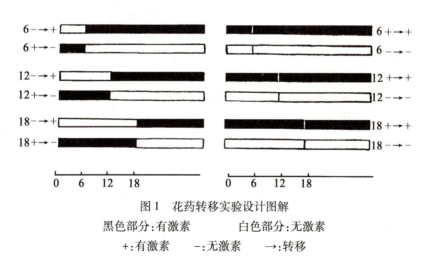

图 1　花药转移实验设计图解
黑色部分:有激素　　白色部分:无激素
+:有激素　-:无激素　→:转移

为了研究激素在花药培养不同阶段所起的作用,以景洪 2 号等品种为材料,设计了专门的花药转移实验(图 1)。即在接种后 6d、12d、

18d,分别用接种针将花药由有激素培养基转到无激素培养基上(代号：+→-)或者相反(代号：-→+)；而以由有激素到有激素(+→+)与由无激素到无激素(-→-)的转移作为对照。为保证实验的准确性,转移操作力求仔细,尽可能不改变花药在培养基上的位置关系,并由一个工作者自始至终进行一组花药转移实验,以求技术与培养条件的一致。此外,在培养6d、12d、18d后分别用醋酸甲醇(1∶3)固定花药,用爱氏苏木精整体染色制片技术观察了当时雄核发育的进度,作为分析实验结果的参考。

结　　果

一、有激素培养与无激素培养的比较实验

诱导率的增长速度(图2) 接种20～25d后,愈伤组织开始突破药壁。以后,随着培养时间的增加,肉眼可以看到愈伤组织的花药数继续增多,因而诱导率逐渐提高。这种诱导率的增长速率,在有激素与无激素条件下有明显的区别：无激素时增长缓慢甚至停滞；有激素时在相当长一段时期内保持持续上升的趋势。

不同品种的诱导率(图3) 共比较了7个试验材料的有激素培养与无激素培养的诱导率。除籼稻品种IR_{26}外,所有粳稻品种在无激素条件下均能产生胚状体或愈伤组织,但品种间差异很大,一般诱导率很低,只有个别品种如景洪2号可达到10%以上。加入外源激素无例外地使诱导率成倍地大幅度提高。而且,在无激素培养时诱导率愈高的品种,在有激素培养时诱导率也愈高,反之亦然,表现出明显的相关性。这说明外源激素通过品种的内在遗传与生理基础而起作用。

愈伤组织的生长势(表1) 培养后一定时期,统计当时愈伤组织达到转移大小的花药数占当时产生愈伤组织的花药总数的百分率(简称转移率)作为衡量愈伤组织生长势的相对指标。结果表明,在无激素条件下产生的愈伤组织(还有一部分是胚状体),只有极少数长大到

能转移的程度,多数生长缓慢,长期停滞甚至枯死。而在有激素条件下愈伤组织生长迅速,很快有相当部分长大达到转移分化的大小。

愈伤组织的分化能力(表2) 通常在无激素培养基上可以发生个别胚状体直接出苗的现象,而在有激素培养基上则不能,说明这类外源激素有抑制分化的效果。但是,如果把愈伤组织转移到第二培养基(无激素)上,就可以看出:第一培养基有激素者,愈伤组织多数能分化成健壮的苗,分化率很高;第一培养基无激素者,分化能力很差,即使个别成苗,也多是瘦小的单苗。由此可见,外源激素具有改良愈伤组织质量和提高愈伤组织分化能力的明显成效。

从上述四个方面的比较,可以初步肯定外源激素对水稻愈伤组织的诱导、持续生长及其以后的分化有多方面的促进作用。为了进一步研究这些作用是否在整个培养过程中都表现出来,以及在培养的不同阶段其作用的特点如何,进行了以下花药转移实验。

二、花药转移实验

花药转移的实验设计如图1所示。为叙述方便起见,后文一律用+号表示有激素,-号表示无激素,→号表示转移。例如,6+→-表示接种6d后将花药由有激素培养基转移到无激素培养基,余类推。

接种后6d的转移实验(图4、表3、表4) 根据雄核发育的观察,无论有激素培养或无激素培养,接种后6d大多为二核花粉阶段,仅极个别花粉粒达到三核时期。看来这6d大体相当于在烟草[11]和小麦[3]上所描述的培养初期的"延滞期"。转移实验表明,这一时期有无外源激素,对培养结果都有一定影响。首先比较6-→+与6+→+:前者比后者仅缺少前6d的激素,但诱导率即有明显的降低,而对愈伤组织的生长与在分化培养基上的分化影响不大。其次比较6+→-与6-→-:前者比后者仅多了前6d的激素,但诱导率即有一定提高,对愈伤组织在分化培养基上的分化也有一定促进作用。可见,在雄核发育高峰到来以前最初6d的培养,外源激素的作用就不可忽视。不过,和以后培养过程中激素较长期的作用相比,这6d的作用是不大的,这表现在图4中6+→-的诱导率普遍低于6-→+以及远远低

15

于 6+→+的事实上。

接种后 12d 的转移实验(图 5、表 5,表 6) 接种后 12d,雄核发育逐渐达到多细胞花粉阶段,但尚无突破花粉壁的愈伤组织。此时进行花药转移,结果表明:缺乏前 12d 的激素使诱导率进一步降低,但对愈伤组织的生长与分化能力影响不大(12-→+和 12+→+相比)。另一方面也可以看出:前 12d 有激素比始终无激素诱导率有明显提高,对愈伤组织的生长与分化也有促进作用(12+→-和 12-→-相比)。此外,三个品种的实验中,有两个品种 12+→-的诱导率高于 12-→+,表明就诱导愈伤组织的作用来说,培养前期 12d 的外源激素的重要性已开始超过 12d 以后的激素。但就愈伤组织的生长与分化能力来说,则前 12d 的激素作用不如 12d 以后的激素作用。

接种后 18d 的转移实验(图 6,表 7,表 8) 通过显微镜观察,接种后 18d,雄核发育的进度正开始达到形成幼小的愈伤组织阶段。此时,将花药由无激素培养基转移到有激素培养基(18-→+),诱导率仅略高于对照 18-→-,而显著低于对照 18+→+,表明缺乏前 18d 的激素对诱导愈伤组织有严重不利影响,18d 以后的激素只有微弱的促进作用。但这样产生的少数愈伤组织和 18-→-的情况迥然不同,能够生长而且部分具有分化能力,则又说明 18d 以后的激素对愈伤组织生长与分化能力的重要作用。再看由有激素到无激素的转移(18+→-),其诱导率远高于 18-→-,接近甚至赶上 18+→+。愈伤组织的生长也很好,能达到转移的大小。但分化的效果不如 18+→+。其中很多不能分化的愈伤组织往往是在分化培养基上生长停滞的类型。可能这是由于 18d 激素对愈伤组织生长的促进作用不能较长期地持续下去所致。可见就愈伤组织的诱导而言,有前 18d 的激素已可基本满足需要,对愈伤组织生长有相当的促进作用;而就愈伤组织的分化能力而言,18d 后的激素仍有重要作用。

总起来看,可以把花药转移实验的结果归纳成四类:第一类是由有激素向有激素的转移(+→+),其愈伤组织的诱导、生长和分化能力都最高。第二类是由无激素向无激素的转移(-→-),在各方面的表现都最差,这是两组对照。第三类是由无激素向有激素的转移(-→+),总的特

第一章　雄性器官与细胞的实验研究

点是对诱导率有程度不同的不利影响,无激素的时间愈长,影响愈严重,而对愈伤组织生长及分化能力影响不大。第四类是由有激素向无激素的转移(+→-),总的特点是有了前 18d 激素可以基本满足诱导愈伤组织的需要,不到 18d 则不足;但由于后期缺乏激素,愈伤组织的生长与分化能力明显减弱,尽管前期的激素在这方面仍有一定的后效。这些结果,为我们认识外源激素在培养过程中的作用提供了一定的依据。

讨　论

在花药培养中是否需要外源激素,仍是一个有争论的问题。Sunderland(1974)[11]在涉及这一问题时曾将各种植物分为两类:一类植物只需要基本培养基就够了;另一类(包括所有形成愈伤组织的种和某些形成胚状体的种)则需补加激素与其它生长因素。但实际情况往往相当复杂。即以烟草这样易起反应的植物来说,一般不需外源激素,但补加少量生长素对花粉植株的形成也有好处(Nitsch 1969[9],郭仲琛等 1973[6])。Sopory 等(1976)[10]在曼陀罗的花药培养中发现:虽然在 NB 基本培养基上也能产生胚状体,但频率很低,而适宜浓度的激素总是能提高诱导频率,其中细胞分裂素的效果最好,这与 Nitsch 等[9]早先作出的激动素有抑制作用的结论恰恰相反。至于禾本科植物,一般都需要应用外源生长素诱导愈伤组织。然而朱至清等(1976)[2]在无激素培养基上成功地诱导了水稻和小麦的花粉植株。玉米花粉植株的诱导也不一定需要外源激素[1,4]。甚至加入抗生长素 TIBA 还可以促进大麦(Clapham 1973[8])和玉米(谷明光等 1978[4])愈伤组织的形成(当然,TIBA 在这里是否起抗生长素的作用,还值得研究)。另一方面,潘景丽等(1978)[7]通过小麦花药无激素和有激素培养的对比实验认为外源激素对花粉胚的建成有重要作用。以上各种资料启示我们:关于各类外源激素在各种植物花药培养中的作用大小与性质,仍然有必要进一步研究。

在本实验中,我们研究了一种与 2,4-D 性质相近而诱导力很强的人工合成生长素 MCPA[5]对水稻花粉植株形成的作用,现提出下列看

法供大家讨论:

1. 在我们试验的 7 个水稻品种中,6 个粳稻品种都可在无激素的 N_6 基本培养基上产生愈伤组织或胚状体,其中某些可在此培养基上直接出苗或转移到 MS 分化培养基上分化成苗,这证实了朱至清等(1976)[2]在粳稻品种廉江密早上所获得的结果,表明外源激素对于水稻花粉植株的形成并非必需。然而无激素培养时一般诱导率很低,胚状体瘦弱,愈伤组织小而成苗少,而相应的有激素培养却无一例外成倍地大幅度提高了诱导率,显著促进了愈伤组织的生长和以后的分化。这又表明外源激素在水稻花药培养中确实具有多方面十分重要的促进作用。在无外源激素条件下虽然极少数的花粉粒能够顺利通过雄核发育的各个阶段形成花粉植株,但对多数有孢子体发育潜能的花粉来说,缺少外源激素毕竟很难使这种发育潜能顺利表达。

2. 关于激素在培养过程中所起的作用,迄今研究不多。Sunderland(1974)[11]认为现在还不知道激素对诱导过程本身有什么贡献,很可能其主要贡献是在诱导后的时期。朱至清等(1978)[3]基于小麦有激素培养和无激素培养雄核发育的观察,认为外源激素对雄核发育的启动并不是必需的,它的作用主要是防止多细胞花粉的败育,使较多的花粉粒长成愈伤组织。在我们的实验中,激素在诱导前期和后期的作用都很明显。无论前期缺少激素或是后期缺少激素,都给培养带来不良的后果。即使接种后最初 6d,缺少激素就有不利影响,而给以激素就有促进作用。这说明就水稻花药培养而论,诱导期的激素也是有作用的。不过,诱导期激素的作用究竟是否也对诱导过程本身(即雄核发育的启动)起作用,或只是对后继的过程发生某种后效,还需要进一步研究。

从花药转移实验也可以看出:激素在培养前期和后期的作用各有所侧重。一般讲,在前期侧重对愈伤组织诱导的影响,对愈伤组织后期的生长与分化也有某些作用;在后期则侧重对愈伤组织生长和分化能力的影响,对继续提高诱导率仅有微弱的效果。由于离体花药中花粉发育的不同步性与激素作用的后效性等复杂因素交织在一起,不可能显示出每个特定时期激素作用的专一性,但以上大致趋势仍然是明显的。看来,外源激素的有利作用在去分化培养的全过程中都有所表现,而在各个培养阶段又各有不同的作用效果。

(作者:周嫦、杨弘远。原载:武汉大学学报(自然科学版),1979(2):86~95。图6幅,表8幅,参考文献11篇,仅选录图1。)

3 水稻花粉两条发育途径的实验研究

Experimental researches on the two pathways of pollen development in *Oryza Sativa* L.

Abstract

The present paper deals with the experimental researches on the gametophytic and sporophytic pathways of pollen development in *Oryza sativa* L. Subsp. *Keng*, Cultivar Jinghong No. 2. Three methods of culture were used:(1) The lemma, palea and pistil of excised spikelets were removed and the pedicel was inserted vertically into the medium with the intact stamens standing freely above the medium surface (vertical culture). (2) The spikelets were manipulated similarly but placed horizontally on the medium so that their anthers were directly contacted with the latter (horizontal culture). (3) The anthers were excised and inoculated separately (anther culture). In all cases the pollen stage at inoculation was in late uninucleate. N_6 basic medium supplemented with or without MCPA (2ppm) was used. After inoculation the samples were collected periodically for cytological observation.

In all cases the pollen passed a short stage of gametophytic development, forming a vegetative and a generative cell, then various pathways commenced in different cultures. In 'vertical culture', most of the pollen went on along the gametophytic pathway up to normal 3-celled stage, but some showed anomalous divisions of vegetative or/and generative nuclei, indicating an initiation of sporophytic development. In 'horizontal culture', the sporophytic development went on further, producing some calli, though the main pollen population remained as gametophyte. In anther culture, the gametophytic pathway to a mature 3-celled pollen was blocked, the unique path-way being sporophytic. In rice, the pollen developed along sporophytic pathway mainly via A route.

These comparative investigations indicate that there are two chief factors concerning the switch of pollen development from one pathway to another: first, to be freed from the *in vivo* restrictions, which, as suggested by Sunderland and as supported by the results of vertical culture in our experiments, is sufficient to trigger the first sporophytic division, and second, direct contact with the medium, which is necessary to support the successive growth of multicellular grains and calli. As to the exogenous hormone, rather than functioning as an agent triggering sporophytic development, it plays an important role in increasing eventual induction frequency, growth rates and differentiating ability of calli.

花粉具有配子体发育(形成雄配子体,产生雄配子)和孢子体发育(通过胚状体或愈伤组织形成花粉植株)两条发育途径。迄今花药培养中的细胞学工作,大多集中于研究孢子体发育中的雄核发育途径,而较少着眼于用实验方法进行两条发育途径的对比研究。本文在后一方面作一些初步探讨。

材 料 与 方 法

实验材料是粳稻品种"景洪2号"。设计了三种培养方式:(1)剪下小穗,在解剖镜下去掉内、外颖与雌蕊,将小穗柄插入培养基,花药不接触培养基（简称直插培养）。(2)将经过同样手术处理的小穗平置于培养基上，使花药紧贴培养基（简称平贴培养）。(3)一般的花药培养。这三种培养方式的特点是：直插培养初步摆脱了整体的制约性，但保留了雄蕊的完整性及其自然营养通道。平贴培养则进一步强迫花药通过药壁从培养基吸取养料。花药培养更进一步，既切断了花药与小穗的联系，又强迫它通过药壁吸收养料。

解剖、接种与培养均在无菌条件下进行。实验材料均采自新鲜稻穗（个别实验用低温处理，文中注明）。接种花药一律为单核靠边期。采用 N_6 基本培养基，加5%蔗糖，不补加外源激素或补加2mg/L 2-甲基-4-氯苯氧乙酸(MCPA,抚顺农药厂产品，纯度99%以上)。后者是一种性质与2,4-D相近、诱导力很强的生长素[3]。培养在25℃左右与黑暗条件下。

接种后定期取样，用醋酸甲醇（1:3）固定，70%甲醇保存。爱氏苏木精整体染色[5]，经脱水、透明后整体封藏或制成石蜡切片（厚20μm）。用此法染色，营养核与生殖核容易区分，生殖细胞膜常清晰可见（图版Ⅰ，1、7、13、19）。

结　　果

一、三种培养方式的诱导效果

用新鲜的、经过接种前低温处理(7℃,5d)或接种后低温处理[7](7℃,5d)的材料进行上述三种方式的培养实验，接种40d后统计，肉眼可见愈伤组织的花药数占总花药数的百分率（诱导频率）（表1）。直插培养在一切场合均不诱导愈伤组织，只是当个别花药下垂接触培养基时偶尔可以产生愈伤组织。平贴培养能诱导愈伤组织，但频率远低于花药培养。低温处理使平贴培养与花药培养诱导频率相应提高。

表 1 三种培养方式诱导频率的比较*

Table 1 Comparison of induction frequency in three cultural methods

培养方式 Culture methods	不预处理 Without pretreatment			低温前处理 Cold-pretreatment before inoculation			低温后处理 Cold-pretreatment after inoculation		
	培养花药数 No. of anthers cultured	诱导花药数 No. of anthers induced	诱导频率(%) Induction frequency (%)	培养花药数 No. of anthers cultured	诱导花药数 No. of anthers induced	诱导频率(%) Induction frequency (%)	培养花药数 No. of anthers cultured	诱导花药数 No. of anthers induced	诱导频率(%) Induction frequency (%)
直插培养 Vertical culture	163	0	0	369	0	0	200	0	0
平贴培养 Horizontal culture	184	11	6.0	728	68	9.3	230	48	20.9
花药培养 Anther culture	228	55	24.1	495	169	34.1	205	84	40.7

* 全部实验都是用有激素的培养基进行的。

All experiments were carried out on hormone-containing medium.

接种后 5~7d,直插培养的花药颜色变黄。用碘液染色镜检,其中花粉大多富含淀粉。若将此种花粉接于新鲜柱头,用乳酚棉蓝液染色镜检,看到有些萌发成花粉管(图版Ⅰ,9)。平贴培养亦含类似成熟花粉。花药培养则否。以上实验结果促使我们进一步作了以下的细胞学观察。

二、直插培养时的花粉发育

图1上表示直插培养的结果。接种后2d,多数花药中已形成由营养细胞与生殖细胞组成的二胞花粉(图版Ⅰ,1)。第4d以后,特别是

第 6～8d,营养细胞的胞质变浓,生殖细胞分裂为一对开始呈圆形以后变长的精子(图版Ⅰ,2、3、4)组成三胞花粉。此后花丝萎蔫,花药干枯,花粉衰死。

花粉发育是不同步的,当多数花粉达到成熟配子体阶段时,少数尚处于液泡化状态。看来花粉二型现象(dimorphism)[14,15]在这里得到了加强。正是在6～10d的液泡化花粉中,有的显示营养核或生殖核的异常分裂。据145粒这类花粉的统计,其中生殖核异常分裂的占65%(图版Ⅰ,5、6),营养核异常分裂的占20%(图版Ⅰ,7),其余15%未能判明来源(图版Ⅰ,8)。在直插培养中出现的这类异常分裂,可以认为是孢子体发育的起始步骤。

三、平贴培养时的花粉发育

如图1中排曲线所示,平贴培养时同样出现和体内相似的配子体发育途径,直至形成成熟的三胞花粉。所不同的,是平贴培养时花药败育严重。这是因为花药不仅不能由输导系统获取养分,反而被迫将药壁吸进的部分养料供给行将枯萎的小穗柄组织所致。平贴培养和直插培养显著不同的另一个特点是:不仅出现类似的异常分裂,而且能进一步形成多细胞花粉粒与愈伤组织(图版Ⅰ,10)。

四、花药培养时的花粉发育

花药培养的结果见图1下排曲线。接种后1～2d,多数花粉通过第一次有丝分裂,形成具典型营养细胞与生殖细胞结构的二胞花粉(图版Ⅰ,11～13)。少数花粉的分裂轴向发生改变,形成两个均等的核(图版Ⅰ,14)。此后,配子体发育途径中止。从第5d起,少数花粉中显示孢子体发育途径。多数情况是营养核分裂,同时伴随生殖细胞的退化(图版Ⅰ,15～18)。如果生殖细胞退化较早,观察不仔细时容易把营养核误认为小孢子核,而把由它产生的子核误认为小孢子均等分裂的子核。Sunderland 等[16]也曾提醒注意不要把 A 途径误判为 B 途径。除营养核分裂外,还观察到生殖核分裂而营养核退化(图版Ⅰ,19、20)或二者都分裂的情况。

关于水稻花药培养时的雄核发育途径，前人报道中有的强调 B 途径[6,8,10]，有的强调 A 途径[9,11]，有的指出 A、B 途径都存在而以前者为主[13]。在我们的材料中，A 途径有充足的证据而 B 途径则尚不能确凿证实。

我们还观察到，水稻多细胞花粉的形成有几种方式：一种情况是最初几次分裂后即形成胞壁将花粉分隔成若干细胞（图版 I，21）。但更常见的方式是在早期分裂时花粉粒中仍然保持一个大液泡；分裂过程通常由花粉一侧开始，向另一侧推进，一边分裂一边形成细胞，直至将液泡的空间逐渐填满（图版 I，22～26）。有时，在液泡另一端还有一至几个子核与主要细胞群遥相对应，暗示它们可能有不同的起源。

五、总的比较

三种培养方式下的花粉发育途径有同有异。相同的是：多数花粉在培养初期都能通过第一次分裂形成营养细胞与生殖细胞，即进行一段配子体发育。以后则显示发育的分歧。直插培养与平贴培养时多数花粉继续沿配子体发育途径前进，形成成熟的雄配子体；少数花粉发生异常分裂，显示向孢子体发育途径的偏移。其中平贴培养时孢子体发育能继续进行下去，形成愈伤组织。花药培养时配子体发育途径在第二次有丝分裂前被阻断，只有少数花粉的孢子体发育成为唯一的发育途径。三种培养方式表现出由配子体发育途径向孢子体发育途径偏移的三个级别。

讨　论

1. Sunderland 等[16]认为花粉由配子体发育途径向孢子体发育途径的转变是由于解除了整体对花药的制约，强调仅仅"离体"这一因素即可启动花粉的胚胎发生，而"培养"的作用不过在于促进与维持胚胎发生花粉的继续生长[15]。上述判断主要是根据烟草低温预处理等实验的结果作出的。本文实验结果支持与补充了这一观点。实验中的三种

第一章 雄性器官与细胞的实验研究

图版 I   plate I

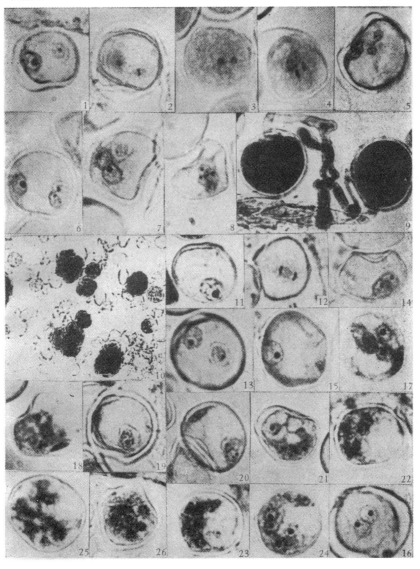

培养方式，"离体"的程度（即摆脱整体制约的程度）是不同的，在发育途径上显示重大的差别。直插与平贴培养的离体程度较轻，多数花粉能顺利地实现配子体发育。花药培养的"离体"程度较重，只能进行一段配子体发育。但就孢子体发育途径来看，只要有了前一种"离体"程度，即足以启动花粉的孢子体发育，而无需花药的彻底分离。实际上，即使在非离体条件下也可以出现作为孢子体发育途径起始标志的异常分裂现象[15]。离体显然加剧了这一倾向。

2. 如果说本文直插培养实验表明仅仅"离体"即可启动孢子体发育，则平贴培养实验表明孢子体发育的继续进行还需"培养"这一条件。Sunderland[16]强调花药直接接触培养基的重要性，指出花丝不能传递对于（胚胎发生）生长的刺激。我们推测这是由于胚状体或愈伤组织的生长要求有充足的水分、养料供应和适当的渗透压值等条件，而短命的花丝输导组织不能承担这些任务。设想今后可以建立一个包括两个阶段的实验系统（第一阶段是直插培养，第二阶段是花药培养），来验证孢子体发育的启动与继续维持是先后两个过程的基本观点，并以此作为进一步开展有关细胞生物学研究的基础。

3. 外源激素在花药培养中的作用也是一个重要的理论问题，我们有另文对此进行专门讨论[4]，这里只作一点补充。外源激素可以显著提高诱导频率是众所周知的事实，也为严格的对比实验所证明[4]。然而在水稻、小麦、天仙子等的实验中得出这样的看法：外源激素对孢子体发育途径的启动并非必需的因素，其作用主要是促进愈伤组织生长与防止败育[1,2,4,12]。本实验的三种培养方式都作了有激素与无激素的对比观察（图 1 左、右两行），结果表明二者在发育途径的频率与速率方面没有显著差异，这对上述观点是一个进一步的证明。很可能，在我们所采用的品种中，内源激素的水平已足够启动花粉的初始孢子体发育，因而无需提供外源激素。但对愈伤组织的生长来说，外源激素则是十分重要的。

图　版　I

除图 9、10 外，均为爱氏苏木精整体染色，石蜡切片。各图放大倍数均为×

800,图10为×300。

1~9. 直插培养。1. 二胞花粉,示营养核(左)与生殖细胞(右)。2. 生殖细胞分裂中期(右)。3. 刚由生殖细胞分裂而成的一对圆形精子。4. 三胞花粉,含一对长形精子。5、6. 生殖核异常分裂而成的一对子核(右)。7. 营养核异常分裂而成的一对子核(左)。8. 异常分裂所产生的四个子核。9. 直插培养6d的花粉人工授于新鲜柱头,示萌发成花粉管。10. 平贴培养16d,示多细胞花粉与愈伤组织。11~26. 花药培养。11. 小孢子分裂前期,细胞核在萌发孔对侧。12. 小孢子分裂后期。13. 二胞花粉,示营养核(左)与生殖细胞(右)。14. 小孢子核均等分裂产生的两个近似的核。15. 二胞花粉,生殖细胞退化(右下),营养核在萌发孔近侧。16. 营养核分裂成一对子核(萌发孔近侧),生殖细胞退化(右下)。17. 营养核分裂成三个子核,生殖细胞退化(右下)。18. 多细胞花粉粒,示退化的生殖细胞(右下)。19. 营养核退化(左),生殖核异常分裂前期(下)。20. 生殖核异常分裂成一对子核(右下),营养核退化(左上)。21. 多细胞花粉粒,其内部空间很早被细胞填满。22~26. 多细胞花粉粒,示细胞由花粉一侧向另一侧增殖,液泡逐渐缩小,最后花粉内腔被细胞填满。

(作者:杨弘远、周嫦。原载:植物学报,1979,12(4):345~351。图版1幅,图1幅,表1幅,参考文献16篇,仅选录表1、图版Ⅰ)

4 水稻幼花直插培养诱导花粉孢子体发育的实验

Induction of sporophytic development in rice pollen by flower cutting

在离体培养条件下,花粉可以由正常的配子体发育途径转向孢子体发育途径。所谓离体培养,其实包含"离体"与"培养"两种因素的作用,它们在花粉发育途径的转变中各起何种作用,是一个值得研究的问题。由于在一般花药培养中此二者互相交织,难以区分,因此有必要设计新的实验。本文在前一篇报告[1]的基础上,通过一套专门的实验对这一问题作了进一步的探讨。

材 料 与 方 法

粳稻品种"景洪2号",在无菌条件下剪下单核靠边花粉期的幼花,去掉内、外颖,保留雌、雄蕊(雌蕊的去留对实验结果影响不大),将花柄垂直地插入固体 N_6 培养基(不加外源激素,含5%蔗糖),注意不使花药接触培养基("幼花直插培养"或简称"插花")。同时,在成分相同的培养基上进行一般的花药培养作为对照。全部接种工作在同一天进行。接种后2d、4d、6d、8d、10d、12d、14d 分别取花药再接种到加入2ppm 2甲基-4-氯苯氧乙酸(MCPA)与5%蔗糖的 N_6 固体培养基上("后继花药培养")。培养于25℃左右的黑暗条件下。插花后每隔1d与后继花药培养后4d、8d、12d 分别取花药以醋酸甲醇(1∶3)固定,保存于70%甲醇(4℃)中以备观察,另留一部分材料最后统计愈伤组织诱导频率。细胞学研究仍用前文的爱氏苏木精整体染色程序[1],共计观察32个处理,每处理约25～40粒花药。全部实验设计与结果见图1。

结 果

一、幼花直插培养

在插花期间,除了沿正常发育途径直至形成成熟雄配子体的花粉外,有一部分花粉能够启动和一般花药培养一样的第一次孢子体分裂(图版Ⅰ,2～6),从而转向孢子体发育途径。插花后4d已有8.3%的花药中出现第一次孢子体分裂后的花粉,这一频率以后逐渐提高,至第10d时达到26.5%(图1)。

已完成第一次孢子体分裂的花粉,随后进行连续的分裂,形成多细胞或多核花粉(图版Ⅰ,7～9)。插花后8d,有3%的花药中出现多细胞花粉;其频率迅速提高,至第12d时达54.3%。如果我们把多细胞

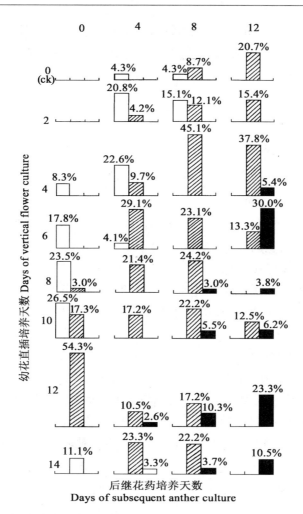

图1 水稻花粉在幼花直插培养与后继花药培养时的孢子体发育（以发育领先的花粉作为判断花药发育的标准）

Fig. 1 Sporophytic development of rice pollen in vertical flower culture and subsequent anther culture (Using the development ally more advanced pollen grains as criterion of anther development)

□第一次孢子体分裂的花粉
　　Pollen passed the first Sporophytic division
▨多细胞或多核花粉
　　Multicellular or multinucleare pollem
■幼小时愈伤组织 Young calli

花粉看成类似原胚的构造,则可认为插花后 4~12d 期间相当于花粉的原胚发育阶段。幼花直插培养所诱导的花粉原胚发育进度与一般花药培养(图1)基本一致,其诱导频率甚至高于后者。不同的是,在插花条件下,后期花丝萎蔫,花药干枯,致使多细胞花粉不能继续生长而形成愈伤组织。

二、后继花药培养

插花后定期将花药取下作后继花药培养,结果如图1各横行所示。很明显,在插花期间已启动了第一次孢子体分裂或已进行了一段原胚发育的花粉,可以在后继花药培养阶段继续其原有的发育过程,并可形成显微镜下可见的愈伤组织(图版Ⅰ,12~14)。而且,插花阶段愈长,后继花药培养时愈伤组织出现愈早。例如,插花4d需要后继花药培养12d才形成愈伤组织;而插花8d只需后继花药培养8d、插花12d只需后继花药培养4d即可形成愈伤组织。其总时间均为16d左右。由此可见,花粉在后继花药培养期间的孢子体发育无疑是其在幼花直插培养期间的发育的延续。

在插花后 50d 统计肉眼可见愈伤组织的花药诱导频率,证明幼花直插培养加后继花药培养确实可以诱导出有活力的愈伤组织。例如,6d 幼花直插培养加后继花药培养的处理,诱导频率为 32.2%,与一般花药培养对照(39.2%)相差不远。12d 幼花直插培养加后继花药培养的诱导频率(15.9%)则显著降低,这是由于插花时间过长引起花粉原胚及由之产生的愈伤组织生活机能衰退所致。

讨 论

花粉由配子体发育途径转向孢子体发育途径要经历一个诱导花粉去分化的过程。Sunderland 学派认为这一诱导主要是由于花药摆脱了整体的制约[6],仅仅"离体"本身即足以诱发这一变化,而"培养"的作用仅在于促进和维持预定单位的生长[7],因此主张把研究的重点由培养基的研制移到培养前的时期上来[3]。这一系列论点的提出主要依据体内花粉二型性的观察与低温预处理实验。看来还需要有更多的实验来为上述观点提供更充分的细胞学方面的证据。

前一篇报告已经肯定水稻花粉在幼花直插培养条件下可以通过第一次孢子体分裂[1]。本文证实在这种条件下还可以进一步完成原胚发育。我们在进行本实验的同时曾观察过大量"景洪2号"品种的成熟花粉,并没有发现多细胞花粉。因此在幼花直插条件下的花粉孢子体发育,显然是"离体"的效应。我们的实验在这一点上支持了Sunderland的观点。

幼花直插培养的突出特点是花药不接触培养基,花粉仅仅通过花柄、花丝中的输导系统获取营养。而且在本实验中培养基不含外源激素。这种条件能够诱导花粉的原胚发育在文献中尚未见报道。过去一直认为花药直接接触培养基是花粉胚胎发生的必要条件。Sunderland等强调指出:"接种时必须使花药直接接触培养基,如果花丝把花药升举到离开培养基,则花粉不能生长。显然,对生长的刺激作用不能通过花丝传给花药[6]。"国内的有关著作也引用了这一看法[2]。本实验结果所表明的实际情况并非如此。因此上述看法需要重新考虑。

(作者:杨弘远、周嫦。原载:植物学报,1981,23(5):414~415。图1幅,图版1幅,参考文献7篇。仅选录图1。)

5 大麦的花药培养与雄核发育

Anther culture and androgenesis of *Hordeum vulgare* L.

摘 要

本实验初步研究了影响大麦花药培养的各种因素,比较了不同类型品种诱导频率的变化规律,并对花粉雄核发育的途径与类型进行了细胞学观察。最后,作者就大麦花药培养与雄核发育的某些特点与问题提出了见解。

Abstract

This paper reports our preliminary experimental results of barley anther culture. The results so far obtained suggested that N_6 (or MS) medium supplemented with 12% sucrose, 1 ppm IAA, 1ppm BAP and 0.5 ppm MCPA (or 2,4-D) is suitable for barley anther culture. Hormone-free culture allows callus and embryoid production. However, exogenous hormones have a significant beneficial effect.

Anthers of 54 cultivars were cultured and their induction frequency was compared. In most of cultivars the value was low, but several highly responsive cultivars were found out. In general, the more responsive cultivars belong to 2-rowed types and the 4- or 6-rowed types are less responsive (Table 1). Microscopical observations made on 36 cultivars revealed that percentage of anthers containing multinucleate and/or multicellular grains are higher than the final induction-frequency, indicating the potentiality of callus production in most of cultivars (Table 2).

There are three pathways of pollen androgenesis in barley, i. e. division of vegetative nucleus (Plate Ⅰ,2-8), division of vegetative and generative nucleus (Plate Ⅰ,9-12) and equal division of the microspore (Plate Ⅰ,13-18). Each pathway involves free-nucleate and cellular types. There are two patterns in cellular types of development: (1) cells remain meristematic for a long time; (2) cells vacuolate early. The free-nucleate type and the "vacuolate" cellular patterns, in comparison with the "meristematic" cellular pattern, generally fail in their further development. The developmental rate and the ploidy status in anther culture of barley were also mentioned.

The problem of relatively low induction-frequency in most of cultivars is discussed. Question is focussed on the cause of abortion of multinucleate and multicellular grains. It is believed that, the initial developmental pattern may play an important role in determining their future fate.

(作者:周嫦、杨弘远。原载:植物学报,1980,22(3):211~215。图版1幅,表2幅,参考文献20篇,均删去。)

第二节　花粉萌发与花粉管生长

提要

在本节中,我们首先介绍两项关于花粉与花粉管细胞学观察的研究结果。一项是用荧光素二醋酸酯(FDA)标记生活花粉管的方法。FDA是广泛用于鉴定细胞生活力的荧光染料。我们则试图以之作为追踪花粉管生长的一种荧光标记。实验设计是先以FDA染花粉,使之呈现荧光,然后洗脱花粉的多余染液,任其在无染料的培养基中萌发花粉管。结果表明由此产生的花粉管在生长期间始终呈现荧光,直至死亡。由此推测此法有可能用于离体授粉或其它细胞融合实验,以追踪花粉管或融合细胞的行为。

另一项方法是用DNA荧光染料H33258染花粉,再进行冬青油透明,以显示花粉细胞核。由于花粉壁的遮掩,常规染色法较难显示花粉中的细胞核。花粉壁的自发荧光也严重干扰单独使用H33258、DAPI等荧光染色法的效果。我们所提出的荧光染色结合冬青油透明的技术则达到了预期目的,减弱甚至消除了花粉壁荧光,凸现了内部细胞核荧光。甚至像向日葵那样外壁很厚且具刺状突起的花粉,此法亦可清晰显示其细长弯曲的精核(参看书末的彩色照片)。

花粉管中生殖细胞的行为及其与营养核的关系曾是研究"雄性生殖单位"时所关心的问题。我们用DAPI荧光染色法研究了玉帘生活花粉管中生殖核分裂时营养核的形态变化及二者的相互关系,重点研究了低温预处理诱导生殖核无丝分裂与异常有丝分裂的现象。无丝分裂在细胞发育中的地位与作用历来是一个有争议的问题,而花粉管中的无丝分裂现象则为我们首次报道。

芸苔属花粉人工萌发具有独特的要求,它不像一般花粉要求偏弱酸性的培养基,而要求偏碱性的条件。此外,前人试验发现用聚乙二醇(PEG)代替蔗糖培养芸苔属花粉有较好效果。我们对此进行了重复实验,提出了一种改良的培养基,同时与化学系研究者合作,以微电极测定花粉与柱头表面的 pH 值,试图解释芸苔属花粉在高 pH 值条件下人工萌发的机理。这一改良的配方在我们以后芸苔属脱外壁花粉的实验中发挥了重要作用(参看本章第四节)。

本节中的另一组系列论文是关于钙及其相关因素对烟草花粉萌发与花粉管生长的影响。所涉及的研究方法包括:在花粉培养基中分别加入 Ca^{2+}、Ca^{2+} 载体、Ca^{2+} 阻滞剂、外源钙调素等进行药理学实验。尽管这些多属重复前人的工作,但仍具有自身的特色:一是我们不仅着眼于培养的最终结果,而且着眼于培养不同阶段的动态影响,并对观测结果进行了统计学处理。另一特色是我们不仅着眼于花粉萌发与花粉管生长,而且着重研究了对生殖细胞核分裂的影响,这是前人未曾注意的方面。

1 Fluorescein diacetate used as a vital stain for labeling living pollen tubes

In recent years methods of using fluorescent dyes for vital staining of protoplasts have provided an interesting new means in cell engineering studies[1-5]. In these experiments the fluorescent dyes, mainly fluorescein isothiocyanate and rhodamine isothiocyanate, were used as labels for identifying heterokaryons. Strangely, we have not yet seen similar reports on the pollen tube, which is the vehicle of male gametes in higher plants and plays an important role in sexual processes. Leuco-aniline blue, a fluorescent dye routinely employed in demonstrating pollen tubes, stains only the tube wall (for a review see Ref. 6). For vital staining, it is more important to mark the tube protoplasm. For this purpose, we have tested fluorescein isothiocyanate, rhodamine B and fluorescein diacetate (FDA). The former 2 dyes, though also showing positive effects, had some drawbacks (unpublished data). The latter resulted in good and interesting effects.

Materials and methods

Experiments were done in 1985 with snapdragon (*Antirrhinum majus*) and tobacco (*Nicotiana tabacum*) as plant materials. Vital staining was performed on artificially germinated pollen. The liquid medium used was according to Brewbaker and Kwack's formula[7] with a few modifications: H_2BO_3(100 mg/L), $CaCl_2 \cdot 2H_2O$ (200 mg/L), $MgSO_2$(200 mg/L), KNO_3(100 mg/L) and sucrose (100g/L). For fluorescence staining, FDA (Sigma) stock solution (5 mg/ml in acetone) was diluted with the medium before use to make $10\mu g/ml$ solution.

A drop of the staining solution was placed in the center of a square cover slip. With a stainless steel needle, fresh pollen was stirred into the medium. The cover slip was immediately put upside down on a glass ring of microchamber; the ring edge was smeared with tetracycline ointment beforehand. After 40 min, the cover slip was taken off and reversed to previous state. With a filter paper strip, the stain solution was absorbed away. Then a drop of the same medium, but without FDA, was put to the same place. This procedure was repeated three times to wash off the excess dye. Care has been taken to avoid removing too much pollen. The hanging drop was then remade on the microchamber and incubated at 28-30℃ for incubation.

For observation and photomicrography, an AO FLUORESTAR 110 microscope with vertical fluorescence illumination equipment and AO 2073 filter combination (exciter filter 436/dichroic mirror 450/barrier filter 475) was used.

Results

In our preliminary experiments with snapdragon we chose a stronger solution (100 μg/ml) to stain the pollen, because this concentration has widely been adopted for viability test since Widholm's work[8]. However, it caused some inhibition of subsequent pollen germination and growth. Experiments on tobacco pollen showed a similar result. Therefore the concentration was lowered to 10μg/ml. Normal germination and growth of the pollen in both species was obtained repeatedly with this weaker staining solution. Meanwhile the pollen tubes maintained good fluorescence after staining of the pollen. As a representative example, a series of photographs taken at intervals during one experiment on tobacco pollen tubes are presented.

It is well known that viable pollen grains treated with FDA show bright green fluorescence, a fluorochromatic reaction caused by intracellular esterase which converts FDA to fluorescein. The fluorescence becomes stronger within 40 min[9]. Taking account of this, in our experiments pollen was treated in FDA solution (10 μg/ml) for 40 min, and then washed with FDA-free medium. When viewed under a fluorescence microscope, all the pollen grains fluoresced brightly (Fig. 1). During subsequent incubation in FDA-free medium, pollen germination took place normally. The pollen grains as well as the tubes all fluoresced strongly (Fig. 2). The pollen tubes kept vigorous growth, elongating significantly and twining with each other. Fluorescence of the pollen grains weakened and vanished gradually as the protoplasm was transferred into the tubes; meanwhile the fluorescence of the tubes maintained, especially at their front part where the protoplasm was centering (Figs. 3 and 4). The pollen tubes showed rapid cytoplasmic streaming when viewed by brightfield illumination, coinciding with the fluorochromasia (Figs. 5-7). This means that they grew quite normally at least

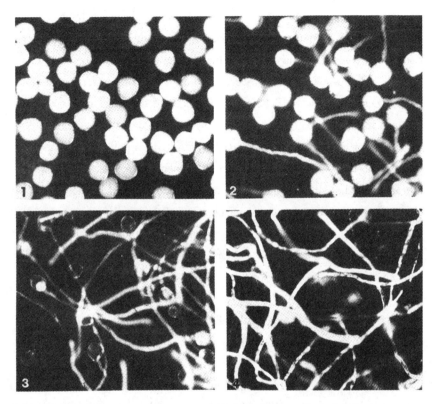

Fig. 1. Pollen grains soon after staining with FDA (10μg/ml) for 40 min and washing with FDA free medium, showing bright fluorescence (×200).

Fig. 2. After 40 min of FDA treatment and 3 h of subsequent incubation in FDA-free medium, normal germination occurred. Both the pollen grains and the tubes fluoresced (×200).

Fig. 3. The pollen tubes elongated significantly and showed fluorescence after 6h of subsequent incubation, meanwhile the pollen grains began losing their fluorescence (×150).

Fig. 4. After 9 h of incubation, most of the pollen grains did not fluoresce. The tubes kept fluorescing, especially at their front part (×150).

up to 9h after FDA treatment. Indeed we have observed similar behavior when the preparations had stayed overnight. After 24h of incubation, however, most of the pollen tubes lost their fluorescence with only a few exceptions (Fig. 8). It should be noted that these tubes that were fluorescing still showed slow cytoplasmic streaming, while the non-fluorescing tubes had already ceased in growth. Besides, in the case of bursting, the discharged contents from the tube tips always failed in fluorescing.

In order to know whether or not the FDA stained pollen can be used in pollination studies, we have done two experiments with tobacco. One was to put an excised stigma by the side of the incubated pollen tubes. The tubes grew towards the stigma, showing a chemotropic effect. However, reaching a certain distance, the tubes fell into a bright halo around the stigma tissue and lost the contrast of their own fluorescence due to the interference of the halo. In another experiment we have attempted to use FDA stained pollen for *in vitro* pollination. A piece of placenta with attached ovules was placed by the side of the pollen tubes. After 8-9h, many tubes were observed growing on the surface of the placenta, but entry of pollen tube into the micropyle has not been confirmed yet. From these preliminary observations we can conclude that FDA-treated pollen retains the ability of chemotropic growth, though the possibility of fertilization with such pollen should be clarified by further experiments.

Discussion

Since the first discovery of fluorochromasia phenomenon in mammalian cells[10], FDA staining has been used as an effective and convenient method for testing cell viability. In plant cell biology, it has been found widely applicable to pollen[9], cultured cells[8], microspore tetrads[11], stigma surface[12], protoplasts[13], and more recently, enzymatically isolated embryo sacs[14]. A late appraisal of the fluorochromatic test on pollen viability is given by Heslop-Harrison et al.[15]. However, in all these and other reports, cells treated with FDA were discarded after microscopic examination. It means that this technique is so far limited, in the exact sense, to a supravital rather than a vital staining. Recently, in a study of plant-bacterial interactions by Millmann and Lurquin, the FDA-labeled spheroplasts of *Escherichia coli*, *Agrobacterium tumefaciens* and *A. radiobacter* were bound to plant protoplasts and examined with flow cytometry[16], thus widening the usage of this technique.

In the present paper, we use FDA as a vital stain for pollen tubes. A short treatment at the pre-germination period can keep fluorescence label at the post-germination period for a rather long time. The fluorescence label maintains so long as the pollen tubes were living. We cannot yet explain the precise mechanism of this phenomenon. Probably the fluorescein converted from FDA by pollen esterase is so abundant that it moves from the pollen grain into the tube. Alternatively, an excess of FDA unexhausted at the pre-germination period moves into the tube and is converted there into fluorescein. Another possibility that cannot be excluded is that a trace of FDA may remain in the incubation medium as a result of incomplete washing after staining. Anyway, that FDA can be used as a vital stain is beyong doubt. It may provide a new means for studying the behavior of living pollen tubes and probably the *in vitro* pollination-fertilization process. Moreover, this report may also arouse the interest of using similar method for vital staining of other living cells of higher plants such as cultured cells and protoplasts.

(Author: Yang HY. Published in: Plant Science, 1986, 44:59-63. with 8 figures and 16 references. Only Figs. 1-4 are retained here.)

2 用荧光染色与冬青油透明技术显示花粉细胞核

A fluorescence staining-methyl salicylate clearing technique for demonstrating pollen nuclei

Abstract

Recently several DNA-binding fluorochromes have been used for demonstrating pollen nuclei. However, the autofluorescence of pollen wall often obscured the fluorescence of nuclei, thus limited the use of this method. Methyl salicylate (MS) as a clearing agent has shown excellent effect for observing embryo sac in whole-mounted ovules. This aroused me to try a combination of fluorescent staining with MS clearing in order to make a better demonstration of the pollen nuclei.

Mature 2-celled or 3-celled pollen of several angiosperm species stained with Hoechst 33258 (H33258) and cleared (via ethanol dehydration) with MS showed clear-cut fluorescence of their generative or sperm nuclei and vegetative nucleus. MS greatly decreased the wall fluorescence and increased the transparency of the pollen contents, meanwhile maintained the H33258 stained fluorescence, consequently made the nuclei brighter under a darkened background. For example, in sunflower pollen a pair of elongated and winding sperm nuclei which could not be identified after simple H33258 staining were quite visible after MS clearing. In artificially germinated pollen tubes, the locomotion of nuclei from pollen grain into the tube, the sequence of generative and vegetative nuclei travelling along the tube and the division of generative nucleus into two sperm nuclei could be well followed by this method.

The present technique may be adoptable for observations on the processes of microsporogenesis and male gametophyte development, androgenesis in cultured anthers, and also possibly for tracing the nuclear events during pollination-fertilization.

研究花粉和花粉管中细胞核的状态与行为,是当前花粉生物学所关心的问题之一。在这方面,除了借助复杂的电镜观察与计算机辅助三维重建技术外,现代光镜技术也可以发挥一定的作用。例如 Heslop-Harrison 等(1986)用"光学切片"结合图像分析的方法研究了 *Helleborus foetidus* 花粉与花粉管中营养核和生殖细胞的相互关系[9]。这就对花粉的光镜观察技术提出了新的、更高的要求。由于常规染色技术对于观察花粉细胞核只能起到有限的作用,近年不少研究者开始把眼光转向荧光显微术,应用与 DNA 结合的荧光染料(如 mithramycin、DAPI、ethidium bromide、Hoechst33258 等)显示花粉与花粉管中的细胞核[1,7,10]。但是花粉壁的强烈自发荧光常常干扰细胞核的观察,在厚壁花粉中这种干扰尤为严重。因而提出了一个如何减少花粉壁自发荧光,增加花粉透明度,从而提高细胞核观察效果的问题。作者在用苏木精染色、冬青油透明技术观察整体胚珠或子房中的胚囊取得成功以后[2],设想将荧光染色与冬青油透明结合起来可能是解决上述困难的途径,故而开展了本试验。

材 料 与 方 法

一、供试植物

石蒜(*Lycoris radiata*)、蚕豆(*Vicia faba*)、虞美人(*Papaver rhoeas*)、玉米(*Zea mays*)、紫菜苔(*Brassica campestris* var. *purpurea*)、向日葵(*Helianthus annuus*)等。

二、花粉粒的制样方法

成熟花药用Carnoy(3∶1)液固定,70%乙醇冷藏。临用前捣碎花药,将散出的花粉逐步下行至蒸馏水并彻底换洗。用一定pH值的缓冲液(见后文)预处理数小时。将事先配制存于冰箱中的H33258贮备液(1mg/ml水溶液)用相应的缓冲液稀释为一定浓度的染液(见后文)。花粉在此染液中于25℃、黑暗条件下染色14~24h。一部分花粉在染液中封藏或加入等量甘油封藏,荧光镜检。一部分花粉经50%、70%、95%、100%乙醇逐级脱水,无水乙醇与冬青油(水杨酸甲酯)等量混合液过渡,冬青油透明14h以上。

三、花粉管的制样方法

将新鲜的成熟花粉接种在盛有浅层培养液的小培养皿中,在25℃、黑暗条件下人工萌发。供试的蚕豆、虞美人、紫菜苔三种植物的花粉,分别采用不同的培养基配方(见后文)。在所有培养基中均含有H33258 20μg/ml。培养期间定期取样活检花粉管生长及染色情况,同时定期取样用Carnoy(3∶1)液固定约1h,然后经逐级乙醇脱水、无水乙醇与冬青油等量混合液过渡、冬青油透明14h以上。

四、荧光镜检

由冬青油中吸取样品滴于载玻片上,加盖玻片,置Olympus NEW VANOX AHBS-514显微镜下,用落射荧光装置的U或V激发滤光组合观察与摄影。

结　　果

一、花粉粒的观察

染色与透明方法的研究　(1)染液浓度:Laloue等(1980)用1μg/ml H33258染Carnoy液固定的植物体细胞,能清楚地显示细胞核与染色体[12]。杨弘远、周嫦(1985)在观察由固定胚珠人工分离的胚囊时也使用同样的染液浓度[4]。Hough 等(1985)则用 20μg/ml H33258 染固定与生活的花粉[10]。在本试验中,比较了 5μg/ml、10μg/ml、20μg/ml 三种浓度,结果表明就多数供试植物的花粉而言,以 20μg/ml H33258 的染色效果为佳;较淡的染液只适用于壁薄的花粉。看来,花粉比体细胞或胚囊要求较高浓度的染液。

(2)缓冲液的种类与 pH 值:Laloue 等(1980)采用 pH5 的柠檬酸-磷酸氢二钠缓冲液作预处理和配制 H33258 染液[12]。Hough 等(1985)则用蒸馏水配制染液[10]。我们试验了柠檬酸-磷酸氢二钠(pH5、pH7.2)、Tris-HCl(pH7.2)三种情况,结果表明以柠檬酸-磷酸氢二钠(pH5)缓冲液效果最稳定。

(3)透明的方法:用染液直接封藏花粉荧光镜检,通常很难透过强烈的花粉壁自发荧光看清内部的细胞核。加入等量甘油透明后,效果有所改进。而冬青油透明则显著提高了观察的效果。透明的时间延长,效果更好。花粉在冬青油中保存于黑暗、低温条件下,细胞核的荧光更为突出。

根据上述研究,确定用 Carnoy 液固定的花粉的染色、透明基本程序为:柠檬酸-磷酸氢二钠缓冲液(0.1mol/L,pH5)预处理→20μg/ml H33258(溶于上述缓冲液)染色→乙醇脱水→冬青油透明。

观察效果 供试的 3 种二细胞型花粉与 3 种三细胞型花粉经 H33258 染色、冬青油透明后,荧光镜检效果良好。

图版Ⅰ,1 示石蒜花粉。当调焦到花粉中部的光学切面时,花粉壁仅呈一圈很薄的自发荧光;在无荧光的花粉内含物背景衬托下,很清楚地看见一个荧光明亮的椭圆形的生殖核和一个荧光较弱的圆形的营养核。蚕豆花粉(图版Ⅰ,2)和虞美人花粉(图版Ⅱ,14)亦属于二细胞型,其观察效果与石蒜相近,即冬青油透明大大减弱了花粉壁荧光,突出了内部的生殖核和营养核。

图版Ⅰ,4、5 示玉米花粉。其花粉壁荧光基本上被消除,花粉内含物呈弱荧光,而一对楔形的精核荧光十分明亮,惟营养核较模糊。紫菜苔花粉的壁仅局部较厚区域尚有荧光,其余部位荧光被消除。一对椭圆形精核荧光明亮,营养核稍淡(图版Ⅰ,20)。向日葵花粉是最有趣的:由于它有很厚的壁且具刺状突起,未经冬青油透明只看到一片强烈的自发荧光,除非将花粉压破无法显示细胞核。但经透明后,花粉壁荧光大为减弱,从而显示出一对精核与营养核。通过调焦可以清楚地看到精核呈细线状盘绕于花粉粒中(图版Ⅰ,6)。图版Ⅰ,7、8示两粒花粉的不同光学切面:其中上方一粒花粉在图7中显示一对弧形的精核,在图8中显示营养核,下方一粒花粉在图7中显示壁的表面观,在图8中则显示精核和营养核的部分结构。图版Ⅰ,5为向日葵的幼嫩花粉,此时尚处于二细胞初期,生殖核尚呈贴壁状态。

二、花粉管的观察

染色与透明方法的研究

(1)培养基的成分与pH值:各种植物的花粉应有适宜的人工萌发培养基。在本试验中,蚕豆花粉采用Brewbaker与Kwack培养基[6],含20%蔗糖;虞美人花粉用不含糖的Brewbaker与Kwack培养基或仅用0.01%硼酸水溶液[11];紫菜苔花粉则用Roberts等(1983)提出的甘蓝花粉培养基[13]。但Roberts等的原配方是用Tris将培养基调至pH8~9。我们在紫菜苔上作了三种pH值的比较:用Tris将pH值调至8.5时,花粉萌发顺利,但荧光染色不佳,不加Tris(pH5.5)时,染色良好而萌发不良;只有加Tris将pH值调至7时,才能兼顾人工萌发与荧光染色二者的要求。

(2)活体观察与固定、透明后观察的效果:三种植物的花粉在含20μg/ml H33258的相应培养基中保温1~2h后,萌发与生长旺盛。如此时取样直接用荧光显微镜活检,无论在花粉粒或花粉管中均看不清细胞核。这一结果与Hougn等(1985)报道的不同:他们用同样浓度的H33258染人工萌发的花粉,据说仅需5~10min染色时间即可[10]。我们多次观察表明,如此短时间染色作活体观察是看不到细胞核的;我们推测上述报道中所涉及的可能是已经死亡的花粉。因为我们在观察染色1~2h的花粉管时,如果由盖玻片边缘加入70%乙醇,很快就显示细胞核。如果再经过脱水与冬青油透明,则细胞核荧光更为明亮。

根据上述研究,确定花粉管的染色、透明基本程序为:在含 20μg/ml H33258 的适宜的培养基中人工萌发与染色→Carnoy 短时间固定或 70%乙醇固定→乙醇脱水→冬青油透明。

观察效果 用上述方法可以观察到在萌发过程中细胞核由花粉粒向花粉管的转移,生殖核与营养核在花粉管生长过程中的行进顺序及其空间距离,生殖核在花粉管中分裂成精核的动态等。

蚕豆花粉萌发后,生殖细胞一般率先进入花粉管。此时生殖核变得比萌发前更为细长,以适应通过狭窄的萌发孔。当生殖细胞已进入花粉管中时,营养核尚留在花粉粒中(图版Ⅰ,9、10)。生殖核在花粉管中又逐渐缩短,恢复到和原来在花粉粒中相似的形态(图版Ⅰ,11)。在少部分花粉管中看到营养核先于生殖细胞进入花粉管的情况,如(图版Ⅰ,12)中,营养核已进入花粉管一段距离,而生殖核尚未通过萌发孔。生殖细胞与营养核最终行进到花粉管尖端(图版Ⅰ,13)。看来,蚕豆的生殖细胞与营养核至少在萌发初期可能拉开很长的距离,二者间似乎缺乏密切的联系。

虞美人花粉在萌发与生长过程中通常是营养核领先于生殖细胞,并且二者互相紧靠(图版Ⅰ,16)。生殖核的分裂很快,一般在培养 3h 即已见到生殖核分裂时出现染色体(图版Ⅰ,17、18)。培养 5h,许多花粉管中已有一对精子(图版Ⅰ,19)。这也说明了培养基中的 H33258 并不抑制生殖核的分裂。另一个有趣现象是在培养过程中观察到多例尚未萌发的花粉粒中已经产生一对精子,好似三细胞型花粉的情况(图版Ⅰ,15)。看来在该种植物中,精子形成不一定非依赖花粉管形成不可。

紫菜苔花粉属三细胞型。萌发后,一对精子常领先进入花粉管,营养核紧随其后(图版Ⅰ,21)。在培养过程中,可以看到一对精子由花粉管基部移向尖端的各个图像,但营养核则荧光褪淡(图版Ⅰ,22~24)。

讨 论

应用与 DNA 结合的荧光染料观察植物细胞核已有较多的工作,其中 H33258 是一种常用的荧光染料。它具有荧光明亮、不易衰退、毒性低、适于活体染色等优点,曾被用于显示藻类的细胞核和细胞质中的含 DNA 小体、高等植物的细胞核与染色体、原生质体培养中的 DNA 含量变化、原生质体活染与细胞分拣、酶法分离胚囊的细胞核等[3]。迄今只有一篇关于用 H33258 显示花粉细胞核的报道[10]。本文结果也证明用它作花粉细胞核的荧光染色效果是良好的,它和 DNA 的结合力很强,甚至能经受染色后的固定、脱水与透明等程序而不褪色。正是由于这一稳定的性能,才能将它和冬青油透明结合起来而收到更好的效果。

冬青油作为一种透明剂早就用于植物显微技术,但主要限于制作显示器官中维管束系统的整体透明标本[5]。Crane(1978)第一次将它引入胚胎学研究,观察整体胚珠中的胚囊[8]。某些禾本科植物的子房也可以用冬青油透明观察其中的无融合生殖[16,17]。Stelly 等(1984、1986)将梅氏苏木精明矾染色和冬清油透明结合起来,改进了观察效果[14,15]。杨弘远进一步发展了爱氏苏木精染色-冬青油透明技术,扩大了它在观察胚囊中的用途[2]。本文则进一步从两方面扩大了它的用途:由与常规染色结合扩大到与荧光染色结合;由胚囊的观察扩大到花粉的观察。

荧光染色-冬青油透明技术具有如下优点:(1)显著提高荧光染色的效果,使细胞核荧光得以突出表现。其原因可能是冬青油溶去了花粉壁上与花粉内一部分发射自发荧光的物质,同时又增加了花粉物质的折射率这样两重的作用。像向日葵花粉这类原来完全看不到细胞核的情况,经过透明后也能显示出清楚的精核和营养核。(2)较长久地保存 H33258 染色的荧光。花粉在冬青油中浸泡一月以上,细胞核荧光明亮如故。这一特点在临时观察、时间匆促的场合更为有利。作者也试验过另一种与 DNA 结合的荧光染料 DAPI,用它染花粉后经冬青油透明也取得类似的效果,但 DAPI 比较不耐保存,一个月后细胞核荧光有所减退,而细胞质中出现扩散的荧光。

本文在几种二细胞型与三细胞型花粉上的试验表明:这一技术可用于研究成熟花粉中细胞核的状态和人工萌发花粉管中细胞核的行为。显然,它也应能适用于观察小孢子发生与雄配子体发育过程,以及离体培养花药中的雄核发育过程。在这些研究中,经冬青油透明后的花粉是否适于作细胞核 DNA 的荧光光度测定,是需要进一步证明的。至于这一方法能否在传粉、受精的研究中发挥作用,也还需要继续探讨。

(作者:杨弘远。原载:植物学报,1988,30(3):242~247。图版1幅,参考文献17篇,均删去。参看书末彩色图版Ⅰ。)

3 玉帘离体萌发花粉管中生殖细胞核的有丝分裂和无丝分裂

Mitosis and amitosis of generative cell nuclei in artificially germinated pollen tubes in *Zephyranthes candida* (lindl.) Herb.

摘 要

应用 DAPI-DNA 荧光显微术观察了玉帘 [*Zephyranthes candida* (Lindl.) Herb.] 人工萌发花粉管中生殖细胞核有丝分裂前后营养核的形态变化及其与前者的空间联系,着重研究了低温预处理诱导生殖细胞核无丝分裂和异常有丝分裂的现象。正常条件下,生殖细胞和营养核贴合后开始进行有丝分裂,这期间营养核弥散、延伸。精子形成后,营养核逐渐恢复原来状态。花粉经低温预处理后萌发的花粉管中,部分生殖细胞核进行无丝分裂,此时观察不到前述营养核的形态变化及其与生殖细胞的贴合。形成的精细胞核也可进一步发生无丝分裂以及形成微核的现象。低温预处理亦诱导了部分生殖细胞核进行异常有丝分裂。

Abstract

This paper deals with the comportment of the vegetative nucleus and its spatial association with the generative cell and sperm cells in the artificially germinated pollen tubes of *Zephyranthes candida* (Lindl.) Herb. before and after generative cell mitosis with the use of DNA-specific fluochrome 4′, 6-diamidino-2-phenylindole (DAPI). The induction of amitosis and abnormal mitosis of generative cell nuclei by cold-pretreatment of the pollen prior to germination was studied in particular. In normal case, the generative cell, after appressing to the vegetative nucleus for certain time, underwent mitosis to form two sperms, while the vegetative nucleus became markedly elongated, diffused, and exhibited blurring of its fluorescence. After division, a pair of sperms remained shortly in close connexion with the vegetative nucleus. Then the vegetative nucleus returned to its original state. In the pollen tubes germinated from cold-pretreated pollen, amitosis of some generative cell nuclei were frequently observed. Amitosis took place via either equal or unequal division with a mode of constriction. During amitosis, the dynamic change of vegetative nucleus and its intimate association with generative cell described above did not occur. Sperm nuclei produced from amitosis could further undergo amitisis resulting in micronuclei. Factors affecting the amitotic rate of generative cells, such as pollen developmental stage, temperature and duration of cold-pretreatment, were studied. Besides amitosis, cold-pretreatment also induced some abnormal mitotic behavior leading to the formation of micronuclei. Based on our observations and previously reported facts in other plant materials, it is inferred that the vegetative nucleus plays an important role in normal mitosis of generative cell and development of sperms.

(作者:莫永胜、杨弘远。原载:植物学报,1992,34(7):485~490。
图版2幅,表1个,参考文献18篇,均删去。)

4 pH 值与聚乙二醇对芸苔属花粉人工萌发的影响

Influence of medium pH and polyethylene glycol on the artificial pollen germination in *Brassica*

摘 要

紫菜苔(*Brassica campestris* var. *purpurea*)花粉在 pH8.5,并以 20% PEG 取代蔗糖的 R 培养基中显著提高了萌发率和改善了花粉管的生长。该改良的 R 培养基亦适用于白菜型油菜(*B. campestris*)与青菜(*B. chinensis*)花粉的人工萌发。将改良 R 培养基水合或萌发的花粉授于花柱切面上,花粉管可在花柱中顺利生长。为了比较花粉离体与活体萌发的 pH 值条件,应用微电极测定了紫菜苔花粉与柱头表面微环境的 pH 值,二者均呈弱酸性,表明花粉在碱性条件下的人工萌发并非对自然萌发时 pH 条件的简单模拟。讨论了 pH 值与 PEG 影响花粉人工萌发的可能机理。

Abstract

Pollen of *Brassica campestris* var. *purpurea* (a vegetable crop) was cultured in a medium developed by Roberts et al. (1983) containing 15% sucrose. The effect of pH value of the medium ranging from 5.16 to 8.97 was studied. The results showed that the best germination was obtained at pH8.47. Complete substitution of sucrose by 20% PEG exhibited excellent effect on both pollen germination (up to 70.8%) and tube growth (prolonged for 20 hours). Then this modified R medium composed of Roberts's salts and 20% PEG, pH8.5, was adopted to the pollen of other *Brassica* species. Similar positive results were obtained in *B. campestris* (a rapeseed plant) and in *B. chinensis*, but were not ideal in *B. napus*. Pollen incubated in this modified R medium, if pollinated onto the cut surface of style, could germinate well and their pollen tubes could grow down the style into the ovary. This might provide a biotechnological means for *Brassica* species using hydrated pollen as the carrier of foreign genes. In order to compare the artificial with natural condition of pollen germination, a microelectrode was used to test the pH value of pollen and stigma surface. The mean pH value was 6.12 in the former case and 5.87 in the latter. This means that the optimal alkaline condition for artificial pollen germination was not a simple mimic of natural condition. The mechanisms of the influence of pH value and PEG on pollen germination are discussed.

(作者:徐秉芳、周嫦、杨弘远、张学记、周性尧。原载:武汉大学学报(自然科学版),1996,42(4):453~458。图版1幅,表3幅,参考文献12篇,均删去。)

5 外源 Ca^{2+} 对烟草花粉管生长和生殖核分裂的调节

Exogenous Ca^{2+} regulation of pollen tube growth and division of generative nucleus in *Nicotiana tabacum*

摘 要

用细胞学和统计学方法研究了外源 Ca^{2+} 对烟草(*Nicotiana tabacum* L.)离体花粉管生长和生殖核分裂的影响。正常培养条件下,花粉管群体内的生殖核分裂率大致呈对数增长,10～18h为其分裂高峰期。所用 Ca^{2+} 浓度中以 10^{-3} mol/L 最适于花粉管生长,与之相比,其它浓度随时间延长愈益明显地表现出抑制效应。生殖核分裂则以 10^{-2} mol/L 与 10^{-3} mol/L 较为适宜,且 10^{-2} mol/L 可相对提前分裂高峰。在含 10^{-3} mol/L Ca^{2+} 培养基中培养 10h 后用不同方法处理,发现高钙抑制花粉管生长,尤以 10^{-1} mol/L Ca^{2+} 抑制最强烈,导致花粉管顶端壁加厚及生殖核的无丝分裂。而 10^{-2} mol/L Ca^{2+} 在处理早期(10～12 h)促进生殖核分裂。EGTA 处理则同时抑制花粉管生长和生殖核分裂。

Abstract

Cytological and statistical studies on the effects of exogenous Ca^{2+} on *in vitro* pollen tube growth and generative nucleus (GN) division of tobacco (*Nicotiana tabacum* L.) were conducted in an artificial experimental system. Under normal cultured conditions, the rate of GN division increased logarithmically in general, and reached the climax at about 10-18h. Among the treatments with various Ca^{2+} concentrations, 10^{-3} mol/L was the optimal concentration for pollen tube growth, whereas other Ca^{2+} concentrations showed increasing inhibitory effect with the time of culture. Generally, Ca^{2+} concentrations at 10^{-2} mol/L and 10^{-3} mol/L favored GN division more than the others. Compared with 10^{-3} mol/L, Ca^{2+} concentration at 10^{-2} mol/L benefitiated GN division at earlier stage of the treatment, but afterwards showed inhibitory effect gradually. Besides, the authors designed another series of experiments, in which 10^{-2}, 10^{-1} mol/L Ca^{2+} (final concentrations) of 2, 10 mmol/L EGTA were respectively added to the medium containing 10^{-3} mol/L Ca^{2+} at 10 h of culture. Pollen tube growth was inhibited by the high Ca^{2+} treatments, especially being severely effected by 10^{-1} mol/L Ca^{2+}. At this concentration thickening of the tube tip, amitotic division of GN leading to micronucleus formation occurred. 10^{-2} mol/L Ca^{2+} treatment, however, promoted GN division at the earlier stage of treatment (10-12h). EGTA treatments inhibited both pollen tube growth and

GN division.

（作者：范六民、杨弘远、周嫦。原载：植物学报，1997，39（10）：899～904。图版1幅，图4幅，表1幅，参考文献12篇，均删去。）

6 Ca^{2+}载体A23187对烟草生殖核分裂的调节

Regulation of generative nucleus division by calcium ionophore A23187 in *Nicotiana tabacum* pollen tubes

摘 要

用Ca^{2+}载体A23187研究了高钙水平对烟草离体花粉管内生殖核分裂的影响。结果表明，A23187抑制花粉管生长，影响其正常形态，调节生殖核分裂。较高浓度（10^{-6} mol/L，10^{-5} mol/L）A23187可阻断生殖核分裂，而较低浓度（10^{-7} mol/L）的效应因处理时间而异：萌发6h处理显著抑制生殖核分裂；而10h、14h处理早期促进分裂，随后逐渐转为抑制。较高浓度还改变生殖核的正常形态，使之趋于圆形。着重讨论了高钙调节生殖核分裂的作用过程。

Abstract

The effects of high calcium on tabacco generative nucleus (GN) division were studied using calcium ionophore A23187. Mature pollen grains were artificially germinated in a medium containing 10^{-3} mol/L calcium ions. A23187 of various concentrations was added at 2, 6, 10 and 14h respectively. Then pollen tubes (PT) were sampled at intervals, measured in length and stained with fluorescent dye DAPI for investigating the dynamics of GN division rate. A23187 could inhibit PT growth, affect PT form and regulate GN division. The influence of A23187 on GN division was related to the mitotic phase of GN and the drug concentration. The higher concentrations (10^{-6} mol/L and 10^{-5} mol/L) blocked GN division completely no matter when the treatment was carried out, whereas the lower concentration (10^{-7} mol/L) showed different effects on GN division depending on the time when A23187 was applied. The application of 10^{-7} mol/L A23187 at 6 h of culture inhibited the increase of GN division rate significantly, but showed changeable effects in 10h and 14h treatments from initial promotion to later inhibition. Moreover, the drug at the higher concentrations also had a dramatic effect on GN form, resulting in its rounding. The mechanism of how high calcium regulates GN division was discussed.

(作者:范六民、杨弘远、周嫦。原载:西北植物学报,1997,17(2):158~162。图版1幅,表2幅,参考文献13篇,均删去。)

7 Nifedipine 对烟草花粉萌发、花粉管生长及生殖核分裂的影响

Effects of nifedipine on pollen germination, pollen tube growth and division of generative nucleus in *Nicotiana tabacum*

摘 要

用细胞学和统计学方法研究了 Ca^{2+} 通道专一性阻滞剂 Nifedipine

(Nif)对烟草(*Nicotiana tabacum* L.)离体花粉萌发、花粉管生长及生殖核分裂的影响。10^{-4} mol/L Nif 可抑制花粉萌发。Nif 对花粉管生长的影响与其浓度和处理持续时间有关,10^{-4} mol/L Nif 始终抑制花粉管生长;而 10^{-7} ~ 10^{-5} mol/L Nif 在较短时间内起不同程度的促进作用,之后逐渐过渡为抑制花粉管生长。较高浓度处理可使花粉管形态趋向异常,细胞质流动趋于停滞。Nif 抑制生殖核的有丝分裂,相对推迟分裂高峰。Nif 使花粉管中的金霉素(CTC)荧光趋于减弱,表明 Nif 通过抑制 Ca^{2+} 通道活性产生生理效应。着重讨论了 Nif 抑制生殖核分裂的可能原因。

Abstract

This paper deals with the effects of nifedipine (Nif), a Ca^{2+} channel blocker of rather high specificity, on pollen germination, pollen tube growth and division of generative nucleus (GN) in experimentally germinated pollen tubes of *Nicotiana tabacum* L. Pollen germination was inhibited by the addition of 10^{-4} mol/L Nif whereas no significant inhibition by 10^{-7}-10^{-5} mol/L Nif was observed. The effects of Nif on pollen tube growth were related to its concentration and duration of treatment. At the earlier stage, tube growth was promoted at the lower concentrations (10^{-7}-10^{-5} mol/L), but was significantly inhibited at a concentration of 10^{-4} mol/L Nif. With increasing time of culture, even the lower concentrations also became harmful; the stronger the concentration, the earlier the transition from promotion to inhibition. Generally, inhibition of tube growth occurred within 24 hours of culture with different extent in various concentrations. Moreover, higher concentrations also tended to disturb tube morphology and cytoplasmic streaming. Nif was observed to perturb GN division at various concentrations, either blocked it completely at 10^{-4} mol/L, or only delayed it at 10^{-7}-10^{-5} mol/L. The dynamics of membrane-associated calcium in pollen tubes was tested with chlorotetracycline (CTC). With increasing time of culture and escalating Nif concentration, CTC fluorescence weakened gradually, indicating that the physiological effects of Nif is mediated by its inhibition on Ca^{2+} channel activities.

(作者:范六民、杨弘远、周嫦。原载:植物学报,1996,38(9):686~691。图版1幅,图2幅,表2幅,参考文献16篇,均删去。)

8 外源钙调素和钙调素拮抗剂对烟草离体花粉管生长和生殖核分裂的调节

Regulation of *in vitro* pollen tube growth and generative nucleus division by exogenous calmodulins and calmodulin antagonist in *Nicotiana tabacum* L.

Abstract

The effects of exogenous calmodulins and the calmodulin antagonist trifluoperazine on *in vitro* tobacco pollen tube growth and generative nucleus (GN) division were investigated using cytological methods. The effects of these agents on pollen tube growth depended on the stage of growth.

At an earlier stage (0-5h of culture), cauliflower calmodulin in the concentration range of 10^{-9}-10^{-6} mol/L stimulated pollen tube growth, and the stimulation increased with increasing concentration. At later stage (5-8h), 10^{-6} mol/L changed to inhibit tube growth whereas 10^{-9}-10^{-7} mol/L still stimulated tube growth. 10^{-8} and 10^{-7} mol/L promoted GN division at an earlier stage (8-10h) whereas the concentrations beyond this range showed no obvious promotion. With increasing duration of treatment, higher concentrations (10^{-7}、10^{-6} mol/L) gradually changed to inhibit GN division (Figs. 2,3). The effects of bovine brain calmodulin-agarose on tube growth and GN division were similar to and even stronger than cauliflower calmodulins. At early stage (0-2h), 10^{-11}-10^{-8} mol/L stimulated tube growth, and the promotion became weak at concentrations higher than 10^{-9} mol/L (Figs. 5,6). In general, the influence on GN division showed a tendency from promotion to inhibition with the prolongation of treatment. TFP at early stage (0-2h), 10^{-4} mol/L inhibited tube growth, while at later stages (2-5h、5-8h), all the concentrations used exhibited inhibition. TFP inhibited GN division at earlier stage (8-12h) and exhibited no significant effect within 12-24h culture. Besides, its inhibition could be partially overcome by bovine brain CaM-agarose at a concentration of 10^{-9} mol/L (Figs. 8,9).

The above results indicate that both extra- and intra-cellular endogenous calmodulins are involved in pollen tube growth and generative nucleus division in *Nicotiana tabacum* L.

钙调素(CaM)在植物钙信号途径中扮演关键的角色。CaM 的荧光标记(Tirlapur 和 Cresti 1992)和免疫细胞化学定位(Tirlapur 等 1994)以及 CaM 拮抗剂的抑制实验(Polito 1983, Gong 等 1994)均提示,内源(包括胞内和胞外)CaM 参与花粉萌发和花粉管生长的调控。最近,CaM 拮抗剂 W7-agarose、CaM 抗血清处理及消除实验证明,胞外内源 CaM 参与启动并调节朱顶红花粉萌发和花粉管伸长,而外源 CaM 可促进上述效应(马力耕和孙大业,1996)。

关于 CaM 与植物细胞分裂和增殖的关系是研究较少的一个领域。已知外源 CaM 可促进植物细胞的增殖和原生质体初生壁的再生(Sun 等 1994,1995)。目前尚未见有关 CaM 与植物生殖细胞分裂关系的报告。本工作对 CaM 在烟草离体花粉管生长不同阶段和生殖细胞分裂中的作用进行了研究。

材料与方法

一、供试材料与药品

供试材料普通烟草(*Nicotiana tabacum* L.)品种 G-80 种植于华中农业大学温室。花椰菜 CaM 由河北师范大学生物系孙大业教授和白娟老师惠赠,牛脑 CaM-agarose 和 CaM 拮抗剂 TFP(trifluoperazine)购自 Sigma。

二、花粉人工萌发

按范六民等(1996)的方法进行花粉萌发。培养起始即加入不同浓度的花椰菜 CaM、牛脑 CaM-agarose 或 TFP 进行处理。

三、外源 CaM 浓度实验

以基本培养基作为对照,基本培养基附加花椰菜 CaM 作为处理,CaM 浓度设置为 10^{-9} mol/L、10^{-8} mol/L、10^{-7} mol/L 和 10^{-6} mol/L。按范六民等(1996)的方法,分时段统计花粉管平均生长速率,定时统计生殖核分裂率,并定时观察花粉管形态和细胞质流动。

四、外源 CaM-agarose 浓度实验

以基本培养基作为对照,基本培养基附加牛脑 CaM-agarose 作为处理,处理的浓度设置为 10^{-11} mol/L、10^{-10} mol/L、10^{-9} mol/L、10^{-8} mol/L、10^{-7} mol/L 和 10^{-6} mol/L。进行各项指标的统计和观察。

五、TFP 处理及其 CaM-agarose 恢复实验

以基本培养基作为对照,处理设置为基本培养基分别附加 TFP 10^{-6} mol/L、10^{-5} mol/L、10^{-4} mol/L、10^{-3} mol/L 以及 10^{-5} mol/L 的 TFP+10^{-9} mol/L 的 CaM-agarose,进行各项指标的统计和观察。

六、统计方法

花粉管平均生长速率的统计分 3 个时段(0~2 h、2~5h 和 5~8h),对照和各处理分别统计 60 个花粉管。定时(8h、10h、12h 和 24h)统计生殖核分裂率,对照和处理分别统计约 300 个花粉管。上述指标的统计方法参考范六民等(1996)。用 t 测验检查处理与对照间的差异显著性程度。

结　果

一、外源花椰菜 CaM 对花粉管生长和生殖核分裂的调节

分时段对花粉管平均生长速率的统计表明(图1),外源花椰菜 CaM 对花粉管生长的影响具有浓度效应和时间效应。花粉管生长较早时期(0~5h),CaM 10^{-9}~10^{-6}mol/L 均促进花粉管生长,且促进作用随浓度提高而增强;但较后时期(5~8h)、较高浓度(10^{-6}mol/L)转而起抑制作用,其余试验浓度虽仍促进花粉管生长,但促进作用随浓度提高而渐减。

定时统计生殖核分裂率表明(图2),较早时期(培养 10h、12h),适当浓度(10^{-8}mol/L、10^{-7}mol/L)可促进生殖核分裂,而随后其促进作用趋于减弱,甚至转而起抑制作用。进一步分析不同时段生殖核分裂率增加量(图3)可更清楚地看出,8~10h,10^{-8}mol/L、10^{-7}mol/L 明显促进生殖核分裂,而其他试验浓度无明显影响;10~12h,各种浓度(10^{-9}~10^{-6}mol/L)对生殖核分裂均无明显影响;12~24h,10^{-9}mol/L、10^{-8}mol/L 无明显影响,而 10^{-7}mol/L、10^{-6}mol/L 却表现出明显的抑制效应。总之,在较早时期(8~10h),适当浓度(10^{-8}mol/L、10^{-7}mol/L)促进生殖核分裂,高于或低于该浓度均无明显的促进作用。随着时间延长,较高浓度 10^{-7}mol/L、10^{-6}mol/L)处理转而抑制生殖核分裂。适当浓度的 CaM 处理使花粉管群体生殖核分裂高峰提前。

二、外源牛脑 CaM-agarose 对花粉管生长和生殖核分裂的调节

分时段统计花粉管平均生长速率表明（图4），外源牛脑 CaM-agarose 对花粉管生长的影响亦具有浓度效应和时间效应。花粉管生长早期（0~2h），较低浓度（10^{-11}~10^{-8} mol/L）CaM-agarose 促进花粉管生长，但超过 10^{-9} mol/L 促进作用减弱，而更高浓度（10^{-7} mol/L）则几乎完全抑制花粉管生长；较后的时期（2~8h），各种浓度的 CaM-agarose 均逐渐转而表现出抑制效应，且浓度越高，作用的转变越早。与花椰菜 CaM 相比，牛脑 CaM-agarose 对花粉萌发和花粉管生长的影响更强烈。

定时统计生殖核分裂率表明，外源牛脑 CaM-agarose 亦可影响生殖核分裂（图5）。培养 10h 之内，CaM-agarose 10^{-10} mol/L、10^{-9} mol/L 明显促进生殖核分裂，而在此两种浓度以外无明显的影响；12h 之内，各浓度均促进生殖核分裂，且浓度大于 10^{-10} mol/L 时促进作用减弱；24h 之内，各浓度均抑制生殖核分裂。进一步考察不同时段生殖核分裂率增加量（图6）可知，CaM-agarose 8~10h，10^{-10} mol/L、10^{-9} mol/L 明显促进生殖核分裂，大于或小于这两个浓度均不明显影响生殖核分裂；10~12h，各浓度均明显促进生殖核分裂，且其作用与浓度大致呈负相关；12~24h，各浓度均表现明显的抑制作用。综上可知，牛脑 CaM-agarose 对生殖核分裂的影响亦具浓度效应和时间效应，总体上表现出从促进到抑制的渐变过程。

三、TFP 对花粉管生长和生殖核分裂的调节

分时段统计花粉管平均生长速率表明（图7），TFP 对花粉管生长的影响具有时间效应。花粉管生长早期（0~2h），较低浓度（10^{-6} mol/L、10^{-5} mol/L）TFP 对其生长无明显影响，较高浓度（10^{-4} mol/L）则表现抑制效应；稍后（2~5h），在所用的低浓度下也表现抑制作用；再后（5~8h），抑制作用更加明显。CaM-agarose 10^{-9} mol/L 不能消除 TFP 对花粉管生长的抑制。

定时统计生殖核分裂率表明（图8），培养 10h、12h 和 24h 之内，TFP 10^{-6} mol/L、10^{-5} mol/L 处理均明显抑制生殖核分裂，而 CaM-agarose 10^{-9} mol/L 可部分消除 TFP 10^{-5} mol/L 的抑制效应。分析不同时段生殖核分裂率增加量的变化(图9)可更清楚地看出,8~10h、10~12h 两个时段,TFP 10^{-6} mol/L、10^{-5} mol/L 处理均明显抑制生殖核分裂,而 CaM-agarose 10^{-9} mol/L 可部分消除 TFP 10^{-5} mol/L 的抑制效应,但后一时段内 TFP 10^{-6} mol/L 处理的抑制作用有所减弱;12~24h,各处理均不表现抑制效应。综上可知,TFP 对生殖核分裂的影响具有时间效应,从较早期的抑制逐渐转变为后期的无明显影响。

讨 论

前人已有报告,一定浓度的外源 CaM 可促进花粉萌发和花粉管生长,而 CaM 拮抗剂则表现抑制效应(Polito 1983,Gong 等 1994,马力耕和孙大业 1996)。本实验取得类似结果,外源花椰菜 CaM 和牛脑 CaM-agarose 以及 CaM 拮抗剂 TFP 均可影响花粉管生长,并且发现它们的作用具有时间效应:较高浓度的 CaM-agarose 强烈抑制花粉萌发和花粉管生长;随处理时间延长,即使早期起促进作用的浓度亦逐渐转变为抑制作用。可见,CaM 并非一味地促进生长。TFP 对花粉管生长的抑制随时间的延长而趋于增强。

以往亦有外源 CaM 促进细胞增殖的报告(Sun 等 1994,1995)。本文首次证明外源 CaM 和 TFP 可调节烟草花粉管中生殖细胞的有丝分裂。与对花粉管生长的作用特点相似,外源 CaM 和 TFP 对生殖细胞分裂的调节亦具有时间效应,而且发现花椰菜 CaM 与牛脑 CaM-agarose 作用有所差异:在培养 24h 之内,整体上前者对生殖核分裂无明显影响,而后者则表现出强烈的抑制效应。TFP 对生殖核分裂的作用随处理时间的延长,抑制逐渐转变为后期的无影响,且抑制效应可被 CaM-agarose 部分消除。

为了鉴别外源 CaM 调节的作用途径,我们在实验中还特别应用了牛脑 CaM-agarose。由于 agarose 颗粒一般大于花粉粒,因此 CaM-agarose 只能通过胞外的作用位点将信号转导到胞内而起作用。这与马力耕和孙大业(1996)在实验中采用 W7-agarose 的原理相同。许多研究证明,植物细胞壁中存在 CaM,即胞外内源 CaM(Biro 等 1984,叶正华等 1989,Li 等 1993)。本文从另一个角度证明了胞外内源 CaM 在花粉萌发和花粉管生长的调节中起重要作用。

CaM 结合蛋白(CaMBP)是 Ca^{2+}-CaM 信号通路中的一个重要环节。现已证明植物细胞外存在 CaMBP,且主要存在于细胞壁中(叶正华等 1989,Tang 等 1996),说明细胞壁上有 CaM 结合及作用位点(孙大业等 1995),但目前尚不清楚 CaMBP 的下游事件。根据以往的报告,植物细胞质膜上的许多酶和蛋白受 CaM 或 Ca^{2+}-CaM 调节,包括钙泵(Ca^{2+}-ATPase)(Robinson 等 1988)、蛋白激酶(Ranjeva 和 Boudt 1987)、磷脂酶 C(Melin 等 1987)等。动物细胞以及植物细胞液泡膜上的离子通道(包括钙通道)(Weiser 等 1991,Blatt 和 Thiel 1993)亦受 Ca^{2+}-CaM 或 CaM 调节。有研究表明,CaM 拮抗剂处理可降低麝香百合花粉管中的[Ca^{2+}]c,且使其正常梯度丧失(Obermeyer 和 Weisenseel 1991)。由以上结果可以推测,胞外内源 CaM 可能通过作用于花粉粒和花粉管质膜上与[Ca^{2+}]c 调控有关的受体而间接调节花粉粒和花粉管中的[Ca^{2+}]c 水平,而外源 TFP 除进入胞内抑制胞内 CaM 外,还抑制胞外内源 CaM 的作用。[Ca^{2+}]c 的变化进而影响花粉萌发、花粉管生长及生殖核分裂。

(作者:范六民、杨弘远、周嫦。原载:植物生理学报,1988,24(3):240~246。图 9 幅,参考文献 19 篇,均删去。)

第三节 花粉原生质体操作

提要

花粉原生质体操作是我们从 20 世纪 80 年代末到 90 年代末的一项重点研究,包括花粉原生质体的分离、培养、融合、转化及有关细胞生物学研究。这项研究在国际上起步较早，也较为系统和深入。

首先面临的是如何分离大量的生活花粉原生质体。由于花粉外壁无法酶解，而化学去壁又会杀死原生质体，所以花粉原生质体的分离自20世纪70年代以来一直是一个难题。我们通过自己摸索，在鸢尾（鸢尾科）、风雨花（石蒜科）、萱草（百合科）三种植物中取得了成功。这几种植物花粉的共同特点是外壁较薄、萌发沟较大。通过吸水膨胀，外壁沿萌发沟裂开，内壁直接暴露在酶液中而得以降解，花粉原生质体从而逸出。这一针对花粉本身的生物学特点而建立的"水合-酶解"技术，以后又在唐菖蒲中获得成功，并由单子叶植物进一步推广到双子叶植物中的芸苔属。芸苔属的花粉原生质体分离较前几种植物困难，我们将水合-酶解法改为二步操作，即先经水合脱去外壁（由此又派生出本章第四节脱外壁花粉研究系列），然后再酶解内壁分离花粉原生质体。以后，我们在酶液中加入强效的pectolyase Y23，使水合与酶解合为一步，亦可分离芸苔属花粉原生质体，并且不仅由新鲜花粉，而且由超低温保存的花粉分离原生质体也取得满意的结果。烟草的情况又有所不同，它的花粉没有容易裂开的萌发沟，因而需要另辟思路。我们的对策是将花粉萌发成短花粉管，此处没有外壁障碍，可以酶解出原生质体，但不可任花粉管太长，否则会产生大量亚原生质体。

成熟花粉可以分离出原生质体，幼嫩花粉是否可能呢？如果可能，幼嫩花粉原生质体可否像花粉（小孢子）培养那样诱导孢子体发育呢？这些都是前人未曾尝试的。我们在萱草的实验中，不仅分离出成熟花粉原生质体，也由单核晚期与二核早期的花粉分离出原生质体，为幼嫩花粉原生质体研究开了先河。以后，又在芸苔属与烟草中获得成功，幼嫩花粉原生质体的分离迄今仍是国际上唯一的记录。

为了探讨成熟的与幼嫩的花粉原生质体在离体培养中的发育规律，我们提出了研究花粉原生质体两条发育途径的设想。实验结果表明这一设想是可行的：以唐菖蒲为代表的成熟原生质体培养萌发出接近正常的花粉管；以萱草为代表的幼嫩花粉原生质体培养则启动了细胞分裂,形成多细胞构造,其中有的类似早期原胚。幼嫩花粉原生质体启动孢子体发育是国际上首次成功。以后,在烟草中也由幼嫩花粉原生质体诱导了第一次分裂;而最近的一项与荷兰的国际合作研究更将油菜幼嫩花粉原生质体培养推进到微愈伤组织形成的新水平(见本章其它论文目录)。

在唐菖蒲成熟花粉原生质体和萱草幼嫩花粉原生质体培养取得阶段性结果之后,紧接着开展了对以上两个过程的超微结构研究。特别对于花粉原生质体再生新壁的方式与特点、幼嫩花粉原生质体脱分化和细胞分裂的特点等进行较为细微的观察。此外,还研究出单个花粉原生质体的电镜样品制备程序,便于选择特定材料进行有针对性的超微结构研究。

20世纪80年代中期国外研究者曾将小孢子四分体原生质体和体细胞原生质体融合,培养出三倍体杂种植株,称为"配子-体细胞杂交"("配子"一词不妥,应为"孢子"——作者按)。四分体被胼胝质壁包围,以蜗牛酶很易分离出原生质体。但四分体为期短暂,体形较小,取材与操作不甚方便。花粉原生质体的分离成功,使我们提出以之替代四分体进行融合的"花粉-体细胞杂交"的设想。在芸苔属中开展的实验首次在国际上证明了这一设想,细胞学与分子生物学鉴定为获得的异源三倍体与四倍体植株提供了证据。继而在烟草属中也获得了类似的结果。

利用花粉原生质体无壁的特点导入外源基因,是另一个思路。日本研究者在百合中、我们在芸苔属中以电激法于1995年同年报道了肯定的结果。GUS基因导入花粉原生质体后,瞬间表达频率远高于完整具壁花粉,表明它是很好的外源基因受体。

利用花粉原生质体无壁的特点开展离子通道的研究,是中国农业大学应用我们的芸苔属花粉原生质体系统合作的又一成果。可参看章末所附其它论文目录。

1 三种植物花粉原生质体的大量分离与初步培养

Mass isolation and culture of pollen protoplasts from three plant species

Abstract

The exine which contains chemically resistant sporopollenin makes a severe obstacle for obtaining pollen protoplasts. A technique was established for protoplast isolation in large number from mature pollen grains in three plant species. *Iris tectorum*, *Zephyranthes grandiflora* and *Hemerocallis fulva*. Dehiscence of exine by the force of pollen hydration and consequently exposure of intine to the direct action of enzymatic maceration led to successful release of protoplasts from pollen grains. The highest isolation rate up to 90% protoplasts was obtained from mature pollen grains of *Z. grandiflora*. Pollen protoplasts from tetrads to mature pollen, except early uninucleate pollen stage, were prepared in *H. fulva*. Purified pollen protoplasts were obtained by repeated centrifugation and density gradient centrifugation and proved to be viable by fluorochromatic reaction with FDA. When cultured *in vitro*, the pollen protoplasts of *H. fulva* underwent regeneration of cell wall, formation of various tube-, nodule-, bead-and dumbbell-shaped structures and sometimes division of the generative nucleus into 2-4 nuclei. Exogenous hormone showed promotive effect on the growth of above mentioned structures.

在现代植物生物学和细胞工程中具有重要研究与应用价值的原生

质体迄今主要取材于体细胞,而来源于生殖系统者尚未充分开发利用。花粉原生质体具有单倍性、结构简单、发育同步和群体数量大等优点,被认为是"遗传操作和突变研究的卓越体系",提供了"研究花粉生理学和个体发育的新途径"(Bajaj,1983)[5]。

早在20世纪70年代,Power(1973)[12],Bajaj和Davey(1974)[6]分别在烟草和几种茄科及禾本科植物中制备花粉原生质体,但效果不佳。以后这方面工作长期停滞。1985年Loewus等(1985)报道用MMNO·H_2O(4-methyl morpholine N-oxide monohydrate)分离出大量麝香百合花粉原生质体,但药物与高温使原生质体严重失活[10]。翌年他们报道技术的改进,然而生活力问题仍未完全解决[7]。最近我们在几种植物中建立的技术能得到大量较纯的生活的花粉原生质体,并初步开展了培养实验。

材 料 与 方 法

一、供试材料

包括鸢尾科的鸢尾(*Iris tectorum* Maxim)、石蒜科的风雨花(*Zephyranthes grandiflora* Lindl)和百合科的萱草(*Hemerocallis fulva* L.)。

二、原生质体分离

花蕾与花药用70%乙醇消毒,花粉置于酶液中25℃黑暗条件下酶解。酶液包括2%果胶酶、2%纤维素酶、0.5%葡聚糖硫酸钾与K_3培养基[11]的大量元素以及甘露醇或蔗糖。鸢尾与萱草一般用10% ~ 12%甘露醇,风雨花用5%蔗糖。根据倒置显微镜检查酶解情况和荧光增白剂鉴定脱壁效果,确定酶解时间,一般约2h。

三、原生质体纯化

经等渗培养基洗涤与离心2~3次后,将富集的原生质体悬浮液铺在含30%蔗糖的培养基液面上,500r/min离心8~10min。吸取富含原生质体的上部液层,培养液洗涤与离心 2~3 次,获得较纯的原生质体。如纯化效果不够好,可反复处理。

四、原生质体培养

用 K_3 培养基附加 0.4~0.5mol/L 蔗糖,分有激素与无激素两种处理。激素为 2,4-D 0.1mg/L,NAA 1 mg/L,6-BA 0.2mg/L。培养的原生质体密度为 $10^4 \sim 10^5$ 个/ml。

五、超声波处理

将新鲜的或 FPA 固定的花粉经 70%、50% 乙醇下行至水。置于盛蒸馏水的小管中,在盛水的超声清洗机(H 00005 型,无锡超声电子设备厂)中处理。根据镜检结果,确定处理时间,一般为 20~30min。

六、显微观察

(1)纤维素鉴定 0.1% 荧光增白剂(Calcofluor white ST)染色。(2)果胶质鉴定 钌红(1∶5000)水溶液染色。(3)细胞核染色 荧光染料 Hoechst 33258 20μg/ml 与冬青油透明技术[3]观察原生质体的细胞核。(4)生活力测定 荧光素二醋酸酯(40 μg/ml)染色。用 AO 1820 BIOSTAR 型倒置显微镜明视野或 Olympus NEW VANOX AHBS-514 型万能显微镜明视野与荧光显微装置观察与摄影。

结　果

一、原生质体分离

鸢尾 花粉具萌发沟。在 12% 甘露醇溶液中花粉因水合而膨胀,导致外壁由萌发沟处裂开,露出内壁(图版Ⅰ,1)。荧光增白剂鉴定暴露的内壁部分,表明其内层含纤维素,外层不含纤维素(图版Ⅰ,2)。超声波处理生活花粉,使原生质体破损,外壁与内壁分开。可清楚地看出内壁有两层,其内层被荧光增白剂染色,而外层不染色(图版Ⅰ,3)。

超声波处理固定花粉,则将内壁外层与具内壁内层的原生质体分开。钌红染色表明:内壁的内、外两层均含果胶质(图版Ⅰ,4)。由此可见,鸢尾花粉内壁分为含果胶质的外层和兼含果胶质与纤维素的内层两部分。这一观察与 Heslop-Harrison(1987)总结的花粉内壁分层规律[8]相符。

花粉置入酶液中,外壁同样由于水合作用而裂开,内壁直接暴露于酶液作用下。此时可见到含果胶质的内壁外层分离开来并逐渐溶解(图版Ⅰ,5)。继而含纤维素的内壁内层溶解,释放出原生质体。图版Ⅰ,5~8示原生质体释放情况。当外壁裂口较宽时,正释放的原生质体呈椭圆形,一半尚留在外壁内,一半已经逸出(图版Ⅰ,5)。当外壁裂口较窄时,释放中的原生质体呈哑铃状,可透过外壁看到尚包围在内的原生质体轮廓(图版Ⅰ,6,7)。酶解 2h 后,大量原生质体逸出。荧光增白剂鉴定其外表已无纤维素。原生质体呈典型的球形,有明显的质膜包围,内含物很丰富(图版Ⅰ,8)。

风雨花 花粉近椭圆形,透过花粉壁隐约看见其内的生殖细胞(图版Ⅰ,9)。采用上述技术,可获得大量原生质体。花粉粒置于酶液约 10min 后即已吸水膨胀,撑开外壁,露出椭圆形的尚具内壁的原生质体(图版Ⅰ,10)。20min 后原生质体逐渐变圆(图版Ⅰ,11,12)。30min 后相当多的花粉内壁已降解,原生质体呈球形,进一步逸出外壁。原生质体富含细胞质。生殖细胞呈纺锤形,清晰可见,但营养核较难分辨(图版Ⅰ,13)。风雨花花粉原生质体分离率高达 90%以上,图版Ⅱ,20示酶解 2h 的群体中包含大量完整的花粉原生质体与空外壁,基本上没有亚原生质体,很少破损花粉粒。

萱草 由萱草成熟花粉亦分离出原生质体。图版Ⅱ,22 示正在酶解的花粉群体,部分原生质体已经逸出,残留空的外壁;另一部分正在分离过程。萱草不同发育时期的花粉,由四分体至成熟花粉,除单核早期外,均能分离出原生质体。由于生活花粉及其原生质体中细胞核不易辨认,故均以相应时期的固定材料经丙酸洋红染色鉴定其发育阶段。图版Ⅰ,14~17分别示四分体、单核晚期、二核早期与成熟花粉的原生质体。

二、原生质体纯化

酶解后的花粉悬浮液中含原生质体、亚原生质体、未充分酶解的花粉粒、破损花粉内含物以及空外壁,具体情况因物种与酶解效果而异。纯化的主要困难是未充分酶解的花粉粒与原生质体的大小和重量相近,不易分开。通过多次离心和间断梯度离心,初步获得了较纯的原生质体群体。图版Ⅱ,18,21 和 23 分别示鸢尾、风雨花和萱草纯化后的群体。后二者的原生质体大小均匀,由完整花粉粒分离而成;前者的群体中除完整的花粉原生质体外,尚有较小的亚原生质体。用荧光素二醋酸酯鉴定了纯化后群体的生活力。图版Ⅱ,19 示鸢尾花粉原生质体发射明亮荧光,证实它们是生活的。下文所述培养结果更为原生质体生活力提供直接证明。

三、离体培养

在 K_3 培养基中,许多花粉原生质体再生了细胞壁。以后形成多种结构:有类似花粉管的管状结构(图版Ⅲ,24);有明显凹陷或隔壁的结节状结构(图版Ⅲ,25);有的类似出芽生殖,形成各种大小的球体;有的连接成串,成为念珠状或哑铃状结构(图版Ⅲ,26,27)。各种形态之间无严格界限,有混合和过渡类型,呈现出变化多端的复杂景象。总的看来,都属于以花粉管生长为基本格局的各种畸态,不属于正常的细胞分裂。与无激素培养相比,外源激素有加快各种结构生长的作用。

用荧光显微术观察了细胞核的行为。萱草成熟花粉经 Hoechst 33258 染色与冬青油透明后,生殖核荧光明亮,呈椭圆形;营养核荧光微弱,形状较模糊。在培养 12d 的材料中,观察到生殖核通过 1~2 次分裂形成 2~4 个细胞核。图版Ⅲ,28 示由花粉原生质体形成结节状结构的明视野显微图像。图 29 是同一材料的荧光显微图像。在细胞质聚集部位,有一对明亮的圆形细胞核,是生殖核分裂一次的产物。侧边有一荧光微弱的营养核,不易分辨。图版Ⅲ,30 示另一花粉原生质体的结节状结构。在荧光显微镜下可看到该材料生殖核分裂两次所形成的三个细胞核(图版Ⅲ,31)。

讨 论

1.10多年来对花粉各个发育时期均开展过原生质体分离的研究[4,5]。整个工作发展是不平衡的：小孢子母细胞至四分体时期，细胞壁主要由胼胝质等构成；花粉管时期，管壁由果胶质、纤维素与胼胝质组成；它们可以被胼胝质酶、果胶酶或纤维素酶所降解，分离出原生质体。唯独花粉时期（小孢子至成熟花粉），外壁主要由孢粉素组成，迄今缺乏相应的酶使之降解。因此，重要的问题是解决外壁障碍。Southworth(1974)试用30多种强酸、强碱、强氧化剂、洗涤剂、有机溶剂等药品分解外壁，仅极少数药品有效，但却同时杀死了花粉[13]。显然，企图用化学药剂分解外壁而又获得生活的原生质体是相当困难的。本文的思路是根据植物材料本身特点，寻找适宜方法避开外壁的障碍：利用花粉水合作用，在适宜渗透压条件下，使外壁破裂，内壁暴露，在酶的作用之下，达到分离大量生活原生质体的目的。关于花粉在水合过程中结构与功能状态的变化，前人有详细的研究[8]，对更好地认识花粉原生质体分离的机理很有参考意义。这一技术能否在更广泛的植物材料中推行，还需要试验与改进。

2. Bajaj与Davey(1974)首先试探了花粉原生质体的培养，观察到它们能有限生长，但未能明确证实有细胞分裂[6]。朱澂等(1984，1985)在花粉管亚原生质体培养中看到细胞壁再生和花粉管状结构的生长[1,2]。我们在萱草花粉原生质体培养中亦观察到细胞壁再生和各种形态的结构，以及生殖核的一二次分裂。总之，迄今花粉原生质体培养尚未有重大突破。如何诱导花粉原生质体脱分化、持续细胞分裂以至最终再生植株，是今后需要解决的重要课题。利用花粉原生质体作为实验体系开展细胞生物学研究也是很有意义的。

3. 关于花粉壁分层的研究，一般采用切片技术。Heslop-Harrison(1982)介绍一种内壁"ghosts"技术，用二乙醇胺、NaOH等较复杂的处理，溶解外壁、果胶质以及花粉内含物，最后仅留下内壁纤维素层，借以研究该层结构的特点及其与花粉粒形状、萌发孔结构的关系[9]。本文在研究原生质体分离的同时,用超声波技术分离出单独的花粉外壁、内壁、内壁外层以及具内壁内层的原生质体等单位,今后可作为花粉壁分层研究的一种新手段。其优点在于分离出花粉壁的不同层次,有利于观察用切片技术得不到的各层内、外表面,以及可用于对各层结构与生化特性的分析与研究。

(作者:周嫦。原载:植物学报,1988,30(4):362~367。图版3幅,参考文献13篇,均删去。参看书末彩色图版Ⅱ。)

2　Cell divisions in pollen protoplast culture of *Hemerocallis fulva* L.

The pollen protoplast became an attractive research subject following the success of somatic protoplast culture since it offered the possibility of a useful haploid system for cell fusion, genetic manipulation and mutation studies as well as a new approach to study pollen biology. However, the isolation of protoplasts from pollen grains faced technical difficulties due to the enzyme-resis-tant exine consisting of sporopollenin, thus limiting the quantities of pollen protoplasts obtained in early efforts[1,2]. In recent years, fresh progress has been gained by successful isolation of large quantities of pollen protoplasts of several species by either chemical[3,4] or enzymatic[5,6] techniques. These approaches enable us to put pollen protoplast culture on the agenda.

Although culture of pollen protoplasts was pioneered by Bajaj and Davey[2], not much progress was made. Recently renewed interest comes from the work of Tanaka et al. who inoculated protoplasts from pre-anthesis binucleate pollen in *Lilium longiflorum*[5] and Zhou who cultured protoplasts from mature pollen of *Hemerocallis fulva*[6]. The results obtained by both authors were similar, e. g. regeneration of cell wall, formation of pollen tube-like and other abnormal structures, division of generative nucleus, etc. Besides, other investigators have tried to culture protoplasts from tetrads[7-9] or pollen tubes[10,11]. In short, although some gametophytic development or other types of growth have occurred, so far there is still no breakthrough in pollen protoplast culture in respect to a cell division pattern leading to in vitro embryogenesis.

Based on the previous work with isolation of protoplasts from *H. fulva* pollen[6], experiments on its culture were carried out and, for the first time, cell divisions were induced. For convenience, the term 'pollenplast' is suggested to simplify 'pollen protoplast' in the following text.

Materials and methods

Plants of *Hemerocallis fulva* were grown in pots. Pollenplast isolation was carried out as described earlier[6] with some minor modification. Buds of 2.0-2.5 cm in length containing late uninucleate microspores were used. For cold-pretreatment, the buds were put in petri dishes and chilled in a refrigerator at 5-6℃ for 4 or 7 days. The buds, either fresh or being chilled, were surface-sterilized with 70% ethanol. Anthers were excised and gently crushed in a washing solution to release the pollen grains. This solution consisted of K_3 medium [12], 1 mg/L 2,4-dichlorophenoxyacetic acid(2,4-D), 2 mg/L naphthalene acetic acid (NAA), 18% mannitol and 1% polybuffer 74 (Pharmacia Fine Chemicals, used by Koop et al. for protoplast culture)[13] (pH5.8). After centrifugation at $80 \times g$ for 2 min, the pollen grains were macerated in an enzyme solution containing 0.5% cellulase (Onozuka R-10), 0.5% pectinase (Serva, Feinbiochemica, Heidelberg, FRG), 0.5% potassium dextran sulfate, 1% calf serum and 18% mannitol dissolved in K_3 medium. The enzymatic maceration was carried on for 7 h at 29-30℃ under stationary condition. After that the resulting pollenplast suspension was centrifuged and washed 3 times. The final precipitate was resuspended with a culture medium containing K_3 medium plus 1 mg/L 2,4-D, 2 mg/L NAA, 1% calf serum, 1% polybuffer 74, 9% mannitol and 17% sucrose (pH5.8) and centrifuged again at 80 × g for 5 min. The pollenplasts usually floated in the upper part of the medium and the debris was precipitated on the bottom. Purified pollenplast population was carefully collected with a pipette. Then the pollenplasts were prepared at a density of 2-5 × 10^4 protoplasts/ml of the medium and cultured at 25 ± 1℃ in darkness.

The cultured pollenplasts were observed periodically with an AO 1820 BIOSTAR inverted microscope. A portion of them were fixed in FPA (formalin/propionic acid/50% ethanol, 5 : 5 : 90, by vol.) and stored in 70% ethanol at 4-5℃. After being gradually hydrated to water, they were stained with propiono-carmine for 30 min and mounted in propionic acid and glycerine (9 : 1, v/v), or cleared and mounted in lactophenol. The preparations were observed with an Olympus NEW VANOX AHBS-514 microscope with Nomarski differential interference contrast equipment.

Results

In the present paper protoplasts isolated from uninucleate pollen were used as starting materials of culture, which were different from the pollenplasts of mature or nearly mature stage used previously[5,6]. Successful isolation of pollenplasts is shown in Fig. 1 where a great deal of pollenplasts have already separated from the empty pollen walls. Purification removed the pollen wall debris and provided a pollenplast population for culture.

During the course of the culture, pollenplasts from both fresh and chilled buds could regenerate wall and be induced to cell division. The percentage of divided pollenplasts after 0, 4 and 7 days of cold-pretreatment and 15 days of culture were 0.8%, 3.5% and 7.2%, respectively, indicating a remarkable promotive effect of cold-pretreatment on the induction of cell division. Moreover, without cold-pretreatment divided units remained in the 2-celled stage and showed signs of degeneration; in contrast, after cold-pretreatment cell divisions could be sustained, leading to formation of multicellular units.

Observations on living materials

Figures 2-10 show various developmental stages of living pollenplasts during culture as observed under an inverted microscope. Figure 2 is a pollenplast at the time of inoculation, in which a large vacuole is seen but the nucleus can not be visualized. After 4-5 days of culture the first cell division occurred to produce either two daughter cells of equal size with equally dense cytoplasm (Fig. 3) or with a big vacuole in one of the cells (Fig. 4), or alternatively two unequal cells (Figs. 5 and 6). In the latter case the vacuole was usually located in the large cell (Fig. 5) or there was no prominent vacuole (Fig. 6). Later, the second cell division took place and 3-4 celled units appeared. In the case of unequal first cell division the small cell divided into two cells and the large cell remained undivided (Fig. 7); otherwise, both the small and large cells entered into division resulting in 4 cells of different sizes (Fig. 8). In the case of equal divisions, four cells of approximately the same size were formed as is shown in Fig. 9 (one of the cells is out of focus). As divisions proceeded, multicellular clusters can be found (Fig. 10).

Observations on fixed materials

After fixation and staining or clearing, the cell wall and nuclei could be clearly identified (Figs. 11-29). Figure 11 shows a typical pollenplast at the time of inoculation with a prominent nucleus and a big vacuole. Obviously it was a protoplast isolated from a uninucleate pollen grain. However, during the course of cold-pretreatment approximately one fourth of the uninucleate pollen had divided into binucleate grains consisting of a vegetative cell and a generative cell. After enzymatic maceration some of the genera-

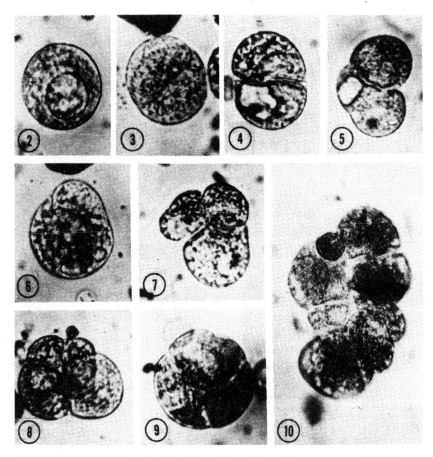

Figs. 2-10. Cell divisions in pollenplast culture after cold-pretreatment of buds for 7 days; Observations on living materials.

Fig. 2. A freshly inoculated pollenplast. (×700).

Figs. 3-6. Two-celled units produced by division from pollenplasts after 5-7 days of culture. (×700).

Figs. 7-9. 3-4-celled units after 10 days of culture. (×700).

Fig. 10. A multicellular cluster after 15 days of culture. (×600).

tive cells might be expelled from the pollenplasts, which were then composed of the vegetative cell only (Fig. 12). Hence, cell division might be initiated besides mainly from the protoplasts of uninucleate pollen, also possibly from the protoplasts of the binucleate pollen or its vegetative cell only.

The first division patterns were similar to those observed on fresh materials, including equal (Figs. 13 and 14) and unequal (Figs. 15 and 16) divisions. Based on the data of 152 divided units, there was almost equal chance of their occurrence, i. e. ,50.7% equal and 49.3% unequal division. The second division resulted in 3- or 4-celled units with preferable occurrence of the latter. Morphology of the 4-celled units was quite divergent: isobilateral (Fig. 17), tetrahedral (Figs. 18 and 19, where the mitotic figure can be seen), decussate (Fig. 20), and T-shaped (Fig. 21, where one of the upper cells is out of focus). At later stages of development, multicellular clusters were formed (Figs. 22-25). A proliferated cell cluster is shown in Fig. 25, where some mitotic figures can be seen.

Besides cell divisions there were many other morphological changes such as swelling, budding, formation of tube- or bead-like structures and so on, which have been reported in the previous paper[6]. In such cases, karyokinesis might take place without subsequent cytokinesis, resulting in coenocytic units. Here several examples are demonstrated, such as oval- and dumbbell-shaped cells with two nuclei (Figs. 26 and 27), a big multinucleate mass with prominent nucleoli (Fig. 28), and a tube-like structure with several nuclei (Fig. 29). Undoubtedly, these structures, though numerous, lack significance regards as embryogenic development.

Discussion

The present paper shows evidence that isolated pollen protoplasts can be triggered *in vitro* to cell divisions which may lead to sporophytic development, a process not realized previously. This represents a starting point to-

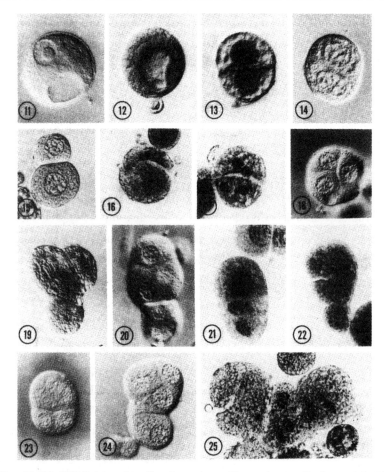

Figs. 11-25. Cell divisions in pollenplast culture: Observations on fixed materials. The materials were either stained with propiono-carmine (Figs. 12,13,16,17,21,22 and 25) or cleared with lactophenol (Figs. 11,14,15,18-20,23 and 24).

Fig. 11. A freshly isolated uninucleate pollenplast. (×700).

Fig. 12. A freshly isolated binucleate pollenplast where the generative cell is seen expelled from the pollenplast. (×700).

Figs. 13-16. Two-celled units produced by first cell division after 7 days of culture. (13, ×700; 14, ×700; 15, ×650; 16, ×650).

Figs. 17-21. Formation of 4-celled units with different morphologies after 11 days of culture. (17, ×650; 18, ×650; 19, ×650; 20, ×700; 21, ×650).

Figs. 22-24. Formation of multicellular proembryos after 11 days (Figs. 22 and 23) and 15 days (Fig. 24) of culture. (×650).

Fig. 25. A multicellular cluster after 15 days of culture. (×650).

ward further regeneration as plantlets.

In this study effort has focused on forcing pollenplasts to develop along a sporophytic pathway other than that *in vivo* programmed, and also on maintaining their survival and development. The success may be due to many factors involved in both preculture and culture stages, such as suitable pollen stage, cold-pretreatment, and culture medium. The two former factors have been widely adopted in anther/pollen culture for initiating androgenesis. Concerning the latter factor, it was an integration of the experience from somatic protoplast culture, e. g. K_3 medium, calf serum etc., with certain important modifications to fit the pollen biology, e. g. high concentration of osmoticum and sufficient level of exogenous hormones. Of course, to push development further to reach regeneration of plantlets, more effort is needed on improvement of the culture procedure.

The embryogenesis in anther/pollen culture as well as somatic protoplast culture has been studied comprehensively. Preliminary observations in this paper indicate that cell divisions in pollenplast culture may possibly be initiated from either uninucleate or binucleate pollen stages; in the latter case the vegetative cell may probably make greater contributions in comparison with the generative cell. The first cell division may be equal or unequal, resulting in similar or different daughter cells. The subsequent growth may lead to various types of cell clusters. To clarify the origin and developmental patterns, more detailed investigations are left to the future.

(Author: Zhou C. Published in: Plant Science, 1989, 62: 229-235, with 29 figures, 13 references. Figs. 2-25 are retained here. See color plate III in the end of this book.)

3 唐菖蒲花粉原生质体及其萌发花粉管的超微结构研究

An ultrastructural study on pollen protoplasts and pollen tubes germinated from them in *Gladiolus gandavensis*

Abstract

Large quantities of protoplasts were isolated enzymatically from the mature pollen grains in *Gladiolus gandavensis*. Regeneration of cell wall and germination of pollen tubes were performed during culture of purified pollen protoplasts in K_3 medium supplemented with 32% sucrose, 0.1 mg/L 2,4-D, 1 mg/L NAA and 0.2 mg/L 6-BA, with a germination rate up to 47.7%. The materials were fixed gently with gradually increasing concentration of glutaraldehyde, followed by osmium, then preembedded in a thin layer of agar and surveyed under an inverted microscope so as to select desired specimens for subsequent procedure. Small agar blocks containing specimens were dehydrated through ethanol-propylene oxide series, embedded in araldite and ultratomed. Electron microscopic observations showed that the pollen protoplasts were surrounded by a smooth plasma membrane and with ultrastructurally intact cytoplasm, a vegetative nucleus and a generative cell. After 8h of culture, wall regeneration commenced resulting in a multilayered, fibrillar wall structure which was different from the intine. No exine was formed. Numerous vesicles participated actively in the wall formation. The wall was uneven in thickness around its periphery; a thickened area somewhat resembling to germ furrow was formed, from which pollen tube emerged. The tubes contained abundant plastids, mitochondria and dictyosomes. Vesicles were released out of the plasma membrane and involved in tube wall formation. After 18h of culture, the vegetative nucleus and generative cell had migrated into the tube. Technical points of preparing pollen protoplast specimens for ultastructural studies and the features of wall regeneration in pollen protoplast culture are discussed.

20世纪70年代以来,先后从四分体、小孢子、花粉粒和花粉管中分离出原生质体,并作了初步的培养工作[1~5,7~9,18,20,21]。这为花粉生物学研究提供了一个新的研究系统[8]。但迄今有关研究甚少,其中花粉原生质体培养的超微结构仅有两篇报道:Bajaj等(1975)首先对烟草四分体原生质体培养条件下的再生壁作了电镜观察[8]。最近,Miki-Hirosige等(1988)对麝香百合花粉原生质体培养时细胞壁的再生作了较详细的研究[18]。近年来,我们在鸢尾、风雨花、萱草三种植物上建立了大量分离花粉原生质体技术,并在萱草花粉原生质体培养中形成了花粉管状等结构[3,4]。本文进一步从唐菖蒲成熟花粉粒中分离出原生质体,经过培养,诱导了大量原生质体形成细胞壁和萌发出花粉管。在此基础上,进行了超微结构的研究。

材 料 与 方 法

一、花粉原生质体分离与培养

供试材料为鸢尾科的唐菖蒲(*Gladiolus gandavensis*)。取开花当天的未开裂花药,用75%乙醇表面消毒。取出花粉,置于酶液中,在(25±1)℃的黑暗条件下酶解。酶液含0.8%纤维素酶("ONOZUKA"R-10)、0.8%果胶酶(SERVA)、0.5%葡聚糖硫酸钾、15%甘露醇与K_3培养基[19]的大量元素。酶解时间一般为3~4h。然后将酶液以500r/min离心8min,弃去上清液,加入含32%蔗糖的K_3培养基再离心一次。吸取漂浮在上层的原生质体,加入等量的加浓1.5倍的$W_5^{[17]}$洗液(NaCl 231mmol、$CaCl_2$2HO 187.5mmol、KCl 7.5mmol、葡萄糖7.5mmol)洗涤一次,获得较纯的原生质体。并接种在附加32%蔗糖,0.1mg/L 2,4-D, 1mg/L NAA, 0.2mg/L 6-BA 的 K_3 培养基中。原生质体密度为 $10^4 \sim 10^5$ 个/ml。在(25 ± 1)℃的黑暗条件下培养。

二、电镜制样方法

用培养基配制戊二醛固定液,以0.5%、1%和2%三种浓度依次固定(但花粉粒直接用2%戊二醛固定),在室温下总共固定3~4h。培养基洗涤一次,转入0.05mol磷酸钠缓冲液(pH值约6.8)洗涤三次,再用同一缓冲液配制的1%锇酸固定,置冰箱过夜。次日用同一缓冲液洗三次。固定后的材料用1%琼脂进行预包埋。即将熔化的35~40℃的琼脂与材料轻轻混匀,用吸管吸出平铺在载玻片上,厚度约1mm。在倒置显微镜下挑选所需要的材料,切成小块。然后将含材料的琼脂小块经乙醇逐级脱水,环氧丙烷过渡,Araldite(Durcupan ACM, Fluka)渗透与包埋。在 Sorvall MT-6000 型超薄切片机上用玻璃刀切片。经醋酸双氧铀与柠檬酸铅先后染色,置于JEM-100CX/Ⅱ型透射电子显微镜下观察和摄影。

结 果

唐菖蒲花粉粒在酶液中吸水膨胀,外壁在萌发沟处裂开,露出内壁。内壁经酶液作用完全降解,释放出圆球形的花粉原生质体。在K_3培养基中培养8h,在光镜下可见再生一层透明的细胞壁,并伴有再生壁的局部增厚。培养到18h,萌发花粉管。一般一个花粉原生质体只萌发一根花粉管,少数情况产生2~3根花粉管,萌发率达47.7%。花粉管长短不一,其中有的长达280μm,外形比较正常,内部有细胞质流动。电镜观察结果分述如下:

一、花粉粒

成熟花粉粒呈椭圆形。花粉壁包括外壁与内壁。外壁覆盖层不连续,基粒棒之间的空隙充满嗜锇物质。基足层厚薄不匀。萌发沟处外壁变薄并消失,露出内壁。内壁连续完整,萌发沟处较厚,从萌发沟向两侧延伸逐渐变薄。内壁分为两层,在萌发沟处加上较厚的小管状结构的外层而分为三层。在质膜和内壁之间的电子透明区域有较多的油滴,特别是在萌发沟油滴更多。花粉粒的细胞质具有较高的电子密度,细胞器丰富,有生殖细胞与营养核。

二、花粉原生质体

花粉原生质体呈圆球形,脱壁彻底,外表仅为质膜包围。细胞质丰富,细胞器保存完好。由于吸水膨胀,液泡变得较为发达(图版Ⅰ,1)。线粒体众多,一般呈椭圆形或长形,内嵴发达。质体均匀分布在细胞质中,不含淀粉,呈电子不透明状态。内质网为粗糙型,在生殖细胞和营养核周围比较集中,有时呈环状分布在细胞质中。高尔基体分泌许多小泡。细胞质中含大量游离核糖体(图版Ⅰ,2~4)。

营养核与生殖细胞相连位于原生质体的一侧。营养核呈不规则形状。生殖细胞为纺锤形,细胞质较少,内含线粒体、游离核糖体和小液泡,但不含质体。细胞核很大,有分散的染色质(图版Ⅰ,2、4)。

三、细胞壁再生

花粉原生质体在培养条件下有旺盛的细胞壁再生能力。培养 8h,已开始再生细胞壁,形成较松散的纤丝结构,其间夹杂着不少含纤丝物质的电子密度很高的小泡(图版Ⅱ,5)。其中,有的小泡较小,似乎直接从高尔基体分泌而来;另一些较大,可能由细胞质中的小液泡组成。小泡通过胞吐作用排出质膜外。因此,在质膜边缘可以见到很多这类小泡(图版Ⅱ,6)。显然,小泡积极地参与了壁的合成。随着培养时间增加,新壁逐渐加厚,至 18h,再生壁虽仍为纤丝结构,但明显分为疏密相间的多个层次,在靠近质膜处的一层集中地含有大量的小泡(图版Ⅱ,7)。

在整个原生质体表面,再生壁厚薄不均,有局部增厚现象,类似花粉的萌发沟。但增厚部分仍为纤丝结构,而无萌发沟的管状和分层的状况(图版Ⅲ,9)。在培养 8h 的材料中,在局部增厚处,往往有小泡排出质膜外(图版Ⅱ,6),有时看到有一团电子密度极高的物质,其中含有纤丝物质的小泡、同心环状的膜状结构、以及解体的细胞器等。许多泡状物质从中释放到再生壁中(图版Ⅱ,8)。培养 18h,这类物质已变成电子密度较低的纤丝结构(图版Ⅲ,10),它们可能与细胞壁的局部增厚有关。

在培养条件下,还可看到,线粒体多分布在细胞的边缘,这似乎与细胞壁的合成有关。质体仍然很多,有些已有淀粉合成。粗糙内质网分散在细胞质中,油滴明显增多。有时观察到膨大的内质网包裹小泡、核糖体和油滴等物质的现象。

四、由花粉原生质体萌发花粉管

花粉原生质体培养到 18h,萌发出花粉管,花粉管往往从局部增厚的再生壁处突出(图版Ⅲ,12)。管尖端没有帽区结构。管基部的细胞壁较厚,从横切面可以观察到明显的分层:外层为纤丝结构,中间为电子透明带,靠近质膜有许多通过胞吐作用溢出的光滑小泡和小液泡(图版Ⅲ,13)。从花粉管基部到尖端,纤丝结构的花粉管壁逐渐变薄,看不出明显的分层。不少小泡加入到花粉管壁的合成中(图版Ⅲ,13)。

随着花粉管的伸长,细胞质逐渐由原生质体转移到花粉管中,质体较为集中在花粉管基部(图版Ⅲ,12),有些质体内形成平行排列的片层。线粒体较均匀分布在花粉管中,内嵴十分发达。高尔基体遍布于花粉管中,两端膨大,分泌出许多小泡(图版Ⅲ,11),后者通过质膜运输到管壁中。相对来说,花粉管顶端区域小泡较多,基部细胞质较密。营养核已经进入花粉管中,但尚未观察到生殖细胞的分裂。

讨　　论

一、花粉原生质体电镜制样技术

常规的电镜制样技术对于柔弱的原生质体是不适合的[6]。Fowke 等(1972)发展了适于体细胞原生质体的电镜制样方法[6,11]。在此基础上,我们还参考了用琼脂进行两栖动物卵和被子植物雄蕊毛细胞预包埋的方法[14,16],建立了一种花粉原生质体的电镜制样技术。它们的特点是:(1)用培养基配制固定剂,从低浓度到高浓度逐级用戊二醛缓慢固定,从而保持了花粉原生质体完好的内部结构;(2)进行琼脂预包埋,减少了材料因多次离心造成的损失;(3)在倒置显微镜下观察预包埋的薄层琼脂,有目的地挑选需要的材料,提高了切片和观察的命中率。结果表明,本技术用于花粉原生质体的超微结构研究是可行的。

二、花粉原生质体细胞壁再生的特点

利用花粉原生质体进行超微结构观察,为研究花粉细胞生物学,特别是壁的再生提供了一条新途径。不少工作显示,四分体、花粉粒和花粉管原生质体都能再生细胞壁[1~5,7,8,18,20,21]。但是,花粉原生质体再生细胞壁的电镜观察,迄今仅有两篇报道[8,18]。Bajaj(1975)最早在烟草四分体原生质体培养中,观察到再生壁有两种类型:一种类似于体细胞原生质体形成的纤维素组成的简单细胞壁,另一种是电子不透明物质(可能是孢粉素)沉积的复杂细胞壁。他们认为,后一情况似乎意味着外壁物质不仅起源于绒毡层,而且还可能部分地起源于小孢子原生质体的活动[7,8]。Miki-Hirosige等(1988)在麝香百合花粉原生质体培养中对再生壁的过程作了详细的观察。培养初期产生无定形的纤维素性质的细胞壁。以后逐渐增厚,形成双层,但其成分与花粉内壁不同,更无类似外壁的结构。而由花粉原生质体与花粉粒萌发的花粉管壁则是相似的。他们还详细研究了衣被小泡和光滑小泡以及小窝(pits)参加壁形成的活动。以上两篇报道均未提到形成类似萌发孔或沟的结构[8,18]。在唐菖蒲中,我们观察到大量花粉原生质体再生了细胞壁并形成了类似萌发沟的局部增厚结构。再生壁不含花粉外壁的孢粉素物质,主要由纤丝组成多层状态,与内壁的结构亦不尽相同。高尔基体的分泌小泡活跃地参与了新壁形成,看来,花粉原生质体再生壁的特点及其形成过程与体细胞的初生壁以及体细胞原生质体的再生壁[10,12]比较相似,而与花粉壁[13,15]有所不同。但是,再生壁表现出厚薄不匀的生长方式,而且局部明显增厚成类似萌发沟的特点,又与花粉内壁在萌发沟的局部增厚[13,15]较为相近。这似乎表明,花粉原生质体的遗传生理基础对于壁的生长具有某些预定作用。

图版 I   plate I

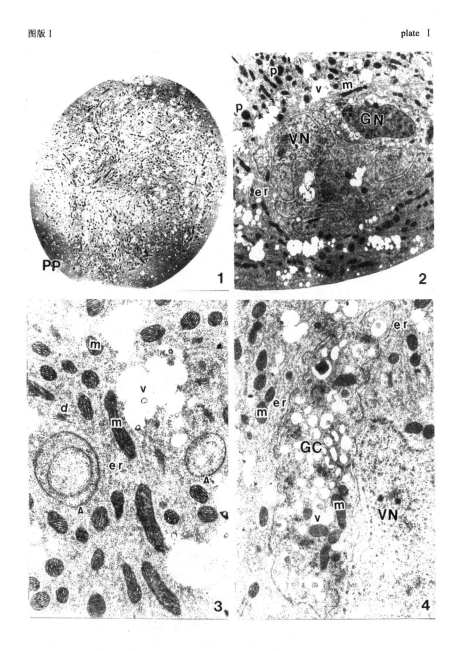

第一章 雄性器官与细胞的实验研究

图版 II   plate II

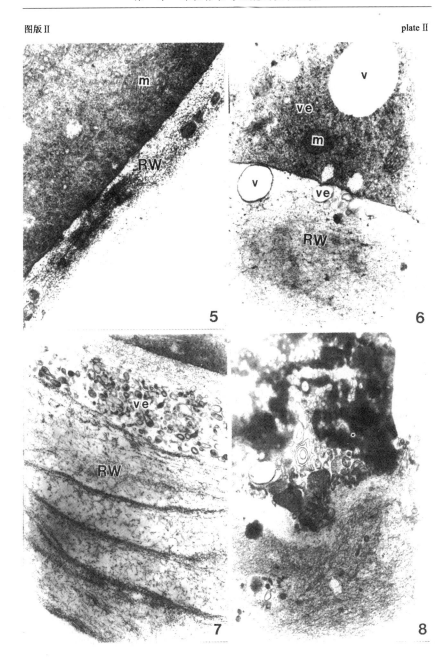

图 版 说 明

E. 外壁　GC. 生殖细胞　GN. 生殖核　I. 内壁　PP. 花粉原生质体　PT. 花粉管　RW. 再生壁　VN. 营养核　d. 高尔基体　er. 内质网　l. 油滴　m. 线粒体　p. 质体　v. 液泡　ve. 小泡

图 版 Ⅰ

1. 成熟花粉原生质体整体观。×1450　**2.** 花粉原生质体中的生殖细胞与营养核。×5400　**3.** 花粉原生质体细胞质。三角号示环状内质网。×21000　**4.** 花粉原生质体中的生殖细胞。×15000

图 版 Ⅱ

5. 花粉原生质体培养8h再生细胞壁。×43500　**6.** 培养8h形成的类似萌发沟处，小泡参与壁的合成。×28500　**7.** 培养18h，再生壁呈纤丝结构分层。×21000　**8.** 培养8h，类似萌发沟处高电子密度物质。　×28500

Explanation of plates

E. Exine　GC. Generative cell　GN. Generative nucleus　I. Intine　PP. Pollen protoplast　PT. Pollen tube　RW. Regenerated wall　VN. Vegetative nucleus　d. Dictyosome　er. Endoplasmic reticulum　l. Lipid　m. Mitochondrium　p. Plastid　v. Vacuole　ve. Vesicle

Plate Ⅰ

Fig. 1. Whole view of a pollen protoplast. ×1450　**Fig. 2.** The generative cell and vegetative nucleus in a pollen protoplast. ×5400　**Fig. 3.** Cytoplasm of a mature pollen protoplast. Triangles show circular endoplasmic reticulum. ×21000　**Fig. 4.** The generative cell in a pollen protoplast. ×15000

Plate Ⅱ

Fig. 5. wall regeneration after eight hours of pollen protoplast culture. ×43500
Fig. 6. Formation of the germ furrow-like structure after eight hours culture, showing vesicles participating wall synthesis. ×28500　**Fig. 7.** Fibrill layers of regenerated wall

after eighteen hours of culture ×21000　**Fig. 8.** Electron dense mass in a germ furrow-like structure. ×28500

(作者:吴燕、周嫦。原载:植物学报,1990,32(7):493~498。
图版3幅,参考文献20篇,仅选录图版Ⅰ、Ⅱ。)

4　萱草幼嫩花粉原生质体培养启动细胞分裂的超微结构研究

An ultrastructural study on the triggering of cell division in young pollen protoplast culture of *Hemerocallis fulva* L.

Abstract

The ultrastructural changes of young pollen protoplasts under culture condition in *Hemerocallis fulva* were studied. In comparison with the original pollen grains, the pollen protoplasts had been completely deprived of pollen wall, but kept the internal structure intact, including a large vacuole, a thin layer of cytoplasm and a peripherally located nucleus. After 8 days of culture a few pollen protoplasts were triggered to cell division: some of them were just undergoing mitosis with clearly visible chromosomes and spindle fibers; the others already divided into 2-celled units. The two daughter cells were equal or unequal in size but with similar distribution of organelles inside. Besides cell division, there were also free nuclear division, amitosis and formation of micronuclei indicating a diversity of division modes in pollen protoplast culture. A series of changes occurred during the process of induction of cell division, such as locomotion of the nucleus toward the central position, disappearence of the large vacuole, increase of electron density of cytoplasm, increase and activation of organelles, diminishing of starch granules in plastids, etc. However, the regeneration of surface wall was not sufficient: it contained mostly vesicles with only a few microfibrils. The wall separating the two daughter cells were either complete or incomplete. The weak capability of wall formation is supposed to be one of the major obstacles which has so far restricted sustained cell divisions of young pollen protoplasts under current culture condition.

花粉原生质体在花粉生物学与植物细胞工程研究中有特殊的价值。随着花粉原生质体分离与培养工作的进展,从超微结构水平上研究其在离体条件下变化的特点就自然提上日程。Bajaj 等(1975)早期曾作了烟草四分体原生质体再生细胞壁的电镜观察[5]。最近,Miki-Hirosige等(1988a,1988b)[13,14]、吴燕和周嫦(1990)[2]分别研究了麝香百合和唐菖蒲成熟花粉原生质体培养再生新壁与萌发出花粉管时的超微结构变化。但迄今幼嫩花粉原生质体培养尚无类似的工作。这里所说的幼嫩花粉是指单核后期至二核初期的花粉,即通常花药与花粉粒培养中诱导雄核发育最适合时期的花粉。近来,萱草幼嫩花粉原生质体培养首次诱导了细胞分裂[3],本文在此工作基础上,进一步研究其超微结构的变化与特点。

材 料 与 方 法

一、分离与培养

供试材料为萱草(*Hemerocallis fulva* L.)。花粉原生质体培养方法基本上参照前文[3]。选取含单核后期花粉的花蕾,低温(5~6℃)预处理6d。接种时,花蕾经75%乙醇表面消毒,取出花粉粒置于酶液中。酶液含0.5%纤维素酶、0.5%果胶酶、0.5%葡聚糖硫酸钾、18%甘露醇和 K_3 基本培养基成分。酶解5~6h后,用含18%甘露醇的 K_3 基本培养基作为洗液洗涤 3 次,再换入下列培养基中离心。培养基成分为 K_3 基本培养基附加 1mg/L2,4-D、2mg/L NAA、1% 小牛血清、1% Poly-buffer 74、9% 甘露醇和 17% 蔗糖。收集漂浮在溶液上层的原生质体置于培养皿中培养。

二、电镜制样

电镜制样与前文[3]描述的方法基本相同。花粉原生质体培养 8d 后,用 0.5%、1%、2% 三种浓度的戊二醛依次固定,0.05mol/L 磷酸缓冲液洗涤 3 次,1% 锇酸后固定,再经缓冲液洗涤 3 次,1% 琼脂包埋,在倒置显微镜下选出所需要的材料切成小块。琼脂块经乙醇脱水、环氧丙烷过渡、Araldite(Durcupan Fluka)渗透与包埋。在 Sorvall MT-6 000 型超薄切片机上用玻璃刀切片。经醋酸双氧铀与柠檬酸铅双染后,在 JEM-100CX/Ⅱ型透射电镜下观察和摄影。

结 果

一、幼嫩花粉粒

萱草单核后期的花粉粒呈椭圆形。花粉壁包括外壁与内壁。外壁覆盖层不连续,在萌发沟处基本消失,基足层厚薄不匀(图版Ⅰ,1)。内壁分为两层,外层的纤丝结构明显(图版Ⅰ,2),内层厚薄不一,主要是在萌发沟处增厚。细胞内含大液泡,仅四周有一薄层细胞质。细胞核位于花粉粒的一侧(图版Ⅰ,1)。

二、接种前的花粉原生质体

如图版Ⅰ图 3 所示,花粉原生质体脱壁彻底,外表仅为质膜包围。细胞中仍有大液泡,细胞质稀薄,细胞器保存完好。线粒体较多,其内嵴不发达而被基质充满。质体内有少量淀粉,内质网不发达,高尔基体稀少,游离核糖体较多。细胞核仍位于原生质体的一侧,内含核仁。

三、花粉原生质体的细胞分裂

培养 8 d 时,少数花粉原生质体已经启动细胞分裂。观察到有丝分裂的图像,如有的花粉原生质体正进入分裂状态。此时,大液泡已经消失,只有许多小液泡。细胞核从原生质体边缘移到中央,核膜消失。染色体排列在细胞中部的纺锤体上,纺锤丝很清晰(图版Ⅰ,4、5)。细胞质的电子密度明显加强,细胞器丰富且呈活跃状态。线粒体的数量增加,内嵴增多。质体中不含淀粉。内质网为粗糙型。高尔基体增多。核糖体有不少为串珠状多聚核糖体。细胞边缘有包裹线粒体和内质网的自体吞噬泡(图版Ⅱ,6)。花粉原生质体表面有一些含有或不含深色物质的小泡,未能形成完整的细胞壁(图版Ⅰ,4;图版Ⅱ,6)。有的花粉原生质体已分裂形成二个细胞。在电镜下共观察了 7 个二细胞单位,两个子细胞大小相近或有所区别。每个子细胞一般各具一个细胞核,其大小、形状基本相似。细胞核均较小,形状规则或不规则。核仁中有小液泡(图版Ⅱ,7、8)。少数情况下一个子细胞内含两个核。偶尔观察到微核,图版Ⅱ图 9 示一个夹在隔壁中的微核。在图版Ⅱ图 8 所示刚完成分裂的二细胞中右侧的子细胞内,核物质基本已进入细胞,仅少部分物质尚处在两个细胞的隔壁间隙,很像无丝分裂的结果。原胚内两个子细胞的细胞质没有明显区别。它们的电子密度均很高,有十分丰富、活跃的细胞器。其明显特征是:线粒体数量增多,分布均匀,内嵴发达,有的正在分裂。核糖体丰富,多呈串珠状。高尔基体囊泡两端膨大并产生小泡。内质网上附着核糖体颗粒。质体一般不含淀粉(图版Ⅱ,6)。仅在少数二细胞单位中,质体积累少量淀粉(图版Ⅱ,10)。值得注意的是,再生的细胞壁不够典型。细胞周围和细胞之间有大量小泡。大部分小泡含有深色纤丝物质。它们有的贴近质膜,有的填充在两个子细胞之间。细胞间壁宽窄不等,含有极少的纤丝物质(图版Ⅱ,9、10;图版Ⅲ,11)。其中,有的细胞壁从原生质体两侧作相对地向心自由生长(图版Ⅱ,9、10),还未完全将细胞分开。

四、花粉原生质体的游离核分裂

少数花粉原生质体在培养中发生游离核分裂。这类原生质体的外形或保持圆球形,或变为梨形,或伸长成管状。游离核多为两个,有的正在形成三个核。图版Ⅲ图12所示可能是无丝分裂形成的细胞核。核的形状往往不规则,有一个或几个核仁,内含核仁液泡。细胞质丰富,中央大液泡缩小或消失,仅有分散的小液泡。线粒体较多。质体中有少量淀粉。高尔基体分泌小泡。多聚核糖体分布较均匀(图版Ⅲ,13)。外围只有小泡,细胞壁亦不典型。

五、未分裂的花粉原生质体

许多花粉原生质体虽未分裂,形态却有若干变化。有的保持圆球形、有的变为哑铃形、结节状等。随着原生质体的形状发生改变,细胞核也有一定的变化,或从原生质体的一侧移向中央,或变成不规则形状。细胞质内大液泡也往往消失,细胞质密度增加,细胞器丰富。线粒体内嵴不甚发达。高尔基体增多。游离核糖体较多。这些现象和分裂的原生质体的情况相似。但在未分裂的原生质体中有两个明显的特点:大部分质体中含一个或多个淀粉粒。出现明显的成束的纤丝状物质,从形态上判断,可能是肌动蛋白微丝(图版Ⅲ,14、15)。这些在分裂的原生质体中是很少见的。这类原生质体一般也没有再生细胞壁,只是质膜表面有少量小泡。总的看来,到培养8d后仍未分裂的花粉原生质体,可能是没有发育前途的。

讨 论

萱草幼嫩花粉原生质体培养中的变化已经作过光镜观察[3],本文进一步从超微结构水平上进行了研究。结果表明,萱草花粉原生质体能够启动细胞有丝分裂,形成二个细胞。两个子细胞大小相似或不同,但细胞器没有明显的极性分布。除细胞分裂外,还观察到游离核分裂、无丝分裂、微核形成等现象,显示了花粉原生质体分裂方式的多样性。

萱草花粉原生质体启动分裂时发生多方面的深刻变化,呈现出活跃的状态:包括大液泡消失;细胞质电子密度增加;细胞核移位与活化;细胞器数量增加,结构复杂化;与未分裂的花粉原生质体相比,启动分裂的花粉原生质体的质体一般不含淀粉,没有微丝的成束排列。由于目前还缺乏同类的研究资料,我们还不能准确地判断其中哪些变化在启动分裂中具有特殊的作用。花药培养与花粉粒培养的超微结构研究的资料虽然不少,但不同植物的雄核发育具有不同的特点,甚至同一种植物上也得出不同的结论,似乎也还难以总结出明确的规律。例如,有的研究者认为烟草花粉在脱分化时有细胞质改组(细胞器脱分化、RNA减少的现象)[7]。但另外的研究者则认为烟草花粉胚胎发生的特点是细胞密度和核糖体的增加以及线粒体结构的变化[16]。在其它植物如曼陀罗[8]、小麦[1]、油菜[17]等的雄核发育均未发现细胞质的改组。最近有研究指出,油菜花粉启动分裂时的主要变化是中央液泡消失、核移向中部、淀粉和颗粒物质积累[17]。

  萱草幼嫩花粉原生质体在离体培养条件下细胞壁再生能力薄弱。这与麝香百合[13,14]、唐菖蒲[2]的成熟花粉原生质体以及烟草四分体[5]具有旺盛的壁再生能力颇不相同。壁的再生与细胞分裂显然有密切关系。许多植物的合子先形成完整细胞壁,然后再分裂[15]。花粉雄核发育时也是首先在质膜与内壁之间形成纤维素壁才发生细胞分裂[7,8,11,16,17]。特别是体细胞原生质体培养的大量资料表明壁的再生对细胞分裂有重要作用[6,9,10]。然而,亦有一些实验显示,当壁的再生并不完全时,细胞亦可启动分裂。例如,向日葵合子分裂发生在细胞壁尚未完全形成之际。烟草叶肉原生质体形成一种"不坚固的假壁"(non-rigid pseudo-wall)时,发生了2~3次细胞分裂,不过以后不能进行持续的分裂[12]。本文所描述的现象与之颇相类似。

  迄今,萱草花粉原生质体培养已经诱导了细胞分裂与多细胞结构的形成[3]。但是,如何促使发育继续下去直到植株再生仍是面临的重要研究课题。目前的培养条件能够诱导细胞核与细胞质转向活跃状态,从而启动细胞分裂。存在的困难之一是在现有培养条件下细胞壁再生能力薄弱,妨碍了细胞的持续分裂。这是今后在改进培养技术时

第一章 雄性器官与细胞的实验研究

需要特别注意的。

图版 I   plate I

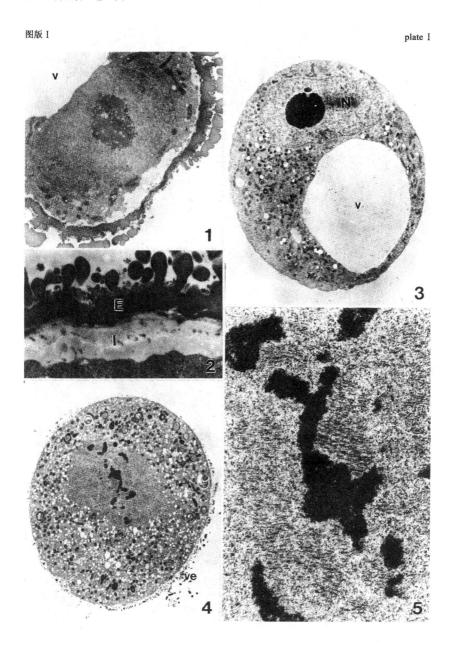

图版 II
plate II

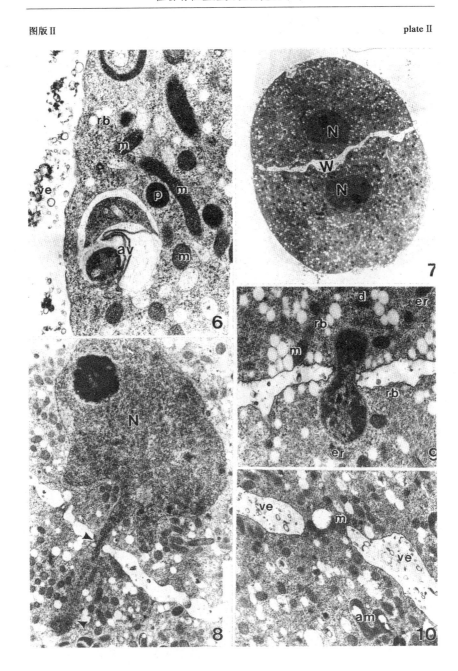

第一章 雄性器官与细胞的实验研究

图 版 说 明

E. 外壁　I. 内壁　N. 细胞核　W. 细胞壁　am. 造粉体　av. 自体吞噬泡　d. 高尔基体　er. 内质网　m. 线粒体　mf. 微丝　p. 质体　rb. 核糖体　v. 液泡　ve. 小泡

图 版 Ⅰ

1. 花粉粒局部,示花粉壁和细胞质。×5400　**2.** 花粉壁。注意内壁的分层。×15000　**3.** 花粉原生质体整体观。×2850　**4.** 启动分裂的花粉原生质体。细胞核已进入分裂状态。×2100　**5.** 上图放大。示染色体和纺锤丝。×15000

图 版 Ⅱ

6. 启动分裂的花粉原生质体。示细胞质和质膜表面的小泡。×15000　**7.** 二细胞单位。×2100　**8.** 二细胞单位中的一个细胞。少量核物质还在两个细胞交界处。×7200　**9.** 示两个细胞之间的微核。×15000　**10.** 示壁的自由生长。注意自由生长壁中的小泡和少量纤丝,以及质体中的淀粉。×10800

Explanation of plates

E. Exine　I. Intine　N. Nucleus　W. Wall　am. Amyloplast　av. Autophagic vacuole　d. Dictyosome　er. Endoplasmic reticulum　m. Mitochodrium　mf. Microfilament.　p. Plastid　rb. Ribosome　v. Vacuole　ve. Vesicle

Plate Ⅰ

Fig. 1. A pollen grain. ×5400　**Fig. 2.** Pollen wall showing exine, intine with 2 layers. ×15000　**Fig. 3.** A pollen protoplast just inoculated. ×2850　**Fig. 4.** A dividing pollen protoplast. ×2100　**Fig. 5.** Magnification of Fig. 4, showing chromosomes and spindle fibers. ×15000

Plate Ⅱ

Fig. 6. A triggered pollen protoplast showing cytoplasm and vesicles on the surface of plasma membrane. ×15000　**Fig. 7.** A two-celled unit. ×2100　**Fig. 8.** Nuclear division of a two-celled unit showing a small part of nucleus still remained between the two cells. ×7200 **Fig. 9.** A micronucleus remaining between the two cells. ×15000 **Fig. 10.** A centripatally growing wall containing vesicles, a few microfibril and starch-containing plastids in cytoplasm. ×10800

(作者:吴燕、周嫦。原载:植物学报,1992,34(1):20~25。
图版3幅,参考文献17篇,仅选录图版Ⅰ、Ⅱ。)

5 玉米花粉与唐菖蒲花粉原生质体中肌动蛋白微丝的荧光显微观察

Fluorescence microscopic observations on actin filament distribution in corn pollen and gladiolus pollen protoplasts

Abstract

Actin filament (AF) distribution in *Zea mays* pollen and *Gladiolus gandavensis* pollen protoplasts was localized by FITC conjugated phalloidin fluorescence microprobe. The pollen was incubated in Brewbaker and Kwack (BK) medium, and the pollen protoplasts were isolated enzymatically and cultured in K_3 medium containing various supplements by a previously reported method. Samples were fixed for 30 min with 1.5% paraformaldehyde dissolved in 0.1 mol/L phosphate buffer (pH 7), half strength of BK elements, 1 mol/L EGTA and sucrose, stained for 30-60 min with 1 μg/ml FITC-phalloidin in the buffer solution, and observed by fluorescence microscopy. In hydrated corn pollen grains, the AFs constituted an irregular network. Prior to germination a part of the pollen grains showed polarized pattern of AFs. At the opposite pole to the germ pore, there was a center from which AF bundles radiated and converged toward the pore, often making a spindle-shaped configuration. In just isolated gladiolus pollen protoplasts, the AFs appeared as irregular fine network. After 4-7h of culture, the AF distribution coincided in some cases with the unevenly regenerated new wall area as exhibited by FITC-phalloidin and Calcofluor White ST double staining, indicating a possible involvement of AF in wall synthesis. After 17-18h of culture, a part of the pollen protoplasts went on germination. The AFs became polarized in such protoplasts and converged into the tubes produced, and ran longitudinally along the tubes just like in the tubes germinated from pollen grains. However, in ungerminated pollen protoplasts, the AFs behaved abnormally, showing various irregular arrangements. When protoplasts bursted, the actin aggregates often located at the protrusion site from which the protoplasts would burst, and were discharged into the medium. In neither corn pollen nor gladiolus pollen protoplasts AFs were observed within the generative or sperm cells.

高等植物细胞中微丝骨架的研究是从20世纪70年代开始的[2,3]，但早期的研究是根据超薄切片和电镜观察，只能了解微丝的细节而不能认识它在细胞中的整体分布情况。20世纪80年代以来，由于应用了以荧光染料标记的、与肌动蛋白专一结合的鬼笔碱(phalloidin)作为荧光探针，使得在广泛的植物细胞中能够进行微丝的三维格局的观察[7,8,14]。其中，关于花粉与花粉管中微丝的研究涉及许多方面：较早的工作有Perdue与Parthasarathy(1985)[9]和Heslop-Harrison等(1986)[5]分别关于花粉管中和花粉萌发前后微丝形态的观察。Kohno与Shimmen(1987)研究了高浓度钙离子诱致花粉管中微丝的碎裂和细胞质流动的停止[6]。Tiwari与Polito(1988)详细观察了梨花粉在水合与萌发过程中微丝的变化过程[15]。另一方面，Pierson等(1986)不仅观察了离体生长的花粉管，而且成功地观察了由麝香百合花柱道中剥离的花粉管中的微丝。该作者还采用二甲基亚砜(DMSO)处理代替醛固定的技术,促进花粉对荧光染料的通透性,获得了比常规方法更清晰和细致的微丝图像[11]。类似的方法在研究不同发育时期花粉中微丝变化方面亦获成功[16]。最近,Pierson 等(1989)又报道了同时显示花粉管中微管与微丝的技术[12]。另一方面,关于应用荧光探针研究植物原生质体中微丝的资料却很少见。Hasezawa 等(1989)研究了烟草原生质体培养初期微管、微丝、纤维素纤丝的分布状态,以探讨三者在决定细胞形态变化上的相互关系[4]。Rutten(1988)摘要报道了由烟草花粉管分离的亚原生质体中微管与微丝的变化。

综上所述,用荧光显微术研究花粉和原生质体中的微丝骨架是近年突起的课题,无论在研究的广度和深度方面都还有不少空白。本文报道我们在玉米花粉和唐菖蒲花粉原生质体两方面所作的初步结果。

材料与方法

实验在武汉大学生物系植物细胞胚胎学研究室完成。材料为玉米(*Zeamays* L.)和唐菖蒲(*Gladiolus gandavensis* L.)的成熟花粉。

花粉水合与萌发　玉米与唐菖蒲花粉分别用含 20% 与 12% 蔗糖的 Brewbaker 与 Kwack(BK)培养基,在 26℃下保温培养 1~2h。

花粉原生质体分离与培养　唐菖蒲花粉原生质体分离方法与周嫦(1988)在其它植物上提出的方法[1]基本相同:花粉在 0.5% 纤维素酶与 0.5% 果胶酶(溶于 K_3 培养基,附加0.5% 葡聚糖硫酸钾与 16% 甘露醇)中处理 3.5~4h,分离出原生质体,经过滤、离心纯化后,用不含酶的培养基洗涤,转入 K_3 培养基(附加 1% 小牛血清、1% Polybuffer 74、1mg/L 2,4-D、2mg/L NAA、1mg/L BA、22% 蔗糖、3% 甘露醇)中,于26℃下培养。

制样方法　花粉或花粉原生质体用 1.5% 多聚甲醛(溶于 0.1 mol/L pH7 的磷酸缓冲液、1mmol/L EGTA、20%~33% 蔗糖、1/2 BK 培养基成分)固定 30min。用不含多聚甲醛与蔗糖的缓冲液与培养基洗 3~4 次。然后用 1μg/ml 异硫氰酸荧光素-鬼笔碱(FITC-phalloidin, Sigma)在黑暗下染色 30~60min。缓冲液洗 3 次,离心富集材料,加入等量甘油,取样封藏于载玻片上以备观察。部分花粉原生质体在上述染色后,用荧光增白剂 Calcofluor White ST 复染 1min,以显示再生壁。

观察与摄影　用 Olympus NEW VANOX AHBS-514 显微镜落射荧光装置。观察微丝采用 B 激发系统;观察再生壁采用 U 激发系统。

结　　果

一、玉米花粉

玉米花粉由于壁薄而自发荧光弱,比较容易透过花粉壁观察到内部被 FITC-phalloidin 染色的肌动蛋白荧光。在未经水合的、直接固定的花粉中,肌动蛋白呈均匀荧光,看不出微丝结构。水合后,出现下列各种变化:多数情况是在花粉细胞质中出现由较短的微丝组成的、不规则的网络(图版Ⅰ,1、2)。有些花粉中微丝细长而盘绕(图版Ⅰ,3)。进一步出现微丝极性分布的格局,通常是以萌发孔为一极,以相对一端为另一极,在两极之间有众多微丝束平行排列,微丝的整体结构呈纺锤体状(图版Ⅰ,4)。在有些花粉粒中,与萌发孔相对的一极可以看到一个微丝聚集的中心,在极面观上呈星芒状,微丝由此中心向对极辐射(图版Ⅰ,5),并汇集于萌发孔(图版Ⅰ,6、7)。这种状态似乎是花粉萌发的前兆。在正在生长的花粉管中,微丝束通常与花粉管纵轴平行。无论在花粉粒或花粉管中均未看到精细胞中有微丝。

二、唐菖蒲花粉原生质体

刚从花粉粒中分离出来的花粉原生质体,肌动蛋白以很细的丝状作不规则的排列(图版Ⅰ,8)。培养 4~7h 后,花粉原生质体正值新壁再生时期,壁的再生是不均衡的,有些部位较快,有些部位较慢。这时,通过 FITC-phalloidin 与 Calcofluor White ST 双重染色,可以看到一部分原生质体中微丝的分布恰好和再生壁的部位互相吻合(图版Ⅱ,15~18)。似乎暗示微丝的活动与壁的形成之间有一定的关联。但并非在所有场合均出现这种规律性的联系,说明情况的复杂性。

培养 17～18h 后,部分原生质体萌发出花粉管。图版Ⅰ图 9 示一个正在萌动的原生质体,其肌动蛋白呈密布的网络状。有些原生质体中微丝呈极性分布,其排列方向和花粉管伸出的方向一致(图版Ⅰ,10)。图版Ⅱ图 14 示一个花粉原生质体已产生较长的花粉管,此时原生质体及其花粉管均已再生新壁,在原生质体中尚可看到极性排列的微丝向花粉管中汇集(图版Ⅱ,13)。由原生质体产生的花粉管,有不少是具有正常的外形和生长行为,在这类花粉管中微丝的排列(图版Ⅰ,11)与由花粉粒萌发的花粉管中微丝的排列(图版Ⅰ,12)并无二致。

在培养中也有不少原生质体表现不正常,其中有些在培养 18h 后仍无萌发迹象。这类原生质体中的微丝常常是异常的:有的呈长索状,在原生质体表层盘绕成环带(图版Ⅱ,19、20);有的在原生质体较深层处呈漩涡状分布(图版Ⅱ,21);有的则呈粗短的条杆状(图版Ⅱ,22)。原生质体破裂时肌动蛋白的分布也往往是异常的,它们常聚集成浓密的团块,分布在原生质体即将破裂的凸起处(图版Ⅱ,23～26),最后随着凸起的破裂喷泄而出,在泄出物中呈现荧光明亮的团块(图版Ⅱ,27、28)。在花粉管尖端破裂时,肌动蛋白也有类似的表现。

在唐菖蒲花粉原生质体中,同样没有看见生殖细胞中有被 FITC-phalloidin 显示的微丝。

讨 论

1. 花粉管是研究微丝骨架的重要材料,这主要是由于花粉管中有活跃的细胞质流动,同时花粉管尖端的多糖分泌小泡积极参与管尖新壁合成,二者均与微丝的功能有关。关于花粉萌发前的微丝动态,研究相对较少。Heslop-Harrison 等(1986)发现:英地百合未水合的花粉粒中存在着被 thodamine 或 FITC 标记的 phalloidin 染色的众多的纺锤针状小体,认为它们是花粉中肌动蛋白的贮藏形式。花粉水合以后,肌动蛋白转变为丝状网络形态。萌发时,微丝向萌发孔汇集并进入花粉管中[5]。Tiwari 与 Polito(1988)报道:梨的未活化花粉中,肌动蛋白以粗颗粒状围绕营养核并以环形小体状分布在周质中。花粉水合活化后，它们转变成丝状，在三个萌发孔之间定向排列。萌发时，微丝束向萌发孔汇集并进入花粉管[15]。Pierson 在麝香百合花粉中同样观察到微丝由不规则的网络状转变为向萌发孔汇集的极性分布状态。我们在玉米花粉和唐菖蒲花粉原生质体萌发花粉管过程中也看到微丝的极性化现象，在玉米花粉粒中有时还发现与萌发孔相对的一极存在一个微丝聚集的中心。在其它植物中未曾报道这样一个中心，我们目前对其意义尚不能作出合理的解释。根据以上几项研究结果，可以认为肌动蛋白的分布动态是花粉萌发过程中一系列细胞生物学变化的一个重要方面，值得进一步深入研究。

2. 关于原生质体分离与培养过程中细胞骨架的变化，在微管方面已有许多资料，但在微丝方面可以说刚刚起步。据 Hasezawa 等(1989)的观察，刚分离的烟草原生质体中微管和微丝均是随机分布的。当原生质体培养后再生细胞壁并伸长时，它们发生有序的重排，并与细胞壁纤丝的排列表现一定的相关。微丝的变化常追随微管的变化，似乎被后者的格局所决定[4]。Rutten(1988)报道，刚分离的烟草花粉管亚原生质体尚无有组织的细胞骨架，培养后转变为有组织的状态。亚原生质体出芽时，在芽处出现一环微丝。由芽伸长为管状结构时，管中微丝的排列与管的生长方向垂直。在不生长的亚原生质体中则有短的微丝随机分布[13]。根据本文对唐菖蒲花粉原生质体的观察，刚分离的原生质体中微丝同样是不规则分布状态。培养后有两点值得注意：第一，至少在一部分原生质体中，微丝分布与新壁再生的部位是吻合的，暗示二者有一定的关联。第二，当原生质体产生正常花粉管时，微丝的行为和由花粉粒萌发花粉管时是一致的，而当原生质体不生长或表现异常时，微丝也表现出其它的排列方式与异常行为。这说明微丝的变化和原生质体的发育前途有密切关系。

图版 I

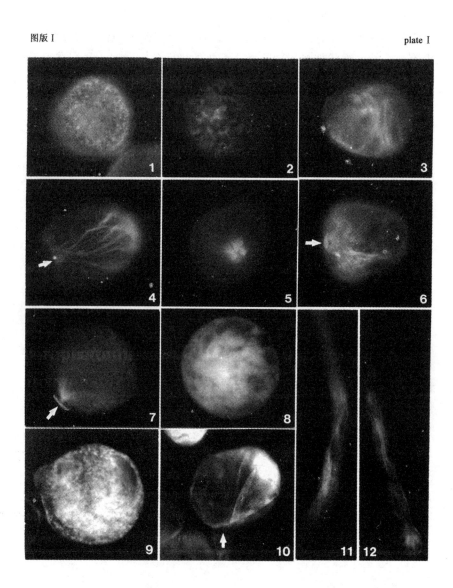

plate I

图 版 说 明

图 版 Ⅰ

FITC-phalloidin 荧光染色显示肌动蛋白微丝。**1~7.**玉米;**8~12.**唐菖蒲。**1.** 玉米水合花粉中的微丝网络。×480　**2.** 微丝网络的表面观。×480　**3.** 水合花粉中的细长微丝。×480　**4.** 极性排列的微丝束由萌发孔对侧横贯花粉粒向萌发孔(箭头)汇集。×560　**5.** 萌发孔对侧星芒状微丝中心的极面观。×480　**6.** 微丝向萌发孔(箭头)汇集。×480　**7.** 萌发孔(箭头)的侧面观,示微丝汇集状况。×480　**8.** 唐菖蒲刚分离的花粉原生质体中的肌动蛋白分布状况。×350　**9.** 一个正在开始萌发的原生质体中的微丝网络。×350　**10.** 微丝束在原生质体中呈极性排列并向花粉管(箭头)汇集。×270　**11.** 由花粉原生质体产生的花粉管,示微丝纵向排列。×300　**12.** 由花粉粒萌发的花粉管,示微丝纵向排列。×200

Explanation of plate

Plate Ⅰ

Actin filament (AF) visualized by FITC-phalloidin fluorescence probe. **Figs. 1-7.** *Zea mays*. **Figs. 8-12.** *Gladiolus gandavensis*.

Fig. 1. AF network in a hydrated corn pollen grain. ×480　**Fig. 2.** Surface view of the AF network in a hydrated pollen grain. ×480　**Fig. 3.** Elongated AFs in a pollen grain. ×480　**Fig. 4.** Polarized AF bundles penetrating a pollen grain from the opposite site of the germ pore toward the pore (arrow). ×560　**Fig. 5.** Polar view of an AF center located at the opposite pole of the germ pore. ×480　**Fig. 6.** The AFs converging to the germ pore (arrow). ×480　**Fig. 7.** Side view of a germ pore (arrow), showing the converging AFs. ×480　**Fig. 8.** Actin distribution in a just isolated gladiolus pollen protoplast. ×350　**Fig. 9.** AF network in a pollen protoplast which began to germinate. ×350　**Fig. 10.** Polarized AF bundles converging to the tube (arrow) of a pollen protoplast. ×270　**Fig. 11.** Longitudinal arrangement of AFs in the pollen tube germinated from a pollen protoplast. ×300　**Fig. 12.** Longitudinal arrangement of AFs in the pollen tube germinated from a pollen grain. Compare Fig. 11 with it. ×200

(作者:周嫦、杨弘远、徐是雄。原载:植物学报.1990,32(9):657~662。图版2幅、参考文献16篇,仅选录图版Ⅰ。)

6 甘蓝型油菜和紫菜苔花粉原生质体的大量分离

Release of pollen protoplasts in large quantities in *Brassica napus* and *B. campestris* var. *purpurea*

Abstract

Protoplasts were released in large quantities from mature pollen of *Brassica napus* L. and *B. campestris* var. *purpurea* for the first time, with a yield up to 66.7% and 70.4% respectively. Most of the pollen protoplasts were viable as tested by fluorochromatic reaction with fluorescein diacetate. The success of isolation of pollen protoplasts in these two *Brassica* species relied on a technique modified from the previous method developed for several monocotyledonous flowers. The procedure included two steps: First, the pollen was hydrated in 1mol/L of sucrose solution at 28-30℃ for ca. 9h. During this process, the exine of most pollen dehisced and was detached from the pollen grains which were then covered by intine alone. Second, the hydrated pollen was transferred into an enzyme solution containing 1% cellulase, 1% pectinase, 1mol/L mannitol, 0.5% potassium dextran sulphate and K_3 medium salts. After 4-6 of enzymatic maceration, the intine was degradated resulting in the release of protoplasts. Factors affecting the two steps have been investigated.

近几年来,花粉原生质体的研究有很大的进展[4],不仅从好几种植物的花粉中分离出大量原生质体,而且开展了有关培养[1,3,5]、融合[12]、超微结构[1,9]、微丝骨架[5]等多方面的研究工作。但是,迄今能够大量分离花粉原生质体的,主要是几种单子叶植物,如百合科的麝香百合[7,11]、萱草[2]、黄花菜[14]，石蒜科的风雨花[2]，鸢尾科的鸢尾[2]、朱顶红[14]等。至于双子叶植物（例如烟草）花粉原生质体的分离，虽然早期曾有报道，但并不能获得大量生活的花粉原生质体[6]。从另一角度看，上述几种植物均为花卉植物，而在主要农作物中，无论双子叶植物或单子叶植物，至今没有一种能够获得大量的花粉原生质体。这种现状对于本项技术的进一步应用有所限制。甘蓝型油菜和紫菜苔既属于双子叶植物中的十字花科芸苔属，又是重要的油料或蔬菜作物。我们改进了原有技术，建立了一种新的分离方法，成功地由这两种植物的花粉中分离出大量原生质体。

材 料 与 方 法

一、材料

甘蓝型油菜（*Brassica napus* L.）和紫菜苔（*B. campestris* var. *purpurea*）。

二、花粉原生质体分离

第一步：水合。由刚刚开放的花朵中摘取花药，置于水合液（成分见后文）中，用平头玻璃棒轻压挤出花粉。花粉悬浮液经400目不锈钢网过滤除去药壁残渣，盛于培养皿中，在28~30℃下保温。定时取样观察计数。第二步：酶解。将水合后的花粉悬浮液离心沉淀，除去上清液。将沉淀物转入含1%纤维素酶（Onozuka R-10）和1%果胶酶的介质（成分见后文）中，在28~30℃下酶解。定时取样观察计数。

三、观察与计数

用 AO 1820 Biostar 型倒置显微镜和 Olympus BH-2 型显微镜观察花粉原生质体的分离过程并摄影。用荧光素二醋酸酯（FDA，50μg/ml）鉴定生活力，在 AO H110 Fluorestar 型荧光显微镜（滤光组合为2073，即激发滤光片436，分色镜450，阻挡滤光片475）下观察与摄影。计算通过水合所获得的去外壁花粉分离率（去外壁花粉数/观察花粉总数×100%)和通过酶解后所获得的花粉原生质体分离率(花粉原生质体数/观察花粉总数×100%)。

结 果

实验初期,采用原有的水合与酶解同时进行的方法(简称"一步法")[2]尝试分离甘蓝型油菜和紫菜苔花粉原生质体。结果只有极少数花粉能够释放原生质体。多数花粉不能释放原生质体,在酶液中最终破裂。以后经过改进,建立了"二步法",在两种植物中均分离出大量花粉原生质体。此法的特点是:首先将花粉经过一段时间水合,通过吸胀作用促使外壁脱离,得到仅有内壁覆盖的花粉(简称"去外壁花粉"),然后再将后者酶解除去内壁,得到原生质体。分别研究了这两个步骤中影响分离效果的关键因素。照片仅以油菜为例。

一、第一步:水合脱去外壁

油菜与紫菜苔花粉均为三沟型。干燥的花粉呈长椭球形,置入水合液后很快膨胀为椭球形(图版Ⅰ,1)。水合1h以后,部分花粉的外壁沿萌发沟一端开裂,露出内壁的一部分(图版Ⅰ,2)。水合继续进行,外壁完全脱离,仅由内壁包围的花粉游离出来,此种去外壁的花粉仍保持椭球形(图版Ⅰ,3~5)。FDA检测表明它们是生活的。

去外壁花粉的分离率,因下列因素的变化而异:

水合液成分 比较了0.8mol/L的蔗糖水溶液、甘露醇水溶液以及在二者中分别加入K_3培养基的无机盐的效果。结果表明,在加有K_3培养基的水合液中花粉难以脱去外壁,而在单独的蔗糖或甘露醇水溶液中脱壁良好,其中甘露醇比蔗糖效果更好(表1)。但是进一步酶解的结果,来源于甘露醇液的去外壁花粉很少产生原生质体;而来源于蔗糖液的去外壁花粉则易于产生原生质体。因此以后的实验均采用蔗糖水溶液作为水合介质。

蔗糖浓度 蔗糖浓度对脱外壁的效果有很大影响。浓度过低(0.5mol/L)或过高(1.2mol/L)均很难获得去外壁花粉。在油菜与紫

菜苔两种材料中,均以 1.0mol/L 蔗糖溶液最为适宜(表2)。

水合时间　花粉的脱外壁过程相当缓慢。在 1mol/L 蔗糖水溶液中,从水合2h后开始有少数花粉脱去外壁。随着时间延长,去外壁花粉数递增,至9h达最高值。以后再延长水合时间作用不大。在保温过程中,间常轻轻摇动培养皿有利于促进外壁脱离。水合一定时间后加水冲击可加速脱壁过程,但FDA测定表明,这不利于保存去外壁花粉的生活力。

二、第二步:酶解获得花粉原生质体

将水合后的花粉转入酶液后,由于花粉的内壁暴露于酶液中而降解,原来椭球形的去外壁花粉逐渐变圆,成为典型的原生质体(图版Ⅰ,6)。酶解的效果与水合时间有关:当水合不充足时即转入酶液,由于相当多的花粉尚未完全脱去外壁,致使酶解所得原生质体不是很多。充分水合后,酶解所得原生质体产量较高(图版Ⅰ,7),90%以上具有生活力(图版Ⅰ,8)。酶解的时间以 4~6h 为宜。

酶解的效果因下列因素的变化而异:

酶浓度　比较了 0.5%~2% 范围内不同配比的纤维素酶与果胶酶的浓度,结果表明原生质体分离率没有显著差异。单独的纤维素酶亦有效,但单独的果胶酶则无效。以后实验均采用1%纤维素酶与1%果胶酶的配比。

渗透压调节剂　比较了 1mol/L 蔗糖或甘露醇作为酶液中渗透压调节剂的效果。结果表明,在油菜与紫菜苔两种材料中均以甘露醇的效果较好。

盐离子　在酶液中加入盐离子对原生质体分离率有重要作用。加入 K_3 培养基的大量元素与微量元素[10]或 CPW13M[8] 均有很好的效果,而不加盐离子的酶液中原生质体分离率很低(表4),原生质大多破损死亡。

保护剂　在酶液中分别添加牛血清白蛋白(BSA)、聚乙烯吡咯烷酮(PVP)、葡聚糖硫酸钾(PDS),与不含以上保护剂的对照相比,没有明显的有利作用,惟PDS略有改进效果(表5)。

总结上述实验结果,"二步法"的主要技术要点如下:

(1) 水合:将花粉悬浮于 1mol/L 蔗糖水溶液中,在 28～30℃下保温 9h,可有约 61%(油菜)或 75%(紫菜苔)的花粉脱去外壁。

(2) 酶解:将水合后的花粉悬浮液离心沉淀,转入含 1% 纤维素酶、1% 果胶酶、1mol/L 甘露醇、0.5% 葡聚糖硫酸钾的 K_3 培养基或在 CPW13M 中重新悬浮,在 28～30℃下酶解 4～6h,可获得高达 67%(油菜)或 70%(紫菜苔)的花粉原生质体分离率。

讨 论

由花粉分离原生质体的主要困难在于花粉外壁的基本成分孢粉素迄今无法酶解。近年大量分离花粉原生质体的成功事例均是利用花粉水合后的吸胀作用撑破外壁,使含果胶质与纤维素的内壁大面积地处于酶的作用之下而被降解[4]。因此,如何促进外壁裂开是技术的关键所在。外壁结构的种属特点具有决定的作用。周嫦与杨弘远(1989)曾指出:当前取得成功的几种百合科、石蒜科与鸢尾科植物,其外壁较薄、萌发沟宽阔等生物学特点是它容易开裂的重要因素[4]。这类花粉置于酶液以后很快吸水膨胀、外壁开裂,可以说,其水合与酶解几乎是同时进行的,采用"一步法"即可顺利地分离大量花粉原生质体。本文所研究的两种芸苔属花粉同样具有萌发沟,但外壁较厚而坚硬,用"一步法"只能分离出很少的原生质体。如将水合与酶解两个步骤分开,使花粉先经较长时间水合,脱去外壁而获得仅有内壁的花粉;然后再进行酶解,除去内壁而获得大量原生质体。实验结果表明,在"二步法"中,第一步所用的水合液只可含有蔗糖,不可含有培养基中的盐离子,否则外壁不能开裂和脱落;而第二步所用的酶液则非含有盐离子不可,否则原生质体不能保护完好。这种将水合和酶解分为先后两步的方法,也许能在更多的植物中应用于分离花粉原生质体,尚需今后试验。

甘蓝型油菜与紫菜苔是有重要经济价值的作物。它们的花粉原生质体大量分离取得成功,使这一技术今后应用于作物改良的距离更为缩短了。周嫦与杨弘远(1989)[4]、Yang and Zhou(1992)[13]曾提出利用花粉原生质体两条发育途径开展遗传工程的设想。目前在配子体途径方面已由唐菖蒲成熟花粉原生质体产生大量花粉管[1];在孢子体途径方面已由萱草幼嫩花粉原生质体培养出多细胞团[3]。Ueda等(1990)用电融合法使麝香百合的成熟花粉原生质体互相融合,经过短期培养后,融合体中来自双方的两个生殖细胞均可分裂成精子[12]。由于花粉原生质体具有单倍性的优点,今后在基因转入和转化实验中可以加以利用。下一步可以试验用摄入外源遗传物质的油菜或紫菜苔成熟花粉原生质体进行授粉、受精。在这两种植物中,幼嫩花粉原生质体的分离尚待突破,如亦能成功,则借鉴目前国际上已经相当成熟的油菜花粉粒培养技术,促使幼嫩花粉原生质体细胞分裂以至再生植株,也是很有前途的。

(作者:李仕琼、杨弘远、周嫦。原载:植物学报,1992,34(5):339~345。图版1幅,表5幅,参考文献14篇,均删去。)

7 紫菜苔花粉超低温保存及其原生质体分离

Pollen cryopreservation and pollen protoplast isolation in *Brassica campestris* var. *purpurea*

Abstract

The authors have investigated the factors affecting pollen cryopreservation in *Brassica campestris* var. *purpurea*, such as pollen development stages, cryoprotectant and the process of freezing. A suitable procedure was established as follows: Pollen grains suspended in B_5 medium containing 10% DMSO and 15% sucrose were frozen by a three-step method (0℃ $\xrightarrow{-1℃/min}$ -10℃, standing for 15 min $\xrightarrow{-1℃/min}$ -40℃, standing for 1 hr→

liquid nitrogen) and later thawed in 40℃ water bath. During a period of 60,90 days' preservation, the relative survival percentage of mature (at the day of anthesis) and nearly mature (2 days before anthesis, trinucleate stage) pollens maintained at ca. 91%; that of young pollens (7-8 days before anthesis, late uninucleate stage to early binucleate stage) slightly declined from the original 91.6% to 84.3%. Culture experiment showed that the cryopreserved young pollen could be induced to cell division just as well as the fresh pollen. The method of isolating protoplasts from fresh mature pollen developed previously was improved and simplified. As a result, protoplasts were isolated more conveniently from mature pollen and young pollen for the first time. The protoplasts from cryopreserved mature and young pollen could be obtained as well with an isolation rate of 77.4% and 35.9% respectively. However, for isolation of protoplasts from preserved young pollen, an incubation in NLN medium at 35℃ after thawing was necessary.

超低温保存作为现代种质保存技术,在植物的多种营养器官、组织、细胞和原生质体方面获得成功[5,12]。在花粉方面,已对30余种植物作了成熟花粉超低温保存的研究[4,8,11]。随着花药与花粉培养及其在育种中应用日益发展,幼嫩花粉的保存亦已开始。Bajaj(1978)[6]作了几种植物花药的超低温保存。Charne 等(1988)研究了甘蓝型油菜(*Brassica napus*)游离小孢子的超低温保存[7]。当前,花粉原生质体分离取得重要进展[1~3,10,13],这使花粉的保存有了新的研究与应用价值。本文的目的是研究紫菜苔不同发育时期花粉的超低温保存以及由新鲜和保存的花粉分离原生质体。

材 料 与 方 法

一、试验材料

供试材料为紫菜苔(*Brassica campestris* var. *purpurea*)。选用3个

发育时期的花粉:幼嫩花粉(开花前 7~8d,单核后期至二核早期),近成熟花粉(开花前 2 天,三核期)和成熟花粉(开花当天)。

二、超低温保存

冰冻保护剂预处理　在含 15% 蔗糖的 B_5 培养基中压出花粉,400目不锈钢网过滤,离心,去上清液,置 4℃ 冰箱中,待温度恒定,加入预冷至 4℃ 的冰冻保护剂(成分见实验结果)。

降温　将花粉悬浮液转入塑料小管(Bio-Freeze™ Vials Costar Co. USA),置程序降温仪(Kryo 10-1.7,Planer Products Ltd.,England)内,由微机控制降温速度与程序进行降温,然后将小管置液氮中保存。试验了两种降温程序:二步法(由 0℃ 以 -1℃/min 的速度降至 -40℃,停留 60min,入液氮);三步法(由 0℃ 以 -1℃/min 的速度先降至 -10℃,停留 15min,以相同速度降至 -40℃,停留 60min,入液氮)。

化冻与洗涤　由液氮取出塑料小管,在 40℃ 恒温水浴内迅速化冻。用 B_5 培养基离心洗涤 2 次。

三、花粉生活力测定

用荧光素二醋酸酯(FDA 50μg/ml)染色,统计生活花粉的比率。用相对存活率(保存后与保存前生活花粉的百分比)分析保存效果。用 Olympus IMT-2 倒置显微镜的荧光显微装置摄影。

四、幼嫩花粉培养

参考 Huang 等(1989)的油菜小孢子培养方法[9],将花粉在含 13% 蔗糖的 NLN 培养基中,35℃、黑暗条件下培养 4d,再转入 25℃ 黑暗培养。

五、花粉原生质体分离

花粉置于酶液(1% 纤维素酶、1% 果胶酶、0.1% Pectolyase Y-23、0.5% 葡聚糖硫酸钾、0.8mol/L 甘露醇、0.4mol/L 山梨醇与 CPW 液)中,在 25~30℃ 下酶解 2~4h。保存花粉分离原生质体前先经化冻、洗涤。保存的幼嫩花粉试验了3种处理,详见"结果"。

结 果

一、花粉超低温保存

花粉发育时期的影响 作了3种不同发育时期花粉的比较。总的看来,成熟花粉的保存效果较好,近成熟花粉与之相近,幼嫩花粉较差。不仅如此,幼嫩花粉对保存条件比前两者敏感,只在适宜条件下才有好的保存效果。如8种不同处理中,保存的成熟与近成熟花粉的相对存活率变动在75.8%～100.0%之间,而幼嫩花粉则变动在46.2%～96.4%之间。图版Ⅰ图1~4分别示保存花粉群体的生活状态。

冰冻保护剂中蔗糖浓度的影响 冰冻保护剂的基本成分为B_5培养基、10% DMSO和蔗糖。其中蔗糖浓度对保存效果有明显影响。在"二步法"降温程序中,比较了5种蔗糖浓度:10%、15%、20%、25%和30%。花粉保存效果随蔗糖浓度的提高达到一个最高值后下降。成熟与近成熟花粉以15%蔗糖最好;幼嫩花粉以20%蔗糖最好。在"三步法"降温程序中,进一步比较15%、20%和25%三种蔗糖浓度的效果,结果3个发育时期的花粉均以15%蔗糖为佳(表1)。

降温程序的影响 比较了两种降温程序的效果。结果表明,"三步法"比"二步法"的效果显著(表1)。以幼嫩花粉为例,同为15%～25%蔗糖条件下,"三步法"的相对存活率为87.4%～96.4%,而"二步法"仅为65.2%～79.4%。成熟与近成熟花粉也是"三步法"优于"二步法"。

保存时间 根据上述实验结果,以10% DMSO、15%蔗糖,B_5培养基作保护剂,采取"三步法"程序降温,在液氮中长期保存,然后定期取样检查(表2)。成熟与近成熟花粉在保存60d、90d期间,生活力保持不变,相对存活率均达到90%以上。幼嫩花粉在保存30d后生活力有所下降,但此后变化不大,90d时的相对存活率仍达84.3%。

幼嫩花粉培养 将新鲜与保存的幼嫩花粉进行初步培养试验。培养2～3d后,有些花粉能分裂1～2次,形成2～4个细胞(图版Ⅰ,7)。

新鲜花粉与保存花粉培养结果相近,表明幼嫩花粉经超低温保存后仍具有脱分化能力。

二、花粉原生质体分离

成熟花粉原生质体分离 用本文技术将花粉直接酶解,一步完成原生质体分离,取得很好的结果。新鲜花粉原生质体分离率可达90.3%;超低温保存花粉原生质体分离率为77.4%(图版Ⅰ,5、8～11),比新鲜花粉的效果有所下降(表3)。

幼嫩花粉原生质体分离 幼嫩花粉原生质体的分离比较困难。用同上方法酶解新鲜的幼嫩花粉,原生质体分离率为32.2%。保存花粉经化冻、洗涤后直接酶解,分离原生质体的效果不佳。为改进技术,做了3种试验。花粉化冻、洗涤后:(1)直接酶解(对照);(2)在含13%蔗糖的 B_5 培养基中25℃温育20h 酶解;(3)在13%蔗糖的NLN培养基中35℃温育9h 后酶解。结果表明,试验(3)的花粉原生质体分离率达35.9%(图版Ⅰ,6),与新鲜花粉的分离效果类同(表3)。而试验(1)与(2)的分离率仅10.1%和14.9%。此外,用试验(2)、(3)的方法处理新鲜花粉,对原生质体分离则无明显影响。

表3　　　　新鲜花粉与保存花粉原生质体的分离

Table 3　Isolation of protoplasts from fresh and cryopreserved pollen grains

材料 Material	幼嫩花粉 Young pollen			成熟花粉 Mature pollen		
	观察花粉数 Pollen grains observed	花粉原生质体 Pollen protoplast		观察花粉数 Pollen grains observed	花粉原生质体 Pollen protoplast	
		数量 Number	分离率 Percentage (%)		数量 Number	分离率 Percentage (%)
新鲜花粉 Fresh pollen	441	142	32.2	236	213	90.3
保存花粉 Cryopreserved pollen	276	99	35.9	248	192	77.4

讨 论

花粉超低温保存的研究集中在成熟花粉方面[5,11]，而在未成熟花粉方面的工作几近空白。就我们所知，迄今仅有 Charne 等(1988)[7]关于甘蓝型油菜小孢子超低温保存的报道。其结果表明，保存花粉能保持胚胎发生与植株再生能力，但生活力显著下降，相对存活率仅4%～10%。本文研究了紫菜苔花粉的超低温保存及其影响因素，建立了较适宜的保存流程。不同发育时期的花粉经保存后，相对存活率可达90%以上。

多种植物材料的研究表明，影响因素的调节是超低温保存成败的关键[5,12]。就紫菜苔花粉而言，花粉发育时期是重要的内在因素，成熟与近成熟花粉对保存条件的适应能力较宽，存活率高；幼嫩花粉对保存条件敏感，存活率因条件变化而波动。幼嫩花粉处于花粉第一次有丝分裂前后时期，与成熟花粉相比，其代谢较活跃，液泡化程度较高，含水量较大。这种生理状态显然是它们对超低温保存条件要求更为严格的内在原因。冰冻保护剂、降温程序以及化冻方式是重要的外在因素。许多植物成熟花粉超低温保存时，一般不需冰冻保护剂，干燥处理后即可保存[5,11]。本文为了在相同条件下进行比较，一律采用由 DMSO、蔗糖与 B_5 培养基组成的冰冻保护剂。在此条件下，成熟花粉也能很好保存。本文采用程序降温仪，由微机严格控制三步，降温程序提供了平稳、逐步降温的条件，使花粉排水较为充分，冰冻伤害较小。此外，常用的快速化冻方法对紫菜苔花粉也是适宜的。

花粉原生质体分离已在若干植物中获得成功[1~3、10、13]，但分离技术仍有待完善。李仕琼等(1992)用"水合-酶解二步法"分离出大量紫菜苔与甘蓝型油菜成熟花粉原生质体[1]，但有水合时间过长的缺点。本文减去水合步骤直接酶解，分离效果也很好。而且首次分离出部分幼嫩花粉原生质体。主要改进措施是酶的组成中增加了 pectolyase Y-23。Tanaka 等(1987)在麝香百合近成熟花粉原生质体分离的研究中，比较了酶的种类，认为纤维素酶与果胶酶效果好，添加 pectolyase 无益[10]。周嫦(1988)分离萱草、鸢尾、风雨花时也是用的这两种酶[2]。

此后,在其进一步的研究中,发现有的只需上述两种酶即可[3]。

保存花粉能否分离原生质体及其技术上有无特点是需要研究的问题。从本试验结果看,保存的成熟花粉经化冻、洗涤后,和新鲜花粉一样,可以直接分离原生质体。保存的幼嫩花粉则需在酶解前增加"温育"处理。Withers(1985)认为保存材料在降温与化冻过程中对环境有适应性变化,往往需要经过一段时间才能恢复保存前状态[12]。简令成(1990)指出,测定保存的成熟花粉生活力或萌发率时,需经过一定时间使花粉恢复其含水量[5]。显然,紫菜苔幼嫩花粉原生质体分离的保温处理是必要的技术环节,具有恢复保存前状态的作用。

(作者:梁立、徐秉芳、郑从义、周嫦。原载:植物学报,1993,35(10):733~738。图版1幅,表3幅,参考文献13篇,仅选录表3。)

8　烟草花粉原生质体的分离

Isolation of pollen protoplasts in *Nicotiana tabacum*

Abstract

A new method for isolation of quantities of mature pollen protoplasts in *Nicotiana tabacum* has been established. The first step was to germinate mature pollen in Brewbaker and Kwack medium containing 20% sucrose. When most of the pollen grains had just germinated short pollen tubes, they were transferred to an enzymatic solution for the second step. The enzymatic solution contained 1% pectinase, 1% cellulase, 0.5% potassium dextran sulfate, 1 mol/L mannitol, 0.4 mol/L sorbitol in D_2 medium with or without 15% Ficoll. The enzymes firstly degraded the pollen tube wall and then the intine. As a result, intact pollen protoplasts were released with the isolation rate up to 50% ~70%. Factors affecting pollen protoplast isolation during the germination and maceration of pollen grains were studied. The success depended on two key points: pollen germination duration and osmoticum concentration. The optimal germination duration was 30 min at 30℃. When it was too long, long pollen tubes formed and subsequently large number of subprotoplasts instead of whole protoplasts were yielded as the case reported by previous investigators. The optimal concentration of mannitol and sorbitol in enzyme solution was as high as 1.4 mol/L in total. Lowering of the osmoticum concentration resulted in decrease of percentage of pollen protoplasts.

近年来,在若干植物中陆续有花粉原生质体分离成功的报道[1~5]。烟草作为植物组织培养与细胞工程的重要模式植物,其花粉原生质体的分离虽有报道[6],但是迄今未能分离大量花粉原生质体。前人用烟草花粉管分离的多为亚原生质体而非完整的原生质体[7~9]。本文报道用改进的方法分离大量烟草成熟花粉原生质体的实验结果。

材 料 与 方 法

供试材料是烟草(*Nicotiana tabacum* L.)品种"G-80"。将成熟花粉置于含20%蔗糖的Brewbaker和Kwack培养基中在30℃和黑暗条件下培养,用含20%蔗糖、D_2培养基[10]大量元素的培养基洗涤2次,转至酶液中在25℃和黑暗条件下酶解3~4h。酶液含D_2培养基大量元素,1%纤维素酶(Cellulase R-10),1%果胶酶(Pectinase, Serva),0.5%葡聚糖硫酸钾,1.0% mol/L甘露醇,0.4mol/L山梨醇,加或不加15% Ficoll。酶解后由培养皿中吸去酶液,用仅含1.0mol/L甘露醇和0.4 mol/L山梨醇的D_2培养基洗涤,用吸管轻轻吸去上层花粉外壁等漂浮杂质,保留沉在底部的原生质体,反复洗涤2次,得到较纯净的花粉原生质体群体。

结 果 和 讨 论

本法包括花粉离体萌发和酶解两个步骤。当萌发出短花粉管(图1)时及时转入酶液。酶液首先降解花粉管壁,然后沿花粉管伸向花粉粒,降解其内壁,从而释放出完整的花粉原生质体。经过纯化步骤,可获得较纯净的花粉原生质体群体(图2)。研究了影响花粉原生质体分离的若干重要因素。

一、花粉萌发阶段的因素

培养基、蔗糖浓度、培养温度、萌发时间等通过影响萌发状态,最终影响花粉原生质体的分离效果。例如,培养基的蔗糖浓度在10%~30%范围内,以20%时花粉萌发率高,且萌发速度适中,便于控制花粉管生长长度,因而花粉原生质体分离率最高。萌发时间在20~70min内,以30min时花粉原生质体分离率最高(表1)。此时,多数花粉已经萌发而花粉管很短(图1)。正是这种萌发状态能够较好地分离出完整的花粉原生质体。若萌发时间过短,许多花粉尚未萌发,就不能分离出大量原生质体;若萌发时间过长,花粉管很长,则减少完整花粉原生质

表1 花粉萌发时间对分离花粉原生质体的影响

Table 1 Effect of pollen germination duration on isolation of pollen protoplasts

萌发时间 Germination duration (min)	观察花粉数 No. of pollen grains observed	花粉原生质体 Pollen protoplast	
		数量 No.	分离率 (%)
20	352	128	36.4
30	303	215	71.0
40	301	202	67.1
50	311	144	46.3
60	315	163	51.7
70	341	160	46.8

体而增多亚原生质体。因此,控制萌发时间是分离烟草完整花粉原生质体的技术关键。前人分离烟草花粉管亚原生质体正是利用长的花粉管酶解的。例如,Power 采用萌发 5h 的花粉管[8];朱澂等[7]采用萌发 4h、长达 0.3～1.4 mm 的花粉管,所以获得大量亚原生质体而非完整的原生质体。

二、酶解阶段的因素

培养基、酶的种类、渗透压调节剂、Ficoll 等均有影响。其中渗透压调节剂浓度很重要。表 2 表明,甘露醇与山梨醇浓度总和在 0.8～1.4 mol/L 范围内,花粉原生质体分离率随浓度增加而大幅度提高。以 1.4 mol/L(甘露醇 1.0 mol/L,山梨醇 0.4 mol/L)效果最好。因此,采用高渗溶液是本法另一技术关键。如加入 Ficoll,渗透压调节剂的适宜浓度可下降至 1.2 mol/L,同时还有减少亚原生质体的效应。

综上所述,本文提出的"萌动-酶解法"能够较顺利地分离出烟草成熟花粉原生质体。不仅如此,对于许多植物而言,上述技术路线也可能有借鉴的价值。

(作者:汪泳、周嫦。原载:植物学报,1995,37(5):413～416。
图 2 幅,表 2 幅,参考文献 10 篇,仅选录表 1。)

9 烟草幼嫩花粉原生质体分离与早期离体发育

Isolation and early *in vitro* development of young pollen protoplasts in *Nicotiana tabacum*

Abstract

For isolating young pollen protoplasts in *Nicotiana tabacum*, the authors established two efficient enzymatic methods via anther preculture or pollen starvation pretreatment. Procedure of the first method included the following steps: 1. Cold pretreatment of flower buds with pollen at late unicellular to early bicellular stage; 2. Anther floating culture for pollen shedding into the culture medium followed by dehiscence of exine; 3. Enzymatic maceration of exine-dehisced pollen resulting in degradation of intine and release of pollen protoplasts in large quantity. Procedure of the second method involved the following steps: 1. Culture of pollen at middle bicellular in Kyo and Harada's B medium for starvation; 2. Enzymatic maceration of starvated pollen resulting in release of pollen protoplasts and subprotoplasts. Factors affecting the results of both methods as well as early *in vitro* developmental events of young pollen protoplasts were studied. The protoplasts could be induced either to trigger the first sporophytic division or to continue the gametophytic pathway leading germinatation of pollen tubes indicating their potentiality of inducing both sporophytic and gametophytic development of pathway. In rare instance a quite interesting phenomenon was observed that a pollen protoplast first divided into two daughter cells and one of which then germinated a pollen tube. It may insinuate that such pollen protoplasts initially started a sporophytic pathway could reverse to a gametophytic pathway.

花粉原生质体兼具花粉与单倍体特性,被认为是一种特殊的原生质体实验体系而受到广泛重视[1,2]。与成熟花粉相比,幼嫩花粉(单胞至二胞中期)的原生质体分离更加困难,而幼嫩花粉正是被诱导脱分化的适宜时期。因此,幼嫩花粉原生质体有望成为开展外源基因导入以获得纯合二倍体转基因植株的理想实验体系。烟草作为植物组织培养的模式植物,早在20世纪70年代即开始试探其花粉原生质体的分离,但进展甚微[1,3]。最近,本实验室以萌发-酶解法分离出较大量的烟草成熟花粉原生质体[4],但对于烟草幼嫩花粉并不适用。本文采用花药预培养与花粉粒饥饿预处理两种方法,尤其是前者高频率地分离出烟草幼嫩花粉原生质体,并初步进行了培养试验。

材料与方法

一、材料

烟草(*Nicotiana tabacum* L.)品种为"G-80",种植于湖北省农业科学院温室与本实验室网室内。

二、花粉原生质体分离

采用以下两种方法分离原生质体:

花药预培养法 取含单胞晚期与二胞早期花粉的花蕾,在4~6℃下预处理11~14d。将花药接种于Sunderland和Roberts的A培养基[5]或0.3mol/L蔗糖或0.3 mol/L甘露醇。待大部分花粉粒从花药中释放1d后,收集花粉粒转入酶液。酶液组成为:1%纤维素酶R-10、1%果胶酶(Serva)、0.1% pectolyase Y-23、0.5%葡聚糖硫酸钾、K_3培养基[6]的大量元素、1 mol/L甘露醇、0.3 mol/L山梨醇,pH5.8~6.0。28℃黑暗条件下酶解10h。分离的花粉原生质体用洗液(K_3大量元素、1 mol/L甘露醇、0.3 mol/L山梨醇,pH5.8~6.0)洗2次,K_3培养基洗1次。

花粉饥饿预处理法 取含二胞中期花粉的新鲜花蕾。将花粉接种于Kyo和Harada的B培养基[7]中。25℃黑暗条件下培养2~4d。收集花粉粒置入酶液(成分同上),28℃黑暗条件下酶解5h,转入4℃条件下过夜。离心收集原生质体,洗液洗2次,K_3培养基洗1次。

三、花粉原生质体培养

将花药预培养法分离的花粉原生质体接种于附加0.4 mol/L蔗糖和0.8 mol/L甘露醇的K_3基本培养基,pH5.8~6.0。原生质体密度调至10^4~10^5个/ml。

四、显微观察与渗透压测定

用 Olympus IMT-2 倒置显微镜观察并摄影。DAPI(10mg/L)鉴定花粉发育时期。荧光素二醋酸酯(FDA,10 mg/L)鉴定花粉原生质体的生活力。Calcoflower white(0.1%)鉴定细胞壁。用渗透压测定仪(The Advanced™ Micro-osmometer Model 3MO, U.S.A)测定酶液渗透压。

结 果

一、幼嫩花粉原生质体分离

花药预培养法 从低温处理的花蕾中取出花药,置入蔗糖或 A 培养基[5]中漂浮培养。在 1d 时间内,花药裂开,陆续释放出花粉。花粉落入培养基中继续培养 1d,大量花粉的外壁裂开,以至仅与花粉一侧相连,裸露出内壁的绝大部分(图版Ⅰ,1、2、14),此时可称为"裂外壁花粉"。后者转入酶液后,外壁迅速脱落,花粉仅存内壁,在光镜下呈折光性强的圆圈,包围着原生质体(图版Ⅰ,3),此时可称为"脱外壁花粉"。在酶液中,内壁逐步降解。当内壁局部降解形成缺口时,原生质体由缺口处逸出(图版Ⅰ,4);当内壁均匀降解时,它逐渐变得松散、模糊,最终消失,原生质体自然释放出来(图版Ⅰ,5、6)。总之,花药预培养阶段,释放出花粉,促使外壁裂开;酶解阶段,外壁脱落,内壁降解,释放原生质体。在上述两个阶段均有一些关键因素影响最终结果。在花药预培养阶段,释放花粉的数量与质量十分重要。花粉数量愈多,生活力愈强,分离的原生质体频率也愈高。而释放花粉的能力又受多方面的因素影响:植物生长季节、生理状态、花粉发育时期、花蕾低温处理的温度与时间等均有影响。外壁开裂是另一关键。刚从花药中释放出的花粉,外壁尚未开裂。释放 1d 后,"裂外壁花粉"的频率可高达 86.5%(表1)。在酶解阶段,酶液的渗透压值具有重要影响。甘露醇与山梨醇作渗透压调节剂,分别将渗透压调至 1281、1614mosm/kg 和

1846mosm/kg。其中以 1614mosm/kg 条件下酶解原生质体的分离率最高，状态最好（表2）。酶的种类与酶解时间也有一定影响。仅用纤维素酶与果胶酶可以分离出原生质体，添加 pectolyase Y-23，能促使外壁裂口较小的花粉亦分离出原生质体，从而有一定的改善效果。酶解时间要充分，时间太短，分离率偏低，3h 的分离率为 60%，延至 6h，分离率可高达 90%。

花药预培养法可以分离出大量花粉原生质体，状态良好，大小均匀，无亚原生质体（图版Ⅰ，15）。FDA 荧光鉴定表明 97.9% 是有生活力的（图版Ⅰ，16）。

花粉饥饿预处理法　此法亦能分离出一定数量的花粉原生质体。将幼嫩花粉由花药中直接压出，此时花粉呈近球形。在不含蔗糖仅含甘露醇和无机盐的 B 培养基[7]中进行饥饿预处理。转入高渗酶液中，花粉很快收缩成长椭圆形。在酶液作用下，原生质体从一个萌发沟处逐渐向外逸出，形成完整的原生质体（图版Ⅰ，7）。有时，同时从 2 个或 3 个萌发沟处逸出，产生 2 个或 3 个亚原生质体，结果分离产物由大小不等的完整原生质体与亚原生质体群体组成。在饥饿阶段，处理时间对花粉原生质体分离有重要作用。未经饥饿处理，花粉原生质体分离率极低，仅 3.3%。处理 2~4d，分离率由 12.8% 提高至 34.0%；处理 6d，分离率又降至 5.2%（表3）。可见一定天数的饥饿处理是必需的。但时间过长，花粉长期处于饥饿条件下导致大量死亡，也不利于原生质体分离。酶的种类对分离有显著影响。在纤维素酶和果胶酶的基础上，附加 pectolyase Y-23，花粉原生质体分离率比无该酶时明显提高。这一点和花药预培养法是不同的。

二、幼嫩花粉原生质体培养

用花药预培养法分离出幼嫩花粉原生质体，在 K_3 培养基中培养 12h，用荧光增白剂鉴定已再生细胞壁（图版Ⅰ，8）。培养 1d 后，在花粉原生质体群体中出现发育途径的分歧：少数花粉原生质体脱分化，发生第一次细胞分裂，产生 2 个子细胞（图版Ⅰ，9）；有的花粉原生质体则萌发出花粉管，内有细胞质环流（图版Ⅰ，10）。此外，还观察到出芽、结节状结构(图版Ⅰ,12)和其它异常生长行为(图版Ⅰ,13)。值得注意的是,观察到4例原生质体进行第一次细胞分裂产生2个子细胞后,其中1个子细胞又萌发出花粉管,另1个子细胞既未萌发花粉管,亦未继续分裂(图版Ⅰ,11)。

讨 论

在迄今各种植物花粉原生质体分离的研究中,一个共同的关键技术问题是如何促使抗酶解的外壁裂开,以便内壁更易受到酶的作用而释放出原生质体[2,8]。根据不同植物的特点,已经建立了直接酶解[9,10]、水合-酶解[11]、萌发-酶解[4]等几种分离花粉原生质体的方法。但上述方法对烟草幼嫩花粉原生质体分离均不适用。本实验借鉴Sunderland和Roberts[5]的花药漂浮培养和Kyo和Harada[7]的花粉饥饿处理[7]两种用于诱导烟草花粉脱分化的方法,作为酶解前的预处理,改变了花粉外壁、内壁与原生质体三者间的联系状态,解决了分离的困难。尤其是花药预培养法导致大量花粉外壁开裂、脱落,从而能够顺利地酶解出大量的花粉原生质体,避免了亚原生质体,便于纯化,具有更大的优越性。花粉饥饿预处理法则具有需时较短、操作简便的特点。以上方法不仅成功地解决了烟草幼嫩花粉原生质体分离的难题,而且可考虑试用于其它植物材料,从方法学上为花粉原生质体分离技术增添了新的思路。

上述方法的另一重要特点是:用于分离花粉原生质体的两种预处理程序均是和诱导花粉启动脱分化、走向孢子体发育途径的程序一致的[5,7]。因而从理论上推测,对于诱导花粉原生质体在培养中脱分化也应该是有利的。初步实验表明,由花药预培养法分离的花粉原生质体,在培养过程中,既可继续配子体发育萌发花粉管,又可表现出进行细胞分裂走向孢子体发育的趋势。之所以表现出两种不同的发育趋向,很可能是由于群体中的花粉原生质体在发育时期上有差异,或它们所受预处理影响程度有所不同所致。进一步研究有望认识两条发育途径的调控机理。一个以往从未观察过的有趣现象是,由1个花粉原生质体分裂成2个子细胞后,其中1个子细胞又萌发花粉管,表明这个子细胞在启动了孢子体发育之后,又重新恢复了原有的配子体发育。而且,这种配子体发育仅表现于1个子细胞,是否暗示其母细胞在培养中发生了某种极性变化,导致分裂产生的两个子细胞在发育上的异质性?值得今后从细胞生物学角度加以研究,作出合理解释。此外,如何促进花粉原生质体在第一次孢子体分裂基础上持续分裂直至再生植株,是今后研究的重要任务。

(作者:夏惠君、周嫦、杨弘远。原载:植物学报,1996,38(2):113~117。图版1幅,表3幅,参考文献11篇,均删去。)

10　烟草单个花粉原生质体透射电镜样品制备及其超微结构观察

Specimen preparation of single pollen protoplasts for TEM and their ultrastructural observation in *Nicotiana tabacum*

Abstract

The single mature pollen protoplasts of *Nicotiana tabacum* were observed under a transmission electron microscope. Their preparation for electron microscopy was performed by selecting single target pollen protoplasts with a microcapillary and, after short prefixation in glutaraldehyde droplets, embedding them in low gelling agarose or gelatin as preliminary steps, As a comparison, the specimen preparation of mass pollen protoplasts was carried out following the method of Wu and Zhou (1991) with minor modification. The results from both methods were comparable with each other pretty well. The pollen protoplasts observed were completely deprived of pollen wall, but kept the internal structure quite intact. This technique for specimen preparation might be used in the ultrastructural research on sexual plant cell manipulation not only for pollen protoplasts but also for gametoplasts (e. g. isolated egg cells).

近年植物实验生殖生物学与性细胞工程的发展趋势之一,是其实验操作技术的日益精密化,已由器官、组织的操作提高到原生质体操作的水平[1]。花粉原生质体在植物性细胞工程中有着重要的潜在价值。随着花粉原生质体分离与培养工作的进展,研究其在离体条件下超微结构的变化是非常必要的。由于迄今花粉原生质体分裂频率偏低,且分裂后发育状况各异,需要在现有花粉原生质体群体制样法的基础上更加精细化、准确化,建立一项适用于研究单个原生质体的电镜制样技术。我们参考了吴燕与周嫦关于花粉原生质体电镜制样技术[2],以及Faure等对分离的单个卵细胞的超微结构的研究[3],根据Paul and Bell关于缓冲液及固定液与渗透压的关系的论述[4],借鉴了孙蒙祥等用微吸管手工吸取单个原生质体进行"一对一"融合的方法[5],以烟草成熟花粉原生质体为试验材料,设计了以手工挑取单个原生质体、戊二醛微滴固定、低熔点琼脂糖或明胶预包埋、小塑料管顶扣包埋[6]为关键步骤的流程,初步建立了适用于研究单个花粉原生质体的电镜制样技术。我们分别对单个包埋的花粉原生质体和群体包埋的原生质体进行了超薄切片,比较了其超微结构,结果是满意的。

材 料 与 方 法

实验材料为烟草(Nicotiana tabacum L.)。花粉原生质体的制备系采用汪泳与周嫦的酶解法分离成熟花粉原生质体[7],酶液由 R_3 培养基大量元素、1%纤维素酶、1%果胶酶组成,渗透压调节剂为 1mol/L 甘露醇、0.4mol/L 山梨醇。

一、花粉原生质体的单个样品制备

单个花粉原生质体的挑取与前固定 在倒置显微镜下,用自制的微吸管将所选目标花粉原生质体吸出,置于不含酶而其它成分与酶液相同的介质中,洗去原生质体表面附着的酶液。预先滴一滴1%戊二醛(用0.2mol/L磷酸缓冲液配制,含1mol/L山梨醇)在盖玻片上,将洗过的单个花粉原生质体转入此滴中。翻转盖玻片,倒扣在自制的玻璃小室上,以防在固定过程中原生质体沉底粘着。在小室内加入相同的戊二醛固定液以防干燥。置4℃冰箱内固定1h。

预包埋与后固定 在倒置显微镜下,找到固定液中的单个原生质体,用微吸管吸出,置于盖玻片上,立即吸一小滴预先熔化的2%低熔点琼脂糖(Agarose A-2790,Sigma)或4%明胶覆盖其上,使原生质体贴着盖玻片,置4℃冰箱中约10min,待其凝固后,用2%戊二醛继续固定1h,洗涤二次,1%锇酸固定3~4h。以上固定均在4℃下进行。

脱水 10%乙醇、30%乙醇脱水各10min,50%乙醇以后每次15min,70%乙醇4℃冰箱内过夜,90%乙醇以后,可在室温下操作。100%乙醇脱水2次,环氧丙烷过渡2次,每次20min。每次更换溶液时,用吸管轻轻将原溶液吸出,将新溶液沿皿壁缓缓加入。否则,溶液的冲力加上脱水剂的作用会使预包埋块与盖玻片分离,而导致实验夭折。

渗透与包埋 用环氧树脂Araldite(Durcupan ACM,Fluka)包埋,一般现配现用。附着在盖玻片上的预包埋块经环氧丙烷与环氧树脂按3∶1、1∶1、1∶3(V/V)的混合液依次渗透各2~3h,纯树脂渗透过夜。次日,吸去预包埋块上多余的渗透液,置于倒置显微镜下寻找目标,在每个花粉原生质体上滴一滴环氧树脂,然后用镊子轻轻扣上一段长约1.5mm、直径0.5~0.7mm的塑料小管。由于毛细管作用,小管内逐渐充满环氧树脂。将扣好小管的预包埋块连同盖玻片放入烘箱聚合。先在45℃下聚合12h,再升温至60℃,保持36h。自然冷却后,将包埋块连同盖玻片投入液氮,利用环氧树脂与玻璃的热膨胀系数不同,在骤然降温中使包埋块与盖玻片分离。将脱落后的包埋块置入30~40℃烘箱中2h,烘干其上的冷凝水。取出置干燥器中冷却至室温。围绕小管修整包埋块,将贴着盖玻片的光面朝上,重新聚合到用过的废包埋块头上。

切片与电镜观察　用 Sorvall MT6000 型超薄切片机上的体视显微镜寻找小管的位置,在小管内径范围内修整包埋块。按常规方法对超薄切片进行醋酸铀—柠檬酸铅双染色。置 H-7000FA 透射电镜下观察。

二、花粉原生质体的群体样品制备

固定与预包埋　将离心收集分离的花粉原生质体,加入离心管中的 1ml 不含酶的前述介质中,然后沿管壁缓缓加入用相同介质配制的 2% 戊二醛,直至与原体积相等,固定 1h 后,换入 2% 戊二醛继续固定 1h,用上述介质洗涤一次,0.1mol/L 磷酸缓冲液洗涤二次,1% 锇酸后固定 3~4h,磷酸缓冲液洗涤三次,离心收集于离心管中。将预先熔化的 2% 纯化琼脂冷却到 40℃ 左右,迅速与材料混匀,立即平铺在载玻片上,琼脂厚度约为 1mm。冷却后用刀片划成 $1mm^2$ 大小的方块,置于磷酸缓冲液中。以上固定均在 4℃ 下进行。

脱水　从 10% 乙醇开始,每 15min 递增 10%,100% 乙醇脱水两次,每次 30min,用 100% 乙醇与环氧丙烷的等量混合液过渡一次,再换二次纯环氧丙烷,每次 15min。80% 乙醇之前的各步骤均在 4℃ 下进行。

渗透与包埋　所有包埋剂同上。将小琼脂块转移到环氧丙烷与环氧树脂 3∶1(V/V)混合液中,渗透 3h,再在 1∶1、1∶3 的混合液中依次渗透 2h,纯树脂渗透时间过夜或稍长一点。用牙签将小琼脂块挑入橡胶平板中包埋,置入 45℃ 烘箱中聚合 12h,再将温度升至 60℃,继续聚合 6h。

切片与电镜观察　在切片机的体视显微镜下找到包埋块中小琼脂块的位置,手工修整后,先在切片机上作半薄切片,经甲苯胺蓝-O 染色,在显微镜下找到所需目标,在其周围重新修整出梯形平面,尔后进行超薄切片。

结　果

烟草成熟花粉属二细胞型,由营养细胞和其中包藏的生殖细胞组成。花粉原生质体外形基本上呈圆球形,外表光滑,脱壁彻底(图1)。单个制样方法所得样品(图4~6),和作为对照的群体制样结果(图1~3)相似,两者均图像清晰,结构完好,尤其是细胞膜、内质网、线粒体等膜结构保存较佳。

营养细胞的胞质浓厚,有许多小而分散的液泡,有的小液泡中含有电子密度高的物质(图1)。线粒体较多,内嵴发达(图2)。质体较大,多为椭圆形,不含淀粉。其内片层结构不发达,一般不超过三层,基质的电子密度很高(图2、5)。内质网丰富,几乎全为粗糙型,常与质体相伴(图5)。高尔基体稀少,游离核糖体丰富,营养核靠近生殖细胞(图4)。生殖细胞略呈椭圆形,有一明显伸长弯曲的尾部(图3)。生殖细胞与营养细胞间有一宽窄不等的电子透明带,其中含有一些电子密度较高的纤丝状物质(图3、6)。生殖细胞的细胞核很大,含丰富的异染色质,细胞质在靠近质膜的周边成一薄层(图3、6),微管常集合成束沿长轴平行排列(图3),没有看到质体。

讨　论

吴燕和周嫦(1991)建立的唐菖蒲花粉原生质体制样方法[2],其技术关键之一为分次固定,特别是以等渗基质配制低浓度戊二醛作初始固定剂,以减少原生质体收缩与破损,这对保持内部结构完整非常重要;关键之二为应用琼脂预包埋法,有目的地挑选有用的材料,减少了工作中的盲目性。但此法是以有大量的供试原生质体为前提,且在操作过程即离心收集过程中,常易使目标材料丢失。本文所述单个花粉原生质体制样技术对此作了重要改进:首先,挑选单个目标材料,以低浓度戊二醛初始固定后,立即预包埋。这样,一方面防止了原生质体在操作过程中发生破损,另一方面从开始就置目标原生质体于人工控制中,自始至终不易丢失,给操作也带来了很大的方便。在单个细胞的电镜制样方面,Faure 等[3]已有成功的经验,他们用显微操作系统挑选分离的玉米卵细胞,先用等渗的低温琼脂糖预包埋,然后再进行常规处理。而本文所述方法是用自制微吸管手工挑取花粉原生质体,无需特殊设备,达到了异曲同工的效果。挑取的花粉原生质体先经等渗的低浓度戊二醛固定后再进行预包埋,因此,无需考虑以后各步骤所用溶液的渗透压,使操作简单化。预包埋时将目标材料先置于玻片上,将琼脂糖覆盖其上,克服了植物细胞在培养中不能贴壁生长而给包埋切片工作带来的困难。此外,小塑料管的应用也给切片工作带来了便利,免去了切片工作中的盲目性。

这一制样技术的成功,不仅适用于单个花粉原生质体进行细致的研究,而且还可望用于配子原生质体(如卵细胞)、细胞融合、融合体早期变化等的超微结构研究。根据它们各自的特性,应对所需的具体条件或步骤细节作适当的调整。

(作者:梁世平、余凡立、孙蒙祥、周嫱、杨弘远。原载:电子显微学报,1995(4):321~325。图版1幅,参考文献7篇,均删去。)

11 紫菜苔花粉原生质体的电激转化及 Zm13-260-GUS-NOS 融合基因的时序表达

Transient transformation of pollen protoplasts via electroporation and temporal expression of Zm 13-260-GUS-NOS chimeric gene in *Brassica campestris* var. *purpurea*

利用花粉作为外源基因的媒介进行植物遗传转化是一个活跃的研究方向[1],而将外源基因导入花粉是这一转化体系的重要环节。但因花粉壁厚,导入外源 DNA 比较困难,而脱壁后的花粉原生质体则理应相对优越。近年花粉原生质体的分离与培养已取得了一定进展[2],以花粉原生质体作为转化受体已成为可能。为此,我们以花粉特异启动子 Zm13-260 控制的 GUS 基因作为报告基因,用电激法分别转化紫菜苔花粉原生质体和花粉粒,通过瞬间表达检测,比较了二者的转化效果,并探讨了 GUS 基因在不同发育时期花粉原生质体中的时序表达特性。

材 料 与 方 法

一、供试材料

紫菜苔(*Brassica campestris* var. *purpurea*)。

二、质粒

pBS-260 gn 质粒由 J. P. Mascarenhas 教授惠赠,它是带有由玉米花粉特异启动子 Zm13-260,GUS 编码区和 NOS 终止子组成的融合基因的表达载体[3]。质粒 DNA 的提取鉴定按本文参考文献[4]所述方法进行。

三、成熟花粉

取当天开花的花药,置于15%的蔗糖溶液中压出花粉,45μm 不锈钢网过滤,离心收集后用15%蔗糖溶液洗涤3次,重悬于花粉电激液(含15%蔗糖的 CPW 溶液,pH5.8)中,调整密度至 3×10^5/ml 待用。

四、花粉原生质体

取单核、二核及三核成熟花粉按梁立等[5]的方法制备花粉原生质体。幼嫩花粉原生质体按李昌功等[6]的过滤离心法纯化;成熟花粉原生质体以沉降法纯化。纯化的花粉原生质体最后重悬于原生质体电激液(附加 0.8mol/L 蔗糖、0.4mol/L 甘露醇、0.2mol/L 葡萄糖的 D_2 培养基,pH5.8)中。

五、电激处理

用天津理工大学研制的 LN-101 基因脉冲导入仪,将 0.8ml 样品置于电极间距为 0.4cm 的电激槽中,加入终浓度为 100μg/ml 的小牛胸腺 DNA 混匀,再加入质粒 DNA 使其终浓度达 100μg/ml,充分混匀后电激处理。电容为 20μF,脉冲时间常数约为 3.2ms,根据实验要求调整电压。电激完成后,室温下静置 30min,电激液洗涤 2 次。花粉粒最后培养于附加 5% 蔗糖和 20% PEG 的 Roberts 培养基中,pH8.5,25℃,80r/min 振荡培养。花粉原生质体则培养于上述原生质体电激液中。培养 16h 后,进行生活力的 FDA 鉴定与 GUS 活性检测。

六、GUS 活性检测

按 Jefferson[7] 的荧光法检测 GUS 活性。GUS 反应底物为 MUG (Molecular Probes Inc.),以不同浓度的 MU(Molecular Probes Lnc.)作为荧光强度标准,荧光分光光度计为 Shimadzu RF-540,激发波长为 360nm,检测波长为 450nm。按 Bradford[8] 的方法测定蛋白质含量。GUS 比活性以 nmol MU/min · mg^{-1} 蛋白质表示。

结　　果

一、电场强度对成熟花粉原生质体和成熟花粉粒生活力的影响

原生质体由于缺少壁的保护,在高场强电脉冲作用后易破碎死亡,因此通常利用较低场强和较长脉冲时间进行电激转化。花粉原生质体同样如此。如表 1 所示,当脉冲时间常数约为 3.2ms 时,随着电场强度的增加,花粉原生质体生活力逐渐降低,其半致死电场强度约为 500V/cm。花粉粒对电场强度的耐受性较大,其半致死电场强度约为 1 000V/cm。为兼顾一定的通透性和生活力,以下实验中均采用半致死电场强度进行电激转化。

表1　电场强度对成熟花粉原生质体和成熟花粉粒生活力的影响

电场强度/V·cm^{-1}		250	500	750	1 000	1 250
生活力/%	花粉原生质体	83	55	31	14	6
	花粉粒	95	90	65	46	18

二、花粉原生质体和花粉粒电激转化效果的比较

成熟花粉原生质体与成熟花粉粒的电激转化效果有很大差异(表2)。由于成熟花粉原生质体消除了花粉壁的障碍,电激处理后外源DNA更易进入细胞,因而GUS基因的瞬间表达水平远较花粉粒者为高。前者的GUS活性为后者的100倍以上。这说明花粉原生质体是一个更好的电激转化受体。

表2　花粉原生质体和花粉粒中GUS基因表达水平的比较

材料	花粉原生质体			花粉粒(三核)
	三核	二核	单核	
GUS活性/nmol MU·min^{-1}·mg^{-1}蛋白质	1.12	0.26	0.04	0.01

三、GUS基因在不同发育时期花粉原生质体中的时序表达

Zm13-260是花粉特异表达启动子,其启动活性具有时空特点。将Zm13-260-GUS-NOS融合基因导入不同发育时期的花粉原生质体,GUS基因的表达水平存在很大差异。在成熟花粉原生质体中有高水平表达,在二核期花粉原生质体中表达水平较低,而在单核期花粉原生质体中极低(表2)。说明Zm13-260是在小孢子有丝分裂后的花粉成熟过程中方才启动基因表达。

讨 论

以花粉原生质体作为转化受体的设想早已提出[9]。本实验证实了这一设想的可行性。与花粉粒相比,花粉原生质体的电激导入效果更高。随着今后花粉原生质体实验体系的建立和完善,以成熟花粉原生质体作用外源基因的媒介,经授粉受精获得转基因植株,可能成为有潜力的遗传转化新体系。

Hamilton 等[4]利用基因枪法将含 Zm13 不同酶切片段的 GUS 基因导入玉米和紫露草成熟花粉以及紫露草叶片,分析了 Zm13 不同片段的功能,并证实了该花粉特异启动子的组织特异性。Guerrero 等[10]在转基因烟草中证实 Zm13 仅能启动 GUS 基因在小孢子有丝分裂后的花粉中表达。我们在紫菜苔花粉原生质体的电激转化及瞬间表达的分析也得到了类似结果,花粉愈幼嫩,Zm13 的启动活性愈低。

(作者:施华中、徐秉芳、杨弘远、周嫦。原载:科学通报,1995,40(18):1704~1706。表 2 幅,参考文献 10 篇。仅选录表 1、2。)

12 芸苔属花粉-下胚轴原生质体融合再生杂种小植株

Regeneration of hybrid plantlets via pollen-hypocotyl protoplast fusion in *Brassica* spp.

Abstract

It has been reported that "gameto-somatic hybridization" was performed by fusion of microspore tetrad protoplasts with somatic protoplasts in *Nicotiana* and *Petunia*. However, since the success of isolation of pollen protoplasts in recent years, the use of protoplasts at pollen stage as one of the fusion partners in such hybridization is a novel experimentation. Young pollen protoplasts were isolated from the pollen grains of *Brassica chinensis* at mid-late unicellular to early bicellular stage by treating the pollen for 1.5-2.5 h at 25℃ in a CPW solution containing 0.8% of cellulase, 0.5% pectinase, 0.1% pectolyase, 13% mannitol, 10% glucose, 0.3% potassium dextran sulphate and 3 mmol/L MES. The purified pollen protoplasts were then fused with the hypocotyl protoplasts of *B. napus* by PEG method. Heterokaryons were identified by means of visualization of the fluorescence from FITC-prelabeled pollen protoplasts. In order to increase heterokaryons and reduce hypocotyl homokaryons, the density of hypocotyl protoplasts were lowered and the ratio of the number of hypocotyl vs. pollen protoplasts were adjusted from 1 : 3 to 1 : 6. The fusion products were cultured in a liquid KM8p medium supplemented with 0.4 mol/L glucose, 0.8 mg/L 2,4-D, 0.25mg/L NAA, 0.5mg/L BA, 500 mg/L glutamine and 3 mmol/L MES where cell division and callus formation took place. The calli, after being transferred to a MS medium supplemented with 2.0 mg/L BA, 3% sucrose and 0.4% agarose, differentiated into a few shoots. The shoots were transferred onto a half-strength MS medium supplemented with 2% sucrose, 0.1-0.2 mg/L NAA, 0.5 mg/L IBA and 20% potato juice for root formation. Finally, three plantlets were regenerated. Chromosome counts by roottip squash method revealed that one plantlet was $2n=48$, corresponding to an allotriploid resulted from a fusion between one pollen protoplast of *B. chinensis* ($2n=20$) and one hypocotyl protoplast of *B. napus* ($2n=38$), and the other two plantlets were $2n=58$, which might be an allotetraploid originated from a fusion between two pollen protoplasts and one hypocotyl protoplast. The isozyme patterns of leaf esterases showed that all the three plantlets had bands characteristic of both parents. This is the first case of success in "gameto-somatic hybridization" by using pollen protoplasts instead of tetrad protoplasts as the haploid partner.

植物性细胞(广义的概念)与体细胞的融合,最初是从小孢子四分体的实验开始的。Pirrie 和 Power 首次将 *Nicotiana glutinosa* 四分体原生质体与 *N. tabacum* 叶肉原生质体融合,获得了种间杂种植株,并称为"配子-体细胞杂交"(gameto-somatic hybridization)[1]。类似的实验相继在烟草属与矮牵牛属的种间与种内杂交中获得成功[2~4]。但是,迄今还没有关于游离花粉的原生质体与体细胞原生质体融合成功的报道。这是因为游离花粉外壁富含孢粉素,无法酶解。近年,花粉原生质体分离技术的突破为其融合奠定了技术基础。本文以青菜幼嫩花粉原生质体与甘蓝型油菜下胚轴原生质体进行融合与培养,再生了小植株,初步证明是种间杂种。

材 料 与 方 法

一、花粉原生质体的制备

将青菜(*Brassica chinensis* L.)品种"南京白"含单胞中后期至二胞早期的花蕾,置入分离液(CPW[5]附加 13% 甘露醇,10% 葡萄糖,0.3% 葡聚糖硫酸钾,3mmol/L MES,pH5.6),压出花粉。经过滤、离心后,在酶液(上述分离液附加 0.8% 纤维素酶,0.5% 果胶酶,0.1% pectolyase)内于 25℃ 酶解 1.5~2.5h。过滤、离心与洗涤后,得到纯化的花粉原生质体,调整其密度至 3×10^5~6×10^5 个/ml。

二、下胚轴原生质体的制备

纵剖甘蓝型油菜(*Brassica napus* L.)品种"4312"无菌苗的下胚轴,置入质壁分离液(CPW 附加 13% 甘露醇)0.5~1.0h,转入酶液(CPW 附加 0.8% 纤维素酶,0.8% 果胶酶,6% 甘露醇,4% 蔗糖,3 mmol/L MES,pH5.6),25℃ 酶解 14~18h。过滤、离心、洗涤后收集原生质体,调整至所需密度。

三、融合

将花粉原生质体与下胚轴原生质体悬浮液混合,用融合液[40% 聚乙二醇(PEG,MW6000),6% 葡萄糖,50 mmol/L $CaCl_2·2H_2O$]诱导融合。参照Terada等[6]的方法并作修改:先滴数滴融合液于培养皿底,再将原生质体混合液轻滴于其上,15~30min 后加入 W_5 洗液[7],静置0.5~1h,再用 W_5 洗液洗涤1~2次后转入培养基进行培养。

四、培养

采用液体浅层培养法,培养基为 KM8p 附加 0.8mg/L 2,4-D, 0.25mg/L NAA, 0.5mg/L BA, 500 mg/L 谷氨酰胺,0.4mol/L 葡萄糖, 3mmol/L MES, pH5.6。两周后,每周加1次稀释培养基(K_3 附加 0.3mg/L 2,4-D, 0.5mg/L BA, 3% 蔗糖,pH5.6)。在扩增培养基(MS 附加 0.2mg/L 2,4-D,0.5mg/L BA, 3% 蔗糖,0.4% 琼脂糖,pH5.6)上增殖愈伤组织。在分化培养基(MS 附加 0~0.2 mg/L 2,4-D,1~3 mg/L BA,3% 蔗糖,0.4% 琼脂糖,pH5.6)上诱导愈伤组织分化出芽。在生根培养基(1/2 MS 附加 0.1~0.2 mg/L NAA,0.5 mg/L IBA,2% 蔗糖,20% 马铃薯汁,pH5.6)上促进生根。

五、显微观察

用 Olympus IMT-2 倒置显微镜观察。部分花粉原生质体用异硫氰酸荧光素(FITC)5~10 μg/ml 活染,充分洗涤后再进行融合试验,以观察荧光标记的融合体。融合率计算方法:异源融合率=(与花粉原生质体融合的下胚轴原生质体数/下胚轴原生质体总数)×100%。同源融合率=(参加同源融合的下胚轴原生质体数/下胚轴原生质体总数)×100%。

六、染色体计数

再生植株根尖用饱和对二氯苯水溶液处理3h,甲醇:乙酸(3:1)

固定 2h 以上,1 mol/L HCl 60℃水解 10min,改良卡宝品红染色压片。

七、酯酶同工酶酶谱分析

用聚丙烯酰胺不连续体系垂直板凝胶电泳法。浓缩胶 4%,分离胶 10%,电极缓冲液为 Tris-甘氨酸缓冲液,pH8.7。样品提取液为 pH8.2 的磷酸缓冲液。稳定电压 20 V/cm。0.1%醋酸-α-萘酯,0.1%醋酸-β-萘酯,0.1%坚牢蓝 RR 盐中染色。

结　果

一、青菜幼嫩花粉原生质体的分离

幼嫩花粉酶解 1.5~2.5h 后释放出原生质体,分离率为 20%~40%(图版Ⅰ,1)。延长酶解时间不能提高分离率,反而引起原生质体破损。酶解后的悬浮液中除花粉原生质体外,尚有花粉粒,花粉外壁等杂质。采用 10μm 孔径尼龙筛网过滤可得到较纯净的花粉原生质体群体(图版Ⅰ,2)。

二、花粉原生质体与下胚轴原生质体的融合

青菜幼嫩花粉原生质体与甘蓝型油菜下胚轴原生质体形态不同,大小悬殊,容易辨别(图版Ⅰ,2,3)。所用的融合液能有效地诱导这两种原生质体融合,但形成的异源融合体在形态上很难与下胚轴原生质体区分。用 FITC 预先标记的花粉原生质体进行融合,可清楚地观察到花粉原生质体与其异源融合体均呈特异的绿色荧光,使之明显地区别于下胚轴原生质体(图版Ⅰ,3~6)。

为了提高异源融合效果,试验了双亲原生质体密度与数量比率的影响。表 1 显示,当双亲原生质体的密度均为 15×10^4 个/ml、比率为 1∶1 时,总的异源融合率为 26.8%,下胚轴原生质体同源融合率为 24.7%。降低下胚轴原生质体的密度,提高花粉原生质体的密度,使二者的比率改变为 1∶3~1∶6 时,异源融合率提高到 40.8%~52.7%,

同源融合率则相应下降。但如果下胚轴原生质体密度过低,虽双亲数量比率达1∶9,异源融合率却不进一步提高。

表1 原生质体密度与比率对花粉-下胚轴原生质体融合效果的影响
Table 1 Effect of density and ratio of pollen-hypocotyl protoplasts on fusion results

下胚轴原生质体 Hypocotyl protoplast (10^4/ml)	花粉原生质体 Pollen protoplast (10^4/ml)	比率 Ratio	统计的下胚轴原生质体数 Hypocotyl protoplast counted	异源融合体 Heterokaryons					总计 Total (%)		下胚轴同源融合体 Hypocotyl homokaryons	
				一对一融合① "One to One" fusion		一对二以上融合② "One to two or more" fusion					
				No.	%	No.	%	No.	%	No.	%
15	15	1∶1	535	99	18.5	44	8.3	26.8		132	24.7
10	30	1∶3	517	127	24.6	83	16.2	40.8		106	20.5
7	42	1∶6	296	102	34.5	54	18.2	52.7		37	12.5
5	45	1∶9	339	112	33.0	58	17.1	50.1		24	7.0

①一对一融合:一个下胚轴原生质体与一个花粉原生质体之间的融合。②一对二以上融合:一个下胚轴原生质体和两个及两个以上花粉原生质体之间的融合。

① "One to one" fusion means a fusion between one hypocotyl protoplast and one pollen protoplast. ② "One to two or more" fusion means the fusion between one hypocotyl protoplast and two or more pollen protoplasts.

三、融合后的发育与植株再生

融合后原生质体在适合的培养基上能持续发育。培养2~3d开始第一次细胞分裂(图版Ⅰ,7);4~5d第二次分裂(图版Ⅰ,8);两周左右形成肉眼可见的多细胞团(图版Ⅰ,9)。愈伤组织经增殖后转到分化培养基上(图版Ⅰ,10),有显著的褐化死亡现象,存活较好的31块愈伤组织中仅有4块出芽,分化率为12.9%。以后进一步诱导生根,形成完整小植株。在存活的3株(图版Ⅱ,11~13)中,叶片不同程度皱缩,生长亦较缓慢,与油菜原生质体再生植株相比,在形态与发育上有一定区别。

四、再生植株的鉴定

1. 染色体计数　青菜 2n=20;甘蓝型油菜 2n=38((图版Ⅱ,18、19)。对融合后再生的 1 号苗 5 个根尖,2 号苗 8 个根尖,3 号苗 5 个根尖进行染色体计数,结果表明,1 号苗 2n=48,2 号与 3 号苗 2n=58(图版Ⅱ,15~17),均与双亲染色体数及油菜下胚轴原生质体同源融合产物的染色体数(应为 2n=76)不同。据此,初步推断 1 号苗可能源于 1 个青菜花粉原生质体与 1 个油菜下胚轴原生质体的融合,为异源三倍体;2 号、3 号苗可能源于 2 个青菜花粉原生质体和 1 个油菜下胚轴原生质体的融合,为异源四倍体。

2. 酯酶同工酶分析　亲本不同个体的酯酶酶谱是稳定的。融合后原生质体再生的 3 个小植株既有青菜特有的酶带(Rf 0.43,0.48),又有甘蓝型油菜所特有的酶带(Rf 0.39,0.55),显示出杂种酶谱的特征(图版Ⅱ,14)。

讨　　论

花粉原生质体兼具花粉、单倍体、原生质体三方面的特性,是植物细胞工程中有特色的实验体系。幼嫩花粉原生质体具有脱分化的潜力,但其离体培养迄今仅在萱草中取得一定进展[8],尚未能再生植株。本文作者曾试探过青菜、甘蓝型油菜与紫菜苔 3 种芸苔属植物的幼嫩花粉原生质体培养,仅在青菜中诱导了 1 次细胞分裂。另一途径则是利用花粉原生质体与具有再生能力的体细胞原生质体融合,开展"配子-体细胞杂交"。本实验证实了这一思路的可行性。

杂种细胞筛选是原生质体融合工作中的重要环节。花粉原生质体迄今不能单独再生植株,因而在融合实验中可以简化筛选程序,只需考虑体细胞亲本一方再生植株。前人在四分体与体细胞融合研究中,先后利用营养缺陷型[1]、白化突变体[2,3]、显性的抗性基因[4]等特殊材料作亲本,成功地筛选出杂种。本实验没有采用这类材料,而是借鉴以前大、小两种原生质体融合实验的经验[9],降低大原生质体的密度以及调整两种原生质体的数量比率,以提高异源融合率。本实验获得的3个小植株,可能分别为异源三倍体和异源四倍体,均不可能为体细胞原生质体单独发育或其同源融合的产物。酯酶同工酶酶谱分析为其杂种特性提供了佐证。

本实验所采用的幼嫩花粉原生质体为单胞中、后期至二胞早期的。从理论上推测,参加融合的应为单胞时期的小孢子核或二胞时期的营养核;生殖细胞因有原生质膜包围,似不可能参与核的融合。前人在四分体与体细胞原生质体融合的研究中,仅着重于遗传学鉴定。我们认为,今后在"配子-体细胞杂交"工作中,对于融合后的细胞学行为仍有深入研究的必要。

图 版 说 明

图 版 Ⅰ

1. 刚分离的幼嫩花粉原生质体群体。×400 **2.** 纯化的花粉原生质体群体。×400 **3.** 纯化的下胚轴原生质体群体。×260 **4~6.** FITC标记的幼嫩花粉原生质体与下胚轴原生质体融合。**4.** 融合前的荧光-明视野图像。×550 **5.** 融合后异核体的明视野图像。×550 **6.** 图5的相应荧光图像。×550 **7~10.** 培养的融合后原生质体进行细胞分裂与形成愈伤组织。**7.** 第1次分裂(培养3d)。×320 **8.** 第2次分裂(培养5d)。×320 **9.** 形成的多细胞团(培养11d)。×100 **10.** 形成的愈伤组织。

图 版 Ⅱ

11~13. 再生的杂种小植株。**11.** 1号苗。**12.** 2号苗。**13.** 3号苗。**14.** 杂种小植株及其亲本的叶片酯酶同工酶酶谱,从左至右依次为:A. 青菜;B. 1号苗;C. 2号苗;D. 3号苗;E. 甘蓝型油菜;F. 青菜与甘蓝型油菜的混合样品的酶谱。箭头示杂种植株的互补酶带位置。**15~19.** 杂种小植株及其亲本的根尖染色体。**15.** 1号苗(2n=48); **16.** 2号苗(2n=58); **17.** 3号苗(2n=58)。**18.** 青菜(2n=20)。**19.** 甘蓝型油菜(2n=38)。

第一章 雄性器官与细胞的实验研究

图版 I plate I

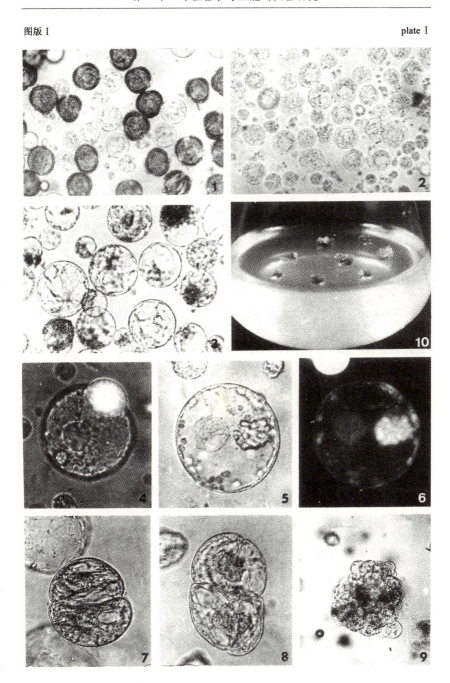

图版Ⅱ plate Ⅱ

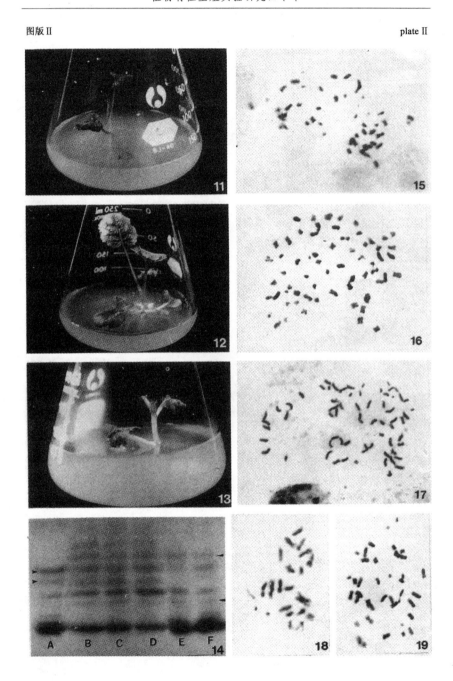

第一章 雄性器官与细胞的实验研究

Explanation of plates

Plate Ⅰ

Fig. 1. A population of young pollen protoplasts just released from the pollen grains. ×400 **Fig. 2.** A population of purified pollen protoplasts. ×400 **Fig. 3.** A population of purified hypocotyl protoplasts. ×260 **Figs. 4 ~ 6.** Fusion between an FITC-prelabeled pollen protoplast and a hypocotyl protoplast. **Fig. 4.** Fluorescence-brightfield image of a pair of protoplasts just before fusion. ×550 **Fig. 5.** Brightfield image of a heterokaryon just after fusion. ×550 **Fig. 6.** Fluorescence image corresponding to Fig. 5. ×550 **Figs. 7-10.** Cell division and callus formation of cultured PEG-treated protoplasts. **Fig. 7.** First division after 3 days of culture. ×320 **Fig. 8.** Second division after **5** days of culture. **×320** **Fig. 9.** Multicellular cluster after 11 days of culture. ×100 **Fig. 10.** Formed callus.

Plate Ⅱ

Figs. 11-13. Regenerated plantlets. **Fig. 11.** No. 1 plantlet. **Fig. 12.** No. 2 plantlet. **Fig. 13.** No. 3 plantlet. **Fig. 14.** Leaf esterases isozyme patterns of the pollen-somatic hybrids and their parents. From left to right: A. *Brassica chinensis*. B. No. 1 plantlet. C. No. 2 plantlet. D. No. 3 plantlet. E. *B. napus*. F. Extract mixture of *B. chinensis* and *B. napus* (1∶1). Arrows show complementary band sites of the hybrid plantlets. **Figs. 15-19.** Root-tip chromosomes of the pollen-somatic hybrids and their parents. **Fig. 15.** No. 1 plantlet (2n=48). **Fig. 16.** No. 2 plantlet (2n=58). **Fig. 17.** No. 3 plantlet (2n=58). **Fig. 18.** *B. chinensis* (2n=20). **Fig. 19.** *B. napus* (2n=38).

（作者：李昌功、周嫦、杨弘远。原载：植物学报，1994，36（12）：905～910。图版2幅，表1幅，参考文献9篇，选录全部图表。）

13 Development and molecular identification of pollen-somatic hybrid plants in *Brassica* spp.

As an important branch of sexual plant cell manipulation studies, gameto-somatic hybridization has made great progress in last decade. Fusion of protoplasts isolated from microspore tetrads, young and mature pollen grains with somatic protoplast achieved successes[1-3]. However, the studies so far mainly concentrated on the plant species in *Solanaceae*[1,3,4], and there has been no clear report on the fertility of hybrid plants. In the previous reports by Li et al. (1994), three plantlets were regenerated from young pollen-somatic protoplast fusion in *Brassica* and were preliminarily identified to be one allotriploid and two allotetraploids by means of chromosome counts and esterase isozyme analysis[2]. In this note we report growth and development of the three plantlets and further characterization of their hybrid nature with randomly amplified polymorphic DNA (RAPD) markers[5].

Materials and methods

Materials and plant morphological observation

The three plantlets regenerated from the fusion between young pollen protoplasts of *Brassica chinensis* and hypocotyl protoplasts of *B. napus*[2] were multiplied by vegetative propagation, and the plantlets higher than 5cm were transferred to soil and maintained under natural conditions. The developmental status of hybrid plants was compared with that of their parents. The viability of mature pollen grains was examined by fluorescein diacetate (FDA) staining. Fertility was evaluated by means of ovary development and seed formation.

Total DNA extraction of plants

Following the CTAB method modified by Zhang et al.[6], DNA was extracted from fresh leaves of five hybrid plants, two allotriploids (Plantlet 1) and three allotetraploids (Plantlets 2 and 3), as well as their parents, and was purified with a mixture of equal volume of phenol and chloroform.

RAPD *analysis*

Ten 10-mer primers synthesized by the State Key Laboratory in Genetic Improvements of Crops in Huazhong Agricultural University were selected arbitrarily for PCR amplification following the protocol established by Wang[1]: the reaction components were 10 mmol/L Tris-HCl buffer (pH8.3), 50 mmol/L KCl, 2.5 mmol/L $MgCl_2$, 10μmol/L primer, 1mmol/L dNTPs, 20 ng of genomic DNA and 1.5 units of Taq DNA polymerase (the product of the same laboratory as 10-mer primers). The final volume per tube for amplification was 25 μl. The PCR machine (Perkin Elmer cetus 9600) was programmed for an initial denaturation step of 95℃ for 3 min, then 40 cycles of 94℃ for 1 min, 38℃ for 1min and 75℃ for 1.5 min after the cycles a final step of 72℃ for 8 min and then 4℃ for store. The products were loaded in 1.4% agarose gel in 1×TAE buffer and run at 1 V/cm for 10h. A 1-kb DNA ladder (BRL) was used as molecular standard. The gel was stained with ethidium bromide and photographed under UV light.

Results

Developmental characteristics of hybrids

Compared with both parents, pollen-somatic hybrid plants were short and thin. They had almost no branches or two nearly equal length branches (Fig.1(a)). The leaves of the hybrids were intermediate in shape and size between the two parents; the buds were bigger than those of both parents. The viabilities of mature pollen as examined by FDA staining were 62.4% for allotriploids, 68.7% for allotetraploids, which were lower than those of the two parents during the same season (77.8% for *B. chinensis* and 81.6% for *B. napus*). All the ovaries of the hybrids developed into fruits. The allotriploid plants had short and thick siliques without seed but with the vestigial ovules. The allotetraploid plants formed normal siliques with mature seeds, but the number of seeds was less than that of the parents (Fig. 1(b),(c)).

Analysis of RAPD *markers*

Among the ten 10-mer primers, apart from RA04, the other nine primers produced 1-8 bands. The band number varied depending on the primer and genomic DNA. The size of amplified DNA fragments ranged from about 250 to 2500 bp. The DNA fragments amplified by primers RA32 and RA58 showed no polymorphism between the two parents. The primers RA08, RA18, RA42 and RA53 produced several detectable markers and some of them were polymorphic between both parents, but the DNA fragment sizes of *B. chinensis* were the same as those of *B. napus*, and the profiles of the five pollen-somatic hybrids were basically unanimous to those of *B. napus*. The products of primers RA34, RA39 and RA59 gave significantly different banding patterns between the two parents. The five hybrid plants had a combination of the main DNA fragments of the two parents banding patterns and some weak bands were lost. The results of RA34 are given in Fig. 2. Each hybrid had a 900-bp DNA fragment of *B. chinensis* and three DNA fragments (350, 650 and 1 300 bp) of *B. napus*. A 700-bp DNA was not polymorphic among all the plants. The results of three repetitions were the same.

Discussion

Close attention should be paid to how the fertility of allopolyploid plants would be. In the previous gameto-somatic hybridizations, only the hybrid plants reported by Pirrie and Power (1986) had lower viable pollen grains than their parents[1]. Other researchers mentioned that the pollen of gameto-somatic plants were sterile or inviable[3,4]. There was no positive report on seed fertility. In this work, the allotriploid plants had viable pollen grains and formed siliques without seed; the allotetraploids produced normal siliques. These results were corresponding to those theoretically expected.

RAPD marker is a molecular technique, in which the total genomic DNA sequences are amplified using 10-bp primers to generate randomly polymorphic DNA fragments by PCR[5]. In recent years, it has been used to identify somatic hybrids[7]. The results of RAPD markers reported here provided direct DNA molecular evidence for the characterization of hybrid nature of the plants originated from pollen-somatic protoplast fusion. To sum up, the chromosome counts, the isozyme patterns of leaf esterases reported previously[2] as well as the above morphological and molecular proofs confirmed that the genetic materials of pollen protoplast have been delivered to the descendants by means of gametosomatic hybridization. This is the first success that the hybrid nature of gametosomatic plants was identified by using RAPD markers.

(Authors:Li CG,Zhou C,Yang HY,Li XH ZhangQF. Published in:Chinese Science Bulletin,1996,41(18):1564-1567. with 2 figures and 7 references, all omitted here.)

14 烟草属花粉-叶肉原生质体的融合及杂种植株再生

Pollen-mesophyll protoplast fusion and hybrid plant regeneration in *Nicotiana*

Abstract

Isolated pollen protoplasts of *Nicotiana tabacum* L. N364 Km$^+$ were fused with mesophyll protoplasts of *N. rustica* L. using PEG-high Ca^{2+} pH method. The cells resulted from fusion between immature (early-middle bi-cellular) pollen protoplasts and mesophyll protoplasts could divide to produce microcalli and regenerated plantlets when cultured in a selection KM8p medium containing 50 g/L kanamycin. Four plantlets were regenerated. The isonzyme patterns of leaf peroxidases of these plantlets had bands characteristic of both parents. Root-tip squash showed that the gameto-somatic hybrids had the expected triploid chromosome number. Aside from these kanamycin-selected plantlets, six of the twenty-one plantlets that had not undergone selection were also evidenced to be gameto-somatic hybrids.

Pirrie 和 Power 首次报道烟草小孢子四分体原生质体和叶肉原生质体融合，再生了配子-体细胞(gameto-somatic)种间杂种植株[1]。类似的实验以后在烟草属与矮牵牛属的种内与种间杂交中获得成功[2-4]。随着游离时期花粉原生质体分离技术的突破，以四分体以后时期的花粉原生质体作为融合供体被提上日程，从而拓宽了配子-体细胞杂交的范围。最近，李昌功等在芸苔属[5]和 Desprez 等[6]在烟草属中用游离花粉原生质体与体细胞原生质体融合获得种间杂种植株。作为生殖细胞工程的一项技术，配子-体细胞融合可望成为作物改良的一个有力工具。同时，它也成为探讨生殖生物学与发育生物学有关理论问题的一种新系统而受到关注。本文以烟草幼嫩花粉原生质体与黄花烟草叶肉原生质体进行融合与培养，再生了小植株，经鉴定证明获得了种间三倍体杂种。

材 料 与 方 法

一、原生质体分离

用饥饿预处理法分离烟草(*Nicotiana tabacum* L.)品系 N364Km⁺二核早中期的花粉原生质体[7]。用常规方法分离黄花烟草(*N. rustica* L.)的叶肉原生质体。

二、融合

采用 Kao 等的 PEG-高钙高 pH 法[8]。将花粉原生质体(1×10^6)和叶肉原生质(1×10^6)等量混合,滴 3~4 滴于小培养皿中,随即加入 2 滴 20% PEG(MW6000)溶液(附加 0.3 mol/L 葡萄糖、3.5 mol/L $CaCl_2 \cdot 2H_2O$、0.7mmol/L KH_2PO_4),静置 5~10min,然后每间隔 5min 依次加入 0.5ml、1 ml 和 1 ml 的高钙高 pH 值溶液(50 mmol/L $CaCl_2 \cdot 2H_2O$、50 mmol/L 甘氨酸钠,pH10.5)。最后加入 2ml 含 13% 甘露醇的 CPW 洗液[9],静置 30min,吸掉上清液,用 KM8p 培养基[10]洗 3 次。

三、培养

采用液体浅层培养法,培养基为 KM8p 附加 0.5mol/L 甘露醇、0.1mol/L 葡萄糖、2% 蔗糖、0.4mg/L 6-BA、0.4mg/L 2,4-D、0.6mg/L NAA,pH5.8。培养 2 周后添加渗透压降低的新鲜培养基(KM8p 附加 0.2mol/L 甘露醇、0.1 mol/L 葡萄糖、2% 蔗糖,激素同上),同时添加 50 mg/L 卡那霉素(Km)。4~5 周后,加入无激素、附加 2% 蔗糖和 50 mg/L Km 的 MS 培养基。继续培养 2 周,将小细胞团转移到固体培养基(MS,2% 蔗糖,0.7% 琼脂,pH5.8)。大约 9 周,小细胞团长至 0.5cm 大小,移入分化培养基(MS 附加 2 mg/L 6-BA 和 50 mg/L Km)。

四、过氧化物酶同工酶分析

采用不连续垂直板聚丙烯酰胺凝胶电泳法。分离胶 10%,浓缩胶 4%,核黄素作光聚合催化剂。电极缓冲液为 Tris-甘氨酸,pH8.3。样品抽提液为 0.2 mol/L Tris-HCl、20% 蔗糖,pH8.5。稳压 20 V/cm 电泳 1.5~2h。0.4% 联苯胺染色。

五、染色体计数

用 0.05% 秋水仙素处理 3~4mm 长的根尖 3h,水洗后用 Carnoy 液(醋酸∶酒精=1∶3)固定 2h 以上,60℃、1 mol/L HCl 水解 10min,45% 醋酸软化 1~2min。卡宝品红染色 2h,压片观察。

结　果

一、N364 品系抗 Km 特性的证实

烟草 N364 品系的叶肉原生质体培养 2 周后，加入 50 mg/L Km。4 周后再加入 50 mg/L Km。愈伤组织转入分化培养基时再加入 100 mg/L Km。最终能旺盛地再生植株，证明该品系确实具有抗 Km 的能力。与此相反，黄花烟草叶肉原生质体培养 2 周后加入 50 mg/L Km，很快死亡，表明后者不具抗 Km 的能力。

二、花粉原生质体与叶肉原生质体的融合

烟草 N364 品系幼嫩花粉原生质体和黄花烟草叶肉原生质体大小相近(图版Ⅰ,1、2)，加入 PEG 后很快粘连，在高钙高 pH 值溶液洗涤稀释过程中，一对原生质体逐渐各自变为半球形，粘合成一个圆球体状。变圆后其间的界限很快消失，叶肉原生质体中的叶绿体逐渐扩散到整个融合体的胞质中(图版Ⅰ,3~5)。

三、培养与筛选

融合后的原生质体培养 3d 左右开始变成椭圆形，5d 左右发生第一次分裂，7d 左右发生第二次分裂(图版Ⅰ,6、7)。2 周后，在培养物中加入 50 mg/L Km。4 周后，再加入 50 mg/L Km，此时培养物已长成肉眼可见的白色多细胞团(图版Ⅰ,8)。继续培养 2 周，将小愈伤组织转移到固体培养基上。培养 2 周，获得 25 块愈伤组织，转入含有 50 mg/L Km 的分化培养基上，获得 4 株小植株(图版Ⅰ,9)。

另有部分融合后的原生质体未经筛选处理，培养后再生了 21 株小植株。

四、杂种植株的鉴定

过氧化物酶同工酶分析　过氧化物酶同工酶酶谱(图版Ⅰ,13)表明:黄花烟草有5条酶谱带:Rf 0.80、0.76、0.72、0.68 和 Rf 0.64;烟草 N364 有 6 条酶谱带:Rf 0.76、0.71、0.68、0.64、Rf 0.59 和 Rf 0.55。经抗性筛选后再生的 4 株小植株,其过氧化物酶同工酶酶谱均具双亲的特异酶带。其中黄花烟草的 Rf 0.80 和烟草 N364 的 Rf 0.59 及 Rf 0.55 是特异酶带。

未经筛选处理再生的 21 株小植株,有 6 株的过氧化物酶同工酶具双亲的特异酶带。

染色体计数 烟草花粉原生质体染色体数 n=24(图版Ⅰ,11),黄花烟草体细胞染色体数 2n=48(图版Ⅰ,12)。普通烟草和黄花烟草的配子-体细胞三倍体杂种植株的染色体数理论上应为 2n=3x=72。经筛选后获得的 4 株小植株,其根尖染色体数变幅为 66~72(表1;图版Ⅰ,10)。未经筛选的上述 6 株杂种小植株染色体数变幅为 58~72。

讨 论

配子-体细胞融合不仅具有一般原生质体融合技术的优点,而且还有其特点。首先,在异核体筛选上选择方法简单[1~4,6]。迄今,四分体与花粉原生质体还不能再生植株,因此仅需一方亲本具有选择标记性状;而一般体细胞杂交则需双亲均有选择标记性状。其次,配子-体细胞融合可获得细胞器的重组。例如,Pental 等用抗 Km 和链霉素的烟草四分体原生质体与黄花烟草体细胞原生质体融合,结果配子-体细胞杂种只抗 Km,不抗链霉素(由叶绿体基因组控制的性状)。DNA 分子杂交结果也证实了杂种植株中缺乏四分体原生质体亲本的叶绿体基因组[11]。

采用游离花粉时期的原生质体与体细胞原生质体融合,是配子-体细胞杂交技术的新发展。以往以小孢子四分体原生质体进行融合,由于四分体时期十分短暂,且体形很小,取材不甚方便。而花粉时期较长,体形较大,可以提供更广阔的选择。李昌功等首次将青菜(*Brassica chinensis*)的幼嫩花粉原生质体与甘蓝型油菜(*B. napus*)的下胚轴原生质体融合,通过控制双亲原生质体的数量比率,提高了异源融合率,虽然在培养过程中未经筛选程序,但最终再生了 3 株种间杂种[5]。最近,

Desprez 等用抗 Km 的烟草(*Nicotiana tabacum*)的成熟花粉原生质体与蓝茉莉叶烟草(*N. plumbaginifolia*)叶肉原生质体融合,经抗性筛选后,再生了 20 株杂种或胞质杂种植株[6]。本实验再一次成功地将花粉原生质体用于配子-体细胞杂交。和 Desprez 等的方法不同的是,本文是用二核早中期的幼嫩花粉原生质体与叶肉原生质体融合获得成功;而用成熟花粉原生质体作为融合供体则只能产生管状结构,不能发生细胞分裂[12]。这似乎表明成熟花粉原生质体具有强大的配子体发育能力,超过并抑制了融合体中体细胞原生质体的细胞分裂能力。幼嫩花粉原生质体的分化程度不如成熟花粉原生质体高,与叶肉原生质体融合后,融合体能够发生细胞分裂形成细胞团。这种分裂能力应该主要来自于叶肉原生质体,但至少幼嫩花粉原生质体的分化程度并不阻碍叶肉原生质体的分裂,最终能形成三倍体植株。至于这一观点如何解释 Desprez 等的实验结果,我们推测有两种可能:一是该文所用的成熟花粉原生质体取自 5cm 长的开花前花蕾,也许实际上尚未达到充分成熟的阶段;二是融合组合不同,该文所用的体细胞原生质体亲本蓝茉莉叶烟草是个原生质体再生能力极强且再生植株周期短的烟草种,也许其细胞分裂能力不易被花粉原生质体所抑制。

(作者:卢萍、周嫦、杨弘远。原载:植物学报,1996,38(5):342~346。图版 1 幅,表 1 幅,参考文献 12 篇,均删去。)

第四节 脱外壁花粉操作

提要

"脱外壁花粉"(exine-detached pollen, de-exined pollen)是我们提出的一个新名词,指介于完整花粉与花粉原生质体间的一种结构单位,其特点是花粉除去外壁仅留内壁与原生质体。

前节已经叙及,芸苔属花粉原生质体的分离包括两步,其中第一步就是通过水合脱去外壁。实际上以前我们在其它材料中也观察到这个现象。后来,我们认识到脱外壁花粉有其独特的研究价值,于是便由花粉原生质体研究中衍生出一个新的课题。脱外壁花粉的价值有两方面:其一,它提供了一种独特的花粉形态学与生理学研究材料。脱去外壁的花粉仅被内壁覆盖,由此可以研究整个内壁的表面结构,也便于透视花粉的内部结构,还可以研究剥离的外壁结构与成分。更重要的是:通过脱外壁花粉在萌发与授粉方面的表现,可以研究外壁在这些过程中的生物学功能。其二,脱外壁花粉在外源基因导入方面也可能有其优点,它比完整花粉少一层外壁的障碍,又比花粉原生质体更坚固,可能成为遗传转化的有用受体。我们在芸苔属与烟草中进行了以上两方面的探索。

首先是在以上两种植物中建立了脱外壁花粉的分离技术。针对两种花粉的生物学特点,分别建立了分离芸苔属脱外壁花粉的水合-热激-渗激程序和分离烟草脱外壁花粉的低温预处理-花药漂浮培养-花粉酶解程序,二者均达到60%以上的分离频率,并可加以纯化。

然后是建立脱外壁花粉的人工萌发实验系统。在芸苔属中,采用改良的花粉培养基,脱外壁花粉萌发率可达30%~40%,花粉管生长正常。利用这一实验系统进行了萌发过程的细胞生物学研究,重点研究了极性形成、萌发位点预定、新壁合成等问题。在烟草中,脱外壁花粉的人工萌发率达48%,并在半数花粉管中观察到生殖细胞分裂成正常精子,表明尽管脱去外壁,并没有根本影响花粉的正常萌发与花粉管生长。

在人工萌发成功的基础上,进一步试验脱外壁花粉的授粉与受精能力。在芸苔属中没有取得这方面的成功,似乎暗示其外壁或含油层在授粉中有不可缺少的作用。烟草则不同,脱外壁花粉的离体授粉取得圆满的成功。为了提高研究的精确性,我们采取了过去限量授粉的经验(参看第三章第一节),在一个柱头上授以事先人工萌发的30~40粒脱外壁花粉,后者产生的花粉管能顺利进入花柱,显微观察计数证明约有半数花粉管在花柱中生长。将授粉后的子房进行离体培养,可产生正常的种子；统计表明每授4粒花粉可结一粒种子。种子可以出苗长成植株。这样就建成了烟草脱外壁花粉离体授粉的完整实验体系。

　　为了探索脱外壁花粉的转化效果，我们先后应用了电激法与基因枪法，均获得肯定的结果。在基因枪实验中，同时采用了 GUS 基因与绿色荧光蛋白（GFP）基因作为报告基因。二者的结果是一致的：脱外壁花粉的导入频率高于作为对照的完整花粉。显微观察证实基因枪轰击的金粒有些可射进花粉生殖细胞。已导入了 GFP 基因的花粉授粉，在花柱内有些花粉管也显示绿色荧光。离体授粉获得了种子与幼苗。这一切预示脱外壁花粉具有将外源基因携带给后代的潜力。

1　芸苔属脱外壁花粉的分离与人工萌发

Isolation and artificial germination of de-exined pollen in *Brassica*

Abstract

In view of the fact that pollen deprived of exine has been shown by electroporation to be a superior receptor of foreign genes in tobacco, the authors developed a technique to isolate and artificially germinate the de-exined pollen in *Brassica*. Pollen grains of *B. chinensis* and *B. campestris* var. *purpurea* were first hydrated at 4℃ in a medium containing Roberts' salts, 24% sucrose, 7.3% mannitol, 0.05% lactoalbumin hydrolysate (pH8.5) for 18 h, followed by heat shock at 34℃ for 40-60 min and osmotic shock for 20 min subsequently. Through this three-step procedure the exine could dehisce along the germination furrows and completely detached off, resulting in the release of pollen coated with naked intine with an isolation rate as high as over 60%. The de-exined pollen were germinated in a modified medium containing 15%-20% polyethylene glycol (MW 6000), 5% sucrose, 0.05% lactoalbumin hydrolysate and Roberts' salts (pH8.5) with a rate of 41% in *B. chinensis* and 33% in the other species. However, in sharp contrast with the behavior of the intact pollen, the de-exined pollen did not germinate in the carbon and nitrogen sources free media. Scanning electron microscopic and light microscopic observations on the processes of isolation and germination of de-exined pollen were carried out. The phenomenon of exine detachment in a few plant species is reviewed. There might be some corresponding association between pollen wall characteristics and methods of isolation. The de-exined pollen may be a useful experimental system not only for introducing and transferring foreign genes, but also for investigating the role of exine in pollen germination and in sporophytic self-incompatibility.

最近,在烟草中研究出制备脱外壁花粉的方法[1]。电激基因转移实验表明,脱去外壁的花粉较完整的花粉粒更有利于外源基因的导入[2]。在芸苔属中,脱外壁花粉曾被作为分离花粉原生质体的中间环节[3],但当时并未认识到其本身独立的价值,因而缺乏深入的研究。我们发现,用前法制备的芸苔属脱外壁花粉丧失了萌发力,因此需要建立新的分离程序,并着重研究其人工萌发的条件。

材 料 与 方 法

一、材料

芸苔属青菜(*Brassica chinensis*)与紫菜苔(*B. campestris* var. *purpurea*)种植于露天。

二、脱外壁花粉的分离

取开花当天的花药,在含15%蔗糖的Roberts培养基[4](pH5.8)中压出花粉,过滤、离心去残渣,将花粉重新悬浮于含 24% 蔗糖、7.3% 甘露醇、0.05% 水解乳蛋白(LH)、Roberts 培养基盐成分的水合液(pH8.5)中,在 4℃下低温水合不同时间,转入 30~38℃热激不同时间,弃去水合液,加入含 24% 蔗糖的 Roberts 培养基(pH8.5)渗激,在 25℃下振荡(100r/min) 20 min。水合液与渗激液的渗透压值用 Micro-Osmometer Model 3MO 测定。

三、脱外壁花粉的人工萌发

将脱外壁花粉转入含不同有机成分的 Roberts 无机盐培养基中(pH8.5),在 25℃下,振荡培养(80r/min)。以未脱外壁的完整花粉作为对照。

四、光镜观察

在 Olympus IMT-2 倒置显微镜下进行 Nomarski 干涉差与荧光显微观察。以含 0.1% Triton 的 DAPI(10μg/L)染色观察精核与营养核。花粉的脱外壁率和萌发率随机取样计数,重复 4 次,计算平均数和标准误差。

五、扫描电镜观察

样品在 1% 与 2% 戊二醛(以含 7.5% 蔗糖的 0.2 mol/L 磷酸盐缓冲液 pH7.2 配制)中依次固定 1h,磷酸缓冲液洗 3 次,1% 锇酸 4℃固定 3h,缓冲液洗后,经乙醇逐级脱水,乙腈真空干燥法干燥[5]。喷金后在 H-450 扫描电镜下观察摄影。

结　　果

一、脱外壁花粉分离中的关键因素

低温水合　花粉在水合液中冷浸一段时间,再进行热激是脱外壁的必要条件。不冷浸或冷浸 3h,热激后脱外壁率极低;10h 冷浸再热激可获 10% 左右的脱外壁花粉;18h 则脱外壁率显著提高至近 40%;进一步延长至24h,提高不显著(表1)。

热激 低温水合后,短时间热激对脱外壁有重要影响。花粉经18h冷浸后,在30℃、34℃、38℃三种条件下处理1h,结果34℃比30℃下脱外壁率提高了约1.5倍。38℃与34℃相比脱外壁效果差异不显著(表1)。

进一步试验34℃热激的时间,在1h之内,随热激时间的延长,脱外壁率提高;1h以上未见进一步提高(图1)。

渗激 在低温水合与热激过程中,花粉水合液的渗透压值为1689 mosm/kg。在此基础上再转入1055 mosm/kg进行低渗冲击,可使脱外壁率进一步提高20~30个百分点,达到60%以上(图1)。并且,经此过渡,脱外壁花粉在转入更为低渗的萌发液中不易破裂。

二、脱外壁花粉的人工萌发

分离过程对脱外壁花粉人工萌发的影响 脱外壁花粉分离过程中的各个环节,不仅影响脱外壁的效果,而且也影响其后人工萌发的效果。首先是水合液的成分与水合温度。李仕琼等曾用单纯的蔗糖溶液28~30℃下浸泡芸苔属花粉9h以上,获得较高的脱外壁率[3]。但据我们试验,该法分离的脱外壁花粉人工萌发率极低,只有用本文所述接近于芸苔属花粉人工萌发碱性培养基的,但又较萌发培养基高渗的水合液,于4℃低温水合,方能取得较高的萌发率。其次是热激的时间。在34℃下热激不同时间的实验结果表明:热激40 min,萌发率为38.7%;延长至60 min,萌发率略有下降;进一步延长则导致萌发力急剧降低(图2),因此热激时间不宜超过1h。再次是渗激液的渗透压值,如过分降低,亦导致脱外壁花粉萌发力下降,甚至破裂。

培养基成分对脱外壁花粉萌发的影响 以Roberts培养基的无机盐成分为基础,分别以(A)15%蔗糖;(B)20%聚乙二醇(PEG 6000);(C)15%~20% PEG+5%蔗糖+0.05% LH作为添加成分,调pH至8.5,用以作为完整花粉和脱外壁花粉的萌发培养基。实验结果(表2)表明,完整花粉在A液中有一定的萌发力;在B液中萌发力有所提高;在C液中萌发率大幅度提高,达80%以上,且萌发速度与花粉管长度亦大有改善。脱外壁花粉在 A、B 液中基本不能萌发,只有在 C 液中可以正常萌发,青菜的萌发率约 41%,紫菜苔的萌发率约 33%。尽管和对照(完整花粉)相比,萌发率还有较大差距,但已基本上达到开展进一步人工萌发实验的要求。而添加 5% 蔗糖与 0.05% LH 竟使脱外壁花粉由基本不萌发变为萌发率较高,其效果是极其明显的。

总结前述研究结果,芸苔属脱外壁花粉的分离与人工萌发的技术程序如下:花粉在含 24% 蔗糖、7.3% 甘露醇、0.05% LH、Roberts 培养基盐成分,pH8.5 的水合液中,于 4℃ 低温水合 18h,转入 34℃ 热激 40~60 min,再经含 24% 蔗糖的 Roberts 培养基(pH8.5)渗激 20 min。分离的脱外壁花粉在含 15%~20% PEG(6000)、5% 蔗糖、0.05% LH 的 Roberts 培养基(pH8.5)中,于 25℃ 振荡培养。

表 2　　培养基成分对花粉与脱外壁花粉萌发的影响

Table 2　Effect of germination media on germination of intact pollen and de-exined pollen

培养基 Germination media	青菜 *Brassica chinensis*		紫菜苔 *B. campestris var. purpurea*	
	完整花粉萌发率 Germination rate of intact pollen ($\bar{X}\pm SE$)(%)	脱外壁花粉萌发率 Germination rate of de-exined pollen ($\bar{X}\pm SE$)(%)	完整花粉萌发率 Germination rate of intact pollen ($\bar{X}\pm SE$)(%)	脱外壁花粉萌发率 Germination rate of de-exined pollen ($\bar{X}\pm SE$)(%)
A 15% sucrose	50.0±1.3	1.1±0.57	46.0±1.1	0.82±0.45
B 20% PEG	57.6±3.2	0	54.5±3.9	0
C 15%~20% PEG+5% sucrose+0.05% LH	82.3±3.0	41.0±3.1	80.2±3.4	32.6±2.3

三、脱外壁花粉分离与萌发过程的细胞形态学观察

扫描电镜观察　青菜花粉粒呈椭圆形,外壁很厚,表面分泌油滴,具 3 条萌发沟(图版 I,1),经冷浸后,在热激和渗激两个阶段外壁开

裂,暴露出内壁。在热激过程中,萌发沟扩大,内壁从 1 条(图版Ⅰ,2)或 2 条(图版Ⅰ,3)萌发沟露出,逐渐分离出脱外壁花粉。在渗激过程中,进一步促进尚未开裂外壁的花粉,由 3 条萌发沟的会合处同时裂开,迅速逸出脱外壁花粉(图版Ⅰ,4)。彻底脱去了外壁的花粉,仍保持原有的椭圆形,其整个内壁表面全部暴露(图版Ⅰ,5)。与外壁相比,内壁表面较光滑,在纵行曲折的凹凸纹络看不出原来萌发沟处有何特异形态。由脱外壁花粉萌发的花粉管,与一般花粉管形态相近(图版Ⅰ,6)。

光镜观察　图版Ⅰ,7 示青菜脱外壁花粉群体的 Nomarski 干涉差显微图像。脱外壁花粉萌发之初,细胞质开始极性化,在一极产生花粉管,另一极出现液泡(图版Ⅰ,8),花粉管继续延长,细胞质开始向管中流动(图版Ⅰ,9),原来位于花粉粒中的两个精子和营养核移入花粉管中(图版Ⅰ,10、11)。总的状况和完整花粉的萌发一致,只是较后者萌发速率慢,需 2 h 方能大量萌发,而后者仅需 30 min。

讨　　论

一、脱外壁花粉的分离

迄今,脱外壁花粉的分离仅在极少数植物中取得成功,其中只有烟草是有目的的研究[1]。其余均为分离花粉原生质体中的附带发现[3,6,7]。总结前人与本文研究结果,可以将已有方法分为 3 种,并且似可看出它们和各种植物花粉壁结构特点有某种对应关系:(A)低渗水合。用于外壁薄而内壁厚的花粉,如南美扁柏(*Cupressus arizonica*)花粉在水中浸泡 3~4 min 即可脱去外壁[6]。风雨花(*Zephyranthes grandiflora*)花粉外壁很薄,在水合液中亦易脱去外壁[7]。(B)水合-酶解。烟草花粉也是外壁较薄而内壁厚,但外壁裂开后附着于花粉一极,需经短时酶解方可完全脱离[1]。(C)多步冲击。芸苔属花粉外壁厚而内壁薄,二者厚度相差数倍,需要长期水合方可脱去外壁[3],但又会导致萌发力丧失。本文提出的低温水合-热激-渗激程序,利用长时间冷浸削弱内外壁的联系,然后通过短时间热激与渗激使外壁裂开脱落,且在接近萌发培养基、但更为高渗的水合液中低温水合得以使脱外壁花粉有一定的萌发力。总之,针对花粉的生物学特点设计不同的方法,是今后在其它植物中制备脱外壁花粉时应遵循的原则。

二、脱外壁花粉的萌发

脱外壁花粉较完整花粉难于萌发,表现在:第一,对萌发介质成分有较高的要求。芸苔属花粉在仅含 PEG 和 Roberts 无机盐的碱性培养基中即可顺利萌发[6],而脱外壁花粉必须添加蔗糖和 LH。第二,萌发率较低,萌发所需时间较长。原因可能有二:一是分离过程中的冲击导致的损害;二是失去外壁导致的不利影响。已经知道,外壁及含油层中的类黄酮素对花粉萌发有重要作用[9];外壁束缚造成的花粉膨压也有利于花粉萌发,且外壁只在萌发沟处的不连续对花粉萌发所需的极性建立有重要意义[10]。由于目前尚难设计出区分上述两个因素的实验,所以关于外壁在花粉萌发中的作用还有待进一步探讨。

芸苔属花粉外壁含有孢子体自交不亲和的识别蛋白[11],含油层对花粉在柱头上的水合与萌发起重要作用[12]。失去外壁的花粉如何与柱头相互作用是饶有兴趣的问题。通过脱外壁花粉授粉,研究孢子体自交不亲和反应,可能是一个新的思路。

图 版 说 明

图 版 Ⅰ

1~6. 扫描电镜下青菜花粉脱外壁与萌发的过程。1. 水合的花粉粒。×3000 2、3. 热激后1条或2条萌发沟裂开,内壁暴露。×2500 4. 渗激后3条萌发沟从汇合处裂开,内壁暴露。×3000 5. 完全脱去外壁的花粉。×3000 6. 脱外壁花粉萌发的花粉管。×3000 7~9. Nomarski 干涉差显微图像。 7. 脱外壁花粉群体。×200 8. 脱外壁花粉萌发初期。×400 9. 花粉管的延伸。×250 10、11. DAPI 荧光染色,示脱外壁花粉(10)中及其花粉管(11)中的一对精子和营养核。×400

第一章 雄性器官与细胞的实验研究

图版 I

plate I

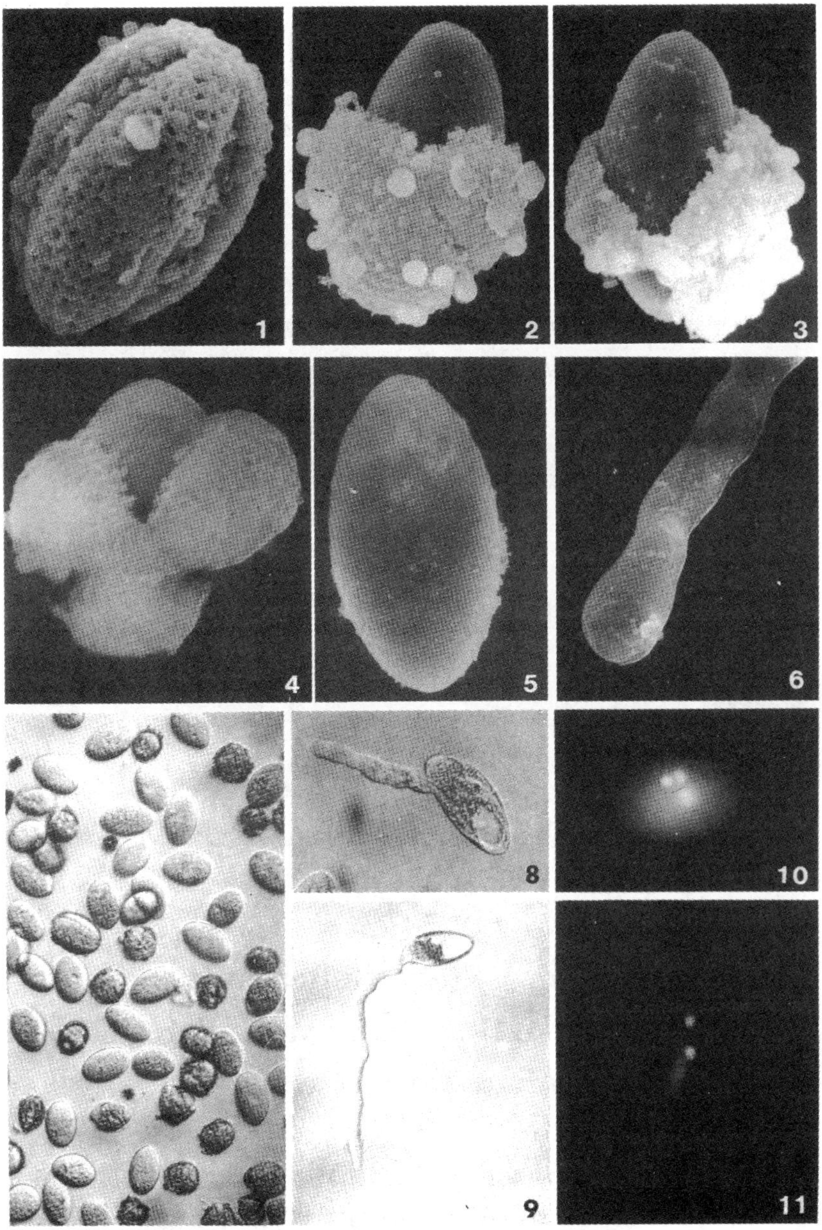

161

Explanation of plate

plate Ⅰ

Figs. 1-6. Processes of isolation and germination of de-exined pollen in *Brassica chinensis*, viewed by scanning electron microscopy. **Fig. 1.** A hydrated pollen. ×3000 **Figs. 2、3.** Exposure of intine from one or two dehisced furrows of exine after heat shock. ×2500 **Fig. 4.** Exposure of intine from the dehisced exine at the joint of three furrows after osmotic shock. ×3000 **Fig. 5.** A completely naked de-exined pollen. ×3000 **Fig. 6.** A pollen tube germinated from the de-exined pollen. ×3000 **Figs. 7-9.** Nomarski interference contrast microphotographys. **Fig. 7.** A population of de-exined pollen. ×200 **Fig. 8.** Early germination stage of de-exined pollen. ×400 **Fig. 9.** A growing pollen tube from the de-exined pollen. ×250 **Figs. 10、11.** DAPI fluorescence microphoto graphs, showing two sperms and a vegetative nucleus in a de-exined pollen (10), and a pollen tube of it (11). ×400

（作者：徐秉芳、梁世平、周嫦、杨弘远。原载：植物学报，1996，38 (12)：963~968。图版1幅，图2幅，表2幅，参考文献12篇，仅选录图版Ⅰ与表2)

2 芸苔属脱外壁花粉作为研究花粉萌发的新实验系统

Brassica de-exined pollen as a new experimental system for studying pollen germination

The de-exined pollen, i. e. pollen deprived of exine and only coated with intine, may become a new experimental system, applicable not only in the study of genetic transformation but also in the fundamental research on pollen biology. Up to date, an experimental system for isolation, artificial germination and *in vitro* pollination of the de-exined pollen has been established in *Nicotiana*[1,2]. Once the barrier of exine is removed, the de-exined pollen has been shown by electroporation to be a superior receptor for foreign genes[3]. The de-exined pollen has been successfully isolated in *Brassica*, and normally germinated in a modified polyethylene glycol (PEG) based medium[4]. In this paper, we report further studies on the processes of activation and polarization, predetermination of germination sites and new wall synthesis during isolation and germination of the de-exined pollen in *Brassica*.

Materials and methods

The methods of isolation and artificial germination of the de-exined pollen of *Brassica chinensis* L. were reported previously[4]. Briefly, the exine was detached from the pollen via a 3-step procedure including cold hydration at 4℃ for 18 h, heat shock at 34℃ for 40 min, and then osmotic shock from 1689 to 1055 mosm/kg. Subsequently the de-exined pollen was germinated in Roberts' medium with 15% PEG, 5% sucrose, 0.05% lactoalbumin hydrolysate (pH8.5).

In the light microscopic studies, the distribution of membrane-associated Ca^{2+} was demonstrated under fluorescence microscope stained with 10^{-4} mol/L chlorotetracycline (CTC). Wall components were identified using 0.1% Calcofluor White (CW) for cellulose, 0.02% decolourized water-soluble aniline blue (DAB) for callose, and 0.5% ruthenium red (RR) for pectin. Olympus microscopes IMT-2 and BH2-RFCA were used for observation and photopraphy.

For transmitting electron microscopic (TEM) observation, the samples were pre-fixed successively in 0.5%, 1%, 2% glutaraldehyde (dissolved in 0.2 mol/L phosphate buffer containing 7.5% sucrose, pH7.2) each for

1 h. After that they were post-fixed in 1% osmium tetraoxide, dehydrated through ethanol series, then embedded in Araldite. The ultrathin sections were stained with uranyl acetate and lead citrate and were examined under a JEM-100CX/II electron microscope. For scanning electron microscopic (SEM) observation, the dehydrated samples were dried using the acetonitrile vaouum drying method[5], coated with gold and examined under a H-450 scanning electron microscope.

Results

Activation and polarization of the de-exined pollen

Ultrastructural observations The mature tricellular pollen grain of *B. chinensis* was ellipsoid in appearance, with two sperm cells and one vegetative nucleus. Its electron-dense cytoplasm contained abundant mitochondria and lipid bodies. Polarity was not present at this time.

Table 1 Relationship between polarization of Ca^{2+} distribution and germination rate of the de-exined pollen*

Isolation treatments	Ca^{2+} distribution of the de-exined pollen(%)			Germination rate(%)
	Polarized	Unpolarized	Unclear	
Heat shock for 40 min, osmotic shock from 1689 to 1055 mosm/kg	54.9	11.7	33.4	41.0
Heat shock for 100 min	16.8	42.2	41.0	8.9
Osmotic shock from 1689 to 589 mosm/kg	15.2	52.9	31.9	6.6

* More than 300 de-exined pollen grains were examined in each treatment.

During the isolation procedure, the pollen was activated and underwent profound changes. Thus the newly isolated de-exined pollen possessed one or rarely two activated vesicle-rich sites at its peripheral region. New wall materials deposited outside these sites (Pl. I ,1). Vesicle aggregation associated with new wall deposition has been considered as a characteristic of pollen germination sites[6], indicating that polarization of the pollen and predetermination of the germination sites had already been accomplished during the process of isolation.

Further polarization was observed during the subsequent germination process. The cytoplasm, vegetative nucleus and sperm cells moved into the newly germinated tube, and a large vacuole occupied the major portion of the de-exined pollen (Pl. I ,2).

CTC observation Membrane-associated Ca^{2+} illustrated by CTC could not be observed in the intact pollen grains due to the strong autofluorescence of the exine. Removal of the exine made it possible for CTC fluorescence visualization. As seen in Figs. 3 and 4 of Plate I , the activated pollen whose exine had dehisced but still partially attached had already exhibited a polarity of the membrane-associated Ca^{2+}. A completely naked de-exined pollen had one or rarely two intensive fluorescence sites (Pl, I ,5、6) which were affirmed as the germination sites[7]. Thus, at the light microscopic level, we also proved that the polarity and germination sites had been established in advance before exine detachment and pollen germination.

There was an evident relationship between the polarization of Ca^{2+} distribution and the germination rate of the de-exined pollen. As seen in Table 1, during the isolation of the de-exined pollen, a prolonged heat shock or too strong an osmotic shock led to a sharp decline of the percentage of de-exined pollen with polarized Ca^{2+} distribution. Correlatively, the germination rate was also remarkably decreased.

New wall deposition on the de-exined pollen

Ultrastructural observations The structure of the *Brassica* mature pollen is characterized by having three germ furrows and a thick sculptural exine (Pl. I,7). The intine is thinner, with an additional fibrillar outer layer at the furrow sites (Pl. I,8).

During the course of isolation, the fibrillar outer layer of intine disappeared. Meanwhile, new wall deposition at the inner side of intine was always localized under the furrows (Pl. II,9). Viewing under a scanning electron microscope the freshly isolated de-exined pollen maintained its original ellipsoidal shape. Its surface was coated with a uniform intine without any trace of the former furrows (Pl. II,10). However, at the inner side of the intine, new wall deposition at the germination sites was visible under a transmitting electron microscope. The small vesicles adjacent to this new wall indicated their role in wall regeneration (Pl. I,1).

1 to 2 h after culturing in the artificial germination medium, the new wall was prominently thickened and composed of two layers: an electron-dense outer layer and an electron-translucent inner one. Some unidentifiable electron-dense materials dispersed among the new wall and outside the intine (Pl. II,11).

For investigating the whereabouts of the above-mentioned new wall, in comparing with the wall of the germinated de-exined pollen tube, several features in common with the wall at the tube base were found, such as their thickness, electron density, laminated feature, as well as some unidentifiable electron-dense materials (Pl. II,12). Therefore, we assumed that the new wall was likely to be the progenitor of the tube base wall.

Identification of wall components The freshly isolated de-exined pollen could only be stained with CW(Pl. II,13) showing uniform cellulosic fluorescence covering the whole surface of the de-exined pollen. RR staining for pectin and DAB staining for callose were not evident. After 1 to 2 h of culture, the wall at the germination sites became remarkably thickened, and was well stained with DAB and RR (Pl. II,14、15),indicating its callose and pectin nature. Concurrently CW fluorescence still appeared all around the cell.

Discussion

As reported previously, pollen deprived of exine without germ furrows can still germinate in *Nicotiana*[2] as well as in *Brassica*[4]. Thus two questions may arise: (1) Since the de-exined pollen lacks germ furrows, how could the germination site be determined or could it still be located at the original furrow? (2) The tube germinated from the de-exined pollen is not limited by the dimension of the germ furrow, then why is it still so slender as the ordinary pollen tube?

The hypothesis supported by Heslop-Harrison and Heslop-Harrison[8], and by Feijó et al.[9] could explain why the germination site is usually established at the site of the germ furrow. It is interpreted as if the lack of exine at the furrow causes preferentially stretching of its plasma membrane by the turgor pressure and opening of its stretch-activated Ca^{2+} channel, and in turn increases Ca^{2+} concentration at the region of the furrow. Thus the germination site is established on this activated region. According to our observation, the germination site of the de-exined pollen is predetermined and located at the region of the previous furrow prior to exine detachment. This conclusion is confirmed by the fact that the new wall as a predisposition of germination site always deposits at the region of the furrow during isolation of the de-exined pollen.

The phenomenon that pollen synthesizes new wall during hydration has been mentioned by Sedley[10] and Gresti et al.[11]. Miki-Hirosige et al.[12] pointed out that the new wall was composed of two layers, an outer electron-dense pectin layer, and an inner electron-translucent callose layer. In our fluorescent microscopic observation, the new wall regenerated from the de-exined pollen also contained pectin and callose, corresponding respectively to the outer electron-dense layer and the inner electron-translucent layer of our ultrastructural observation. The function of the new wall is unclear. Heslop-Harrison and Heslop-Harrison[6] suggested that the new wall forms an annular structure to define the diameter of the pollen tube. The de-exined pollen is an ideal system to support this view, since there is already no germ furrow to limit the tube size. As a progenitor of the tube base wall, the new wall reasonably plays a role in determining the diameter of the tube base.

The de-exined pollen facilitates fluorescent microscopic researches in pollen biology because the strong fluorescence exhibited by exine is eliminated. Nuclei, membrane-associated Ca^{2+}, and wall components of intine and the newly formed wall can thus be well visualized after fluorescent staining. Owing to all these merits, the de-exined pollen appears to be a useful experimental system for further deepening our knowledge on pollen biology.

Explanation of plates

Plate I

Fig. 1. A newly isolated de-exined pollen with a vesicle-enriched region (arrowhead) and a new wall site (arrow). ×5000 **Fig. 2.** A germinated de-exined pollen, showing the sperm cells (SC) and a vegetative nucleus (VN) moving into the tube, and a vacuole (V) occupying a large portion of the de-exined pollen. ×4000 **Figs. 3、4.** The pollen partially deprived of exine, showing polarity of membrane-associated Ca^{2+} stained with chlorotetracycline (CTC). ×500 **Figs. 5、6.** The freshly isolated de-exined pollen, with one (Fig. 5) or rarely two (Fig. 6) CTC intensive fluorescence sites. ×500

Fig. 7. Surface view of an intact pollen grain under scanning electron microscopic (SEM) observation. ×3000 **Fig. 8.** The pollen wall composed of a thick exine (E), a thin intine (I) and a fibrillar outer layer (arrow) covering the intine at the furrow. ×20000

Plate II

Fig. 9. The new wall (arrowhead) deposited at the region of the furrow during the isolation procedure. ×20000 **Fig. 10.** Surface view of a de-exined pollen under SEM observation. ×3000 **Fig. 11.** The thickened new wall inside the intine (I) composed of two layers: an electron-dense outer layer (arrowhead), and an electron-translucent

第一章 雄性器官与细胞的实验研究

图版 I

Plate I

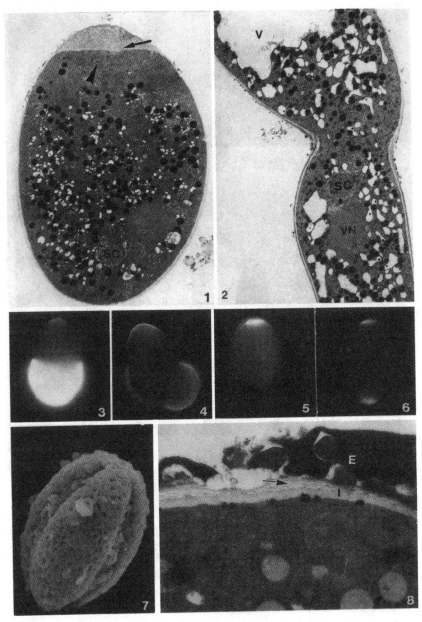

图版 II　　　　　　　　　　　　　　　　　　　　Plate II

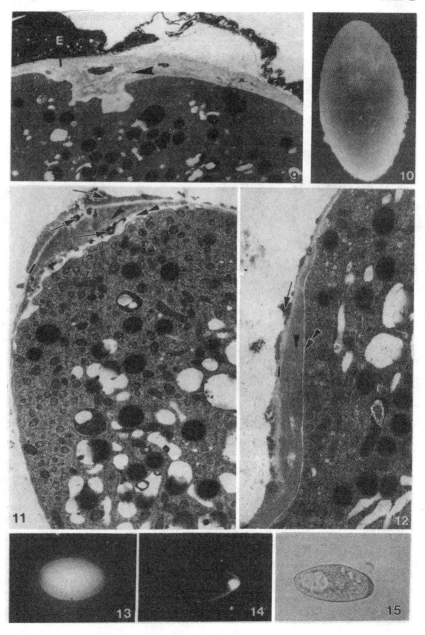

inner one (double arrowhead), with some unidentifiable electron-dense materials (arrow) dispersed among the new wall and outside the intine. ×11500 **Fig. 12.** The tube base wall of a germinated de-exined pollen, showing the same structure as in Fig. 11. ×10000 **Fig. 13.** Whole surface of a de-exined pollen stained with Calcofluor White (CW). ×450 **Figs. 14、15.** Thickened wall at the germination site stained with decolourized water-soluble aniline blue (DAB) (Fig. 14) and ruthenium red (RR) (Fig. 15). ×450

(Authors: Xu BF, Liang SP. Zhou C, Yang HY, Published in:
Acta Botanica Sinica, 1997,39(6):489-493, with 2 plates,
1 table and 12 references, all retained here.)

3 Preparation of exine-detached pollen in *Nicotiana tabacum*

A pollen grain is composed of three parts: exine, intine and protoplast. There have been many publications on the biology of intact pollen (Mulcahy and Ottaviano 1982, Russell and Dumas 1992). Great progress has been made about studying of pollen protoplasts in recent years (Zhou and Yang 1989, Yang and Zhou 1992). To date, no studies have been reported on exine-detached pollen, a unit between pollen grain and pollen protoplast. The influence that pollen has only intine on germination, pollination and fertilization is now unknown. Pollen without covering of exine should be easier to accept foreign DNA materials in genetic transformation. Therefore, the exine-detached pollen as an experimental system may not only be helpful in studying the biological function of exine and intine, but also provides a new receptor system for gene engineering. We describe in this paper an efficient protocol for preparing large populations of exine-detached pollen and the main factors affecting its preparation. The main results are as follows.

Materials and methods

Plant material

The experimental material was anthers from *Nicotiana tabacum* L. cultivar G-80.

Preparation of exine-detached pollen.

Anthers containing early-middle binucleate pollen were kept at 4-6℃ for 7-14 days. After surface-sterilization, the anthers were floated in 0.3 mol · L^{-1} sucrose solution and cultured in the dark at 28℃. After 2 days, pollen grains shed from the anthers were collected by centrifugation, then transferred into an enzyme solution containing 1% cellulase R-10, 1% pectinase, 0.1% pectolyase Y-23, 0.5% potassium dextran sulphate, K_3 medium macro elements, 1 mol · L^{-1} mannitol and 0.3 mol · L^{-1} sorbitol, pH 5.8-6.0 (Xia et al. 1995). The enzyme solution was incubated in the dark at 28℃ for 15-20 min, shaking with hand at the same time. Sediment was collected by centrifugation. Rinsed with K_3 medium (supplemented with 1 mol · L^{-1} mannitol, 0.2 mol · L^{-1} sorbitol) and then resuspended in the same medium.

Results and discussion

Process of exine-detachment of the pollen

When anthers were cultured in liquid medium, a large number of pollen grains were shed into the medium (Fig. 4). The exine of most pollen grains gradually dehisced during the next day (Fig. 1). Finally, they only attached to one side of the pollen grains and almost all the intine was exposed (Fig. 2,5). After they were transferred into an enzyme solution, the exine was detached immediately resulting in the release of exine-detached pollen (Fig. 3). Purified populations of exine (Fig. 6) and exine-detached pollen (Fig. 7) were obtained by centrifugation. The exine-detached pollen had not a completely spherical shape and a clearly-visible intine.

Main factors affecting isolation of the exine-detached pollen

Cold-pretreatment The anthers that had been cold-pretreated released numerous pollen grains. It was estimated that the exine of 86.5% pollen grains dehisced after 1 day. By enzymatic maceration, the exine-detached pollen was obtained in large numbers. The anthers that were not cold-pretreated did not release pollen grains to the medium. The exine of only a few pollen grains that were directly squashed from anthers dehisced. When they were placed into enzyme solution, only a limited number of exine-detached pollen was produced.

Osmoticum concentration To define the optimal conditions, we tested various concentrations of mannitol and sorbitol, such as $0.3, 0.5, 0.8, 1.0, 1.2 \text{ mol} \cdot \text{L}^{-1}$. The results showed that osmotic pressure below $0.5 \text{ mol} \cdot \text{L}^{-1}$ gave rise to rupture of exine-detached pollen. When osmoticum concentration was $1 \text{ mol} \cdot \text{L}^{-1}$, the exine-detached pollen still swelled. The optimal condition for the production of exine-detached pollen was $1.2 \text{ mol} \cdot \text{L}^{-1}$. Under this condition, the exine-detached pollen kept normal shape.

Enzymes used Effects of enzymes used on the isolation of exine-detached pollen are shown in Table 1. Without enzyme, only a few pollen grains detached off their exine. Cellulase alone had not any effect on detachment of exine. However, pectinase had a prominent effect. Addition of pectolyase Y-23 caused improvement of the yield of the exine-detached pollen. Mixture of the three enzymes proved to be the best for the yield. It was not the same as the report on *Brasica napus* and *Brassica campestris* var. *purpurea* (Li et al. 1992), whose pollen could detach exine only by hydration.

Exine-detached pollen as a new experimental system will be of great significance in both fundamental research and practical applications. First, by comparison of intact pollen with exine-detached pollen, the biological function of exine will be known deeper; second, by comparison of the exine-detached pollen with pollen protoplast, the biological function of intine will be also well understood. Further, it may deepen our knowledge about the function and mutual relations between exine, intine and protoplast in the process of pollination and fertilization; third, it will be a good system for introducing foreign genes due to removing the barrier of the exine. It enables us to find a new approach to transfer foreign genes into the fertilized egg cell via pollination of exine-detached pollen.

(Authors: Xia HJ, Zhou C, Yang HY. Published in Wuhan University Journal of Natural Sciences, 1996, 1(1):116-118, with 7 figures, 1 table and 7 references, all omitted here.)

4 烟草脱外壁花粉人工萌发与离体授粉实验系统的建立

Establishment of an experimental system for artificial germination and *in vitro* pollination with de-exined pollen in *Nicotiana tabacum*

Abstract

Tobacco flower buds at mid-binucleate pollen stage were cold-treated and their anthers were then float-cultured to release the pollen, which were subsequently macerated in enzyme solution for a short time. De-exined pollen grains were eventually isolated. The main factors affecting artificial germination of the de-exined pollen were investigated, including osmotic pressure of the enzyme solution during the isolation process, polyethylene glycol (PEG) or sucrose in the culture medium as well as supplementation of lactoalbumin hydrolysate (LH). Finally, a medium containing D_2 macroelements, 30% PEG-6000 and 0.1% LH was established, which supported the de-exined pollen to germinate well with a frequency up to 57.8%. After 24 h of culture, the generative cell in more than half of the pollen tubes, already divided into two sperms. Using a method of micro-suspension droplets with the aid of a small piece of filter paper, 30 to 40 de-exined pollen grains were pollinated onto the stigma, resulting in nearly half of the pollen tubes growing in the style and approximately a yield of one seed out of four de-exined pollen grains after subsequent ovary culture. The seeds were germinated into seedlings. The artificial germination of de-exined pollen can be further used as a tool for understanding the role of exine in pollen germination. The *in vitro* pollination with such exine-free pollen might become a new means for introducing foreign genes into the seeds and offsprings.

花粉粒由外壁与内壁及其所包围的原生质体所组成。用人工方法脱去双重壁分离出花粉原生质体；如仅脱去外壁而保留内壁与原生质体，则得到介于花粉粒与花粉原生质体之间的结构单位，可命名为脱外壁花粉。李仕琼等[1]研究了芸苔属脱外壁花粉的制备方法，但当时只是把它作为获得花粉原生质体的一个中间环节。Desprez 等[2]亦在分离烟草成熟花粉原生质体的过程中获得了脱外壁花粉。夏惠君等[3]专文报道了由烟草幼嫩花粉制备脱外壁花粉的技术。脱外壁花粉由于排除了外壁对生物大分子的屏障作用，应有利于外源基因的导入。这一推测已被施华中等[4]的实验所证实：烟草脱外壁花粉的外源 GUS 基因瞬间表达水平约为花粉粒的 30 倍，这表明脱外壁花粉可能成为一种有潜力的遗传转化受体。本文的目的是研究烟草脱外壁花粉人工萌发的最适条件,创建脱外壁花粉离体授粉实验系统,为利用这一实验系统开拓转基因技术新途径奠定基础。同时,通过这一研究,也可以对外壁在花粉萌发与授粉中的生物学意义获得某些新的认识。

材 料 与 方 法

供试材料为烟草(*Nicotiana tabacum* L.)品种"G-80"。

一、脱外壁花粉的制备

参照夏惠君等[3]的方法,取含二核中期花粉的花蕾,在 4~6℃下贮存 7~14d。在 25℃下将花药漂浮于 0.2 mol/L 蔗糖、0.2mol/L 甘露醇溶液中培养 36h 左右。离心收集从花药中释放出的、外壁大部分裂开的花粉,转入 1% 纤维素酶 R-10、1% 果胶酶、0.1% pectolyase、D_2 培养基大量元素[5]、1.0 mol/L 甘露醇、0.2 mol/L 山梨醇,pH5.8~6.0 的酶液(渗透压值用 Micro-Osmometer Model 3 MO)测定,在 25℃、弱光下酶解 20~30 min。离心收集完全脱去外壁的花粉,用洗液(D_2 大量元素、1.0mol/L 甘露醇、0.2 mol/L 山梨醇、pH5.8~6.0)充分洗涤。

二、人工萌发

采用附加 30% 聚乙二醇(PEG)6000、0.1% 水解乳蛋白,pH5.6~5.8 的 D_2 基本培养基,在 25℃下液体浅层、静止、弱光培养,使之萌发花粉管。培养 24h 后,用 10 mg/L DAPI 染色,荧光显微镜观察统计花粉管中的精子形成。

三、离体授粉

开花前 1d 去雄,切下 1~1.5 cm 的柱头与花柱,插入附加 10% 蔗糖、0.1% 水解乳蛋白的 D_2 培养基中。授粉方法:(1)用微吸管在倒置显微镜下挑取少量在前述培养基中温育 1h 的脱外壁花粉,授于柱头;

(2)吸取脱外壁花粉置于小滤纸片上,稍干,倒贴于柱头上;(3)以上两法相结合。授粉20h后,按Martin[6]的方法,用0.1%脱色水溶性苯胺蓝染色,荧光显微镜观察花粉管在花柱中的生长。

另取开花前1d的花蕾,去雄后自花柄基部切下,插入含4%蔗糖的Nitsch培养基的三角瓶中。开花当天,用上述方法授粉,继续培养3d后,取下子房,经消毒后培养在含4%蔗糖、500 mg/L水解乳蛋白、10 mg/L抗坏血酸、0.8%琼脂,pH6.0的Nitsch培养基中。40d后,剥取种子在含3%蔗糖、0.8琼脂的MS培养基中萌发幼苗。

结　果

一、影响脱外壁花粉离体萌发的关键因素

制备过程中的酶解条件　在脱外壁花粉的制备过程中,酶液的渗透压对于其以后的离体萌发有重要影响(表1)。以甘露醇和山梨醇作渗透压调节剂,将酶液分别调到900、1281、1614、1846 mosm/kg H_2O 4种渗透压。其中以1614 mosm/kg H_2O 条件下所获得的脱外壁花粉萌发率最高,达57.8%(图版Ⅰ,3)。渗透压过高,质壁分离严重;过低则导致膨胀,均不利于萌发。此外,酶解时间也有一定影响。时间过长,花粉内壁也被损伤,对萌发不利;时间太短,外壁不能脱去。一般以酶解20～30 min为宜。

PEG和蔗糖　在D_2培养基中附加不同浓度的蔗糖或PEG,结果表明(表2):脱外壁花粉在含20%与30%蔗糖的培养基中完全不能萌发;在含20% PEG时几乎全部破裂;PEG为30%时,可正常萌发,频率达32.7%。对照为未脱壁的完整花粉粒,在上述4种培养基中都能萌发,但也以含PEG的培养基中萌发率较高,花粉管生长较快且长,培养20h后,花粉管生长仍然正常,有胞质流动。

水解乳蛋白　在不同PEG浓度的D_2培养基中添加0.1%水解乳蛋白,对提高脱外壁花粉萌发率均有作用(表3)。同时也改善了花粉管的生长:管细直而长,生长速度加快,管尖不易破裂。30%和35%

PEG 差别不大,但 PEG 太浓,以后操作不便,故以 30% PEG、0.1% 水解乳蛋白为宜。

表2 聚乙二醇和蔗糖浓度对脱外壁花粉萌发率的影响
Table 2 Influence of polyethylene glycol and sucrose concentration on germination of the de-exined pollen

处理 Treatment		观察的脱外壁花粉数 No. of the de-exined pollen grains observed	萌发的脱外壁花粉数 No. of the de-exined pollen grains germinated		观察的完整花粉数 No. of the intact pollen grains observed	萌发的完整花粉数 No. of the intact pollen grains germinated	
			数量 No.	频率 Frequency (%)		数量 No.	频率 Frequency (%)
蔗糖 Sucrose	20%	312	0	0	316	217	68.7
	30%	308	0	0	322	177	55.0
PEG	20%	322	0	0	324	254	78.4
	30%	339	111	32.7	318	270	84.9

二、花粉管中的精子形成

烟草具 2-细胞型花粉,只有在花粉管中生殖细胞分裂成精子,才能进行正常受精。在本实验条件下,脱外壁花粉在培养 24h 后,花粉管中能形成正常的精子(图版Ⅰ,4)。研究了不同培养基中精子形成的情况(表4)。结果表明,改变 PEG 的浓度对精子形成没有明显影响。但添加 0.1% 水解乳蛋白后,精子形成率有大幅度提高,这一结果与 Read 等[7]报道的相似。高于 50% 的精子形成率应能满足受精要求。

表4　不同培养基对脱外壁花粉萌发花粉管中精子形成的影响
Table 4　Influence of culture media on the sperm formation in pollen tubes from the de-exined pollen

处　　理 Treatment	观察的脱外壁花粉管数 No. of the de-exined pollen tubes observed	形成精子的花粉管数 No. of the pollen tubes containing sperms	
		数量 No.	频率 Frequency (%)
30% PEG	56	21	37.5
30% PEG+0.1% LH	58	32	55.2
35% PEG	50	16	32.0
35% PEG+0.1% LH	50	27	54.0

PEG. Polyethylene glycol;LH. Lactoalbumin hydrolysate.

综合上述实验结果,可确定烟草脱外壁花粉的较佳人工萌发条件是在 D_2 基本培养基中附加 30% PEG-6000、0.1% 水解乳蛋白,pH5.6~5.8,于25℃下液体浅层、弱光、静止培养。

三、授粉后花粉管在花柱中的生长

在上述培养基中温育1h的脱外壁花粉进行人工授粉实验(表5)。在花粉预培养过程中,花粉外壁裂开的频率可达86.6%(图版Ⅰ,1),经过其后的酶解处理及多次离心之后,脱外壁花粉的纯化率可达90%以上(图版Ⅰ,2)。在倒置显微镜下用微吸管挑取状态良好的脱外壁花粉,每次挑取一至数个,其纯化率可高达97.4%,基本上排除可能混杂的未脱壁花粉。然后以微滴授在柱头上,花粉能萌发,花粉管伸入花柱中正常生长(图版Ⅰ,5),并能由花柱基部切口长出。但如授于柱头上的液滴过大,则不利于花粉管进入花柱。如柱头上同时覆盖带有脱外壁花粉的小滤纸片,花柱中花粉管的生长频率提高20%左右。若将脱外壁花粉授在切去柱头的花柱切面上,未观察到花粉管的生长。从表3和表5可以看出,脱外壁花粉在花柱中和在培养基中的花粉管生长频率十分相近,均有近一半的花粉能萌发正常花粉管。

表5 授粉技术对花粉管在花柱中生长的影响
Table 5 Influence of the pollination technique on pollen tube growth in styles

授粉技术 Pollination technique	观察的花柱数 No. of styles observed	每一柱头上的脱外壁花粉数 No. of the de-exined pollen grains per stigma ($\bar{X}\pm SE$)	每一花柱中的花粉管数 No. of pollen tubes per style	
			数量 No. ($\bar{X}\pm SE$)	频率 Frequency（%）
液滴 Droplet	30	17.3±0.5	5.3±0.4	30.6±2.4
滤纸片 Filter paper	28	23.8±0.6	6.8±0.5	28.6±2.2
液滴附加滤纸片 Droplet plus filter paper	33	21.7±0.4	10.3±0.5	47.3±1.9**

两种处理差异极显著。Significance difference between the two treatments at $P \leqslant 0.01$.

四、离体授粉后的结实与出苗

在自然条件下,烟草"G-80"从授粉到受精的时间为36~48h。脱外壁花粉授粉48h后子房没有种子形成,72h则有,表明脱外壁花粉从授粉到受精的时间较正常花粉授粉长,大约在48h以后。培养5d后,子房开始膨大。20d后,可见子房不同部位膨胀,子房壁逐渐变褐。40d后,种子成熟(图版Ⅰ,6)。在所观察的37个子房中,22个发育成含种子的蒴果,占总数的59.5%。每一柱头上平均授34.8粒脱外壁花粉,每一蒴果中平均含8.8粒种子,结籽率为25.3%。即约1/4的花粉可以完成受精,形成种子。种子转入MS培养基中,5d后开始发芽,长成幼苗(图版Ⅰ,7)。

讨 论

一、脱外壁花粉的萌发需求

脱外壁花粉是介于花粉粒和花粉原生质体之间的独特结构。这三者的萌发需求是不同的。烟草花粉粒在附加一定蔗糖的简单的BK培养基中即可获得很高的萌发率;脱外壁花粉在 BK 培养基中难以萌发,只有在较复杂的、适用于原生质体的 D_2 培养基中才能正常萌发,并且需要以 PEG 代替蔗糖,以及附加水解乳蛋白;而烟草花粉原生质体迄今在各种试验中还不能正常萌发,只能得到少量类似花粉管的结构(未发表资料)。从完整的花粉粒到脱外壁花粉再到花粉原生质体,对萌发条件的要求愈来愈苛刻,是否可以认为这是由于外壁与内壁在花粉萌发的生理过程中各自起着重要的作用呢?以往关于外壁的结构、成分与生物学功能(如识别)积累了丰富的研究资料[8,9]。但是,外壁对花粉萌发是否也有重要作用,迄今文献中却缺乏论证。这主要是由于过去缺少相应的实验方法来探讨这一问题。本文为此提供了一项有用的实验系统,在排除外壁的情况下研究花粉人工萌发的需求。问题在于,制备脱外壁花粉过程中的低温贮存、漂浮培养、酶处理等环节,都可能会对花粉萌发力产生不利的影响,从而干扰实验结果的分析。因此,在进一步设计更严密的实验之前,还不足以就外壁对花粉萌发的生物学功能作出肯定的结论。

二、PEG 作为花粉人工萌发介质的优越性

Zhang 和 Croes[10]在矮牵牛花粉的人工萌发中比较了 PEG 和蔗糖的作用,指出高浓度的蔗糖对花粉管生长是有害的。这是由于蔗糖改变了细胞膜的通透性,促进花粉活性物质外泄;部分蔗糖被水解,从而导致培养基渗透压值与 pH 值的改变。PEG 则不被分解,可以保持培养基渗透压的稳定。这一结论被 Jahnen 等[11]在花烟草上的实验所证实。本文发现,与完整花粉粒相比,脱外壁花粉对蔗糖的不良反应更为强烈,基本不能萌发,而在含 PEG 的培养基中萌发良好,花粉管生长可以维持 24h 之久,并且一半以上的花粉管中可以形成精子,再次证明 PEG 作为花粉人工萌发介质的优越性。

三、脱外壁花粉离体授粉受精实验系统建立的意义

烟草的离体授粉早有报道。近年采用离体成熟的烟草花粉,在活体柱头上授以约含 5 000 个花粉粒的 4μl 液滴,获得 500～1 000 个种

子/子房[12,13]。本实验采用微滴附加滤纸片的技术,在离体柱头上授以微量脱外壁花粉,经过子房培养,获得一定数量有萌发力的种子。用于脱外壁花粉人工萌发的培养基同样适用于离体授粉。至此,已经建成了脱外壁花粉离体授粉实验系统。脱外壁花粉具有比完整花粉粒更易导入外源基因的特点已被施华中等[4]的实验所证明,今后需要进一步研究通过脱外壁花粉离体授粉技术将外源基因导入种子后代以获得转基因植株的可行性。

图 版 说 明

图 版 I

1. 外壁裂开而尚未脱落的花粉。×220　**2.** 外壁完全脱去的脱外壁花粉。×260　**3.** 脱外壁花粉萌发的花粉管群体。×120　**4.** 脱外壁花粉培养24h后,DAPI荧光染色,示花粉管中已形成一对精子。×150　**5.** 脱外壁花粉授粉后,脱色水溶性苯胺蓝荧光染色,示在花柱中生长的花粉管。×80　**6.** 脱外壁花粉授粉的子房培养40d后,剥开果皮,种子成熟。**7.** 由上述种子萌发的幼苗。

Explanation of plate

Plate I

Fig. 1. Pollen grains with dehisced but not completely detached exine. ×220　**Fig. 2.** De-exined pollen grains. ×260　**Fig. 3.** Population of pollen tubes germinated from the de-exined pollen. ×120　**Fig. 4.** A pollen tube from the de-exined pollen after 24 h of culture. DAPI fluorescent staining, showing two sperms in the tube. ×150　**Fig. 5.** De-colorized aniline blue fluorescent staining, showing pollen tubes grown in the style after pollination with de-exined pollen. ×80　**Fig. 6.** A dissected ovary after pollination with de-exined pollen at 40 d of culture, showing matured seeds. **Fig. 7.** Seedlings germinated from pollination with de-exined pollen.

(作者:王劲、夏惠君、周嫦、杨弘远。原载:植物学报,1997,39(5):405~410。图版1幅,表5幅,参考文献13篇,选录图版I和表2、表4。)

第一章 雄性器官与细胞的实验研究

图版 I Plate I

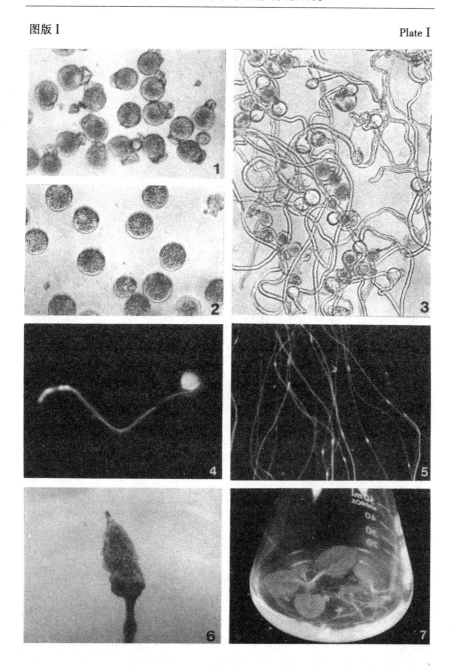

5 烟草花粉发育过程及不同组织中的内源 GUS 活性

Intrinsic GUS activity in various tissues and during pollen development of *Nicotiana Tabacum*

摘 要

以 X-gluc 作为葡糖苷酸酶(GUS)酶反应底物,用酶组织化学方法检测了烟草(*Nicotiana tabacum* L.)不同组织细胞的内源 GUS。在幼苗的根、茎、叶不同发育时期的花药壁、柱头、胚珠分离的生殖细胞和胚囊中,均未检测到内源 GUS 活性。不同发育时期的烟草花粉内源 GUS 活性存在差异,在小孢子分裂前后和成熟花粉萌发前后有两个活性高峰。检测花粉内源 GUS 活性的适宜 pH 值为 5.0; pH7.0 的条件下未检测到内源 GUS。用20%甲醇或0.2mmol/L葡糖二酸内酯处理不能完全抑制内源 GUS 活性。

Abstract

The β-glucuronidase (GUS) gene has been widely used as a reporter gene in the study of plant molecular biology and genetic engineering. One of the major reasons leading to the popularity of GUS-fusion system was the belief that there was no detectable intrinsic GUS activity in plant tissues. However, investigators have been troubled by the "false positive" results or "background" activities when GUS assays were performed. In the present experiment, histochemical observations of intrinsic GUS activity in various tissues and during pollen development of tobacco (*Nicotiana tabacum* L.) were carried out using 5-bromo-4-chloro-3-indolyl-β-D-glucuronic acid (X-gluc) as a substrate for overnight incubation of the treated tissues at 37℃. No detectable intrinsic GUS activity was found in seedling root, stem, leaf, anther wall and stigma of different stages, ovule as well as isolated generative cell and embryo sac. During pollen development, two peaks of intrinsic GUS activity appeared, one, close to the microspore mitosis and the other from the full maturation of pollen lasting to the post-germination pollen tube stage, no or weak activity was found at other pollen developmental stages. GUS was located in the cytoplasm of the pollen. The pH value of staining solution strongly influenced the experimental results. Blue color was visualized at pH 5, even when 20% methanol or 0.2 mmol/L glucaric acid-1-4-lactone (GAL, a specific GUS inhibitor) were added. At pH 7, no detectable reaction was found at all. The aforementioned results indicate that when using tobacco pollen as the target of GUS gene transformation, the assay should be strictly controlled to neutral condition for avoiding false positive results.

(作者：施华中、杨弘远、周嫦。原载：植物学报，1995，37（2）：134～139。图版1幅，表2幅，参考文献16篇，均删去。)

6 烟草脱外壁花粉的电激基因转移

Exine-detached pollen of *Nicotiana tabacum* as an electroporation target for gene transfer

摘 要

以β-葡糖苷酸酶（GUS）基因作为报告基因，通过瞬间表达的检测，比较了烟草（*Nicotiana tabacum* L.）脱外壁花粉、未萌发与萌发花粉的电激导入效果,探讨了不同电激条件及启动子对外源基因瞬间表达的影响。结果表明:当脉冲时间常数为 13ms 时,导致脱外壁花粉和萌发花粉生活力下降约 50% 的电场强度分别为 750V/cm 和 1250V/cm,在此条件下电激,二者的导入效果最好。脱外壁花粉的 GUS 基因表达水平约为萌发花粉的 5 倍、花粉粒的 30 倍。玉米花粉特异启动子(Zm13-260)能启动 GUS 基因在脱外壁花粉和萌发花粉中高效表达,而 CaMV 35S 的启动活性很低。

Abstract

In developing alternative systems for plant transformation the authors investigated the use of male gametophyte as the foreign gene receptor. However, delivery of foreign DNA into pollen is difficult because of the existence of a thick exine, therefore a new experimental system was developed using exine-detached pollen (EDP) of *Nicotiana tabacum* as an electroporation target which was also compared with germinating pollen (GP) and pollen grains (P). A transient GUS expression assay was conducted to analyze the effects of different electroporation conditions and promoter activity. The pollen-specific promoter Zm13 from *Zea mays* mediated high level of GUS gene expression but CaMV 35S only had very low activity in both EDP and GP. The optimal field strength for gene transfer was obtained at 750 V/cm for EDP and 1250 V/cm for GP when the time constant of pulse was 13 ms. The GUS activity in EDP had a 5-fold increase as compared with GP and P respectively. The level of GUS gene expression was slightly increased when adding 10% PEG into the electroporation buffer. This result indicates that pollen deprived of exine responds much better to foreign gene transfer than the previously used intact pollen grains and may be a better vector to introduce, via pollen tube, genes into the egg cell and offsprings.

(作者:施华中、王劲、杨弘远、周嫦。原载:植物学报,1996,
38(8):626~630。表 2 幅,参考文献 18 篇,均删去。)

7 β-glucuronidase gene and green fluorescent protein gene expression in de-exined pollen of *Nicotiana tabacum* by microprojectile bombardment

Pollen transformation has been proposed as a convenient method of gene transfer in higher plants (Heberle-Bors et al. 1990). It may give vital information about the molecular and cellular biology of pollination with pollen grains carrying foreign genes (Morikawa et al. 1994). To date, there have been several investigations on gene transfer into pollen by particle bombardment (Twell et al. 1989; Hamilton et al. 1992; Plegt et al. 1992; Stöger et al. 1992; Nishihara et al. 1993). Expression of the β-glucuronidase (GUS) gene in embryogenic pollen of tobacco and formation of stable transformants after regeneration of haploid plants have been reported recently (Stöger et al. 1995). However, gene transfer into mature pollen was not so easy, due mainly to the thick exine. In our lab, de-exined pollen deprived of exine and coated only with intine has been successfully isolated in *Nicotiana* (Xia et al. 1996) as well as in *Brassica* (Xu et al. 1997). We then established an experimental system for artifical germination and *in vitro* pollination with de-exined pollen in *Nicotiana tabacum* (Wang et al. 1997). Electroporation experiments verified that de-exined pollen responded much better to foreign genes than intact pollen grains (Shi et al. 1996). However, since electroporation resulted in rather low expression frequency of the tested GUS gene, we should search for a more efficient method that can be adapted to this experimental system.

Materials and methods

Nicotiana tabacum cv. G-80 was grown in the greenhouse. Intact pollen was collected from anthers at different developmental stages and cultured in BK medium (Brewbaker and Kwack 1963) supplemented with 20% sucrose. Flower buds 25-30 mm in length containing middle-late binucleate pollen were collected for isolation of de-exined pollen as previously described (Wang et al. 1997). The de-exined pollen was cultured in a medium containing D_2 macroelements (Li 1981), 30% polyethyleneglycol 6000 and 0.1% lactoalbumin hydrolysate.

The plasmid DNAs used for microprojectile bombardment were pBI121 containing the GUS-coding region flanked by the *CaMV35S* promoter, and pBS-260gn and mGFP4 containing the GUS- or GFP-coding region, respectively, flanked by the *Zm*13-260 pollen-specific promoter from maize (Hamilton et al. 1992). All plasmids contained the nopaline synthase terminator.

Purified plasmid (10 μg) was absorbed to 3 mg of gold particles (1.0 μm diameter, Bio-Rad) using the procedure described by Klein et al. (1987). DNA transfer was carried out using the helium-driven PDS-1000/He system (Bio-Rad) following the manufacturer's recommendations. In each experiment, uninucleate, mid-binucleate, mature pollen and de-exined pollen were bombarded with the three different chimeric genes. Pollen bombarded with gold particles free of DNA and pollen without bombardment were used as controls. After bombardment the plates were incubated at 25 ℃ for 24 h. Histochemical GUS assays were based on methods described in Jefferson et al. (1987). The samples were incubated in GUS histochemical buffer at 37 ℃ for 24 h. Pollen grains with a blue color, indicating GUS expression, were counted under a microscope. Fluorescent GFP assays were carried out under an Olympus IMT-2 fluorescent microscope with B excitation. Pollen grains with green fluorescence were counted. Bombarded pollen was either stained with 10 mg/L 4,6-Diamidino-2-phenylindole (DAPI) or left unstained. Intracellular localization of gold particles in pollen was determined microscopically with bright-field or fluorescent illumination. Using the method described by Wang et al. (1997), the pollen bombarded with plasmid mGFP4 was placed onto the stigma with style cuts 1-1.5 cm long which were inserted in the MS medium. After 24h, pollen tubes grew from the style cut. The pollen tubes were then observed by fluorescent microscopy.

Results and discussion

The exine of tobacco pollen shed from float-cultured anthers (Fig. 1a) ruptured when incubated in 0.2 mol/L sucrose and 0.2 mol/L mannitol, and the intine was gradually exposed (Fig. 1b). After incubation for 30 min in an enzyme solution purified de-exined pollen was obtained, as shown in Fig. 1c. This provided the material for the experiments on microprojectile bombardment.

Gold particles introduced by bombardment were visible in both de-exined pollen (Fig. 1d) and intact pollen (Fig. 1e). After pollen germination, gold particles could be moved from the pollen grains into the pollen tubes, conveyed by the vegetative cell cytoplasm (Fig. 1f). Introduced gold particles were observed by fluorescent microscopy after DAPI staining not only in the cytoplasm and vegetative nucleus (Fig. 1g) but also at the generative nucleus (Fig. 1h). Blue color was not detected in pollen bombarded with uncoated gold particles or in unbombarded pollen. Blue-colored, GUS-expressing pollen grains could be detected after addition of the substrate. GFP-expressing pollen could be detected directly by fluorescent microscopy without any substrate. Transient expression of the GUS and GFP genes in de-exined pollen is shown in Fig. 2a–c.

To establish the optimal conditions required to support transient gene expression in de-exined and intact pollen, frequencies of GUS and GFP

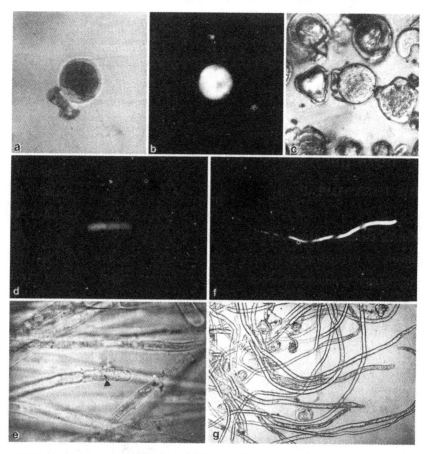

Fig. 2a-g Transient expression of the GUS and GFP genes. **a.** A blue-colored, GUS-expressing de-exined pollen grain. ×280. **b.** A green fluorescent, GFP-expressing de-exined pollen grain. ×260. **c.** Bright-field image of b. ×260. **d.** A green fluorescent, GFP-expressing pollen tube growing from the style cut. ×320. **e.** Bright-field image of d, showing pollen tubes growing from the style cut. *Arrowhead*: gold particles at one of the tube tip. ×320. **f.** A green fluorescent, GFP-expressing *in vitro* germinated pollen tube. ×300. **g.** Bright-field image of f. ×300

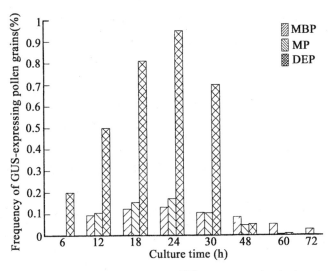

Fig. 3. Influence of culture time on transient GUS gene expression in de-exined pollen compared with intact pollen of different developmental stages. MBP Middle-binucleate pollen, MP mature pollen, DEP de-exined pollen

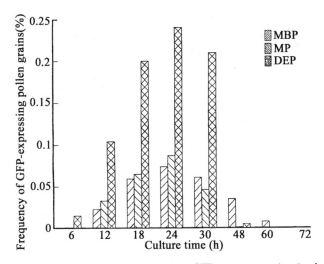

Fig. 4. Influence of culture time on transient GFP gene expression in de-exined pollen compared with intact pollen of different developmental stages (*Abbreviations* see Fig. 3)

gene expression were assayed at various time points following bombardment (Figs. 3 and 4). In general, gene expression was observed in a few de-exined pollen grains after 6h of incubation, whereas in intact pollen gene expression was not observed until after 12h of incubation. Frequencies of gene expression in pollen increased to a peak at 24h and then declined to a low level 60h after bombardment. Note that in all cases expression frequency was much higher in de-exined pollen than in intact pollen. At 24h (peak time), the frequency of GUS-expressing de-exined pollen was 0.95%, nearly 6 times than that of the intact pollen, while the frequency of GFP-expressing de-exined pollen was 0.24%, approximately 3 times that of the intact pollen. Mascarenhas and Hamilton (1992) indicated that diffusion of GUS reaction products can occur from stained tissues to otherwise unstained pollen. These would appear as diffusion artifacts and there would be false staining of pollen over a long incubation time with the substrate. In addition, blue photobleaching over time would result in loss of GFP-expressing individual grains. Therefore, this might explain the higher expression frequency of GUS than of GFP, although both constructs employed the same promoter. The efficiency of transient gene expression in pollen by particle bombardment varies in previous reports. The frequency of tobacco pollen expressing the GUS gene under the control of anther-specific promoter LAT52 was 0.1% (Twell et al. 1989) and 0.7% (Nishihara et al. 1993). However, the frequency of GUS-expressing tobacco pollen under the control of a pollen-specific promoter PA2 was merely 0.025% (Stöger et al. 1992). In the present research, for the first time, we used de-exined tobacco pollen as the target for gene transfer by particle bombardment and obtained high frequencies of transient GUS and GFP gene expression, 0.95% and 0.24% respectively, much higher than with intact pollen.

To determine whether the expression of the introduced gene could be assayed throughout the pollination process, we used an *in vivo/in vitro* technique; for pollination to occur, the de-exined and intact pollen bombarded with gold particles carrying the GFP gene were placed onto the stigma surface. The styles were cut and inserted into a medium; 24 h later, when pollen tubes grew out of the cut surfaces, GFP expression was detected in a few tubes (Fig. 2d,e). This was quite similar to the case of the *in-vitro* germinated pollen tubes (Fig. 2f,g). A large number of seeds have been obtained after pollination with de-exined and intact pollen bombarded with plasmid pBI121.

(Authors: Wang J, Shi HZ, Zhou C, Yang HY, Zhang XL, Zhang RD.
Published in Sexual Plant Reproduction, 1998, 11:159-162.
with 4 figures and 17 references. Figures 2-4 are retained here.)

第五节 生殖细胞操作

提要

中文的"生殖细胞"有两种不同的含义：一为 reproductive cell，泛指所有具有生殖功能的细胞；二为 generative cell，特指花粉中精细胞的前身。本节涉及的是后一种含义。生殖细胞是包藏于花粉粒内的"细胞中的细胞"。它的分离与操作意味着由性器官操作到性细胞操作的更进一步发展。

我们与美国研究者合作，于1986年首次报道以压片法由绣球百合花粉中分离出生活的生殖细胞，并研究了其离体后的细胞学变化。这一研究随后在国内继续进行，提出"二步渗透压冲击法"，从蚕豆与玉帘花粉中分离出大量生殖细胞，并通过过滤与离心获得了纯化的群体。在此基础上，对属于6科的11种植物的生殖细胞分离技术展开了大规模的研究，根据它们的生物学特点摸索出适于不同花粉的分离方法，即

"一步渗击法"、"二步渗击法"与"低酶法"。在国外,只有一种由花粉原生质体间接地分离百合生殖细胞的方法。我们实验室建立方法之多、覆盖植物种类之广,均大大超过了国际同类研究。

生殖细胞的离体培养是前人从未尝试的课题。我们将分离的黄花菜生殖细胞培养在含多种附加物和以花药为饲养物的琼脂糖培养基中。在近 6 000 个生殖细胞中,发现其中少数进行了一次核分裂,个别进行了二次核分裂。这一结果表明生殖细胞在离体条件下可以存活并完成有限的核分裂。

生殖细胞的离体培养潜能使我们可以利用它开展细胞融合实验。应用 PEG 诱导技术,进行了同种生殖细胞间、异种生殖细胞间、生殖细胞与体细胞原生质体间各种组合的融合,均获得了生活的同核体或异核体,证明在技术上是可行的。利用这一实验系统同时进行了荧光活染的实验。DAPI、H33258、mithramycin 等细胞核荧光染料通常用于显示生活细胞的核。但染色后的细胞是否仍然存活还没有人去求证,因而是否真正意义的"活体染色"仍未肯定。我们以花粉管中的细胞质流动、分离生殖细胞的 FDA 生活力鉴定以及生殖细胞的融合能力作为三项判断标准,证明上述染料尤其是其中的 DAPI 确实可用于活体染色,在性细胞操作中可作为鉴定、示踪与筛选异核体的有效方法。

分离的生殖细胞用于细胞生物学研究也获得不少有趣的结果。在一项与美国的国际合作研究中,对绣球百合的分离生殖细胞进行了 Nomarski 干涉差与扫描电镜观察,以视频增差显微术追踪了生殖细胞离体后的细胞形态学变化,还以免疫荧光显微术首次观察了生殖细胞中微管的整体分布。在此基础上,以后继续重点研究了生殖细胞形态变化与微管骨架的关系。在与香港大学研究者合作的风雨花研究中,发现随着生殖细胞离体后由纺锤形变为圆球形,微管骨架也由平行束变为网络状,显示明显的相关性。在朱顶红中,我们建立了一种边分离边固定的方法,可以保持分离生殖细胞形态与微管格局的体内原有状态。以此研究了生殖细胞体内发育过程中细胞形态与微管骨架变化的关系,实验规模之大足以进行翔实的观察统计。大体上,由幼嫩的圆球形生殖细胞发育为成熟的梭形生殖细胞过程中,微管格局也由网络状演变为平行束状;期间出现各种中间形态。以往研究生殖细胞中微管的分布多依靠超微结构观察,而分离生殖细胞容许对整个生殖细胞进行大规模观察是其一大优点。我们的研究为此后国际上同类研究开辟了道路。

1 Direct observations on generative cells isolated from pollen grains of *Haemanthus katherinae* Baker

Approximately two thirds of all pollen types in angiosperms are bicellular (Brewbaker 1957), with a generative cell embedded within a vegetative cell forming a unique "cell within a cell" structure. The thick pollen wall and concentrated organelles and storage products associated with the vegetative cell make it very difficult to study the generative cell directly. Unimpeded observations of generative cells became possible when a squash method was used to isolate them from mature pollen grains of *H. katherinae*.

Materials and methods

Plants of *Haemanthus* were grown in the Murdough Greenhouse at Dartmouth College and used as experimental material. Plants were maintained at a temperature of 25-30℃ (days) and 18-20℃ (nights) in natural daylengths. Mature undehisced anthers were collected from opened flowers for these experiments.

Generative cell isolation

Pollen grains from 1-2 anthers were suspended in 2-3 drops of Brewbaker and Kwack medium (1963) with varying concentrations of sucrose (0%, 6%, 9%, 12%, 18%, 24% and 48%). For some experiments a large number of anthers was used. Pressure was applied to the coverslip by gently tapping with the finger tip. In this manner pollen grains were squashed to release cytoplasmic contents along with the generative cells. Effectiveness of release was checked microscopically. Cells were routinely examined under the light microscope with or without aceto-orcein staining.

Cell wall detection

Four different methods were utilized to determine the existence of a generative cell wall and the chemical nature of the wall.

(1) Unfixed generative cells were examined by polarizing microscopy.

(2) Unfixed cells were stained for 10 minutes with 0.1% fluorescent brightener 28 (i. e. calcofluor white M2R, Sigma Chemical Co., St. Louis, MO., USA) and studied by epi-fluorescence microscopy (exciter G405, splitter FT 460, barrier LP 495) according to Nagata and Takebe (1970).

(3) To identify wall polysaccharides cells were collected, fixed in Carnoy's solution and stained with PAS according to the basic procedure described by Feder and O'Brien (1968).

(4) To identify pectin unfixed cells were stained in a ruthenium red solution (1 : 5000) according to Jensen (1962).

Viability testing

The viability of the generative cells was examined by the fluorescein diacetate (FDA) method according to Larkin (1976) and Shivanna and Heslop-Harrison (1981) with modifications. A stock solution of 2 mg/ml FDA (Sigma Chemical Co., St. Louis, MO, USA), diluted 50 times with Brewbaker and Kwack medium (1963), was used to treat the generative cells. After 10 minutes, the treated cells were observed by epi-fluorescence microscopy (exciter filter BP 450-490, beam splitter FT 510, barrier LP 520). Cells retained their viability for approximately 1h under the conditions employed in these investigations.

Microscopy

Microscopic observations were made using an inverted Zeiss Axiomat microscope with an internal zoom lens to 3.2X. Planapochromatic objectives of 100X, 50X and 25X were used with an achromatic condenser. For DIC and Polarized light observations a Je Sernarmont compensator consisting of a quarter wave plate and a rotating analyzer was used with a 50-W AC mercury burner for illumination. Fluorescence images were illuminated with a 100W DC mercury-burner.

DIC and Polarized light images were recorded on Kodak TP-2415 35 mm film and processed in Diafine (Acufine, Inc., Chicago, IL, USA) according to their recommended procedure. The fluorescence image was projected through a 1.6X extension tube and detected with a Hamamatsu C1000-12 silicon-intensified target (SIT) camera head connected to a Hamamatsu C1966 image processor (Photonics Microscopy, Inc. Oak Brook, IL, USA). The image was then recorded onto a 3/4 inch U-Matic recorder. Kodak Plus-X 35 mm film processed in Kodak Microdol-X (1 : 3), was used to record the image from a television monitor.

Results and discussion

A mature pollen grain of *H. katherinae* consists of a generative cell within a vegetative cell, both of which can be observed through the pollen wall after aceto-orcein staining (Fig. 1).

Generative cell isolation

When pressure was applied to the pollen grains, cytoplasm flowed out of the mature grains carrying with their generative cells, vegetative nuclei, vacuoles and other organelles. Many of the isolated generative cells maintained their structural integrity. A typical spindle-shaped generative cell with an elongate, deeply aceto-orcein stained nucleus and a thin layer of cytoplasm is shown in Fig. 2. Observations of unstained cells at high magnification (oil immersion objective 100X) present a much clearer image, especially of the boundary and the cytoplasmic contents (Fig. 4). The adherence of small amounts of vegetative cell cytoplasm to the surface of the generative cells is indicative of the intimate connection between these two cells.

Change in generative cell shape

The shape of the generative cells varied considerably in media with different concentrations of sucrose (0%, 6%, 9%, 12%, 18%, 24%, 48%). In general, at sucrose concentrations of 12% or 18%, the majority of cells were spindle-shaped as described above and illustrated in Figs. 2 and 4. When the concentrations of sucrose were raised above 18%, the cells retained their spindle shape, but the cytoplasm frequently contracted. At concentrations below 12%, many generative cells lost their spindle shape and became oval or spherical (Fig. 5-7), resembling protoplasts. Figure 6 presents a clear image of an entire cell, whereas Fig. 7 focuses on the nucleus. Brownian movement of cytoplasmic particles was observed in the spherical and oval cells at sucrose concentrations of 6%-9%. When water was used during the process of releasing the generative cells instead of a sucrose medium, isolated cells broke down within seconds. In many cases, the isolated spindle-shaped cells demonstrated a tendency to change to spherical shape with the passage of time. It is possible that the mechanical pressure involved in the squash technique influences the cell shape. The experiments indicate that when *Haemanthus* generative cells are observed in isolation, certain important morphological changes in their usual structure becomes evident. When isolated, cells tend to change shape rapidly and show great sensitivity to different concentrations of sucrose in the medium. Similar changes in the shape a generative cells, from spindle to oval or spherical, were also observed by Sanger and Jackson (1971b) when anthers of *Haemanthus* were exposed to colchicine or isopropyl N-phenyl carbamate solution for a period of 24 hours.

Cell wall detection

The presence or absence of a cell wall in generative cells of mature pollen grains has been the subject of considerable controversy in the literature (Owens and Westmuckett 1983; Souvre et al. 1985; Vijayaraghavan and Bhatia 1985). In order to investigate whether or not a cell wall surrounds the generative cell, isolated cells were studied using the following techniques.

(1) Unfixed generative cells of different shapes were observed by polarizing microscopy. When viewed between crossed polarizers, the obvious birefringence associated with cellulosic cell wall was not seen.

(2) Cells treated with fluorescent brightener, did not display the fluorescence typical of cellulose.

(3) The boundary layer of PAS treated cells exhibited no positive reaction.

(4) A negative result for the presence of pectin was obtained using ruthenium red.

The observations at the lightmicroscopic level indicate that isolated generative cells from mature *Haemanthus* pollen grains are unlikely to possess a typical cellulosic wall and might even prove to be naked cells.

Viability testing

One of the important questions concerning the isolated generative cells was whether or not they were viable. A test using fluorescein diacetate (FDA) has been widely adapted for examining the viability of pollen grains, plant cells, protoplasts and isolated embryo sacs (Shivanna and Heslop-Harrison 1981; Larkin 1976; Zhou and Yang 1985). It was used in our experiments. Generative cells isolated in medium with sucrose concentrations from 9 to 18% had shapes varying from spindle to spherical. When observed by fluorescence microscopy many cells showed the strong fluorochromasia caused by intracellular esterase (Fig. 3). Uniform fluorescence indicated the compact nature of the generative cells, but at times a very weak fluorochromasia appeared at the two poles of elongate cells where a small amount of cytoplasm adhered. In this case, the video intensified microscope (Allen 1985) proved advantageous in strengthening the fluorescent intensity.

Summary

As the precursor of the two male gametes, generative cells have been studied intensively. In addition, isolated nuclei of generative and vegetative cells have been analyzed biochemically by several authors among them Sheridan (1972) and Wever and Takats (1971). Also, isolated sperms of barley have been observed by Cass (1973). To our knowledge isolated generative cells have not been previously studied. Using the squash method we isolated generative cells from mature pollen grains and then proved their viability. Our experimental results complement the ultrastructural study on pollen grains of *H. katherinae* by Sanger and Jackson (1971a,b). The isolation technique described and utilized for *Haemanthus* has been successfully applied to studies of pollen from other species including *Narcissus pseudonarcissus*, *Hippeastrum hybrid*, *Hyacinthus orientalis* and *Tulipa* sp. This technique appears to provide certain advantages.

(1) The method is simple and requires no elaborate or time consuming techniques.

(2) Naked and living generative cells of various shapes could be observed clearly and without interference.

(3) The three dimensional cell structure could be examined microscopically simply by adjusting the focus.

(4) Chemicals could be applied directly to the generative cells and

their reactions to those chemicals observed directly and without delay.

We believe the isolation technique described here is an approach to studying generative cells that could provide information that will complement studies using intact pollen grains. For future progress to be made in studying generative cells in the living state more attention must be given to the couditions necessary for isolating and maintaining the cells *in vitro*.

Generative cells differ from somatic cells in that they appear to be naked cells with a large nucleus and a thin layer of cytoplasm. They are haploids and divide only once to produce two sperm cells. With further improvement in the isolating technique, generative cells could provide a unique experimental system for *in vitro* culture, genetic manipulation and other biotechnical applications.

(Authors: Zhou C, Orndorff, K, Allen D, Demaggio AE. Published in: Plant Cell Reports, 1986, 5:306-309, with 7 figures and 17 references, all omitted here.)

2　Isolation and purification of generative cells from fresh pollen of *Vicia faba* L.

The generative cell (GC) of angiosperm pollen is embedded within the vegetative cell which in turn is surrounded by the pollen wall. Recently GCs were isolated from mature pollen grains by means of a squash method and observed microscopically (Zhou et al. 1986). The squash method could offer GCs of high quality for cytological studies owing to absence of interference by the vegetative cell and pollen wall. However, for further manipulation in biotechnical studies the technique should be improved to obtain GCs in sufficiently large numbers. For this purpose Russell's osmotic shock method for isolating sperm cells of *Plumbago* was modified for isolating GCs of *Vicia faba*. The critical point was to use a two-step procedure instead of Russell's one-step technique.

Materials and methods

Plants of broad bean (*Vicia faba* L.) grown in the field were used as experimental material. Flower buds were collected from the field in the morning just before anthesis. Undehisced mature anthers were put into sucrose solution (10%, 20%, 30% and 40% were tried respectively and 20% finally chosen) in petri dishes, usually 25-30 anthers per ml solution. Pollen grains were released from the anthers by gentle squashing with smooth bottom of a small beaker and anther walls were removed with forceps. The Petri dishes were sealed with parafilm and incubated in darkness at 25℃ for 30-45 min. Then distilled water of equal volume was added to the sucrose solution. Within 40-50 min the majority of pollen grains burst and GCs were released in large quantities.

For purification of GCs such a mixture including isolated GCs and other pollen debris was collected into centrifuge tubes and centrifuged at a speed of 500 rpm for 6 min. The supernatant and precipitate were then filtered by different ways: the supernatant was filtered through nylon screens with 10 μm pores. The filtrate was discarded. The screens were rinsed with isotonic sucrose solution (usually 10% sucrose) and this solution was collected for further use. The precipitate was filtered through nylon screens with 25 μm pores. In this case the filtrate was collected. The two collected suspensions were mixed together and centrifuged again with a speed of 1000 rpm for 8-10 min to obtain enriched and purified GCs.

The freshly isolated GCs in sucrose solution either before or after purification were observed and photographed under AO 1820 BIOSTAR inverted microscope. The materials stained with carbol-fuchsin (Miller et al. 1971) were observed under Olympus NEW VANOX AHBS-514 microscope.

The viability of GCs was determined by fluorochromatic reaction test mainly according to Heslop-Harrison (1984). Stock solution of fluorescein diacetate (FDA, 10mg/ml) was diluted 50-100 times with isotonic sucrose solution before use. The cell suspension and FDA solution were mixed 1 ∶ 1 (vol ∶ vol) and then dropped on a slide and mounted with coverslip. After 5-10 min, it was observed under AO FLUORESTAR 110 epi-fluorescence microscope with a filter combination of AO 2073 (exciter filter 436, dichroic mirror 450, barrier filter 475).

Results

Isolation

Following Russell's method the first experiment was designed to incubate pollen grains of *V. faba* in sucrose solution of various concentration (0%, 10%, 20%, 30%, 40%). After 1 hour, pollen grains burst in distilled water, but the released GCs could not keep well. On the other hand, only a few of pollen grains in 10%-40% sucrose burst and the others showed a tendency of germination. Therefore, one-step osmotic shock appeared not so effective to isolate GCs in *V. faba* as sperm cells in *Plumbago*.

The second experiment was carried out and as a result an improved technique of two-step osmotic shock was established. The first step was to incubate pollen grains in a sucrose solution of suitable concentration for a proper time. Hydration led to their swelling and protrusion of intine from the germ pores (Fig. 1). Then the second step was carried out by adding distilled water. A sudden osmotic shock made them to burst rapidly and their contents carrying GCs spurted out from the pores. Within one hour most of the pollen grains burst and released their GCs. It is important to determine the time for adding water; if it was too early, the pollen grains were not ready to accept the shock. It seems that the success of two-step osmotic shock depends on a proper interaction between pollen physiological state and external osmotic change.

The GCs in mature pollen of *V. faba* are spindle-shaped. Just released GCs maintained their shape. An oval nucleus occupying the center of the cell could be seen after staining with carbol-fucshin (Fig. 2). With passage of time the cytoplasm gradually contracted. If the contraction force was equal at two poles, the cell turned oval-shaped; otherwise it might form one "tail" (Fig. 3). However, finally most of them became spherical (Figs. 4, 5). Sometimes an elongated nucleus curved within a spherical cell could be seen, as shown in Fig. 5. These morphological changes of isolated GCs were quite similar to those shown in other species by previously used squash method (Zhou et al. 1986). The isolated sperm cells also had a tendency to change from spindle-shaped to spherical form (Russell 1986, Matthys-Rochon 1987).

Purification

The suspension obtained by osmotic shock was a complex mixture of various components: besides GCs there were also empty pollen walls, unburst pollen grains, vegetative nuclei and other cytoplasmic particles. Microscopic observation revealed difference of size between these components. The diameter of spherical GC was 18-20 μm, much smaller than the pollen (70×35 μm) and bigger than the cytoplasmic particles. Therefore, it would be possible to separate them by filtration. For this purpose the suspension was first centrifuged for a short time. As a result, the supernatant contained a part of GCs which could be separated from the smaller particles by filtration with 10 μm nylon screens, and the precipitate contained more GCs which could be separated from the bigger pollen grains and pollen wall debris by filtration with 25 μm nylon screens. Fig. 6 shows the purified GC population made up these two components after many of the contaminants were removed.

Viability test

The GCs after release were examined by FDA test and showed fluorochromasia. Some of them maintained the reaction even after 4 hours on the slides. The purified GCs also demonstrated strong fluorescence (Fig. 7). In this photograph the cells were not as crowded as in Fig. 6 due to dilution with FDA solution and also because of the coverslip. From these tests it could be concluded that the isolated GCs before or after purification were viable.

Discussion

In 1970s sperm cells were first isolated and observed by Cass in barley. In the meantime generative and vegetative nuclei have been isolated for biochemical analysis by several authors (La Fountain and Mascarenhas 1972, Sheridan 1973, Weber and Takats 1971). There was no progress until recently when a renewed interest in isolation of male reproductive cells (including sperm and GC) appeared. Dumas et al. (1984) pointed out that "The isolation of the male germ unit is an essential step that will provide new opportunities to control and modify the male genetic material". Russell (1986) also emphasized: "The isolation of living sperm cells of an angiosperm in numbers large enough to permit physiological characterization has been regarded as a technically challenging task." He suggested a new term "gametoplast" to name the isolated sperm cell or GC. We have noticed that the isolation technique provides a new approach not only for cytological observations on GC but also for their *in vitro* culture, genetic manipulation and other biotechnical applications (Zhou et al. 1986).

Russell (1986) has obtained mass-isolated sperm cells of *Plumbago*. Dupuis et al. (1987) reported the isolation of viable sperm cells from *Zea mays* pollen. Matthys-Rochon et al. (1987) isolated male germ units and sperm cells from three tricellular pollen species by osmotic shock with grinding. Knox et al. obtained released sperm cells of *Gerbera* (personal communication). As for isolation of GCs, so far only we have succeeded first by squash (Zhou et al. 1986) and now by osmotic shock. The two-step osmotic shock method developed in the present paper may be applied in a wider range of species which are recalcitrant to simple mechanical or osmotic pressure.

(Author: Zhou C. Published in: Plant Cell Reports, 1988, 7:107-110.
with 7 figures and 11 references, all omitted here.)

3 花粉生殖细胞的大量分离与纯化

Mass isolation and purification of generative cells from pollen grains

Abstract

A technique named "two-step osmotic shock" was developed for the isolation of large number of generative cells from pollen grains in *Vicia faba* and *Zephyranthes candida*. The procedure included: 1. Pollen hydration: mature pollen grains were incubated in adequate sucrose solution for 30 ~ 50 min. 2. Osmotic shock: when the pollen grains swelled and showed the tendency of germination, a sudden osmotic shock by adding equal volume of water made them bursting. The contents carrying generative cells spurted from the germ pores. By subsequent filtration and centrifugation, purified generative cell population was obtained. FDA test proved that the isolated and purified generative cells were viable.

第一章 雄性器官与细胞的实验研究

被子植物的花粉生殖细胞是精子的前体,在生殖过程中占有重要地位。由于它被营养细胞与花粉壁包围,一般只能以花粉粒为单位进行研究。1986 年以来,我们用压片法分离出花粉生殖细胞,作了有关细胞生物学研究,同时指出分离的生殖细胞用于离体培养与遗传操作的意义[10]。作者参考了 Russell 分离精子的一步"渗透压冲击法"[7],建立了适于大量分离与纯化生殖细胞的"二步渗透压冲击法"(以下简称"二步法")。此外,还作了从花粉原生质体分离生殖细胞的试验。

材 料 与 方 法

一、分离方法

供试材料主要为蚕豆(*Vicia faba*)和玉帘(*Zephyranthes candida*)。成熟花粉在适当浓度的蔗糖液中温育 30～45min。当花粉充分水合,开始萌动时,加入与蔗糖液等量蒸馏水,使花粉粒破裂,释放出生殖细胞。此外,以风雨花(*Z. grandiflora*)为材料,首先制备花粉原生质体,具体方法详见另文[2],然后加蒸馏水冲击,使原生质体破裂,亦可分离出生殖细胞。

二、纯化方法

蚕豆生殖细胞从花粉粒中逸出后,将悬浮液离心(500r/min)6min,分上清液和沉淀两部分:上清液经 10μm 孔尼龙筛网过滤,收集滤网上的细胞;沉淀部分用等渗液冲洗,经 25μm 尼龙筛网过滤,弃去网上残渣,收集滤液。将收集液混合,离心后,可获得较纯的生殖细胞群体。如仍不够纯化,尚需反复处理。玉帘生殖细胞分离后用 400 目不锈钢筛网过滤悬浮液。滤液用 10μm 尼龙筛网过滤,除去较细的杂质,再将收集的滤液铺在 15% 蔗糖液面上,离心 10min。富集纯化的生殖细胞群体。

三、生活力测定与显微观察

生活力测定是用荧光素二醋酸酯(FDA)处理,FDA 浓度为 50~200μg/ml。用 AO FLUORESTAR 110 型荧光显微镜的 2073 滤光组合(激光滤光片 436,分色镜 450,阻挡滤光片 475)观察与摄影,明视野显微观察用 AO 1820 BIOSTAR 型倒置显微镜或 Olympus 顺置显微镜。

结 果 与 讨 论

蚕豆 用蔗糖液处理花粉,在浓度为 0%、5%、10%、20%、30%、40% 范围内,10%~20% 蔗糖液能使花粉破裂,但破裂花粉数不多,又不同步,无实际应用价值。以后改进技术,建立了"二步法"。第一步将花粉置蔗糖液中保温培养,当少数花粉开始萌动时(图版Ⅰ,1),进行第二步:加入蒸馏水使蔗糖浓度突然下降,渗透压冲击导致花粉破裂,内含物从萌发孔迅速喷出,生殖细胞随之流出。蔗糖浓度与加水时间很关键。蔗糖浓度对生殖细胞分离的速度、数量以及细胞形状均有明显影响,其中蔗糖浓度为 20%,分离效果较好,适时加水亦颇重要。加水过早,花粉尚未充分吸水,不易破裂;加水过晚,花粉萌发成长短不等的花粉管,导致生殖细胞分离不同步。此外,培养基中各种离子特别是 Ca^{2+} 起保护质膜、阻止花粉破裂作用。所以,单纯的蔗糖液比较适宜。

玉帘 用玉帘进一步比较了"一步法"与"二步法"的效果。在"一步法"的 5%、10%、15% 蔗糖液中,花粉虽能破裂,但破裂时间参差不齐,在 20% 以上蔗糖液中,花粉往往皱缩,不易破裂。而用"二步法",花粉在较短时间内破裂,生殖细胞大量逸出。蔗糖浓度以 15% 为宜。保温培养 40 min,花粉开始萌动(图版Ⅰ,2),加入蒸馏水后,花粉破裂,生殖细胞随细胞质流出(图版Ⅰ,3)。

风雨花 最近,风雨花花粉原生质体分离成功[2],在此基础上,利用渗透压冲击,使花粉原生质膜破裂,内含物泄出,从而获得生殖细胞。此法与 Tanaka 等(1988)分离麝香百合花粉生殖细胞的方法相似[8]。

分离的生殖细胞悬浮液中还包括未破裂的花粉粒、空花粉壁、营养核、液泡和细胞质等各种颗粒。我们采用不同孔径筛网过滤与离心方法,初步达到纯化生殖细胞的目的。一般多次纯化处理,可以提高纯化程度,但细胞得率会明显下降。在蚕豆纯化工作的基础上,用玉帘作了技术改进,相比而言,后者生殖细胞纯化效果较好。图版Ⅰ,4~6分别示蚕豆和玉帘纯化后的生殖细胞群体。经纯化之后,蚕豆生殖细胞基本上变成圆球形,玉帘的生殖细胞一般不能保持原有的长纺锤形,不同程度地发生收缩,呈现短纺锤形、椭圆形以及圆球形等多种形态。

对蚕豆和玉帘生殖细胞进行了生活力测定。实验结果表明,经过分离与纯化的许多步骤之后,生殖细胞有可能保持其生活状态,无明显不利的影响。

生殖细胞与精子的实验操作是当前国内外一个重要研究方向。它们的分离是实验操作的必要技术前提。精子的分离有较长的探索历史,最近有明显的进展[3~7,9]。与精子分离工作相比,生殖细胞分离尚属新近的尝试[1,8,10]。从现有资料分析,二者分离方法有许多可以互相借鉴之处。迄今建立的技术包括一步法、二步法、花粉原生质体冲击法以及研磨法,前三者都利用了渗透压冲击作用,而辅以其它不同措施,各具特色,可以根据实验材料和研究目的加以选用。

图 版 说 明

1. 蚕豆花粉水合后开始萌动。×130 **2.** 玉帘花粉水合后开始萌动。×200 **3.** 玉帘花粉水合后,经渗透压冲击而破裂,释放出生殖细胞。×160 **4.** 蚕豆纯化的生殖细胞群体。×260 **5、6.** 玉帘纯化的生殖细胞群体。 **5.** ×200 **6.** ×800。

Explanation of plate

Fig. 1. Pollen grains of *Vicia faba* hydrated in sucrose solution and prior to germination. ×130 **Fig. 2.** Pollen grains of *Zephyranthes candida* hydrated in sucrose solution and prior to germination. ×200 **Fig. 3.** Pollen grains of *Z. candida* bursting and releasing generative cells after hydration and osmotic shock. ×160 **Fig. 4.** Purified generative cell population of *V. faba*. ×260 **Figs 5、6.** Purified generative cell population of *Z. candida*. **Fig. 5.** ×200 **Fig. 6.** ×800

图版 I

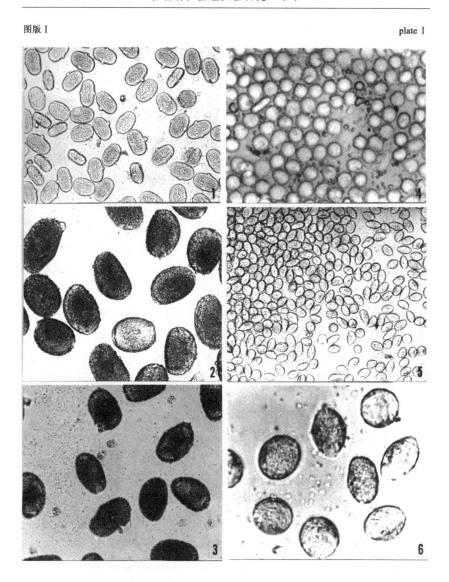

plate I

(作者:周嫦、吴新莉。原载:植物学报,1990,32(5):404~406。图版1幅,参考文献10篇。选录图版 I。参看书末彩色图版Ⅳ。)

4 多种被子植物花粉生殖细胞大量分离技术的比较研究

A comparative study on methods for isolation of generative cells in various angiosperm species

Abstract

Four methods for isolating generative cells from pollen grains, named one-step osmotic shock, two-step osmotic shock, weak enzyme treatment and isolation via pollen protoplast preparation, were compared among 11 angiosperm species belonging to 6 families. The result revealed that each method was effective only for a group of species. The method of one-step osmotic shock is adaptable to species as *Gossypium hirsutum* whose pollen grains are easy to burst in a low osmotic sucrose solution. The two-step osmotic shock method is suitable for species as *Nicotiana tabacum*, *Zephyranthes candida*, *Z. grandiflora*, *Lycoris radiata* and *Impatiens balsamina* whose pollen grains are characterized by easy germination in sucrose solution. A sudden osmotic shock by addition of water just at the beginning of germination causes the pollen to burst. In *Iris tectorium*, *Gladiolus gandavensis*, *Hemerocallis minor*, *Allium tuberosum* and *Hippeastrum vittatum* pollen grains neither burst nor germinate well in simple sucrose solution, but weak enzyme treatment promotes bursting of pollen grains in large quantity. Based on this result, it is assumed that the responce of pollen grains in sucrose solution could be used as a criterion for choosing suitable method for isolation of generative cells in a given plant species. Besides, the fourth method of isolating via pollen protoplast preparation was examined. Although it is adaptable to certain species, the isolation efficiency is not high enough. Factors affecting the isolation result, such as pollen developmental stage, medium components, sucrose and enzyme concentration, duration of pollen incubation, quantity of water added to, were studied as well. Viability test by fluorescence microscopy showed over 80% cells were viable after purification procedure.

被子植物的生殖细胞和精子包藏在花粉粒或花粉管中,过去对它们的研究一般是在原位进行的。最近,由于受精理论研究的深入和生殖系统细胞工程技术的发展,分离精子的研究迅猛发展[1],已从9科14种植物的花粉粒或花粉管中分离出大量精子[1~5]。与精子相比,作为精子前体的生殖细胞的分离工作目前却比较少。周嫦等(1986,1987)首先用压片法分离出生殖细胞[6,7]。随后,参考 Russell 分离精子的渗透压冲击法[8](简称一步法)加以改进,建立了大量分离生殖细胞的二步渗透压冲击法(简称二步法)[9]。Tanaka[10]还提出了通过花粉原生质体释放生殖细胞的方法(简称花粉原生质体释放法)。在此基础上,本文进一步用多种植物为材料,研究不同的方法和各种因素对分离效果的影响以及纯化技术等。通过比较研究,总结出了适用于不同植物特点的生殖细胞分离技术。

材 料 与 方 法

一、实验材料

供试材料包括6科11种植物,它们是:石蒜科的玉帘(*Zephyranthes candida* Herb.)、风雨花(*Z. grandiflora* Lindl.)、石蒜(*Lycoris radiata* Herb.)、朱顶红(*Hippeastrum vittatum* Herb.);鸢尾科的鸢尾(*Iris tectorium* Maxim.)、唐菖蒲(*Gladiolus gandavensis* Van Houtte);百合科的黄花菜(*Hemerocallis minor* Mill.)、韭菜(*Allium tuberosum* Rottl. ex Spreng.);锦葵科的棉花(*Gossypium hirsutum* L.);茄科的烟草(*Nicotiana tabacum* L.);凤仙花科的凤仙花(*Impatiens balsamina* L.)。共计8

种单子叶植物和3种双子叶植物。

二、分离方法

先后采用四种分离方法:(1)一步法[8]:将花粉在低渗蔗糖溶液中保温一段时间使其自行破裂。(2)二步法[9]:将花粉在等渗蔗糖溶液中培养,当部分花粉开始萌动时加入蒸馏水,使渗透压突然下降,花粉破裂。(3)低酶法:花粉在含少量纤维素酶与果胶酶(一般各0.1%)的低渗蔗糖溶液中短时保温而破裂。(4)花粉原生质体释放法[10]:花粉在酶液(含0.5%纤维素酶、0.5%果胶酶和10%甘露醇)中酶解,待分离出原生质体后,离心除去酶液,加入少许培养基用吸管轻轻吸打,原生质体破裂而释放生殖细胞。以上各种处理均在25℃和黑暗条件下进行。

三、纯化步骤

将分离生殖细胞的悬浮液经过 $25\mu m$ 和 $10\mu m$ 孔径的尼龙网二次过滤,除去花粉壁等残渣,然后将细胞悬浮液铺于相对高浓度的蔗糖溶液上,进行间断密度梯度离心(500r/min)12 min,使生殖细胞降至界面下,其它花粉的颗粒漂于上层。除去上层的溶液,将界面以下溶液再离心 5 min,使生殖细胞富集于底部,即可得到纯化的生殖细胞。在采用低酶法或花粉原生质体释放法时,经尼龙网过滤后,需要离心除去酶液。

四、显微观察和统计

生殖细胞分离率:一般按分离的生殖细胞占花粉总数的百分率统计。经实验检验,破裂花粉数与分离生殖细胞数相当一致,所以,有时用破裂花粉数代表生殖细胞统计分离率。纯化得率:按纯化后生殖细胞数占花粉总数的百分率计算。生活力测定:将荧光素二醋酸酯(FDA)加入待测细胞悬浮液,使 FDA 最终浓度为 $10\mu g/ml$。5min 后,在 AO BIOSTAR 110 型荧光显微镜下统计生殖细胞总数和其中生活细胞数。细胞形态学观察:经丙酸洋红染色或未染色的生殖细胞用

Olympus BH-2 型显微镜观察与摄影。

结　　果

一、不同植物生殖细胞分离方法的比较

最初的试验是比较当时已经建立的两种分离方法即一步法[8]与二步法[9]。它们均是利用渗透压冲击使花粉破裂,释放出生殖细胞。实验结果表明,11 种植物大致有三种反应:第一类植物仅有棉花。采用一步法,花粉在 5% ~35% 蔗糖溶液中能够大量破裂,生殖细胞分离率高达 85% 以上,不需二步法即可获得满意的效果。第二类植物有烟草、玉帘、风雨花、石蒜与凤仙花 5 种植物。一步法分离效果较差,二步法可明显提高分离率。如烟草由一步法改为二步法时,分离率由 37% 左右提高到 78.8% 。第三类植物包括鸢尾、黄花菜、唐菖蒲、朱顶红与韭菜 5 种植物,无论一步法或二步法,生殖细胞的分离率均很低,如鸢尾的分离率在 50% 以下(表 1)。

以后的试验着重改进第三类植物的分离技术。经过研究,建立了低酶法,即在蔗糖溶液中加入少量细胞壁降解酶。经短时处理后,花粉在酶液中大量破裂,大幅度提高了分离率。如表 1 所示鸢尾低酶法分离率高达 85% ~90% 。需要指出,低酶法主要适用于第三类植物,而在第二类植物中,低酶法的分离效果与二步法相比,并无优越性。

值得注意的是各类植物的花粉在蔗糖溶液中的反应是不同的。第一类植物花粉容易破裂;第二类植物花粉不易破裂,但能够大量萌发;第三类植物花粉既不易破裂,又难萌发。显然,根据花粉本身的特性,采用与之适合的方法,是生殖细胞分离成败的关键。

此外,还试验了花粉原生质体释放法,即先分离花粉原生质体,再促使原生质体破裂释放出生殖细胞。结果在 6 种植物(鸢尾、唐菖蒲、黄花菜、韭菜、朱顶红和风雨花)中取得成功。但是,此法步骤较繁,分离率偏低。

表1　　　　　　　　　　　　　　分离方法的比较

材料	烟草			玉帘			鸢尾			黄花菜			
分离方法	一步法	二步法		一步法	二步法		一步法	二步法	低酶法	一步法	二步法		低酶法
蔗糖浓度(%)	15	30	30→15	7.5	15	15→7.5	5	10	10→5	5	10	10→5	5	10
花粉总数	626	582	590	623	575	637	601	606	612	627	607	634	659	586	612	574
生殖细胞数	231	216	465	322	196	422	233	223	310	564	521	264	55	302	560	83
分离率(%)	36.9	37.1	78.8	51.7	34.2	66.3	38.8	36.8	50.7	90.0	85.8	41.6	8.35	51.5	91.5	14.5

注：表1~3中→表示二步法加水前后蔗糖浓度的变化。

供试的11种植物的生殖细胞在分离条件下发生的形态变化，与文献报道的结果[6,7,9,10]类似：生殖细胞在花粉粒内一般呈纺锤形，分离后往往不易保持原有形状，而转变为椭圆形和圆球形，只有少数植物如烟草，一部分生殖细胞能够长时间保持纺锤形（图版Ⅰ,1~10）。细胞形状的变化还与分离方法、蔗糖浓度以及花粉发育时期等因素有关。

二、影响生殖细胞分离效果的因素

为了更好地掌握分离技术,进一步研究了若干影响生殖细胞分离效果的因素。

花粉发育时期　花粉发育时期对分离效果的影响因植物类型和分离方法而异。第一、二类植物用一步法和二步法,只有开花当天的成熟花粉才能分离出生殖细胞。如棉花、烟草、玉帘开花当天的分离率分别为85%、66%、78%,而开花前一天的花粉则很难破裂,分离率均低于5%。第三类植物用低酶法和花粉原生质体释放法时,由开花前三天到开花当天的花粉均能分离生殖细胞。随着花粉发育成熟,分离率逐步提高,开花当天分离率最高(图1)。

溶液成分　分离生殖细胞一般宜用单纯的蔗糖溶液。如加入无机盐成分,往往阻碍花粉的破裂而降低分离率;即使花粉破裂,流出的内含物聚集成团,生殖细胞很难纯化。只有用花粉原生质体释放法时,加入无机盐离子对分离率无明显的不利影响,而对生殖细胞的存活可能有利。

蔗糖浓度　蔗糖浓度对分离效果有重要影响。原则上,一步法和低酶法必须低渗条件花粉才能破裂。随着蔗糖浓度提高,分离率逐渐降低。但是,棉花和烟草在10%~30%的蔗糖溶液中,分离率大致相似,分别为85%和35%左右(图2)。二步法中起始蔗糖浓度是重要因素。浓度太低,生殖细胞易破。浓度过高,分离率下降。例如,玉帘起始浓度为15%时,效果最好,分离率达66%;浓度提高到20%,分离率就猛降为23%。其它几种植物的起始蔗糖浓度也以15%为宜,只有烟草宜用更高(30%)的浓度。

保温时间　采用一步法和低酶法时,保温时间大约15min。采用二步法时,以部分花粉开始萌动为最适时间,此时加水冲击,分离率最高。

加水量　采用二步法分离生殖细胞时,一般以加入与原有蔗糖液等量的蒸馏水分离效果最好,但烟草加入较少量的水亦可。

酶的浓度　低酶法试验了四种浓度:1%、0.5%、0.1%和0.05%。浓度过高(0.5%~1%),生殖细胞分离率较低,并且花粉释放物不易分散,生殖细胞很难纯化。浓度为0.1%时,分离率最高,生殖细胞生活力也不受影响。浓度为0.05%时,分离率有所下降。采用花粉原生质体释放法时,则酶浓度一般宜高于0.5%,花粉原生质体与生殖细胞分离较好。

三、生殖细胞的纯化

花粉破裂释出生殖细胞后,溶液中尚有破裂和未破裂花粉以及大量花粉内含物。通过筛网过滤与蔗糖密度梯度离心两个步骤,先后弃除大、小残渣,获得纯化的生殖细胞群体。玉帘、烟草、鸢尾、黄花菜和朱顶红5种植物均获得较满意的结果(图版Ⅰ,11~13)。经统计,纯化后的玉帘生殖细胞密度达10^6个/ml,纯化得率为30.3%;烟草的纯化得率为11.6%。

四、生殖细胞的生活力

根据FDA荧光显微测定法,刚分离和纯化后的生殖细胞在适宜条件下有很强的生活力。从表2可见,玉帘、鸢尾、黄花菜生殖细胞纯化后2h仍有80%以上细胞具有生活力。图版Ⅰ,14~16示生活的生殖细胞群体在FDA作用下发射出明亮荧光。最近,本文作者进行黄花菜分离生殖细胞的培养,诱导了细胞核分裂(见下一篇论文),更进一步证实了分离生殖细胞的确是生活的。

表2　　　　　　　　生殖细胞纯化后2h生活力测定

材　料	玉　帘	鸢　尾	黄花菜
分离方法	二步法	低酶法	低酶法
蔗糖浓度(%)	15→7.5	10	5.0
生殖细胞总数	214	200	205
生活生殖细胞数	206	168	173
生活生殖细胞比率(%)	96.3	84.0	84.3

从初步试验看来,分离方法对生殖细胞生活力有一定影响,如玉帘用一步法(7.5%蔗糖)分离的生殖细胞只有75.6%是生活的,而二步法(最终浓度也是7.5%)分离的细胞则有93.6%是生活的。分离液最终浓度对生殖细胞的生活力有较大影响:浓度过低,细胞生活力很快下降;如果将细胞迅速转入适宜的培养基中,可以存活较长时间。玉帘的分离生殖细胞在15%蔗糖溶液中保存在4℃条件下,48h后仍有70%是生活的。

综合实验结果,总结出各种植物具体的分离方法和条件(表3)。

表 3　花粉生殖细胞较适宜的分离方法和条件

分离方法	植物名称	花粉发育时期	蔗糖浓度(%)	保温时间(min)	加水量
一步法	棉花	开花当天	20	15	不加水
二步法	玉帝、风雨花、石蒜、凤仙花	同上	15→7.5	40	等量
	烟草	同上	30→15(或 20)	50	等量或较少量
低酶法	鸢尾、唐菖蒲、韭菜、朱顶红	开花前三天至开花当天	10	15	不加水
	黄花菜	同上	5	15	不加水

讨　论

从花粉中分离精子和生殖细胞是当前植物生殖系统细胞操作的研究热点之一。与分离精子的广泛工作[1]相比，分离生殖细胞的工作相对薄弱，迄今只有周嫦等[6,7,9,11]和 Tanaka[10]有过研究，并且其中只在蚕豆、玉帝和麝香百合上建立了大量分离与纯化技术。因此，需要在更广泛的植物种类中加以检验和改进。本文通过大规模试验，在 6 科 11 种单子叶和双子叶植物中成功地分离出大量生殖细胞，并在此基础上比较了四种不同方法在不同植物中的效果，摸索出适合于每一种植物的分离方法，得到了一定的规律性认识。第一类植物的花粉在蔗糖溶液中容易破裂，可用一步法分离生殖细胞。在供试植物中只有棉花属于此类。但在精子分离工作中很多植物采用这种方法[1,8]。第二类植物花粉的特点是在蔗糖溶液中不易破裂而能顺利萌发，可以采用二步法，先经过培养促使花粉萌动，然后加水造成渗透压骤降，使刚萌动的花粉管尖端薄弱区域破裂。有 5 种供试植物适用这一方法。第三类植物的花粉既难萌发也不易破裂。本文试验出一种低酶法，即在一步法中辅以低浓度酶处理，利用短时酶解作用促使花粉萌发孔处内壁部分降解而释放内含物。有另 5 种供试植物宜用此方法。此外,在几种植物中还试验了花粉原生质体释放法[10],技术上改用机械力量使原生质体破裂。优点是生殖细胞分离时不受渗透压剧变的影响,而且用培养基作介质比单纯的蔗糖溶液对细胞有较好的保护作用。但是此法必须以分离花粉原生质体为前提,技术繁冗而分离率并不高。总之,研究者可以根据材料的特点和研究目的,灵活选用以上任一方法。

生殖细胞和精子类似,缺乏典型的细胞壁,Russell 称之为"配子原生质体"(gametoplasts)[8]。和体细胞相比,它们具有单倍体的遗传组成、较大的核质比。作为雄配子或其前身,它们应具有与雌性细胞识别的特殊机制。此外,占被子植物多数的二细胞型花粉,在授粉前只含有生殖细胞而尚未形成精子。这些特点,使生殖细胞的分离在植物生殖工程和植物生殖生物学基础研究两方面均有重要的意义和应用前景。

(作者:吴新莉、周嫦。原载:实验生物学报,1991,24(1):15~23。图版 1 幅,表 3 个,参考文献 11 篇,仅选录表 1~3。参看书末彩色图版 V。)

5 分离的黄花菜花粉生殖细胞在培养条件下的核分裂

Nuclear divisions of isolated, *in vitro* cultured generative cells in *Hemerocallis minor* Mill.

Abstract

The generative cells isolated from pollen grains of *Hemerocallis minor* Mill. were cultured in modified K_3 and MS agarose media surrounded by liquid media containing anthers as a feeder. Microscopical observations on a total of 5961 cells revealed that nuclear divisions occured leading to the formation of 2-4 nuclei within one cell. The average induction frequency among 14 culture experiments was 3.27% and the 11.46% as the highest. The generative cells varied considerably from spindle, oval to spherical shape. Among them the spherical cells appeared easier to be triggered. There were equal and unequal first nuclear divisions with almost equal chance of their occurence. A few of the cells passed through the second nuclear division, producing 3-4 nucleate cells. Cytokinesis resulting in 2-celled structures seemed to be induced in rare cases. This is the first experimental attempt to culture isolated generative cells and the results show the potential of free generative cells surviving and developing under *in vitro* conditions.

被子植物的花粉生殖细胞是雄配子(精子)的前身,同时它又是一种独特的细胞类型,具有单倍体的遗传组成,较大的核质比,缺乏典型形态的细胞壁等特点。因此,如果将生殖细胞从花粉粒中分离出来,实现游离生殖细胞的离体培养,便有可能把它用于细胞工程,成为一种有价值的实验体系。近年来,已经研究和建立了若干分离技术,包括压片法、渗透压冲击法和花粉原生质体释放法[1,2,6,8],已能由花粉粒中大量地分离生殖细胞。而生殖细胞的离体培养迄今尚无实验报道。本文是这方面的首次尝试。

材 料 与 方 法

实验材料为黄花菜(*Hemerocallis minor* Mill.)。生殖细胞分离的主要方法为首先分离花粉原生质体,然后释放生殖细胞[2,6]。开花前一天的新鲜花蕾或经 4~5℃ 低温预处理 1~2d 花蕾的花粉粒悬浮于酶液(0.5% 纤维素酶、0.5% 果胶酶和 10% 甘露醇)中,在 25±1℃ 和黑暗条件下保温 4h,分离出原生质体。换入培养基,轻轻摇荡促使原生质体破裂,释放出生殖细胞,经过滤、离心后获得纯化的生殖细胞群体。少数实验是将花粉粒悬浮于0.1%纤维素酶、0.1%果胶酶和5%蔗糖组成的稀酶液中保温15 min,促使花粉粒破裂,直接释放出生殖细胞。经过滤、离心后,悬浮于10%蔗糖液中,铺于15%蔗糖液面上,进行梯度离心,亦可获得纯化的生殖细胞。

将分离的生殖细胞与融化的琼脂糖培养基混匀,在培养皿中央铺一薄层。周围加入液体培养基,并置放二枚单核晚期至二核早期的花药作为饲养物。一般在25±1℃黑暗条件下培养。液体培养基为 K_3 或 MS 基本培养基附加1.2%小牛血清,1.5%椰乳,1% Polybuffer74（Pharmacia Fine Chemicals）[4],14%蔗糖,不同浓度的甘露醇和激素。激素有2,4-D 和 NAA 或者2,4-D,NAA 和 6-BA 二个组合的6种不同浓度与比例。固体培养基的组成与上述液体培养基相似,但小牛血清浓度为0.8%,椰乳为1%,琼脂糖为0.3%。

培养前后的生殖细胞用 FPA 固定剂固定6h,转入70%乙醇保存于冰箱中。临用时下行至水,用2%丙酸洋红染色。Olympus BH-2 型显微镜观察与摄影。每个处理统计的细胞数在400个以上。

结　　果

一、核分裂的诱导

黄花菜生殖细胞在接种时少数尚停留在间期,多数已进入分裂前期,但从未见有二核细胞。从新鲜花粉中分离的生殖细胞处于间期与前期者分别占30.61%和69.39%。从低温预处理1~2d后的花粉粒中分离的生殖细胞进入前期的数目略有增加（表1）。

离体培养4~20d,生殖细胞发生了核分裂。在观察的5 961个生殖细胞中,有195个细胞通过核分裂,形成2~4个核的细胞。平均诱导频率为3.27%。其中一次和二次核分裂的诱导频率分别为2.94%和0.33%（表2）。先后共做过14次实验,每次都观察到启动了核分裂的细胞,但诱导频率有很大波动,最高达11.46%,最低仅0.22%。这是因为每次实验采用不同的培养基,用以广泛探寻适宜的培养条件。由于未做严格的对比处理,目前,我们尚不能准确肯定各种因素的具体效果。故仅选出5次结果较好的实验作为代表列入表2。总的看来,黄花菜生殖细胞对培养条件有较广泛的适应能力,而从较好的几次实验结果(表2)分析,采用新鲜材料、K_3培养基、小牛血清、椰乳、Polybuffer 74、花药饲养、激素的组合以及短时高温刺激可能是比较有利的因素。

二、显微观察

  由于生殖细胞体形很小,又是生活材料,在倒置显微镜下很难分辨内部结构的变化,所以本实验采用固定材料进行显微观察。黄花菜生殖细胞在花粉粒内呈纺锤形,分离的生殖细胞从形态上大致可分为三类:纺锤形、椭圆形与圆球形。细胞核很大,细胞质稀薄(图版Ⅰ,1~3)。培养后,生殖细胞仍具上述三种形态。它们均能启动核分裂。表3统计结果显示,圆球形细胞数虽然仅占细胞总数的44.91%,但是,其诱导频率占总数的58.12%。由此可见,圆球形细胞相对易于启动核分裂。在部分细胞中观察到细胞核的分裂相,图版Ⅰ,4、5分别示纺锤形和圆球形细胞在培养过程中进入分裂中期,此时核膜消失,染色体清晰可见。表明至少在部分细胞中发生的是有丝分裂。没有证据表明是否也发生了无丝分裂。第一次分裂产生的二核细胞中,有的含有大小相近的二核(图版Ⅰ,8~11);有的含有大小悬殊的二核(图版Ⅰ,6、7)。也就是说存在着均等与非均等分裂二种方式。根据172个核分裂细胞的统计,二种方式出现的几率相近:均等分裂型为47.67%;非均等分裂型为52.33%。此外,细胞核的形态亦有差异,表现出圆球形或椭圆形等变化。以后在少数细胞中继续进行第二次分裂,形成三核或四核细胞(图版Ⅰ,14~16)。在个别细胞中似乎还发生了胞质分裂,二核之间产生隔壁,形成二细胞结构(图版Ⅰ,12、13)。由于该类细胞数量少,染色方法不够适宜,有关胞质分裂的结果尚待进一步确证。

第一章 雄性器官与细胞的实验研究

图版 I   plate I

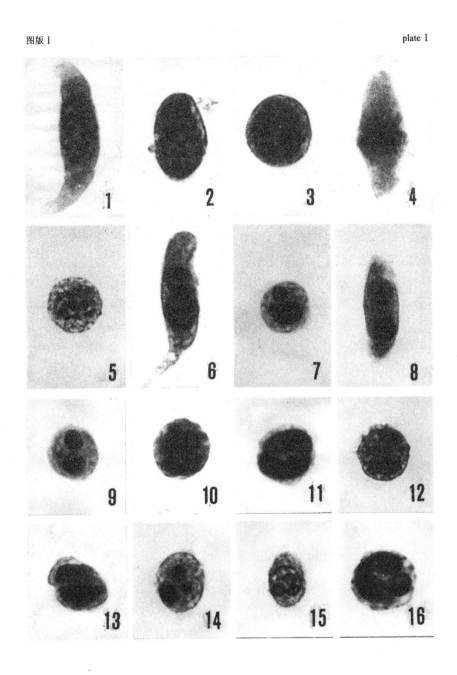

讨 论

生殖细胞在体内条件下被包围在营养细胞内,带有半寄生性质。它本身细胞质稀少,而且缺乏一般细胞所具有的典型形态的细胞壁。它从花粉粒中分离出来,能否在体外存活和发育,是迄今尚未研究的新课题。本文提供的实验证据初步表明,分离的生殖细胞在培养条件下可以存活并进行核分裂,因此,它有发育潜能,有希望用于细胞工程实验。实验取得初步成功的因素可能包括:第一,黄花菜的多数生殖细胞在接种时已达到分裂前期。这种状态有利于它们在离体条件下完成核分裂的后续过程。第二,参考原生质体培养积累的经验,设置的培养条件(如基本培养基、各种添加成分,饲养法)等[3~5,7]为生殖细胞的体外发育提供了较为适宜的条件。在某些实验中,核分裂频率高达8%和10%以上。当然,目前仅是初步实验,还存在着许多有待解决的问题,特别是如何进一步促使胞质分裂和细胞壁形成,达到诱导细胞分裂的目的,就体形微小、细胞质贫乏的生殖细胞而言,是一个难题。显微观察显示了生殖细胞及其细胞核在体外的发育与变化。值得注意的是,三种形态的生殖细胞中以圆球形细胞最易启动核分裂。这是否暗示更多地利用圆球形细胞是进一步提高诱导频率的一条出路?本文还揭示了生殖细胞核分裂具有均等与非均等分裂方式以及细胞核形态上的变化。这与生殖细胞在体内分裂形成一对精子的情况相比,显然有颇大的区别,似乎暗示离体培养已经改变了体内预定的发育程序。沿着这一方向开展实验研究,将有助于认识生殖细胞分化与脱分化机理等理论问题。

图 版 说 明

丙酸洋红染色。×1650　**1~3.** 刚接种的生殖细胞。**4、5.** 分裂中期的生殖细胞。**6、7.** 非均等核分裂形成的二核细胞。**8~11.** 均等分裂形成的二核细胞。**12、13.** 细胞分裂形成的二个细胞。**14、15.** 二次核分裂形成的三核细胞。**16.** 二次核分裂形成的四核细胞。

Explanation of plate

All materials were stained by propiono-carmine. ×1650 **Figs. 1 ~ 3.** Freshly inoculated generative cells. **Figs. 4、5.** Generative cells in metaphase. **Figs. 6 ,7.** Binucleate generative cells after unequal nuclear division. **Figs. 8 ~ 11.** Binucleate generative cells after equal nuclear division. **Figs. 12、13.** 2-celled units after first cell division. **Figs. 14、15.** Trinucleate generative cells after second nuclear division. **Fig. 16.** Tetranucleate generative cell after second nuclear division.

（作者：吴新莉、周嫦。原载：植物学报,1990,32(8):577 ~ 581。图版1幅,表3幅,参考文献8篇,仅选录图版Ⅰ。参看书末彩色图版Ⅴ。）

6 几种植物花粉生殖细胞的融合实验

Fusion experiments of isolated generative cells in several angiosperm species

摘 要

用 PEG 法诱导了玉帘(*Zephyranthes candida*)、石蒜(*Lycoris radiata*)、黄花菜(*Hemerocallis minor*)、朱顶红(*Hippeastrum vittata*)和鸢尾(*lris tectorum*)五种植物的同种生殖细胞之间、玉帘与石蒜异种生殖细胞之间、石蒜生殖细胞与其花瓣原生质体之间、石蒜生殖细胞与油菜下胚轴原生质体之间的融合,获得了含二至多个细胞核的同核体和异核体。其中石蒜生殖细胞的同核体还进一步发生了细胞核的融合。用与 DNA 特异结合的荧光染料 DAPI 预先标记的方法,可以在融合体中显示生殖细胞核。FDA 荧光鉴定表明融合体是生活的。还研究了若干因素对玉帘生殖细胞之间融合的影响。

Abstract

The generative cells used for fusion experiments were isolated from pollen grains of *Zephyranthes candida* and *Lycoris radiata* by "2-step osmotic shock" and from those of *Hippeastrum vittate*, *Hemerocallis minor* and *Iris tectorum* by "weak enzyme treatment" as reported previously. Using PEG method, fusions have been successfully induced between generative cells of the same species mentioned above, between generative cells of *Z. candida* and *L. radiata*, between generative cells and petal protoplasts in *L. radiata*, and between generative cells of *L. radiata* and hypocotyl protoplasts of *Brassica napus*. In all cases either homokaryons or heterokaryons could be obtained. Fusion of nuclei was observed sometimes in homokaryons of generative cells in *L. radiata*. The generative nuclei in fusion products could be well identified by labelling the generative cells before fusion with DAPI. FDA test demonstrated that most of the fusion products were viable. Factors affecting fusion efficiency including cell density, PEG concentration, duration of PEG treatment and effect of calcium ions were studied in fusion of generative cells in *Z. candida*. Our experiments indicate that isolated generative cells are likely to be deprived of cell walls and may be regarded as a special kind of protoplasts for direct fusion experiments.

(作者:吴新莉、周嫦。原载:植物学报,1991,33(12):897~904。
图版2幅,表5个,参考文献15篇,均删去。)

7 Fluorescent vital staining of plant sexual cell nuclei with DNA-specific fluorochromes and its application in gametoplast fusion

Vital staining of nucleus is a useful method in plant cell culture and protoplast fusion studies. The DNA-specific fluorochromes such as 4,6-diamidino-2-phenylindole (DAPI), Hoechst 33258 (H33258) and mithramycin have usually been considered as vital nuclear stains. In regard to sexual cells, DAPI or mithramycin staining has been used for visualization of generative and vegetative nuclei in artificially germinated, fresh pollen tubes[1-4], and pollination with H33258-or DAPI-prestained pollen has been used for tracing sperm nuclei in style or ovule tissues[5-8]. However, not in all these cases were direct evidences given for the real viability of stained cells at the moment of fluorescence observation; nor has there been reported such a vital nuclear staining technique adaptable to the in vitro fusion studies with gametoplast, i.e. isolated sperm or generative cell[9], as one of the fused partners.

The present investigation attempted to determine if the fluorescent staining of sexual cell nuclei was really capable of maintaining the viability of cells for further manipulation. Three criteria were adopted to estimate the viability of stained cells (a) cytoplasmic streaming in pollen tubes whose generative and vegetative nuclei showed fluorescence of tested fluorochrome, (b) fluorochromatic reaction (FCR) of isolated generative cells which were double-stained with fluorescein diacetate (FDA) and nuclear fluorochrome, and (c) the capability of fusion of isolated, prestained gametoplasts with other protoplasts.

Materials and methods

Plant material, fluorochromes and microscopy

Plant species used as materials were *Zephyranthes grandiflora* Lindle, *Hemerocallis minor* Mill. and *Lycoris radiata* Herb. Fluorescent dyes used were DAPI (Mannheim), H33258 (Mannheim), mithramycin (Sigma) and FDA (Sigma). Microscopic observations were carried out with an Olympus NEW VANOX AHBS-514 research microscope or an Olympus

IMT-2 inverted microscope, both equipped with epifluorescence attachments, U excitation for DAPI and H33258, B excitation for mithramycin and FDA.

Staining of nuclei in pollen tubes

In one experiment, *Z. grandiflora* pollen was germinated in hanging drops of BK medium[10] containing 12% sucrose. After incubation at 25℃ for 3h, when the vegetative nucleus and generative cell had been located in the pollen tubes, the original drops were absorbed with a piece of filter paper and replaced by new drops of fluorochrome-supplemented medium. Then samples were monitored at intervals to examine the nuclear fluorescence and cytoplasmic streaming in the same pollen tubes.

In another experiment, the pollen was stained prior to germination in fluorochrome-containing BK medium in Petri dishes. Samples were collected periodically and washed thoroughly with fluorochrome-free medium. Then the prestained pollen was cultured in hanging drops for 3 h and the pollen tubes thus germinated were examined for viewing the nucleus fluorescence and cytoplasmic streaming.

Staining of isolated generative cells

Generative cells of *H. minor* were isolated in quantities by a "weak enzymatic method"[11]: The pollen was incubated in 5% sucrose solution containing 0.1% cellulase (Onozuka R-10) and 0.1% pectinase (Serva) at 25℃ for 20-25 min to release the generative cells. After filtration and centrifugation, the purified generative cells were stained with nuclear fluorochrome (20 μg/ml in 19% sucrose solution) for 20 min. Then FDA stock solution was added to, a final concentration of 10 μg/ml. Through centrifugation, the precipitated generative cells were observed to determine the nucleus staining and FCR simultaneously.

Fusion of isolated, prestained gametoplasts with other protoplasts

Two sets of experiments were carried out. In the first experiment, isolated generative cells were fused with petal protoplasts in *L. radiata*. The generative cells were isolated by a "two-step osmotic shock method"[12] and stained with 5 μg/ml DAPI for 30 min, followed by thorough washing with fluorochrome-free solution. The petal protoplasts were prepared by maceration of petal strips with 1% cellulase and 1% pectinase solution at 28℃ for 6h. Fusion was induced by polyethylene glycol (PEG, M. W. 6000) according to a "small-scale fusion procedure"[13].

The second experiment dealt with fusion between isolated sperm cells and microspore protoplasts in *H. minor*. The sperms were isolated by an *in vivo/in vitro* technique[14]: Cut styles of 1 cm were inserted into BK medium containing 8% sucrose. After 20h of pollination, when sperm cells were observed in the pollen tube tips growing out from the cut ends, the styles were transferred to BK medium containing 0.5% cellulase, 0.5% pectinase and 4% sucrose to release sperms. The isolated sperms were then stained with 10 μg/ml DAPI for 20 min. The microspore protoplasts were prepared basically according to the method developed by Zhou[15]. Fusion was induced by the "small-scale fusion procedure"[13].

Results

Visualization of stained nuclei in living pollen tubes

The vegetative nucleus in pollen tubes began to show fluorescence after stained with mithramycin (20 or 50 μg/ml) for 30 min, whereas the generative nucleus fluoresced much later and weaker, usually after 90 min of staining (Fig 1 and 2). Cytoplasmic streaming was evident in all the tubes with fluorescing nuclei. In many cases the vegetative nucleus changing its shape and moving slowly along the tubes was observed. Supplementation of dimethyl sulphoxide (DMSO,2 or 5%) into the medium could enhance the penetration of fluorochrome, but did not show effect on the relative fluorescence intensity of vegetative and generative nuclei (Fig. 3). When pollen prestained with mithramycin (20 μg/ml) was germinated in dye-free medium, the pollen tubes grew well with vigorous cytoplasmic streaming. Prestaining for 15 min did not result in nucleus fluorescence; 30-45 min prestaining only labeled vegetative nuclei; 60 min prestaining could label generative nuclei as well.

Direct pollen tube staining or, alternatively, pollen prestaining with DAPI (20 μg/ml) showed similar results of simultaneous nucleus fluorescence and cytoplasmic streaming in the pollen tubes. However, there were two distinct points different from mithramycin staining. First, the fluorescence of DAPI maintained much longer than that of mithramycin which faded rapidly under excitation. Second, in the case of DAPI staining, the generative nucleus usually fluoresced brighter than the vegetative one did (Fig. 4). This was considered normal since the DNA of generative nucleus is usually more condensed than that of vegetative nucleus.

H33258 failed to work in either pollen tube staining or pollen prestaining. Attempts to improve its effect including raising the dye concentration (from 20 to 200 μg/ml) or the pH value of the medium (from pH 7 to 10) as well as supplementation of DMSO (2% or 5%) could not lead to positive results (Fig 5 and 6). The nuclei did not fluoresce unless the cytoplasmic streaming had ceased or the nuclei had been released from bursted tubes into the medium.

Visualization of stained nucleus in isolated, living generative cells

Isolated generative cells of *H. minor* sequentially stained with DAPI (20μg/ml) and FDA (10 μg/ml) showed simultaneous nucleus fluorescence and FCR on observation with U and B excitation respectively (Figs. 7-10). Of the 124 generative cells demonstrating FCR, 115 cells exhibited also DAPI fluorescence, denoting that 92.7% of the viable cells allowed DAPI staining. Alternatively, of the 240 generative cells demonstrating DAPI fluorescence, 213 cells showed FCR, indicating that 88.8% of the DAPI-stained cells were viable. By any means of calculation, a high efficiency of vital staining was confirmed.

Generative cells could be double-stained similarly with H33258 (20 μg/ml) and FDA (10 μg/ml). Here, the nucleus fluorescence was somewhat weaker than that stained with DAPI under the same condition of dye concentration and staining duration (Fig. 11, 12).

Double staining did not succeed with mithramycin (20 μg/ml). It was not clear yet why negative results were obtained. One possible reason was that mithramycin molecules could hardly penetrate the plasma membrane of generative cells, since in the experiments mentioned above, the staining process of generative cell in pollen tubes with mithramycin was very slow and weak. Another possibility was that mithramycin fluorescence and FCR were viewed with the same B excitation, that caused a masking of the former by the latter.

Visualization of prestained gametoplast nuclei during protoplast fusion

In the experiment on fusion between generative cells and petal protoplasts in *L. radiata*, the generative cells were prestained with DAPI before fusion while the petal protoplasts were unstained. During the fusion process, bright-field observation could distinguish the outline of a generative cell by its yellow-brown pigmentation of the cytoplasm from the petal protoplast which was characterized by red-colored anthocyanin in the vacuole (Fig. 13). Fluorescence observation on the same specimen could detect a strongly fluorescing generative nucleus in the fusion products. The petal protoplast nucleus in the fusion products sometimes also showed faint fluorescence, though the nucleus of unfused petal protoplasts never did so (Fig. 14).

In the experiment dealt with fusion between sperm cells and microspore protoplasts in *H. minor*, similar behavior was observed. The sperm nuclei prestained with DAPI showed bright fluorescence whereas the unstained microspore protoplast nuclei did not (Fig 15 and 16). When a sperm cell had adhered to a microspore protoplast, still only the former demonstrated fluorescence (Fig 17 and 18). After uptake of a sperm by a petal protoplast, both nuclei were not discernible by bright-field observation (Fig. 19), but by fluorescence observation the sperm nucleus could be detected and sometimes the microspore nucleus as well (Fig. 20).

Discussion

Though DNA-specific fluorochromes have often been used to detect the nucleus of fresh cells, there have been reports concerning limitations in their utilization. For instance, H33258 can not well stain the living protoplasts except in the presence of Triton X-100[16]; so are DAPI and ethidium bromide[17]. H33342, an analog of H33258, can stain the living protoplasts only when the medium is adjusted to pH 10[18,19]. In extreme cases, H33258[20] and propidium iodide[21] are used as probes for dead cells instead of living ones. Even in recent years, there has been a dilemma about whether DAPI-stained cells are "alive, or dead, or a bit of both" (see letters in Trends in Genetics, Vol 5, No. 9, 1989). Therefore, it is now still worthwhile to reexamine this problem.

If we consider the concept of "vital staining" as a staining of real living cells, we should exclude such cases as the formerly living cells are dying at the moment of observation. Since DNA-binding fluorochromes stain dead cells much easier than living ones, such possibility should always be kept in mind. As an example, when using H33258-prestained pollen for *semi-vitro* pollination in *Nicotiana alata*, Mulcahy and Mulcahy (1986) saw fluorescing sperm nuclei in the pollen tubes growing along the style and beyond the stylar cut ends. However, they noticed that the nuclei were visible only after cytoplasmic streaming in the tube had ceased[6]. Our present observation on H33258-stained pollen tubes confirmed this point.

In this paper we used three criteria to estimate the validity of fluorescent vital staining of sexual cell nuclei. Judged by cytoplasmic streaming in pollen tubes, both DAPI and mithramycin were confirmed to have the ability of vital staining (though mithramycin stained the generative nucleus less prominently), but H33258 did not. When FCR was used as a criterion in an isolated generative cell system, the nuclei were vitally stained by either DAPI or H33258, but not by mithramycin. The success of simultaneous visualization of nucleus fluorescence and FCR in isolated generative cells was in contrast to such cases as in pollen grains[2] and aleurone protoplasts[21], where nucleus fluorescence could not be visualized together with FCR. The last evidence of vital staining was the ability of stained gametoplasts to fuse with other protoplasts as has been suggested[22]. Since the results from the first two experiments had shown that DAPI was the most superior one among the three tested fluorochromes, we used only DAPI for gametoplast fusion experiments, and positive results were obtained. Thus through three sets of experiments we can conclude that for the tested fluorochromes, DAPI has its really vital staining ability to a certain degree.

Recently there have been attempts to utilize isolated gametoplasts in plant biotechnology[23-26]. With regard to fusion studies, there were reports on fusion of generative cell with generative cell[27,28], pollen protoplast[27] and somatic protoplast[28]. In all these cases homokaryons or heterokaryons have been obtained. Kranz et al. have succeeded in fusion between isolated single sperm and egg cell in maize and cultured the fusion products to multicellular structures[29]. In the present paper, DAPI-labeled gametoplasts were fused with petal or microspore protoplasts. The fluorescence of gametoplast nuclei could be clearly visualized in the heterokaryon. It is interesing that sometimes the unstained somatic or microspore protoplast nucleus in the fusion products showed fluorescence too. This can be interpreted as a consequence of diffusion of the fluorochrome molecules from the donor nucleus during the fusion process. Similarly, Williams and Keijzer noticed that DAPI-labeled pollen tube could stain nuclei of female tissues after its entering the ovule and discharging into the embryo sac[8]. To sum up, the fluorescence vital staining technique is useful for tracinggametoplast nucleus in fusion process and for selecting desirable heterokaryons in gameto-gametic or gameto-somatic hybridization studies.

(Authors: Yang HY, Wu XL, MO YS, Zhou C. Published in: Cell Research, 1993, 3: 121-130. with 20 figures and 29 references all omitted here.)

8 Isolated generative cells in some angiosperms: a further study

In a previous paper the isolation of generative cells of *Haemanthus katherinae* from mature pollen grains by means of a simple squash method was described. This technique appears to provide certain advantages in the study of generative cells (Zhou et al. 1986). In order to extend its application to more plant species and to investigate the changes in morphology that take place in generative cells, the present study was carried out.

Materials and methods

The experimental materials included the following species: *Narcissus pseudonarcissus* L., *Crinum moorei* Hook. f. (Amarylidaceae) and *Tulipa gesneriana* L. (Liliaceae). Plants were grown either in the Murdough greenhouse, Dartmouth College, maintained at 25-30℃ (days) and 18-20℃ (nights) under natural daylength, or on the college campus. The isolation of generative cells and the microscopic observations were performed as reported previously (Zhou et al. 1986). In summary, mature pollen grains, mounted in Brewbaker and Kwack's medium with 5%-20% sucrose (Brewbaker and Kwack 1963), were squashed under a coverslip and generative cells were subsequently released from the pollen grains. The generative cells, fresh or fixed, were observed under an inverted Zeiss Aximat microscope equipped with Nomarski differential interference contrast accessories. The morphological changes occurring after isolation were recorded by Allen video-enhanced microscopy (Allen 1985).

For scanning electron microscopy, the generative cells were isolated in 100 mM PIPES buffer, with 3 mM EGTA added for *T. gesneriana* and 25 mM phosphate buffer for *N. pseudonarcissus*, fixed in 2% (v/v) glutaraldehyde in the same buffer for 1.5h, then washed with the buffer and centrifuged three times. The concentrated cell suspensions were placed on coverslips that had been coated with 1.5% poly-*l*-lysine. They were then postfixed in 1% osmium tetroxide for 2h, washed again with the buffer and distilled water, and either stained with 4% uranyl acetate (*T. gesneriana*), or not stained (*N. pseudonarcissus*). The fixed material was dehydrated in a graded series of ethanol and critical-point-dried with carbon dioxide. Dried samples were then coated with approximately 100 Å of gold/palladium in a Technics Hummer V sputter coater and examined at 20 kV under an AMR 1000 scanning electron microscope.

The procedure for the immunofluorescence study of microtubules followed those of Wicks (Wicks et al. 1981) and Derkson (Derkson et al. 1985) with some modifications. EGTA and $MgSO_4$ were added to the pollen grains suspended in Brewbaker and Kwack's medium supplemented with 10% sucrose to give final concentrations of 5.0 and 2.5mM, respectively. Generative cell obtained by squashing in this medium were fixed for 20 min in a 20 mM phosphate buffer containing 4% paraformaldehyde, 10% sucrose and 5 mM $MgSO_4$, then transferred to the same fixative without sucrose for 3 h, washed, and centrifuged three or four times. The concentrated cells suspension was attached to coverslips previously coated with 1.5% poly-l-lysine, treated with methanol (-20℃ for 8 min) and washed with PBS (0.13M NaCl, 5.1 mM Na_2HPO_4, 1.56 mM KH_2PO_4, pH7.0) for 15 min. The cells were then treated with the Kilmartin tubulin antibody diluted 100 times in PBS (monoclonal anti-α-tubulin YOL 34, a gift from Dr. Kilmartin) (Kilmartin et al. 1982). The preparation was incubated in a moisture chamber at 37℃ for 50 min and washed three times with PBS before being treated with the secondary antibody. The secondary antibody was FITC antimouse IgG (Miles, Yeda, Canada) diluted 500 times with PBS. Preparations were incubated as for the primary antibody, washed three times in PBS and mounted in buffered glycerol with p-phenylenediamine (Johnson and de Araujo 1981). The preparations were observed by epi-fluorescence microscopy (exciter filter BP450-490, beam splitter FI510, barrier LP520).

Results

Nomarski differential interference contrast microscopic observations

Generative cells were isolated from mature pollen grains of the three plant species used in this study. The isolated generative cells from each of these species were similar morphologically to those of *H. katherinae* (Zhou et al. 1986). Figures 1-5 show the generative cells of *T. gesneriana* that were isolated and fixed in glutaraldehyde. Various cell shapes were observed: typical spindle-shaped cells with two long "tails" (Fig. 1), spherical cells (Fig. 5) and some transitional oval shapes with gradually shortened tails (Figs. 2-4). The cell structure was quite discernible: the cells had a large nucleus occupying the center of the cytoplasm and one or two prominent nucleoli. The cytoplasm appeared dense and evenly distributed without obvious vacuoles (Figs. 1-5).

第一章 雄性器官与细胞的实验研究

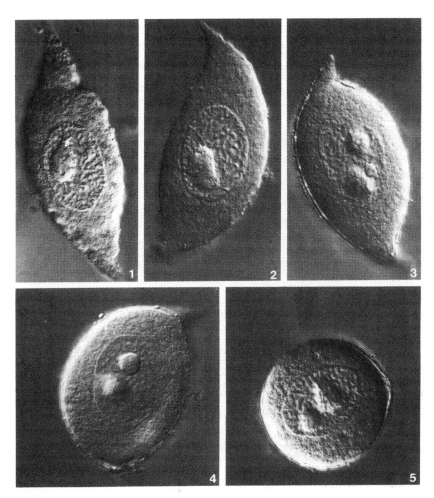

Figs. 1-5. Isolated generative cells of *Tulipa gesneriana* fixed in glutaraldehyde and observed with Nomarski interference contrast microscopy. ×1500

Scanning electron microscopic observations

In order to study the external features of the generative cells, scanning electron-microscopic observations were carried out in *T. gesneriana*. The same cell shapes as seen with the light microscope were observed at high resolution and high magnification. These were spindle-shaped cells (Fig. 6), cells with shortened tails (Fig. 7), spherical cells with a "tail" remnant (Fig. 8), and spherical cells (Fig. 9). The sizes of these cells were 61.3×20.6; 38.8×20.6; 26.5×20.9; and 20.3×20.3 μm, respectively. The measurements indicate that the cells are quite different in their length but similar in width. Large or small folds could be seen on the cell surface but no special or unusual structures were noted. Sometimes the cytoplasmic material of the vegetative cell was seen adhering to the surface of isolated generative cells.

The scanning electron-microscopic features of generative cells in *N. pseudonarcissus* appeared to be comparable to those seen in *T. gesneriana*. They might be spindle-shaped, oval or spherical. Figure 10 shows one oval cell.

Video-enhanced microscopic observations

Generative cells *in situ* are usually spindle-shaped, but following isolation they attained quite different forms. A common tendency was a transitory change to an oval and finally to a spherical shape over time. Using video-enhanced microscopy we traced the changes in shape of the generative cells in *N. pseudonarcissus* (Figs. 11-14). After isolation, the generative cell soon became oval. However, a separation between the membrane and cytoplasm on one side of the cell was observed. The cytoplasm appeared to gradually condense from the two poles towards the center, causing the whole cell to contract significantly (Figs. 12、13). Within approximately 40 min the cell had assumed a spherical shape (Fig. 14).

Immunofluorescence observations for microtubules

Using a monoclonal anti-α-tubulin antibody (Kilmartin et al. 1982) and FITC anti-mouse IgG as the secondary antibody, an immunofluores-

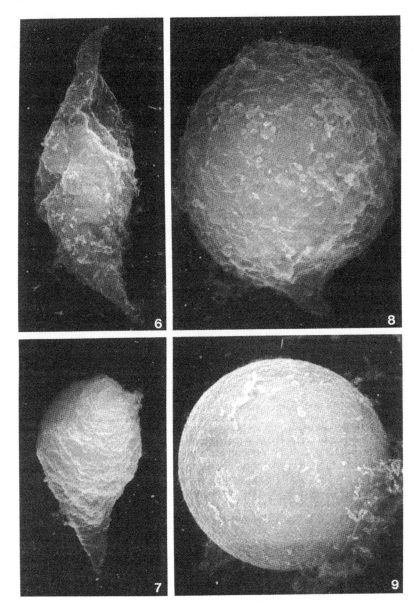

Figs. 6-9. Isolated generative cells of *T. gesneriana* seen by means of scanning electron microscopy. ×1600(**Fig. 6**); ×1800 (**Fig. 7**); ×3500 (**Figs. 8,9**)

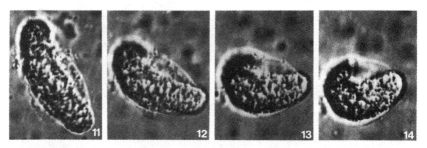

Figs. 11-14. Time-lapse observation of an isolated generative cell of *N. pseudonarcissus* viewed with video-enhanced microscopy. ×1000

cence study on the isolated generative cells of *C. moorei* was carried out. Figures 15-17 show different optical sections of one cell from the surface to the center. When focused on the surface region, long parallel strands of microtubules could be seen throughout the cell; their orientation was mainly axial with some irregularities (Fig. 15). When focused nearer the center, microtubules were seen to be distributed at the periphery of the cytoplasm; the nucleus and a large part of the cytoplasm occupied the dark region (Figs. 16, 17). Other cells presented a similar arrangement of microtubules, as shown in Figs. 18 and 19. These observations provide us with an overall surface/cortical distribution pattern of microtubules in generative cells.

Discussion

Employing a simple squash method, we have so far isolated generative cells from mature pollen grains in four species, including *H. katherinae*. As noted in both a previous (Zhou et al. 1986) and the present report, this isolation technique has distinct advantages when observing both the external and the internal structure of generative cells. Emphasis has been given to the cell surface, the three-dimensional structure of the cell and to the changes in cell morphology that occur *in vitro*. There is no interference from the pollen wall and the vegetative cell. The isolation technique permits unobstructed observations of living cells. For example, using this method we were able to observe scanning electron microscopic images of naked generative cells for the first time. In addition, we provide immunological evidence for the distribution of microtubules in intact, whole generative cells. The existence and arrangement of microtubules in generative cells has been previously documented only in ultrastructural studies (Sanger and Jackson 1971; Cresti et al. 1984). We have utilized an immunofluorescence technique as a means of visualizing the overall three-dimensional distribution of microtubules in the cell. Derkson et al. (1985) and Van Lammeren et al. (1985) identified tubulin in the vegetative and generative cells of tobacco and lily pollen tubes, and in developing pollen of *Gasteria*. However, attempts to observe immunofluorescence of a generative cell within an intact pollen grain encounter obvious technical difficulties. The pollen wall is not only a serious barrier to antibody permeability; in addition, its strong autofluorescence may mask the immunofluorescence. Using isolated generative cells, we overcame these problems and obtained three-dimensional images of microtubule immunofluorescence. We were also able to observe the dynamic alternations that take place in the shape of living generative cells after release from pollen grains. Our preliminary study by video-enhanced microscopy permitted us to record the processes of contraction of the generative cell after isolation. Using the methods we have developed, it will be possible to study the internal changes (e. g. , microtubule distribution) occurring during this contraction. Also, it is now feasible to submit the isolated generative cells to various environmental conditions and to monitor the response.

The potential utility of isolated generative cells for *in vitro* study and possible genetic manipulation was pointed out in our previous paper (Zhou et al. 1986). Other workers have recently contributed new information concerning the isolation of protoplasts and sperm cells from pollen. Loewus et al. (1985) have isolated protoplasts from lily pollen grains using a technique which chemically dissolves the pollen wall. Russell (1986) described a technique for releasing sperm cells from the pollen grains of *Plumbago* using osmotic shock. He suggests the term "gametoplast" for the isolated sperm and generative cells. Matthys-Rochon et al. (1987) reported the isolation of sperm cells from three tricellular pollen species. More recently, one of us (Zhou) has effectively isolated protoplasts from mature pollen grains of *Iris* in large numbers.

All this research signals a growing interest in efforts to manipulate the male reproductive system. However, many problems remain to be solved before isolated generative cells and/or sperm cells can be conveniently utilized in experimental work. Improvements are needed in techniques for cell isolation, in the separation and purification of cells from the cytoplasmic content of the pollen grains, and in maintaining the viability of isolated cells *in vitro*. Our own efforts are currently directed towards these problems. A method named the two-step osmotic shock technique has been developed by Zhou for isolating generative cells in large quantities (to be reported elsewhere).

(Authors:Zhou C, Orndorff K,Daghlian CP, DeMaggio AE. Published in:Sexual Plant Reproduction, 1988 1:97-102,with 19 figures and 13 references. Figs 10,15-19 are omitted here.)

9 Microtubule organization of *in situ* and isolated generative cells in *Zephyranthes grandiflora* Lindl

Although electron microscopy has shown the presence of microtubules in generative cells (Sanger and Jackson 1971; Cresti et al. 1984), immunofluorescence microscopy has received a considerable degree of attention in recent years as it facilitates large-scale investigations on the spatial and temporal organization of microtubules in cells. With the latter technique, a basket-like structure formed by microtubules was found to be characteristic of *in situ* generative cells in several angiosperm species (Derksen et al. 1985; Pierson et al. 1986; Lancelle et al. 1987; Tiezzi et al. 1980; Pierson 1989; Taylor et al. 1989). Heslop-Harrison et al. (1988) showed that the microtubule cytoskeleton of generative cells corresponds closely to cell shaping and movement in the pollen tube. Palevitz and Cresti (1989) showed that the microtubule pattern could undergo structural changes during sperm formation and revealed some unusual characteristics of generative cell mitosis in *Tradescantia virginiana*. All of these researchers used generative cells *in situ* as experimental material, namely, pollen grains or growing pollen tubes. Zhou et al. (1988), however, made some preliminary observations on the microtubule cytoskeleton using artificially isolated generative cells. In the study presented here, further observations were made on the microtubule cytoskeleton of generative cells of *Zephyranthes grandiflora* under both *in situ* and isolated conditions.

Materials and methods

Materials

The experimental material was an ornamental, *Zephyranthes grandiflora* Lindl., grown in pots. In order to study the microtubule patterns of *in situ* generative cells, fresh mature pollen grains were used, some of which were subjected to a short-time enzymatic treatment, others not. Pollen tubes grown in a culture medium were also used.

Pollen grains subjected to enzyme treatment were incubated in the macroelement solution of the K_3 medium (Nagy and Maliga 1976) containing 12% mannitol, 0.5% cellulase (Onozuka R-10, Japan), 0.5% pectinase (Serva, Finebiochemica, Heidelberg, FRG) and 0.1% pectolyase Y-23 (Sheishin Pharmaceutical, Tokyo, Japan) for 8 min, rinsed three times with the same medium free of enzymes, and then incubated for about 1h.

For collection of the pollen tubes, pollen grains were artificially germinated at 26℃ in BK medium (Brewbaker and Kwack 1963) with 12% sucrose for 1-3h.

Two methods were used to isolate the generative cells: osmotic shock and grinding. In the osmotic shock method (Zhou 1988), pollen grains were incubated in 24% sucrose solution for 1h and then shocked by adding water. One hour later the generative cells released from the bursting pollen grains were collected for fixation. In the grinding method (Yang and Zhou 1989), pollen grains in 20% sucrose solution or BK medium containing 20% sucrose and 4 mM EGTA (ethylene glycol-bis(-aminoethyl-ether)-N, N,N',N'-tetracetic acid) were ground with a glass homogenizer to release the generative cells. The samples were fixed immediately after grinding.

Immunofluorescence

Sample processing was basically according to the method of Derksen et al. (1985) and Zhou et al. (1988) with some minor modifications. The samples were fixed with freshly prepared fixative containing 3.7% paraformaldehyde, 100 mM phosphate buffer, half-strength BK medium, 5 mM EGTA, and 12% or 20% sucrose for 3h, rinsed three times with a washing solution (100 mM phosphate buffer, half-strength BK medium and 1 mM EGTA), treated in methanol at −20℃ for 6-8 min, rinsed three times with the washing solution, digested with 0.5% cellulase and 0.5% pectinase (dissolved in the washing solution) for 10 min, and rinsed thoroughly. For the isolated generative cells the steps involving methanol and enzyme treatments were omitted. Following a final rinse in PBS (0.13M

NaCl, 5.1mM Na$_2$HPO$_4$, 1.5mM KH$_2$PO$_4$, pH7.0), the samples were incubated at 37℃ for 50 min in a monoclonal anti-α-tubulin (Sigma, diluted 500× with PBS) as the primary antibody, rinsed three times with PBS, and then incubated in FITC-labelled anti-mouse IgG (Sigma, diluted 60× with PBS) as the secondary antibody at 37℃ for 50 min. After three rinses the materials were resuspended in a PBS/glycerol mixture, mounted on slides and examined under a microscope (Olympus, NEW VANOX AHBS-514) equipped for epifluorescence viewing with B excitation system. In the controls without the primary antibody treatment microtubules could not be located.

All the experiments were carried out in the Laboratory of Plant Cytoembryology, Department of Biology, Wuhan University.

Results

Microtubule patterns of generative cells in pollen grains

It was only possible to visualize the microtubules of generative cells in the pollen grains in some unhydrated pollen grains, but this could be greatly improved by a short-time enzymatic treatment followed by hydration, which caused the dehiscence and removal of the exine, leaving the pollen protoplasts enclosed in the intine alone. The vegetative cell in the pollen grain does not seem to have any organized microtubules and shows only a weak generalized fluorescence. The generative cell, on the other hand, contains longitudinally oriented microtubule bundles with strong fluorescence. The generative cell is spindle-shaped and often has a bent tail-like extension at each pole. The microtubule bundles usually run parallel to the long axis and are present throughout the cell, giving a spindle-shaped configuration of microtubule cytoskeleton (Figs. 1, 2). Microtubules are also visible in the tail-like extensions (Figs. 3, 4). Sometimes microtubule bundles are seen crossing each other (Fig. 2) and occasionally obliquely-oriented microtubules running in opposite directions, one above the other, are present in the same cell (Figs. 5,6). In polar views, the microtubule bundles of the generative cell appear as a circular structure at the periphery of the cell (Figs. 7,8).

Microtubule patterns of generative cells in pollen tubes

After pollen germination microtubules oriented parallel to the long axis of the pollen tube are visible in the vegetative cell (Fig. 9); in the generative cell the microtubules show more diverse patterns and brighter fluorescence. The main type in spindle-shaped cells is also of longitudinally oriented microtubule bundles. However, the generative cell in the pollen tube often appears more elongated and the tail-like extensions are no longer bending backward (Fig. 10). Sometimes microtubule bundles appear to twist around each other at the tail or near the end of the cell (Figs. 12,13). Occasionally longitudinally oriented microtubules also occur in the ellipsoidal cells (Fig. 11). In rare cases when the cell turns into an ellipsoidal form, microtubules may even organize into spirals (Fig. 14). It is interesting to note that in some cells two different microtubule patterns may co-exist within the same cell: longitudinally oriented bundles at one pole and meshes at the other (Fig. 15). Such a state seems to be a transitional form of microtubule rearrangement, from the typical longitudinal pattern towards a new, mesh pattern (see also Figs. 16,17).

Microtubule patterns of isolated generative cells

For the labelling of microtubules in isolated generative cells, three different methods were compared: (1) treatment with methanol and enzyme, (2) treatment with methanol alone, and (3) no treatments at all before antibody labelling. No significant differences were found among them. This indicates that in contrast to the *in situ* condition, where the generative cell is enclosed by either the pollen or the pollen-tube wall, the permeability of the naked generative cell to antibody molecules must be quite sufficient to allow a procedure without the need of any pretreatment.

Usually the generative cells isolated by osmotic shock soon lose the typical spindle shape and become spherical (Zhou 1988). The present study showed that more spindle-shaped cells can be maintained if the pollen grains are ground in BK medium supplement with the microtubule stabilizer EGTA followed by immediate fixation. Under such conditions a small number of the spindle-shaped cells are able to maintain their microtubules in a longitudinal orientation (Figs. 18, 19). The majority, however, show a basket-like structure of generalized fluorescence around the nucleus (Figs. 20, 21). Sometimes a certain amount of mesh structure can also be seen through the generalized fluorescence, giving an impression that a new mesh structure is beginning to form from the dispersed tubulin (Fig. 21). As the cells become ellipsoldal and finally spherical in shape, microtubule bundles present in these cells disappear and are replaced by the mesh as a main pattern. Figures 22 and 23 demonstrate the transition form in some ellipsoidal cells in which the mesh structure is in coexistence with a few longitudinal bundles; Figs. 24 and 25 show some ellipsoidal cells with mesh structure alone. In the spherical cells, bundles are no longer visible and all microtubules are rearranged in the mesh form (Figs. 26-29).

Discussion

In recent years the cytoskeleton organization of the generative cells has received considerable attention. Interest in it has been prompted by the role of the generative cell in male gamete formation and its migration along the pollen tube. A basket-like structure consisting of longitudinally oriented microtubule bundles is believed to be a widespread property of generative cells of most angiosperm species under *in situ* condition (Derksen et al. 1985; Tiezzi et al. 1986, 1988; Heslop-Harrison et al. 1988; Pierson 1989; Taylor et al. 1989). Nevertheless, deviations from this typical pattern have

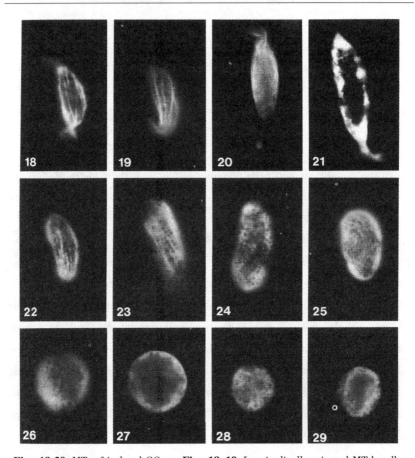

Figs. 18-29. MTs of isolated GCs. **Figs. 18,19.** Longitudinally oriented MT bundles persisting of a spindle-shaped (Fig. 18) and a one-tailed (Fig. 19) GC. ×800. **Fig. 20.** Generalized fluorescence of tubulin exhibited in a spindle-shaped GC. ×800. **Fig. 21.** Generalized fluorescence with some mesh structure appearing in a spindle-shaped GC. ×1000. **Figs. 22,23.** Mesh MTs mixed with MTs arranged in parallel in ellipsoidal GCs. **Fig. 22** ×800, **Fig. 23** ×1000. **Figs. 24,25.** Mesh pattern of MTs in ellipsoidal GCs. ×1000. **Figs. 26~29.** Mesh pattern of MTs in spherical GCs, focused on the surface region (Figs. 26,28) and focused on the center region (Figs. 27,29). **Figs. 26,27** ×1000, **Figs. 28,29** ×800

been reported. For example, in the pollen grains of *Tradescantia virginiana* the generative cell contains a transverse or reticulate microtubule structure; after pollen germination, the microtubule orientation changes to a longitudinal or helical arrangement (Palevitz and Cresti 1989). Our results demonstrate that the microtubule patterns in generative cells of *Z. grandiflora* are quite variable. Besides the typical longitudinal arrangement, various other patterns are found to be present under the *in situ* condition. More diverse forms occur in the generative cells in pollen tubes than in those in pollen grains. This is probably because in the pollen tube the generative cell is actively undergoing reshaping. The problem of generative cell migration down the tube has been discussed in detail by a number of researchers (Derksen et al. 1985; Tiezzi et al. 1986, 1988; Heslop-Harrison 1988; Heslop-Harrison et al. 1988; Palevitz and Cresti 1989). Tiezzi et al. (1988) revealed that the microtubules present in the generative cells of some species are linked together by lateral projections. They assumed that the microtubules with the associated lateral structures could be cooperating together in generative cell reshaping during its migration. According to Heslop-Harrison et al. (1988) the microtubules also seem to be involved in shaping the generative cell and adapting it to its passage along the pollen tube, whereas the microfilaments provide the motive force for movement. Recently, using anti-actin antibody, Taylor et al. (1989) demonstrated that there are abundant actin microfilaments in generative cells and that actin corresponds closely in distribution to tubulin. Obviously, they may be involved in generative cell movement. Palevitz and Cresti (1989) are making ongoing efforts to more fully elucidate the mechanism of movement and shaping of generative cells by pharmacological, molecular and microscopic methods.

As for microtubule organization in isolated generative cells, our previous work has demon-strated the axial arrangement of microtubules in *Crinum moorei* (Zhou et al. 1988). The present paper offers a more detailed study on isolated generative cells of *Z. grandiflora* and reveals the alterations in microtubule patterns in relation to cell shaping. Microtubule orientation usually changes from an axial arrangement to a mesh structure as the cell changes from a spindle to an ellipsoidal and finally to a spherical shape. It is obvious that isolated generative cells can be used as an experimental system for studying the relationship between changes in microtubule organization and cell shape. There are some advantages: firstly, observations can be made on naked generative cells without interference from the vegetative cell and pollen wall; secondly; the use of the isolated cell-system can make studies on the effect of *in vitro* conditions quite different from that of the *in situ* environment.

(Authors:Zhou C,Zee SY,Yang HY. Published in: Sexual Plant Reproduction,1990,3:213-218. with 29 figures and 17 references. Only Figs 18-29 are retained here.)

10 Microtubule changes during the development of generative cells in *Hippeastrum vittatum* pollen

The microtubule (MT) cytoskeleton of generative cells (GC) within pollen tubes has been investigated extensively by means of immunofluorescence microscopy (Derksen et al. 1985; Pierson et al. 1986; Tiezzi et al. 1986; Heslop-Harrison et al. 1988; Tiezzi et al. 1988; Palevitz and Cresti 1989; Pierson 1989; Taylor et al. 1989; Xu et al. 1990; Zhou et al. 1990; Zhu and Liu 1990). More recently, microtubular changes in GCs artificially isolated from mature pollen grains have been dealt with (Zhou et al. 1988; Tanaka et al. 1989; Zhou et al. 1990). In some of these investigations, GCs inside mature pollen grains were also observed for MT arrangement.

However, the study of changes in the MT cytoskeleton of GCs during their development inside pollen grains at various developmental stages remains difficult due to methodological difficulties; these cells are enclosed by the pollen wall, which barricades antibody penetration and also masks immuno-fluorescence observation. We have established a simple and effective method for isolating GCs at various developmental stages while still preserving the original MT organization. This technique enables us to carry out a large-scale investigation on the MT changes that occur in the young, spherical cell stage up to their full maturation.

Materials and methods

Hippeastrum vittatum Herb. plants growing in pots under natural conditions were used. The pollen grains were collected separately from each individual bud of a certain length and immediately fixed in a freshly prepared fixation solution containing 4% paraformaldehyde, 2.5 mM EGTA, half-strength BK medium (Brewbaker and Kwack 1963) and 50 mM phosphate buffer, pH 6.9. After 5-10 min of fixation, the pollen grains were ground gently in the fixative with a glass homogenizer to release the GCs. The fixation was then continued for another 2.5-3 h. The samples were then rinsed 3 times with a washing solution containing 1.5 mM EGTA, half strength BK medium and 50 mM phosphate buffer, pH 6.9, followed by one rinse of phosphate-buffered saline (PBS), pH 7.0.

The procedure used for immunofluorescence was similar to that described previously (Zhou et al. 1990). The samples were first incubated with a monoclonal anti-α-tubulin (Sigma, diluted 500 times with PBS) as the primary antibody at 37℃ for 50 min. After three rinses with PBS, they were then incubated with FITC-labelled anti-mouse IgG (Sigma, diluted 60 times with PBS) as the secondary antibody at 37℃ for 50 min. Then after three rinses with PBS and resuspension in a PBS/glycerol mixture, the samples were finally mounted on slides and examined under a microscope (Olympus NEW VANOX AHBS-514) equipped for epifluorescence observation with B excitation system. The control without primary antibody treatment did not show specific fluorescence.

Results

Two aspects of the MT cytoskeleton of GCs were studied: the three-dimensional configuration of the MT framework and the detailed arrangement of MTs making up this framework. It was necessary to discriminate between these two concepts since we have observed more than a few cases where the framework of the same configuration might bear different MT arrangements and *vice versa*.

Changes in MT framework configuration during GC development

The GCs of each individual bud of a certain length were examined separately. GCs from a 7.4-cm-long bud were the youngest that could be isolated from the pollen grains; those from shorter buds were not yet able to be liberated, probably because they were still attached to the pollen wall. The GCs in buds with a length of 14.5 and 15.0 cm were fully mature and were released at the day of anthesis. We observed that with the increase in the bud length during this period the configuration of the MT framework changed from a spherical form to a spindle-shaped form via varous transitional forms. For the convenience of description, they were classified into five forms: spherical (Figs. 1-5), ellipsoidal (Figs. 6-8), pera-shaped with an extension at one pole (Figs. 9-14), short spindle-shaped (Figs. 15-17) and long spindle-shaped (Figs. 18-20). Since the MT frameworks were always located at the periphery of the GC cytoplasm, they probable represent the full shape of GCs at various stages. The distribution data of these forms in individual buds were summarized into five stages (Table 1). It was evi-

第一章 雄性器官与细胞的实验研究

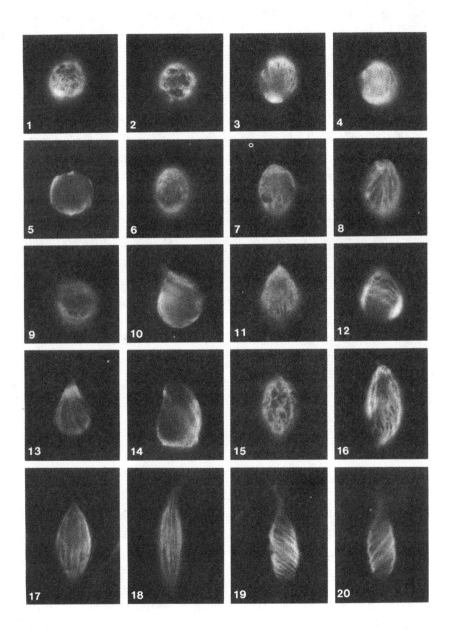

253

Figs. 1-20. ×1000.   **Figs. 1-5.** MTs in spherical cells.   **Fig. 1.** Network pattern with evenly distributed MTs.   **Fig. 2.** Network pattern with unevenly distributed MTs.
Fig. 3. Intermediate pattern of MTs in a GC. The reticulate MTs were changing into bundles.
Fig. 4. Long MT bundles running parallelly across the cell.   **Fig. 5.** MTs in a long bundle pattern in a GC viewed on the central optical section. Note the pointed tips that have emerged at the poles.
Figs. 6-8. MTs in ellipsoidal GCs.   **Fig. 6.** Network pattern of MTs.   **Fig. 7.** Intermediate pattern showing some short bundles just organized from randomly distributed MTs.
Fig. 8. MTs in a long bundle pattern.
Figs. 9-14. MTs in pear-shaped GCs.   **Figs. 9** and **10.** Network patern of MTs in a GC viewed on the surface and central optical sections, respectively.   **Fig. 11.** Intermediate pattern showing long MT bundles just organized.   **Fig. 12.** MT bundles oriented obliquely across a GC.   **Fig. 13.** A GC with typically axially arranged MT bundles.   **Fig. 14.** MT bundles running obliquely around a GC.
Figs. 15-17. MTs in short spindle-shaped GCs.   **Fig. 15.** Network pattern of MTs.
Fig. 16. Intermediate pattern consisting of long bundles and reticulate MTs.   **Fig. 17.** Typical long bundle pattern of MTs.
Figs. 18-20. MTs in long spindle-shaped GCs.   **Fig. 18.** Typical long bundle pattern.
Fig. 19 and **20.** A GC with MT bundles running helically around the cell viewed on two diffenert surfaces.

dent that GCs in the same bud were heterogeneous with respect to MT configuration. However, each bud of a certain stage usually had one predominant form. ***Stage* 1.** Bud length: 7.4cm. More than 90% of the GCs had a spherical form. Afew cells were ellipsoidal or pear-shaped. ***Stage* 2.** Bud length: 8.0 cm. This stage was characterized by various transitional forms, such as ellipsoidal (approximately 10%), pearshaped (approximately 20%) and short spindle-shaped (approximately 50%). ***Stage* 3.** Bud length: 8.4-9.4cm. All the forms from spherical to long spindle-shaped could be seen, with the short spindleshaped being predominant (approximately 70%-80%). ***Stage* 4.** Bud length: 11.5 cm. The spherical and ellipsoidal forms no longer existed; the long spindle-shaped form was present in a largerproportion (approximately 60%). ***Stage 5***. Bud length: 14.5-15.0 cm. 85%-90% of the GCspossessed the long spindle-shaped form.

MT arrangement patterns in relation to MT framework forms

A great diversity in MT arrangement occurred during GC development. The MT framework of the same form might be made up of MTs of different arrangements, which could be roughly classified into three patterns: the network pattern composed of randomly distributed MTs, the organizedlong bundle pattern, and an intermediate pattern with a mixture of both. The relationship betweenthe MT arrangement patterns and the frame-work forms are summarized in Table. 2.

MT arrangements in spherical GCs

The network pattern appeared to be the standard type in approximately 97% of the spherical GCs. The MTs were scattered either evenly (Fig. 1) or unevenly (Fig. 2). There was also a small proportion of other patterns. For example, Fig. 3 shows an intermediate pattern with both a network and bundles. In Fig. 4, long bundles running parallelly are seen. In some GCs the MT bundles converged into two opposite poles and protruded out as if it would form the two forthcoming extensions(Fig. 5)

MT arrangements in ellipsoidal GCs

The network was still the predominant pattern(Fig. 6). In a few GCs the MTs were beginning to organize into some short bundles (Fig. 7); in others they had already gathered into long bundles(Fig. 8).

Table 1. Changes of MT framework configuration during GC development

Stage	Bud length (cm)	Number of GCs observed	Spherical		Ellipsoidal		Pear-shaped		Short spindle-shaped		Long spindle-shaped	
			No.	%	No.	%	No.	%	No.	%	No.	%
1	7.4	294	273	92.8	14	4.8	7	2.4				
2	8.0	217	38	17.5	22	10.1	47	21.7	110	50.7		
3	8.4	240	25	10.4	18	7.5	19	7.9	174	72.5	4	1.7
	9.4	220	4	1.8	12	5.4	18	8.2	181	82.3	5	2.3
4	11.5	204					5	2.4	75	36.8	124	60.8
5	14.5	104					2	1.9	14	13.5	88	84.6
	15.0	235					3	1.3	23	9.8	209	88.9

* Short spindle-shaped: length/width ratio < 3 : 1; long spindle-shaped: length/width ratio > 3 : 1

Table 2. MT arrangement patterns in relation to MT framework forms of GCs

MT framework from	Number of GCs observed	Network		Intermediate		Long bundle	
		No.	%	No.	%	No.	%
Spherical	328	319	97.3	3	0.9	6	1.8
Ellipsoidal	60	54	90.0	2	3.3	4	6.7
Pear-shaped	91	33	36.2	18	19.8	40	44.0
Short spindle-shaped	570	33	5.8	35	6.1	502	88.1
Long spindle-shaped	328					328	100.0

MT arrangements in pear-shaped GCs

Here various MT arrangements existed, none of which predominated. A general tendency was apparent however: the network pattern cells had decreased and the long bundle pattern cells increased significantly. Figs. 9

and 10 show a GC in which most of the MTs were distributed randomly, although a few bundles had stretched to the tail-like extension. In Figs. 11 and 12, it can be seen that a transition towards the long bundle pattern has become more evident. Finally, in 44% of the cells the MTs had already organized into long bundles (Fig. 13). Usually the bundles were oriented along the long axis of the cells, but sometimes they ran at an oblique angle (Fig. 14) or even nearly perpendicular to the long axis (Fig. 12).

MT arrangements in short spindle-shaped GCs

The long bundle pattern had become predominant (Fig. 17), although the network pattern remained in a few cells (Fig. 5). In Fig. 16 the intermediate pattern is clear.

MT arrangements tn long spindle-shaped GCs

All the MTs were organized into long bundles. As usual, the bundles were oriented parallel to the long axis and stretched to the very end of the two poles (Fig. 18). Sometimes they ran obliquely as a helical around the cell (Figs. 19 and 20). Such a helical arrangement was present in 11.5% of the 208 GCs calculated. A similar phenomenon was also observed in less mature, short spindle-shaped cells, at a frequency of 7.2% (306 cells calculated).

Discussion

Since a series of morphogenetic events occur during GC development in pollen grains (Knox 1984), it is meaningful to know what changes take place in MT organization during this process. Sanger and Jackson (1971) and Brighigna et al. (1981), using electron microscopy, reported that MTs were scattered throughout the cytoplasm when the GC was spherical and were aligned in longitudinal parallel bundles when the cell was elongated. The electron microscope observations provided detailed but spatially restricted and quantitatively limited information on MT organization. The method presented in this paper is promising for large-scale investigations on MT organization of GCs in pollen grains by immunofluorescence microscopy. The clarity of the MTs in the GCs observed was greatly improved by removing the pollen wall and the vegetative cell. This enabled us to trace the spatial and temporal changes of the MT cytoskeleton of GCs during their development and to see many transitional, transiently occurring and rarely observed behavior.

This investigation confirms that the three-dimensional configuration of the MT framework is always a basket or cage-like structure at the periphery of the GC cytoplasm, even when it is still spherical. A basket-like MT framework has been described previously in spindle-shaped GCs (Heslop-Harrison 1988). In the case of typical long spindle-shaped GCs, there is a high correspondence between the MT framework form and the long bundle arrangement pattern, as reported by previous investigators (Cresti et al. 1984; Derksen et al. 1985; Heslop-Harrison 1988). However, in the case of young, developing GCs, the situation appears more complicated. On one hand, when the framework remains spherical, the MT arrangement may possibly change from disordered to polarized bundles in a few cells; on the other hand, when the framework has already changed to short spindle-shaped, the MT arrangement may still remain partially in a disordered state with many cells showing an intermediate pattern. This implies that the change in MT arrangement may go either ahead or behind the change in the MT framework in some cases. It should be mentioned that the change in MT arrangement is a process of rearrangement rather than via depolymerization and repolymerization: among the thousands of cells examined we have not seen any case of a dispersed fluorescence that can be considered as depolymerized tubulin.

The MT changes taking place in GCs freshly isolated from pollen grains have been reported by Zhou et al. (1990) in *Zephyranthes*. It is interesting now that the changes in MT organization in these two processes, from young spherical GCs to mature spindle-shaped GCs and from *in vivo* spindle-shaped GCs to *in vitro* spherical GCs, are quite similar to each other if viewed in a reverse order. For instance, the isolated GCs also show MT rearrangement from axial bundles to the network pattern through an intermediate form. The difference is that the *in vivo* development is a natural process with gradual and smooth changes, whereas artificial isolation causing a sudden change of the environment may result in a certain degree of MT depolymerization (Tanaka et al. 1989; Zhou et al. 1990).

(Authors: Zhou C, Yang HY. Published in: Sexual Plant Reproduction, 1991, 4: 293-297. with 20 figures, 2 tables and 19 references. All figures and tables are retained here. See also color plate V in the end of this book.)

第六节 精细胞操作

提要

在生殖细胞分离成功之后,我们进一步研究了精细胞的分离。首先,在油菜、玉米、黑麦三种植物中以渗透压冲击或匀浆器轻度研磨的方法分离出生活的精细胞。生活力鉴定仍用 FDA 荧光染色技术,但有所改进。在本章第二节所述 FDA 标记生活花粉管方法的启示下,转用于精细胞的标记,即在分离精细胞的同时,在介质中加入 FDA。然后将分离的精细胞冷藏,以后每天从中取样直接荧光镜检。这一改良方法的优点是无需在每次取样后进行 FDA 染色,既节约了时间,又避免了每次染色时精细胞的损失,大大方便了精细胞保存期间生活力的鉴定。此法在以后的研究中继续沿用。

在分离精细胞基本成功之后,我们对以紫菜苔为代表的三细胞型花粉和多种二细胞型花粉的精细胞分离分别进行了更大规模的研究。在紫菜苔实验中,对渗击法与研磨法作了进一步改良,分离率分别达34%与86%。研磨法不仅分离率高,而且在4℃条件下存活最长可达一周,考虑到实验是在有菌条件下操作的,如用无菌操作,存活应当更久。对影响分离率与生活力的几种因素进行了研究,其中特别要提到的是,在分离与保存介质中加入葡聚糖硫酸钾、牛血清白蛋白、聚乙烯吡咯烷酮(PVP)等保护剂,使效果大为提高。此外,还有趣地观察到刚释放的一对精细胞被 FDA 荧光显示了二型性:其中一个精细胞具长尾状延伸物,尾部亦呈荧光。这与文献中通过电镜观察到的情况一致。

二细胞型花粉的精细胞是在花粉管中形成的,其分离方法亦异于三细胞型花粉精细胞的分离。前人采用两种方法:一种方法是由人工萌发的花粉管中分离精细胞。此法虽然比较简便,但对于难以人工萌发的花粉则不适用。更须注意的是,人工萌发往往是不同步的,有些花粉管中已经产生了精细胞,有些则尚未。这样分离的精细胞群体中难免混杂生殖细胞。另一种方法是借助"活体-离体技术",将花粉授于柱头上,将花柱切段插入培养基中,经一定时间后,花粉管由花柱切口长出,此时精子已经形成。然后以渗击或酶处理从花粉管中分离精细胞。由于由花柱切口长出的花粉管在发育上是同步的,从而可以保证精细胞的相对纯洁。应用这一技术,我们在属于5科的8种植物中分离出精细胞,其中多数是新的纪录。

精细胞的分离成功为开展其融合实验创造了条件。进行了同种精细胞间、异种精细胞间、精细胞与小孢子原生质体间、与体细胞原生质体间各种组合的融合,均获得了同核体或异核体,表明用于融合的精细胞是生活的,诱导融合的 PEC 法原则上是可行的,为几年以后我们开展精、卵融合(即离体受精)实验奠定了初步基础(参看第三章第二节)。

第一章 雄性器官与细胞的实验研究

1 Isolation of viable sperms from pollen of *Brassica napus*, *Zea mays* and *Secale cereale*

The progress of plant reproductive biology and reproductive cell engineering requires isolation of sperm cells from the pollen. Although microscopic observations on isolated sperm cells were pioneered early by Cass (1973) and subsequently by Russell and Cass (1981), the techniques for isolating sperms in large quantities have been highlighted and refined not until recent years. So far sperm populations of various degree of purity have been obtained in *Plumbago zeylanica* (Russell, 1986), *Brassica campestris* (Cass et al., 1986), *Brassica oleracea* (Matthys-Rochon et al., 1987), *Zea mays* (Cass et al., 1986; Matthys-Rochon et al., 1987; Dupuis et al., 1987; Cass & Fabi, 1988), *Triticum aestivum* (Matthys-Rochon et al., 1987), *Gerbera jamsonii* (Southworth & Knox, 1987), *Rhododendron* spp. and *Gladiolus gandavensis* (Shivanna et al., 1988). However, in only a few of these reports the viability of isolated sperms has been confirmed by fluorochromatic reaction (FCR) (Dupuis et al., 1987; Matthys-Rochon et al., 1987), the others either did not make viability test or use negative reaction of Evans blue staining only. Inorder to obtain sperm cells available for manipulation work, proof of viability is an indispensable prerequisite. Here we briefly report our preliminary results on sperm isolation with emphasis on obtaining viable cells.

Materials and methods

Plants of *Brassica napus*, *Zea mays* and *Secale cereale* were grown in the field. Fresh pollen was collected from the mature, nearly dehiscing anthers. Various media for isolating sperms were tested, including sucrose solution, medium special for pollen germination of *Brassica oleracea* (Roberts et al., 1983), Brewbaker and Kwack medium (1963), and macroelements of K_3 medium developed by Kao for protoplast culture (Nagy & Maliga, 1976). Two methods for liberation of sperms were tried: (A) Grinding. The pollen was incubated in the medium for 15-20 min. and, after hydration, gently ground with a glass homogenizer. (B) Osmotic shock. A two step procedure developed by Zhou (1988) for isolation of generative cells was adopted. The pollen was first hydrated in sucrose solution of certain concentration, then a sudden osmotic shock was made by adding approximately equal volume of distilled water, causing the pollen to burst rapidly. The mixture was filtered through stainless or nylon screens of various pore sizes to remove the pollen wall debris, and centrifuged for making the sperm cells concentrated. The resulted suspension contained large quantities of sperms together with other small particles. For labeling living sperms, fluorescein diacetate (FDA) was used as vital stain according to Yang (1986): Stock solution of FDA (5mg/ml in acetone) was diluted with the isolation medium to a final concentration of 10 μg/ml. The labeled sperm suspension was stored at 4℃ in a refrigerator. Samples were taken at intervals and monitored with the fluorescence illumination of an Olympus NEW VANOX AHBS-514 Microscope using B excitation system. The viable sperms fluoresced green light. Sperm nuclei were visualized by Hoechst 33258 staining (1 mg/ml stock water solution diluted with the medium to a final concentration of 20 μg/ml). The nuclei fluoresced blue light with U excitation.

Results

Isolation of sperms

Brassica napus Both methods could cause pollen to burst and release sperms. With osmotic shock method, the pollen bursted well when 20% sucrose solution was used as incubation medium before addition of water. 30% sucrose resulted in bursting of only a part of pollen grains and 40% sucrose failed in pollen bursting. After filtration and centrifugation, sperm cell population without pollen wall debris was resuspended in Roberts et al. medium containing 5% sucrose, 8% sorbitol, 0.3% potassium dextran sulphate and 10 μg/ml FDA. In case of grinding, the pollen was first hydrated in Roberts et al. medium with the same supplements and then ground with a glass homogenizer. Pollen wall debris was removed by filtration and concentrated sperm population was obtained after centrifugation.

Zea mays Large quantity of sperms could be obtained by either osmotic shock or grinding. The former method was worth in convenience of filtrating off the pollen wall debris thoroughly. The pollen was first hydrated in 30% sucrose solution for 30 min. After that addition of water caused most of pollen grains to release sperms. The following procedure was similar to that with *Brassica napus*. K_3 medium appeared better than Brewbaker and Kwack medium in maintaining sperm viability. So far as the experiment proceeded, the optimal medium for sperm storage was K_3 macro-elements supplemented with 5% sucrose, 8% sorbitol, 0.5% potassium dextran sulphate and 10 μg/ml FDA.

Secale cereale Only the method of grinding has been tried. In Brewbaker and kwack medium containing 5% sucrose, 8% or 13% sorbitol, 0.3% potassium dextran sulphate and 10 μg/ml FDA, large number of sperms have been obtained. however, little success has been won in sprm storage.

Fluorescence micorscopic observations

Using FCR as a mark of viable cells, it was reliable that large number of living sperms have been isolated from the pollen of all three species. We did not take photographs of *Secale cereale* sperms owing to their rapid decrease of viabil ity.

In intact pollen grains, the sperms possessed their own shapes characteristic of each species. The sperms just discharged from the pollen grains could maintain their original shape for merely a short time and then turned to spherical bodies. Plate Ⅰ, 1 shows a pair of ovoid *B. napus* sperms near the pollen grain releasing them. After filtration and centrifugation, most of sperms rounded as in Plate Ⅰ, 3. Nevertheless, a few of isolated sperms maintained their original shape even through the whole isolation procedure (Plate Ⅰ, 2). In *Z. mays* intact pollen the sperms were elongated form (Plate Ⅰ, 4). After isolation procedure most of the sperms rounded (Plate Ⅰ, 6) with still a few exceptions in which the original form was maintained (Plate Ⅰ, 5). Maintenance of shape was better in more concentrated medium (e. g. with 13% sorbitol) than in less concentrated one (e. g. with 8% sorbitol), indicating that osmotic pressure of the medium played a role, though limited, in sperm shape maintenance.

The two sperms just liberated from one pollen grain connected with each other for a while. Then the connection was cut off and the sperms became scattered singly in the medium. It was interesting that some sperms still kept in pairs even when they had turned to spherical bodies (Plate Ⅰ, 3 and 6).

After being mounted under a coverslip, the isolated sperms could no longer be viable. In contrast, when stored in suspension at low temperature, the viability was maintained much longer. The sperms of *B. napus* survived after 24,48 and 72 hours of storage. The sperms of *Z. mays* survived for at least 24 hours. Plate Ⅰ, 7 and 8 show FCR of isolated sperm cell population of these two species respectively. Plate Ⅰ, 9 and 10 show nuclei of sperm cell population of Z. *mays* visualized by Hoechst 33258 staining. The sperms of S. *cereale* could only survive no longer than 30 min.

Discussion

Methods for isolation of sperms from tricellular pollen include osmotic shock (Cass & Fabi, 1988; Dupuis et al., 1987; Russell & Cass, 1981) and grinding (Cass et al., 1986; Southworth & Knox, 1987). However, to date only a few of the reports confirmed the viability of isolated sperms by FCR test (Dupuis et al., 1987; Matthys-Rochon et al., 1987). It seems that failure of FCR in other works may be due to technical causes rather than species specificities. In previously reported works the medium in which sperms were isolated was either sucrose solution or Brewbaker and Kwack medium. It is doubtful that such medium developed for pollen germination offers optimal conditions also for isolated sperms. For instance, high concentration of boric acid contained in Brewbaker and Kwack medium favoring pollen tube growth is not likely to be suitable for sperms. Logically, more elaborated medium for protoplast culture may yield better result. In our experiments, Roberts et al. medium used for isolation of B. *napus* sperms contained lower concentration of boric acid as compared with Brewbaker and Kwack's. For isolation of Z. *mays* sperms, K_3 medium developed for protoplast culture showed better effect than Brewbaker and Kwack's did. The conditions for sperm storage will be a further subject worth research, just as it has been done for several decades with pollen storage, but on a new level.

Technique of FDA test may be another critical problem. In the present paper, a method formerly developed for labeling pollen tubes (Yang, 1986) was adopted: FDA was added into the medium for grinding and sperm storage. The sperm cells labeled in such manner were stored as a pool. Samples were taken at intervals from the pool and monitored microscopically. This not only greatly facilitated the test procedure but might also minimize the damage of sperm viability. Based on FCR, only living sperms show typical fluorescence; the dead sperms and other particles do not. Hence FCR can be used as an excellent probe for quick distinguishing viable sperms, which is difficult by using the negative reaction of Evans blue staining.

(Authors: Yang HY, Zhou C. Published in: Chinese Journal of Botany, 1989, 1:80-84. with 10 figures and 14 references, all omitted here. See color plate IV in the end of this book.)

2 紫菜苔精细胞的大量分离和生活力保存

Mass isolation and preservation of viable sperm cells in *Brassica campestris* var. *purpurea*

Abstract

The two-step osmotic shock and grinding methods reported by Yang and Zhou (1989) were modified for isolation of viable sperm cells in large quantities from pollen grains of *Brassica campestris* var. *purpurea*. Factors affecting the yield and survival of isolated sperm cells have been investigated. These included physiological status of donor flowers, sucrose concentration used for pollen hydration, basic media, protectants and osmotica supplemented in the medium etc. As a result, two procedures have been developed. For osmotic shock method, pollen grains at the day of anthesis were hydrated in 25% sucrose solution for 30 min and, after centrifugation and removal of the supernatant, the pellet was shocked by a medium containing 12.5% sucrose, 0.1g/L KNO_3, 0.36 g/L $CaCl_2 \cdot 2H_2O$, 0.3% potassium dextran sulphate (PDS), 0.6% bovine serum albumin (BSA), and 0.3% polyvinylpyrrolidone (PVP). The viable sperm yield was 34%. After removal of pollen wall debris by filtration and centrifugation, the sperm cell-rich pellets were resuspended in a medium containing 20% sucrose, 5% sorbitol, 0.1g/L KNO_3, 0.36g/L $CaCl_2 \cdot 2H_2O$, 0.6% BSA and 0.3% PDS, and preserved at 4℃ for two days. For grinding method, the pollen grains hydrated in 30% sucrose solution for 30 min. were resuspended in a medium containing 20% sucrose, 5% sorbitol, 0.1g/L KNO_3, 0.36g/L $CaCl_2 \cdot 2H_2O$, 0.3% PDS, 0.6% BSA, 0.3% PVP and 20 μg/ml fluorescein diacetate, then ground with a glass homogenizer to release the sperm cells. The viable sperm yield was up to 86%. Following filtration and centrifugation for removal of pollen wall debris, the sperm cells were stored at 4℃ in the same medium but without supplementation of PVP. Tested by fluorochromatic reaction, the sperm cells could survive up to one week with a gradual decline of viability. Cytological observations revealed that pairs of ellipsoidal sperm cells just released were linked together; one of the pair had a long tail-like extension which also show fluorochromasia. Soon after, the sperm cells separated and turned to be spherical. The present results open a prospect to use isolated viable sperm cells for further experimental manipulations.

精细胞作为植物的雄配子在生殖过程中占有重要地位。以往被子植物精细胞的研究仅局限于体内原位观察。从花粉粒或花粒管中分离精细胞的成功,在受精作用基础理论研究和生物工程应用两方面均具有重要的意义[4,9,18]。最近几年,精细胞的分离有了长足的进展。迄今已从9科16种植物的花粉粒或花粉管中分离出精细胞[4,10,14,19~24]。然而,在这些分离成功的物种中,很多缺乏关于分离率的报道,并且,仅在部分研究中采用FDA染色技术证实分离的精细胞具生活力,而在其它研究中仅用了伊凡斯蓝负反应或未做生活力测定[4]。至于分离的精细胞的离体保存,虽有初步尝试[7,22,24],但至今尚未见到有关保存条件的准确研究资料。因此,精细胞分离率的提高和生活力保存仍是亟待进一步研究的重要课题。

本工作在以前研究[24]的基础上进一步建立了紫菜苔(*Brassica campestris* var. *purpurea*)精细胞大量分离及其生活力保存的适宜技术,并研究了影响分离率和生活力的若干因素。

材 料 与 方 法

一、试验材料

紫菜苔9月播种于武汉大学生物系试验地,1月开花。

二、分离和保存方法

将新鲜花粉置于一定浓度(见后文)的蔗糖液中制成悬浮液,在28℃、黑暗条件下温育20~40min。然后按下述两种方法释放精细胞:(1)渗透压冲击法。参考原有的"一步法"[17]和"二步法"[24]建立了下列改良方法:先将经过适时温育的花粉悬浮液离心去上清液,再倾入含保护剂的低渗培养基进行冲击,使渗透压骤降,导致花粉破裂释放精细胞。然后通过离心沉淀,将精细胞转移到含保护剂的等渗培养基中。所用的保护剂包括葡聚糖硫酸钾(potassium dextran sulphate,简称PDS)、牛血清白蛋白(bovine serum albumin,简称BSA)和聚乙烯吡咯烷酮(polyvinylpyrrolidone,简称PVP)。(2)研磨法。参考以前的方法[24]加以改进,即将温育后的花粉离心沉淀,悬浮于含保护剂的培养基中,用玻璃匀浆器轻轻研磨。将得到的含大量精细胞的混合物直接置于冰箱(4℃)保存(便于统计),或先依次通过400目不锈钢网和20μm、10μm孔径尼龙筛网过滤,以除去未破裂的花粉和花粉壁残渣,收集滤液,再置于冰箱(4℃)保存。定时取样检测生活力,统计保存率变化。

三、显微观察

在上述分离和保存试验的每个环节中均按前法[24]采用荧光素二醋酸酯(FDA,20 μg/ml)测定生活力;采用 DNA 荧光染料 Hoechst 33258(20μg/ml)结合冬青油透明技术[2]观察花粉粒中的细胞核。用 AO Fluorestar 110 型落射荧光显微镜的 2073 滤光组合(激发滤光片 436,分色镜 450,阻挡滤光片 475)观察与摄影。

四、统计方法

在荧光显微镜下随机观察 5 个视野中的花粉与精细胞,按下列公式计算各项指标:

$$花粉破裂率 = \frac{破裂花粉数}{总花粉数} \times 100\%$$

$$生活精细胞分离率 = \frac{生活精细胞数}{总花粉数 \times 2} \times 100\%$$

$$精细胞相对存活率 = \frac{\frac{存活精细胞数}{破裂花粉数 \times 2}(保存 t 时间后)}{\frac{生活精细胞数}{破裂花粉数 \times 2}(刚分离时)} \times 100\%$$

以刚分离时的精细胞存活率为 100%。

结　果

一、分离方法的比较

渗透压冲击法　"一步法"[17]难以导致花粉破裂释放精细胞。应用"二步法"[24],试验了 10%~40% 范围内的起始蔗糖浓度,表明 25% 浓度下的分离效果略好,花粉破裂率为 27%,生活精细胞分离率最高为 11%,分离的精细胞在低渗的蔗糖溶液中存活不到 1h;即使分离后马上转移到下述含保护剂的等渗培养基中低温保存,10h 后存活 34%, 28h 后仅为 8%,36h 后全部死亡。采用本文的改良方法,成熟花粉的破裂率提高到39%,生活精细胞分离率提高到34%。精细胞相对存活率在保存10h后为82%,28h为64%,保存时间最长可达48h。以上结果表明,改良方法中生活精细胞的分离率和保存率均有明显提高。

研磨法 首先比较了三种方法:一是花粉不经水合直接研磨;二是花粉在培养基中水合后研磨;三是花粉先在等渗蔗糖液中水合后再在培养基中研磨。结果表明生活精细胞的分离和保存效果均以第三种方法最佳。在此基础上,应用第三种方法再比较了 Roberts(1983)[15]、$BK^{[16]}$ 和 $K_3^{[13]}$ 等三种基本培养基对精细胞分离和保存的影响。结果表明它们之间差异不显著,而以不含硼酸的 Roberts 培养基效果略好,说明盐离子成分的差异对粗细胞生活力影响不大。因此,以后的实验均采用去硼酸的 Roberts 培养基。

二、影响精细胞分离和保存的因素

花粉生理状况 花粉生理状况对分离和保存影响很大。以盛花期的晴天早晨取即将开放的新鲜花朵试验效果为好,阴雨天取样不利于分离;在开花尾期,花粉粒生活力降低,精细胞得率亦低,其生活力亦易丧失。花粉发育时期亦是关键因素,充分成熟的花粉效果好,而较幼嫩的花粉较差。例如,开花前一天的花粉,应用前述改良的渗透压冲击法,在25%蔗糖液中需水合1h,花粉破裂率最高为21%,生活精细胞分离率仅为13%,分离保存10h后仅存活11%,15h后全部死亡。应用改良的研磨法,花粉破裂率最高为53%,生活精细胞分离率亦仅为26%,生活力最长只能维持一天。

水合液中的蔗糖浓度 水合液中的起始蔗糖浓度对精细胞的分离也有影响,尤以应用渗透压冲击法时更为明显。从表1中可看出,起始蔗糖浓度过低(如10%),虽然花粉破裂率高,但生活精细胞分离率却很低。这可能由于精细胞刚释放时处于过于低渗的溶液中易于遭受损害之故。起始蔗糖浓度过高(如40%),花粉不易破裂,因而释放的精细胞亦少。以25%的起始蔗糖浓度水合效果最佳。若冲击时蔗糖浓

度的骤降幅度低于一半,则30%的起始蔗糖浓度也有效,但过剧地降低浓度对精细胞生活力有不利影响。应用改良研磨法时则以30%蔗糖液水合较理想。两种方法的水合时间一般均为30min。

保护剂 在两种方法中均试验了三种保护剂对分离率和生活力的影响,结果大致相同。例如,应用改良的渗透压冲击法(表2),添加保护剂对生活精细胞的分离和保存均有一定的促进作用。三种保护剂单独使用时,其效果有一定差异,依次为PVP>BSA>PDS,而以三者混合使用效果最好。但如在分离介质中使用三种保护剂而在保存液中去掉PVP,其保存效果比保存液中含PVP者更好。添加保护剂的促进作用在改良研磨法中表现更为明显。总结表2和表3结果,可以认为分离紫菜苔生活精细胞以在研磨介质中添加PDS、BSA和PVP三种保护剂而保存液中去掉PVP效果最好。

渗透压调节剂 不同渗透压调节剂对生活精细胞分离率和生活力保存的效果有明显差异。在渗透压冲击法和研磨法中结果相近,现仅以改良研磨法为例(表4)。在0.6mol/L左右的浓度下,试验了蔗糖、山梨醇、甘露醇三种成分的5种组合。结果表明,生活精细胞的分离率高低顺序依次为21%蔗糖>5%蔗糖+8%山梨醇>11%山梨醇>5%蔗糖+8%甘露醇>11%甘露醇,生活力保存时间长短顺序为11%山梨醇>21%蔗糖>5%蔗糖+8%山梨醇>5%蔗糖+8%甘露醇>11%甘露醇。在0.88mol/L左右的浓度下,试验了上述三种成分的三种组合。结果表明,30%蔗糖和20%蔗糖+5%山梨醇组合对生活精细胞的分离和保存效果基本相近,以后者略好,它们均明显高于20%蔗糖+5%甘露醇组合。综合上述两种渗透压条件下的对比试验结果,说明蔗糖和山梨醇作为渗透压调节剂既有利于生活精细胞的分离也有利于生活力的保存,并以20%蔗糖+5%山梨醇或30%蔗糖两种条件最佳。生活精细胞的分离率最高达86%,生活力保存最长可达一周。甘露醇作为渗透压调节剂不利于紫菜苔生活精细胞的分离和保存。

表3 保护剂对精细胞分离率和保存率的影响(改良研磨法)*

Table 3 Effects of protectants on isolation and preservation of sperm cells (modified grinding method)*

保护剂 Protectant	观察花粉粒数 No. of pollen grains observed	生活精细胞数 No. of viable sperms	生活精细胞分离率 Viable sperm yield (%)	精细胞相对存活率 Viable sperm relative survival (%) 保存天数 Storage duration (days)							
				0	1	2	3	4	5	6	7
对照① Control	419	226	27	100	67	31	7	0	0	0	0
0.3% PDS	433	294	34	100	73	44	11	0	0	0	0
0.6% BSA	471	367	49	100	76	59	37	8	0	0	0
0.3% PVP	449	521	62	100	79	68	51	37	16	0	0
0.3% PDS+ 0.6% BSA+ 0.3% PVP	458	760	83	100 (100)	83 (87)	72 (79)	63 (71)	47 (62)	24 (45)	3 (21)	0 (4)

* 30%蔗糖液中水合30min后研磨。

①对照组合 0.1g/L KNO$_3$、0.36 g/L CaCl$_2$·2H$_2$O 和 30%蔗糖。括号中数据为保存液中不含 PVP 时的结果

* The pollen were hydrated for 30 min in 30% sucrose solution before grinding in the medium.

①The control contained 0.1 g/L KNO$_3$, 0.36 g/L CaCl$_2$·2H$_2$O and 30% sucrose. The data in brackets were obtained when PVP was deleted from the preservation medium.

通过上述比较研究,总结出紫菜苔生活精细胞大量分离和生活力保存的技术流程如下:

渗透压冲击法:成熟花粉在25%蔗糖液中于28℃下水合30min ——→ 离心沉淀——→用含12.5%蔗糖、0.1g/L KNO$_3$、0.36g/L CaCl$_2$·2H$_2$O、0.3% PDS、0.6% BSA 和0.3% PVP 的低渗液冲击——→将释放的精细胞及

时转移到含 20% 蔗糖、5% 山梨醇、0.1g/L KNO$_3$、0.36g/L CaCl$_2$·2H$_2$O、0.3% PDS 和 0.6% BSA 的培养基中,置于冰箱(4℃)冷藏。

研磨法:成熟花粉在 30% 蔗糖液中于 28℃下水合 30min ——→离心沉淀——→在含 20% 蔗糖、5% 山梨醇、0.1g/L KNO$_3$、0.36g/L CaCl$_2$·2H$_2$O、0.3% PDS、0.6% BSA 和 0.3% PVP 的介质中研磨——→将释放的精细胞转移到不含 PVP 的上述培养基中,置于冰箱(4℃)冷藏。

三、细胞学观察

成熟花粉经 Hoechst 33258 染色和冬青油透明后,可以显示其中一对精核(图版Ⅰ,1)。采用 FDA 荧光反应作为生活细胞的标志,可以清楚地观察到生活精细胞从花粉粒中释放出来的过程及其随后的形态学变化。刚由一粒花粉中分离出来的两个姊妹精细胞是成对相连的。精细胞形状略长,其中一个精细胞的一端具长尾状延伸物(图版Ⅰ,2)。经 FDA 处理后,分离的精细胞呈均匀的绿色荧光,而且直达尾部,说明尾部的细胞质亦呈生活状态(图版Ⅰ,3)。以后,它们逐渐变为椭圆球形,仍维持成对状态(图版Ⅰ,4)。然后联系中断而单个地分散于介质中(图版Ⅰ,5),最终变为彼此分开的圆球形(图版Ⅰ,6)。图版Ⅰ,7示经全部分离程序后的生活精细胞群体。

讨 论

精细胞分离与保存技术的研究是当今植物生殖生物学和生殖细胞工程的前沿课题之一[4,9]。迄今已建立的由三细胞型花粉中分离精细胞的方法包括渗透压冲击法和研磨法[4,18]。从细胞工程的角度看,分离精细胞的关键在于创造一种最佳条件以达到提高分离率和保存生活力的目的。已有的渗透压冲击法(无论是"一步法"[17]还是"二步法"[24])中都使刚释放出来的精细胞处于缺乏保护剂的仅含蔗糖的低渗溶液中,看来这不是精细胞在离体条件下维持生活状态的最佳条件。基于上述考虑,我们改良了现有的二步渗透压冲击法,即将经过蔗糖液水合的花粉转入含多种保护剂的低渗培养基中冲击,再将释放出来的精细胞及时转移到含保护剂的等渗培养基中保存。试验表明,分离条件不仅关系到精细胞分离率的高低,而且也关系到以后的生活力保存。这一方法具有两个优点:一是使精细胞刚一释放就立即处于含多种保护剂的培养基中,从而减少损伤;二是低渗培养基的直接加入可能比以前加等量蒸馏水的方法能够在花粉群体内造成更均匀的渗透压骤降。在此基础上,我们对原先的研磨法[24]也进行了适当的改进,建立了先在等渗蔗糖液中水合再在含保护剂的培养基中研磨的改良方法。

保护剂的加入对生活精细胞的分离和保存有显著的促进作用。Southworth 和 Knox(1988)[18]曾经指出保护剂在精细胞分离中可能具有重要的作用。他们通过在研磨介质中加入 PVP 和 BSA 等保护剂成功地分离出非洲菊的精细胞[19]。杨弘远与周嫦(1989)在油菜与玉米精细胞分离和保存中应用 PDS 作为保护剂[24]。PDS 在原生质体分离中是常用的保护剂,其作用被认为是降低酶液中的核糖核酸酶活性,增强膜的稳定性[1]。在精细胞分离方面,PDS 可能会降低营养细胞破裂后释放的核糖核酸酶等活性而对精细胞起一定的保护作用。BSA 和 PVP 在细胞器分离等生理生化研究中早已得到广泛应用,一般认为 PVP 有助于对抗破碎组织释放出来的多酚类物质的有害作用[3]。Southworth 和 Knox(1988)[18]认为 PVP 和 BSA 均作为脂类结合物而对精细胞起保护作用。本试验中精细胞保存液中以不加 PVP 为宜,其原因之一可能是由于 PVP 是一种芳香环化合物,它的长期作用可能对精细胞具有一定的毒害。

关于精细胞具尾状延伸物的现象,已有较多报道[5]。最近在非洲菊[19]等植物的精细胞分离过程中也观察到。但用 FDA 检测证明精细胞具生活状态的长尾,则是本文首次报道。在紫菜苔中,来自同一粒花粉的一对姊妹精细胞中只是一个精细胞具尾,这同 McConchie 等(1985)[12]对同种植物(不同变种)成熟花粉粒中的一对精细胞的计算机辅助三维重构研究结果是一致的。若应用前文[24]方法,在紫菜苔精细胞分离中难于观察到具尾状延伸物现象。这也表明本文建立的分离条件更为合适,能使分离的精细胞在更长时间内维持体内原状。

关于分离的精细胞的生活力保存,现有文献中只有零星报道而缺乏翔实的实验资料[7,8,11,16,17,22,24]。我们的研究结果在迄今用冰箱(4℃)保存精细胞的研究中,保存效果较佳。考虑到本实验中没有采用无菌操作,精细胞在冰箱中保存一周可能会有菌类滋生,推测如果在无菌条件下保存,可能还会延缓生活力的下降速度。此外,应用超低温方法保存精细胞,也值得进一步试验。

图 版 说 明

除图 2 为明视野摄影外,其余均为荧光显微摄影。其中,图 1 为 Hoechst 33258 染色和冬青油透明,其余为 FDA 染色。

1. 一粒花粉中的一对精核(营养核不在此焦面上)。×800　**2、3.** 刚分离的一对精细胞,其中一个精细胞具长尾状延伸物(2 和 3 分别为同一材料的明视野和荧光摄影)。×2250　**4.** 一对椭圆球形精细胞。×2250　**5.** 另一对椭圆球形精细胞。×2250　**6.** 一对圆球形精细胞。×2250　**7.** 生活精细胞群体。×250

Explanation of plate

All photos were taken by fluorescence microscopy except Fig. 2 which was a bright-field image. The sperm cells were stained with Hoechst 33258 and cleared with methyl salicylate in Fig. 1 and stained with FDA to show fluorochromacia in Figs. 3-7.

Fig. 1. Two sperm nuclei within a mature pollen grain (vegetative nucleus out of focus). ×800　**Figs. 2、3.** A pair of sperm cells just released and still linked together visualized by bright-field and fluorescence microscopy respectively. One of them has a long tail-like extension. ×2250　**Fig. 4** A pair of ellipsoidal sperm cells still linked together. ×2250　**Fig. 5.** A pair of ellipsoidal sperm cells detached from each other. ×2250　**Fig. 6.** A pair of spherical-shaped sperm cells. ×2250　**Fig. 7.** A population of isolated viable sperm cells. ×250

(作者:莫永胜、杨弘远。原载:植物学报,1991,33(9):649~657。

图版1幅,表4幅,参考文献24篇,仅选录图版Ⅰ,表3。)

图版 I　　　　　　　　　　　　　　　　　　　　　　　　　　　　　　　Plate I

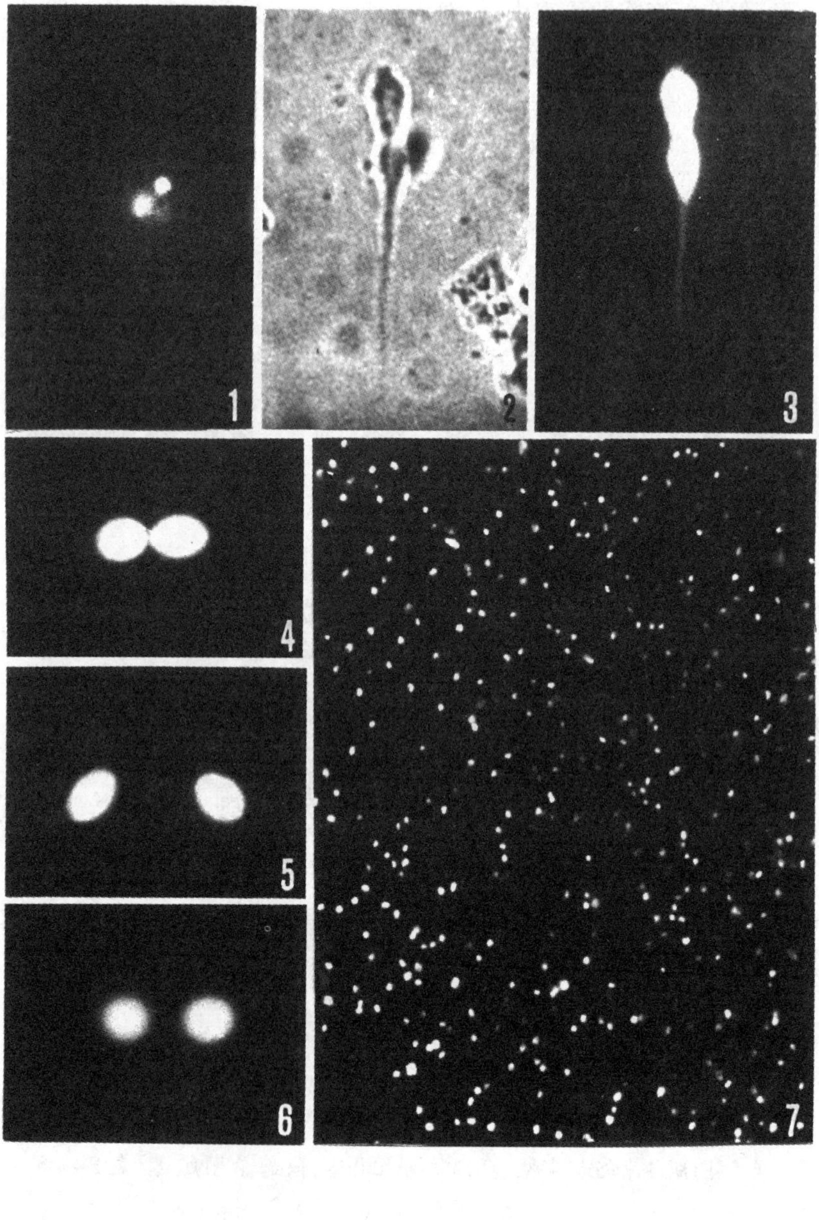

3 几种具二细胞型花粉植物精细胞的分离和融合

Isolation and fusion of sperm cells in several bicellular pollen species

Abstract

Living sperm cells were isolated in large quantities from the pollen tubes, grown by the "*in vivo-in vitro* technique" in 8 bicellular pollen species belonging to 5 families. An "osmotic shock-weak enzyme treatment" method could effectively release sperms from pollen tubes and favor subsequent purification. The viable sperm yields were up to 82.9% in *Zephyranthes candida* and 78.2% in *Hemerocallis minor*. Fusions were successfully induced by polyethylene glycol (PEG) according to the "small-scale fusion" procedure in various combinations, viz., between the same sperm cells in 5 species, between sperm cells of *Gladiolus gandavensis* and *Hippeastrun vittatum*, between sperm cells and microspore protoplasts in *Hemerocallis minor*, and between sperm cells of *H. vittatum* and microspore protoplasts of *Hemerocallis fulva*. Test with fluorochromatic reaction, more than 85% of the fusion products of sperm cells in *Z. candida* were viable. The yield of viable fusion products between sperm cells and microspore protoplasts in *Hemerocallis minor* was about 75% and half of them could survive after culture for 24h. The induction of fusion between sperm cells and petal protoplasts in *G. gandavensis* by a combined PEG-dimethyl sulfoxide (DMSO) treatment was investigated in detail. About 90% of the fusion products thus obtained were viable. Several critical factors affecting the fusion efficiency were studied. These included the ratio of sperm cell number to petal protoplast number in the mixture, concentrations of PEG and DMSO, and duration of incubation in the inducing solution. It appeared that addition of DMSO could significantly increase the fusion frequency, and that there may be a synergistic effect between PEG and DMSO. This is the first attempt to use isolated sperm cells for fusion studies in bicellular pollen species.

精细胞的离体实验操作是当前植物性细胞生物学和细胞工程研究的前沿课题[6,12,22]。迄今已从7科13种具三细胞型花粉的植物中分离出大量精细胞[6,22]。在二细胞型花粉植物方面,分离精细胞的工作相对薄弱,已建立了两种分离方法:一种是由人工萌发的花粉管中分离,如百合[21]和蓝猪耳[13];一种是"活性-离体技术(*in vivo-in vitro* technique),如杜鹃花和唐菖蒲[20]。后法在获得均一的精细胞群体等方面比较优越。精细胞的大量分离成功为进一步的实验操作奠定了基础。关于分离的精细胞的融合研究,直至最近仅在三细胞型花粉植物玉米中有精卵体外融合等报道[14,15]。本文报道我们在5科8种具二细胞型花粉的植物中,开展的生活精细胞分离和分离的精细胞间及其与其它原生质体间融合的实验结果。

材料与方法

一、实验材料

石蒜科的玉帘(*Zephyranthes candida* Herb.)和朱顶红(*Hippeastrum vittatum* Herb.),鸢尾科的鸢尾(*Iris tectorum* Maxim.)和唐菖蒲(*Gladiolus gandavensis* Van Houtte),百合科的黄花菜(*Hemerocallis minor* Mill.)和萱草(*H. fulva* L.),茄科的烟草(*Nicotiana tabacum* L.)和锦葵科的陆地棉(*Gossypium hirsutum* L.)。

二、精细胞的分离和纯化

精细胞分离按"活体-离体技术"[20]进行。先在整朵花上对柱头进行人工授粉,当花粉管在花柱中生长到一定程度后,切下1cm左右的花柱,将切口插入液体培养基,使花粉管从切口处长出,在培养基中继续生长(图版Ⅰ,1)。不同植物有各自适合花粉管生长的人工培养基:萱草、黄花菜、玉帘、朱顶红和鸢尾分别为含7%、8%、15%、15%和30%蔗糖的BK培养基[10],唐菖蒲沿用Shivanna等(1988)的培养基[20],烟草为含10%蔗糖和0.01%硼酸的水溶液[2],棉花为Barrow培养基[9]。当花粉管中已形成精细胞后,将众多的花粉管从基部切断,然后采用渗透压冲击、酶解或两者结合的方法从花粉管中释放精细胞。渗透压冲击液和酶解液中均添加葡聚糖硫酸钾(PDS)、牛血清白蛋白(BSA)和聚乙烯吡咯烷酮(PVP)等保护剂[8]。将含精细胞的混合物通过筛网过滤,除去破裂的花粉管残渣,再进行间断密度梯度离心[3]而得到相对纯化的精细胞群体。

三、精细胞的融合

萱草和黄花菜小孢子原生质体的分离和纯化基本上按周嫦(1989)的方法[5],取材为单核期的花粉。唐菖蒲花瓣原生质体的制备按Power和Chapman(1985)[18]所述方法,但酶组分改为1%纤维素酶和1%果胶酶。按"小规模融合法"[18]用聚乙二醇(PEG,分子量6000)诱导融合。唐菖蒲精细胞与花瓣原生质体的融合试验中,在融合液中添加适宜浓度的二甲基亚砜(DMSO)。

四、显微观察

花粉管、分离的小孢子原生质体和花瓣原生质体经FPA或2%戊二醛固定后,用DAPI荧光染色鉴定细胞核[7]。精细胞和融合体的生活力测定均用荧光素二醋酸酯(FDA,20μg/ml)处理。精细胞先用DAPI(10μg/ml)活染标记,充分洗涤后再进行融合,便于观察融合过程中的核动态和判断融合体。所有材料均用Olympus IMT-2倒置显微镜或Olympus NEW VANOX AHBS-514显微镜的荧光(FDA用B激发,DAPI用U激发)、干涉差或明视野装置观察和摄影。

五、统计

$$\text{生活精细胞分离率} = \frac{\text{生活精细胞数}}{\text{花粉管总数} \times 2} \times 100\%$$

$$\text{精细胞融合率} = \frac{\text{融合的精细胞数}}{\text{精细胞总数}} \times 100\%$$

$$\text{精细胞与小孢子或花瓣原生质体的融合率} = \frac{\text{摄取精细胞的原生质体数}}{\text{原生质体总数}} \times 100\%$$

结　　果

一、精细胞分离

应用"活性-离体技术"在所试验的 8 种植物中,均能从花柱切口顺利地长出花粉管,并在各自适宜的培养基中继续生长(图版Ⅰ,2)。不同植物从授粉到由切口长出花粉管所需的时间大约为:鸢尾 8h、唐菖蒲 12h、棉花 15h、朱顶红和烟草 20h、黄花菜和萱草 24h、玉帘 32h。精细胞的分离方法起初采用 Shivanna 等(1988)[20]的渗透压冲击法和酶解法。观察到精细胞以两种方式从花粉管中释放:一是花粉管顶端或亚顶端侧面破裂而释出精细胞(图版Ⅰ,3);二是花粉管细胞质与精细胞向花粉管切口倒流逸出(图版Ⅰ,4)。在渗透压冲击法中,两种方式的频率相近,而在酶解法中以第一种方式为主。渗透压冲击法仅在唐菖蒲和鸢尾中获得较高的生活精细胞分离率,而在其余植物中效果不好。酶解法有三个问题:一是花粉管释放精细胞不够同步;二是释放的精细胞易与花粉管细胞质粘结,不利于纯化[3];三是会释放少量花粉管亚原生质体[2]而与精细胞混淆。在两种方法中,添加 PDS、BSA 和 PVP 等保护剂[8]有利于保持精细胞的生活力,从而提高生活精细胞分离率,如在唐菖蒲中应用渗透压冲击法分离时,能从 46.3% 提高到 71.9%。

以后,参考分离生殖细胞的低酶法[3],建立了低渗-低酶法:将蔗糖浓度降为相应培养基中的一半,并采用较低的酶浓度(0.2%纤维素酶和0.2%果胶酶)。在5种植物中比较了各种分离方法的效果(表1),从总体上看,低渗-低酶法效果较好,尤其在玉帘和黄花菜中能大幅度提高生活精细胞的分离率。

经纯化步骤后,在玉帘、鸢尾、唐菖蒲、朱顶红和黄花菜中均获得了相对纯化的精细胞群体(图版Ⅰ,5~7;图版Ⅱ,11),而在萱草、棉花和烟草中纯化未获成功。FDA检测表明上述5种植物纯化后的精细胞具有生活力(图版Ⅰ,8,9;图版Ⅱ,10)。卡宝品红染色显示,精细胞的胞质贫乏,核质比大,类似核质体(图版Ⅱ,12)。

二、精细胞的融合

在融合试验中,除唐菖蒲精细胞按渗透压冲击法分离外,其余植物均按低渗-低酶法分离。

同种植物精细胞之间的融合 玉帘、鸢尾、唐菖蒲、黄花菜和朱顶红的同种精细胞之间均易融合形成同核体。在PEG的诱导下,精细胞逐渐接触和粘连(图版Ⅱ,13~15)。加洗涤液后,质膜很快融合而形成一个完整细胞(图版Ⅱ,16),在融合体中可清楚地观察到事先被DAPI标记的两个精核(图版Ⅱ,17)。FDA检测证明,玉帘中约80%以上的融合体均有生活力。以玉帘为材料,探讨了影响精细胞融合的因素。结果表明,细胞密度对融合率有很大影响:密度过低时(1×10^4个细胞/ml),细胞粘连少,二胞融合率低于15%;密度过高时(1×10^6个细胞/ml),细胞聚集快,粘连多,融合率高,但易形成多胞融合体,二胞融合率最高仅18%;适宜的细胞密度为1×10^5个细胞/ml,二胞融合率最高,可达26%。比较了5%、10%、20%、30%、40%、50%6种PEG浓度对融合率的影响:浓度为5%时,细胞聚集慢,粘连少,融合率低于5%;浓度高于40%时,则易引起许多精细胞粘连成团而形成多胞融合体,并且对精细胞及其融合体的生活力也有损害。从提高二胞融合率及融合体生活力的角度考虑,以10%~30%的PEG浓度较适宜。在PEG中的保温时间对融合效果也有影响。以15%和25%两种PEG浓度比较了不同保温时间对融合率的影响。结果表明:在15%浓度下,

表 1　　不同分离方法对生活精细胞分离率的影响

Table 1　Effect of isolation methods on viable sperm yield

材料 Materials	唐菖蒲 Gladiolus gandavensis			鸢尾 Iris tectorum			朱顶红 Hippeastrum vittatum			玉帘 Zephyranthes candida			黄花菜 Hemerocallis minor		
分离方法[①] Isolation methods	A	B	C	A	B	C	A	B	C	A	B	C	A	B	C
花粉管数 No. of pollen tubes	374	346	417	507	441	539	437	529	495	546	469	490	332	434	401
生活精细胞数 No. of viable sperms	538	261	362	518	485	511	228	335	566	253	520	812	256	393	627
生活精细胞分离率 Viable sperm yield (%)	71.9	37.7	43.4	51.1	55.0	47.4	26.1	31.7	57.2	23.1	55.4	82.9	38.6	45.3	78.2

①A. 渗透压冲击法(含降低一半浓度的蔗糖);

B. 酶解法(1%纤维素酶+1%果胶酶);

C. 低渗-低酶法(含降低一半浓度的蔗糖,0.2%纤维素酶+0.2%果胶酶)。三种方法中均添加相同的保护剂[8]。

A. Osmotic shock (half strength sucrose of the correspondent germination medium);

B. Enzyme treatment (1% cellulase + 1% pectinase);

C. Osmotic shock/weak enzyme treatment (half strength sucrose of the correspondent germination medium, 0.2% cellulase + 0.2% pectinase). Protectants[8] were supplemented in all the three methods.

精细胞粘连较慢,保温 30 min 后,二胞融合率达最高值(27.9%);在 25% 浓度下,精细胞粘连较快,保温 20min 后二胞融合率即达到最高值 (26.1%)。保温时间继续延长,二胞融合率反而降低。

异种植物精细胞之间的融合 试验了朱顶红和唐菖蒲异种精细胞之间的融合。朱顶红精细胞的直径约为唐菖蒲的 1.5 倍,且细胞质较淡。在 PEG 诱导下,二者粘连(图版Ⅱ,18)。加入洗涤液后,二者紧密相贴,但细胞质差异仍清楚易辨(图版Ⅱ,19)。融合继续进行,二个精细胞间仍可看到有一接触线(图版Ⅱ,20)。最后接触线消失而融合完成(图版Ⅱ,21)。异核体中可看到大小不同的两个异种精核(图版Ⅱ,22)。双体异核融合率最高达 7.8%。

精细胞与小孢子原生质体的融合 在黄花菜精细胞与同种小孢子原生质体和朱顶红精细胞与萱草小孢子原生质体之间的二组融合试验中,精细胞均能被小孢子原生质体摄取。用 DAPI 标记精细胞,与未染色的小孢子原生质体融合,可追踪融合过程中核的动态。如在黄花菜中,融合前精核呈现明亮荧光而小孢子原生质体核无荧光(图版Ⅲ,27、30)。精细胞同小孢子原生质体粘附时,也只有精核具荧光(图版Ⅲ,28、31)。当精细胞被摄入小孢子原生质体后,在异核体中可看到呈明亮荧光的精核,在很多场合下也观察到原来未染色的小孢子原生质体核有荧光(图版Ⅲ,29、32)。未融合的小孢子原生质体的核从不显示荧光。因而可以推测异核体中的小孢子核之所以显示荧光,是由于融合过程中被精细胞扩散出来的 DAPI 染色所致。因此,在融合后,根据小孢子原生质体是否具荧光核即可判断它是否摄取了精细胞。这些现象在 DAPI 标记的朱顶红精细胞与未染色的萱草小孢子原生质体的融合中亦观察到(图版Ⅱ,23~26)。在黄花菜中,小孢子原生质体摄取精细胞的总频率接近 3%。FDA 检测表明,约占 75% 的融合体具生活力。融合产物在小孢子原生质体培养基[5]中培养 24h 后,约有 50% 的异核体仍然生活,但尚未观察到异核体中发生核融合的迹象。

精细胞与花瓣原生质体的融合 唐菖蒲精细胞与花瓣原生质体二者差异显著。和精细胞相比,花瓣原生质体不仅细胞与核大得多,并且具有含花色素的大液泡。加入 PEG 溶液后,花瓣原生质体略有收缩,事先用DAPI标记的精细胞核呈明亮荧光而花瓣原生质体核没有荧光(图版Ⅲ,33、36)。洗涤时,花瓣原生质体摄取所粘附的精细胞,此时也只观察到精核具荧光(图版Ⅲ,34、37)。但在融合完成后的异核体中,除精核呈现明亮荧光外,花瓣原生质体核往往也显示荧光(图版Ⅲ,35、38)。没有摄取精细胞的花瓣原生质体的核均不呈现荧光。

进一步研究了影响花瓣原生质体摄取精细胞频率的若干因素:(1)精细胞与花瓣原生质体的比例:增大精细胞的比例(如30∶1或50∶1),可以明显地提高花瓣原生质体摄取精细胞的频率。就摄取一个精细胞(即双体异核体)而言,以30∶1的比例为适宜(表2)。(2)PEG与DMSO组合:在PEG融合液中添加一定的DMSO(表3),能明显地促进花瓣原生质体对精细胞的摄取,但浓度过高时,精细胞的生活力受损,异核体频率反而降低。PEG和DMSO似乎存在协同作用:PEG浓度较低(如10%)时,以较高的DMSO浓度(如10%)为宜;PEG浓度较高(如20%)时,则以较低的DMSO浓度(如2.5%)为宜。总的看来以15% PEG+5% DMSO的组合效果最佳。FDA检测表明约90%摄取了精细胞的花瓣原生质体有生活力。(3)保温时间:保温时间也影响融合效果(表4)。在15% PEG+5% DMSO条件下,保温1min时即已见到精细胞粘附花瓣原生质体。5min时异核体频率已有1.9%。15min时频率最高。继续延长保温反而使融合率下降,融合体存活时间缩短。

讨 论

关于二细胞型花粉植物的精细胞分离技术,迄今虽有若干有意义的研究[13,21],但就获得用于细胞工程的大量、均一的生活精细胞而言,分离技术仍待改进。Shivanna等(1988)[20]应用"活体-离体技术"在唐菖蒲和杜鹃花中分离出大量精细胞用于扫描电镜观察,但未作生活力测定。本文在前人的基础上,将植物种类扩大到5科8种,并应用了比较适合生活精细胞分离和随后纯化的低渗-低酶法。推测此法亦可在某些用渗透压冲击法和研磨法难以奏效的三细胞型花粉的精细胞分离中应用。本试验中也证明在酶液或渗透压冲击液中添加PDS、BSA和

PVP 等保护剂有利于维持精细胞的生活力,这同前文[8]在三细胞型花粉植物中的研究结果是一致的。

分离的配子原生质体(gametoplast)能否应用于细胞工程,已日益受到关注[6,12,22]。就融合研究而言,迄今已开展了生殖细胞之间[4,23]、生殖细胞与花粉原生质体[23]和生殖细胞与体细胞原生质体[4]之间的融合试验。关于精细胞的融合,最近 Kranz 等(1991)[14,15]应用电融合方法在三细胞型植物玉米中首次成功地开展了精细胞间、精细胞与卵细胞及其它胚囊组成细胞的原生质体和胞质体间的融合,并使精卵的融合体("人工合子")在培养条件下分裂成微愈伤组织。我们应用 PEG 融合法首次在二细胞型花粉植物中成功地开展了同种或异种精细胞间、精细胞与小孢子原生质体以及与花瓣原生质体间的融合研究,结果表明:植物精细胞具有广泛的融合能力,在细胞工程中具有很大的潜力。其中在精细胞与小孢子原生质体间、精细胞与花瓣原生质体间都成功地获得了异核体。考虑到小孢子原生质体已可启动胚胎发生分裂[5],而体细胞原生质体原则上亦可被诱导再生植株,则利用精细胞与这类原生质体融合开展诸如"配子-体细胞杂交"(gameto-somatic hybridization),或利用精细胞携带外源基因导入原生质体是富有前景的。

在精细胞与花瓣原生质体这样大小悬殊的材料融合中,提高精细胞的比例是成功的关键因素之一。前人关于原生质体摄取叶绿体[1]、体细胞原生质体摄取细胞核[19]以及植物原生质体摄取真菌原生质体[16]等实验中,亦均指出必须提高体积小的一方的比例。在 PEG 融合液中添加适量的 DMSO 能促进花瓣原生质体对精细胞的摄取,这和前人关于 DMSO 提高细胞融合率的报道[11,17,19]是一致的。其可能的机理据认为是 DMSO 通过改变细胞膜特性而提高了细胞对 PEG 作用的感受性[11,19]。我们的试验结果还表明 PEG 和 DMSO 之间似存在协同性,它们共同协调地制约着融合率的高低。这一点尚未见诸文献,值得进一步探讨。

图版 I
Plate I

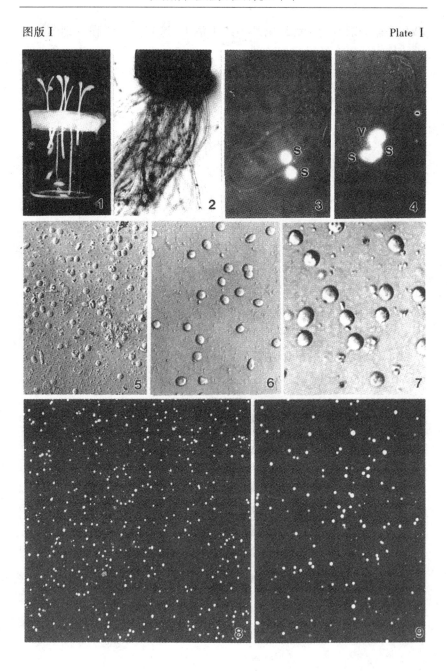

第一章　雄性器官与细胞的实验研究

图 版 说 明

s. 精细胞或精核　v. 营养核

图　版　Ⅰ

1. "活体-离体技术"的实验装置。烧杯用 parafilm 封口,带柱头的花柱切断穿过 parafilm 插入液体培养基。　**2.** 从黄花菜花柱切口长出的大量花粉管。×10　**3.** 从烟草花粉管亚顶端释出的一对精细胞,FDA 荧光染色。×880　**4.** 从萱草花粉管切端释出的一对精细胞的核和营养核,DAPI 荧光染色。×650
5~7. 分离的精细胞群体,干涉差显微观察。　**5.** 鸢尾。×320　**6.** 黄花菜。×440
7. 玉帘。×960　**8、9.** 生活精细胞群体,FDA 荧光染色。**8.** 玉帘。×160　**9.** 黄花菜。×200

Explanation of plate

s. Sperm cell or its nucleus　v. Vegetative nucleus

Plate Ⅰ

Fig. 1. Experimental design for "*in vivo-in vitro* technique". The beaker is sealed with parafilm. The cut styles are inserted through the parafilm into liquid medium. **Fig. 2.** Bundles of pollen tubes growing out from the stylar cut end in *Hemerocallis minor*. ×10　**Fig. 3.** A pair of sperms released from the sub-tip of a pollen tube in *Nicotiana tabacum* after FDA staining. ×880　**Fig. 4.** A pair of sperms and vegetative nucleus released from the cut base of pollen tube in *Hemerocallis fulva*, stained with DAPI. ×650

Figs. 5-7. Populations of isolated sperm cells, observed by Nomarski interference contrast microscopy.　**Fig. 5.** *Iris tectorium.* ×320　**Fig. 6.** *Hemerocallis minor.* ×440　**Fig. 7.** *Zephyranthes candida.* ×960　**Figs. 8,9.** Populations of viable sperm cells after FDA staining. **Fig. 8.** *Z. candida.* ×160　**Fig. 9.** *Hemerocallis minor.* ×200

（作者:莫永胜、杨弘远。原载:植物学报,1992,34(9):688~697。
图版3幅,表4幅,参考文献23篇,仅选录图版Ⅰ。）

其它论文目录

1. 湖北省水稻单倍体育种协作组,武汉大学遗传研究室花培组、水稻花药简化培养基的研究.武汉大学学报(自然科学版),1977.(1):33~34.(周嫦撰)
2. 武汉大学遗传研究室花培组、培养加入多菌录防止霉菌污染析试验简报.湖北农业科学,1978.(4):28~29.(周嫦撰)
3. Chou C.(Zhou. C). Yu T. C. Chang C Y. Cheng C C. 1978：An investigation on callus induction hormones in rice anther culture. In：Proceedings of Symposium on Plant Tissue Culture, Sceince Press, 247.
4. 杨弘远,周嫦.水稻花药的一种整体染色制片技术.遗传,1979.1(3):36.
5. 周嫦,杨弘远.大麦雄核发育的特点及其与诱导频率的关系.武汉大学学报(自然科学版),1981.(1):89~92.
6. 蔡得田,周嫦.胡萝卜花药培养诱导出单倍体小植株.植物学通报,1983.(2):36~37.
7. Yang H Y. Zhou C. 1983. Sporophytic development of rice pollen. In：Cell and Tissue Culture Techniques for Cereal Crop Improvement, Science Press, Beijing, Gordon and Breach Science Publishers New York, 159-164.
8. Zhou C. Wu Y. 1990. Two pathways in pollen protoplast culture：Cell divisions and tube growth. In：Progress in Plant Cellular and Molecular Biology (ed. by H. J. J. Nijkamp, L. H. W. Van der plas and J. Van Aartrijk), Kluwer Acad. Publ., Amsterdam,222-227.
9. 吴燕,周嫦.一种花粉原生质体电镜制样技术.植物学通报,1991.8(3):62~64.
10. 梁立,徐秉芳,郑从义,周嫦.油菜花粉超低温保存及花粉原生质体分离.武汉大学学报(自然科学版),1993.(6):133~136.
11. 施华中,杨弘远,周嫦,陈正华.根癌农杆菌的高效电激转化.武汉大学学报(自然科学版),1995.41(6):724~728.

12. 卢萍,周嫦,杨弘远.烟草花粉原生质体与叶肉原生质体的融合及培养早期变化.植物研究,1996.16(1):96~99.

13. 李昌功,周嫦,杨弘远,李香花,张启发.青菜与芥菜——体细胞原生质体融合的研究.武汉植物学研究,1996.14(4):289~293.

14. 李昌功,孙海鹏,周嫦,杨弘远.甘蓝型油菜胞质体大量制备技术的研究.武汉植物学研究,1998.16(1):77~81.

15. 王劲,施华中,周嫦,杨弘远.烟草脱外壁花粉与完整花粉电激导入效果的比较研究.武汉大学学报(自然科学版),1998.44(2):206~208.

16. 王劲,施华中,周嫦,杨弘远,张献龙,章荣德.用基因枪法将GUS基因导入烟草脱外壁花粉及其瞬间表达.植物研究,1998.18(4):422~427.

17. Fan L M. Wu WH. Yang H Y. 1999. Identification and characterization of the inward K^+ channel in the plasma membrane of *Brassica* pollen protoplasts. Plant Cell Physiology,40:859-865.

18. Sun MX. Kieft H. Zhou C. van Lammeren A. 1999. A co-culture system leads to the formation of microcalli derived from microspore protoplasts of *Brassica napus* L. cv. Topas. Protoplasma,208.265-274.

第二章 雌性器官与细胞的实验研究

提要

20世纪70年代末，我们正在从事花药培养，一次意外的发现促使我们放下手边的工作，毅然投入到未传粉子房与胚珠培养的研究，不久又开始了胚囊操作的研究，并且连续工作了10年。这一由雄性系统操作到雌性系统操作的战略转移是基于当时国内外客观形势的分析：经过20世纪70年代花药培养的热潮，国内这方面的研究已经是"过饱和"了，而未传粉子房与胚珠培养无论国内或国外还仅仅开始，很少受到人们的注意。我们由水稻子房培育出单倍体植株，是起步最早的研究者之一。既然与"雄核发育"对等的"雌核发育"也能产生单倍体植株，且它又是一片未开垦的处女地，我们又幸运地捷足先登，于是抓住机遇耕耘了一番。更有甚者，研究雌核发育，不仅需要进行大量有关培养操作的实验，还需要进行系统深入的胚胎学研究，这正是我们自己的优势所在。以后的发展证明，从实验胚胎学的角度开拓这一新研究领域的战略决策是正确的，使我们在相对落后的工作条件下迅速步入国际前沿。1981年我们在国际刊物上发表的研究论文，是国际上在这一领域内第一次用英文发表的正式论文，此前只有法文或中文的报道。翌年我们又在另一国际刊物上发表了首篇这一领域的综述论文。这两篇文章，引起国际学术界对这个研究方向的重视，推动了研究的深入。我们最后一篇综述论文见于1990年由外国学者主编出版的一部专著，概括了整个20世纪80年代国际上有关研究的成果(参见第五章)。

我们最初的胚胎学观察证明，由未传粉子房与胚珠培养诱导的单倍体植株确系起源于胚囊内未受精的卵细胞或助细胞。这就萌发一个新的思路：可否将胚囊由子房与胚珠中分离出来直接操作呢？这是一个大胆的设想。借助原生质体分离技术的经验，用酶法分散胚珠细胞是不成问题的，问题在于酶解的结果是否能保持胚囊的整体结构，以及如何在分离的大量胚珠细胞中寻找为数很少的胚囊细胞。在缺乏前人经验的条件下，我们自行设计了一个装置，在微型混合器上安装了微型试管以代替原来的多孔板，将悬浮于酶液中的胚珠置入试管，在微型混合器上进行振荡,再辅之以小吸管人工吸打。经过反复试验,终于分离出完整的胚囊。当我们准备撰写第一篇论文时,在查阅文献中发现原来有一个苏联研究小组此前已经成功地以酶法分离出胚囊。尽管他们的研究结果只在会议上以俄文作了摘要报道,鲜为世人所知,但我们还是以科学家应有的客观与公正的态度引证了他们的工作。1982年我们发表了国内首篇论文,1985年又在国际刊物上发表了首篇论文。在这些论文中我们都实事求是地承认了他们的首创权。我们在胚囊操作方面的系列研究,推动了此后国际上同类研究的发展,并且成为后来雌性细胞分离、离体受精和合子培养的先导(参见第三章第二、三节)。

第一节 未传粉子房与胚珠培养和离体雌核发育

提要

在一次花药培养实验中,我们将水稻去颖的幼花(连同雄蕊与雌蕊)进行液体漂浮培养,除由花药诱导出愈伤组织外,还发现子房膨大的现象。剥开子房与胚珠,内有白色块状物。将后者转移到分化培养上再生了小植株。根尖染色体观察表明有单倍体。研究结果与法国科学家于1980年各自在本国刊物上发表。由此揭开了水稻未传粉子房培养诱导单倍体植株及离体雌核发育研究的新篇章。

以后工作朝两个方向发展:一方面,以水稻为重点继续进行系统深入的研究;另一方面,将研究对象扩大到其它植物,以求取得新的成功。无论水稻或其它材料,一旦初步成功,就接着进行三方面的研究:一是通过培养实验了解影响诱导效果的关键因素,寻求最佳的培养方法;二是通过胚胎学观察揭示雌核发育的起源与发育规律;三是通过染色体检查与形态性状观察认识再生植株的倍性与性状特点。以上三方面往往同时展开、互相交错。例如,诱导率高并不意味雌核发育率高,还需要通过倍性鉴定或胚胎学观察,甄别其中是否有体细胞起源的二倍体植株。这种甄别反过来又促进培养技术的改进和关键因素的阐明。有的材料(如向日葵)较难分化根、芽,在这种情况下胚胎学观察就代替染色体检查起主要作用。向日葵胚珠培养时,愈伤组织有三种可能的来源:卵细胞、珠被绒毡层、外层珠被组织。培养实验结合胚胎学观察揭示了以上三者对外界条件的不同要求,特别是在有外源激素时易诱导胚珠体细胞组织增生,而卵细胞孤雌生殖则意外地需要无外源激素的培养条件。

由于子房或胚珠培养时雌核发育诱导率一般在10%以下,因此胚胎学切片的工作量必须很大才能命中有用的样品。为了节省人力与物力,我们采用在花药培养中曾经行之有效的爱氏苏木精整体染色方法(参看第一章第一节),先染色再包埋切片。这使我们得以完成成千上万个样品的切片观察,从大量的观察中总结出可靠的结论。以后进一步在方法学上加以简化,提出爱氏苏木精整体染色——冬青油透明技术,可以透视整个胚珠中的胚囊,十分便于诱导频率的统计;当看到有用的样品后,还可以再转入包埋切片作精细的观察。这些方法上的简化,在此项研究中起到了重要的作用。本节第10与第18两篇英文论文中,对这些方法作了文字与照片的介绍。

在简略介绍了我们的研究技术路线的特点以后,下面再简略介绍主要的研究结果:

在水稻子房培养研究中,通过多种因素的比较实验,确定了"幼花低激素液体漂浮培养"技术流程,在多个品种(主要指粳稻品种)和多年的实验中,均取得了可重复的效果。这一严格控制的条件不仅有效地诱导了雌核发育、排除了因花药存在而可能怀疑会发生的受精,而且抑制了体细胞的发育。染色体检查表明再生植株多数为单倍体;亦有二倍体和非整倍体(可能来自培养过程中的自我加倍)。再生植株中绿苗占多数,白化苗较少,与水稻花粉植株中白化苗比例很高的现象形成鲜明的反差。胚胎学研究通过多年逐步深入的观察揭示了雌核发育的起源与发育规律:首先,肯定了雌核发育原胚与愈伤组织均位于胚囊的珠孔端,从而判断其起源于卵器中的细胞。接着,大规模的切片观察查明助细胞是雌核发育的原始细胞,而卵细胞仅进行游离核分裂形成

第二章 雌性器官与细胞的实验研究

多核体。再进一步查明了助细胞分裂的起因与模式,并从细胞化学角度探讨了助细胞和卵细胞在分化过程中功能变化的可能机制。对于助细胞原胚以后的发育方式也做了研究,发现在外源激素 MCPA 的诱导下主要遵循类似"原球茎"的途径,形成愈伤组织,再通过根芽分化再生植株;而以毒莠定(picloram)代替 MCPA 则可促进胚状体途径达到直接出苗(关于 MCPA 与 picloram 的发现,请参看第一章第一节花药培养的研究)。这样,经过一系列逐渐深化的研究,可以肯定水稻未传粉子房培养中单倍体植株的来源是助细胞无配子生殖。在我们自己和国际上关于离体雌核发育的研究中,只发现水稻属于这种特例。

向日葵是我们另一个重点研究的对象。在国际上我们首次报道向日葵离体诱导单倍体小植株。随后的胚胎学观察查明雌核发育起源于卵细胞的分裂,即属于离体诱导的孤雌生殖。基于大规模的培养实验结合胚胎学鉴定,总结出"低温预处理、无激素胚珠培养"技术,有效地诱导了孤雌生殖,抑制了珠被绒毡层与外层珠被组织的增生。在光镜观察的基础上,又进行了电镜观察,揭示了卵细胞启动离体孤雌生殖的超微结构特征。这是国际上首次对离体孤雌生殖进行超微结构的研究。与此同时,对向日葵及水稻体内受精前后胚囊的发育变化也进行了配套的超微结构观察(参看第四章第一节)。

大麦是法国研究者首次在国际上进行未传粉子房培养成功的材料,但他们没有提供胚胎学证据。我们没有作为重点研究对象,主要着重以胚胎学观察证实了大麦子房培养中卵细胞的离体孤雌生殖现象。

韭菜是一种具有自然孤雌生殖与无配子生殖现象的植物,但在自然条件下并不产生单倍体植株。我们利用它的这种特性,通过未传粉子房培养由卵细胞与反足细胞诱导出单倍体植株,并对此进行了光镜与电镜下的观察,对这一很有兴趣的材料进行了较深入细致的研究。

总之,我们在 4 种植物中取得了未传粉子房或胚珠培养诱导单倍体的成功,其中水稻、向日葵、韭菜属国际首次成功。对上述几种植物,分别研究了有效的培养技术,开展了翔实的胚胎学观察。研究结果表明:所谓"离体雌核发育"(*in vitro* gynogenesis)实际上包括卵细胞孤雌生殖、助细胞无配子生殖、反足细胞无配子生殖几种不同的类型。水稻中卵细胞游离核分裂形成多核体构造是一种在自然界中不存在的奇特现象。此外,还观察到未受精的中央细胞进行游离核分裂形成类似胚乳的现象。本书第五章第3篇文章对我们和国际上离体雌核发育的研究作了总结性的评述。

本节编入的论文数目较多,但当时原期刊的图版有的质量不佳,且篇幅过大,因此收录图版数目较其它章节为少。其中两篇英文文章分别总结了我们在水稻与向日葵两种植物中的离体雌核发育研究结果,并有质量较高的代表性照片作为插图,可供读者参阅。

1　从水稻未授粉的幼嫩子房培养出单倍体小植株

In vitro induction of haploid plantlets from unpollinated young ovaries of *Oryza sativa* L.

通过子房或胚珠的离体培养,诱导大孢子或雌配子体产生单倍体植株是国际、国内正在探索的一项新课题[1],但由于技术原因迄今未取得如同花药培养那样广泛的进展[5]。Nishi 等最先从水稻子房培养产生了二倍体与四倍体植株,但未获得单倍体[6]。Uchimiya 等在茄子未受精胚珠的培养中观察到单倍体细胞的分裂;在玉米的子房培养中由愈伤组织分化了根[7]。Mullins 等进行葡萄未受精胚珠的液体培养,由珠心产生的愈伤组织再生了小植株[8]。Hsu 等用 2-chloroethylphos-phonic acid (ethyphon) 诱导棉花离体胚珠的珠孔端产生愈伤组织[9]。Jensen 等用电镜研究棉花未受精胚珠在培养过程中的变化,发现极核可进行有限的游离核分裂与细胞形成[10]。上海植物生理研究所细胞分化组在油菜子房培养中诱导出一个小植株,又从胚珠培养获得若干愈伤组织与类似胚状体的构造[2]。颜昌敬等在进行水稻叶鞘和枝梗愈伤组织再生植株的研究时,也由子房培养出小植株[3]。但以上工作

第二章 雌性器官与细胞的实验研究

均未能最终获得单倍体植株。最近,祝仲纯等从小麦与烟草的子房培养中成功地得到单倍体植株[4],为我国在这一领域中的研究取得一项突破。本文报道我们由水稻子房培养诱导出单倍体小植株的实验结果。

材料与方法

1979年秋季进行实验。将粳稻品种景洪2号的幼穗经0.1%升汞消毒与无菌水洗后,取花粉单核靠边期的小花,除去内、外颖,将连接于花柄上的雌、雄蕊一并接种于液体培养基上进行漂浮培养(保留雄蕊的本来目的是进行花药培养实验,结果子房中也产生了愈伤组织)。每试管盛培养液25～30ml,每管接种5～6朵小花。去分化培养基成分为 N_6+蔗糖(3%或5%)+MCPA(0.125ppm)或NAA(0.25ppm),有的还补加水解乳蛋白(500ppm)。接种后有的材料在12～13℃下进行了6d低温处理,效果较好。培养在25℃与黑暗条件下。分化培养基成分为 N_6+蔗糖(3%或5%)+MCPA(0.033ppm)或NAA(0.25ppm)+琼脂(0.8%),有的还补加水解乳蛋白(500ppm)。分化培养除采用固体培养基外,也试验了液体漂浮培养技术,以厚约5毫米的泡沫塑料小片作为愈伤组织的载体以防沉没(图版Ⅰ,5)。根据对花粉愈伤组织的试验,后一方法分化率较高,分化速度较快。小植株的根尖染色体鉴定是用铁矾苏木精压片法进行的。

结　果

接种后7～8d,约90%的子房显著膨大。以后多数呈半透明状,仅含水液,显然是无子结实类型。但约有20%的接种子房变成不透明状,其中有的基部凸出。接种后42天,在解剖镜下观察了少数后一类子房,发现有的子房壁内尚有一层浅绿色包膜(估计是珠被),膜内含一至数个白色的愈伤组织小块(图版Ⅰ,1)。接种后45～60d,在培养的269个子房中,有12个子房(占4.5%)的愈伤组织自然地突破子房壁(图版Ⅰ,2),其余的子房逐渐变褐死亡。

将12个出现愈伤组织的子房转移到分化培养基上,一段时间后,1个产生绿苗,4个产生白苗,3个只产生根(图版Ⅰ,3~5)。

对部分再生小植株进行了根尖染色体检查,发现其中有单倍体,$n=12$(图版Ⅰ,6),但也有是二倍体或多倍体的。

根据愈伤组织最初是孤立地被包含在胚珠内和再生的小植株中有单倍体这两点,可以断定至少一部分小植株是起源于大孢子或雌配子体。对培养条件、雌雄发育过程和染色体倍性的详细研究将在以后继续进行。

过去我们在水稻花药培养工作中,很少看到附带接种的子房发生膨大。而本实验中子房膨大是普遍现象,其中一部分诱导出愈伤组织与单倍体小植株。也许本实验采用的液体漂浮培养技术与培养基成分为水稻大孢子的孢子体发育途径提供了比较适宜的条件。

1. 培养42天的子房,子房壁已去掉,示包膜(可能是珠被)内含有的两枚幼小愈伤组织。 2. 愈伤组织突破子房壁。 3. 子房转移到分化培养基上以后,由愈伤组织分化根和苗。 4. 再生的绿苗。 5. 再生的白苗,示以泡沫塑料小片作为载体的液体漂浮培养。 6. 再生小植株根尖压片,铁矾苏木精染色,示单倍体数($n=12$)染色体。

1. A rice ovary after 42 days' culture, showing two young calli dissected out from an envelope (which is possibly the integuments of ovule), the ovary wall being removed. 2. Calli protruding from the ovary wall. 3. Differentiation of roots and shoots from calli transferred on medium for differentiation. 4. A regenerated green plantlet. 5. Regenerated albinos supported by a piece of spongy plastics floating on liquid medium for differentiation. 6. Iron-hematoxylin squash of a root-tip from regenerated plantlets, showing haploid number ($n=12$) of chromosomes.

(作者:周嫦、杨弘远。原载:遗传学报,1980,7(3):287~288。

图版1幅,参考文献10篇,选录图版Ⅰ。)

第二章 雌性器官与细胞的实验研究

图版 I　　　　　　　　　　　　　　　　　　　　　　　　Plate I

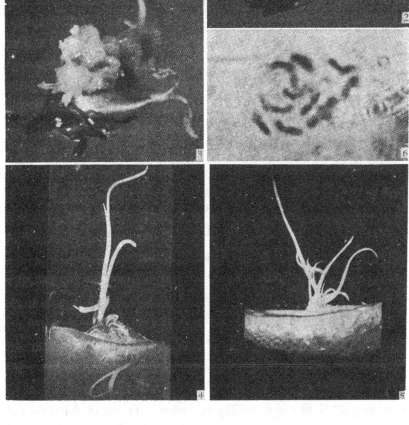

299

2 Induction of haploid rice plantlets by ovary culture

The production of haploid plants by anther culture has become routine practice in some plant species. More attention is now being given to the haploid cells of the ovule as an alternative source. San Noeum[1,2] has obtained haploid barley plants from unfertilized ovaries cultured *in vitro*. Zhu and Wu[3] have also induced haploid plants by ovary culture of wheat and tobacco. These studies are a distinct advance on earlier ones which were either abortive or resulted only in the formation of calli or plantlets of unknown origin[4-7].

We have induced haploid plantlets from cultured rice ovaries[8]. Here a more detailed report is offered.

Materials and methods

Experiments were carried out in 1979 on the *japonica* cultivar of *Oryza sativa*, Jing Hong No. 2, and in 1980 on eight *japonica* and *indica* cultivars. Young panicles were surface-sterilized in a 0.1% solution of mercuric chloride for 12 min and washed thoroughly in several changes of sterile distilled water. Florets at late uninucleate pollen stage were dissected aseptically and the excised ovaries with attached receptacles were inoculated into glass tubes (30mm×200 mm), 5-6 ovaries per tube, on liquid N_6 medium[9] (25 ml per tube) supplemented with 2-methyl-4-chlorophenoxyacetic acid (MCPA)[10] (0.125ppm) and sucrose (3%). Cultures were incubated at about 25℃ in darkness. The ovaries were floated on the surface of the medium as previously described for anthers by Sunderland and

Roberts[11].

For plantlet regeneration, calli derived from the cultured ovaries were transferred to the same medium (except that the level of MCPA was reduced to 0.033ppm). The medium was either solidified with agar or remained as liquid. In latter case the calli were supported on pieces of plastic foam floating on the liquid (Fig. 1E). The ploidy of regenerated plantlets was checked by root-tip squashes stained in iron-hematoxylin.

For embryological observations, ovaries of two *japonica* cultivars, Zao Geng No. 19 and Jing Hong No. 2, were sampled at 2-4 day intervals after inoculation, fixed in aceto-methanol (1 : 3) for 20-24h, stored in 70% methanol at 4℃, stained in toto for 2 days in Ehrlich's hematoxylin which was diluted with equal volume of aceto-ethanol (1 part of glacial acetic acid plus 1 part of 50% ethanol), washed thoroughly in several changes of distilled water and then tap water, each for 3-4h. The stained materials were passed through a routine procedure of paraffin method, sectioned to 8-15 μm and, after being dewaxed with xylol, mounted directly in Canada balsam.

Results

Culture experiments

After 7-8 days of culture, about 90% of the ovaries had enlarged markedly. Most of them contained only liquid within their cavities. These ovaries eventually turned brown and degenerated. However, some ovaries swelled at the base. When dissected, after about a month of culture, some opaque ovaries were found to contain one or several small white calli enclosed in a thin greenish envelope of ovular tissues, inside the ovary wall (Fig. 1A). Five *japonica* and two *indica* cultivars were induced to callus formation; the average induction-frequency of 1192 ovaries cultured was 5.2% (Table I). With further culture, the calli emerged from the ovary wall (Fig. 1B).

In 1979, among 11 calli transferred to fresh medium, one gave green haploid plantlets (Figs. 1D, F) and four gave albinos. Chromosome counts made on part of the albino plantlets showed that there were not only haploid but also some diploid and polyploid roots. In 1980, among 104 calli transferred, 24 gave green plantlets, 6 gave albinos and the others remained undifferentiated. Plantlets from one ovary were identical in respect of their color. Chromosome counts on these materials are still to be conducted.

Embryological observations

The florets inoculated at late uninucleate pollen stage were found to contain embryo sacs of different stages. Of 35 florets sectioned, 23 were uninucleate, 7 were binucleate and 5 were 4-nucleate. After 4 days of culture, many embryo sacs grew older with polarized differentiation of egg apparatus, central cell and antipodals. From about 7 days on, embryogenesis took place in some embryo sacs. Figure 2A shows mitosis of a cell of egg apparatus, but it is uncertain whether it is egg cell or a synergid. Proembryos of various sizes and shapes appeared in the micropylar regions; some consisted of normal meristematic cells, others were highly vacuolated (Figs. 2 B-I). A partitioned proembryo consisting of different cell types was observed (Fig. 2 H). Long unicellular or multicellular suspensors atypical of rice zygotic proembryos were found in some cases (Figs. 2 G, I). Antipodals generally persisted a period and degenerated at last, but some proliferated significantly (Fig. 2 L). However, there was no evidence so far for the origin of proembryos from antipodals. In some cases, the unfertilized polar nuclei were induced to division, forming structures similar to endosperm nuclei (Fig. 2M). However, embryo development was not necessarily accompanied by these structures as it takes place *in vivo*.

After 12-15 days, the proembryos might grow into microscopical calli, one or several in an embryo sac, usually at micropylar end (Figs. 2J, K). The origin of multiple calli in one embryo sac remained unknown.

Discussion

There are some distinguishing features in the techniques adopted in the present experiments as compared with those of many previous works[1-7], namely, the inoculation of intact ovaries attached to the receptacles, the use of liquid medium for float culture, and the less supply of exogenous hormone. The success in this case over many previous failures, may be attributed to the effectiveness of these technical points. Detailed comparative studies are under way and will be reported elsewhere.

Embryological data presented in this paper reveal that embryogenesis, followed by callus formation, may be initiated from the female gametophyte rather than directly from the megaspore. This finding confirms the experiments of San Noeum[1,2], in which plantlets were obtained from ovaries inoculated at binucleate or early trinucleate pollen stages and younger ovaries gave no results. It appears that switch to sporophytic pathway can occur over a period of gametophytic development.

According to San Noeum[2], barley proembryos were found to originate from egg and other cells of embryo sac in various proportion. Since observations were made merely on dissected ovaries under binocular microscope, this conclusion seems vague to us. In our work, evidences are given to the origin of proembryos from the egg apparatus and to the morphogenetic divergences of proembryos from normal development. Initiation of endosperm like structure from unfertilized polar nuclei, similar to that found in cultured cotton ovules[12], is described too. These may offer opportunity for further study on embryogenesis and endosperm development from *in vitro* female gametophyte in the absence of fertilization.

Our findings conflict with those of San Noeum[1,2] and Zhu and Wu[3] who reported the exclusive formation of green haploid plants in ovary culture of barley, wheat and tobacco. Our results on rice suggest that the plantlets regenerated may vary in ploidy and may be green or albino.

(Authors: Zhou C, Yang HY. Published in: Plant Science Letters, 1981, 20：231-237. with 19 figures, I table and 12 references, all omitted here.)

3　水稻未受精胚囊的离体胚胎发生

In vitro embryogenesis in unfertilized embryo sacs of *Oryza sativa* L.

Abstract

Haploid rice plantlets were induced from cultured ovaries in our previously reported experiment. The present paper is an embryological study on this subject.

Young flowers of two *japonica* cultivars were excised and cultured just in the same manner as before. Liquid medium used for float culture was N_6 + 3% sucrose + 0.125 ppm MCPA. The inoculated materials were checked to be at late uninucleate pollen stage which corresponded mainly to the uninucleate embryo sac stage, but as well as some 2-or 4-nucleate embryo sacs. Samples were fixed at 2-4 days' intervals in aceto-methanol (1：3), stained *in toto* with diluted Ehrlich's hematoxylin and sectioned by paraffin method for microscopical observation.

4 days after inoculation most of the embryo sacs developed up to 8-nucleate stage with polarized differentiation of the egg apparatus, central cell and antipodals. From 7th day on, proembryos of various sizes and shapes appeared in the micropylar region of some embryo sacs; some consisted of

meristematic cells, others were highly vacuolated. One-celled as well as linear multicellular suspensors atypical of in vivo zygote proembryos were observed. However, it was uncertain whether the proembryos originated from the egg cell, the synergids, or the differentiating egg apparatus as a whole. Another peculiar event occured during culture was the formation of endosperm-like free nuclei from the unfertilized polar nuclei in some embryo sacs. Sometimes the free nuclei were numerous and showed a tendency of cell formation in localized areas. 12-15 days after inoculation, the proembryos developed into microscopical calli with globular or pearlike shape, which continued enlarging to visible size with naked eyes at about 18-24 th day. Further growth eventually led the calli protruding out the ovary wall beyond 32-35 th day.

These observations indicate that the embryo sac, similarly as the pollen, can be induced to embryogenesis in vitro. This may open a new way to study the mechanism controlling gametophytic and sporophytic developmental pathways of embryo sac and provide means for large-scale production of "embryo sac plants" in future.

在此前,我们报道了通过液体漂浮培养,诱导水稻子房产生愈伤组织,并再生单倍体小植株的实验结果[2]。为了阐明单倍体的来源及其发生特点,进行了本文的胚胎学研究。

材 料 与 方 法

实验于1980年夏季进行。以两个粳稻品种("早粳19"、"景洪2号")作为材料。在无菌条件下将处于花粉单核靠边期的幼花,除去内、外稃,保留雌、雄蕊,连同花托一起接种于 N_6+3%蔗糖+0.125ppm MCPA(2甲基-4氯苯氧乙酸)的液体培养基上,进行漂浮培养。接种后先在12~14℃冰箱中放置6天,然后转至25℃左右、黑暗条件下继续培养。接种后每隔2~4天取样,用醋酸甲醇(1∶3)固定,70%甲醇保存，经爱氏苏木精整体染色[4]后制成 8~15μm 的石蜡切片进行观察。

结　　果

一、接种时的胚囊及其在培养初期的发育

接种材料一律处于花粉单核靠边期。根据 35 朵花的观察，其中 65.7% 为单核胚囊，20.0% 为二核胚囊，14.3% 为四核胚囊（图版 I，1~3）。前人的工作也指出水稻胚囊的发育较花粉为快，当胚囊进入二核、四核甚至八核时，花粉仍在单胞阶段[1]。由于胚囊游离核分裂过程迅速，我们根据花粉确定发育时期，目前只能做到使接种时期相对一致。

接种后 4 天，胚囊发育到四核、八核与接近成熟的阶段。后者一般已分化出典型的卵器、中央细胞与反足器，看不出与体内发育的胚囊的区别（图版 I，4）。但有个别胚囊，珠孔端三个核的排列方向发生变化，呈与胚囊纵轴平行的线形排列（图版 I，5）。这种分裂轴向的变化是否与以后的原胚发生有关，尚需更多研究。

二、原胚的发生

从接种后 7 天开始，在某些胚囊中即有原胚发生。原胚都位于胚囊的珠孔端。在一个材料中看到珠孔端三个细胞中的一个正在发生横分裂（图版 I，6），但从形态上还不能肯定它是否卵细胞。图版 I，9 所示的原胚呈奇特的分节状态，联系上述珠孔端三个核呈线形排列的现象，是否有可能当胚囊发育到八核阶段时由珠孔端一群未分化的细胞共同组成原胚？因此关于胚囊原胚发生的确切方式还需继续探讨。

原胚在形态上呈现很大的多样性。有些原胚的细胞很小，液泡化程度较低，显示分生的能力；另一些原胚细胞很大而高度液泡化。在同一放大倍数下比较其大小，相差十分悬殊（图版 I，7、9；图版 II，10~12）。还看到一个原胚包括显著不同的三个部分：顶部为分生细胞，中

部为薄壁细胞,基部为一个巨大的液泡化胚柄细胞,各部之间界限分明,呈嵌合状态(图版Ⅱ,14)。就原胚的外形而言,除接近体内合子原胚的梨形外,尚有各种非典型的形态。胚柄的变异尤其引人注目:有的没有明显的胚柄;有的只有一个胚柄细胞;有的则由一个或一列细胞构成一个长形的胚柄而将胚向胚囊中央推进,表现出与水稻正常合子原胚迥异的形态特征(图版Ⅱ,13、15)。

培养早期的胚囊中常有类似体内的反足器,维持一段时期后大多逐渐退化,少数持续发育。在一个培养 15 天的胚囊中,看到一个巨大的反足细胞群,分布在 9 张连续切片上,其形状、大小与细胞数目不同于水稻正常的反足器(图版Ⅱ,19)。这种特殊增生的反足器是否有可能发育成原胚与愈伤组织尚难肯定。

三、愈伤组织的形成

培养 12～15 天,原胚发育成球形、梨形或不规则形状的愈伤组织,一般位于胚囊的珠孔端,经常一个胚囊中只有一个,也有两个以上的情况(图版Ⅱ,16～18)。在某些情况下几个愈伤组织互相密接,它们可能是由同一原胚以后分裂而来的。在我们的材料中,一个胚囊中只看到一个原胚,似乎支持上述单元起源的假设。但因观察的材料不够多,愈伤组织多元起源的可能也是存在的。

培养后 18～24 天,愈伤组织已长大到肉眼可以辨认的程度,有些已开始芽的分化。从 32～35 天起,它们开始突破子房壁。将达到肉眼可以辨认大小的愈伤组织,由子房中分离出来,或将已突破子房壁的愈伤组织取出,转移到分化培养基上,可再生小植株。根据我们对 8 个品种的实验,除一个品种未产生愈伤组织外,其余品种的诱导频率为 1.1%～12.0%。已初步由 30 个子房分化出绿苗 24 丛、白苗 6 丛。

四、类似胚乳构造的发生

培养过程中另一值得注意的现象是在某些胚囊的中央细胞中产生类似胚乳游离核的构造。游离核的数目有的较少,有的较多并成片分布,甚至有的在局部区域开始形成细胞(图版Ⅱ,20、21)。但没有看到像体内胚乳那样大规模形成细胞与积累淀粉的情况,也没有看到由它们形成原胚的迹象。

类似胚乳的游离核显然是由未受精的极核分裂产生的。然而由于没有看到极核初始分裂的图像以及游离核的染色体倍性,因此它们究竟是来自两个极核或其中之一的独立分裂,或系二者融合后的分裂产物,尚属疑问。至于它们与原胚的关系,有三种情况:同一胚囊中既有原胚又有"胚乳";有原胚而无"胚乳";有"胚乳"而无原胚。从这一点来看,很难认为它们和体内的胚乳一样是原胚发育的必要条件。

讨 论

在被子植物的生殖过程中,胚囊的离体研究是一个空白。Raghavan 分析其原因在于胚囊分离培养技术的困难使人望而却步[10]。尽管如此,企图通过未受精的子房或胚珠培养诱导单倍体的工作仍然时有所见,并于最近在大麦[11,12]、小麦、烟草[5]与水稻[2]上相继突破。本文以胚胎学观察,证明了在离体条件下水稻未受精胚囊中的胚胎发生,表明胚囊和花粉一样能够由配子体发育途径转向孢子体发育途径,通过原胚发生、愈伤组织形成、根芽分化再生植株。根据初步观察结果提出以下问题讨论。

1. 在我们采用的培养方式下,胚囊在培养早期继续沿配子体途径发育,由单核、二核或四核的幼年阶段直至形成成熟的胚囊,并在发育过程中显示与体内基本一致的极性分化。发育途径的转变是在配子体发育的过程中发生的。凡有原胚发生的胚囊均已显示极性分化,原胚几乎均由珠孔端发生,没有看到由大孢子直接分裂形成原胚的情况。因此,孢子体发育的起点是雌配子体而不是大孢子。在 San Noeum 关于大麦子房培养的工作中,接种时花粉处于二核期或三核早期(此时胚囊已分化)有诱导效果,而较幼嫩的子房培养则不成功[11,12]。显然原胚也不是由大孢子直接产生的。此外,据报道,大麦子房培养时一个子房中含一至数个原胚,其来源多数是卵细胞,也可能是助细胞或反足细胞[12]。但作者没有做切片观察,而仅仅借助子房解剖作出推断与统计,说服力是不充足的。我们的工作肯定原胚主要发生于胚囊的珠孔端。至于其确切来源及其比例,我们认为有必要在更大规模的胚胎学研究后才能解决。

2. "胚囊原胚"与合子原胚相比,形态发生上有同有异。相同的是:二者都有明显的极性分化,即胚朝合点端、胚柄朝珠孔端(如果"胚囊原胚"有胚柄)。不同的是:二者的形态特征可以有很大的差异。"胚囊原胚"表现出明显的多样性,其中有些特点与花粉原胚或体细胞原胚颇相类似。例如,很早就出现由液泡化细胞组成的原胚,与我们在大麦花粉中所见到的十分类似[3]。又如由几部分细胞"嵌合"而成的原胚,有些像 Sunderland 在花粉中所看到的"间隔单位"(partitioned unit)[13]。再如与合子胚柄迥异的变态胚柄的出现也是很有趣的。这些似乎都反映出离体培养的效应。在胡萝卜上曾观察到体细胞原胚与合子原胚最初分隔方式的明显差异,以及培养中产生的种种异常的胚与胚柄[9]。玉米花药培养中也发现各种非典型的双子叶型与多子叶型的胚状体[6]。进一步研究离体胚囊中的胚胎发生,并与其它来源的胚胎发生相互比较,将会为探索离体胚胎发生的控制机理增添新的知识。

3. 在人工诱导体内孤雌生殖的试验中,胚乳的诱导比胚的诱导往往更为困难。Maheshwari 指出,缺乏胚乳是妨碍孤雌生殖胚继续发育的真正困难所在[8]。然而 Jensen 等观察棉花未受精胚珠在培养过程中的变化时,发现某些情况下两个极核互相融合并进行有限的游离核分裂,以后早熟地形成细胞[7]。在本文中又一次报道了由水稻未受精极核产生类似胚乳游离核的现象,而且并不罕见。看来在离体条件下诱导极核的增生并不像过去设想的那样困难。这对于进一步研究胚乳发生的机理将提供一个有用的实验系统。

(作者:周嫦、杨弘远。原载:植物学报,1981,23(3):176~179。
图版2幅,参考文献12篇,均删去。)

4 水稻子房培养时助细胞的无配子生殖和卵细胞的异常分裂

Synergid apogamy and egg cell anomalous division in cultured ovaries of *Oryza sativa* L.

Abstract

This paper is a further research on the origin of gynogenic structures of rice which have been induced by the method of unpollinated ovary culture in our laboratory for several years. Three cultivars were used as materials. The culture medium and technique were basically the same as in previously reported experiments [1-3,11]. Samples were fixed, stained and sectioned by paraffin method.

All the four components of embryo sac could be induced to division, however, with different frequency and results (Tab. 1, 2). Most of the gynogenic proembryos were originated from the synergids. The processes of synergid apogamy are described in detail (Plate Ⅰ, 1-16; Plate Ⅱ, 1-5). The egg cell, in contrast, usually vacuolated considerably and underwent free nuclear division, leading to coenocyte formation (Plate Ⅱ, 6-15).

The problem of why and how synergid apogamy takes place in such high frequency in ovary culture is discussed. The anomalous behavior of the egg cell is discussed in relation to the physiological as well as the phylogenetic aspect.

关于水稻未传粉子房培养,周嫦、杨弘远[2,11]与郭仲琛[6]已分别作过胚胎学研究,证明雌核发育的胚状体或愈伤组织系起源于胚囊内,多

第二章 雌性器官与细胞的实验研究

数位于胚囊的珠孔端。但正如前文所指出的,关于雌核发育的确切起源与发生过程尚有待在更大规模的研究后加以阐明。本文将就这一问题作出回答,并同时报道观察到的若干有趣的现象。

材料与方法

实验于 1980～1982 年进行。以"农垦 4 号"(早稻)、"景洪 2 号"(中稻)和"鄂晚 3 号"(晚稻)为材料。由于"农垦 4 号"子房培养有较高的诱导率[3],在本实验中被选作主要材料。采用幼花液体漂浮培养技术,其接种时期与方式、培养基与培养方法和过去报道的相同[1～3],惟接种后不经低温处理,直接在 25～27℃的遮光条件下培养。接种后每天或每两天取样,用醋酸甲醇(1∶3)固定,经爱氏苏木精整体染色后制作石蜡切片。总共观察子房 1 万个以上,本文仅围绕雌核发育的起源问题报道其中部分结果,其它问题将另文发表。

结　果

一、总的描述

在本实验中,接种材料大多处于二核至八核胚囊期。接种后的培养初期,配子体发育继续顺利进行,三天内全部达到胚囊细胞形成的阶段。雌核发育是在配子体形成的基础上发生的,这和以前在水稻[2]与大麦[7]上所得到的结论一致。

在培养条件下形成的雌配子体,其四种组成细胞均可启动分裂发育,但分裂的频率不同,发育的特点与结局亦各异。根据"农垦 4 号"培养 3～7d 期间 1155 个胚囊的观察结果(表 1),有 359 个胚囊中具有各种情况的分裂,占观察胚囊总数的 31.1%。其中,有些胚囊中只有一种细胞发生分裂;有些胚囊中几种细胞发生分裂。助细胞发生分裂的频率最高。

表1　水稻子房培养初期胚囊中各种细胞的分裂频率*

Table 1　Division frequency of embryo sac cells during early stage of rice ovary culture*

观察胚囊数 No. of embryo sacs observed	发生分裂的细胞类型 Kinds of dividing cells	发生分裂的胚囊数 No. of embryo sacs with dividing cell(s)	发生分裂的胚囊百分率(%) Percentage of embryo sacs with dividing cell(s)(%)
1155	卵细胞 E	31	2.7
	助细胞 S	179	15.5
	中央细胞 C	17	1.5
	反足细胞 A	10	0.9
	助细胞+卵细胞 S+E	42	3.6
	助细胞+助细胞 S+S	9	0.8
	助细胞+中央细胞 S+C	5	0.4
	助细胞+反足细胞 S+A	6	0.5
	不详 Unknown	60	5.2
	总计 Total	359	31.1

* "农垦4号"培养3~7d的材料。

* Cultivar "Nong-Keng No.4" after 3-7 days of culture.

　　Abbreviations: E. Egg cell, S. Synergid, C. Central cell, A. Antipodal cell.

通常是一个助细胞分裂；也有两个助细胞均分裂或助细胞与另一种细胞同时分裂的情况。助细胞发生分裂的胚囊占观察胚囊总数的20.9%。在发生分裂的361个细胞中(表2)，助细胞占69.3%；其次是卵细胞，占20.2%；中央细胞和反足细胞只占很小的比例。助细胞不仅分裂频率高，而且以分生细胞型的分裂方式为主，导致原胚或愈伤组织的形成。少数助细胞表现液泡化细胞型或游离核型的分裂方式，但比重不大。卵细胞则相反，主要是游离核分裂，少数为液泡化细胞型，未观察到通过分生细胞型分裂形成原胚的情况。中央细胞的极核可以发生游离核分裂形成类似幼期胚乳的构造，再一次证实了过去报道的结果[2,11]。反足细胞一般为类似活体内的腺型多细胞构造，在很少的

第二章 雌性器官与细胞的实验研究

表 2　水稻子房培养初期胚囊中各种细胞的分裂特点及比例*

Table 2　Types and percentages of dividing embryo sac cells during early stage of rice ovary culture*

分裂的细胞总数 Total number of dividing cells	分裂的细胞类型 Kinds of dividing cells	个数 Number	占分裂细胞总数的百分率 Percentage (%)	分裂特点 Types of division	个数 Number	比例 (%) Percentage (%)
361	卵细胞 Egg cell	73	20.2	游离核型 Free nuclear division	67	91.8
				液泡化细胞型 Vacuolated cell division	6	8.2
	助细胞 Synergid	250	69.3	分生细胞型 Meristematic cell division	216	86.4
				液泡化细胞型 Vacuolated cell division	21	8.4
				游离核型 Free nuclear division	13	5.2
	中央细胞 Central cell	22	6.1	游离核型 Free nuclear division	22	100
	反足细胞 Antipodal cell	16	4.4	分生细胞型 Meristematic cell division	16	100

*"农垦 4 号"培养 3～7 d 的材料。

* Cultivar "Nong-Keng No. 4" after 3-7 days of culture.

场合能通过分生细胞型的分裂形成类似原胚或愈伤组织的构造,显然在数量上不能和由助细胞所形成的相比拟。

因此,在探讨雌核发育的起源问题时,注意力自然就集中到助细胞身上。同时,对具有一定分裂频率和特殊研究价值的卵细胞也予以足够的重视。而中央细胞与反足细胞的分裂则在本文中不再赘述。

二、助细胞的无配子生殖

助细胞与卵细胞同处胚囊的珠孔端且又互相紧贴。因此在判断培养材料中位于胚囊珠孔端的原胚的确切来源时,需要仔细推敲。我们主要依据以下两点来辨别:第一,水稻助细胞的核与核仁比卵细胞的小得多;第二,当原胚附近有明显可认的卵细胞存在时,后者可作为该原胚起源于助细胞的旁证。图版Ⅰ、Ⅱ的各张照片上显示了以上两点依据。

助细胞的第一次分裂通常在培养后数天,即雌配子体形成后不久即可发生(图版Ⅰ,1、2)。以后通过连续的分裂,形成二胞、三胞(图版Ⅰ,3、4)以至多细胞的原胚。助细胞原胚的形成方式有二。一种方式是:助细胞在卵细胞的珠孔方向分裂,形成始为三角锥形或球形、终为类似体内合子梨形原胚的构造(图版Ⅰ,5~16)。而卵细胞则位于助细胞的合点方向。造成这一位置变化的原因,是由于助细胞在分裂前就居于卵细胞的侧旁而相对地偏于珠孔端的位置,其细胞核的位置差异就更大。卵细胞向合点方向的膨大更加强了这一趋势。

另一种方式是:助细胞在卵细胞的侧旁增生,沿着卵细胞表面向合点方向延伸,越过卵细胞合点端,并向两侧扩展,将后者部分地环抱起来。这样一种助细胞原胚,在切面上呈钩形,其立体形态应是"风雪帽"式的结构(图版Ⅱ,17~21)。助细胞原胚环抱卵细胞的奇特现象如何解释尚属疑问。在蚕豆中曾发现胚柄细胞环抱胚体的现象[8]。很可能在两种不同的细胞之间存在着某种生理上的相互作用。"风雪帽"式的助细胞原胚发育前途如何也还不清楚。

三、卵细胞的异常分裂

卵细胞在培养条件下的变化颇为特殊。卵细胞往往异常膨大,形成高度液泡化的巨大泡状物(图版Ⅱ,22),其直径有时可增长 10 倍以上。如果不是由于大体保持了原有的位置与轮廓,加之卵核的特有标志,很难相信它是由卵细胞变成的。在某些液泡化的卵细胞中发生核的分裂(图版Ⅱ,23),形成二核以至多核细胞(图版Ⅱ,24、25)。游离核多以有丝分裂方式产生,也有少数无丝分裂的情况(图版Ⅱ,26)。游离核分裂继续进行,通常在卵细胞的外周布成一层。在某些胚囊中,可看到由卵细胞形成的多核体与助细胞原胚并存的现象(图版Ⅱ,27~29)。

一部分多核体的进一步发育与体内胚乳的发育颇相类似:游离核一面分裂,一面由外周部分开始产生细胞壁,形成多细胞的结构。此时卵细胞已丧失原有外形,并增生很大的体积。判断这种构造来源于卵细胞而非中央细胞的根据,一是它的位置在胚囊的珠孔端,二是在它的合点方向常有原封未动的极核存在(图版Ⅱ,27、30、31)。卵细胞多核体及其衍生的多细胞构造,很少有产生雌核发育原胚与愈伤组织的可能。在个别情况下,卵细胞亦可不经游离核分裂直接形成多细胞原胚,但其液泡化程度很高,看来难以继续进行雌核发育(图版Ⅱ,32)。

讨 论

未传粉子房培养胚胎学研究中的一个重要问题是雌核发育的起源。迄今关于这一问题已在一篇综述中作了简略的介绍[4]。由于诱导频率尚低,加之培养过程中可能发生各种反常的变化,要解决雌核发育的确切起源是不容易的。为此,在本实验中注意了从两方面获得可供研究的大量图像:一是选择高诱导率的材料;二是增加制片的规模。本室多年应用的整体染色技术容许我们在不太长的时间内制作大量的切片。此外,还要加上观察时的反复斟酌与慎重判断。这些方法上的考虑被证明是有关键意义的。

通过观察,我们可以作出如下结论:(1)在培养条件下水稻胚囊中的四种组成细胞均可分裂发育,但分裂的频率与发育的特点各不相同。(2)雌核发育主要起源于助细胞的无配子生殖;反足细胞亦可发生无配子生殖,但频率不高;卵细胞的分裂尚未证明有雌核发育的前途。现就助细胞与卵细胞的行为作如下讨论。

在胚囊中,助细胞的位置和形态、生理特征比其它细胞更接近卵细胞,即更具有胚性细胞的潜能。Raghavan 指出:"在被子植物中,助细胞是自发产生胚状构造的最普通的来源。助细胞变成和卵细胞相似,然后经过或不经过受精发育为胚。"[10]因此,在特殊条件下,诱导助细胞代替卵细胞行使生殖功能是不足为奇的。由于自然发生频率低,而体内诱导的方法并不十分有效。文献中很少有关助细胞无配子生殖过程的详细胚胎学与细胞学资料,但在本实验中,助细胞的无配子生殖频率相当高,从而为这方面的研究提供了基础。本文观察记载的水稻助细胞分裂发育的各种现象具有进一步研究的价值。今后可应用其它细胞学技术尤其是从亚显微水平深入研究其内在的机理。

卵细胞理应是胚胎发生的当然起点,但在本实验中却让位于助细胞,这是另一个值得注意的问题。卵细胞的分裂频率并不太低,主要问题是它表现液泡化与游离核分裂的特点,从而丧失了胚胎发生的前途。在体内条件下,卵细胞(受精或不受精)通过分生细胞型的分裂方式形成胚,而中央细胞受精后则通过游离核分裂或液泡化细胞型的分裂方式形成胚乳。为什么离体培养子房中的卵细胞在某种程度上仿效了中央细胞的发育方式?这究竟是水稻子房培养的普遍规律,还是在特殊实验条件下(如液体漂浮培养、异常的激素与蔗糖浓度、异常的温度等)的一种特殊表达?这些都是需要继续研究的问题。我们还可以提醒注意:在花粉雄核发育中同样包括分生细胞型、液泡化细胞型、游离核型等各种发育方式,它们对雄核发育的前途有重要作用[5]。由此可见这是子房培养与花药培养中一个带共性的问题。

卵细胞的异常分裂还可以从另一个角度加以讨论。裸子植物合子的最初分裂是游离核分裂。被子植物中,迄今只发现芍药属(*Paeonia*)的合子行游离核分裂。Яковлев 与 Иоффе[12~14]曾描述了其合子形成多核体、游离核分布于多核体外周并形成细胞以及由多核体的外周细胞产生胚的过程。Murgai重复了这一研究,认为芍药属合子的第一次分裂仍为细胞分裂,然后由基细胞单独形成多核体。至于由多核体产生胚的过程则与前面作者观察的一致[9]。Яковлев等曾根据芍药属胚胎发生的特殊情况提出被子植物新胚型(芍药型)的建议,以及被子植物与裸子植物胚胎发生统一性的观点。现在我们在水稻未传粉子房培养中也观察到未受精卵细胞分裂成多核体及其后细胞形成的现象,除了尚未证明有胚胎发生前途这一点外,和芍药属的情况颇为近似,是很有趣的。

（作者：田惠桥、杨弘远。原载：植物学报,1983,25(5):403~408。图版2幅。表2幅,参考文献14篇。仅选录表1、2。）

5 水稻子房培养中的胚状体与愈伤组织形态发生特点

Morphogenetic aspects of gynogenetic embryoid and callus in ovary culture of *Oryza sativa* L.

摘 要

在液体漂浮培养条件下,水稻未传粉子房培养中雌核发育的主要途径是由原胚转变为类似愈伤组织的构造。极少数情况下原胚可以分化为胚状体。描述并讨论了以上过程的形态发生特点。此外还研究了多胚现象和原胚的败育问题。

Abstract

This paper is a further study on the morphogenetic processes of gynogenetic embryoid and callus formation in rice unpollinated ovary culture. Techniques for culture and for specimen preparation were basically the same as in previously reported experiments in our lab. During early stage of culture, gynogenetic structures within the embryo sacs were predominantly young pear-shaped proembryos. A few of them developed later along a more or less normal pathway into differentiated embryoids. However, most proembryos grew much larger without embryo differentiation and finally turned to callusing. A discussion is given to the morphogenetic aspects of such kind of callus, which is considered to be a structure somewhat similar to the "protocorm" described by Nostog in barley proembryo culture. In this paper, the process of proembryo abortion and the occurence of polyembryony are reported as well.

(作者:田惠桥、杨弘远。原载:植物学报,1984,26(4):372~375。
图版1幅,参考文献14篇,均删去。)

6 水稻离体无配子生殖的进一步胚胎学研究

Further embryological studies on the in vitro apogamy in *Oryza sativa* L.

Abstract

In the cultured young ovaries of rice, the processes of megagametophyte development could be switched to the formation of various abnormally organized embryo sacs and then to the initiation of synergid apogamy. The main pathway leading to apogamy was found to go via a linearly oriented 4-nucleate embryo sac to the formation of a linearly oriented egg apparatus, from which it was usually the chalazal synergid giving rise to an apogamous proembryo, and the micropylar synergid degenerated. The proembryo thus produced was located at the base of a vacuolated egg cell (Plate I, 1-7). The second pathway went through a nonlinearly oriented 4-nucleate embryo sac to the formation of an egg apparatus in which the two synergids were located at one side of the egg and oriented longitudinally. In this case it was often the chalazal synergid that could be triggered to apogamy, resulting in a hook-shaped proembryo embracing the egg cell from one side (Plate I, 8-11). When ovaries with nearly matured embryo sac were cultured, in a few cases where apogamy was induced, the proembryos observed were all situated at one side of the egg and were hook-shaped (Plate I, 12). All these pathways are summarized in a diagram (Fig. 23).

Some interesting changes were observed in the synergid and the egg cell of the cultured ovaries by PAS reaction and mercuric bromophenol blue staining. The egg cells, in contrast to *in vivo* condition, often contained abundant starch grains. The synergids and synergid proembryos were rich of cytoplasmic protein (Plate II, 13, 14). We supposed that the egg may supply some nutrients as well as stimulants to the developing synergid in the course of apogamy. The distribution of starch and protein in apogamous embryoids during subsequent development was also described in this paper (Plate, 15-22).

近几年来我们实验室的研究工作已经证明:由未传粉的水稻子房培养的单倍体植株主要是起源于助细胞的无配子生殖而不是卵细胞的孤雌生殖[2]。但是仍然留下一些未被解决的疑问,如:原胚究竟起源于哪一个助细胞? 在两个助细胞中,是否有一个特殊的助细胞预定启动无配子生殖? 为什么存在着位置与形态上不同的两类助细胞原胚?

助细胞和卵细胞在培养条件下性质与前途发生根本的变化,如何从生理角度解释?和体内合子胚相比,助细胞胚状体在发育过程中有什么代谢方面的特点?回答这些问题需要作更多的基础研究工作。本文仅就其中几点深入一步。

材 料 与 方 法

实验于 1983～1984 年进行。主要实验材料是粳稻品种"鄂晚 3 号"。子房培养沿用本实验室中所采用的方法,即在含 3% 蔗糖与 0.125mg/L MCPA(2 甲基-4 氯苯氧乙酸)的 N_6 液体培养基上进行去颖幼花的漂浮培养[6]。为了研究无配子生殖的不同发生途径,分别采用单核花粉和三核花粉两个时期的幼花接种;后者于接种前除去花药,以免受精。培养期间定期取样,用醋酸甲醇(1:3)固定,70% 甲醇保存。部分样品经爱氏苏木精整体染色[5]后制作石蜡切片;部分样品在石蜡切片后分别用高碘酸-锡夫反应(PAS 反应)[4]显示淀粉与多糖,用汞-溴酚蓝染色[8]显示蛋白质。总共观察了约 6000 多个子房。

结 果

一、无配子生殖的发生途径

单核花粉期接种的材料,胚囊处于单核至四核期而以二核胚囊为主(占53%)。在培养过程中,由于胚囊游离核分裂轴向和细胞空间排列方位的改变,导致各种变态胚囊的形成,并在此基础上启动助细胞的无配子生殖。现已查明其中一条主要途径(暂称 A 途径)是:二核胚囊分裂时,两个纺锤体与胚囊纵轴平行,这样形成的四核沿胚囊纵轴排列成一线(图版 I,1)。再一次分裂形成八核时,珠孔端四核亦排成线形,其中合点端一核以后成为上极核,其余三核组成线形的卵器(图版 I,2、3)。这种情况观察到98 例。我们以前曾经报道过线形排列的卵器,当时曾提出:"这种分裂轴向的变化是否与以后的原胚发生有关尚需更多的研究[7]"。以后的研究中再次观察到很多这种情况,但未将它和无配子生殖联系起来[1]。现在进一步发现线形卵器正是无配子生殖的重要基础。如图版Ⅰ,4~6所示,线形卵器中的一个合点端细胞(即上极核的姊妹细胞)以后发育为液泡化的卵细胞。其余两个为助细胞,其中近合点(即紧靠卵细胞)的一个通常具胚性细胞的特征,可以分裂形成原胚;而近珠孔的一个则常退化。由于线形排列的结构,这样形成的助细胞原胚通常位于卵细胞的基部,而使后者向胚囊中央方向推移(图版Ⅰ,7)。田惠桥与杨弘远曾看到众多的这类原胚[2]。在本实验中这一类型同样占主要地位。

田惠桥与杨弘远还观察到另一类助细胞原胚,其特点是位于卵细胞一侧并向后者环抱,形成"风雪帽"状的构造[2]。本实验也追溯了这类原胚的起因。在单核花粉期接种的材料中观察到胚囊发育的另一条途径(暂称B途径):二核胚囊分裂时,两个纺锤体不与胚囊纵轴平行,结果形成非线形排列的胚囊(图版Ⅰ,8)。珠孔端二核再分裂时,两个纺锤体轴以锐角相交,形成非线形排列的四核(图版Ⅰ,9)。以后,上极核向胚囊中央移动;其余三核组成卵器,两个助细胞均位于卵细胞一侧而作上下方排列。这样的情况共观察到28例,其中有5例近珠孔端的助细胞退化(图版Ⅰ,10)。如果在这种情况下近合点的助细胞启动分裂,则产生的原胚应位于卵细胞的一侧而可能形成"风雪帽"状的结构。实际上我们观察到7例这样的原胚(图版Ⅰ,11)。

三核花粉期接种时,胚囊已经成熟,诱导频率极低,在2000多个子房中,仅看到10多个原胚。值得注意的是,能够认清和卵细胞关系的所有5个助细胞原胚,一律位于卵的侧方而向后者作环抱状(图版Ⅰ,12)。尽管由于数目太少无法追踪详细的发生过程,我们仍然可以认为由体内正常分化的卵器中存在一个助细胞直接启动无配子生殖(暂称C途径)。

以上三条发生途径,以图解方式总结如图23所示。

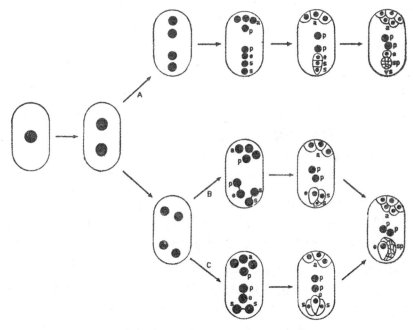

图23　水稻离体无配子生殖发生途径图解
s.助细胞　e.卵细胞　p.极核　a.反足细胞　sp.助细胞原胚

二、无配子生殖过程中的淀粉与蛋白质动态

在体内发育的水稻胚囊中,卵细胞仅含很少的细小淀粉粒,助细胞(尤其是其珠孔端)呈强烈的多糖反应。但在离体培养条件下情况发生显著的变化:无论在线形或非线形排列的卵器中,卵细胞中大多密布淀粉粒,而助细胞则呈弱的多糖反应(图版Ⅱ,13)。我们也再一次肯定了以前所发现的卵细胞游离核分裂现象[2]。由此形成的卵细胞多核体亦常含淀粉粒。另一方面,汞-溴酚蓝染色表明:助细胞及其生成的原胚细胞富含原生质蛋白质,而卵细胞及其分裂产物则高度液泡化,原生质蛋白质很少(图版Ⅱ,14)。这些情况说明,在子房培养过程中分化的卵细胞和助细胞和体内预定分化的产物具有本质的差别。

观察了助细胞胚状体的各个发育阶段的淀粉和蛋白质分布特点。早期的助细胞原胚常呈正常的分生组织特征:富含浓厚的原生质蛋白质(图版Ⅱ,14),而细胞壁很薄,不易被 PAS 反应着色,不含淀粉粒(图版Ⅱ,15)。少数原胚在早期即已出现细胞壁增厚(图版Ⅱ,16)或积累淀粉等早衰特征。原胚生长成为大的球形团块后,内部组织仍保持分生状态;外周几层细胞则胞质变稀薄(图版Ⅱ,17)并积累淀粉(图版Ⅱ,18)而逐渐解体。这样的胚状体由于内方增生新的细胞仍能继续生长。偶尔看到一个特殊的胚状体,内部淀粉粒密集成团,完全变为贮藏组织状态,显然难以继续发育(图版Ⅱ,19)。胚状体进一步长大后,其外围尤其是合点端出现不整齐的凸凹,田惠桥与杨弘远曾称这种胚状体为类似"原球茎"的构造[3]。此时淀粉多分布于其外周与合点端的组织中;在其"分生组织带"的珠孔端组织中也密集大颗粒淀粉(图版Ⅱ,20)。"原球茎"的外周组织中同样会有较多的贮藏性蛋白质粒(图版Ⅱ,21)。"原球茎"分化根、芽时,由富含淀粉粒的组织产生根尖和芽尖。这时淀粉粒在根、芽分生组织中消失(图版Ⅱ,22)。这和 Thorpe 总结烟草等体细胞愈伤组织分化出芽时淀粉动态的规律是一致的[10]。

讨 论

1. 我们实验室以前曾报道水稻未传粉子房培养时有两类位置与形态不同的助细胞原胚:一类是位于卵细胞基部的、始为三角形或球形、终为梨形的原胚;另一类是位于卵细胞侧面而环抱后者的弯钩形或"风雪帽"形原胚[2]。本文对上述两类原胚的起因作了说明。我们发现在离体培养的子房中水稻的胚囊发育与无配子生殖可能遵循多条途径(图版23)。幼嫩子房(以二核胚囊为主)接种时主要是通过 A 途径(二核胚囊——→线形排列的四核胚囊——→线形排列的卵器——→近珠孔助细胞退化,近合点助细胞分裂)形成第一类原胚。也可能通过 B 途径(非线形排列的四核胚囊——→非线形排列的卵器——→近珠孔助细胞退化,近合点助细胞分裂)形成第二类原胚。成熟子房接种时,虽然没有观察到发生过程,但很可能是通过 C 途径(即由体内成熟卵器中的一个助细胞分裂),也是形成第二类原胚。我们推测在胚囊发育过程中至少有两个控制点决定着分裂轴向与细胞的空间排布:一个在二核期和四核期之间,它决定了线形或非线形的四核胚囊的形成;另一个在四核期和八核期之间,它决定了线形或非线形的卵器的形成。只有接种时胚囊尚处于第一控制点之前,才能经由 A 途径形成第一类助细胞原胚。如果接种时胚囊已在体内通过了第一控制点,则仅能通过 B 或 C 途径形成第二类助细胞原胚。周嫦与杨弘远在讨论水稻雌核发育的启动时期问题时曾经推断:"发育途径的转变是在配子体发育的过程中发生的"[7]。本文为这一论断提供了具体的论证。关于从细胞学角度研究无配子生殖的发生途径,在文献中几乎是空白[9]。本文利用离体的实验系统提供了这方面的初步资料。

2. 在离体条件下,助细胞和卵细胞不仅在形态上、而且在生理特征上也表现出明显的变化:前者保持未分化时富含原生质蛋白质的胚性细胞状态;后者积累淀粉而成为贮藏型的细胞。联系到两个助细胞中通常是紧靠卵细胞的那一个启动无配子生殖、助细胞原胚在生长过程中可向卵细胞环抱以及卵细胞可行游离核分裂等现象,似乎可以认为在这种情况下卵细胞有着类似胚乳的功能,可能向启动无配子生殖的助细胞提供营养或(和)活性物质。这样两个相邻细胞相互关系上质的转化,在花粉雄核发育中同样见到:原来预定形成雄配子的生殖细胞停止发育;原来预定起滋养和输送雄配子作用的营养细胞反而成为胚胎发生的起点。这种转化过程中必然包含深刻的细胞生物学变化,值得今后进一步探讨。

(作者:李国民、杨弘远。原载:植物学报,1986,28(3):229~234。图版 2 幅,图 1 幅,参考文献 10 篇。仅选录图 23。)

7 毒莠定作为外源激素促进水稻子房
培养中胚状体的分化

Picloram as an exogenous hormone promotes embryoid differentiation in rice ovary culture

Abstract

In the previous works of our lab MCPA (2-methyl-4-chlorophenoxyacetic acid) has been used as exogenous hormone for induction of synergid apogamy in unpollinated rice ovary culture. The proembryos thus produced were frequently poorly differentiated and had to be transferred to the differentiation medium for regeneration. In the present experiments we used picloram (4-amino-3,5,6-trichloropicolinic acid) instead of MCPA and found a promoting effect on embryoid differentiation. Although its induction-frequency was somewhat lower, the differentiation rate was much higher, and the callusing tendency was weakened as compared with the effects of MCPA. Direct regeneration of plantlets was observed on the induction medium. The best result was obtained from N_6 basic medium plus 3% sucrose, 2 mg/L picloram and 500 mg/L lactalbumin hydrolysate. Histological observations revealed that most of the proembryos developed into bipolar embryoids with plumule, radical and other parts similar to the *in vivo* zygotic embryos but different from the protocorm-like structures induced by MCPA.

自从用未传粉子房培养方法诱导出水稻单倍体植株以来,在我们实验室的一系列研究工作中均采用2-甲基-4-氯苯氧乙酸(MCPA)作为培养基中的外源激素[1]。MCPA具有诱导培养子房中的助细胞分裂形成无配子生殖原胚的培养效果。但其中只有极少数原胚分化成为典型的胚状体;多数原胚仅在愈伤组织上生长并发生不同程度的愈伤组织化,形成"原球茎"状的构造,需要转移到分化培养基上才能出苗[2]。这种状况显然归因于未受粉子房的原胚缺乏胚乳的哺育,同时也说明现行培养基尚有待于进一步改良。为此,我们试验了多种方法,结果表明用毒莠定代替MCPA作为外源激素有促进胚状体分化的显著效果。

材 料 与 方 法

供试水稻(*Oryza sativa* L.)为粳稻品种"景洪2号"与"鄂晚3号"。按本实验室过去常用的漂浮培养方法进行未传粉子房培养[1]。用N_6培养基加入3%的蔗糖作为基本配方,以附加0.2mg/L MCPA作为对照。实验处理包括:用毒莠定(4-氨基-3,5,6-三氯-2-羧基吡啶,Picloram)代替MCPA;在MCPA或/和毒莠定之外再附加水解乳蛋白(500mg/L)。毒莠定的来源,一为河南生物研究所与北京农药二厂生产的农药制品,在本实验室中经过去杂质处理;一为法国科学研究中心植物细胞生理实验室Guern教授赠送的纯制品。接种后,6d、14d、20d、30d、40d、60d分别取样,用醋酸甲醇(1∶3)固定,保存于70%的甲醇中冷藏备用。部分材料用戊二醛(4%)固定。按常规石蜡切片技术用爱氏苏木精(Ehrlich's hematoxylin)整染或片染。根据切片观察统计胚状体诱导率(含无配子生殖胚状体的子房百分率)、胚芽分化率(产生胚芽的胚状体的百分率)、愈伤组织率(愈伤组织化的胚状体的百分率)。光镜切片用Olympus BH-2显微镜摄影。选择少数戊二醛固定的胚状体,经乙醇系列脱水、醋酸异戊酯过渡、CO_2临界点干燥、镀铂后,在Hitachi S-450型扫描电子显微镜下观察摄影。

结 果

一、毒莠定浓度对胚状体诱导和分化的影响

在培养基中加入不同浓度的毒莠定,与加入 0.2mg/L MCPA 的对照进行比较(表1)。结果表明毒莠定在 0.2~4mg/L 浓度范围内均有一定诱导效果,胚状体诱导率低于对照,并与其浓度成反比。但浓度为 1~4mg/L 时,芽与根的分化率都显著高于对照。更高浓度(8mg/L)则对诱导与分化均有不利影响。根据这一结果,我们在以后的各次实验中均将毒莠定的浓度定为 2mg/L。又由于水稻子房培养中再生植株的关键在于芽的分化而非根的分化,因此以后的实验中均主要统计芽的分化率。

二、毒莠定作用下胚状体诱导与分化动态

在表 1 实验结果的基础上,进一步作了三个处理的比较:(1) 0.2mg/L MCPA;(2) 0.2mg/L MCPA+500mg/L 水解乳蛋白(LH);(3) 2mg/L 毒莠定。分别从胚状体诱导率、胚芽分化率、愈伤组织率三方面研究培养后不同时期胚状体的变化动态。结果见图 1。

1. 胚状体诱导率(图 1,上)　培养 6~60d 期间,MCPA 的胚状体诱导率变动在 4.8%~7.5% 之间,MCPA+LH 变动在 5.3%~9.3% 之间,毒莠定变动在 5.0%~7.2% 之间。总的看来,MCPA+LH 略高于对照 MCPA,毒莠定略低于 MCPA。从变化动态看,三个处理的诱导率均有随培养时间逐渐下降的趋势,但毒莠定的后期诱导率相对稳定。镜检表明,毒莠定诱导的胚状体在培养过程中较少发生败育。

2. 胚芽分化率(图 1,中)　培养 6~14d 的子房中原胚尚未分化;培养 20~60d 的各期材料中均有分化的胚状体。MCPA 的分化率很低,最高仅为 20%,并随培养时间的延长而不断降低。MCPA+LH 对胚状体的分化有相当大的促进作用,至培养 60d 时约有 1/3 的胚状体分化出芽。毒莠定的促进作用更显著,并且分化率随培养时间的推移而不断上升,至 60d 时约有 4/5 的胚状体分化出芽,其中还有三株小苗直接由诱导培养基上再生。

3. 愈伤组织率(图1,下)　MCPA 引起胚状体的愈伤组织化,这一趋势在培养过程中不断增强,至60d 时 90% 以上的胚状体发生不同程度的愈伤组织化。附加 LH 后,虽然培养早期的愈伤组织化有所缓和,但以后这一趋势也不断加强,最后达到与单加 MCPA 处理相似的程度。毒莠定的结果则与此不同,愈伤组织率保持在较低的水平上,至培养60d 时仍未超过 50%。

综上所述可以看出:MCPA 虽然在胚状体诱导率方面稍占优势,但胚芽分化率很低,愈伤组织化的程度很高;MCPA 附加 LH 后,诱导率略有提高,分化状况有较大的改善;毒莠定虽然诱导率稍低,但促进分化的效果极为显著,愈伤组织化的倾向大为减弱。

三、毒莠定与 LH 或 MCPA 配合使用的效果

在单因子实验肯定了毒莠定对分化的有利效果之后,进一步作了毒莠定与 LH 或/和 MCPA 配合使用的比较。共做了两次实验,其中一次是用国产的毒莠定制品,另一次采用进口的纯制品(表2)。

两次实验所取得的结果基本一致。无论就胚芽分化率的绝对值而言,或就 40~60d 期间胚芽分化率的消长而言,均可以看出各处理的效果呈由强到弱的趋势:毒莠定+LH>毒莠定>毒莠定+LH+MCPA>毒莠定+MCPA>MCPA。即在毒莠定浓度稳定不变的情况下,附加 LH 可促进分化;附加 MCPA 抑制分化。最佳的培养成分组合是:N_6 基本培养基+3% 蔗糖+2mg/L 毒莠定+500mg/L LH。

四、毒莠定作用下胚状体形态发生的特征

培养条件下诱导的无配子生殖产物表现出各种形态学上的变化,从典型的胚状体到典型的愈伤组织,还有各种中间状态。一般来讲,由 MCPA 诱导的胚状体多数呈不分化状态,后期通常愈伤组织化,这和过去所观察的结果是一致的[2]。由毒莠定所诱导的胚状体,虽然同样有不分化和愈伤组织化的情况,但多数分化为具胚芽、胚根的两极结构。其中有些胚状体的形态与体内合子胚极为相似,胚芽、胚芽鞘、胚根、胚根鞘、盾片、下胚轴、外胚叶各部一应俱全,只是各部的形状、大小略有

变化而已(图版Ⅰ,1)。有些胚状体从外形看似乎是一个未分化的组织块,但在切片上仍可观察到清楚的分化,仅仅由于生长的不平衡,致使各部的位置发生了很大的变动。例如在有些胚状体中,背侧的盾片组织加速生长,而腹侧生长缓慢,发达的背部逐渐环抱腹部,致使胚芽、胚根一并转向珠孔端。(图版Ⅰ,2~4)表示这一转变过程的若干过渡形态。这类胚状体用肉眼观察时和 MCPA 诱导的"原球茎"状构造[2]相似,但前者是有根芽分化的两极结构,后者是缺乏分化而有相当程度愈伤组织化的结构。在扫描电子显微镜下,即使从表面形态上也可看出它们的区别(图版Ⅰ,5、6)。由毒莠定诱导的胚状体后期出芽的情况在切片上相当清楚(图版Ⅰ,7、8)。

讨 论

1. 在水稻未传粉子房培养中,由助细胞分裂产生的原胚后期难以正常分化,可能与使用 MCPA 这一强烈的外源生长刺激剂有关。在本实验中,附加水解乳蛋白能在一定程度上促进原胚的有组织生长和分化,这和前人在胚胎培养方面的实验结果是一致的[3]。如用毒莠定代替 MCPA 作为外源激素,无论是否附加水解乳蛋白,促进分化的效果均更好。可见外源激素在这里起着更为重要的调节作用。

毒莠定本是一种除草剂,在低浓度时有促进生长的作用[4]。Чернова 等(1975)在几种豆科和禾本科植物的组织培养实验中比较了毒莠定和 2,4-D 的效果:在菜豆中,它对愈伤组织的生长有较强的促进作用;在小麦中,它的去分化作用较为柔和,后效期较短,较有利于分化[5]。周嫦等(1978)在水稻的花药培养实验中用毒莠定作为外源激素,能促进花药的去分化与再分化作用[6]。Beyl 与 Sharma(1983)证明它对 *Gasteria* 和 *Haworthia* 体细胞胚胎发生的效果超过 2,4-D[7]。Reddy 等(1985)应用 2,4-D、毒莠定与玉米素的激素组合,使水稻花药培养的愈伤组织诱导率和植株再生率获得最佳效果[3]。在未传粉子房培养方面,刘中来在本实验室中曾应用毒莠定成功地诱导了水稻的雌核发育(研究生论文,未发表资料)。本文结果进一步表明毒莠定在促进胚状体分化方面有显著的效果,从而第一次在水稻未传粉子房培养中获得大量分化的胚状体,并观察到在诱导培养基上直接出苗的现象。在国内,毒莠定作为外源激素尚少使用,从上述各方面的研究报道看,它的应用潜力值得注意。

2. 田惠桥、杨弘远(1984)在水稻子房培养的胚胎学观察中,曾经注意到有些胚状体朝胚囊珠孔端分化芽的现象[2],当时不能解释这种反常的极性是如何形成的。现在通过毒莠定诱导了大量的胚状体,使我们有可能追溯这一形态发生的过程,即背部的盾片组织加速生长,环抱并迫使腹侧的胚芽、胚根转向珠孔端。Shah(1982、1983)在研究水稻活体内的胚胎发育时指出:原胚分化时背侧组织的分裂与生长速度高于腹侧组织,致使以后根、芽发生中心偏离中轴,形成两侧不对称的状态[9,10]。可见,活体已经具有背、腹侧生长不均衡的基础,而离体条件下一部分胚状体更加加强了这一倾向。这样的胚状体,合点端由原来盾片增生的组织十分发达。在培养后期这部分组织通常退化解体,可能其分解产物为珠孔端的根、芽生长所利用。作者推测在这种情况下,高度增生的盾片组织似乎部分地代替了胚乳的功能,作为胚状体在无胚乳环境中生长的一种适应。

(作者:何才平、杨弘远。原载:实验生物学报,1987,20(3):283~288。
图版1幅,图1幅,表2幅,参考文献10篇,均删去。)

8 水稻胚囊植株染色体倍性及其它性状的研究

An investigation on ploidy and other characters of the gynogenic plants in *Oryza sativa* L.

Abstract

This paper deals with the ploidy and characters of the gynogenic plants of rice derived from unpollinated ovaries cultured *in vitro* and gives a comparison between gynogenic and androgenic plants of the same cultivar and in the same culture conditions.

Cultivar "Zao Geng No. 19" was used as a main material. Gynogenic and androgenic calli were induced by culture of unhusked young flowers floating on liquid medium as reported before. Regenerated plants and their progenies were studied cytologically and morphologically. Using a "wall degradation hypotonic method", chromosome counts were carried out.

Chromosome counts on 111 gynogenic and 97 androgenic plants showed that in the former both haploid (77.5%) and diploid (22.5%) types were found, while in the latter, besides haploids (63.9%) and diploids (34.0%), a few polyploids (2.1%) were obtained as well. Thus, ovary culture gave more haploids and less non-haploids than anther culture did. It was noticed that all the above mentioned types included a part of mixploids, that means plants with a small proportion of cells of atypical ploidy.

Among 159 clusters of gynogenic plantlets, 89.3% were exclusive green plantlets, 8.8% were exclusive albinos and 1.9% had both green and albino plantlets; in the meanwhile, among 213 clusters of androgenic plantlets they were 36.4%, 61.0% and 2.6% respectively. The predominance of green plants over albinos is a distinguishing feature of rice ovary culture in sharp contrast with large number of albinos in anther culture.

Some morphological characters of the gynogenic plants were observed and compared with those of the androgenic plants.

由未传粉子房培养诱导单倍体植株成功后,提出了很多亟待研究的基础问题。其中起源于胚囊、通过雌核发育发生的植株(以下简称"胚囊植株")的特性如何,与起源于花粉、通过雄核发育产生的花粉植株有何异同,就是一个具有理论与应用意义而迄今很少研究的问题。两年来,我们在水稻未传粉子房培养成功的基础上,对胚囊植株作了较大规模的染色体分析与初步的性状考查,并与花粉植株进行比较。主要结果报道如下。

材料与方法

以粳稻(*Oryza sativa* subsp. *japonica*)为主要实验材料。子房培养采用幼花液体漂浮培养技术,与前文报道的培养基与培养方法相同[7,21]。当幼花同时产生了胚囊与花粉愈伤组织后,将它们分别转移分化成苗。待小植株生根后,采用去壁低渗、火焰干燥、Giemsa 染色技术[3]制片,鉴定胚囊植株与花粉植株的染色体倍性。在供体品种种子根上所作的对照试验证明了这一技术用于鉴定染色体数目的可靠性。共计观察 111 个胚囊植株和 97 个花粉植株的根尖染色体倍性。对成熟的胚囊植株与花粉植株的观察是在冬季温室中进行的;对其二代株系的观察在翌年田间与盆栽条件下进行。

结　果

一、胚囊植株染色体倍性及其与花粉植株的比较

根据"早粳19"111 个胚囊植株与 97 个花粉植株的根尖染色体考查,胚囊植株有单倍体和二倍体类型(图版Ⅰ,1、2)。两类植株分别占总数的 77.5% 与 22.5%,单倍体类型占大多数。花粉植株则有单倍体、二倍体与多倍体类型(图版Ⅰ,7~9),分别占 63.9%、34.0% 和 2.1%。与花粉植株相比,胚囊植株中单倍体数多,二倍体类型较少(表1)。

根据每个植株所观察的细胞的染色体数目变化,无论胚囊植株或花粉植株其单倍体与二倍体类型实际上包括两种情况:全部细胞倍性一致的个体;以一种倍性细胞为主的混倍体。后者兼有不同倍性的细胞或非整倍体的细胞(图版Ⅰ,3~6;10~12)。胚囊植株与花粉植株两种类型的染色体数目的分布概括于表 2 及表 3。由此可见,虽然染

色体数目有一定范围的变化,但仍以一种倍性细胞为主,故分别划归为单倍体或二倍体类型。

表1 **胚囊植株与花粉植株染色体倍性的比较**
Table 1　A comparison of chromosome ploidy between gynogenic and androgenic plants

植株类型 Kind of plants	观察苗数 No. of observed plantlets	单倍体类型植株 Haploid type		二倍体类型植株 Diploid type		多倍体类型植株 Polyploid type	
		数 Number	%	数 Number	%	数 Number	%
胚囊植株 Gynogenic plants	111	86	77.5	25	22.5	0	0
花粉植株 Androgenic plants	97	62	63.9	33	34.0	2	2.1

二、胚囊植株及其二代株系的性状表现及其与花粉植株的比较

由供试的4个品种共计诱导了胚囊植株240株(丛),其中绿苗占79.2%,白苗占13.8%,绿、白苗(同一块愈伤组织既分化绿苗又分化白苗)占7.0%(表4)。品种之间有一定差异,总的趋势是绿苗占多数。

比较"早粳19"在相同条件下形成的胚囊植株与花粉植株,发现前者绿苗率高达89.3%,后者仅36.4%;前者白苗率仅8.8%,后者则高达61.0%(表4)。二者对比,胚囊植株中显然绿苗多、白苗少,恰与迄今水稻花药培养产生大量白苗现象形成鲜明对照,成为胚囊植株的一个显著的特点。

将"早粳19"绿苗移栽温室,34个胚囊植株和10个花粉植株抽穗开花。它们的若干性状考查结果列于表5。可以看出,倍性相同的胚囊植株与花粉植株表现基本一致。单倍体与二倍体相比,在株高、穗长、剑叶长宽以及颖花大小等指标上都显著小于二倍体(表5),唯着粒密度大于后者(图1)。

表 4　胚囊植株与花粉植株中绿苗与白苗的比例
Table 4　Percentage of green and albino plantlets

植株类型 Kind of plants	品　种 Cultivar	再生植株总丛数 Total No. regenarated plantlet clusters	绿　苗　丛 Clusters with green plantlets		白　苗　丛 Clusters with albinos		绿苗与白苗丛 Clusters with green and albino plantlets	
			数 Number	%	数 Number	%	数 Number	%
花粉植株 Androgenic plants	早粳 19 Zao Geng No.19	231	84	36.4	141	61.0	6	2.6
	早粳 19 Zao Geng No.19	159	142	89.3	14	8.8	3	1.9
	景洪 2 号 Jing Hong No.2	27	15	55.6	8	29.6	4	14.8
胚囊植株 Gynogenic plants	辐晚 3 号 Fu Wan No.3	11	9	81.8	1	9.1	1	9.1
	鄂晚 3 号 E Wan No.3	43	24	55.8	10	23.2	9	20.9
	总　计 Sum	240	190	79.2	33	13.8	17	7.0

多数胚囊植株的形态基本正常,与花粉植株、原始品种相似。但有一单倍体植株的叶脉与少数颖花出现白色条斑;另有3个二倍体不育株,每株多达30~40个稻穗,开花后子房不膨大或膨大而无籽结实。对其中一个不育株作了花粉母细胞减数分裂染色体构型的观察,发现在83个中期Ⅰ细胞中仅27个细胞(占32.5%)具12个二价体,其余56个细胞(占67.5%)具有数目不等的单价体(详细结果另文发表)。

初步观察了"早粳19"、"景洪2号"胚囊二代株系、花粉二代株系与原始品种(图2)。与花粉株系、原始品种相比,胚囊株系植株之间没有明显变异,仅在一个胚囊株系中发现一株白苗。株高等性状的考查结果初步表明胚囊各株系内基本整齐一致。

讨　　论

1. 关于由未受精子房或胚珠培养产生的"胚囊植株"的染色体考查,限于群体的数量,迄今较系统的研究资料尚少,亦缺乏与花粉植株的比较[20]。本文以较大规模的染色体倍性分析表明,水稻胚囊植株中单倍体类型占3/4以上,其余为二倍体类型,两种类型中均有少部分混倍体植株。胚胎学研究证明,在我们的试验条件下,再生植株起源于胚囊细胞[8],主要是助细胞,没有胚珠体细胞增生(另文发表),故可认为产生的二倍体是由染色体自发加倍的结果。与花粉植株相比,胚囊植株的单倍体比例较高,似乎暗示子房培养时较少发生染色体自然加倍。

关于花粉植株的染色体倍性已有较多工作[15,16]。烟草花药培养通过胚状体途径形成的花粉植株几乎全是单倍体[16]。禾本科植物花药培养主要经由愈伤组织再分化植株,倍性情况比较复杂。水稻[1]、小麦[9]花粉植株中除了大量单倍体与二倍体外,还曾发现多倍体、混倍体或非整倍体如缺体等。玉米[4]、大麦[17]花粉植株中多数为混倍体,典型单倍体与二倍体植株却较少。

据子房培养文献报道,小麦[11]、大麦[2,18]、玉米[10]通过未传粉子房培养诱导的植株迄今仅发现单倍体。另一些植物如烟草[5,11]、水稻[6,12,13]、非洲菊[14],再生植株除大多数为单倍体外,尚产生少量二倍体、多倍体或混倍体。从总的情况来看,胚囊植株的倍性似乎与发育途径有关。经由胚状体途径直接出苗往往形成单倍体;通过愈伤组织途径再分化则较易导致染色体发生加倍或其它变化。水稻的雌核发育主要是通过原胚阶段并在后期以愈伤组织化的发育方式实现的[8]。本试验中胚囊植株多数为单倍体、少数为二倍体,并有混倍体的现象,是否与这种发育方式有关,值得进一步研究。

2. 迄今未传粉子房与胚珠培养成功的各个物种中,烟草、黄花烟草[5,11]、非洲菊[14]、玉米[10]等获得的胚囊植株均系绿苗。水稻、小麦以及大麦胚囊植株既有绿苗也有少数白苗,其中 San Noeum[18]、王敬驹[2]在大麦,Asselin de Beauville 在水稻[13]试验中得到的全是绿苗,我们在水稻[6]和大麦(未发表)、郭仲琛在水稻[12]上都发现一些白苗。以上试验中胚囊植株往往数量偏少,需要较大工作规模与较大群体,并与花粉植株进行比较,才能更全面地了解胚囊植株的特点。为此,本试验以一个品种为主要材料,比较了相同条件下诱导的数百个胚囊植株与花粉植株,进一步肯定了水稻未传粉子房培养虽然也可能形成少量白苗,但绿苗占明显优势。与花药培养相比,这是一个重要的特点。白化苗的大量出现是禾本科植物花药培养应用于育种的严重障碍,因而成为一个理论研究课题,吸引了许多工作者的注意[15,16]。由未传粉子房与花药培养分别诱导的再生植株虽然同是起源于单倍体的原始细胞,但是胚囊与花粉具有明显的结构与生理上的区别,雌核发育与雄核发育的形态发生过程亦各有不同特点,这可能导致由它们产生的植株在形成绿苗上的差异。胚囊植株的高绿苗率不仅在育种应用上是有利的特点,而且作为一个对照实验体系,将有助于探讨白化苗的成因。

3. 关于花粉植株及其后代的遗传学研究已有相当丰富的资料[15,16],然而关于胚囊植株及其后代的有关研究却甚为缺乏。一些报道描述了胚囊植株若干典型的单倍体特征。San Noeum 对大麦 2 个胚囊株系、3 个花粉株系和亲本作了某些农艺性状的分析比较,认为胚囊株系比花粉株系更接近原始亲本[19]。我们对 34 个成熟胚囊植株、10 个成熟花粉植株作了观察,它们之间没有明显不同,植株形态特征主要因染色体倍性而异,分别具典型单倍体与二倍体的特点。胚囊二代株系和花粉二代株系间性状上也没有显著差异,株系内相对整齐。这方面所做的工作仅是初步的,还有待进一步研究。

(作者:刘中来、周嫦。原载:遗传学报,1984,11(2):113~119。图版1幅,图2幅,表5幅,参考文献21篇。仅选录表1、4。)

9 水稻的未传粉子房培养

Unpollinated ovary culture in rice

20世纪70年代中期首次突破的未传粉子房培养技术,由于开辟了诱导植物单倍体的一条新途径,现已引起愈来愈多研究者的注意,关于这一领域的研究现状,我们已有专文评述[5,11]。我们于1979年由水稻未传粉子房培养出单倍体植株,1980年与1981年分别在国内[1]和国际[9]刊物上报道。在初次诱导成功以后,我们即认识到这是一个对育种实践和学科发展均有潜在意义的课题;同时也认识到这是一个待开垦的处女地,当务之急是开展多方面的、系统的基本研究。四年来,进行了下述三方面的工作:(1)对可能影响培养效果的各种因素进行了实验研究,以期摸清规律,制定适宜的培养技术。(2)对培养的子房进行系统的胚胎学观察,以期阐明再生植株的起源与发育途径。(3)对再生植株进行细胞学与形态学的考查,以期了解其染色体倍性和性状表现方面的特点。以上几方面的研究结果,已经或即将陆续加以报道[2,3,4,6,7,8,10]。为了将水稻子房培养诸方面的工作概括成一完整的体系,并在若干问题上作一些补充,现加以阶段性的总结。

影响培养的各种因素

在花药培养方面,通过许多研究者多年的工作,对于影响培养效果的各种因素已有相当详细的研究资料。但在子房培养方面的同类工作仍很欠缺。我们在着手这一研究时,主要借鉴花药培养的研究经验,并结合子房培养的特点,对各种可能发生影响的因素逐一开展比较实验,在此基础上加以综合,提出比较适宜的培养技术。

一、品种特性

用附录所载的同一培养技术,先后试验了 19 个水稻品种,统计了其诱导频率(产生雌核发育愈伤组织的子房百分率)。其中粳稻品种 15 个:农垦 4 号(12.0)、石羽(9.1)、晓(8.5)、辐晚 23(8.4)、早粳 19(7.5)、鄂晚 3 号(7.1)、702(5.6)、景洪 2 号(5.1)、红旗 16(4.5)、延粳 5 号(3.0)、桂花黄(2.8)、041(2.5)、黎明(2.5)、大粒 3 号(2.0)、徒稻(1.5);籼稻品种 4 个:5350(2.8)、金南特 43B(1.1)、珍汕 97(0)、二九青(0)。除 2 个籼稻品种外,17 个供试品种均有程度不同的诱导频率(括号中的数字为百分率)。粳稻的诱导频率一般高于籼稻,这和花药培养的总趋势是一致的。但就各个具体品种而言,则花药培养诱导率高者(如景洪 2 号),子房培养不一定高;花药培养诱导率低者(如农垦 4 号),子房培养却可能很高。暗示二者的诱导机理不尽相同。就同一品种而言,子房培养的诱导频率低于花药培养。然而一个花药中有成千花粉粒,而一个子房中仅有一个胚囊。因此以胚囊和花粉相比,诱导的几率显然是不算低的。

二、胚囊发育时期

接种时的子房处于何种胚囊发育时期,对诱导结果有明显影响。根据多个品种上的比较实验,大孢子母细胞至四分体时期的幼小材料,一般难以产生雌核发育愈伤组织。完全发育成熟的胚囊虽然可被诱导,但频率不甚高。最适于接种的时期是由单核至四核胚囊阶段,大致相当于花粉单核靠边至二核初期。以上结果表明:就水稻来说,在大孢子发生过程中尚难通过离体培养诱导雌核发育;只有当发育进入雌配子体阶段后方适于诱导。显微观察对这一点提供了说明:即无论何时接种,也要在培养条件下发育到胚囊细胞形成以后,才由其中某些成员细胞启动胚胎发生。看来,培养初期的配子体发育可能是诱导雌核发育的关键时期。

三、低温处理

我们最初试验成功的水稻子房培养,是在接种后给予6d,12~13℃的低温处理。但后来的比较实验证明:无论接种前或接种后的低温处理,诱导频率均未超过或甚至低于未处理的对照。这和花药培养时低温处理能大幅度提高诱导率的现象有明显的区别。在"红旗16"品种上曾观察到:稻穗在4~8℃下经过24h后,胚囊的发育表现多种丧失极性的异常变化,这类变化显然并不能导致雌核发育。

四、花的部分

在不少植物的传粉后的子房培养实验中已经证明:保留花柄、花萼、颖片等附加部分是促进子房发育为果实及其中胚胎发育的重要因素。这在未传粉子房培养中可以借鉴。我们在水稻子房培养中比较了几种外植体的培养效果。发现除掉颖壳的幼花(包括雌蕊、雄蕊、护颖与花柄)诱导效率最高;如将雄蕊也除掉,仍能培养成功但效果稍差;单独的子房培养则很难膨大和产生雌核发育愈伤组织。可见花托、护颖与雄蕊的存在均起有利的作用。需要指出:根据大量的显微观察,在连同雄蕊接种的情况下,并没有发生花粉萌发和受精现象,因此胚囊内的胚胎发生不可能是传粉受精的结果。花托、护颖与花药的有利作用可能归因于其中所含营养物质的转移与被利用。

五、液体培养基与固体培养基

迄今其他研究者的子房培养实验均系采用固体培养基。只有我们在水稻子房培养和向日葵未受精胚珠培养(待发表)上尝试了液体培养并取得了良好的效果。比较实验证明:虽然固体培养亦能成功,但在液体培养基上进行漂浮培养,水稻子房膨大较好,诱导率较高,体细胞愈伤组织发生较少,优越性是明显的。

六、诱导培养基中的外源激素

水稻子房培养以 N_6 作为基本培养基,以我们在花药培养中多年应用的2-甲基-4-氯苯氧乙酸(MCPA,抚顺农药厂产品)作为外源激素。激素浓度的调节对于促进雌核发育、防止体细胞增生起非常关键的作用。无激素条件下子房通常不膨大,不产生愈伤组织。在0.125~8ppm 浓度范围内,子房膨大率随 MCPA 浓度的提高而递增;但雌核发育的诱导则以较低浓度(0.125~0.5ppm)为宜,过高反而不利。2ppm 以上的 MCPA 常导致子房与花托等处体细胞愈伤组织的增生,而抑制胚囊内的雌核发育。此外,诱导培养基中的激素浓度对愈伤组织转移到分化培养基上以后的表现也有影响。适当提高前者的浓度有利于以后的再分化。除 MCPA 外,萘乙酸(NAA)和4-氯基-3,5,6-三氯-2-羧基吡啶(TCP,Picloram)也能诱导水稻的雌核发育。

七、诱导培养基中的蔗糖浓度

初步试验了蔗糖浓度的影响。在以 N_6 为基本培养基的液体漂浮培养条件下,3%~6%蔗糖较为适宜。浓度过低(1%)或过高(9%)均不利于雌核发育。

八、分化培养基

以 N_6 或 MS 为基本培养基,补加微量的 MCPA(0.033ppm),或补加其它生长素与细胞分裂素(IAA 0.5ppm+NAA0.25ppm+KT2ppm)。蔗糖浓度为3%~5%。采用固体培养基;亦可采用带支持物的液体培养基。愈伤组织的分化率可达37%~57%。

九、壮根培养基

愈伤组织长根后,常需切取幼嫩根尖作染色体观察。为了促进新根发生与生长健壮,研究了壮根培养基。在分化培养基配方中去掉 KT,降低蔗糖浓度(至1.5%),提高 NAA 浓度(至0.5ppm),并加入适量的活性炭(0.1%),有利于壮根。切去了根尖的小植株转入壮根培养基后,一周左右便可旺盛发根与移植。

雌核发育的胚胎学

与花粉的"雄核发育"(androgenesis)相对应,有人把在未传粉子房培养中由胚囊产生单倍体植株的过程称为"雌核发育"(gynogenesis)。尽管这两个名称和它们原来的词义相去甚远,用起来不甚恰当,但出于习惯和使用上的简便,我们仍继续沿用。关于花粉雄核发育,已有不少细致深入的细胞胚胎学工作,而在雌核发育方面则缺乏相应的研究。除研究历史短浅的原因外,雌核发育的诱导率尚低,需要进行繁重的切片工作,因此如欲获得准确的、系统的观察结果,也有一定的难度。我们从两方面解决面临的技术挑战:一是选择诱导率较高的材料和适宜的培养技术;二是采用在本实验室中多年应用的爱氏苏木精整体染色技术以增加制片的规模。从而获得了可供研究的大量显微图像。这些方法上的考虑被证明是有关键意义的。

雌核发育的胚胎学应该回答三方面的问题:(1)雌核发育是在什么时期、由什么细胞起源的?(2)发育遵循何种途径?(3)发育过程中有哪些形态发生特点?下文所述,可以就水稻雌核发育中的以上问题作出基本的答案。

一、助细胞的无配子生殖

水稻的雌核发育原胚一般位于胚囊的珠孔端,显然是起源于卵器中的某一细胞。进一步大规模的观察证实原胚主要来自助细胞的分裂。助细胞的第一次分裂通常在培养后数天,即雌配子体形成后不久即可发生,以后通过连续的分裂形成多细胞原胚。原胚有两种发生方式:一种方式是助细胞在卵细胞的珠孔方向分裂,形成类似体内合子梨形原胚的结构。另一种方式是助细胞在卵的侧旁分裂,增生的细胞沿卵的一侧向合点方向扩展,部分地将卵细胞环抱起来,形成一种类似"风雪帽"状的结构。前一种原胚是有发展前途的,后一种原胚前途如何尚属疑问。总之,水稻子房培养时的雌核发育,实质上属于助细胞的无配子生殖(apogamy)。在自然情况下,无配子生殖是很罕见的现象,在文献中很少关于这一过程的细胞胚胎学报道,水稻方面更无所见。子房培养无疑为研究这一过程提供了具有一定诱导频率的、可以重复的实验体系。

二、卵细胞的异常分裂

在大麦的子房培养中,我们曾初步判断卵细胞为雌核发育原胚的起点[4]。但在水稻方面,卵细胞的行为却很特殊。和助细胞相反,卵细胞往往在培养后异常膨大并高度液泡化。某些液泡化的卵中发生游离核分裂,形成特殊的多核体,以后又可通过形成胞壁,衍生为很大的、液泡化的多细胞结构。这种结构很难有进一步发育的前途。在自然情况下,卵(或受精卵)的游离核分裂是极为罕见的。而在离体培养条件下,却有相当数量的卵分裂为多核体,这是一个有趣而值得注意的问题。

三、极核的分裂

子房培养中另一个有趣的现象是未受精极核的分裂。这一现象首先是在棉花未受精胚珠培养中发现的。我们在水稻和大麦的子房培养中也都观察到未受精的极核分裂成类似胚乳游离核。游离核的数目多寡不定,有的成片分布,甚至在局部区域形成细胞。但从来没有出现像体内胚乳那样大规模形成细胞和积累淀粉的情况。极核的分裂和雌核发育的胚胎发生之间并不相关,很难认为它们具有体内胚乳的类似功能。众所周知,诱发极核的分裂比诱发卵的分裂更为困难。因此,子房培养在这一点上也可作为研究胚乳发生机理的一个手段。

四、反足细胞的分裂

在水稻的培养子房中,反足细胞一般为类似体内的腺型多细胞构造。但偶尔亦可分裂成类似愈伤组织或原胚的结构。在大麦中也观察到同样的情况[4]。但它们的发生频率很低,而且尚难确定有否发育前途,显然在雌核发育的起源上不占重要地位。

五、原胚的生长和愈伤组织化

雌核发育原胚早期在形态、大小上和体内的合子原胚近似。以后，一部分原胚细胞数目不断增多，生长为占据胚囊内很大一部分空间的巨大的梨形或球形结构。其内部常出现弧带形的分生组织区，显示一定程度的组织分化；但在器官发生上仍保持不分化的状态。这和体内合子胚迥然不同。在原胚生长的不同时期，均可局部地或整体地愈伤组织化，尤其是其合点端的组织常发生不规则增生而呈现不同程度的凸凹。培养后期，大部分仍然生活的原胚均趋于愈伤组织化，一般需要经过解剖、转移到分化培养基上才分化根、芽。这样一种构造，在形态发生上颇为特殊，似乎是介于胚状体和愈伤组织之间的一种情况，有点类似于兰科植物的"原球茎"(protocorm)。它的形成，表明胚胎发育偏离正常有组织生长的轨道，而转以无组织生长占优势。导致这一现象的原因尚不清楚，很可能和缺乏胚乳哺育以及外源激素的施用有关。

六、胚状体的分化

极少数原胚可以正常分化。首先由侧面产生胚芽鞘原基的唇状突起，继而出现盾片、胚芽、胚根等构造。其分化方式和体内合子胚基本相似，惟外形不如后者规则。在大麦的子房培养中，我们也看到类似的过程，而没有前述愈伤组织形成的情况[4]。但在水稻方面，分化的胚状体只占微不足道的比重（2%左右）。如何控制原胚的发育，使之走胚状体分化而不走愈伤组织化的途径，是一个很有研究价值的问题。

七、原胚的异常发育与多胚现象

在培养材料中，可看到原胚呈现各种异常的形态：(1)液泡化的原胚。(2)由不同组织构成的"分区"原胚。(3)具长形胚柄的原胚。(4)多胚现象。一般一个胚囊中仅有一个雌核发育原胚，但也有两个以上的情况。我们曾推测多胚现象有"多元起源"（由胚囊中不同细胞产生）和"单元起源"（由一个细胞产生）两种可能。随后在大麦与水稻中均证明了多元起源，在水稻中还证明了单元起源。观察到原胚在生长过程中通过"胚出芽"(embryonal budding)方式形成"分裂多胚"的现象。多胚现象在水稻子房培养中所占比重虽然不大，但若能提高其发生频率，对扩大繁殖是有意义的。

八、原胚的败育

经过二周的培养后，约有半数以上的原胚先后出现败育迹象。败育通常始于原胚外围组织的液泡化，继而组织离散、细胞解体。这一过程逐渐向心发展，扩及全体。在禾本科植物花药培养中，多细胞花粉或愈伤组织的败育是限制花粉植株产量的一个严重不利因素。看来子房培养也有类似的问题，有必要加强对败育原因及防止措施的研究。

九、体细胞增生

如何区分配子体起源和孢子体起源的愈伤组织(或胚状体)，是子房培养中一个不可回避的重要问题。在水稻中，已经证明过高的外源激素浓度导致体细胞愈伤组织的增生，但如激素浓度适宜则无此现象发生。根据大量切片的观察，从未发现在0.125ppm MCPA 条件下由珠心、珠被组织产生愈伤组织或不定胚的现象。可以认为：凡是由子房中解剖出来的愈伤组织，均系起源于胚囊细胞的雌核发育。由于避免了"鱼目混珠"，在估计不同处理的诱导频率和分析再生植株倍性变化的原因时，就能做到有所根据。此外，从培育单倍体及纯合二倍体的目的来看，水稻子房培养技术中的这一优点也是显而易见的。

再生植株的染色体与性状表现

对于花药培养的植株及其后代，已有较多的细胞学与遗传学研究。但在未传粉子房培养方面的工作则较贫乏。这是由于迄今所作的子房培养实验多带探索性质，诱导的植株数偏少，不够满足统计的需要。注意到这一点，我们就力求扩大培养的规模以获得足够的植株群体；同时，通过比较同一品种在相同培养条件下产生的"胚囊植株"与"花粉植株",来认识它们之间的共性与个性。

一、染色体倍性

根据"早粳19"品种111个"胚囊植株"和97个"花粉植株"的根尖染色体观察,"胚囊植株"中包括单倍体(占77.5%)和二倍体(占22.5%)两种类型;"花粉植株"中则包括单倍体(占63.9%)、二倍体(占34.0%)与多倍体(占2.1%)三种类型。在单倍体和二倍体类型中,既包括所有观察细胞均属同一倍性的植株,也包括以某一倍性的细胞为主而兼有其它倍性细胞的混倍体植株,虽然后者所占比例不高。总起来看,与"花粉植株"的倍性分布相比,"胚囊植株"中的单倍体比例较高,二倍体比例较低。由于胚胎学观察基本排除了愈伤组织起源于子房或花药体细胞的情况,因此在这里二倍体类型出现的原因便只能解释为雌核发育与雄核发育过程中染色体自然加倍的结果,即主要为纯合二倍体。如果是这样,则在相同条件下子房培养时染色体的自然加倍频率相对地较低。这可能是由于雌核发育原胚的愈伤组织化程度不及花粉愈伤组织所致。

二、白化现象

白化现象是多数禾本科植物花粉植株的通病。有时白化苗在比例上占压倒优势,给花药培养带来非所预期的后果。在子房培养方面,以往限于观察的群体,往往不能得出准确的结论。第一年我们由5个水稻子房中培养出1株绿苗,4株白化苗。第二年在诱导成功的30个子房中,24个出绿苗,6个出白化苗。第三年由4个供试品种共计诱导了240个子房,其中79.2%产生绿苗,13.8%产生白化苗,7.0%产生绿、白苗丛(即同一愈伤组织既分化绿苗又分化白苗)。不同品种的白化苗率有所差别,但白化苗的比例均偏低。根据同一品种(早粳19)"胚囊植株"与"花粉植株"的比较,231个花粉愈伤组织中,36.4%产生绿苗,61.0%产生白化苗,2.6%产生绿、白苗丛;而159个胚囊愈伤组织中,89.3%产生绿苗,8.8%产生白化苗,1.9%产生绿、白苗丛。由此可见,"胚囊植株"绿苗率高、白化苗率低,恰与"花粉植株"呈明显的对比。这是子房培养中的一个十分突出的特点。

三、其它性状表现

对"早粳19"品种成熟的34个"胚囊植株"和10个"花粉植株"的性状进行了观察与测量。多数二倍体"胚囊植株"的形态性状和二倍体"花粉植株"及原始品种相似,但出现三个分蘖旺盛而无子结实的不育株。对其中一株作了花粉母细胞减数分裂染色体构型的分析:在83个中期I细胞中,仅27个细胞(占32.5%)具12个二价体;其余56个细胞(占67.5%)则具数目不等的单价体。产生不育株的原因不清楚。单倍体的"胚囊植株"与"花粉植株"在株高、穗长、剑叶长宽、颖花大小等性状指标上均显著低于二倍体植株,惟着粒密度较大。

对二倍体"胚囊植株"与"花粉植株"的二代株系亦作了初步考查。除在一个"胚囊株系"中有一株白化苗外,未发现其它性状变异。初步观察表明各株系内的植株基本整齐一致。

阶 段 性 结 论

1. 采用液体漂浮培养技术,可以稳定地由水稻(主要指粳稻)未传粉的子房中诱导出雌核发育的愈伤组织和再生植株。诱导频率的高低受多种因素的影响。已经研制出比较适宜的培养技术流程。其中的关键措施是:(1)选用诱导率较高的品种。(2)选取胚囊发育时期适宜的接种材料。(3)采用去颖、带花托、护颖与雄蕊的幼花作为外植体。(4)采用适宜的培养基,尤其是控制外源激素的浓度。(5)在液体培养基上进行漂浮培养,适时解剖子房,将愈伤组织转移到分化培养基。

2. 在子房培养条件下,水稻胚囊内的四种细胞均可不经受精启动分裂,但其发育前途各异。助细胞无配子生殖是雌核发育的主要来源;卵细胞可行异常的游离核分裂;极核可分裂成类似胚乳游离核的构造;反足细胞偶尔亦可形成愈伤组织状构造。未观察到来自珠心与珠被的体细胞愈伤组织或胚状体。一般一个胚囊中仅含一个雌核发育构造,也有多胚现象。

3. 水稻子房培养时雌核发育的主要途径是:培养初期的雌配子体发育→原胚发生→原胚生长与愈伤组织化→根、芽发生。次要途径是由原胚分化为类似体内合子胚的胚状体。在培养过程中,原胚可呈现各种异常的形态。在发育中、后期,有相当数量的原胚败育。

4. 水稻子房培养产生的"胚囊植株",四分之三以上为单倍体类型,余为二倍体类型;绿苗率可高达 80% ~ 90%,白化苗很少。

附录:水稻未传粉子房培养技术流程
一、取材
1. 选取生长健壮、发育适宜的稻穗,用酒精拭擦剑叶,剥出幼穗,在 0.1% 升汞中表面消毒 10 ~ 12min,无菌水洗四次。

2. 根据花粉发育(单核靠边至二核初期)选取胚囊发育适宜(单核至四核胚囊)的小花。除掉外颖与内颖,保留花托、护颖、雄蕊与雌蕊,作为接种材料。

二、诱导培养
1. 培养基成分为:N_6+MCPA 0.125 ~ 0.5ppm+蔗糖 3%。

2. 将外植体接种于上述液体培养基表面,任其漂浮,不可沉没。每试管(30mm×200mm)盛 20 ~ 25ml 培养基,接种 10 ~ 15 个外植体。

3. 置于 25℃、黑暗或散射光条件下培养。约经一周,大多数子房膨大。

4. 约经一个月培养后,用镊子剖开膨大的子房和胚珠,如肉眼观察到其中包含白色的团块,即为雌核发育的愈伤组织。

三、分化培养
1. 培养基成分为:MS(或 N_6)+IAA 0.5ppm+NAA 0.25ppm+KT1 ~ 2ppm+蔗糖 3%+琼脂 0.8%。或为:N_6+MCPA 0.033ppm+蔗糖 3%+琼脂 0.8%。

2. 将解剖出来的雌核发育愈伤组织转移到上述分化培养基上,置有光条件下培养。

3. 约经数日或一月后,愈伤组织开始分化根、芽,形成小植株。

4. 将小植株移植盆中,进行温室栽培,获得成熟的"胚囊植株"。

(作者:周嫦、杨弘远、田惠桥、刘中来、阎华。原载:武汉大学学报(自然科学版),1983,4:146~153。参考文献11篇,均删去。)

10 *In vitro* production of haploids in rice through ovary culture

The spontaneous occurrence of haploids in rice (*Oryza sativa* L.) is a rare and sporadic phenomenon. Since anther culture in rice was first successfully carried out by Niizeki and Oono (1968), technical improvements have made it easy to produce haploid pollen plants in this crop. Nevertheless, it has remained unknown whether it is possible to obtain haploids also from the female cells by the *in vitro* technique. At the beginning of the 1980's, two laboratories separately reported the induction of haploid rice plants by culture of unpollinated ovaries (Asselin de Beauville 1980; Zhou and Yang 1980, 1981a). Subsequently, Kuo (1982) also obtained similar results. From that time on, a series of fundamental research has been done in order to improve the culture technique and to elucidate the origin and developmental aspects of the ovary-derived haploids (Zhou and Yang 1981b, c; Zhou et al. 1983, 1986; Tian and Yang 1983, 1984; Liu and Zhou 1984; Li and Yang 1986; He and Yang 1987, 1988). Ovary culture now provides a reliable experimental system for haploid production in rice, at least in *japonica* varieties, and is therefore worth pursuing.

Culture techniques

Two procedures have been worked out for rice ovary culture with application of different exogenous hormones: one with 2-methyl-4-chlorophenoxy acetic acid (MCPA), another with 4-amino-3,5,6-trichloropicolinic acid (Picloram):

Procedure I

1. Choose suitable young panicles and sterilize them with 0.1% $HgCl_2$ solution for 10-12 min. Rinse thoroughly with sterile distilled water.

2. Choose suitable spikelets by pollen staging. Late-unicellular until early bicellular pollen stage coinsiding with uninucleate to four-nucleate embryo sac stage is recommended.

3. Remove lemma and palea from the spikelets and keep the remaining parts as a unit for inoculation. This comprises a pistil, six stamens, and a pair of sterile glumes, all attached to a short piece of pedicel.

4. Place the unhusked spikelets on liquid N_6 medium supplemented with 0.125-0.5 mg/L MCPA and 3% sucrose, 10-15 spikelets per culture tube containing 20-25 ml medium.

5. Keep the explants floating on the medium at 25℃ in darkness.

6. After 30-40 days of culture, take out the enlarged ovaries, dissect them, and pick out the apogamous proembryos, if any are inside, under aseptic conditions.

7. Transfer the apogamous proembryos on solid N_6 or MS medium supplemented with 0.5 mg/L indole acetic acid, 0.25 mg/L naphthalene acetic acid, 1-2 mg/L kinetin and 3% sucrose, or alternatively with merely 0.033 mg/L MCPA and 3% sucrose. Keep the culture under illumination.

8. After one or several weeks, the proembryos grow into larger callusing masses. Shoots and roots may emerge and plantlets are regenerated.

9. Transplant the plantlets into pots for greenhouse cultivation.

Procedure II

1. Follow steps 1-3 of procedure I.
2. Float the unhusked spikelets on liquid N_6 medium supplemented

with 2 mg/L picloram, 500 mg/L lactalbumin hydrolysate, and 3% sucrose.

3. Keep the culture at 25℃ in darkness.

4. After 30-40 days of culture, embryoids can be excised from some of the enlarged ovaries and transferred onto differentiation medium. Alternatively, continuing culture for another 20-30 days under illumination, a few plantlets can be regenerated directly from inside the ovaries.

Factors affecting induction of haploids

Donor cultivars

Among 19 rice cultivars tested, all 15 *japonica* cultivars were responsive to ovary culture, but the *indicas* were relatively recalcitrant (Table 1).

Embryo sac stage

The optimal stage for induction of apogamous proembryos ranged from uninucleate to four-nucleate embryo sacs, corresponding to the unicellular to early bicellular pollen stage. As the embryo sacs matured, the induction frequency significantly lowered. On the other hand, inoculated at the megaspore-tetrad stage, the ovaries did not form haploid proembryos at all (Table 2).

Explant

Reports on pollinated ovary culture have shown that floral organs, e. g., calyx or perianth, play important roles in ovary growth and embryo development (Bajaj 1966; Rangan 1984). The same situation was found in rice unpollinated ovary culture. As Table 3 shows, the best result was obtained when intact pistil, stamens, and sterile glumes attached to a short piece of pedicel as a unit was cultured. A positive, yet less effective result was gained when stamens were removed from such a unit. Single, detached pistils, however, did not respond. Microscope observations have excluded the possibility of pollination and fertilization while stamens are involved in culture. Therefore, the increase of induction frequency appeared to be attributed to the supply of nutritive and stimulative factors from the floral parts including stamens.

Table 1 Induction frequency of apogamous proembryos in various rice cultivars

Cultivar	% ovaries producing apogamous proembryo
Nong Ken No. 4	12.0
Shi Yu	9.1
Xiao	8.5
Fu Wan No. 23	8.4
Zao Geng No. 19	7.5
E Wan No. 3	7.1
702	5.6
Da Li No. 3	5.5
Jing Hong No. 2	5.1
Hong Qi No. 16	4.5
Yan Geng No. 5	3.0
Guai Hua Huang	2.8
041	2.5
Li Ming	2.5
Tu Dao	1.5
5350[a]	2.8
Jing Nan Te No. 43B*	1.0
Zhen San No. 97*	0
Er Jiu Qing*	0

* *indica* cultivars; the others are *japonicas*.

Table 2 Effect of embryo sac stage on rice ovary culture

Embryo sac stage	Corresponding pollen stage	% ovaries producing apogamous proembryos		
		Zao Geng No. 19	Jing Hong No. 2	Hong Gi No. 16
Megaspore tetrad	Early unicellular	0	0	0
1-4-nucleate	Late unicellular	10.0	3.2	3.8
8-nucleate to cellular	Bicellular	3.1	2.6	7.8

Table 3 Effect of floral parts on rice ovary culture

Floral parts incoculated as a unit	Ovaries inoculated	% enlarged ovaries	% ovaries producing apogamous proembryos
Pistil + stamens + sterile glumes + piece of pedicel	98	81.6	4.1
Pistil + sterile glumes + piece of pedicel	110	77.3	2.7
Pistil	100	0	0

Culture method

Most investigators working on ovary culture used solid medium. In rice, Asselin de Beauville (1980), and Kuo (1982) succeeded with solid medium, whereas Zhou and Yang (1980, 1981a) worked with liquid medium. A comparison of the effects of the different culture methods shows the advantage of liquid over solid medium (Table 4).

Table 4 Effects of liquid and solid media on rice ovary culture

Culture medium	% ovaries producing apogamous proembryos		
	Nong Keng No. 4	Da Li No. 3	Xiao
Liquid	7.8	5.5	8.5
Solid	4.9	1.8	0

Exogenous hormone

MCPA was found by Chou (Zhou) et al. (1978) to have a strong promotive effect in rice anther culture and since then has been used routinely as an exogenous hormone both for anther and ovary culture in our laboratory. For rice ovary culture, the optimal concentration of MCPA lay between 0.125 and 0.5 mg/L. A higher concentration stimulated unwelcome somatic calli and inhibited haploid products. On the other hand, deletion of exogenous hormone hindered ovary enlargement and embryoid development (Table 5). Besides MCPA, other exogenous hormones could also be used in rice ovary culture, such as naphthalene acetic acid (Asselin de Beauville 1980; Zhou and Yang 1980), 2,4-dichlorophenoxy acetic acid (Kuo 1982), and picloram (Liu and Zhou unpubl.). He and Yang (1987) found that picloram exhibited a much better effect on the differentiation of apogamous proembryos than MCPA did, although its induction frequency was somewhat lower (Table 6).

Table 5 Effect of MCPA concentration on rice ovary culture

MCPA concentration (mg/L)	% ovaries producing apogamous proembryos		
	Nong Keng No. 4	Jing Hong No. 2	Hong Qi No. 16
0	0.9	0	0
0.125	8.2	6.3	3.8
0.5	9.0	1.1	2.0
2	3.1	1.0	2.1
8	0	0	0

Table 6 Effects of picloram and MCPA on rice ovary culture

Exogenous hormone	% ovaries producing apogamous proembryos			% differentiated embryoids		
	Days of culture			Days of culture		
	30	40	60	30	40	60
Picloram (2 mg/L)	5.4	5.0	5.8	50.0	63.6	84.7
MCPA (0.2 mg/L)	7.1	5.9	4.8	14.2	7.1	0

Sucrose concentration

One percent sucrose was not able to support ovary enlargement and proembryo development; 3% was optimum; 6% was the next best; 9% showed an inhibitory effect (Table 7).

Table 7 Effect of sucrose concentration on rice ovary culture

Sucrose concentration (%)	% enlarged ovaries	% ovaries producing apogamous proembryos
1	37.5	0
3	91.7	6.7
6	78.0	5.0
9	80.0	1.4

Dark and light

In contrast with Asselin de Beauville (1980), who carried out her rice ovary culture under 1000 lx illumination, Zhou and Yang (1980, 1981a) kept rice ovaries in darkness during the first month of culture. He and Yang (1988) found that illumination of 800 lx caused the apogamous proembryos to suffer serious degeneration as compared with dark culture (Table 8).

Table 8 Effects of dark and light on rice ovary culture

Condition	% ovaries producing apogamous proembryos			% ovaries with degenerated proembryos		
	Days of culture			Days of culture		
	6	14	35	6	14	35
Dark	10.8	5.8	3.8	7.0	17.0	25.0
Light	10.0	3.5	2.4	36.0	66.7	100.0

Embryological studies

Methodology

Embryological study on unpollinated ovary culture is by no means an easy task due to its relatively low induction frequency and the necessity of preparing a large number of microscopic sections. Methodologically, two points should be considered: first, choose possible high-responsive cultivars; second, simplify specimen preparation. Ehrlich's hematoxylin in toto staining combined with paraffin sectioning greatly facilitates the work. In toto staining combined with methyl salicylate clearing offers a more simplified method (Yang 1986).

Method I: In toto staining-paraffin sectioning

1. Fix the cultured ovaries in Carnoy's fluid (3 : 1) and store in 70% ethanol in a refrigerator.

2. Hydrate the materials gradually to distilled water.

3. Stain the whole ovaries with diluted Ehrlich's hematoxylin. One part of fully ripened Ehrlich's stock solution diluted with one part of acetic acid-50% ethanol (1 : 1) mixture is recommended. Stain for 48h.

4. Wash the ovaries with distilled water several times for 1-2 days.

5. Rinse with tap water for 1-2 days until the materials turn blue.

6. Dehydrate, infiltrate, embed, and section the ovaries according to the routine paraffin method.

7. Deparaffinize the sections with xylene and mount in balsam.

Method II : In toto staining-methyl salicylate clearing (Fig. 1)

1. Follow steps 1-2 of method I.

2. Stain the whole ovaries with diluted Ehrlich's hematoxylin as in step 3 of method I. However, the staining duration is shortened to approximate-

ly 2h (at 20℃).

3. Follow steps 4-5 of method I.

4. Dehydrate with ethanol series and , through anhydrous ethanol-methyl salicylate (1 : 1) mixture, clear with methyl salicylate for at least 1 day.

5. Mount the whole ovaries in methyl salicylate and observe with a bright field microscope. Reduce the aperture diaphragm to obtain a better contrast.

6. After calculating the percentage of ovaries containing apogamous proembryos, these ovaries can be picked out for further embedding and paraffin sectioning, if more detailed observations are needed.

Division in the embryo sacs of cultured ovaries

All four kinds of component cells in rice embryo sac could be triggered to division *in vitro* in the absence of fertilization, but with different frequencies and along divergent pathways. Among 1155 embryo sacs observed after 3-7 days of culture, divisions occurred in 31.1% of the embryo sacs (Table 9). The synergid possessed the highest division frequency and usually resulted in proembryo formation. The egg cell, with the next high division frequency, often underwent free nuclear division giving rise to coenocyte formation. The polar nuclei in the central cell sometimes divided into limited free nuclei similar to early endosperm. Only in rare cases did the antipodal cells divide into proembryo-like structures. Obviously, embryological studies confirmed that in rice ovary culture it was actually an *in vitro* apogamy and not a parthenogenesis that occurred (Tian and Yang 1983). This conclusion was further proved by the failure of attempts aimed to provoke egg cell parthenogenesis by changing culture conditions. In a series of experiments including changing the basic media, kind and concentration of exogenous hormones, organic supplements, and culture methods and conditions, synergid apogamy took place repeatedly (He and Yang 1988).

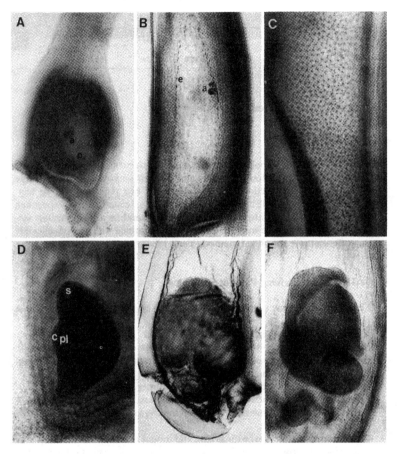

Fig. 1A-F. Observations on rice embryo sac by in toto staining-methyl salicylate clearing technique. **A.** Whole view of a cleared ovary, showing the embryo sac. A pair of polar nuclei(p) and the antipodal cell mass (a) are visible. × 100
B. An ovary after anthesis, showing free endosperm nuclei (e) distributed at the periphery of the embryo sac and the degenerated antipodal cells. ×80 **C.** Large area of endosperm cells filling the embryo sac are visualized by focusing on the suitable optical section. ×100 **D.** A young zygotic embryo. The configuration of the scutellum (s), coteoptile (c) and plumule (pl) is seen. ×100 **E.** In a cultured ovary a protocorm-like embryoid is shown. ×30 **F.** A differentiated embryoid in a cultured ovary. ×80

Table 9 Division of embryo sac cells during 3-7 days of rice ovary culture
(1155 embryo sacs were observed)

Kind of cells triggered to division	No. embryo sacs with division	% embryo sacs with division
Synergid	188	16.3
Synergid + egg cell	42	3.6
Synergid + central cell	5	0.4
Synergid + antipodal cell	6	0.5
Egg cell	31	2.7
Central cell	17	1.5
Antipodal cell	10	0.9
Unknown	60	5.2

Processes of synergid apogamy

Ovary culture offers a stable, reproducible system for studying the detailed processes of synergid apogamy, which is otherwise unavailable since its natural occurrence is of extremely low frequency (Raghavan 1976). The results obtained by Tian and Yang (1983), followed by Li and Yang (1986), can be summarized as follows: In the cultured young ovaries, development could be switched to formation of various abnormally organized embryo sacs. The main pathway leading to apogamy went via a linearly oriented four-nucleate embryo sac to a linearly oriented egg apparatus, which was quite different from the triangular distribution of a normal egg apparatus. It was usually the chalazal synergid (adjacent to the egg cell) that gave rise to the apogamous proembryo, and the micropylar synergid (distant from the egg) degenerated. The proembryo thus formed was located at the base of a highly vacuolated egg cell (Fig. 2A-D). Another pathway went through a nonlinearly oriented four-nucleate embryo sac to the formation of an egg apparatus in which the two synergids were located at one side of the egg and arranged longitudinally. In this case, it was often the chalazal synergid that could be triggered to apogamy, resulting in a peculiar hook-shaped proembryo embracing the egg (Fig. 2E-G). The third, rarely occurring pathway was taking place in ovaries inoculated at nearly matured embryo sacs with normally arranged egg apparatus. The few proembryos observed were all located by the side of the egg and were hookedshaped(Fig. 2H).

Growth and differentiation of the proembryos

In the presence of MCPA, most of the synergid proembryos grew into undifferentiated large masses and often had a varying degree of callusing (Fig. 2I, J). Such structures were called protocorm by Nostog (1977). They indicated a divergence from the normal embryo development to more or less unorganized growth. Consequently, such protocorm-like structures had to be transferred to a differentiation medium for regeneration of plantlets (Tian and Yang 1984). When picloram was used as exogenous hormone instead of MCPA, many proembryos developed into differentiated embryoids with plumule, radicle, coleoptyle, and other parts resembling *in vivo* zygotic embryos (Fig. 2K, L). Direct regeneration of some plantlets on the induction medium was obtained (He and Yang 1987).

Characteristics of regenerated plants

Just as anther culture can induce haploid pollen cells, ovary culture can induce haploid embryo sac cells to develop into sporophytic plants. Whether or not these plants of different origin bear the same features is a question still to be made clear. Liu and Zhou (1984) carried out an experiment by culturing the ovaries and anthers of the same cultivar Zao Geng No. 19 under the same condition and comparing the "embryo sac plants" and "pollen plants" thus regenerated. Chromosome ploidy as well as proportion of green/albino plants were two main aspects to which attention was paid.

植物有性生殖实验研究四十年

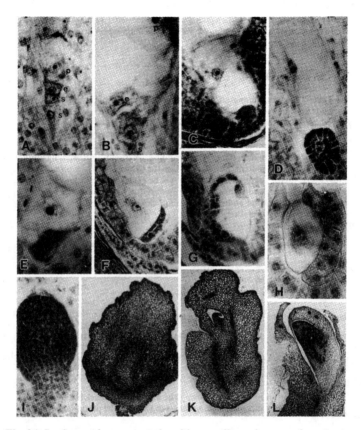

Fig. 2A-L. Synergid apogamy induced by unpollinated ovary culture in rice. The ovaries were stained in toto with Ehrlich's hematoxylin and sectioned by paraffin method. **A.** A linearly oriented egg apparatus induced by ovary culture. *Above to below* upper polar nucleus (p), egg cell (e), and two synergids (s). ×530. **B.** In linearly oriented egg apparatus, the chalazal synergid developed into an embryogenic cell, and the micropylar synergid began to degenerate. ×530. **C,D.** The synergid-derived proembryo located at the base of a highly vacuolated egg cell. ×280. **E.** Three cells divided from a synergid by the side of an egg. ×700. **F.** A synergid proembryo growing along one side of an egg. ×280. **G,H.** Hook-shaped synergid proembryo growing up to the chalazal end of the egg and embracing it. ×280. **I.** A larger synergid proembryo. ×280. **J.** A protocorm-like structure developed from the synergid proembryo in the presence of MCPA. ×150. **K.** A well-differentiated synergid embryo in the presence of picloram. ×40. **L.** Budding of a synergid embryo. ×25

Chromosome ploidy

As Table 10 shows, among all 111 "embryo sac plants" observed, 77.5% were haploids and the remaining 22.5% were diploids; among 97 "pollen plants" observed, 63.9% were haploids, 34.0% were diploids, and 2.1% were polyploids. This means that haploids were predominant among the regenerated plants of either origin, and their proportion was even higher in the former case. As for the occurrence of nonhaploids, it might be attributed to chromosome doubling during the culture rather than to derivation from somatic tissues. The lower proportion of nonhaploids in the "embryo sac plants" might be explained by the lower callusing tendency of the protocorm-like structures induced in ovary culture as compared with the calli induced in anther culture, since chromosome doubling often takes place during callus formation.

Table 10 A comparison of chromosome ploidy between "embryo sac plants" and "pollen plants" in cultivar Zao Geng No. 19

Kind of plant	Observed individuals	Ploidy					
		Haploid		Diploid		Polypoid	
		No.	%	No.	%	No.	%
Embryo sac plants	111	86	77.5	25	22.5	0	0
Pollen plants	97	62	63.9	33	34.0	2	2.1

Green and albino plants

The high proportion of albinos is known to be a serious drawback in anther culture of many cereal crops including rice. As shown in Table 11, the proportion of albinos was more than 60% of the pollen plants; whereas among embryo sac plants this proportion was reduced to negligible. Although the proportion of green plants in ovary culture of other tested rice cultivars was certainly not as high as in Zao Geng No. 19 (approx. 90%), yet its predominance was obvious, since an average proportion of 80% was observed. Other investigators working on ovary culture also reported similar results (for references see Yang and Zhou 1982).

Table 11 A comparison of green/albino ratio between "embryo sac plants" and "pollen plants" in cultivar Zao Geng No. 19

Kind of plant	Observed plant clusters	Green cluster		Albino cluster		Green and albino cluster	
		No.	%	No.	%	No.	%
Embryo sac plants	159	142	89.3	14	8.8	3	1.9
Pollen plants	231	84	36.4	141	61.0	6	2.6

Concluding remarks

During recent years, *in vitro* culture of unpollinated ovaries or ovules has succeeded in induction of haploids at least in 14 species belonging to seven families (Yang et al. 1990). However, rice ovary culture may be the unique system in which a repeatedly confirmed synergid apogamy takes place. The induction frequency of proembryo formation ranges from 1 to 12% depending on the rice cultivar. About two-thirds of the proembryos transferred to differentiation medium can be regenerated to plantlets. Compared with anther culture, this yield is, of course, not high enough. However, taking account of the fact that only one embryo sac exists in an ovary in comparison with thousands of pollen grains per anther, the rate of induced embryo sacs is actually higher than that of induced pollen grains. Moreover, the development is more close to a normal embryogenic way, and the plants regenerated behave more normally in respect of green/albino ratio. Therefore, ovary culture is a new useful way for induction of haploid plants following anther culture. Academically, it offers a stable experimental system for further research on the mechanism of synergid apogamy, an interesting but not yet well-known phenomenon in plant reproductive biology. Ovary culture may also provide a feasible means for gene engineering, since exogenous DNA can be injected into the embryo sacs to obtain transformed plants, as has been shown by some experiments in other species and on *in vivo* ovaries (Hepher et al. 1985), and in this respect an *in vitro* system for haploid production may have certain merits.

(Authors: Zhou C, Yang HY. Published in: Biotecnology in Agriculture and Forestry, vol 14 (ed, by Bajaj YPS), Springer-Verlag, Berlin, 1991, pp. 180-192. with 2 figures, 11 tables and 24 references.
All figures and tables are retained here.)

11 大麦未授粉子房培养的胚胎学观察

Embryological observations on ovary culture of unpollenated young flowers in *Hordeum vulgare* L.

摘 要

在离体条件下诱导了大麦未授粉子房内的胚状体发生。雌核发育的胚主要起源于卵器,也可能起源于反足器。原胚发育与胚分化过程中既有类似体内合子胚的正常类型,也有变态的类型。未受精的极核亦可分裂成类似胚乳的游离核。研究了胚囊发育时期与培养方式对诱导胚胎发生的影响。

Abstract

Young barley flowers of various stages (from megaspore tetrad to mature embryo sac) were used as materials for culture and subsequent embryological observation. Two culture methods, vertical flower culture and horizontal ovary culture, were adopted. The inocula were cultured at nearly 25°C in dark on N_6 medium solidified with agar (0.8%) and supplemented with sucrose (3%-12%), MCPA (0.5-2 ppm), NAA or IAA (1 ppm) and KT or BAP (0.5-1ppm). After inoculation, ovaries were sampled at 2-3 day intervals, fixed in aceto-methanol (1∶3), stained *in toto* in diluted Ehrlich's hematoxylin and sectioned by paraffin method.

In all three cultivars tested, embryogenesis within unfertilized embryo sac was observed. The gynogenetic embryos, totally 59 in number, derived mostly from egg apparatus, but some of them came from antipodals too. Usually only one embryo was located in an embryo sac, but in a few cases, two embryos within one embryo sac were observed. The first embryogenic division was transverse in direction, resulting in a basal cell and a terminal cell. The basal cell elongated strikingly and thus pushed the terminal cell toward the center of the embryo sac. Subsequent divisions often led to the formation of a proembryo with peculiar linear shape. Later, multicellular embryoids with various sizes and shapes were observed. Some of them showed organ differentiation. Most of the differentiating embryoids were similar to the ordinary zygotic embryo of barley, with a terminal scutellum and a lateral coleoptile. However, some of them showed some abnormal appearance.

Ovaries inoculated at megaspore tetrad stage could not be induced to gynogenesis, although in a few cases probable nucellus embryos were observed. Instead, ovaries inoculated at later stages (from uninucleate to mature embryo sacs) did give rise to gynogenetic embryogenisis without the occurence of adventitious embryogeny. The induction-frequency was higher in materials inoculated at 8-nucleate or mature embryo sac stages than at earlier stages. In the latter cases, triggering of embryogenesis could take place only when the embryo sacs were well-differentiated after a period of gametophytic development during culture.

Gynogenetic embryos could be induced by both vertical flower culture or horizontal ovary culture, but the former was superior in providing better conditions for growth of ovaries and embryo sacs and thus yielded more embryoids.

Divisions of unfertilized polar nuclei leading to endosperm-like free nuclei were also found in cultured ovaries. However, such structure was not likely to play a similar role of nurse tissue as *in vivo* for the gynogenetic embryos *in vitro*, since it did not often accompany the occurrence of embryoids within the same embryo sacs.

(作者:黄群飞、杨弘远、周嫦。原载:植物学报,1982,24(4):295～300。图版2幅,表3幅,参考文献11篇,均删去。)

12　由向日葵幼花或胚珠培养出单倍体小植株与胚状体

In vitro induction of haploid embryos and plantlets from unpollinated young florets and ovules of *Helianthus annuus* L.

未传粉子房与胚珠培养是一项新的研究课题。迄今仅有八种植物通过这一方法培育出单倍体小植株[1]。两年来,我们由向日葵幼花与胚珠培养诱导出单倍体小植株与胚状体。

材　料　与　方　法

先后采用10个向日葵品种为材料进行实验。方法大致如下:采用开花前 1~6d 的管状花,用 0.1%升汞消毒 12~15min,无菌水洗 3~4 次。将整朵幼花、单独的子房、胚珠或花药分别接种在液体培养基上静置漂浮培养。每管接种幼花 5 朵或胚珠 10 枚。诱导培养基以 MS 或 N_6 为基本培养基,分别补加 2%或 6%蔗糖以及不同浓度或组合的生长素(MCPA、NAA 或 IAA)和(或)细胞分裂素(KT、BA)等。接种 45~55d 后,在无菌条件下将幼花或胚珠取出,将胚珠内的胚状体小心地挑出,转移到以 MS、N_6 或 White 为基本培养基,附加 KT 与 IAA 或不加激素的固体培养基上,促使继续生长分化。生根之后,取下根尖,用 0.01%秋水仙素+0.001 M 8-羟基喹啉混合液预处理 2.5~3h。卡诺液(3∶1)固定 4h,70%乙醇保存。用孚尔根染色法压片,检查根尖染色体。

结　　果

两年来,进行了幼花、子房、胚珠与花药培养。幼花与胚珠培养诱导单倍体初获成功,子房与花药培养则告失败。以下简述前二者的实验结果。

幼花培养　幼花接种在 MS+MCPA 0.125~2mg/L+蔗糖 6%等较适宜的培养基上,花柱或多或少伸长,子房迅速膨大。3~4 周后,有的子房接近正常瘦果大小,呈现成熟果皮色泽与条纹。接种 45~55d,少数膨大子房内可解剖出形状、大小不一的胚状体。9 个品种中有 6 个品种的 49 朵幼花产生了胚状体。诱导频率变动在 0.5%~16.0%之间。胚状体转移至固体培养基上,长出 4 个小植株(图 1),另有 8 个胚状体生根。对其中 6 个植株与胚状体的根尖压片,鉴定它们是单倍体(2n=x=17)(图 2)。

胚珠培养　胚珠在 N_6+肌醇 100mg/L+MCPA 0.5~2mg/L+蔗糖 2%~6%等较适宜的培养基上亦迅速膨大。开始保持乳白色,以后可能逐渐转黄或变褐色。45~55d 后,剖开珠被,部分胚珠内含发育程度不等的小胚状体(图 3)。8 个品种有 6 个品种的 360 个胚珠内含胚状体,诱导频率为 3.3%~44.8%。有 96 个胚状体长出了根,但未能成苗(图4)。根尖压片表明,8个胚状体是单倍体(图5),1个是二倍体。

体细胞愈伤组织 一般而言,体细胞愈伤组织的产生往往不利于雌核发育进行。在向日葵幼花或胚珠培养试验中,均曾观察到体细胞愈伤组织的形成。在含激动素的培养基上,体细胞愈伤组织大量产生,甚或遍布子房或胚珠表面。如培养基不含激动素,体细胞愈伤组织明显减少或无。这类愈伤组织经转移分化,亦能出苗(图6)。根尖染色体检查为二倍体。这一工作为向日葵体细胞培养分化成苗提供一个新的实验结果。

讨 论

我们曾经指出:未传粉子房与胚珠培养的方法在某些情况下可能具有独特的价值。例如有些植物花药培养至今难以成功,子房与胚珠培养可能提供一条产生单倍体的有效途径[1]。关于向日葵花药或子房、胚珠培养均未见有成功的报道[1,2]。本实验由向日葵幼花培养获得了单倍体小植株,胚珠培养获得了单倍体胚状体。这一初步试验预示了通过雌核发育途径进行向日葵及其它类似情况下的植物单倍体育种的前景。

防止体细胞增生是诱导雌核发育中的一个重要问题[1]。水稻未传粉子房培养试验中,调节外源激素浓度有抑制体细胞愈伤组织的效果[3]。向日葵幼花与胚珠培养亦存在这一问题。初步试验表明,激动素往往促进体细胞愈伤组织的形成。除去激动素,体细胞增生明显降低而雌核发育相对地得到促进。因此,外源激素的种类是影响向日葵幼花与胚珠培养的一个重要因素。此外,我们参考水稻子房培养经验[3,4],在向日葵工作中采用了液体漂浮培养技术,获得良好效果。再次表明液体培养是诱导雌核发育值得研究和利用的一种方法。

在某些实验组合中向日葵幼花与胚珠培养有相当高的诱导频率。但是成苗都很困难。幼花培养虽然产生了小植株,数量却有限,又不易长大。这和向日葵组织培养很少再生植株的现象有类似之处[2,5]。因此,如何促进胚状体充分发育、正常分化与成苗仍是近期必须着重研究解决的问题。

(作者:蔡得田、周嫦。原载:科学通报,1983,(22):1399~1401。图1幅,参考文献,均删去。)

13 向日葵未受精胚珠培养时胚状体发生的显微观察

Microscopical observations on the embryoid formation in cultured unfertilized ovules of *Helianthus annuus* L.

Abstract

A recent advance in plant experimental embryology is the induction of haploid plants via *in vitro* culture of unfertilized ovules. Using float culture method on liquid media, we have raised haploid as well as diploid embryoids in sunflower cultivars by ovule culture. The present investigation was aimed to know the origin and developmental processes of these embryoids. Young flowers 1-4 days before anthesis were dissected and ovules were inoculated on N_6 medium supplemented with 0.5-2 ppm MCPA and 6% sucrose. During culture period, samples were collected at intervals, fixed, stained and sectioned by paraffin method. Fifty one gynogenic embryoids of various sizes were observed among ca. 2000 ovules. They were located at the micropylar end of the embryo sacs and proved to be originated from the unfertilized egg cells. At the early stages, they bore a strong resemblance to the zygotic proembryos *in vivo*, but after a considerable enlargement, they grew into globular, ovoid or elongated big bodies without polarized organ differentiation. Chromosome counts on some mitotic figures in these embyoids revealed their haploid nature.

Embryoids were also produced from the endothelial tissue, which proliferated markedly after inoculation, especially at the chalazal parts, resulting in massive multilayered irregular folds and then degenerated. In some cases, cell divisions at one or several places led to embryoid or callus formation. The problems of how to regulate the growth of *in vitro* ovules in order to promote the gynogenic embryoids and inhibit the somatic embryoids or calli are left for future research.

近年来,关于未传粉子房的培养已有一些研究;而直接由未受精的胚珠培养出单倍体植株只有极少数成功的报道[1,6],并且完全缺乏胚胎学方面的资料。最近,蔡得田与周嫦在向日葵的未受精胚珠培养中诱导出胚状体,生根后经染色体鉴定为单倍体,亦有二倍体[5]。为了弄清在这种情况下胚状体的起源和发育过程,我们进行了较详细的胚胎学研究。

材 料 与 方 法

实验于 1982~1983 年进行。材料为晚熟品种"当阳向日葵"(*Helianthus annuus* L.)。取开花前三、四天和开花前一天的幼花,剥开子房,将胚珠取出接种于含 2 甲基-4 氯苯氧乙酸(MCPA)2(或 0.5)ppm 与蔗糖 6% 的 N_6 液体培养基上,在 27℃ 左右、散射光条件下进行静置的漂浮培养[5]。接种后 40 天内分期取样,用醋酸甲醇(1:3)固定,70% 甲醇保存,经稀释的爱氏苏木精整体染色后,按石蜡法制片,进行显微观察与摄影。共观察了两千多个胚珠。

结　果

一、接种时的胚珠

向日葵的胚珠属单珠被、薄珠心、倒生型。珠心早期退化，胚囊成熟时直接与珠被最内一层特化细胞——珠被绒毡层接触。本实验中用于接种的胚珠，有开花前三、四天和开花前一天两个时期。从内部发育状况看，它们均已达到雌配子体形成阶段。图版 I，1 示开花前三、四天的胚囊，雌配子体的各部分已经形成：一个卵细胞、两个助细胞、一个中央细胞（含一个次生核）、两个反足细胞（含多核）。图版 I，2 示开花前一天胚囊的近珠孔部分：卵细胞已经成熟，具明显的核与核仁，胞质中含丰富的物质；助细胞长而具充分发育的丝状器；中央细胞中胞质染色很浓，次生核很大；反足细胞不在本切片上；珠被绒毡层由一层含单核或二核的细胞构成。考虑到这两类胚珠在培养条件下均有胚状体发生，且其起源和发育过程基本一致，因此下文将它们合并起来描述。

二、雌核发育的胚状体

起源　从培养后 5d 开始，胚囊中出现胚胎发生，至培养后 40d 止，共观察到 51 个处于不同发育时期的胚状体，约占观察胚囊总数的 2.8%。一般一个胚囊中仅有一个胚状体，均位于珠孔端。在培养早期的材料中，根据卵细胞、助细胞和次生核的状况，可以断定胚状体是由卵细胞分裂而成。图版 I，3，4 为这一点提供了明确的证据：早期的原胚不仅具有卵细胞的典型形态和位置，而且附近尚有助细胞和次生核残存可作为旁证。因此向日葵未受精胚珠培养时的雌核发育，实质上是一种离体的孤雌生殖。未观察到助细胞分裂发育的现象。在某些情况下，次生核可发生有限的游离核分裂，与棉花[8]、水稻[3]、大麦[4]中的情况相似。反足细胞有时亦可分裂为松散的多细胞群体。但未看到它们形成有组织的胚状结构。

发育过程与形态发生特点　培养后 5d，观察到卵细胞分裂为二细胞原胚,其胞质中仍含有原来染色很深的物质(图版Ⅰ,3)。以后发育速度很快,7d时已有多细胞的原胚,但仍保持原有梨形,且具明显的胚柄(图版Ⅰ,4、5)。其中有的原胚是经游离核分裂形成的多核体结构,此类原胚没有发展前途,多呈败育状态(图版Ⅰ,6,图版Ⅱ,14)。10~13d的原胚丧失极性生长,成为球形或卵形的构造(图版Ⅰ,7、8、9)。这样的胚状体可以继续细胞分裂而增大体积。培养20d以后,胚状体内部的细胞渐停分裂,外周保持一个在切面上呈圆圈形的分生组织区,表层细胞不断离散脱落而由分生组织区补充新的细胞(图版Ⅱ,10、11)。直至培养40d也没有看到胚状体分化胚芽、胚根、子叶、下胚轴等器官。

染色体倍性 在生长旺盛的胚状体中常可看到正在分裂的细胞。对其中某些分裂相可以进行染色体计数。图版Ⅰ,8的胚状体顶端有一个很清晰的分裂相。图版Ⅱ,12示另一个胚状体的分裂相放大。它们的染色体均为单倍数($n=17$)。用这样的方法,不只一次地证明了胚囊内胚状体的单倍体性质,并可借以区分于孢子体起源的胚状体。直接由切片材料鉴定正在发育的胚状体的倍性在向日葵上被证明是可行的。

三、珠被绒毡层的变化及其胚状体发生

增生 培养后,珠被绒毡层普遍有一个增生的过程:一方面通过垂周分裂扩大其周长;另一方面通过平周分裂增加其细胞层数,由原来的一层细胞变成由多层细胞构成的结构。与此同时,珠被绒毡层在中央细胞和反足细胞之间的位置相向生长,形成葫芦形缢缩,最终互相连接而将胚囊分隔成上下两个腔室,近珠孔腔含卵器与中央细胞,近合点腔含反足细胞。这样,整个珠被绒毡层在切面上呈"8"字形,其近珠孔部分的细胞排列较整齐,近合点部分的细胞则作不规则增生而形成高度折叠的构造(图版Ⅱ,13、14)。约在培养10d以后,增生达到顶峰,退化过程渐占优势,终至完全解体。无论胚囊中有无雌核发育胚状体,均存在上述珠被绒毡层的增生和退化过程。当有胚状体存在时,可以看到正在生长的胚状体逐渐将胚囊的近珠孔腔填满,其周围的珠被绒毡

层处于解体之中,最后在胚状体周围仅留一圈残迹(图版Ⅱ,10)。

胚状体发生 从培养 10d 后开始,在少数胚珠中,珠被绒毡层在增生的过程中可由其局部位置产生胚状体,它们和一般不定胚一样没有胚柄(图版Ⅱ,15)。以后,珠被绒毡层的其它部分相继退化,只有胚状体存活与生长,成为孤立的构造,其形状或者与雌核发育胚状体类似,或者不太整齐而呈愈伤组织状。一个胚珠中的珠被绒毡层胚状体数目,可能是一个,也可能是多个(图版Ⅱ,16、17)。它们虽然发生较迟,但生长速率较快,后期可以达到和雌核发育胚状体相仿的形状与大小,有时仅从外形不易区分二者。但雌核发育胚状体位于胚囊的近珠孔腔中,和周围的组织没有联系,而珠被绒毡层胚状体与退化的珠被绒毡层残迹直接相连,且常位于胚囊腔外(图版Ⅱ,16、17),因此经仔细判断仍是可以辨别的。

四、珠被组织的变化与愈伤组织发生

向日葵的珠被组织很发达,其厚度约有 30 多层细胞。培养后,随着珠被绒毡层的增生,其周围特别是合点端的珠被组织逐渐解体。至第 7d 以后,珠被仅由 7～10 层细胞组成一个空壳,内方形成一个很大的空腔。雌核发育的或珠被绒毡层的胚状体如果存在,就游离于这一空腔中,而与外围的残存珠被组织失去联系。

在某些胚珠中,珠被组织可产生愈伤组织。后者的发生位置与数目不定,既可向胚珠内生长,亦可向外生长而突出于珠被之外(图版Ⅱ,18),但以位于珠孔端、外向生长者居多。

讨 论

1. 胚珠培养的实验虽然早在 20 世纪 30 年代已经开始,但旨在诱导单倍体的未受精胚珠培养则是在 50 年代后期由以 Maheshwari 为首的胚胎学派倡导的[11]。当时,在百合属[16]、葱莲属[17]、罂粟[12]、烟草[18]、洋葱[7]等多种植物上曾试探了未受精和已受精的胚珠培养(同时也试探了未传粉的子房培养),结果后者获得很大的成功,而前者屡遭失败[13]。70年代,少数学者在未受精胚珠培养中观察到一些有趣的现象[19,8]。但直到80年代初才在诱导成单倍体植株方面有所突破。Cagnet-Sitbon 培养非洲菊(*Gerbera jamsonii*)的未受精胚珠,由内部产生愈伤组织,经转移后分化出单倍体植株[6]。冉邦定进行烟草未受精胚珠的培养,获得了形态特征为单倍体的苗(无染色体鉴定)[1]。最近,蔡得田与周嬁在向日葵未受精胚珠漂浮培养的实验中,由6个品种的胚珠内解剖出360个胚状体,其中96个转移后生根,根尖压片证明有单倍体,也有二倍体[5]。本文的胚胎学观察是这一工作的继续,在未受精胚珠培养方面,这还是第一次进行胚胎学的研究。

2. 本实验表明:(1)向日葵胚珠培养时雌核发育是在雌配子体形成以后开始的。更幼嫩的胚珠诱导效果不好(未发表)。这和在水稻[3]、大麦[4]子房培养中得到的结果是一致的。非洲菊也是用超过子房腔空间一半的较大胚珠诱导出单倍体植株;小于子房腔一半的胚珠培养未获成功[6]。(2)向日葵的雌核发育胚状体是起源于未受精的卵细胞分裂,实质上是一种离体条件下的孤雌生殖。和水稻子房培养中助细胞的无配子生殖[2]有所不同,向日葵中未观察到助细胞的分裂发育。次生核或反足细胞亦可进行有限的游离核或细胞分裂,但均未见到产生胚状体。总之,从现有资料来看,在离体条件下胚囊中四种细胞均可被诱导分裂;它们在不同植物中的发育类型与前途有相同之处,也有不同之处。(3)向日葵的雌核发育胚状体,早期经历和体内合子胚相似的原胚发育,以后则只有原胚生长而无器官分化。这种情况和水稻中观察到的颇相近似,不过不像后者原胚长大后愈伤组织化[3]。胚状体的不分化,可能是由于缺少胚乳而培养基成分尚不够适合所致。是今后需要研究解决的问题。

3. 向日葵胚珠培养中珠被绒毡层的行为有两点值得注意。第一是它的增生。据 Newcomb[15] 的观察和我们的证实,在体内条件下,向日葵受精后珠被绒毡层也有一个适度的增生过程。这是它作为兼有腺组织与分生组织双重功能[9]的一种表现。但在培养条件下增生的程度远远超过体内的范围。在文献资料中,在下述各种场合可能发生珠被绒毡层的过度增生:如向日葵的异常发育的小花中[20]、不少植物的不亲和交配时、烟草子房培养以及曼陀罗、番茄子房被注入植物激素等条件下(见 Kapil 与 Tiwani 的综述[9])。在本实验中,珠被绒毡层的初期增生也许对雌核发育胚状体的营养有一定的有利作用;以后的过度增生则可能是有害无益的。第二,是由珠被绒毡层产生胚状体。和上述无组织的增生不同,这是局部细胞中发生的有组织的生长。文献中曾记录过珠被绒毡层产生不定胚的情况[9]。菊科植物 *Melampodium divaricatum* 的多数胚珠中,由珠被绒毡层的某些细胞分裂形成不定胚,代替了合子胚[10]。在本实验中,共观察到二十多例由珠被绒毡层发生的胚状体。从发生顺序来看,它们多在雌核发育胚状体之后。从后期的形成特征来看,与雌核发育胚状体有些近似,易引起"鱼目混珠"。此外,珠被的其它组织也可产生愈伤组织。这样就为今后的研究提出一个问题:如何设法调节雌核发育和体细胞增生之间的平衡,使之有利于前者的发育。

(作者:阎华、吴燕、陈小民、魏正元、周嫦、杨弘远。原载:植物学报,1985,27(1):13~18。图版2幅,参考文献20篇,均删去。)

14 向日葵离体孤雌生殖过程的组织化学研究

Histochemical studies on *in vitro* parthenogenesis in *Helianthus annuus* L.

摘　要

对向日葵未受精胚珠培养所诱导的孤雌生殖过程进行了 DNA、RNA、蛋白质与多糖的组织化学研究,同时也观察了体内胚胎发育的相应变化以作比较。离体孤雌生殖原胚发生初期的组织化学特征和体内合子胚相近；其中，卵细胞质中嗜派洛宁颗粒和卵核中孚尔根反应的动态值得注意。在后期，离体孤雌生殖胚状体和体内合子胚在淀粉动态上呈现显著的差异，这可能是前者不能顺利分化时的一种生理表现。

Abstract

In recent years haploid sunflower embryoids were induced in our lab by culture of unfertilized ovules[6] and these were revealed to be originated from the egg cells, i. e. via *in vitro* parthenogenesis[5]. The present paper reports some preliminary results of our histochemical studies on this process. Ovules 1-2 days prior to anthesis were cultured on liquid N_6 medium supplemented with 2 ppm MCPA, 6% sucrose and 100 ppm inositol. Samples were fixed at intervals in aceto-methanol (1∶3) and processed by paraffin method. *In vivo* ovules were also collected at intervals before and after anthesis and processed in the same way serving as a control system. The sections were stained by Feulgen' reaction for DNA (Plate I,1-6), methyl-green pyronin method for RNA (Plate I,7-12), mercuric bromphenol blue for protein (Plate II,13-18), and PAS reaction for polysaccharide (Plate II,19-24). The nearly matured egg cells were characterized by negative Feulgen reaction of the nucleus, negative pyronin staining of the nucleolus, abundant pyroninophylic granules in the cytoplasm, rich cytoplasmic protein, but without starch grains. After fertilization, or when parthenogenesis was triggered, the cells of the early proembryo maintained weak Feulgen stain and pyroninophylic granules for a period, which indicated that the RNA synthesized formerly in the egg might play a role in early proembryo development until new transcriptional program was established. Although the early status of the *in vitro* parthenogenetic proembryos was similar to that of the *in vivo* zygotic ones, the later development was divergent. In zygotic proembryos, the starch grains were distributed in a polarized decreasing gradient from the suspensor to the embryo proper, and were served as temporary stores, ultimately displaced by oil and protein. In parthenogenetic proembryos, however, they often showed an unpolarized, centripetally distributed gradient and turned to be a long-term storage. This abnormal physiological status may have a relation to the underdevelopment of the parthenogenetic embryoids.

(作者:魏正元、杨弘远。原载:植物学报,1986,28(2):117~122。图版 2 幅,参考文献 12 篇,均删去。)

15 几种因素对向日葵离体孤雌生殖和体细胞增生的调节作用

Regulation of *in vitro* parthenogenesis and somatic proliferation in sunflower by several factors

Abstract

Culture of unfertilized ovules in sunflower (*Helianthus annuus* L.) could be induced to form parthenogenetic embryoids, endothelial embryoids and integumentary calli. Various factors exerted regulatory influences on the production of these structures. The key factors were exogenous hormone, sucrose concentration and cold-pretreatment. In hormone-free condition, parthenogenesis was promoted without any occurrence of somatic embryoids or calli. Addition of exogenous hormone MCPA induced proliferation of sporophytic tissue and meanwhile, reduced the induction frequency of parthenogenesis. Relatively high sucrose level favored the production of parthenogenetic embryoids. Reducing sucrose concentration favored the integumentary calli. The optimal sucrose concentration for endothelial embryoids was intermediate between them. Cold-pretreatment also played an effective role in promoting parthenogenesis and inhibiting integumentary calli. By combining these three factors, we were able to work out a culture technique which significantly enhanced the induction frequency and growth of parthenogenetic embryoids and completely eliminated somatic embryoids and calli.

蔡得田与周嫦(1983)首次由向日葵胚珠中培养出单倍体和二倍体的胚状体[9]。阎华等(1985)通过显微观察发现,向日葵胚珠培养中有3种诱导产物:孤雌生殖胚状体、珠被绒毡层胚状体和珠被愈伤组织[7]。按其起源与性质而言,它们分别属于单倍体和二倍体两类不同的范畴。因研究的目的不同,要求只诱导其中某些产物而排除其他类型。就诱导单倍体的目的而言,当然要求促进孤雌生殖、抑制后两种体细胞增生。这是未传粉子房与胚珠培养的实验胚胎学工作中必须解决的问题之一。本文报道3种重要因素在向日葵未受精胚珠培养中的调节作用。

材 料 与 方 法

实验于1983—1985年进行。材料为向日葵晚熟品种"当阳"和早熟品种"阿尔及利亚"。取开花前1～3d的幼花,剥出胚珠,接种于含肌醇100ppm 的 N_6 液体培养基上进行漂浮培养。每管盛20ml 培养液,接种10个胚珠。着重对2-甲基-4氯苯氧乙酸(MCPA)浓度、蔗糖浓度和低温预处理进行了比较。MCPA 浓度分为 0ppm、0.125ppm、0.5ppm、2ppm、4ppm、8ppm 6个处理;蔗糖浓度分为1%、3%、6%、9%、12%、15%、18%、21% 8个处理;低温预处理是将花序插入盛自来水的烧杯中,外罩塑料袋,置2～4℃冰箱中24h 和48h。所有培养物均置于27℃左右、散射光或黑暗条件下。培养 10d 和 30d 时,分别取样用醋酸甲醇(1∶3)固定,70%甲醇保存,经爱氏苏木精整体染色制成石蜡切片。通过显微观察鉴定不同来源的诱导产物,以便准确地统计实验结果。

结　　果

一、外源生长素的调节作用

在各种培养条件下比较了不加 MCPA 和加入 2ppm MCPA 的实验结果。表 1 所列为培养 10d 时的孤雌生殖诱导频率。可以看出,无论其他条件(培养基成分、接种时期或培养方式)如何改变,8 个处理中除一个例外,全部结果均为无激素较有激素优越。表 2 表示培养 30d 时 3 种诱导产物的频率。除了同样证明无激素培养较有激素培养更有利于提高孤雌生殖诱导频率以外,一个明显的结果是:无论在任何条件下,只要不加外源激素,就完全不产生珠被绒毡层胚状体和珠被愈伤组织,反之,只要附加 MCPA 就诱导这两种体细胞组织的增生(图版 I,7、8、9)。

二、蔗糖浓度的调节作用

3 种诱导产物对蔗糖浓度有不同的要求。从图 1 所示"阿尔及利亚"品种的实验结果可以看出,三者的诱导高峰位于不同的蔗糖浓度。孤雌生殖胚状体所需的蔗糖浓度最高,为 12%;珠被绒毡层胚状体的最适浓度次之,为 3%～9%;珠被愈伤组织的适宜浓度最低,为 1%。

用孤雌生殖诱导频率较高的"当阳"向日葵为材料做实验,以确定诱导孤雌生殖的蔗糖浓度范围及最适浓度。结果表明该品种的孤雌生殖对蔗糖浓度有较宽的(3%～21%)适应范围,但诱导高峰仍位于 12%(图 1),与"阿尔及利亚"的实验结果一致。上述结果均在 2ppm MCPA 时获得。在无激素条件下,比较 12% 和 6% 蔗糖的效果,也是以 12% 蔗糖时孤雌生殖频率较高(表 1)。

表1 MCPA 的有无对孤雌生殖的调节作用(品种:当阳;培养 10d 观察)*

Table 1　Regulation of parthenogenesis by exclusion or inclusion of MCPA in the media (Cultivar: Dang Yang; Observation on 10 days' culture)

培养方式 Mode of culture	MCPA 浓度 MCPA concentration					
	0			2		
	观察胚珠数 Observed ovules (No.)	孤雌生殖胚状体数 Parthenogenetic embryoids (No.)	诱导频率 Induction-frequency (%)	观察胚珠数 Observed ovules (No.)	孤雌生殖胚状体数 Parthenogenetic embryoids (No.)	诱导频率 Induction-frequency (%)
加入 PB74 Addition of PB74	113	18	15.9	114	4	3.5
加入幼果提取物 Addition of achene extract	111	9	8.1	113	4	3.5
浅层培养(6% 蔗糖) Shallow liquid culture (6% sucrose)	123	8	6.5	127	2	1.6
浅层培养 Shallow liquid culture	134	12	9.0	119	4	3.4
漂浮培养① Float liquid culture*	121	9	7.4	120	8	6.7
固体直插培养① Vertical culture on solid medium*	124	23	18.5	126	8	6.3
漂浮培养 Float liquid culture	113	7	6.2	118	12	10.2
固定平放培养 Horizontal culture on solid medium	120	9	7.5	109	6	5.5

*除一个浅层培养处理外,均为 12% 蔗糖条件。

All under 12% sucrose condition except one shallow liquid culture.

① 接种胚珠为开花前 1d,其余均为开花前 3d。

Ovules inoculated 1 day before anthesis, the others all inoculated 3 days before anthesis.

表2　MCPA 的有无对各种诱导产物的调节作用
（品种：当阳；培养 30d 观察）*

Table 2　Regulation of various induced structures by exclusion or inclusion of MCPA (Cultivar: Dang Yang; Observation on 30days' culture)

培养方式 Mode of culture	MCPA 浓度 MCPA concentration (ppm) 诱导频率 Induction-frequency (%)					
	孤雌生殖胚状体 Parthenoogenetic embryoids		珠被绒毡层胚状体 Endothelial embryoids		珠被愈伤组织 Integumentary callus	
	0	2	0	2	0	2
加入 PB74 Addition of PB74	3.5	1.7	0	1.7	0	2.5
加入幼果提取物 Addition of achene extract	3.4	0.9	0	6.0	0	1.7
漂浮培养① Liquid float culture	6.6	2.4	0	2.4	0	8.1
固体直插培养① Vertical culture on solid medium	9.1	3.4	0	5.9	0	12.6
漂浮培养 Liquid float culture	3.5	0.8	0	0	0	2.6
固体平放培养 Horizontal culture on solid medium	8.7	0.8	0	0.8	0	13.8

* 均为 12% 蔗糖条件。

All under 12% sucrose condition.

①接种胚珠为开花前 1d，其余均为开花前 3d。

Ovules inoculated 1 day before anthesis, the others all inoculated 3 days before anthesis.

三、低温预处理的调节作用

在 2ppm MCPA、6% 蔗糖的培养条件下，接种前进行 24h 或 48h 低温预处理，孤雌生殖诱导频率有所提高；珠被绒毡层胚状体的诱导频率

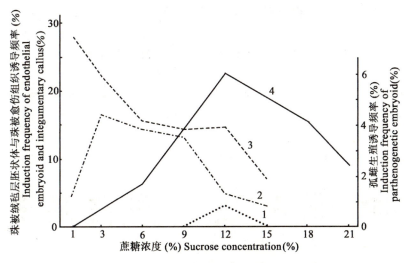

图1 蔗糖浓度对各种诱导产物的作用

Fig. 1 Role of sucrose concentration in inducing various structures

品种：阿尔及利亚(Cultivar："Algeria")
1. 孤雌生殖胚状体(Parthenogenetic embryoid)
2. 珠被绒毡层胚状体(Endothelial embryoid)
3. 珠被愈伤组织(Integumentary callus)

品种：当阳(Cultivar："Dang Yang")
4. 孤雌生殖胚状体(Parthenogenetic embryoid)

变化不大；而珠被愈伤组织则受到明显抑制(表3)。

在上述两种有利条件(即无激素和12%蔗糖)下进行24h低温预处理，取得如下结果：

1. 孤雌生殖诱导频率大幅度提高，由对照的6.2%增加到23.9%(表4)，与过去向日葵未受精胚珠培养中孤雌生殖不足3%[7]相比，提高的幅度更大。

2. 孤雌生殖胚状体的早期发育状况显著改善。培养初期卵细胞的分裂产物有3种类型：分生细胞型、液泡化细胞型和游离核型。其中只有分生细胞型属于正常胚胎发生类型(图版Ⅰ，1、2)，决定着孤雌生殖的最终频率。低温预处理使分生细胞型的诱导频率由对照的2.7%激增为 15.4%,其比例由 42.9% 上升到 64.3%(表4)。

表4 低温预处理对孤雌生殖胚状体诱导频率及早期发育的影响
(品种:当阳;培养 10d 观察)*

Table 4 Effect of cold-pretreatment on the induction-frequency and early development*
of parthenogenetic embryoids (Cultivar: Dang Yang; observation on 10 days' culture)

预处理时间(h) Duration of cold pretreatment(h)	观察胚珠数 Observed ovules (No.)	孤雌生殖胚状体 Parthenogenetic embryoid		比例(%) Proportion(%)		
		数 No.	诱导频率(%) Induction-frequency(%)	分生细胞型 Meristimatic type	液泡化细胞型 Vacuolated type	游离核型 Free-nuclear type
0(对照)(Control)	113	7	6.2	42.9	0	57.1
24	117	28	23.9	64.3	10.7	25.0

*在无激素与 12% 蔗糖条件下。
Under hormone-free and 12% sucrose condition.

3. 孤雌生殖胚状体的后期发育状况也有显著改善。表现在生长较旺盛,平均长度为对照的 2.7 倍,退化减少,并约有一半胚状体显示初步的分化迹象(表5)。与以往实验中,2ppm MCPA、6% 蔗糖、不经低温预处理的结果[7]相比,胚状体质量的改善尤为显著:由原来无极性的、表层组织松散脱落的状态(图版Ⅰ,6)变成有一定极性,有正常的表皮原(图版Ⅰ,3),并开始显示根原基分化的状态(图版Ⅰ,4、5)。

讨 论

在水稻[17]、玉米[6]、非洲菊[11]、杨树[3]和薏苡[2]的子房或胚珠培养中都发现有雌核发育和体细胞增生两类诱导产物。后者有鱼目混珠的作用,因而必须严格鉴别。向日葵胚珠培养中除诱导孤雌生殖外,还

表5　低温预处理对孤雌生殖胚状体后期发育的影响（品种：当阳；培养30d观察）*

Table 5　Effect of cold-pretreatment on the late development of parthenogenetic embryoids
(Cultivar: Dang Yang; observation on 30 days' culture)*

预处理时间(h) Duration of cold pretreatment (h)	观察胚珠数 Observed ovules (No.)	孤雌生殖胚状体 Parthenogenetic embryoid		比例 Proportion(%)			胚状体平均长度 Mean length of embryoids (um)
		数 No.	诱导频率(%) Induction-frequency (%)	分化中的胚状体 Embryoid in differentiation	退化中的胚状体 Embryoid in degeneration	其它 Others	
(对照) 0 (control)	113	4	3.5	0	25.0	75.0	158.1
24	114	8	7.0	50.0	12.5	37.5	431.0

*在无激素与12%蔗糖条件下。
Under hormone-free and 12% sucrose condition.

可产生珠被绒毡层胚状体和珠被愈伤组织[7]。这样3种类型的诱导产物使情况更为复杂。特别是珠被绒毡层胚状体与孤雌生殖胚状体位置相近,形态相似,用肉眼观察无法鉴别。在这种情况下,为了准确地统计诱导频率和评价各种因素的调节效果,只有用显微观察方法严格区分三种来源的诱导产物。本实验就是采用这一方法。

在活体胚珠中,配子体和孢子体组织间维持着精巧的协调关系,从而保障有性生殖过程的顺利进行与新一代孢子体的正常发育。当这种协调关系遭到破坏时,体细胞组织可能过度增生,而不利于合子胚的发育。例如在不亲和交配、人工注射植物激素[14]及异常发育[18]的情况下,珠被绒毡层常过度增生,甚至形成不定胚;与此相伴常常发生合子胚的败育[14]。在离体培养条件下,由于胚珠摆脱了整体的制约,这种不正常现象大大加强。因此如何控制体细胞增生是单倍体诱导中值得注意的问题。正是基于这种看法,我们在本实验中着重分析了3种诱导产物对各种因素的不同反应,而不限于孤立地研究孤雌生殖本身。

在向日葵未受精胚珠培养中有3个重要的因素:外源激素的有无、蔗糖浓度、低温预处理。在未传粉子房和胚珠培养已取得成功的14种植物中,除矮牵牛[12]以外,均采用了外源激素诱导雌核发育[1~3,9,13,16]。在小麦中由无激素和14%蔗糖的培养基上仅获得一株单倍体小苗[4]。向日葵胚珠培养曾沿用外源激素诱导,结果表明:外源激素对诱导孤雌生殖并非必需,反而有害。在无激素培养条件下,孤雌生殖诱导频率提高,完全不产生体细胞胚状体与愈伤组织。蔗糖浓度的影响也很大。孤雌生殖胚状体需较高的蔗糖浓度(12%),这与幼胚培养以及体内的胚胎发育环境是一致的[5]。降低蔗糖浓度则有利于体细胞增生。因此,过去实验中所采用的6%蔗糖浓度[7]是偏低的。在杨树的未传粉子房培养中也发现,当蔗糖由5%增至10%时,雌核发育频率提高,体细胞愈伤组织减少[3]。低温预处理是花药和花粉培养中大幅度提高雄核发育频率的手段之一[15]。在人工诱导体内孤雌生殖的工作中也采用过[10]。在未传粉子房和胚珠培养方面,低温预处理虽然在水稻[16]、非洲菊[11]的实验中未获得预期效果,但在向日葵中则有非常显著的效果。有趣的是低温预处理对孤雌生殖和珠被愈伤组织具有完相反的作用,促进前者而抑制后者。这是一个值得进一步研究的问题。综上所述,在向日葵未受精胚珠培养中把以上三种因素结合起来,可以有效地调节不同的组织的生长发育,大幅度提高孤雌生殖的诱导频率,并排除孢子体组织增生产物的干扰。

(作者:阎华、董健、周嫦、杨弘远。原载:植物学报,1987,29(6):586~587。图版1幅,图1幅,表5幅,参考文献18篇。选录图1、表1、2、4、5。)

16 用整体染色与透明技术观察胚囊、胚、胚乳和胚状体

The use of a whole stain-clearing technique for observations on embryo sac, embryo, endosperm and embryoid

Abstract

A new stain-clearing procedure has been developed for embryological observations on whole mounted specimens. Ovules of *Helianthus annus* and *Nicotiana tabacum* as well as ovaries of *Oryza sativa* were stained with diluted Ehrlich's hematoxylin for a proper short time, followed by steps of washing and dehydration, and finally cleared and mounted in methyl salicylate. When observed by ordinary bright-field microscopy, the embryo sacs before fertilization and the embryos and endosperms after fertilization were clearly visible. The gynogenic embryoids induced in unpollinated rice ovaries *in vitro* were also finely detectable. The Ehrlich's hematoxylin-methyl salicylate technique has the merits of rapidity in specimen preparation, high contrast and three dimensional view, needlessness of phase-or interference-contrast equipment, and the feasibility for a wide range of materials. The special significance of this technique for *in vitro* embryological studies is emphasized.

胚囊处于胚珠组织的层层包裹之中,观察胚囊的传统方法是通过切片。切片的制作过程相当繁冗,并且根据观察连续切片来建立立体概念也需要耐心与经验。为了寻求简便而又富于立体感的观察方法,先后提出过整体解剖、整体透明、酶法分离等研究胚囊的技术。其中,整体解剖法虽然有时可以得到很好的效果,但要求精巧的手工操作,而且难免改变细胞组织原来的位置与形状。酶法分离胚囊的观察效果很佳,但其主要潜力是开展生活胚囊的研究和作为今后实现胚囊培养的技术前提。由于酶法分离的成功率有限,目前还无法用它对单个的胚珠进行指定性的观察。从这个角度看,整体透明法则具备原位地形学(topography)观察和指定性观察这样两个优点。

Herr 首先提出胚珠整体透明技术[9]。他用一种简称为"4½"的复合透明剂(配方见后文)浸泡胚珠,在相差或干涉差显微镜下观察大孢子母细胞、大孢子和胚囊。Herr 和其他研究者先后用这种方法研究了多种植物材料均获成功[5,7,8,10~12,14~17]。Crane 提出用另一种透明剂水杨酸甲酯(methyl salicylate,即冬青油)[6]。后者透明度超过"4½",不仅可用于胚珠,而且也可用于子房的透明[18~20]。然而,单纯的透明法有若干局限性,主要缺点是反差太小,在普通光镜下往往只能显示细胞壁。为了分辨核和核仁,必须借助相差或干涉差装置。即使如此,从已发表的有关论文看,透明法多限于观察幼小材料中的大孢子发生与雌配子体发育,而对较大的材料和受精后的发育则应用有限。我们数年前曾在烟草胚珠和水稻子房上试验了上述两种透明方法,当时也只能观察到幼期的胚囊[4]。最近,Stelly 等又提出:在冬青油透明前,先用梅氏苏木精明矾(Mayer's hemalum)染色,这样增大了反差,用一般明视野观察即可察见 *Solanum* sp. 胚珠中的大孢子发生与胚囊结构[18]。但他们尚未试验子房的透明,亦未观察受精后的构造。考虑到染色-透明比单纯透明有明显的优点,我们在几种植物的胚珠或子房上开展了试验,并建立了一项效果很好的"爱氏苏木精染色-冬青油透明"新程序,使本项技术又推进了一步。

材 料 与 方 法

受精前与受精后的向日葵(*Helianthus annuus*)、烟草(*Nicotiana tabacum*)胚珠和水稻(*Oryza sativa*)子房,固定于醋酸甲醇(1∶3)或FPA(福尔马林、丙酸、乙醇,5∶5∶90)中,保存于70%甲醇或乙醇(4℃)中备用。

水稻的离体培养子房由本室何才平同志提供。按本室过去常用的水稻未传粉子房培养方法[3],于接种后14d将膨大的子房固定于醋酸甲醇(1∶3)中并保存备用。

"4$\frac{1}{2}$"透明剂按 Herr 的配方[9]:乳酸(85%)、水合三氯乙醛、酚、丁香油、二甲苯,2∶2∶2∶2∶1(按重量计),依次混合。材料经95%乙醇约1h后转入"4$\frac{1}{2}$"中1d以上,用"4$\frac{1}{2}$"封藏并观察。

冬青油透明的基本程序为:材料经95%乙醇转入无水乙醇脱水三次并过夜,在无水乙醇与冬青油的等量混合液中约1h,再用冬青油透明,换液三次,每次1至数小时,最后一次1d以上。

试验了冬青油透明以前的各种整体染色方法:丙酸洋红、爱氏苏木精、Feulgen 反应、PAS 反应。其中取得成功的爱氏苏木精染色法为:用爱氏苏木精(Ehrlich's hematoxylin)原液(需充分氧化成熟)1份与45%醋酸和50%乙醇等量混合液2份,配成稀释的染液。材料先下行至蒸馏水,然后在上液中染色不同时间(5min、10min、20min、30min、60min、120min、1d以上)。蒸馏水换洗4~5次(共1~2d),自来水换洗2~3次(共1~2d)。依前述程序脱水、透明。所有步骤均在20℃恒温条件下进行。

经过透明后的胚珠或子房,可在冬青油中保存一个月以上。制片时,由于材料较厚,可采用 Herr 等[9]的 Raj slide 装片,也可用不同深度的凹玻片。材料置凹玻片中央,滴加冬青油,加上盖玻片。不可加压,以免材料变形或破碎。

观察与摄影用 Olympus BH-2 显微镜的明视野照明系统。为了增大反差、获得清晰的图像,需适当缩小孔径光栏,相应地提高照明电压。

结　果

一、染色的方法与时间

用"4½"透明的三种材料均能观察到胚囊的构造,但不甚清晰,更难摄影。单用冬青油透明,透明度有余而反差不足,很难看清内部结构。先染色再用冬青油透明,结果因染色方法而异。丙酸洋红可显示胚囊和幼胚,但分色不佳,核质界线模糊。Feulgen 反应使整个胚珠组织的细胞核染色很深,而胚囊中的卵核与极核反而染色很浅,加以细胞壁不染色,因此难以分辨各部结构。PAS 反应使细胞壁和淀粉粒染色明显,但亦无法显示胚囊构造。长时间（1~2d）的爱氏苏木精染色继之以分色,效果亦不佳。只有短时间的爱氏苏木精染色继之以水洗,方才获得满意的结果。

爱氏苏木精染色的适宜时间,因植物种类与所用材料而异。一般讲,大型胚珠（如向日葵）与小型胚珠（如烟草）相比,老龄材料（如受精后）与幼龄材料（受精前）相比,子房（如水稻）与胚珠相比,前者染色时间需延长一些,但仍属于短时间染色。宁可染色偏浅而采取缩小孔径光栏的方法提高分辨率;不可染色偏深以至妨碍透视。通过比较试验,确立了各种材料的适宜染色时间（表1）。但因染色深度受其它各种因素（特别是苏木精原液的成熟度）所影响,表中数字只能看作近似值。

二、观察效果

向日葵　用本法观察向日葵受精前的胚珠,可以看清内部胚囊的构造（图版Ⅰ,1）。其中位于珠孔端的卵器和中央细胞较位于合点端的反足细胞染色为深,珠被绒毡层中度着色,外围的珠被组织则染色很浅或不染色。除了卵器和中央细胞原生质较浓稠而易染色的原因外,显然染液是由珠孔进入胚囊,然后向合点和周围扩散。所以,尽管珠被很厚,由于染色较浅,并不遮掩胚囊。在高倍物镜下,卵核和次生核染色清晰,核仁尤为显著;通过调焦亦可看到一对助细胞(图版Ⅰ,2)。

受精后,虽然胚珠体积增大很多,但用整体观察法同样可以看到原胚和胚分化的不同阶段:球形胚(图版Ⅰ,3)、晚球形胚(图版Ⅰ,4)、心形胚(图版Ⅰ,5、6)。(图版Ⅰ,6)显示高倍观察时一个心形胚的图像。此时胚细胞层次已经很多,但通过适当染色和充分透明,观察效果仍然极好。通过调焦,不仅可以看出整个胚体的三维形态,而且在一定的光学切面上,胚和胚柄细胞的清晰度接近厚切片的水平,甚至胚柄细胞中的微小淀粉粒亦能辨认。

表1 各种材料的整体染色时间(20℃ F)

Table 1 Duration of whole stain for various materials (at 20℃)

植 物 Species	材 料 Materials	发 育 时 期 Developmental stage	适宜染色时间(分) Optimal stain duration (min.)
烟 草 *Nicotiana tabacum*	胚 珠 Ovule	成 熟 胚 囊 Mature embryo sac	<5
向日葵 *Helianthus annuus*	胚 珠 Ovule	成 熟 胚 囊 Mature embryo sac	10
		球形胚-心形胚 Globular-heart shaped embryo	15
水 稻 *Oryza sativa*	活体子房 *in vivo* ovary	成 熟 胚 囊 Mature embryo sac	30~40
		原胚-胚分化 Proembryo-embryo differentiation	120
	离体培养的 未传粉子房 *in vitro* cultured unpollinated ovary	培养14天的雌核 发育胚状体 Gynogenic embryoid after 14 days' culture	>120

在整体封藏的胚珠中,胚乳外形呈袋状。它的包围着胚的珠孔端

部分比较整齐,合点端则折叠为不规则状态(图版Ⅰ,3、5)。这种不规则的形态可能是因为胚囊空间的限制所致。也可能意味着胚乳通过这种方式扩大其吸收表面。与胚乳毗邻,尤其是其合点端周围的珠被组织染色特别浅,似乎表示这些组织的物质已被胚乳分解吸收。总之,用本法观察胚乳的三维形态,可以获得比切片观察更清晰的印象。

烟草 受精前的烟草胚珠由于体积很大,仅仅经过5min的染色,胚囊和珠被均已着色偏深。但透明后仍可辨别胚囊中的各个成员细胞。烟草胚珠近于球形,因在装片中所处方位不同,可从不同角度观察胚囊内部,获得各种光学切面。(图版Ⅰ,7)表示一个胚囊的光学纵切面:卵细胞、极核、反足细胞分布在大致相同的光学切面上。(图版Ⅰ,8)示卵细胞和一对助细胞的三足鼎立状态;极核分布在另一光学切面上;反足细胞只现阴影。(图版Ⅰ,9)表示一对明显的极核;卵器分布在另一光学切面上;反足细胞仅现阴影;极核和反足细胞周围的淀粉粒呈透明颗粒状。以上各例胚囊,通过调焦均可看到不同光学层面的构造。在烟草上,我们没有用受精后的胚珠做试验。

水稻 水稻子房很小,从中剖出胚珠比较费时。我们用整个子房做试验也取得了成功。由于水稻细胞核与细胞质本来不易分色,加之有很厚的子房壁掩盖,所以观察效果稍逊于前两种植物材料,但用于鉴别胚囊、胚和胚乳的构造是不成问题的。受精前的子房在低倍观察时,可以透过染色很浅的子房壁和染色较深的珠被、珠心组织看到胚囊内的构造,其中以反足细胞群和一对极核比较显著(图版Ⅱ,10)。在高倍观察时,通过调焦,胚囊内的所有结构均可看见(图版Ⅱ,11)。

受精后,子房不断伸长、膨大,但子房壁仍然浅染,同时胚囊内部空间扩大,所以观察效果反而超过受精以前的子房。在开花后2d的子房中可以看到胚乳游离核分布在胚囊的周边区;此时退化的反足细胞残迹尚存(图版Ⅱ,12)。开花后3d的子房,通过调焦可以显示大面积的胚乳核,其中许多正在进行有丝分裂(图版Ⅱ,13),有些部位已经形成细胞。更迟的胚乳细胞充满淀粉粒,无法看清(图版Ⅱ,14、15)。

从原胚到胚分化初期的各种形态均可显示,如:梨形原胚(图版Ⅱ,14);开始出现胚芽鞘原基(图版Ⅱ,15);胚芽鞘已形成帽状,其内产生胚芽原基,盾片开始延伸(图版Ⅱ,16)。我们试验的材料最晚到此为止,没有观察分化程度更高的胚。

未传粉的水稻子房经 14d 离体培养后,已经膨大到在凹玻片上无法平放盖玻片的程度,只能用低倍镜观察。可以看见一部分子房中包藏雌核发育的胚状体(图版Ⅱ,17)和愈伤组织化形成类似"原球茎"的构造(图版Ⅱ,18)。

讨 论

自从 20 世纪 70 年代初期发明胚珠整体透明技术[9]以来,植物胚胎学文献中陆续有对这一技术的应用和改进方面的报道。最近 Johri 主编的《被子植物胚胎学》一书中对此法作了充分的评价[13]。在国内,迄今还没有对此予以足够的重视。几年来,我们实验室在未传粉子房和胚珠培养的胚胎学研究中,为了观察离体雌核发育的发生过程及其在不同培养条件下的诱导频率,常需完成规模极大的石蜡切片任务,而实际上大部分材料中并没有启动雌核发育,人力和物力浪费很大。因此深感需要一种简捷的方法来代替繁重的切片工作。本试验的主要目的即在于此。本文报道的三种材料的整体染色与透明实验结果,首先是在活体内的观察方面获得比较满意的效果,这将为植物胚胎学与教学提供一项简便有效的方法。其次,通过对水稻未传粉子房培养材料的初步试验,第一次表明透明技术对离体培养材料的研究也是适用的,从而将对今后的试管受精、胚珠培养和子房培养(无论未传粉或已传粉的)的胚胎学观察带来相当大的方便。

和"4½"透明法相比,冬青油透明法不仅透明效果更好,而且不会使预先的染色褪掉。正是利用这一特点,Stelly 等最近提出的染色-透明方案[18]是这项技术的一个新突破。沿着这一思路,我们又试验成爱氏苏木精染色-冬青油透明新程序。它有如下显著的优点:(1)兼有透明好和反差强的双重特点,应用一般明视野照明系统即可收到清楚的观察效果,无需应用相差或干涉差装置。(2)不仅可以显示受精前胚囊中的各个成员细胞,而且还能显示受精后发育到相当程度的胚和胚乳;还可用于观察离体培养的材料。(3)对材料可能有较广泛的适应性,不仅适用于小型的胚珠(如烟草),而且适用于大型的胚珠(如向日葵);不仅适用于胚珠的整体观察,而且适用于某些子房(如水稻)的整体观察。从上述后两点来看,我们的结果超过了 Stelly 等的方法,可能具有更大的应用前景。

  本项技术的成功关键在于染色的方法与时间。Stelly 等的梅氏苏木精明矾染色需时很长(1~2d),染色后再用 0.5%~2% 醋酸分色 1~2d[18]。我们实验室多年应用于花药、子房、胚珠培养研究中的爱氏苏木精整体染色-石蜡切片程序,其染色时间也是 1~2d(原始方法见[2])。在本试验中,最初按同样的长时间染色方法不能成功。后来注意到染料进入胚囊的途径,既非首先经由珠柄-合点,亦非通过珠被表皮逐层向内渗透,而是首先从珠孔径直进入胚囊,然后再向外周扩展。因此,利用适当的短时间染色,可以区分胚囊和周围组织的染色深度,经过冬青油透明以后就能透过外层子房,胚珠组织看清胚囊内的构造。浅染的另一优点是可以使细胞层次较多的观察对象(如图版Ⅰ,6中的向日葵心形胚)通过调焦获得近乎厚切片的清晰程度。这对于研究胚和胚状体尤其重要。浅染和观察时与缩小孔径光栏的方法相结合,可以达到较好的反差效果。本技术在推广应用时,不同材料的染色"火候",还需要研究者自行摸索经验。

(作者:杨弘远。原载:植物学报,1986,28(6):575~581. 图版 2 幅,表 1 幅,参考文献 20 篇。仅选录表 1。参看书末彩色图版Ⅵ。)

17 An electron microscope study on *in vitro* parthenogenesis in sunflower

  Spontaneous haploidy has been reported in many higher plant species (Kimber and Riley 1963; Magoon and Khanna 1963). Due to its low frequency, however, cytological data are absent in the majority of cases, although the phenomenon has long been recognized. Recently, the *in vitro* culture of unpollinated ovaries and ovules has provided a useful method for inducing haploid plants from female cells in higher frequencies. This has led to the development of a stable experimental system for conducting cytological studies of parthenogenesis and apogamy (Yang and Zhou 1982). Microscopical observations on embryo formation in cultured unfertilized ovules of sunflower revealed that haploid embryos are derived from egg cells (Yan et al. 1985). Subsequently, to this study, the frequency of inducing parthenogenesis *in vitro* in sunflower has been greatly enhanced by improvements in culture techniques (Yan et al. 1987, 1988). Consequently, we were able to make preliminary electron microscope observations on parthenogenesis induced by the culture of unfertilized ovules of sunflower (Yan and Yang 1989). The present study was undertaken to further the search for distinct ultrastructural features associated with parthenogenesis. In the present paper, we describe the egg cell from before inoculation, through the activated egg, to the parthenogenic proembryo. The ultrastructure of the normal egg, zygote and proembryo of sunflower will be reported in another paper.

Material and methods

Highly responsive sunflower cv "Dang Yang" was used as the test plant and grown in the field at Wuhan University. Ovules at 1 day before anthesis were inoculated on solid N_6 medium that was supplemented with 12% sucrose and 100 mg inositol, but free of exogenous hormone. The culture was maintained at 28°C under dark conditions. Samples were collected prior to inoculation and after 5 and 10 days of culture, respectively. The chalazal portion of the ovules was cut off, and the micropylar portion containing the embryo sac was fixed with glutaraldehyde and osmium. After dehydration through an ethanol series, the samples were infiltrated and embedded in Epon 812 or Poly/bed 812 via propylene oxide. Ultrathin sections were cut with a glass knife on a Sorvall MT-6000 ultramicrotome (Wuhan University) and with a diamond knife on a Reichert ultramicrotome (Ohio State University). The grids were double-stained with uranyl acetate and lead citrate. Ultrastructural observations were made using a JEM-100 cx/II and Zeiss-EM 10 electron microscope at 80 kv.

Results

Egg cell before inoculation

The eggs used for inoculation were collected 1 day before anthesis. As the ultrastructure of the egg 3 days and just before anthesis has been described in another paper, only a summary of important features of the egg at inoculation will be presented here.

The egg cell at 1 day before anthesis is almost mature. It is a strongly polarized cell with the nucleus and most of the cytoplasm accumulating towards the chalazal end and many small vacuoles towards the micropylar end. A large vacuolate nucleolus is present in the egg nucleus, and numerous ribosomes are distributed throughout the cell. The egg contains a large number of mitochondria plastids, both of which are structurally simple, the former having relatively few cristae and the latter only occasional internal lamellae and starch grains. The egg is poor in dictyosomes and endoplasmic reticulum, the latter having a thin cisternal phase. There are a few small lipid bodies randomly distributed in the cytoplasm. The egg cell is only partially surrounded by a cell wall, the chalazal portion being enclosed only by the plasma membrane. In comparison to the zygote, the egg has the appearance of a quiescent cell, and its metabolism appears relatively low.

Activated egg

After 5 days of culture, the eggs contained in most of the ovules are degenerating; in only about 20% of the ovules do the eggs appear to survive and subsequently activated during *in vitro* culture. The ultrastructural changes observed in the activated egg include:

1. The nucleus has moved from its original position close to the chalazal pole to a more central position in the cell (Fig. 1). This repositioning of the nucleus has also been observed in the zygote.

2. Organelles increase both in number and apparent activity: plastids are abundant and peripherally arranged (Figs. 1,2,5), and some of them appear to be dividing (Fig. 2); many mitochondria with developed cristae are distributed throughout the cell (Figs. 1, 7); the number of dictyosomes and the amount of endoplasmic reticulum is much greater than before culture. The dictyosomes have four to six cisternae that have a large number of vesicles at their edges (Fig. 3). The cisternal phase of the endoplasmic reticulum is swollen, and ribosomes are attached to the outer membrane (Fig. 4). Ribosomes that are presumed to be newly synthesized are present near the periphery of the nucleus and on the outer nuclear membrane. They are also found in the cytoplasm (Figs. 1, 6). In common with the zygote, the density of the ribosomes of the activated egg decreases gradually during maturity (Fig. 2). The activated egg also has more lipid bodies than the zygote.

3. In most activated eggs, vacuoles increase in size and are distributed throughout the cytoplasm (Fig. 2); this is contrary to the condition found in the zygote, where many small vacuoles are concentrated at the micropylar end of the cell. This means that the original polarity of the cell decreases or disappears.

4. The cell wall is deposited on the chalazal surface of the cultured activated egg (Figs. 6-8). During the development of the egg cell wall, endoplasmic reticulum of the central cell is arranged in parallel rows adjacent to the wall being formed (Figs. 5, 8). The abundant number of dictyosomes of the central cell are located near the wall (Fig. 8a), and many

vesicles are present between the endoplasmic reticulum and the wall (Figs. 5, 8b). A similar phenomenon has been observed in the zygote. At an early stage of development, the newly synthesized wall of the activated egg only covers part of the cell (Figs. 1,6,7), but as the time of culture increases, the wall eventually completely surrounds the cell (Figs. 2, 8). This is in contrast to the zygote where the chalazal wall is still incomplete when the zygote undergoes division.

Parthenogenic embryo

Parthenogenic embryos can be found in some ovules after 10 days of culture. The different types of embryos found are shown in Figs. 9-13. Parthenogenic proembryos and zygotic proembryos have many features in common-high ribosome density, abundant dictyosomes with many associated vesicles, much endoplasmic reticulum and many mitochondria, small vacuoles and the disappearance of lipid bodies, among others. Plasmodesmata is also found in the internal embryo walls, rather than in the external embryo wall (Fig. 14). A portion of a parthenogenic embryo that developed quickly to the early heart-shaped stage is shown in Fig. 12. This embryo is similar to the zygotic embryo except that it has larger vacuoles.

However, the parthenogenic proembryos also have some distinctive features, and the most significance of these are:

1. In some parthenogenic proembryos, there are two regions that differ markedly in density (Figs. 10,15). Interestingly, one region of the embryo consists of small, deeply staining cells that are densely packed with ribosomes and contain few vacuoles while another region of the embryo comprises large, lighter staining cells that possess a large amount of endoplasmic reticulum and many dictyosomes (Figs. 16a, b). The ultrastructural features of the darker staining cells suggest that they are similar to normal embryo cells, while the cells that are lighter staining are similar to suspensor cells.

2. The polarity of many proembryos changes in various ways. A proembryo consisting of two different parts in terms of the density of its cytoplasm is shown in Fig. 10. The dense portion, which resembles the cells of the proembryo itself, is at the micropylar end, while the lighter portion, which is similar to the suspensor cells, is at the chalazal end-just the reverse of what would normally be found. In another proembryo (Fig. 13), the suspensor cell has dense cytoplasm and small vacuoles, which are concentrated in the chalazal end of the cell. In contrast, the chalazal portion of the embryo contains large vacuoles. This is just the opposite of what is found in normal embryos. In Figs. 9 and 11, ribosome density does not vary in the proembryo shown, but the vacuoles are larger and more numerous in the chalazal part than the micropylar part. Again, the polarity of this proembryo is opposite to that seen in zygotio proembryos.

3. Autophagic vacuoles are often found in the cells of parthenogenic proembryo (Figs. 17,18): plastids, mitochondria, ribosomes and dictyosomes are seen to be encircled by what appears to be autophagic vacuoles. The majority of the organelles within these autophagic vacuoles are plastids and ribosomes. In some cases, the autophagic vacuoles are encircled by several layers of elongate mitochondria, which may be involved in the energy metabolism of the autophagic vacuoles (Fig. 18).

4. Special features are found in the walls of some of the parthenogenic proembryos. First, the wall of the parthenogenic proembryo is generally much thicker than that of the zygotic proembryo (Figs. 9,10,14,17). Second, dense material is visible on the cell walls of some parthenogenic proembryos (Figs. 19,20). The wall of those parthenogenic proembryos adjacent to the central cell usually have an electron-dense layer (Fig. 19) as well as a gap between the proembryo and the central cell (Figs. 9,13, 19,21). In Fig. 21 it appears that various materials from the central cell have penetrated the external wall of the parthenogenic proembryo. Third, in addition to the formation of a phragmoplast (Figs. 22,23), another

mode of wall formation is found in the parthenogenic proembryo cell, i. e., the free growth of wall fragments. This free growth is initiated at the lateral cell wall and then extended centripetally (Figs. 9, 10, 13). Many dictyosomes with abundant vesicles are present near the growing tips of the wall (Fig. 24). The growing tip of the wall is often expanded, and the cell walls formed via the centripetal growth of wall fragments are usually incomplete (Figs. 9, 13). This mode of wall formation is frequently found in those cells that have lighter staining cytoplasm (Fig. 13), whereas phragmoplast formation is often associated with cells that have densely staining cytoplasm (Fig. 22).

5. Some parthenogenic proembryo nuclei undergo free nuclear division (Figs. 25, 26) to give rise to a coenocyte (Fig. 11). In others, free nuclear division occurs in one part of the proembryo, while cell division complete with the formation of cell walls occurs in another part (Fig. 13). Occasionally some free nuclei are formed by amitosis (Fig. 27). In addition, free nuclei are often accompanied by the growth of incomplete walls (Figs. 13, 24).

6. During *in vitro* culture, a few parthenogenic proembryos apparently stop developing and subsequently degenerate. One of these is shown in Fig. 28. This was a proembryo that had an inversion of polarity. The organelle structure appears disorganized, the tonoplast has disappeared in some cells and many lipid bodies are present in the cytoplasm. In contrast, lipid bodies disappear gradually during normal embryogenesis both *in vivo* and *in vitro*. The endothelium in the same *ovule* remains viable.

Discussion

In the egg of sunflower, the cytoplasm is poor in endoplasmic reticulum and dictyosomes, both mitochondria and plastids are structurally simple and the cell is not covered by a wall at its chalazal end. These ultrastructural features can be interpreted as characterizing a dormant cell that will remain undivided *in vivo*. Therefore, it appears that the transition from egg to proembryo will involve an activation stage. In normal *in vivo* development, the egg is activated by fertilization, a process that takes several hours. However, under culture conditions, the egg appears to need several days for induction. Once the process is complete, the egg has apparently been transformed from a dormant gamete into a potentially embryogenic cell that represents the unicellular stage of the new sporophyte generation. At the light microscope level, it is difficult to distinguish the activated egg from an inactive egg as the only distinctive feature that alters in the egg is the staining of the nucleolus-from negative to positive pyronin staining-indicating the beginning of RNA synthesis (Wei and Yang 1986). In contrast, at the electron microscope level, remarkable changes can be seen in the egg embarking on embryogenesis. Those include nuclear migration, an increase in the number of organelles and in their activation, the enlargement and redistribution of vacuoles, a decrease of electron density of the cytoplasm, wall formation at the chalazal end of the cell, among others. In all these features, especially in organelle activation and wall formation, the activated egg is comparable to the zygote, but it is very different from the egg before inoculation. These features may account for the sporophytic potential of the egg, indicating a triggering to embryogenesis-i. e., parthenogenesis-in such a cell.

Activation of the egg is always accompanied by wall formation in the cell. In the activated egg, a complete wall is laid down before the first division; in the sunflower zygote only a partial wall is present when division occurs, although in the zygotes of many other plants, there is a complete cell wall at first division (Natesh and Rau 1984). Rashid et al. (1982) have reported that a fibrillar wall forms around the pollen cytoplasm and within the intine before pollen embryogenesis in *Nicotiana*. This wall is characteristic of embryo formation. In addition, cell wall formation is a prerequisite for nuclear and cell division in tobacco protoplast culture (Schilde-Rentschler 1977): protoplasts are unable to divide and form clonies if wall formation is inhibited by the addition of cellulase to the culture medium; nuclear division also does not occur. In the cultures of unfertilized sunflower ovules, wall formation appears to be necessary for the transformation of the normal egg into an activated one. During wall synthesis, the endoplasmic reticulum of the central cell is arranged parallel to the wall. The same situation has been found in the zygote, except that the endoplasmic reticulum is part of the endosperm. Thus, it appears that the central cell is involved in the wall synthesis of the activated egg.

The parthenogenic proembryo has some features in common with the zygotic proembyro. However, as the culture medium used is different from the *in vivo* condition, the parthenogenic proembryo develops more slowly and less uniformly than the zygotic proembryo. In addition, several unique ultrastructure features have been observed in the parthenogenic proembryo:

1. Polarity inversion occurs in some parthenogenic proembryos. The micropylar portion of the proembryo consists of dense cells, which are similar to an embryo; the chalazal portion consists of light cells, which may be similar in function to a suspensor. The polarity of the parthenogenic proembryo is in complete contrast to what occurs in the zygotic proembryo in which embryo cells are found towards the chalazal end and a suspensor towards the micropylar end (Newcomb 1973b; Schulz and Jensen 1969). This inversion of polarity in the parthenogenic proembryo has been observed with the light microscope (Wei and Yang 1986). In a parthenogenic proembryo, in which different parts show different densities, the micropylar part remains meristematic and the chalazal part becomes vacuolate. It has also been reported that when unpollinated ovaries of rice are cultured, in some cases bud initiation occurs at the micropylar end of the gynogenic proembryo (Tian and Yang 1984; He and Yang 1987).

The polarity inversion of the parthenogenic proembryo may be caused by a change in the food transport pathway within the ovule. In the *in vivo* condition, the embryo sac receives most of its nutrients from the parent plant via a vascular trace that ends at the chalaza of the ovule. In addition, most of the endosperm is located around the chalazal end of the proembryo. Thus, it appears that the zygotic proembryo gets the major portion of its nutrition from the chalazal end. In *Quercus*, the translocation of food within the ovule is from the outer integument to the embryo sac via the chalazal end of the embryo sac and not the micropylar end (Mogensen 1973). In the *in vitro* condition, however, the separation of the ovule from the parent plant and the absence of endosperm in the embryo sac appears to lead to a lack of nutrition in the chalazal end of the embryo sac. It appears that in the *in vitro* cultured ovules, the micropyle provides a more convenient route for the free flow of food from the medium. As a result, in some parthenogenic proembryos, the micropylar part differentiates to form an embryo, while the chalazal part, where the nutrient level is low, differentiates into a suspensor-like structure. However, many parthenogenic proembryos with normal polarity can also be found (Fig. 12), and normal embryos have also been reported elsewhere (Yan et al. 1985; Wei and Yang 1986). These observations indicate that culture conditions are complex and varied, and this makes the observations difficult to interpret.

2. Autophagic vacuoles are often observed in cells of the parthenogenic proembryo. There are several possible explanations for this phenomenon. First, autophagic vacuoles may be involved in the dedifferentiation of the cells. Anther culture of *Nicotiana* (Dunwell and Sunderland 1974; Sunderland and Dunwell 1977) indicated that *Nicotiana* embryogenic pollen is characterized by the presence of multivesiculate bodies resembling lysosomes. These bodies were interpreted to be a means of eliminating gametophytic influence from the cytoplasm of the pollen. In the ovule culture of sunflower, there are more autophagic vacuoles in the early proembryo than in the later one; most of the organelles encircled by autophagic vacuoles are those that were plentiful in the egg before culture. This can be interpre-

ted to mean that the presence of autophagic vacuoles in the parthenogenic proembryo is a continuation of the dedifferentiation process of the egg. Autophagic vacuoles have also been seen in the zygote. A second role for the autophagic vacuoles may be to break down parts of the cytoplasm to provide the energy necessary for growth. A similar function has been ascribed to cytolysomes in long-dormant plant embryo cells (Villiers 1967). Third, the occurrence of autophagic vacuoles may be a symptom of the degeneration of the parthenogenic proembyro. In cultured ovaries and ovules from unpollinated flowers, the abortion of gynogenic proembryos is a common phenomenon (Tian and Yang 1984; Yan et al. 1985). In the zygotic proembryo of sunflower, autophagic vacuoles are present only in the suspensor and not in the embryo.

3. An incomplete wall and the free growth of wall fragments are frequently seen in the parthenogenic proembryo. Fragments of cell walls project from the lateral wall of the proembryo cell and appear to grow centripetally. If two ends of the growing wall fail to meet, or if the wall tip ceases its growth, incomplete walls are formed. The centripetal growth of the cell wall usually takes place in vacuolate and less dense cells and is accompanied by amitosis and by the absence or delay of cytokinesis. These phenomena do not occur in the zygotic proembryo of sunflower, where cell walls form via phragmoplast immediately after each mitosis. The centripetal growth of the cell wall and incomplete walls were seen by Newcomb (1973a, b) in the endosperm, antipodals and endothelium of sunflower-all of which are nutritive tissue or cells. Presumably both the absence of endosperm in cultured ovules and insufficient food in the medium could account for the differentiation of part of the parthenogenic proembryo into nutritive cells. Huang (1986) observed the centripetal growth of incomplete walls in anther culture of barley and wheat; the presence of two or more nuclei in one cell was common during pollen embryogenesis.

Our observations suggest that most of the ultrastructural peculiarities of the parthenogenic proembryos occur in response to culture conditions and are not features of parthenogenesis. We believe this because many parthenogenic proembryos show a developmental pattern that is quite analogous to that of the zygotic proembryo. This includes rapid development, normal polarity, meristematic and highly dense cells, the absence of autophagic vacuoles, a thin cell wall, and wall formation via phragmoplast. Thus, it appears that *in vitro* culture conditions are not as perfect as the *in vivo* condition with respect to embryogenesis and that the occurrence of some abnormal phenomena are inevitable in culture. Clearly, culture conditions should be improved in order to obtain normal embryo development consistently and repeatedly in future studies.

Fig. 1. Unless otherwise indicated, all figures are oriented with the micropylar end towards the bottom of the phage. An activated egg (Eg) after 5 days in culture. Note the position of the cell nucleus (N). CC Central cell, rb ribosome, v vacuole. ×1700

Fig. 2. Portion of an activated egg (Eg) and central cell (CC) after 5 days in culture. er Endoplasmic reticulum, l lipid body, p plastid, v vacuole, FN fusion nucleus. ×2700

Fig. 3. Portion of the activated egg from the same ovule as Fig. 2 showing various cytoplasmic components. The dictyosomes (d) have increased in number and activity. l lipid. ×6700

Fig. 4. Portion of an activated egg (Eg) showing various organelles. Note that the amount of endoplasmic reticulum (er) is greater in the activated egg than in the egg before culture. d Dictyosome, l lipid body, m mitochondrium. ×10200

Fig. 5. Chalazal end of an activated egg (Eg). Note the many vesicles (*arrowheads*) and er of the central cell (CC) near the egg that is oriented parallel to the surface of the egg. d Dictyosome, m mitochondrium, p plastid. ×16000

Fig. 6. Chalazal portion of the activated egg (Eg) from the same ovule as Fig. 1 showing newly synthesized ribosomes (rb). CC Central cell, EN egg nucleus, v vacuole, w cell wall. ×4000

Fig. 7. Chalazal end of an activated egg (Eg) showing its organelles and partial cell wall (w). l Lipid body, m mitochondrium, v vacuole. ×10200

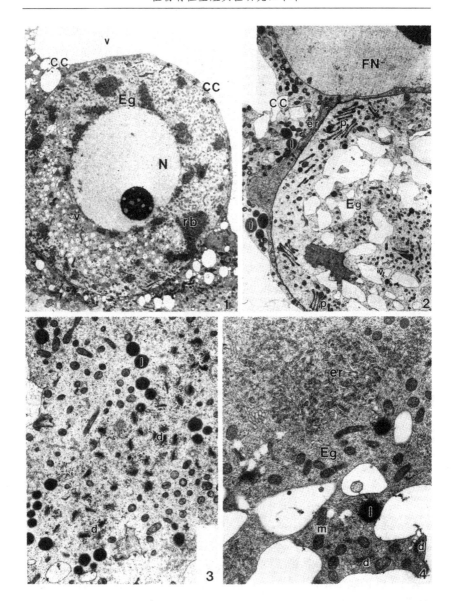

Fig. 8a, b. Chalazal end of an activated egg (*Eg*) showing the newly formed cell wall (*w*). Note the dictyosomes (*d*) and the parallel arrangement of *er* adjacent to the wall on the side of the central cell (*CC*). *Arrowheads* vesicles. a×8700, b×20000

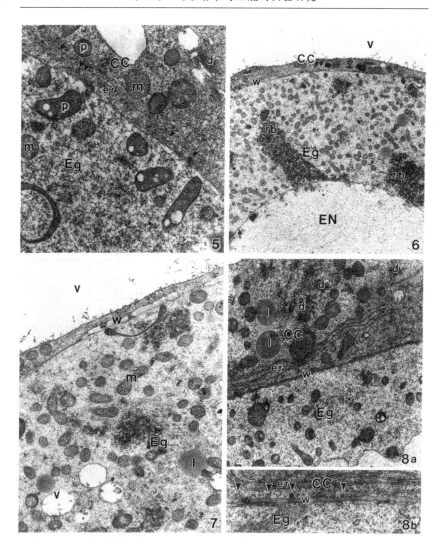

Fig. 9. Parthenogenic proembryo (*Em*) after 10 days of ovule culture. Micropylar end is oriented towards the *right* and *upwards*. The chalazal portion of the proembryo has more vacuoles (*v*) than the micropylar end. Incomplete cell walls (*w*) are present in the chalazal end. Two centripetal growth walls almost meet (*bottom*). Note the space (*arrowheads*) between the proembryo and the central cell (*CC*). ×1500

Fig. 10. Parthenogenic proembryo (*Em*) after 10 days of ovule culture. Note the mi-

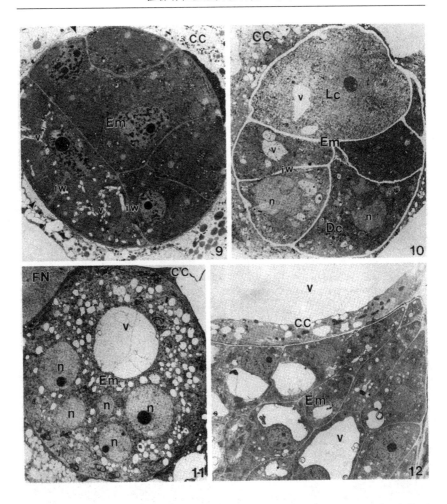

cropylar part of the proembryo has a cytoplasmic density greater than that of the chalazal part. Compare the size of the cells in the two parts of the proembryo. *CC* Central cell, *Dc* dense cell, *iw* incomplete wall, *Lc* light cell, *n* nucleus, *v* vacuole. ×1500

Fig. 11. Parthenogenic proembryo (*Em*) after 10 days of ovule culture. The proembryo is a coenocyte, no cell wall separates the nuclei (*n*). Note that the chalazal part of the proembryo is more vacuolate than the micropylar end. *CC* Central cell, *FN* fusion nucleus, *v* vacuole. ×1500

Fig. 12. Chalazal portion of a parthenogenic embryo (*Em*) after 10 days of ovule culture. The embryo is at the early heartshaped stage. *CC* Central cell, *v* vacuole. ×2000

(Authors: Yan H, Yang HY, Jensen A. Published in: Sexual Plant Reproduction, 1982, 2: 154-166. with 28 figures and 22 references, Figs 1-12 are retained here.)

18 *In vitro* production of haploids in *Helianthus*

Sunflower (*Helianthus annuus* L.) is one of the most important oilseed crops in the world. As a highly cross-pollinated species, homozygotes are especially needed both for genetic studies and breeding programs. However, so far we lack a method of haploid production for shortening the duration of pure line establishment or the breeding cycle of hybrid offspring. *In vitro* techniques may offer a useful means for this purpose. In this chapter, the *in vitro* production of haploids through ovary/ovule culture in sunflower is reviewed. The emphasis is given on the recent work done in our laboratory on the fundamental problems of ovule culture of sunflower which are indispensible for making of a new biotechnology in this crop.

Ovary and Ovule Culture for the Induction of Haploids

The female gamete (egg cell) of angiosperms is harbored in the female gametophyte (embryo sac), which is in turn surrounded by sporophytic tissues of the ovule and ovary. After fertilization, the ovule and ovary serve as a natural "womb" of the developing zygotic embryo. In occasional cases, parthenogenetic or apogamous embryos occur spontaneously in this womb as well. Therefore it is logically possible to induce haploid embryoids by submitting the unpollinated and unfertilized ovaries or ovules to suitable *in vitro* conditions. However, in contrast to anther culture, little attention has been paid to ovary/ovule culture until recent years. We made a preliminary review of this topic several years ago (Yang and Zhou 1982). Up to now, records scored in this field have shown at least 14 species belonging to seven families in which haploids have been induced by this method (Table 1). It is believable that production of haploids via *in vitro* gynogenesis will sooner or later become a practical way following that via *in vitro* androgenesis.

Table 1 Records of induction of haploids by ovary or ovule culture

Family	Species	Explant	Reference
Gramineae	*Hordeum vulgare*	Ovary	San Noeum (1976)
			Wang and Kuang (1981)
			Huang et al. (1982)
			Gu and Zheng (1984)
	Triticum asetivum	Ovary	Zhu and Wu (1979)
	Oryza sativa	Ovary	Asselin de Beauville (1980)
			Zhou and Yang (1980)
			Kuo (1982)
	Zea mays	Ovary	Ao et al. (1982)
			Truong-Andre and Demarly (1984)
	Coix lacryma-jobi	Ovary	Li and Zhang (1984)
Solanaceae	*Nicotiana tabacum* and *N. rustica*	Ovary	Zhu and Wu (1979)
			Wu and Cheng (1982)
	Petunia axillaris	Ovule	De Verna and Collins (1984)
Compositae	*Gerbera jamesonii*	Ovule	Sitbon (1981)
			Meynet and Sibi (1984)
	Helianthus annuus	Ovary	Cai and Zhou (1984)
			Gelebart and San (1987)
Liliaceae	*Lilium davidii*	Ovary	Gu and Cheng (1983)
Salicaceae	*Populus×simonigra*	Ovary	Wu and Xu (1984)
Chenopodiaceae	*Beta vulgaris*	Ovule	Hosemans and Bossoutrot (1983)
			Borman (1985)
Euphorbiaceae	*Hevea brasiliensis*	Ovule	Chen et al. (unpubl.)

Note: Only the first report is listed for works from one laboratory

Induction of Haploids in Sunflower by Ovary/Ovule Culture

Studies on sunflower ovary and ovule culture were begun in our laboratory in 1982. Cai and Zhou (1984) first reported the results obtained by young floret culture and ovule culture with eight cultivars. In these experiments, N_6 medium supplemented with 0.125-2 mg/L 2-methyl-4-chlorphenoxyacetic acid (MCPA), 100mg/L inositol and 2%-6% sucrose was used for float culture of the florets and ovules. After 45-55 days of culture, small globular or rod-shaped embryoids could be dissected from some enlarged ovaries and ovules. When transferred onto solid media, four embryoids from the floret culture grew into small plantlets, eight gave rise only to roots, and 96 embryoids from the ovule culture produced roots without shooting. Chromosome counts on root-tip squashes showed both haploids and diploids. In both culture methods, somatic calli often occurred on the surface of ovaries or ovules, especially when the medium was supplemented with kinetin; a few somatic calluses could regenerate into plantlets after transferring on differentiation medium. This original work led to more detailed studies on the origin and development of the gynogenetic embryoids and the role of various factors in sunflower ovule culture. The advances since then are described below.

Mix (1985) also cultured 3590 unpollinated ovaries of eight cultivars of sunflower on MS medium enriched with auxins and cytokinins. Calli formed in 2 weeks with induction frequency ranging from 10% to 85%. When transferred to differentiation media, 19 plantlets were regenerated from the calli of three cultivars. Some of them grew to flowering in the culture flask. In all of these plantlets diploid metaphases were observed.

Gelebart and San (1987) carried out successful work on sunflower ovary culture. Eight genotypes, including male-sterile and restorer lines and their F_1 hybrids, were used. Ovaries 12-24 h prior to anthesis were placed on a medium with half-strength MS macroelements, full MS microelements,

Morel vitamins, 10^{-4} M Fe-EDTA, 6%-10% sucrose, 9% agar and various combinations of exogenous hormones. During the course of culture, ovules were dissected out of the ovaries and placed on the same medium. This intermediate dissection was emphasized as necessary to avoid nutritional shortage of the ovules enclosed inside the ovaries. After 4-6 weeks from the beginning of ovary culture, the ovules were dissected and the calli or embryoids inside were transferred for regeneration to a medium of the same basic components but with 2% sucrose and free of growth substances. Gynogenesis was observed when either 2,4-D or NAA was included in the induction medium. The best results were obtained with 2 mg/L NAA and 10% sucrose. Among 6562 ovaries cultured, 78 embryoids (1.19%) were formed, from which 37 plants (47%) were regenerated. Observations on squashes or sections revealed that the embryoids or calli were located at the micropylar end of the ovules. Gynogenesis was said to originate from the egg cell or the fused polar nuclei. Root-tip examination at different stages of regenerated plants showed that among 37 plants, 20 had haploid metaphases, 8 had diploid metaphases, and 9 had both haploid and diploid cells, i.e., mixoploids. (In a recent personal communication, the authors told us that no more mixoploids were obtained; they were all haploids.) However, during the course of growth, the frequency of haploid metaphases decreased rapidly and all the cells observed became diploid when flowering occurred.

Embryological Studies

From the above description one can conclude that both haploid and diploid derivatives could be induced in sunflower ovary or ovule culture, so that it is very important to know the precise origin of the products and to distinguish them in order to get precise information on haploid induction. This can only be clarified by embryological observations. In our laboratory embryological observation combined with culture experiments is a routine

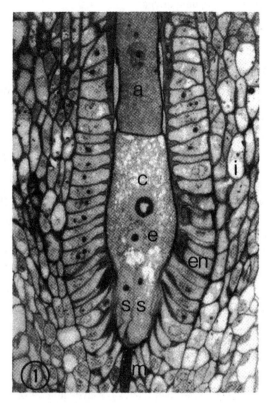

Fig. 1. Semi-thin section of a sunflower ovule 2 days before anthesis, fixed with glutaraldehyde-osmium, embedded in Poly/Bed 812 and stained with toluidine blue O. Note the well-organized embryo sac. *a* antipodal cell; *c* central cell; *e* egg cell; *en* endothelium; *i* integument tissue; *m* micropyle; *s* synergid

task of particular significance in a system as complicated as sunflower ovule culture. Yan et al. (1985) made a microscopic observation on the embryoid formation in cultured sunflower ovules. The main results of this study have been described in detail by Yang et al. (1986). Subsequently, Wei and Yang (1986) reported histochemical studies on the *in vitro* induced gynogenetic embryoids as compared with the *in vivo* zygotic embryos. Here only the main points are summarized.

The sunflower ovule is anatropous, unitegmic, and tenuinucellate. At the time of inoculation, the nearly matured embryo sac is composed of an egg cell, two synergids, a central cell with a secondary nucleus (fused polar nuclei), and two antipodal cells. The embryo sac is surrounded by the endothelium, or integumentary tapetum, a specialized inner cell layer of the integument (Fig. 1).

In cultured ovules, the gynogenetic embryoids were all located at the micropylar end of the embryo sac and undoubtedly derived from the unfertilized egg cells. Process of gynogenesis from the initiation of proembryo till advanced stages has been observed (Figs. 2, 3 A, B, C; also see Figs. 2 and 3 in Yang et al. 1986).

Sporophytic derivatives from sunflower ovule culture showed two types of structures: one was embryoid-like structures originating from the endothelium; the other was callus-like structure produced from the remaining part of the integument. These somatic derivatives usually occurred behind the gynogenetic embryoids, but often grew more vigorously and surpassed the latter in competition for nutrition. At advanced stages, the endothelial embryoids often bore a shape, similar to the gynogenetic ones. They could hardly be distinguished by the naked eye at the time of ovule dissection. Even by microscopic judgement, we could identify them only with great care, mainly according to their different position in relation to the embryo sac (Fig. 3).

Regulation of gynogenesis and somatic proliferation

In sunflower ovule culture, how to regulate the balance between gynogenetic and somatic products in favor of the former is an important task. In order to know the effects of various factors concerned; we carried out a large-scale experiment. The factors involved were genotype, kind of explants, embryo sac stage, cold pretreatment, culture on solid or liquid me-

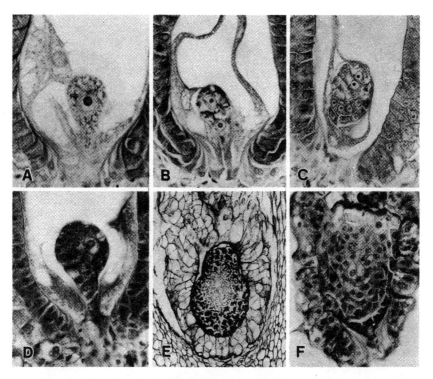

Fig. 2. A-F. Histochemical observations on early stages of gynogenetic embryoids in sunflower ovules. Paraffin sections were stained by methyl green-pyronin (A, B, C), mercuric bromphenol blue (D, F) and periodic acid-Schiff reaction (E). **A.** An egg cell after 2 days of ovule culture with prominent nucleolus and cytoplasmic pyroninophilic substances. **B.** A four-celled proembryo after 4 days of ovule culture, showing cytoplasmic pyroninophilic substances. Only two nuclei were seen in this section. **C.** A "partitioned" proembryo after 4 days of ovule culture. Its basal part remained meristimatic and apical part became vacuolated. **D.** A proembryo after 4 days of ovule culture. The cells were rich in protoplasmic protein. **E.** An ovoid embryoid after 8 days of ovule culture, showing centripetal distribution of starch grains. **F.** A gynogenetic embryoid already filling up the embryo sac after 8 days of ovule culture

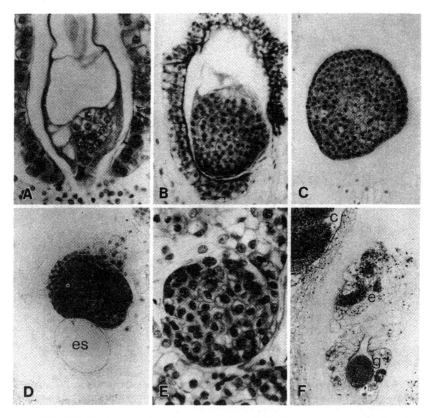

Fig. 3A-F. A comparison between various derivatives in sunflower ovule culture. The ovules were *en bloc* stained with Ehrlich's hematoxylin and sectioned by paraffin method. Notice that the gynogenetic embryoids were located inside the embryo sac (**A, B**); the endothelial embryoids were located outside the embryo sac (*es*, **D**). At advanced stage, the endothelial embryoids (**E**) were morphologically similar to the gynogenetic embryoids (**C**). In F, a gynogenetic embryoid (**g**), some endothelial derivatives (**e**) and a large integument callus (**c**) were seen in one section at low magnification

dium, kind of basal media, concentration of exogenous hormone, concentration of sucrose, culture in light or darkness, etc. More than 20000 ovules were cultured and compared. To identify the induced products, ovules were sampled after 10 and 30 days of culture, stained *en bloc* with Ehrlich's hematoxylin, sectioned by the paraffin method and examined microscopically (Fig. 3). Alternatively, an Ehrlich's hematoxylin staining-methyl salicylate clearing technique developed recently (Yang 1986) was adopted, which greatly facilitated the work. For an example of this method see Fig. 4.

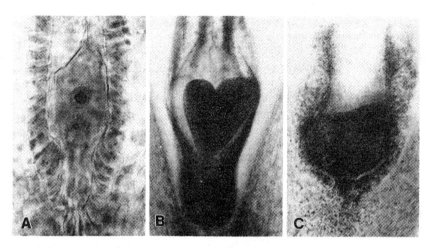

Fig. 4A ~ C. Whole mounts of sunflower ovules stained with Ehrlich's hematoxylin and cleared in methyl salicylate *en bloc*, showing a mature embryo sac (**A**), a zygotic heart-shaped embryo surrounded by endosperm (**B**), and an early heart-shaped gynogenetic emboryoid in contact with the degenerating endothelium (**C**).

Based on these experiments, three factors have been affirmed to play important roles in regulating the balance between gynogenetic and somatic products, i. e., exogenous hormone, cold-pretreatment, and sucrose concentration (Yan et al. 1987). Other factors were, of course, by no means negligible and are summarized elsewhere (Yan et al. 1988).

The presence or absence of exogenous hormone is a key factor. In most of the previous works on unpollinated ovary/ovule culture in various species including sunflower, exogenous hormone has been an indispensable supplement to the induction medium. It appeared to be different with sunflower. A series of comparative experiments have shown that a medium deprived of exogenous hormone gave even better results than its counterpart-hormone-supplemented-regardless of the culture methods used (Table 2). Furthermore, perhaps more importantly and unexpectedly, the hormone-free condition almost fully inhibited the occurrence of somatic products. This situation was in sharp contrast with the hormone supplemented condition, in which the somatic derivatives almost invariably took place (Table 3).

Sucrose concentration is also important. Experiments with three cultivars revealed that each kind of the derivatives responded differently to sucrose concentration of a medium: The optimum concentration was the lowest (1%) for integumentary callus, higher (3%-9%) for endothelial embryoid and the highest (above 12%) for gynogenetic embryoid. In the cultivar Dang Yang, the optimum sucrose concentration for gynogenetic embryoid was certified as 12% (Table 4, also see Table 2).

Cold-pretreatment also plays an important role. When the inflorescence was treated 4℃ for 24-48 h prior to inoculation, the induction frequency of gynogenetic embryoids increased and that of integument calluses decreased. The effect was more evident when cold-pretreatment was coupled with a condition of 12% sucrose and exclusion of exogenous hormone. In this case, not only the yield of dividing egg cells increased, but also cell division proceeded in a more normal, embryogenetic way: The meristimatic cell type predominated over the vacuolated type and the free-nucleate type (Table 5). At an advanced stage, the growth and differentiation of the embryoids was also improved (Fig. 5).

On the basis of these experiments, we now approach to final establishment of a procedure for inducing haploid embryoids and simultaneously excluding somatic derivatives:

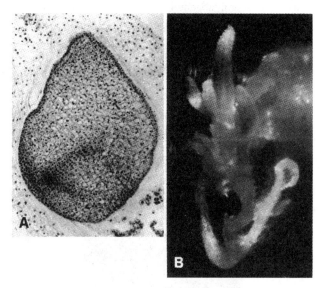

Fig. 5. A. Gynogenetic embryoid after 30 days of culture showing root differentiation. **B.** Shooting of a gynogenetic embryoid

1. Put the capitula in polyethylene bags and at low temperature (−4℃) for 24h.

2. Cut the capitula into several parts, sterilize them in 0.1% $HgCl_2$ solution, and rinse with sterile distilled water several times. Choose suitable tubular florets (1-3 days prior to anthesis) and pick out the ovules aseptically.

3. Inoculate the ovules on the surface of liquid N_6 medium (or alternatively solid medium) supplemented with 100 mg/L inositol and 12% sucrose, 10 ovules per tube containing 20 ml medium. Keep the culture at approximately 27℃ in the dark.

4. After 1-2 months of culture, dissect the enlarged ovules and take out the gynogenetic embryoids, if any induced, for further experiments.

Experiments are now being done for refinement of the regeneration procedure. Preliminary results by Hua and Yang (unpubl.) showed that it seemed better to transfer gynogenetic embroids to a medium originally developed by Paterson and Everett (1985) for regeneration of plants from hypocotyl-derived callus. This was composed of MS salts and vitamins supplemented with 5g/L KNO_3, 100 mg/L inositol, 1 mg/L NAA, 1-2 mg/L BA, 0.1 mg/L GA_3 and 3%-6% sucrose. Under illumination conditions the underdeveloped embryoids cultured on this medium grew vigorousely into calli with green spots. After further transferring and deletion of NAA, green plantlets were regenerated (Fig. 5B, 6).

Fig. 6. A regenerated plantlet of gynogenetic origin

Conclusions

Sunflower has been one of a few crops lacking methods for artificial induction of haploid plants. This situation is now changing, since haploid plants or plantlets have been reared recently via either anther or ovary/ovule culture in several laboratories. Nevertheless, the techniques are still immature due to low induction-frequency as well as usual occurrence of undesirable somatic products. This is perhaps a serious problem in ovary/ovule culture works also in many other plant species, which accounts for the fact that so far this method is still far from practical application. Our experience indicates the necessity of combining culture experimentation with embryological studies in solving these difficulties. To do so, we should first distinguish products of different origin with great care. Next we should know the different responses of these products to various factors involved in the culture. By regulating some key factors, we might find out the best possible means to promote gynogenesis and inhibit or even delete somatic products. It is desirable that the potentiality of unpollinated ovary/ovule culture as a new biotechnology be exploited better in the near future.

(Authors: Yang HY, Yan H, Zhou C. Published in: Biotechnology in Agriculture and Forestry, Vol 10 (ed. by Bajaj YPS). Springer-Verlag, Berlin, 1990, pp. 472-484. with 6 figures, 5 tables and 32 references. Only all figures are retained here.)

19 韭菜未传粉子房培养中单倍体的胚胎发生和植株再生

Haploid embryogeny and plant regeneration in unpollinated ovary culture of *Allium tuberosum*

Abstract

In a cultivar of *Allium tuberosum*, polyembryony comes from zygote and antipodal cell. However, only the zygotic embryo can reach maturity and the antipodal proembryo degenerates during its later development. Using unpollinated ovary culture technique we were able to make the unfertilized egg cell and antipodal cell developing into mature embryoids which finally regenerated into haploid plants. In some treatment, nearly half of the cultured ovules appeared to contain parthenogenetic and/or apogametic proembryos. This method can be used in haploid breading of *Allium tuberosum* for its high frequency of embryoid formation. Embryological observations have revealed the similarity of antipodal complex to egg apparatus in their morphological and developmental patterns. Some problems were discussed on the apogamy of antipodal cells of *Allium tuberosum*.

早在20世纪30年代，Модилевский 发现韭菜的卵细胞、助细胞、反足细胞和珠被细胞都具有胚胎发生能力，在每个胚囊中都有多胚发生，但在种子成熟时，一般只保留合子胚或不定胚[1]。此后，在许多胚胎学文献中多引证其助细胞和反足细胞的胚胎发生作为无配子生殖的典型例证，但很少继续深入研究[2~4]。近年来通过未传粉子房和胚珠培养诱导单倍体已在一些植物上获得成功[5]。将这一实验体系用于韭菜有可能对其多胚现象有更深入的了解，同时也可用于探索其单倍体育种的可能性。我们最近进行了韭菜未传粉子房培养的研究，获得了大量单倍体植株，并作了胚胎学观察，证实了上述的设想。

材 料 与 方 法

供试材料为武汉市蔬菜科研所提供的武汉地区栽培品种"黄格子"(*Allium tuberosum*)将花冠即将张开而尚未传粉的花蕾用75%乙醇浸泡约30s，0.1%升汞消毒12min，无菌水冲洗三次后，在无菌条件下剥去花被和雄蕊，将带花柄的子房置于MS液体培养基上进行漂浮培养。培养基附加成分为3%~12%蔗糖，附加0~2ppm玉米素(Zt)和0~0.4ppm 2-甲基-4氯苯氧乙酸(MCPA)。在27℃左右、散射光条件下培养。接种后5d、10d、20d分别取样并用醋酸甲醇(1:3)固定，冷藏于70%甲醇中。剥出胚珠用爱氏苏木精整体染色后制成石蜡切片作胚胎学观察。用5%改良卡宝品红液染色、压片观察染色体数目。作为对照系,取自然传粉条件下活体内不同发育时期的胚珠,按上述方法石蜡制片。所有制片均用 Olympus BH_2 显微镜观察与摄影。

结　果

一、体内发育的一般状况

韭菜成熟胚囊由卵细胞、两个助细胞、中央细胞(含两个极核)和三个反足细胞组成。卵细胞的细胞质染色很浅,核仁、核膜清楚,核位于细胞合点端。助细胞的细胞质染色很深,核大染色尤深,核仁、核膜难以区分,整个细胞体积也较大。三个反足细胞有明显差异:通常两个反足细胞类似助细胞,细胞质染色很深,核仁、核膜难以区分。另一个反足细胞类似卵细胞,细胞质染色很浅,核仁、核膜清晰,核位于细胞的珠孔端,很容易和前述两个反足细胞区分开(图版Ⅰ,1)。

在开花后的胚珠中,卵细胞状的反足细胞常与合子同时分裂,在胚囊的合点端形成多细胞原胚(图版Ⅰ,2)。在具合子分裂的257个胚囊中,反足细胞进行胚胎发生的有79个(占30.7%)。随着合子胚的生长和胚乳核的增加,反足细胞原胚逐渐退化。在成熟种子中,仅剩合子胚。

在所观察的四百多个活体胚珠中,未见到助细胞胚胎发生,也未看到不定胚的发生。在这方面,这种韭菜与 Модилевский 所用的材料有所不同。

二、培养子房中的原胚发生

子房培养20d时,由爱氏苏木精整体染色的胚珠,脱水后经二甲苯透明即可在双筒解剖镜下清楚看到内部的原胚。有些胚珠具一个原胚,位于合点端或珠孔端,有些胚珠具两个甚至三个原胚,通常是合点端和珠孔端各具一个(图版Ⅰ,3)。经切片观察培养早期的胚珠,证明这些原胚分别起源于卵细胞和反足细胞。未受精极核虽也可分裂,但经多次游离核分裂后逐步消失。与活体内一样,未看到助细胞分裂形成原胚的证据,也未见到不定胚的发生。因此,在韭菜未传粉子房培养中的原胚发生是来源于卵细胞的孤雌生殖和反足细胞的无配子生殖。以下分别详细描述这两个过程:

卵细胞的孤雌生殖　接种时,胚囊发育已近成熟,助细胞和卵细胞的明显形态特征为鉴别珠孔端原胚的起源提供了标准。有些子房培养 4d 时,胚囊珠孔端就已看到二胞原胚。此时卵器三细胞相互紧靠,使鉴定原胚的起源有一定困难。培养 6d 时,有一些令人信服的切片显示:两个助细胞形态完整,中间夹了一个多细胞原胚(图版Ⅰ,4),这就确证这类原胚起源于卵细胞而非助细胞。

卵细胞的早期胚胎发生过程呈现一定规律性。第一次为横分裂,产生一个由顶细胞和基细胞组成的二胞原胚(图版Ⅰ,5)。接着基细胞横分裂成两个胚柄细胞,顶细胞则进行一次纵分裂,共同组成"T"形四胞原胚(图版Ⅰ,6)。以后胚柄细胞不再分裂,而由顶细胞产生的两个细胞进行连续分裂,构成胚体,经梨形胚(图版Ⅰ,7)再进一步发育为大型的胚状体(图版Ⅰ,8)。

反足细胞的无配子生殖　反足细胞一端紧靠中央细胞,三面被胚珠组织包围。为了鉴别合点端原胚的起源,我们根据主要两点:第一,原胚只发生于反足细胞位置并位于中央细胞质中;第二,原胚和退化及宿存反足细胞总数恒定不超过三个(图版Ⅰ,9)。从而断定这些位于合点端的原胚确实起源于反足细胞而非珠心细胞。

反足细胞的胚胎发生过程似有一定预定性。助细胞状的反足细胞一般很少分裂,只有卵细胞状的反足细胞才具胚胎发生能力。有时三个反足细胞形态上非常类似卵器:两个助细胞状的细胞中间夹着一个卵细胞状的细胞,以后原胚就由中间这一细胞产生(图版Ⅱ,10)。反足细胞的早期胚胎发生过程也类似于卵细胞,第一次为横分裂,产生一个二胞原胚(图版Ⅱ,11),接着基细胞横分裂和顶细胞纵分裂,形成倒"T"形四胞原胚(图版Ⅱ,12)。以后来自基细胞的两个胚柄细胞不再分裂,而由顶细胞分裂形成的两个细胞继续分裂(图版Ⅱ,13),经梨形胚(图版Ⅱ,16)发育为大型胚状体(图版Ⅱ,17)。

在极少数胚囊中,两个反足细胞能够同时进行胚胎发生(图版Ⅱ,14),在合点端形成两个并列的原胚(图版Ⅱ,15)。

统计了162个具原胚的胚珠,发现有四种情况(表1):最常见的是一个胚珠中同时含有一个卵细胞原胚和一个反足细胞原胚(占53.7%);其次是仅含反足细胞原胚(占30.2%);再次是仅含卵细胞胚(占13.0%);极少数含两个反足细胞胚和一个卵细胞胚(3.1%)。总起来看,反足细胞的胚胎发生频率稍高于卵细胞。

表1　　　　　　　　不同来源原胚的分布

总数	卵细胞原胚		反足细胞原胚		卵细胞原胚+反足细胞原胚		卵细胞原胚+2个反足细胞原胚	
	个数	%	个数	%	个数	%	个数	%
162	21	13.0	49	30.2	87	53.7	5	3.1

三、外源激素和蔗糖浓度对原胚发生的影响

设计了四种激素浓度和四种蔗糖浓度的配比试验,结果表明(表2):就激素影响而言,培养5d时便可看到绝大多数子房显著膨大,但培养10d时,在无激素处理中培养的子房直径(0.31cm)小于在有激素处理中的(0.46cm)。这时胚囊中已有原胚发生。以胚珠为单位的统计数字表明,虽然无激素培养可以诱导出原胚,但附加低浓度激素0.5(ppm Zt 和 0.1ppm MCPA)使诱导频率有所提高。随着激素浓度的增加,诱导率呈下降趋势。

就蔗糖影响而言,除个别因试验误差导致的例外,在4种处理中均以9%的蔗糖浓度效果最佳。过低的(3%)和过高的(12%)蔗糖浓度诱导效果均不佳。

总结以上两种因素的实验结果,韭菜子房培养的最佳外源激素与蔗糖浓度组合为:Zt 0.5 ppm、MCPA 0.1 ppm、蔗糖9%。

表2　外源激素和蔗糖浓度对原胚发生的影响(诱导频率%)

蔗糖浓度(%) \ 诱导频率(%) \ 激素浓度(ppm)	0	0.5Zt+ 0.1MCPA	1Zt+ 0.2MCPA	2Zt+ 0.4MCPA
3	23.8	14.4	13.3	32.2
6	30.0	42.7	34.6	16.5
9	42.2	48.7	41.9	24.0
12	—	47.5	29.5	18.1

注：每处理观察80~110个胚珠。诱导频率 = $\frac{含原胚胚珠数}{观察胚珠数}$

四、胚的发育与植株再生

培养20d时，有些胚状体已明显分化，外形变为长形，内部分化出胚根原基(图版Ⅰ,8)。两种不同来源的胚状体均继承了原胚的极性。如前所述，进行胚胎发生的卵细胞和反足细胞都具有明显极性，表现在它们的核都位于朝向中央细胞的部位。以后形成的原胚，也都是胚体朝向中央，胚柄朝向外方(图版Ⅱ,16)。原胚分化时胚根原基也都是朝向原胚的胚柄方向，这样，卵细胞胚的胚根恒定朝向珠孔端，反足细胞胚的胚根则恒定朝向合点端。

后期胚状体数目少于原胚数。这是因为有相当数量的原胚中途退化。与水稻未传粉子房培养中的情况相似[6]，退化过程由原胚外围细胞液泡化开始，继而细胞离散、解体，最后扩及整个原胚。

培养40d时，子房直径达0.7cm左右。许多子房的腹缝线裂开，胚状体从里面长出，个别子房里直接长出根。培养50d时，将子房剖开，把胚状体(图版Ⅱ,18)转移到无激素、附加3%蔗糖的MS固体培养基上，再经13d培养即有25%的胚状体长成独苗。如先将胚状体转移到含2ppm BA和0.2ppm NAA的MS固体培养基上，则有70%的胚状体形成芽丛，但根的生长被抑制。经过15d培养后，再将芽丛转移到无激素培养基上，根的生长很快恢复，形成茂盛的丛生植株(图版Ⅱ,

20)。总之,在韭菜子房培养中,孤雌生殖与无配子生殖均是经胚状体途径再生植株。

将再生植株根尖用卡宝品红压片法进行染色体计数,在 11 个植株中,有 10 个的根尖细胞为 2n=x=16,即为单倍体(图版Ⅱ,19),一个植株为二倍体。

讨　　论

我们证明了 Модилевский 关于韭菜多胚现象的发现[1]。本文所用的韭菜,活体内的多胚来自合子和反足细胞;而在未传粉子房培养时,多胚则来自未受精的卵细胞和反足细胞;没有看到如上述作者所报道的由助细胞或珠被细胞产生的多胚。在体内条件下,反足细胞只能分裂到原胚阶段,不能发育成熟,而子房培养则成功地促使卵细胞胚和反足细胞胚发育成熟并再生单倍体植株。这一实验系统在两方面显示其特殊意义。在理论研究方面,在被子植物中由活体内反足细胞形成胚的例子极少,且都只报道最初几次分裂,形成原胚状的结构[7,8]。由于未观察到成熟的、有生活力的反足细胞胚[3,4,8],导致某些学者认为"反足细胞胚的概念是一种误解"[9],并将其从无配子生殖中删除[10]。现在利用子房培养的实验系统,就可以充分肯定反足细胞无配子生殖的价值,并为精确研究韭菜孤雌生殖和无配子生殖提供了可能性。此外,在本实验中无论是来自卵细胞还是反足细胞的原胚,均可在无胚乳、也无特殊因素的简单培养基上生长和分化,这又为研究胚的发育提供了一种实验方法。在应用方面,就花药培养尚未获得成功的韭菜而言[11],由胚囊中诱导出高频率的单倍体植株,已可作为一项相对成熟的技术用于韭菜的单倍体育种,从而显示子房培养技术与韭菜这种特殊材料相结合的优越性。

在体内,虽然约有 30% 的胚珠中反足细胞分裂形成原胚,但一般最大的反足细胞原胚不超过 15 个细胞。这说明反足细胞原胚的发育在体内是受到抑制的。观察表明,当胚乳游离核开始分布于胚囊周缘区时,反足细胞胚即已退化。而在两个胚乳核异常少的胚囊中,反足细胞胚较大。联系到在子房培养条件下没有胚乳发生,反足细胞胚能够发育成熟的现象,似乎可以推测胚乳的发育和反足细胞胚的退化有着某种内在联系。这是一个值得进一步研究的问题。

本试验中的一个特点是反足细胞的特征和胚胎发生过程与卵器非常相似,表现在:第一,在不少情况下,三个反足细胞之中,两个呈助细胞状,一个呈卵细胞状。第二,卵细胞状的反足细胞初期的胚胎发生过程与卵细胞相同。第三,由反足细胞产生的原胚以至成熟胚,其极性与由卵细胞产生的相同。韭菜胚囊中这种反足细胞与卵器在结构与功能上的相似性,是一个非常有趣的现象,需要继续在超微结构和组织化学上进行深入研究。

(作者:田惠桥、杨弘远。原载:实验生物学报,1989,22(2):139~147。图版2幅,表2幅,参考文献11篇。仅选录表1、2。)

20 韭菜孤雌生殖和反足细胞无配子生殖的超微结构观察

Ultrastructural observations on parthenogenesis and antipodal apogamy of *Allium tuberosum* Roxb

Abstract

Allium tubcrosum Roxb is a species characterized by spontaneous parthenogenesis and antipodal apogamy. This paper deals with the ultrastructural changes during these processes. Before pollination, the mature egg cell contained abundant mitochondria with well developed cristae, spherical or ellipsoidal plastids and some polyribosomes, which suggested a relatively high metabolic activity. After fertilization, in zygotes the mitochondria changed to irregular shape and their cristae degenerated, the plastids elongated or became cup-shaped, the polyribosomes decreased and the free ribosomes increased in number. Some unfertilized egg cells, two days after anthesis, showed similar ultrastructural changes as those taking place in zygote, which seemed to be a sign of triggering to parthenogensis. In *Allium tuberosum* Roxb, the three antipodal cells bore a close resemblance to the egg apparatus: Among them two antipodal cells were similar to the synergids with a filiform apparatus-like structure and plentiful organelles at their chalazal end. The other was an egg-like antipodal cell which could undergo similar ultrastructural changes as those happened in zygote, leading to divide into apogamic proembryos two days after anthesis. The problems of parthenogenesis and antipodal apogamy in *Allium tuberosum* Roxb is discussed in view of ultrastructural features of the egg and egg-like antipodal cell.

在被子植物中,韭菜的生殖过程比较特殊。早在1895年Tretjakow就对韭菜作过胚胎学观察,在近一半的胚珠中发现有助细胞原胚或反足细胞原胚(引自Hakansson 1951[5])。Haberlandt(1923,1925)发现在已去雄的花中胚囊可自发地由卵细胞、助细胞、反足细胞和珠被细胞产生原胚,但因极核不受精不能形成胚乳,所有原胚以后均退化(引自Maheshwari 1950[12]。Модилевский(1925,1930,1931)的观察结果亦相近似[17]。Hakansson(1951)还发现,韭菜的三个反足细胞形态上类似卵器,他称为第二卵器。开花后多胚主要由合子胚和反足胚组成[5]。韭菜中的这些特殊现象,曾引起胚胎学家的很大兴趣,但以后很少有人继续深入研究。最近,我们对本地韭菜品种"黄格子"进行了胚胎学观察,证实了其有自发的孤雌生殖和反足细胞无配子生殖倾向,但在体内只能达到原胚阶段;而采用未传粉子房培养方法可以使之在离体条件下继续发育,再生大量单倍体植株[1,2]。在此基础上,我们对韭菜的卵细胞与反足细胞以及由之产生的孤雌生殖与无配子生殖原胚进行超微结构观察。

材 料 与 方 法

供试材料为武汉市蔬菜科研所提供的韭菜（*Allium tuberosum* Roxb）品种"黄格子"。将开花前和开花后1d、2d、3d的子房分别采集，在0.2mol/L、pH7.3的二甲胂酸钠缓冲液中剥出胚珠。用二甲胂酸钠缓冲液配制的3%戊二醛在室温下固定4h，洗涤后再用1%锇酸在4℃冰箱中固定过夜。以后经乙醇逐级脱水，环氧丙烷过渡，用Epon 812树脂包埋。在Sorvall MT-6000型超薄切片机上制作超薄切片，经醋酸双氧铀与柠檬酸铅先后染色各20min，用JEM-100cx/Ⅱ型电子显微镜观察与摄影。

结 果

一、卵细胞及其孤雌生殖

成熟卵细胞 传粉前的成熟卵细胞具明显极性，其珠孔半部被一大液泡占据，核和大部分细胞质位于其合点端。细胞质中，含有较丰富的细胞器。线粒体数量较多，嵴发育良好。质体呈圆形或椭圆形，具电子密度较高的基质，其中含有一些细小的淀粉粒。内质网数量不多，有些是粗糙内质网。高尔基体较少。游离核糖体数量中等，且有一定数量的多聚核糖体（图版Ⅰ，1）。总之，从超微结构特征来看，卵细胞受精前已呈现出一定的代谢活性。韭菜卵细胞的另一特点是其周围具壁，包括其合点端大部分亦被壁包围（图版Ⅰ，1），仅有一范围较小的无壁区域作为以后受精的靶区。

卵细胞受精后的变化 作为研究孤雌生殖和无配子生殖的对照系统，有必要对韭菜受精后的早期超微结构变化作一简略的描述。从受精完成到合子第一次分裂大约需要1d。合子在这段时间内发生了一些明显的结构变化：首先，在其合点端的细胞壁上出现片段的电子浓密物质。与此相关,在细胞外方的胚乳细胞质中,有一些内质网与合子壁平行排列(图版Ⅰ,2)。其次,细胞核中出现二个核仁;核质中的染色质团块变得更小、更多,电子密度有所下降。第三,细胞器的数量与结构发生变化(图版Ⅰ,3):游离核糖体密度增加,多聚核糖体减少。质体拉长呈分裂趋势,或呈杯状。线粒体外形趋于不规则,内部嵴退化,基质中出现一些絮状物。以上变化似乎表示在合子期间发生了某种细胞质改组的过程。开花后2d,合子横分裂,形成由顶细胞和基细胞组成的2胞原胚。顶细胞的细胞质较浓密,含质体较多。基细胞含一大液泡,细胞质中的线粒体和高尔基体较多。在二个细胞的间壁上很少有胞间连丝。从总体上看,2胞原胚中的游离核糖体密度较合子为低,恢复到与受精前卵细胞中核糖体密度相近的水平。一定数量的多聚核糖体重新出现。大部分质体又恢复圆形。线粒体内部的嵴又重新形成。

  孤雌生殖 在自然条件下,开花后2d如果胚囊中两个助细胞之一没有退化,胚乳游离核亦未产生,而卵细胞呈现启动分裂的迹象或已分裂成原胚,可以判定该胚囊中发生了孤雌生殖。在这种情况下,卵细胞尚未分裂,但已显示发育迹象:细胞核中出现电子密度很高的染色质团块,线粒体数量较多,但嵴开始退化,其基质中出现絮状物,质体拉长而有分裂趋势(图版Ⅰ,4、5)。这些特征与合子中的变化很相似,表明该卵细胞可能已经完成了某种激活的过程,显示即将启动孤雌生殖分裂的征兆。唯一与合子有所区别的是,其核糖体的密度较合子中为低,仍有一定数量的多聚核糖体存在,并且不像合子那样在合点端的壁上有片段的电子浓密物质存在。由孤雌生殖产生的2胞原胚,除基细胞的液泡化程度较顶细胞高外,两个细胞之间的细胞器差异不明显,它们之间的公共壁上胞间连丝也较少(图版Ⅱ,6、7)。这些特征与由受精产生的正常2胞合子原胚的特征相似,仅在孤雌生殖原胚中游离核糖体的密度明显较多,而多聚核糖体较少而已。

二、反足细胞及其无配子生殖

  开花前的反足细胞 韭菜的反足细胞在被子植物中非常特殊。我们以前的光镜观察表明:胚囊细胞分化时,与胚囊纵轴垂直的一对姊妹核所形成的两个反足细胞在形态上类似助细胞(助状反足细胞);而与胚囊纵轴平行,即和合点极核同一来源的那个反足细胞则类似卵细胞(卵状反足细胞)。三个反足细胞总体上类似珠孔端的卵器[2]。(图版Ⅱ,8)示卵状反足细胞和一个助状反足细胞的半薄切片,可见其间区别之明显。接近开花时,卵状反足细胞中出现大液泡,细胞核和大部分细胞质被挤到细胞珠孔端,形成极性分明的状态。整个细胞被一完全壁包围。在细胞质中,细胞器的超微结构状态亦类似卵细胞(图版Ⅱ,9)。助状反足细胞中不形成大液泡,细胞核及核仁形状常不规则。细胞质中聚集了许多质体,它们多呈长形,内含数个小淀粉粒。线粒体数量较多,嵴发育良好。核糖体以游离状态为主。这些超微结构特征和助细胞很相似。更有趣的是,两个助状反足细胞的合点端、界壁及与珠心相连的壁较厚,形成类似丝状器的结构(图版Ⅱ,10)。在超微结构上,该种"丝状器"的壁内突由内部的微纤丝骨架和其外围的电子密度较低的壁成分组成。与助细胞丝状器有所不同的是,反足细胞的"丝状器"壁内突范围较小,而且基部扩展到胚囊壁上(图版Ⅱ,11)。

无配子生殖 开花后,助状反足细胞常已退化,部分卵状反足细胞也退化,但有些卵状反足细胞并不退化而显示继续发育的迹象。根据壁的状况,它们可分为两种情况:一种卵状反足细胞的珠孔端有三分之一的区域无壁,其外方常不与中央细胞或胚乳的细胞质接触;另一种卵状反足细胞则有一完全壁,壁的外侧常有发育良好的中央细胞或胚乳的细胞质紧贴。推测中央细胞与胚乳可能与反足细胞的壁代谢有关。这些宿存的卵状反足细胞中,游离核糖体的密度明显增高;线粒体外形不规则,内部嵴退化,基质中出现絮状物;质体拉长呈分裂趋势;核质电子密度降低(图版Ⅲ,12)。这些变化与受精后合子中所发生的变化很相似。通常开花后 2d,卵状反足细胞开始分裂。(图版Ⅲ,13)示一个细胞中出现电子浓密的前期染色体。卵状反足细胞的第一次分裂为横向分裂,由此形成的 2 胞反足原胚其珠孔端一个细胞类似顶细胞,合点端一个细胞类似基细胞。从总体上看,反足原胚和合子原胚结构相似而朝向相反(图版Ⅲ,14)。在反足 2 胞原胚中,质体恢复圆形,线粒体内部重新出现嵴。通常"顶细胞"中含质体稍多,"基细胞"中含线粒体稍多。这些超微结构特征也和合子2胞原胚或孤雌生殖2胞原胚相近。值得注意的是,由珠孔端无壁的卵状反足细胞分裂后形成的2胞原胚,其珠孔端顶细胞外围仍然无壁(图版Ⅲ,15),而由具完全壁的卵状反足细胞分裂成的2胞原胚则被一完全壁包围(图版Ⅲ,16)。受精后,代谢旺盛的胚乳迅速向胚囊合点端发展。观察到硕大的胚乳游离核将反足细胞或反足原胚包围的现象。另外,胚乳还将胚囊合点端的珠心组织分解以扩大胚囊范围,这些现象都不利于反足细胞及其原胚的继续生存。我们曾报道,在体内条件下,反足细胞的无配子生殖只能达到形成原胚阶段,不可能继续发展为成熟的胚。在种子成熟时,只有合子胚存在[2]。

讨 论

一、从超微结构特征看卵细胞孤雌生殖的内在基础

通常被子植物的卵细胞具有类似的超微结构特征:细胞高度液泡化;极性明显(禾本科植物例外);细胞壁从珠孔端到合点端逐渐变薄,在合点端约三分之一的区域仅以质膜与中央细胞相邻;细胞器数量较少且处于不活动状态;很少多聚核糖体。这些特征表明卵细胞受精前是偏于休眠状态,通常只有通过受精才能激活[9,15]。与一般植物相比,韭菜的卵细胞比较特殊:细胞壁比较发达,合点端除一狭小区域外均被壁包围;线粒体数目较多,且嵴发育良好;具有一定数目的多聚核糖体。表明它在受精前即具有相对较高的代谢活性。这也许可以解释为什么韭菜具有较大的自发孤雌生殖潜力。

迄今关于被子植物中孤雌生殖的胚胎学研究主要限于光镜观察,很少超微结构研究。仅最近阎华等(1989)和Yan等(1989)对向日葵未受精胚珠培养中诱导的离体孤雌生殖过程进行了超微结构观察,表明未经受精而被激活的卵细胞中发生了许多与合子相似的变化[4,16]。韭菜由于具有较高的自发孤雌生殖频率,可以作为研究孤雌生殖超微

结构变化的一种良好材料。根据本文观察,韭菜体内启动孤雌生殖的卵细胞中,同样发生许多类似合子中的细胞器结构变化。由于观察数量较少,目前我们还不能对此作全面的评价,有待于今后在更广泛的材料中进行研究。此外,从卵器的整体角度来分析,韭菜卵细胞分裂成孤雌生殖原胚往往伴随着助细胞不退化。Murty 等(1984)在 *Sorghum bicolor* 中同样观察到无融合生殖胚囊中两个助细胞保持完好[13]。目前公认在大多数植物中,两个助细胞中的一个在传粉后、受精前预先退化,花粉管总是进入退化助细胞释放精子[9,15]。因而助细胞的退化是卵细胞受精的前提之一。我们推测,很可能正是由于助细胞不退化,导致花粉管不能进入胚囊使卵细胞受精。而卵细胞本身又具有较高的发育潜能,从而转向孤雌生殖。孤雌生殖不妨被认为是在助细胞丧失接受花粉管功能、受精受阻的条件下的一种补偿性的生殖机能。

二、从超微结构特征看反足细胞无配子生殖的内在基础

在被子植物中,反足细胞在体内分裂形成原胚的例子很少。加之由于从未看到成熟的反足胚,致使学术界对反足细胞是否能够进行无配子生殖产生了怀疑[7,10,11],其焦点在于反足细胞的分裂"仅仅是增殖,还是形成真正的胚"[6]。我们曾经证实韭菜的反足原胚仅局限于从卵状反足细胞产生,并以一种有规律的、类似合子胚的分裂方式形成原胚[2]。这种反足原胚与其它植物中的反足细胞增殖有本质上的区别。在这些植物中,所有反足细胞一起分裂,形成一团无组织结构的不规则细胞团[8]。我们还成功地应用子房培养技术促使韭菜反足原胚形成胚状体并再生单倍体植株[1]。从超微结构看,韭菜助状反足细胞合点端的基部形成了明显的壁内突,细胞质中细胞器丰富并呈代谢活跃状态,这些结构特征具传递细胞性质[3,14]。而卵状反足细胞则完全不具备行使营养功能的结构。它的细胞壁很薄,细胞器的特点和卵细胞相似并发生类似合子中的形态结构变化,由卵状反足细胞分裂产生的 2 胞原胚在超微结构上与合子原胚亦相近。结合上述光镜观察和子房培养的研究结果,我们认为韭菜的卵状反足细胞的分裂属胚胎发生性质,充分体现了其无配子生殖的机能。在其它一些植物中,如 *Aster novae-angliac*, *Cistus laurifolius*, *Ulmus americana*, *U. hollandicabelgica*, *U. glabra*, *Nothoscordum fragrans*, *Allium paradoxicum*, *Rudbeckia bicolor*, *R. sullivantii*,也曾观察到反足细胞类似卵器的现象。其中 *U. americana*, *U. glabra*, *R. sullivantii* 的卵状反足细胞也可以分裂形成原胚。在 *Nicotiana rustica*, *N. paniculata* 中,卵状反足细胞不仅在形态上、而且在组织化学特点上也类似卵细胞(以上均引自 Kapil 与 Bhatnagar[8])。由此可见,像韭菜这样的反足细胞在植物界还是有一定代表性的。

图 版 说 明

各图片均按珠孔端朝下方向排列。A.顶细胞, B.基细胞, CC.中央细胞, E.卵细胞, EA.卵状反足细胞, En.胚乳 er.内质网, m.线粒体, n.细胞核, p.质体, S.助细胞, SA.助状反足细胞, v.液泡, w.细胞壁, Z.合子

1.成熟卵细胞局部,示合点端的壁和细胞质状态。×14000 2.合子合点端的细胞壁。×14000 3.合子细胞质,示质体和线粒体的变化。×10080 4.开花后2d,即将启动孤雌生殖的卵细胞。×1400 5.图4合点端放大,示质体、线粒体的变化和细胞壁状态。×10080

6.孤雌生殖的2胞原胚,注意此时两个助细胞保存完好。×1400 7.图6放大,示孤雌生殖2胞原胚中顶细胞与基细胞的局部细胞质。×10080 8.一个卵状反足细胞和一个助状反足细胞。×1000 9.卵状反足细胞的细胞质状态。×10080

10.二个具"丝状器"的助状反足细胞。×360 11.图10的放大,示"丝状器"。×2660

12.卵状反足细胞中的细胞器变化。×10080 13.卵状反足细胞进入无配子生殖的分裂前期。×1400 14.由卵状反足细胞分裂而成的无配子生殖2胞原胚。×1400 15.一种类型的2胞反足原胚,示其顶细胞珠孔端缺乏细胞壁。×10080 16.另一类型的2胞反足原胚,示其顶细胞珠孔端具有完全壁。×14000

Explanation of plates

All figures were oriented with the micropylar end downwards.

A. Apical cell, B. Basal cell, CC. Central cell, E. Egg cell, EA. Egg-like antipodal cell, er. endoplasmic reticulum, fa. filiform apparatus, m. mitochondrium, n. nucleus, p. plastid, S. Synergid, SA. Synergid-like antipodal cell, v. vacuole, w. cell wall, Z. Zygote.

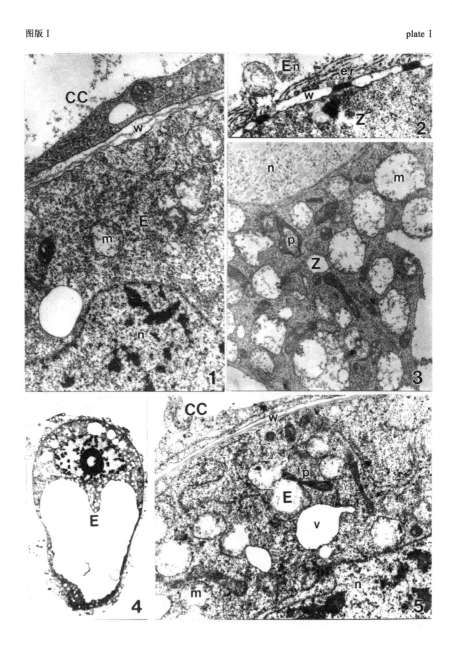

图版 II plate II

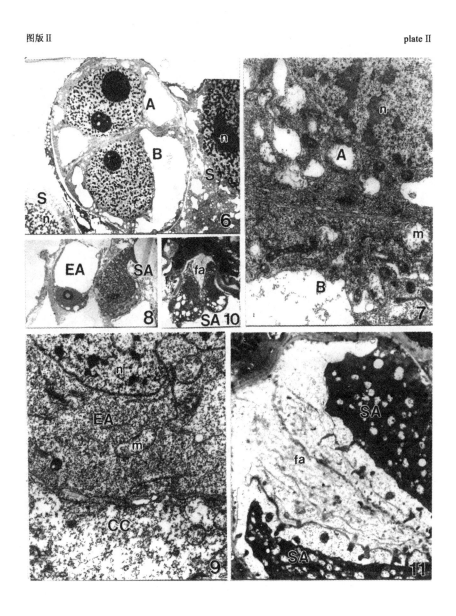

图版III

plate III

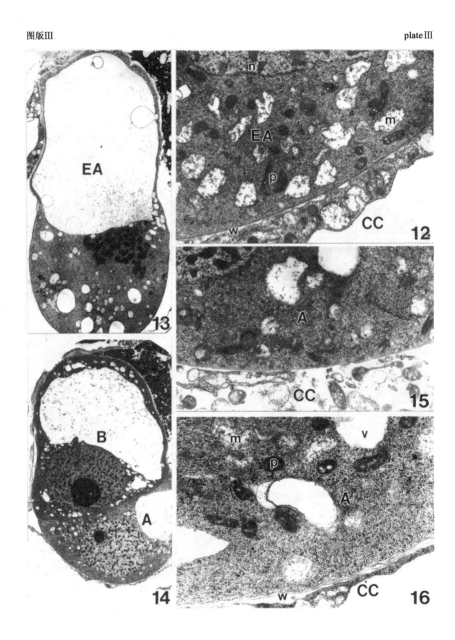

Fig. 1. The cytoplasm and cell wall at the chalazal end of a mature egg cell. ×14000 **Fig. 2.** The cell wall of a zygote at its chalazal end. ×14000 **Fug. 3.** The cytoplasm of a zygote, showing the status of plastids and mitochondria. ×10080 **Fig. 4.** Two days after anthesis, an egg cell just prior to parthenogenetic development. ×1400 **Fig. 5.** Enlargement of the chalazal partion of Fig. 4, showing the status of plastids, mitochondria and cell wall. ×10080

Fig. 6. A parthenogenetic 2-celled proembryo. Note two synergids still maintained intact. ×1400 **Fig. 7.** Enlarged portion of Fig. 6, showing the cytoplasm starus of the parthenogenetic proembryo. ×10080 **Fig. 8.** An egg-like antipodal cell and a synergid-like antipodal cell. ×720 **Fig. 9.** The cytoplasmic status of an egg-like antipodal cell. ×10080 **Fig. 10.** Two synergid-like antipodal cells with a "filiform apparatus". ×360 **Fig. 11.** Enlarged portion of Fig. 10, showing the "filiform apparatus". ×2600 **Fig. 12.** Organelle changes in an egg-like antipodal cell. ×10080 **Fig. 13.** An egg-like antipordal cell entering prophase of apogametic division. ×1400 **Fig. 14.** A 2-celled apogametic proembryo divided from an egg-like antipodal cell. ×1400 **Fig. 15.** One kind of the 2-celled antipodal proembryo, showing absence of a cell wall at the micropylar end of its apical cell. ×10080 **Fig. 16.** Another kind of the 2-celled antipodal proembryo, showing presence of a complete wall of its apical cell. ×14000

(作者:田惠桥、杨弘远。原载:植物学报.1991,33(11):
819~824。图版3幅保留,参考文献17篇已删。)

第二节 胚囊操作

提要

胚囊是被子植物的雌配子体,是双受精和胚与胚乳发生的场所,具有重要的研究价值。然而和对等的雄配子体(花粉)相比,研究胚囊的手段受到很大的限制,一般只能借助切片观察固定的胚囊,而不能像花粉那样很容易地由花药中分离出来进行直接的操作。从历史上看,分离胚囊的最初目的是企图寻找一种简易而又能显示其整体结构的观察方法,以后才发展成以分离生活胚囊为目的。分离的生活胚囊不仅可用于胚囊及其成员细胞生活状态的细胞生理学研究,更重要的是可以由此开辟其离体操作的道路。但是由于胚囊十分柔弱,对孢子体组织又有高度的依赖性,生活胚囊的分离与操作显然相当困难。

我们在未传粉子房与胚珠培养初步成功之后不久,就将生活胚囊的分离与操作列为下一个攻关目标。考虑到其技术难度和没有前人经验可循,我们采取了以下两点由易入难的策略:首先,选择子房内胚珠数目较多的植物(如烟草)作为突破口;其次,先解决固定材料的胚囊分离,再攻生活胚囊的分离。考虑到酶解是分离生活胚囊的唯一途径(当时还没有打算用手工解剖方法),而酸水解等只适于分离固定胚囊,因此从一开始就确定攻克胚囊的酶法分离。为此,我们自行设计了一个简易装置(见本章提要),进行振荡酶解,再辅之以吸管人工振荡,终于第一次在显微镜下看到一个个翻滚着的完整胚囊,与以往根据连续切片在脑中叠加的印象相比,其新鲜感可想而知。1982年,我们报道了烟草、紫菜苔、蚕豆三种材料固定胚囊的分离。接着,又以芝麻与泡桐为材料,分离出由大孢子至雌配子体整个胚囊发育过程中各期的构造,借助Nomarski干涉差与相差显微术甚至能够显示清晰的减数分裂图像。进一步,试验受精后胚囊的分离也取得成功,在向日葵与金鱼草的受精后胚囊中观察到早期胚乳与原胚的发育。

生活胚囊的分离是在金鱼草中首先成功的,于1984年报道,然后又在向日葵与烟草中成功。应用细胞化学与荧光显微术,显示了胚囊中的细胞核,研究了胚囊壁的成分,观察到生活胚囊中存在大量的油脂(在石蜡切片过程中油脂大多被溶去)与淀粉粒;以荧光素二醋酸酯(FDA)显示了分离胚囊细胞的生活力。

在研究金鱼草胚囊壁的过程中发现,胚囊壁可被金胺O与樱草黄荧光染色,表明可能存在孢粉素或角质成分。为了进一步澄清这一点,采用包括乙酰解在内的多种化学处理,破坏胚囊的所有结构,唯独保存细胞壁的囊状结构。这种简称为"壁囊"的结构,起初被认为是胚囊的周壁,但通过石蜡切片与超薄切片的观察,证明其实是包围胚囊的珠被绒毡层的内壁,它全面地包裹受精后胚囊的胚乳与胚,仅在珠孔与合点两端为胚柄与胚乳吸器留下与外界相通的孔道。它可能具有既防止珠被绒毡层分泌的水解酶侵蚀胚囊,又保证胚囊由外界吸取营养物质的双重功能。这是研究分离胚囊的一项"副产品"。

利用分离的生活胚囊,我们以荧光染料 H33258 观察了向日葵的受精。由于卵核与次生核呈弱荧光,而精核则呈强荧光,很容易在分离胚囊中对它们加以鉴别,观察到精子进入助细胞、两个精子分别和卵核与次生核融合、合子与胚乳核分裂等受精环节,还在少数胚囊中观察到附加的精子(顺便指出:在杏叶沙参的分离胚囊中也证实了多精子入胚囊现象,见本章其它论文目录[15])。尝试了向日葵胚囊的离体培养,发现有些胚囊培养 10d 后仍然存活,但没有发育的迹象。

上述工作是 20 世纪 80 年代完成的。到了 90 年代,随着技术水平的提高,在烟草中已达到能够批量分离生活胚囊的地步。在和德国研究者合作完成以及国内完成的两项研究中均取得类似的成功。特别是后一次研究,提出了新的"酶解—渗透压冲击法",不仅能分离数量较多的胚囊,而且进一步由胚囊中分离出其中各个成员细胞,为此后离体受精实验(第三章第二节)奠定了基础。

1 被子植物胚囊酶法分离的研究:固定材料的分离技术与显微观察

Enzymatic isolation of embryo sacs in angiosperms: isolation and microscopical observation on fixed materials

Abstract

An enzymatic technique was developed to isolate the embryo sacs

(ES) of angiosperms. ES could be isolated from the surrounding ovular tissue and remain their structural entireness. Fixed ovules of Nicotiana tabacum, Vicia faba and Brassica campestris var. purpurea were macerated by pectinase-cellulase solution in a microshaker at 28-30℃ for about 5 hours. The resulted cell suspensions containing ES were washed, centrifuged, cleared in lactophenol and then observed by Nomarski interference contrast or negative phase contrast microscopy (Plate Ⅰ). Various methods for observing ES are discussed. The enzymatic technique shows many merits, especially in view of further histochemical and physiological study on viable ES as well as in vitro culture of ES.

作为被子植物的雌配子体世代,胚囊具有重要的研究价值。然而就观察的简易与研究的深入程度而言,胚囊远不如花粉。这是由于胚囊位于胚珠内部,被层层孢子体组织包围,而本身又相当柔嫩,一般只能依靠切片观察,受到很大的限制。几十年来,虽然提出过一些观察胚囊的简易方法,但要把它从胚珠中完整无损地分离出来并保持其生活状态,酶法离解应是最有希望的途径。据我们所知,只有Тырнов等在这方面作出了成功的先例[23,24]。本文报道我们从固定的胚珠材料入手探索胚囊酶法分离的试验结果。

材 料 与 方 法

实验于1981年至1982年进行。材料为烟草(Nicotiana tabacum)、蚕豆(Vicia faba)与紫菜苔(Brassica campestris var. purpurea)。前者取不同发育时期的胚珠做试验;后二者只试验了成熟的胚珠。酶法分离胚囊的实验步骤如下:

一、固定

按花蕾长度、花萼与花冠比例等形态指标区分花蕾的发育时期,根据需要选择适宜的新鲜花蕾,取出雌蕊,剥去子房壁,将胚珠固定于

FPA(配方为福尔马林5ml,丙酸5ml,50%乙醇90ml)中20~24h,换入70%乙醇,贮存于4℃冰箱中备用。

二、酶解

将贮存的胚珠依次经50%、30%乙醇下行至蒸馏水。在水中换洗数次,最好过夜,以彻底脱去乙醇。用果胶酶(Pectinase,Serva Finbiochemica Heidelberg)与纤维素酶(中国科学院生物化学研究所制)的混合液进行酶解。酶液浓度因材料而异:烟草用1.5%果胶酶与1.5%纤维素酶,蚕豆与紫菜苔用2.5%~3%果胶酶与2.5%~3%纤维素酶。在微型混合器(MM-1型,江苏省南通县电子分析仪器厂制)的微型多孔板的小孔中加入酶液,取胚珠置酶液中,在28~30℃保温条件下连续振动约5h,以便促使酶解作用均匀并加速其进程。取一滴酶液在显微镜下检查胚珠酶解程度,如不够离散,可适当补行人工振荡,直至大部分胚珠组织离散成悬浮状态。

三、水洗与透明

将含材料的酶液转入离心管,加蒸馏水稀释,在微型离心机(HE-70型,浙江黄岩医疗器械厂制)中以1500r/min的速率离心约5min。略加静置,用吸管弃去上清液,再换水离心1~2次。弃去上清液,加入乳酚甘油(配方为乳酸20ml、酚20ml、甘油40ml、蒸馏水20ml)。离心弃去乳酚甘油上清液,滴入适量新的乳酚甘油,使材料富集。材料在此液中兼有透明与保存之用。

四、制片与显微观察

吸一滴含悬浮材料的乳酚甘油置清洁的载玻片上。液体不宜过多,以免材料逸出盖玻片外,或不断晃动影响显微摄影。加盖玻片。用Olympus VANOX多用显微镜的Nomarski干涉差装置或相差装置进行观察与摄影。如需制作半永久片,可在盖玻片四周用改正液封边。

结　果

酶法分离胚囊的试验在分属三科的烟草、蚕豆、紫菜苔中均获成功,并有良好的重复性。经酶液处理后,胚珠组织离散很好,而胚囊则可保持完整结构。根据形态特征不难将混在大量体细胞中的胚囊鉴别出来。由于排除了周围细胞的遮掩,胚囊轮廓分明,内部各种细胞构造清晰可见,其相互空间关系亦很清楚。应用干涉差与相差显微术,不需染色即可观察。

一、烟草

图版Ⅰ,1~7为酶法分离的烟草不同时期的胚囊及其母细胞。图1示一个正在减数分裂前期Ⅰ的大孢子母细胞。细胞核占很大比例,细长的染色体盘绕其中,核仁尚存,胞质浓密。图2为一个大孢子(单核胚囊),其周围与珠孔端尚附有少数未剥离的胚珠组织。核居中央,核仁明显,其中有核仁液泡。胞质丰富,液泡开始出现。图3、4表示同一个二核胚囊分别用干涉差与负相差摄影的图像。二者表现方式不同,但均可看出胚囊内的情况:两个细胞核已分别移至两极,胚囊中央为大液泡占据。核的周围及二核之间有胞质联系。图5~7为三个不同的成熟胚囊,均呈梨形,珠孔端狭窄而合点端钝圆。胚囊外周有明显的界壁(这在一般石蜡切片中难以显示)。图5的成熟胚囊包含卵器、中央细胞、反足器三部分。珠孔端有一个卵细胞与一个助细胞,另一助细胞不在此光学平面上。极核在胚囊中央一侧。合点端有一个反足细胞(烟草胚囊成熟时,反足细胞数目可以变动在3~0之间[5])。图6的胚囊着重显示典型的卵器:一个卵细胞与二个助细胞三角鼎立。卵核位于细胞合点端,液泡居珠孔端;助细胞的核与液泡相对位置恰好相反。图7中一对极核已移至卵细胞附近,为受精前的状态。综上所述,由酶法分离的烟草胚囊与通过切片观察的结果是基本一致的[1]。

二、蚕豆与紫菜苔

图8系蚕豆成熟胚囊,其外形略呈椭圆形,可见卵器、含两个极核的

第二章 雌性器官与细胞的实验研究

中央细胞及一个残存的反足细胞。图9为另一蚕豆胚囊的一部分的放大,重点显示极核与卵器间的细胞质丝联系。图10为紫菜苔的成熟胚囊,外形狭长,合点端常不易与胚珠组织分离。胚囊内含物十分丰富,因而透视度不及烟草及蚕豆,但仍可看到卵细胞和两个极核等的构造。

讨 论

1. 为了寻求快捷而又能显示胚囊整体结构的观察技术,研究者们从不同的角度进行探索,先后提出过几类方法。现略加综述如下:第一类是 Поддубная-Арнолъди 的胚珠整体染色与活胚珠观察技术[22],仅适用于像兰科植物那样胚珠微小、珠被薄而透明的特殊材料,没有广泛应用的价值。第二类是胚珠透明技术,如 Herr 提出的"4½"混合液透明法,曾被若干研究者用于观察以豆科为主的多种植物的胚珠发育、大孢子发生与胚囊早期发育[4,8,10~12,14,17~19]。Young 等提出的水杨酸甲酯透明法被用于观察禾本科植物的胚囊与无融合生殖[20,22]。据我们在烟草、紫菜苔和水稻上的试验,这类方法对某些材料的幼期胚囊是适合的,但难以透过多层胚珠组织洞察成熟胚囊的内部结构细节。第三类技术是盐酸水解压片法。Bradley 最早用于观察烟草的胚囊[5]。Murty 等研究高粱的无融合生殖与有性过程[13]。类似方法也被用来作大孢子母细胞减数分裂的压片[2~4,6,9]。第四类是 Forbes 采用的酶解涂抹技术,将雀稗属植物的子房酶解、染色后,在解剖镜下手工剖出胚囊进行观察[7]。以上各类方法,作为研究胚囊的显微技术各有其应用价值,又各有其局限性。后两类方法虽能分离胚囊,但对胚珠进行解剖或压片,费时甚多且需高度手工技巧,更无法分离生活的胚囊。比较理想的途径是利用酶解方法直接将胚囊分离出来。Тырнов 等在这方面进行了多年的工作,取得了显著的进展。根据我们从文摘中了解的简略情况[23,24],他们已由10科25种植物的固定胚珠材料中分离出完整的胚囊进行显微观察。烟草生活胚囊的分离也已成功,并尝试了其离体培养。但我们不知道他们所用的具体方法及详细结果。本文实验结果又一次证明酶法分离胚囊是一项切实可行的新技术,具有操作简便

图版 I

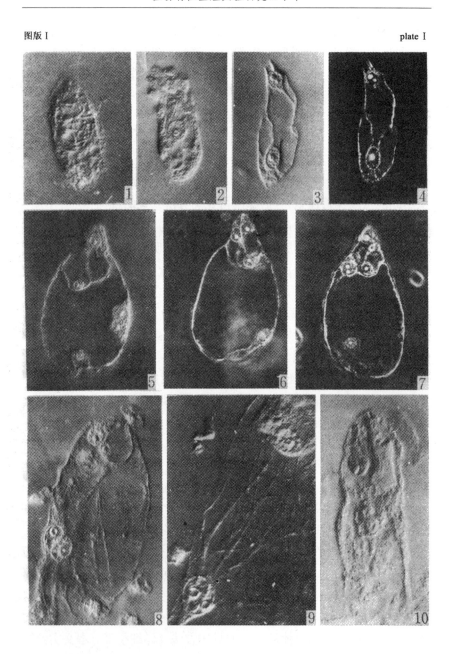

快速、能保持胚囊的完整性、显微观察效果良好等优点。适用于各种特点的植物材料(烟草一个子房约有 2000～4000 个胚珠[1],胚珠体积微小;蚕豆一个子房仅 2～3 个胚珠,体积很大),也适用于分离各个发育时期的胚囊。

2. 酶法分离胚囊为进一步研究被子植物的雌配子体开辟了新途径。一方面,借助这一技术分离生活的胚囊,结合组织化学与细胞生理学研究,将有助于更深入地了解胚囊的生理生化特征,使目前雌、雄配子体研究的不平衡的状况得以逐步改观。另一方面,只有通过这一途径才能逐步实现胚囊及其各组成部分的离体培养。Raghavan 指出:"分离与培养被子植物柔嫩的雌配子体由于技术原因迄今阻碍人们去尝试研究用离体方法由它诱导胚状体"[15]。但是,"通过改善操作技术,有可能表明被子植物的雌配子体细胞即使在没有受精的条件下也能在体外遵循与合子相似的发育途径而再生新植株"[16]。近年来,未传粉子房与胚珠培养的成功已经初步实现了这一预测。现在由于酶法分离胚囊的成功,直接进行胚囊的离体培养也会为期不远了。

图 版 说 明

酶法分离的胚囊图像,除 4、6、7 为负相差摄影外,其余均为干涉差摄影。所有胚囊均为珠孔端朝上、合点端朝下。各图的详细说明见正文。

1. 烟草胚囊母细胞。×900 **2.** 烟草单核胚囊。×450 **3、4.** 烟草同一的二核胚囊的干涉差与负相差摄影。×450 **5、6、7.** 烟草的不同成熟胚囊。×400 **8.** 蚕豆成熟胚囊。×300 **9.** 另一蚕豆成熟胚囊的一部分。×400 **10.** 紫菜苔成熟胚囊。×450

Explanation of plate

Embryo sacs (ES) were enzymatically isolated from fixed ovules of *Nicotiana tabacum* (1-7), *Vicia faba* (8 and 9) and *Brassica campestris* var. *purpurea* (10). Most of the photographs were taken by Nomarski interference contrast micrococpy except 4,6 and 7, which were photographed by negative phase contrast microscopy. In all pictures, the micropylar ends of ES were oriented upwards and the chalazal ends downwards.

Fig. 1. Megaspore mother cell of tobacco, showing its large nucleus in meiotic prophase I. ×900 **Fig. 2.** Megaspore (uninucleate ES) of tobacco with some attached ovular cells around it and at its micropylar end. ×450 **Fig. 3, 4.** Interference contrast (3) and negative phase contrast (4) pictures of a same binucleate ES, showing two nuclei at opposite poles and cytoplasmic connections between them. ×450 **Fig. 5.** A mature ES of tobacco, showing its egg cell, a synergid, two polar nuclei and one antipodal cell. The boundary wall of ES is obvious. ×400 **Fig. 6.** Another mature ES. The structure of egg apparatus is emphasized. Notice the opposite nucleus-vacuole position in the egg and synergids. ×400 **Fig. 7.** Another mature ES. The polar nuclei are near by the egg cell. ×400 **Fig. 8.** Mature ES of *Vicia faba*, showing egg apparatus, central cell and a remaining antipodal cell. ×300 **Fig. 9.** A part of another ES of *Vicia faba*. Notice the cytoplasmic connections between egg apparatus and polar nuclei. ×400 **Fig. 10.** Mature ES of *Brassica campestris* var. *purpurea*. The egg apparatus and the polar nuclei are close by each other. The storage contents are abundant. ×450

（作者：周嫦、杨弘远。原载：植物学报，1982，24(5)：403～407。图版1幅，参考文献24篇。选录图版Ⅰ。参看书末彩色图版Ⅶ。）

2 用酶解技术观察泡桐与芝麻的大孢子发生和雌配子体发育过程

Observations on megasporogenesis and megagametophyte development in *Paulownia* sp. and *Sesamum indicum* by enzymatic maceration technique

Abstract

A technique has been recently developed in our lab to isolate embryo sacs by means of enzymatic maceration of ovules (Zhou and Yang, 1982). In the present paper this method was adopted in observing the whole processes of megasporgensis and megagametophyte development in *Paulownia sp.* and *Sesamum indicum*. FPA fixed ovules were macerated in pectinase-cellulase solution with a microshaker, cleared in lactophenol, and then observed under a microscope with Nomarski interference contrast or phase contrast equipments. In both species, various developmental stages, from megasporocyte till mature embryo sac, were successfully identified and described. As a kind of microtechnique, enzymatic method shows some merits as its rapidness in specimen preparation and convenience for obtaining whole structural image. Several technical points are discussed hereof.

在此前一篇论文中,我们报道了用酶法分离几种被子植物胚囊的初步结果[2]。为了在更广泛的植物材料中试验这一技术,并探讨其在研究大孢子发生和雌配子体发育过程中的应用价值与前景,进一步开展了本实验。

材 料 与 方 法

实验于1982~1983年两年进行。以泡桐(*Paulownia sp.*)与芝麻(*Sesamum indicum*)不同发育时期的胚珠为材料。酶解技术流程与前一报告相同[2],在细节上略有改变。固定剂仍为 FPA。离解剂仍为果胶酶与纤维素酶的混合液。二种酶的各自浓度,泡桐均为3%,芝麻均为1.5%。采用保温振荡方法进行酶解。透明剂用乳酚甘油,或在其中加入丙酸洋红染色。用 Nomarski 干涉差装置或相差装置显微观察和摄影。

结 果

在两种供试植物中,用酶解法均分离出各个发育时期的样品,借以观察了大孢子发生和雌配子体发育的全部过程。

一、泡　桐

泡桐胚珠经酶液作用后,珠被可被彻底除去,珠心则较难离散。在早期的材料中,大孢子母细胞外围常有一层珠心表皮包围,合点端常附着一团未离散的胚珠组织。这样的分离单位很易获得,并在镜检时易与周围离散的胚珠体细胞相区别。经过透明之后,可以清楚地透视内部的大孢子母细胞(图版Ⅰ,1)及其减数分裂的各期图像。图版Ⅰ,2、3为两个母细胞的减数第一分裂中期相,分别以干涉差与负相差摄影,示纵向排列的纺锤体及其上的染色体。图版Ⅰ,4为减数第一分裂后产生的二分体。图版Ⅰ,5、6为两个先后略有不同的减数第二分裂末期相,可以辨认出纺锤体、成膜体和新生的子核。泡桐二分体的上下两个细胞,第二分裂的时间不同步,合点端细胞分裂较早,其纺锤体亦较大。四分体形成后不久,珠孔端三个大孢子相继退化,合点端一个发育为有功能的大孢子。图版Ⅰ,7～9示这一退化与发育过程的不同阶段。图版Ⅰ,10为充分发育的大孢子(单核胚囊),核居中央,液泡已经形成,有明显的细胞质丝贯穿其中。大孢子进行有丝分裂形成二核胚囊。两个游离核起初彼此靠近(图版Ⅰ,11),以后分趋两极(图版Ⅰ,12),中央为大液泡所占。此时胚囊一方面纵向延长,一方面其近珠孔部分横向扩展(图版Ⅰ,13、14)。至四核胚囊时发展成类似蝌蚪形,细胞核两两成对分布于胚囊两极而互相远离,珠孔端二核与胚囊纵轴垂直排列,合点端二核与胚囊纵轴平行排列(图版Ⅰ,15、16)。自二核至四核期,随着胚囊的生长,其外围的珠心表皮逐渐解体。因此经酶解除去珠被后,整个胚囊完全裸露,只有合点端仍然常附有一团孢子体组织。由于四核期以后胚囊高度液泡化且其近合点部分很细长,加之后期珠被的胞壁较难酶解,致使操作过程中常易断裂,分离的成功率降低。图版Ⅰ,17、18示两个成熟的胚囊。卵细胞呈长梨形,卵核偏于合点端而其余空间主要是一个大液泡。助细胞狭长而富含细胞质,核居近中位置。中央细胞很少贮藏物质,极核尚处于远离卵器的部位(不在同一光学平面)。反足细胞已经退化。

二、芝　麻

与泡桐相似,幼嫩的芝麻胚珠经酶处理后,珠被亦可被彻底除去,大孢子母细胞外仅余一层珠心表皮和合点端一团孢子体组织。芝麻的大孢子母细胞较为狭长,核常居于较远离珠心顶端的位置(图版Ⅱ,19),并在此处进行减数分裂(图版Ⅱ,20),形成线形的大孢子四分体(图版Ⅱ,21)。合点大孢子发育成单核胚囊,体积增大,液泡出现,核居中央(图版Ⅱ,22)。此时珠孔端三个大孢子已经退化,单核胚囊在远离珠心顶端的位置进行有丝分裂(图版Ⅱ,23),产生二核胚囊。起初两个游离核比较接近(图版Ⅱ,24)。随着囊体纵向延伸,二核各趋一极(图版Ⅱ,25)。芝麻的四核胚囊十分狭长,细胞核两两成对分布于远离的两极,中央大液泡中有很长的细胞质丝纵向贯穿,将两极的细胞核联系起来(图版Ⅱ,26)。此时很难对两极的核同时摄影,但通过调焦仍易观察。随后,随着胚囊的成熟,其近珠孔部分扩展而近合点部分更细长,外形渐呈蝌蚪状。图版Ⅱ,27 为一个接近成熟的胚囊,示上、下两极核尚未靠拢,卵器不在同一光学平面。图版Ⅱ,28 为一个已成熟胚囊,示狭长的卵细胞、两个紧贴的极核和其周围密布的淀粉粒。图版Ⅱ,29 为另一个成熟胚囊,用盖玻片稍微压扁,以同时显示其各组成部分的位置。一对并列的助细胞,珠孔端锐窄而合点端钝阔,细胞核居近中位置。卵细胞向合点方向延伸很长。两个极核被密集的淀粉粒包围。总的说来,芝麻胚珠较泡桐易于酶解,直至胚囊成熟时分离的成功率仍很高。

讨　　论

1. 用酶法分离被子植物的胚囊是一项新的尝试。20 世纪 70 年代以来,前苏联萨拉托夫大学的研究者在这方面开展了大量的工作[9~13]。在国内,我们近年也已试验成功[2]。从现有资料看来,Еналеева 等[9]和我们各自的方法虽然不尽相同,但在酶解与机械振荡相结合、不用徒手解剖方法分离胚囊这一点上是一致的。此法可以同

图版 I

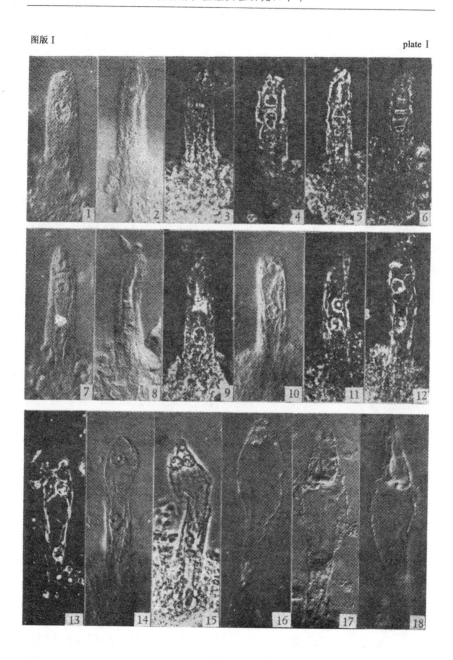

plate I

时制备大批材料,尤其适于分离体积较小或较为幼期的材料。Еналеева等曾据此观察了烟草的大孢子发生与雌配子体发育,惜无照片可供参考[10]。我们曾对烟草不同发育时期的胚囊作了初步的实验与观察[2],在本实验中又在泡桐与芝麻两种植物上作了更系统的工作,收到了肯定的效果。例如在一个单元时间中鉴定过164个泡桐胚珠的发育时期,其中大孢子母细胞前期Ⅰ:133;中期或后期Ⅰ:13;二分体:11;四分体:7。表明这一技术特别适用于多胚珠材料大孢子发生时期的快速鉴定。和常规的切片技术相比,酶解技术除了上述快速性外尚有若干优点。如有些材料在包埋与切片时较难定向,而酶法则不存在这种困难。又如观察酶解的材料立体感特别强,镜检时只需通过调焦即可对同一对象的整体结构一目了然,而无需对连续切片所获得的图像进行费时的拼接。我们在本实验中的观察结果,和前人在芝麻[4,7,8]与泡桐[1]观察的结果基本一致,在细节上还可作一定补充。此外,如欲进行显微光度测定,采用未切破的整体细胞也是有利的。因此,尽管在观察与摄影的精密度方面不如切片,以及仍然存在着其它一些尚待克服的弱点,酶法作为一项显微观察技术是有前途的。

2. 酶法分离胚囊是一项探索中的技术,迄今无成熟经验可循。以下几个问题值得探讨。首先是酶的种类。Forbes最早是用Pectinol 100-D[3]。前苏联研究者多用自制的蜗牛胃液酶[9,11,12]或更辅以果胶酶[13]。我们在已试验成功的几种植物中,一律用果胶酶和纤维素酶混合液。曾在泡桐实验中试加崩溃酶(driselase)、葡糖苷酸酶(β-glucuronidase)等其它酶种,未见改进。看来,只要胚珠组织保持初生壁,果胶酶与纤维素酶即可促其离散而又不致溶解大孢子母细胞的胼胝质壁与含角质等特殊成分的胚囊壁[5]。其次,酶解的效果亦因植物种类、发育时期与组织性质的不同而异。例如就成熟胚珠而言,烟草、芝麻均较易酶解,而泡桐则较难,这可能归因于后者有一定程度的木质化。泡桐和芝麻的珠心表皮和合点端一团胚珠组织常结合紧密不易离散,而烟草却无这种情况。据报道芝麻胚囊合点端有特殊的"垫状组织"[4](即承珠盘)。后者常发生胞壁的木质化或栓质化[6]。因此果胶酶与纤维素酶对其无效。这启示我们:开展胚囊的酶法分离,有必要认识胚珠各个时期、各种组织的成分和性质。

图 版 说 明

图 版 Ⅰ

泡桐的大孢子发生与雌配子体发育。酶法分离,乳酚甘油透明。图1、2、7、8、10、14、16、17、18为干涉差摄影;图3、4、5、6、9、11、12、13为负相差摄影;图15为正相差摄影。所有材料均按珠孔端朝上方向排列。放大倍数相同:×450。

1.大孢子母细胞　**2、3**.减数分裂中期Ⅰ　**4**.二分体　**5、6**.减数分裂末期Ⅱ　**7、8、9**.四分体中珠孔端三个大孢子退化,合点端大孢子发育　**10**.单核胚囊　**11**.二核胚囊(初期)　**12**.二核胚囊(中期)　**13、14**.二核胚囊(后期)　**15**.四核胚囊(初期)　**16**.四核胚囊(后期)　**17、18**.成熟胚囊

Explanation of plates

Plate Ⅰ

Megasporogenesis and megagametophyte development in *Paulownia sp.* through the use of enzymatic macerating. Materials were cleared in lactophenol. Photographs were taken either by Nomarski interference contrast (Figs. 1,2,7,8,10,14,16,17,18) or by negative (Figs. 3,4,5,6,9,11,12,13) or positive (Fig. 15) phase contrast equipments with the same magnification (×450). In all figures, the micropylar ends were oriented upwards and the chalazal ends downwards. **Fig. 1**. Megasporocyte surrounded by a layer of nucellar epidermis and attached to a mass of chalazal ovular tissue. **Figs. 2,3**. Meiotic metaphase I. **Fig. 4**. Dyad. **Figs. 5, 6**. Meiotic telophase II. **Figs. 7, 8,9**. Tetrads, showing the growing chalazal functional spore and the degeneration micropylar spores. **Fig. 10**. Uninucleate embryo sac. **Fig. 11**. Binucleate embryo sac (early stage). **Fig. 12**. Ibid (middle stage). **Figs. 13, 14**. Ibid (late stage). **Fig. 15**. 4-nucleate embryo sac (early stage). **Fig. 16**. Ibid (late stage). **Figs. 17, 18**. Mature embryo sacs.

(作者:杨弘远、周嫦。原载:植物学报,1984,26(4):355~358。
图版2幅,参考文献13篇。选录图版Ⅰ。)

3 金鱼草胚囊的人工分离

The enzymatic isolation of embryo sacs from fixed and fresh ovules of *Antirrhinum majus* L.

Abstract

A technique has recently been developed in our lab for isolation of embryo sacs by enzymatic maceration of fixed ovules. The present paper is an extension of this work. Intact embryo sacs were successfully isolated from fixed as well as fresh ovules of *Antirrhinum majus*. The method used on fixed ovules was similar to that described previously. A series of figures including mature embryo sacs prior to fertilization, entry of pollen tube, and embryo sacs after fertilization with 2, 4, 8 and multicellular endosperm were identified in macerated materials. Isolation of viable embryo sacs was carried out with fresh ovules, which were macerated in a mixture solution of 2% pectinase and 1%-1.5% cellulase with or without mannitol and/or potassium dextran sulphate by a microshaker at 28-30℃ for 2-3 hours. Embryo sacs both before and after fertilization were isolated from fresh materials, though their cellular structure was not as clear as fixed ones. Histochemical reactions revealed that these embryo sacs were rich of ergastic substances, such as lipid drops, polysaccharide and starch grains. The significance of isolation of viable embryo sacs in view of further cytochemical and cytophysiological studies and *in vitro* culture of embryo sacs is discussed hereof.

近年来,我们采用酶解技术从多种植物的固定胚珠中成功地分离出受精前不同发育阶段的完整胚囊[1,2]。这不仅是观察胚囊的一种技术,而且为进一步分离与研究活的胚囊奠定了技术基础。最近,我们在金鱼草上,首先以固定胚珠为材料,分离出受精前、特别是受精后不同时期的胚囊,并在此基础上试探分离活的胚囊,获得成功。现报道如下。

材 料 与 方 法

材料为金鱼草(Antirrhinum majus L.)。固定胚珠的酶解方法与前文报道大致相同[1]。活胚囊的分离是采用受精前后的新鲜花朵,由子房中小心剖出胚珠,置于2%果胶酶与1%~1.5%纤维素酶混合液中,酶液内可加入蔗糖(2%~4%)和葡聚糖硫酸钾(0.2%)。在28~30℃条件下振荡酶解2~3h,并适当补行人工振动,直至大部分胚珠细胞离散成悬浮状况。离心除去酶液后,用蒸馏水或等渗溶液洗涤2次。直接进行显微观察。亦可用0.05%甲苯胺蓝(溶于柠檬酸-磷酸氢二钠缓冲液,pH4.0)[3]染色后观察。对酶解的活胚囊进行了几种显微化学反应:苏丹Ⅲ反应[4]、碘-碘化钾反应[4]与高碘酸-锡夫(PAS)反应[5]。采用Nomarski干涉差或普通光显微镜观察与摄影。

结 果

一、固定胚囊的分离和观察

利用果胶酶与纤维素酶离解固定的胚珠,可以分离金鱼草由大孢子母细胞直至受精后胚乳发育初期各阶段的胚囊。为了与后文所述活胚囊的酶解实验结果相对应,本文仅描述受精后胚囊的发育过程。

图版Ⅰ,1示受精前的成熟胚囊结构:二个狭小的助细胞,核位于细胞中部;一个梨形的卵细胞,比助细胞显著为大,核在细胞的合点端,紧靠细胞壁;中央细胞内的二极核已融合成一显著的次生核;反足细胞

第二章 雌性器官与细胞的实验研究

已经退化。图版Ⅰ,2示传粉后花粉管进入胚囊,卵细胞已经受精而长大(另一光学平面上有初生胚乳核分裂相)。有趣的是,虽经酶解作用,花粉管并未与胚囊脱离,细长的花粉管自外伸入胚囊清晰可见。花粉管浓厚物质可保留到受精后一定时期。当初生胚乳核已分裂为两个互相远离的胚乳核时,珠孔端仍有残存的花粉管(图版Ⅰ,3)。受精以后,助细胞迅速退化。合子有较长的休眠期,形态上看不出明显变化。胚乳属细胞型。初生胚乳核在受精后迅速分裂成二核并在其间产生隔壁,形成二个细胞,珠孔端细胞较大,合点端细胞较小(图版Ⅰ,4)。图版Ⅰ,5、6分别可见含四个与八个胚乳细胞的胚囊,此时合子仍未分裂。以后胚乳继续发育,形成大量的胚乳细胞,这样的胚囊仍可被完整地分离出来(图版Ⅰ,7)。

二、活胚囊的分离和观察

在果胶酶与纤维素酶的作用下,新鲜胚珠的体细胞亦可离散,裸露出完整的胚囊。图版Ⅰ,8示受精前的成熟胚囊,形态与由固定材料分离的胚囊相似。胚囊内含丰富的颗粒物质(它们在固定胚囊中大多被溶解),比较集中地分布于珠孔端与次生核周围,并因之影响胚囊内细胞结构的观察。图版Ⅰ,9示受精后胚乳已相当发育的胚囊,相当于图版Ⅰ,6或稍迟的发育时期。胚囊轮廓清楚完整,内部细胞质均匀分布,惟细胞核与细胞壁不够清晰。用甲苯胺蓝染色,可见胚囊是有细胞结构的,只是由于观察效果尚不够好,未作显微摄影。

对分离的胚囊进行了几种简易的显微化学反应。苏丹Ⅲ反应表明胚囊内含较丰富的油滴。图版Ⅰ,10示受精前的成熟胚囊,内含大小不等的橘红色油滴,分散在胚囊内,较多集中于胚囊中部。胚囊壁呈浅橘红色,说明可能含有脂类成分。PAS反应显示胚囊内有大量紫红色多糖颗粒。图版Ⅰ,11为与图10时期相当的胚囊,多糖颗粒大多集中于次生核周围和珠孔端。胚囊壁和卵器细胞呈浅粉红色,显示它们尚含有多糖成分。碘-碘化钾反应证实胚囊内存在大量淀粉粒。受精前胚囊内淀粉粒的分布与图版Ⅰ,11PAS反应的结果类似。受精后胚囊内仍保留相当多的淀粉粒(图版Ⅰ,12)。

讨 论

胚囊的酶法分离是一项正在探索中的技术。我们曾由固定的胚珠中分离出烟草、蚕豆和紫菜苔的胚囊[1]，并用此技术观察了泡桐和芝麻的大孢子发生和雌配子体发育过程[2]。当前又在金鱼草上获得成功。本实验有两方面的进展：第一，不仅分离出受精前的胚囊，而且也分离出受精后的胚囊（从花粉管进入胚囊直至上百个胚乳细胞时期）。第二，不仅由固定材料，而且也初步由新鲜胚珠中分离出受精前后的胚囊。

Heslop-Harrison 1972 年曾指出："在有花植物所有重要器官中，再没有比胚珠和胚囊的生理学研究更缺乏的了"[6]。1980 年他再次呼吁对"被人们遗忘的世代"——被子植物的配子体——加强研究[7]。纵观胚囊研究的历史与现状不难察觉，尽管由于电镜、细胞化学、荧光显微镜等现代技术的应用，这方面工作取得一定的进展，但仍多局限于对死材料作描述性观察，而缺乏对活材料进行实验性的研究。因而难以更深入地了解胚囊发育的动态变化与机理，更无从施加人工的控制。

用酶解技术分离生活胚囊开拓了研究与控制胚囊发育的新前景：第一，可以直接对裸露的活胚囊进行其它方法比较难作的组织化学与细胞生理学研究。本实验对金鱼草胚囊所作油脂的观察即为一简单的例证。尤其是胚囊内酶的组织化学、细胞渗透压与 pH 值等生理指标的测定等将会因之改观。第二，活胚囊的分离是实现胚囊离体培养这一重要目标的必要技术前提。胚囊是由形态与功能各异的卵细胞、助细胞、中央细胞和反足细胞共同组成的复合结构，是一个"其组成细胞之间有着复杂生理相互作用的特化的功能单位"（Kapil 与 Bhatnagar 1981）[8]。在人工控制条件下培养胚囊，可以观察其整体及各组成部分的反应与变化，研究它们之间的相互关系，进而能动地控制其发育进程。Raghavan（1976）指出："分离与培养被子植物柔嫩的雌配子体由于技术原因迄今阻碍人们去尝试研究用离体方法由它诱导胚状体。"[9]当前，活胚囊分离技术既已开始突破，胚囊的离体培养也就有可能提上日程了。

迄今仅 Тырнов 与 Еналеева[10,11]首先用酶法分离出烟草与列当的活胚囊,但他们未曾报道具体方法。根据我们的经验,活胚囊的分离技术尚存在以下问题:第一,酶解效果不够稳定,容易发生破损或质壁分离;易受胚囊发育时期、酶的种类与浓度、酶液的渗透压与 pH 值、酶解温度与时间等多种因素的影响。第二,观察方法尚待改进。固定材料经透明与染色后有良好的显微观察效果;活胚囊则迄今较难鉴定内部结构。第三,如何保持离体胚囊的生活力亦有待今后研究。

(作者:周嫦、杨弘远。原载:实验生物学报,1984,17(2):141~147。图版1幅,参考文献11篇,均删去。参看书末彩色图版Ⅸ。)

4 酶法分离胚囊的荧光显微观察

Fluorescence microscopic observations on enzymatically isolated embryo sacs

Abstract

The enzymatic isolation of embryo sac combined with various histochemical assays could offer a new means for studying the female gametophyte of angiosperms. In this paper some results of fluorescence microscopic observations on enzymatically isolated embryo sacs are reported. Ovules were macerated in pectinase-cellulase solution in the same manner as previously described and naked embryo sacs were liberated out. Using a DNA-binding fluorescent dye Hoechst 33258, cell nuclei within the embryo sacs of *Antirrhinum majus*, *Vicia faba*, *Sesamum indicum*, *Nicotiana tabacum* and *Solanum nigrum* were stained well. This not only serves as a means for identification of nuclei in embryo sac cells, but also may provide a basis for further fluorimetric measurement of DNA content in these nuclei. Auramine O and primulin staining revealed the cutin, or/and sporopollenin, nature of the boundary wall of snapdragon embryo sac. This may explain the tolerance of this wall to the pectinase-cellulase action, resulting in maintenance of the intactness of the isolated embryo sacs. Fluorescence microscopic observations on the megasporogenesis in enzymatically isolated materials of *Paulownia* stained by leuco aniline blue showed some interesting behavior of callose depositon in this species. These preliminary results indicate that the enzymatic isolation method combined with fluorescence microscopy is useful for studying the biology of embryo sac.

在酶法分离胚囊的技术取得成功以后,我们曾经推测可以应用这一实验系统对胚囊进行组织化学方面的研究[1~3]。作为组织化学领域中的一个重要分支,荧光显微技术受到愈来愈广泛的重视[4]。用荧光显微技术研究酶法分离的裸露的胚囊还没有见到报道。本文是我们在这方面所获得的若干结果。

材 料 与 方 法

实验材料有:金鱼草(*Antirrhinum majus*)受精前和受精后的胚珠;蚕豆(*Vicia faba*)、芝麻(*Sesamum indicum*)、烟草(*Nicotiana tabacum*)、龙葵(*Solanum nigrum*)受精前的成熟胚珠;泡桐(*Paulownia* sp.)大孢子发生时期的幼嫩胚珠。多数材料用福尔马林-丙酸-乙醇(FPA)固定[1],少数情况下是用新鲜胚珠进行实验。胚囊的分离方法与以前所用的基本相同[1~3]。酶解后的材料经过水洗、离心,分别用以下的荧光染料染色:

1. Hoechst 33258(H33258,Guern 教授赠送)按 Laloue 等[5]的方法,用 pH5 的柠檬酸-磷酸氢二钠缓冲液浸泡材料 1~2h,换入 1 μg/ml 的 H33258 染液(配制 1mg/ml H33258 水溶液作为贮备液,临用前用上述缓冲液稀释 1 000 倍),染色数小时或更长时间,观察前加入等量的甘油封藏。

2. 荧光增白剂 Calcofluor White ST（Amer. Cyanamide Co.,以下简称增白剂 ST）按 Nagata 等[6]的方法,用 0.1% 的增白剂 ST 水溶液(pH6～7)染色 10min 或更长,水洗多次除去染液,用蒸馏水封藏观察。

3. 金胺 O(auramine O,北京化工厂)按 Heslop-Harrison[7]的方法,染液浓度减低为 10μg/ml(溶于 pH7.2 的 Tris-HCl 缓冲液),染色数分钟,立即封藏观察。

4. 樱草黄(primulin,上海新中化学厂)用 10μg/ml 水溶液染色数分钟,立即封藏观察。

5. 水溶性苯胺蓝(aniline blue W.S.,北京化工厂):按 Jensen[8]书中引用的方法,用 pH8.2 的磷酸氢二钾缓冲液配制的 0.005% 脱色苯胺蓝溶液染色数小时或过夜,加入少量的甘油封藏观察。

用 AO FLUORESTAR 或 OPTON STANDARD 型的落射荧光显微镜观察。H33258、金胺 O 樱草黄、脱色苯胺蓝的染色标本采用激发滤光片 436、分色镜 450 或 460、阻挡滤光片 470 或 475。增白剂 ST 的染色标本采用激发滤光片 436、分色镜 510、阻挡滤光片 520。用 21°全色 135 胶卷摄影。

结 果 和 讨 论

一、H33258 染色的几种植物胚囊的荧光显微观察

H33258 是与 DNA 重复序列有结合力的荧光染料,曾被用于染色体的荧光显带;近年开始用于显示培养细胞的核和染色体[5]、原生质体的活染[9]、细胞杂种的鉴别[10]以及 DNA 含量的荧光光度测定[11～14]等方面。在本实验中我们用它染分离的胚囊细胞核,在多种植物材料上获得良好的效果。图版Ⅰ,1 示蚕豆成熟胚囊经 H33258 染色后,其中所含两个助细胞核(在不同光学平面上)、卵核和一对极核均发射荧光。芝麻、烟草、龙葵的成熟胚囊中各个组成细胞的核同样可以显示出来(图版Ⅰ,2～4)。金鱼草受精前的胚囊用此法观察,可以清楚地看

到珠孔端一对平行的亮点,即为助细胞核;其下方一个亮点是卵核;胚囊中部一个较大的亮点则是由两极核融合而成的次生核(图版Ⅰ,5)。图6为金鱼草受精后的胚囊,此时中央细胞受精后已分裂为八个胚乳细胞,各含一个细胞核;珠孔端还有一个较小的亮点是合子的核。

以上各个胚囊均系由 FPA 固定的胚珠中分离出来的。由于固定剂中的甲醛有诱导蛋白质产生荧光的作用[4]或由于其它原因,致使细胞中的非染色质成分如核仁、细胞质、细胞膜等有时亦呈现荧光。但甲醛诱导的荧光程度较弱且呈黄色,很容易和 DNA 与 H33258 结合所发射的强绿色荧光区别。在作 DNA 的荧光光度测定时应避免用甲醛作固定剂,但作为一种显示细胞核的方法,有一些非特异荧光的衬托,对辨认胚囊的结构反而比较方便。我们也对金鱼草的新鲜胚囊作了H33258 的荧光观察,结果将另文报道。

以往研究胚囊细胞核 DNA 的荧光染料常用吖啶橙[15]。上述实验结果表明,H33258 作为与 DNA 专一结合的荧光染料,效果是良好的。它还有荧光衰退较慢、毒性较低等优点。不仅适用于观察体细胞,而且也适用于观察分离的胚囊。进一步可考虑用它作胚囊、胚和胚乳发育过程中 DNA 的定量观察,以及研究它们在离体培养条件下的变化。

二、金鱼草胚囊壁成分的荧光显微鉴定

在以前的历次实验中,胚珠经果胶酶和纤维素酶处理后,体细胞组织虽被离解,但胚囊尚可保持完整[1~3]。这就使我们推测胚囊壁中可能含有某种耐上述酶作用的成分。为此,将金鱼草的新鲜胚珠经1.5%的果胶酶与纤维素酶混合液处理后,分离出胚囊,然后分别用荧光增白剂 ST、金胺 O、樱草黄三种荧光染料染色,以图鉴定胚囊壁的成分。为了检验上述三种方法的可靠性,还用其它植物材料设置了相应的对照实验。

1.荧光增白剂对细胞壁中的纤维素与几丁质有亲和力[16],曾被用于高等植物原生质体的细胞壁再生的研究[6,17,18]。我们用增白剂 ST 对白菜叶肉海绵组织细胞和棉纤维染色,细胞壁均呈特有的蓝绿色荧光。但用同样的方法,酶法分离的金鱼草胚囊壁仅呈现微弱的淡黄色自发荧光。这一结果不意味着胚囊壁中缺少纤维素成分,而应解释为其中的纤维素成分在酶的作用下已被降解。

2. 金胺O曾被用于鉴定柱头表面的角质层[7]和花粉壁中的孢粉素[19]。我们用它处理女贞叶的徒手切片,证明其横切面的上、下表皮的角质化外壁与覆盖的角质层呈强绿色荧光,而其它细胞壁不显荧光(木质化壁的自发荧光除外)。用于试验金鱼草花粉时,花粉壁亦呈现强烈荧光。在金鱼草新鲜胚囊上的试验结果是:无论受精前或受精后的胚囊,其囊壁均呈强烈的绿色荧光,而胚囊内部或四周离散的体细胞则仅染成弱黄色。在某些情况下可以看到胚囊壁的外表面呈明亮的网格状(图版Ⅰ,7),这可能是囊壁和珠心细胞的连接处,因后者被离散所留下的痕迹。

3. 樱草黄曾被用于鉴定花粉壁的孢粉素[15],我们没有查到原始方法。而根据我们自己的染色程序(见"材料与方法"),女贞叶徒手切片横切面上的表皮角质壁呈强绿色荧光,其它细胞壁也有较弱的荧光。金鱼草花粉壁亦呈强荧光。金鱼草新鲜胚囊经染色后,胚囊壁呈强绿色荧光(图版Ⅰ,8)。

综上所述可以认为:经果胶酶与纤维素酶处理后的胚囊壁中已不含纤维素,但仍含耐上述酶作用的角质或孢粉素。前人根据切片研究已在少数植物中发现胚囊外有角质壁包围[20]。在本实验中,金胺O与樱草黄对角质和孢粉素均有特异荧光,我们认为含角质的可能性更大,但也不排除含孢粉素的可能性。

三、泡桐大孢子发生过程中胼胝质的观察

在被子植物大、小孢子发生过程中,胼胝质的消长是有规律的。Rodkiewicz等曾研究了多种植物大孢子发生过程中胼胝质的变化[21]。梁汉兴在天麻上也作了胼胝质动态的观察[22]。以上作者均用脱色苯胺蓝荧光染色法鉴定胼胝质,但系采用切片或整体压片法。我们则用酶解分离的材料观察了泡桐大孢子发生期间胼胝质的动态。

泡桐的幼小胚珠经3%果胶酶处理除去珠被,将外面包有一层珠心表皮和下接一团承珠盘组织的大孢子母细胞分离出来[2]。经脱色

苯胺蓝染色后,胼胝质部位呈强烈的黄绿色荧光。图版Ⅰ,9、10显示减数第一次分裂前期的大孢子母细胞,其珠孔端壁出现一帽状的胼胝质构造,侧壁与合点端壁则无特异荧光。形成二分体后,两个子细胞之间的隔壁积累胼胝质,而珠孔端的胼胝质帽则逐渐消失(图版Ⅰ,11、12)。四分体时期,所有三个隔壁均有胼胝质,但第二次分裂形成的隔壁荧光强度较第一次分裂形成的隔壁为弱(图版Ⅰ,13),表明此时胼胝质合成已渐减弱。最后,随着珠孔端三个大孢子退化和合点端功能大孢子长大,所有隔壁上的胼胝质趋于消失(图版Ⅰ,14)。上述观察结果表明泡桐大孢子发生过程中胼胝质的变化有一些特点:在大孢子母细胞的合点端未看到胼胝质壁;侧壁上也没有胼胝质(后一点与梁汉兴在天麻上的观察一致[22]);而珠孔端的胼胝质帽却特别明显。

用酶法分离的材料观察大孢子发生过程中胼胝质的动态,比用通常的切片方法简便得多。但只能用单纯的果胶酶处理;如果酶液中包含纤维素酶或蜗牛酶,则因胼胝质降解而无荧光发生。

(作者:杨弘远、周嫦。原载:实验生物学报,1985,18(2):127~133。图版1幅,参考文献22篇,均删去。)

5 Observations on enzymatically isolated living and fixed embryo sacs in several angiosperm species

The embryo sac (ES) represents the female gametophytic generation of angiosperms. However, its biology is not so well studied as that of its male counterpart, the pollen. The main reason is that the embryo sac is deeply buried in the ovule tissues. It is therefore difficult to observe and handle directly, and microscopic observations on living material, cell-physiological studies, *in vitro* culture and other approaches, which have greatly enhanced our understanding of pollen biology, could so far not be successfully applied to the study of ES biology. As Heslop-Harrison (1972) has pointed out, "Of all the principal organs of the flowering plant, none have received less physiological study than the ovule and embryo sac."

One way out of this situation would be to isolate ESs from their surrounding sporophytic tissues and submit them to various experimental treatments. Various methods for isolating ESs have been tried, including HCl hydrolysis of ovules followed by squashing them under a coverslip (Bradley 1948; Murty et al. 1979) and enzymatic maceration of ovules followed by manual dissection with needles under a binocular microscope (Forbes 1960; Solntseva and Levkovsky 1978). Although these methods do permit the isolation of ESs, they are time-consuming and require high manipulative skills, and are rather a means of isolation for microscopic observation without much concern for keeping the ESs viable. As a way to isolate viable embryo sacs in large quantities, an enzymatic technique not followed by squashing or dissection might be considered as promising. A research group at Saratov University (USSR) has worked along this line for many years (Enaleeva and Dushaeva 1975; Tyrnov et al. 1975; Tyrnov and Enaleeva 1979). However, their results seem to be available only in abstracts, without detailed information.

We have developed a technique for isolation of ESs by enzymatic maceration of the ovules. To start with, we have used fixed ovules as material and have first isolated ESs from *Nicotiana tabacum* L., *Vicia faba* L. and *Brassica campestris* L. var. *purpurea* (Zhou and Yang 1982). Subsequently, this technique has been used for investigations of megasporogenesis and megagametogenesis in *Paulownia* sp. as well as *Sesamum indicum* L. (Yang and Zhou 1984). Finally, we have worked on isolation of ESs from fresh ovules, succeeding first with *Antirrhinum* (Zhou and Yang 1984) and more recently with *Helianthus* and *Nicotiana*. In the present paper we describe this work on the isolation and study of viable ESs in these three species, and for comparison similar work with fixed materials.

Materials and methods

Collection of ovules. Flowers of *Antirrhinum majus* L., *Helianthus annuus* L. and *Nicotiana tabacum* L. were collected before and after anthesis from the field or campus gardens. The ovary walls were removed and the ovules were excised from the placentae. For isolation of fresh ESs the ovules were immediately placed into the enzyme solution. For isolation of fixed ESs they were first put into FPA (formalin-propionic acid-50% ethanol, 5:5:90, by vol.) for 1-3 d, then stored in 70% ethanol at 4℃ until use. Before enzymatic maceration the fixed ovules were rehydrated and washed thoroughly in distilled water.

Enzyme solutions. The enzymes used in our experiments were: pectinase (Serva, Feinbiochemica, Heidelberg, FRG), cellulase (Shanghai Institute of Plant Physiology, Academia Sinica, Shanghai), "snailase" (Institute of Biophysics, Academia Sinica, Beijing) and pectolyase Y-23 (Sheishin Pharmaceutical, Tokyo). For maceration of fixed material, the enzymes were dissolved in distilled water; for fresh material the maceration solutions were prepared with enzymes, sucrose and potassium dextran sulphate. The solution used for snapdragon consisted of pectinase 1.5%-2%, sucrose 2%-8% (depending on the ES stage) and potassium dextran sulphate 0.2%; the solution for the two other species consisted of pectinase 2%-3%, cellulase 2%-3%, snailase 1%-2%, pectolyase Y-23 0.4%-1%, sucrose 7%-14% and potassium dextran sulphate 0.2%. The pH of the solutions was 5-5.5.

Maceration. The enzyme solutions were placed in small centrifuge tubes and ovules were immersed in them. The tubes, sealed with parafilm, were placed on a microshaker and left shaking for 2-6 h at 28-30℃. After incubation, the preparations were agitated manually with a dropping pipette with rubber bulb until the ovular tissues were well macerated. A drop of the suspension was placed on a slide and inspected under a microscope to determine whether or not the ESs were isolated. When isolation had occurred, the suspension was centrifuged at a speed of 1500 rpm for 5 min and the supernatant was discarded. The precipitate was washed twice by resuspension and recentrifugation with a sucrose solution (for fresh materials: 0.06-0.24M for snapdragon and 0.2-0.4M for the other two species) or distilled water (for fixed materials).

Morphological observations. The fixed materials containing isolated ESs were cleared directly in lactophenol (lactic acid, phenol, glycerin, distilled water, 2 : 2 : 4 : 2, by vol.) for one or several days. Alternatively, a 1 : 1 (v/v) mixture of lactophenol and propiono-carmine (2g carmine dissolved in 100 ml propionic acid) was used for simultaneous staining and clearing. A drop of cleared and stained suspension was mounted on a slide with a coverslip and observed with a universal research microscope, model VANOX (Olympus Optical Co.), with Nomarski interference of phase-contrast equipment. The fresh materials were mounted directly after maceration and washing, or temporarily stained with propiono-carmine before observation.

Histochemical reactions. Two histochemical reactions were done on freshly isolated ESs. For identification of lipids, the macerated and washed materials were treated with 50% ethanol for 5 min and then stained in Sudan III solution (0.1g Sudan III dissolved in 10 ml 95% ethanol+10 ml glycerin) for 10 min. For identification of starch, the materials were directly stained in I-KI solution (I 1g, KI 2g, distilled water 300 ml). The stained materials were mounted and observed under a microscope.

Viability test. The viability of ESs was determined by fluorochromasia according to Larkin (1976) and Shivanna and Heslop-Harrison (1981). A stock solution of 2 mg/ml fluorescein diacetate (FDA; Sigma Chemical Co., St. Louis, Mo., USA) was diluted 50 times with isosmotic solution (the same sucrose solutions as used for maceration) before use. Freshly isolated ESs were treated with this solution, mounted on a slide and observed after 5 min.

Fluorescent staining of cell nuclei. Hoechst 33258 (H33258), a bisbenzimidazole derivative (Höchst, Mannheim, FRG; a gift from Professor J. Guern, Génétique et Physiologie du Development des Plantes, CNRS, Gif-sur-Yvette, France), was used for fluorescent staining of nuclei in the ESs. For fixed material the cells were resuspended in 0.1M citric acid-Na_2HPO_4 buffer (pH 5) for 1-2h and then stained with 1μg/ml H33258 (dissolved in the same buffer solution) for 4 h or overnight. An equal volume of glycerin was added before mounting and observation (Laloue et al. 1980). For fresh material the isolated ESs were directly stained with 10 μg/ml H33258 dissolved in isosmotic solution (according to Meadows and Potrykus 1981, without the supplementation of Triton X-100).

Fluorescent staining of the ES wall. Two methods were used to identify the chemical nature of the ES wall. Freshly macerated material was either stained with 0.1% Calcofluor white ST (American Cyanamide Co.) dissolved in isosmotic solution for 10 min to detect cellulose (Nagata and Takebe 1970) or 10 μg/ml auramine O (Beijing Chemical-Industrial Co.) in 0.05 M 2-amino-2 (hydroxymethyl-1, 3-propanediol)(Tris)-HCl buffer (pH 7.2) for 5 min to detect cutin (following Heslop-Harrison 1977, with a decrease in auramine O concentration).

Fluorescence microscopy. The epi-fluorescence system of a Fluorestar 110 (AO Scientific Instruments Co.) or Standard microscope (Opton Feintechnik GmbH,) was used, the former with a filter combination of AO 2073 (exciter filter 436, dichroic mirror 450, barrier filter 475), the latter with Opton V (exciter filter BP 436, beam splitter FT460, barrier filter LP 470).

Results

Isolated ESs of antirrhinum majus.

Using enzymatic maceration, we were able to isolate a series of snapdragon ESs both before and after fertilization. Figures 1-5 show ESs isolated from fixed ovules. The mature ES prior to anthesis consists of two narrow synergids, a pear-shaped egg cell, and a central cell containing a secondary nucleus; the antipodal cells are degenerated (Fig. 1). After penetration of the pollen tube into the ES the egg cell is fertilized and the primary endosperm nucleus divides into two nuclei (Fig. 2) which are then separated by a transverse wall into two endosperm cells (Fig. 3). The endosperm keeps dividing, resulting in four, eight, etc. cells; the zygote remains quiescent for some period of time (Figs. 4,5).

Snapdragon ESs were also isolated in the fresh, living condition. In comparison with the fixed ones they are characterized by abundant contents which would otherwise be largely removed by fixation and subsequent treatments. Histochemical reactions show the presence of many lipid droplets (Fig. 7) and of starch grains (Fig. 8).

When the DNA-binding fluorescent dye H33258 was used, the nuclei within the isolated ESs showed green fluorescence. Figure 9 shows a fixed mature ES. A pair of synergid nuclei near the micropylar end can be seen. To their lower side, the egg nucleus is located. Still lower, at the center of the sac, a large secondary nucleus can be clearly observed. Figure 10 is of a fixed ES after fertilization. The zygote nucleus and the nuclei of eight endosperm cells are evident. Fresh ESs could also be stained effectively. For example, a large, fresh post-fertilization ES with strong H33258 fluorescence of many endosperm nuclei is shown in Fig. 11, and a fresh ES of approximately the same stage photographed by interference contrast microscopy is shown for comparison in Fig. 6. The FDA test provides evidence for the viability of such isolated ES.

In order to obtain some information on the chemical nature of the ES wall, two fluorescent dyes were used on freshly isolated material. Calcofluor white ST did not cause the ES wall to become fluorescent, most probably because of enzymatic degradation of the cellulose in the wall during isolation. In contrast, auramine O-stained material showed strong green fluorescence in the wall (Fig. 12). Auramine O has been used to detect the cuticle on the stigma surface (Heslop-Harrison 1977) and the pollen-wall sporopollenin (Heslop-Harrison and Heslop-Harrison 1982). A cuticle has been found in the ES wall of some species by other means (Kapil and Bhatnagar 1981). Our present results provide evidence of the presence of cutin in the snapdragon ES wall. This may be the reason why ovular tissues treated with cellulase and pectinase are macerated whereas the ES remains intact, the cutin being resistant to the enzyme action.

Isolated ESs of Helianthus annuus and Nicotiana tabacum.

As is snapdragon, ESs both before and after fertilization were isolated in sunflower. Figures 13 and 14 show mature ESs isolated from a fixed and a fresh ovule, respectively. A pearshaped egg cell with a chalazally located nucleus can be clearly observed. One of the synergids is located nearby. The central cell is closely connected with the egg apparatus and its large prominent secondary nucleus is in the proximity of the egg. The antipodal cells are usually disconnected from the central cell by the enzymatic maceration. Embryo sacs isolated from fresh ovules are rich in inclusions. Like the snapdragon ESs they contain lipid drops.

Figure 15 is of a fresh sunflower ES treated with FDA. Notice the strong fluorochromasia throughout the ES; a pair of synergids and the large nucleolus of the secondary nucleus are particularly evident. Similarly, the post-fertilization ESs consisting of proembryo and multicellular endosperm also exhibit fluorochromasia (Fig. 16).

Embryo sacs of tobacco prior to fertilization have been isolated from fixed as well as fresh ovules. Figure 17 is an interference-contrast image of a mature ES. An egg cell and one of the synergids are shown at its micropylar end. The opposite position of nucleus and vacuole can be well seen in these two cells. The central cell is highly vacuolated, with a pair of polar nuclei. The number of antipodal cells varied from three to none. Figure 18 is a negative phase-contrast image of a similar ES, but viewed from another direction. The triangular arrangement of the egg and synergids is very apparent. As in the case of snapdragon, the tobacco ESs stained with H33258 show strong fluorescence of the component cell nuclei (Fig. 19). Figure 20 is of a fresh ES stained with propionocarmine and slightly pressed under the coverslip to have all its component cells in one optical plane.

Discussion

The enzymatic-maceration technique recently developed is a new approach to the isolation of the female gametophytes of angiosperms. It permits direct observation of naked ESs. It is convenient and time-saving compared with complicated sectioning procedures, and a three-dimensional image can be obtained simply by focusing the microscope. Cells kept in their entirety are available for microphotometric measurements. For instance, H33258, which has been used for fluorometric measurement of DNA content in both animal and plant cells (Arnodt-Jovin and Jovin 1977; Howard et al. 1979; Cowell and Franks 1980; Galbraith et al. 1981; Meadows 1983) can be used to stain isolated ESs and has yielded good fluorescence of the nuclei. This may permit quantitative studies on DNA changes during ES as well as early endosperm and embryo development.

The new technique also opens possibilities for histochemical and cell-physiological investigations of fresh and viable ESs which were previously difficult to carry out. As a simple example, it enabled us to observe lipid drops in isolated ESs, which otherwise would mostly be lost in the fixation, dehydration and clearing procedures. Another example is the fluorochromatic reaction. Fluorochromasia of FDA induced by intracellular esterases, which has long been used as a method for testing the viability of pollen and somatic cells (Heslop-Harrison and Heslop-Harrison 1970; Widholm 1972; Ockendon and Gates 1976; Larkin 1976; Shivanna and Heslop-Harrison 1981; Smith et al. 1982) can now be applied to ESs as well. It should generally become possible to study various enzyme activities in the ES and its component cells.

The enzymatic isolation technique may also be an important and perhaps indispensable prerequisite for *in vitro* culture of the ES which so far has not been carried out successfully at all (see Raghavan 1976, 1979). Induction of haploid plants by culture of unpollinated ovaries and ovules has been done with considerable success in recent years (for a review see Yang and Zhou 1982). Therefore, it is reasonable to predict that direct ES culture will be a next logical advance. Further, such an *in vitro* experimental system will be useful not only for possible induction of gynogenesis, but also for understanding the physiological responses and morphogenetic changes during ES development, *in vitro* fertilization, and post-fertilization events.

Enzymatic isolation of ESs is obviously a developing technique, with problems still remaining open. Especially, isolation of fresh viable units is considerably more difficult than that of fixed ones. The results depend on various factors, such as plant species, ES stage, the kind of tissue, combination and concentration of enzymes, the osmotic value of the maceration solution, the method of maceration, etc. Under unfavorable conditions the ESs often burst or suffer from plasmolysis; or the ovular tissues may be recalcitrant to maceration. These and other aspects of the technique need further improvement.

Figs. 1-12. Isolated embryo sacs of *Antirrhinum majus*. Each ES is oriented with its micropylar end upwards and its chalazal end downwards. **Figs. 1-6**, Nomarksi interference contrast microscopy; **Figs. 7, 8**, ordinary light microscopy; **Figs. 9-12**, fluorescence microscopy. All bars=30 μm. **Fig. 1.** A mature ES before fertilization, showing two synergids, an egg cell and a secondary nucleus. FPA-fixed and propiono-carmine-stained. ×500. **Fig. 2.** An ES soon after fertilization. The pollen tube is still visi-

第二章 雌性器官与细胞的实验研究

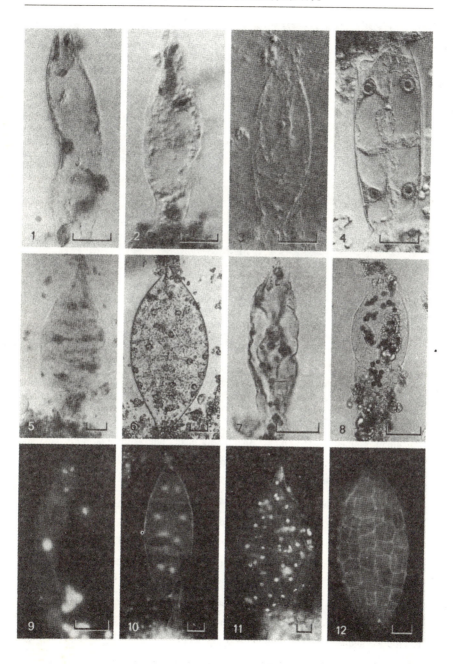

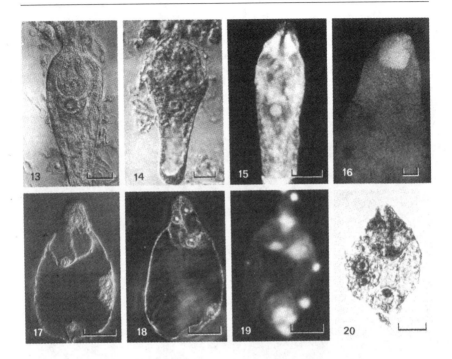

ble. Two endosperm nuclei are located far from each other. FPA-fixed and propiono-carmine-stained. ×500. **Fig. 3.** An ES with two endosperm cells separated by a transverse wall. FPA fixed. ×500. **Fig. 4.** An ES with four endosperm cells. FPA-fixed and propiono-carmine-stained. ×500. **Fig. 5.** An ES with eight endosperm cells. FPA-fixed and propiono-carmine-stained. ×250. **Fig. 6.** An ES isolated from fresh material, with a zygote and multicellular endosperm. ×200. **Fig. 7.** An ES before fertilization, isolated from fresh material and stained with Sudan III, showing lipid drops. ×550. **Fig. 8.** An ES isolated from fresh material and stained with I-KI, showing starch grains. ×400. **Fig. 9.** An ES before fertilization. FPA-fixed and H33258-stained. Notice the fluorescence of a pair of synergid nuclei, an egg nucleus and a large secondary nucleus. ×400. **Fig. 10.** An ES with a zygote and eight endosperm cells. FPA-fixed and H33258-stained. ×200. **Fig. 11.** An ES isolated from fresh material and stained with H33258, showing the nuclei of multicellular endosperm. ×150. **Fig. 12.** An ES isolated from fresh material and stained with auramine O, showing the fluorescence of the ES wall. The network represents sites connected with surrounding ovular tissue. ×250

Figs. 13-20. Isolated embryo sacs of *Helianthus annuus* Figs. 13-16 and of *Nicotiana tabacum* Figs. 17-20. Orientation as in Figs. 1-12. Bars = 30 μm (except Fig. 16). **Fig. 13.** A mature ES of sunflower isolated from FPA-fixed material, showing one of the synergids, the egg cell and a central cell with secondary nucleus. The antipodal cells have been detached because of enzymatic maceration. Nomarski interference contrast. ×350. **Fig. 14.** A mature ES isolated from fresh material, showing the same structure as in Fig. 13. Nomarski interference contrast. ×300. **Fig. 15.** An ES isolated from fresh material and stained with FDA, showing strong fluorochromasia. Fluorescence microscopy. ×400. **Fig. 16.** A post-fertilization ES with proembryo and multicellular endosperm. Fresh material. Stained with FDA, showing fluorochomasia. Bar = 100 μm. ×60. **Fig. 17.** A mature ES of tobacco. FPA-fixed. Nomarski interference contrast. Notice the opposite position of nucleus and vacuole in the synergid and the egg. A pair of polar nuclei and one antipodal cell are seen. ×450. **Fig. 18.** A mature ES FPA-fixed. Negative phase contrast. Notice the triangular topography of the synergids and the egg. ×450. **Fig. 19.** A mature ES FPA-fixed and H33258-stained. Fluorescence microscopy showing two synergid nuclei, an egg nucleus, two polar nuclei and three antipodal nuclei. ×450. **Fig. 20.** A mature ES isolated from fresh material and stained with propiono-carmine, slightly pressed to show simultaneously all its component cells. ×400

(Authors: Zhou C, Yang H Y. Published in: Planta, 1985, 165: 225~231, with 20 figures and 31 references. All figures are retained here. See color plate VII and VIII in the end of this book.)

6 A study of fertilization events in living embryo sacs isolated from sunflower ovules

Introduction

So far the progresses in understanding of fertilization process in angiosperms have been obtained mainly from fixed materials. To study it in living state is beyond doubt of great significance in modern plant embryology. There were only a few such works to date scattered in literature: early in the 1950s Poddubnaya Arnoldi observed fertilization in living ovules of orchids[1]. Recently Erdelska recorded interesting dynamic changes of fertilization as well as early embryo and endosperm development in *Jasione monatana* and *Galanthus nivalis* by microcinematography[2]. Both of them used whole ovules as experimental materials and had to choose special plant species with thin and transparent integuments through which the structures inside the embryo sacs could be seen. Obviously, such a methodology is unavailable for the majority of angiosperm ovules whose integuments and nuclei hinder a direct observation on fertilization. Therefore, our attention must be given to finding a new approach applicable to a wider range of plant species. In recent years we have developed a technique for isolation of living embryo sacs from ovules of several species including *Helianthus annuus* L. [3,4]. This provides the possibility of direct observation on structural changes in the embryo sacs. Combining this technique with fluorescent vital staining, we were able to examine the fertilization events in isolated living embryo sacs of sunflower for the first time.

Materials and methods

Sunflower (*Helianthus annuus* L.) cultivar 'Dang Yang' was used as material. Pollen grains were collected 1 day before pollination and stored in petri dish in refrigerator. At 6-7 o'clock the next morning capitula were picked from the field and their stalks were inserted in water. The tubular florets just at anthesis were soon artificially pollinated with the collected pollen grains. The capitula were kept in the same room conditions under scattered light and 25-30℃. After pollination ovules were dissected out from the ovaries at intervals during 3-24 h and were macerated according to an enzymatic method previously used[3]: the fresh ovules were immediately placed in a mixture of 3% pectinase (Serva, Feinbiochemica), 2% cellulase R10 (YaKult Honska Co.), 2% snailase (Institute of Biophysics, Academia Sinica, Beijing, China), 1% pectolyase Y-23 (Sheishin Pharmaceutical), 10% sucrose, 200 μg/ml KH_2PO_4, 83 μg/ml $CaCl_2 \cdot 2H_2O$ and 20-25 μg/ml Hoechst 33258 (H33258), and incubated on a microshaker at 28-30℃ for about 4-5h. After maceration the cell suspensions were washed and centrifuged 3 times at a speed of 1500 rev./min for 5 min with a solution of KH_2PO_4, $CaCl_2 \cdot 2H_2O$ and sucrose of the same concentration. The enriched cell suspensions were dropped and mounted on slides and observed and photographed under an epi-fluorescence microscope AO FLUORESTAR 110 (AO Scientific Instruments Co.) with filter combination 2073 (exciter filter 436, dichroic mirror 450, barrier filter 475).

Results

The fertilization events could be seen clearly in isolated living embryo sacs of sunflower after H33258 fluorescent staining. Usually only the nuclei showed specific green fluorescence of H33258 and other parts of the cells had no reaction. But the cytoplasm of embryo sac gave yellow autofluorescence, serving as a background for demonstration of sperm position in the embryo sacs.

The isolated sunflower embryo sac unit usually comprised an egg cell, a pair of synergids and a central cell with the secondary nucleus; the antipodals were often disconnected from the unit by maceration[3]. When organized cellular embryo sac was still young, its nuclear DNA could bind with H33258 and give bright fluorescence. Figure 1 demonstrates one of such embryo sacs where an oval nucleus and a spherical secondary nucleus were discernible. Synergid nuclei, though out of focus in this figure, also showed fluorescence. As the embryo sac matured, they grew much bigger than before. Except synergid nuclei, now all the female nuclei (including egg and secondary nuclei) became H33258-negative as shown in Figs. 2 and 3.

The sperm nuclei in mature pollen grains usually appeared as long curved bands. When discharged by a pollen tube into one of the synergids, they already winded themselves into a spherical, oval or spiral shape (Fig. 2). The sperm nucleus was characterized by small size (approx. 4 μm), compact chromatin and strong fluorescence, by which it was quite distinguishable from the synergid nucleus. Figure 3 illustrates a pair of sperms overlapped with each other exhibiting very strong fluorescence. When different optical planes were focused, each individual sperm with characteristic shape could be discernible (Fig. 4).

During the time of nuclear fusion the female nuclei appeared H33258-negative, which made it easy to trace the behavior of the sperms. In Figs. 5 and 6 double fertilization was just taking place; the sperm chromatin was dispersing in one side of the female nuclei and showing diffusive fluorescence. Sometimes a pair of additional sperms might enter into an embryo sac in which the previously discharged sperms had already fused with the female nuclei (Fig. 7). The possible entry of a second set of sperms in an already fertilized embryo sac has been repeatedly reported previously[5] and observed in isolated fixed embryo sacs of *Adenophora* in our lab[6].

After fertilization the zygote nucleus began to show weak green fluorescence of its chromatin, especially near the inner surface of its nuclear membrane (Figs. 8,9). First division of the zygote was transverse, consequently two groups of anaphase chromosomes could be seen (Fig. 10). Later, the nuclei of the proembryo emitted strong H33258 fluorescence as shown in Fig. 12.

The primary endosperm nucleus divided earlier than the zygote did. In Fig. 8 two endosperm nuclei had already started a new division, fluorescing very strongly, while the zygote had just formed with weak fluorescence. Fig. 9 demonstrates two pairs of endosperm nuclei at interphase. It is easy to find the difference of fluorescence pattern and intensity between interphase and mitotic nuclei (Figs. 8, 9). Figure 11 shows that at a later stage more endosperm nuclei occupied in the micropylar part of the central cell and a big vacuole in its chalazal part. Endosperm nuclei kept dividing and eventually filled up the whole embryo sac.

Discussion

Nowadays, new advances in biology of fertilization in angiosperms have been made by using electron microscopy with computer-assisted 3-dimensional reconstruction technique and a series of new concepts have been developed, including sperm heteromorphism, male germ unit, preprogramming of double fertilization, etc.[7,8]. On the other hand, *in vitro* study on fertilization should be considered as another important approach because it would offer the possibility of directly tracing the dynamic process in controlled conditions. Although attractive, it faces a series of technical challenges and still remains in an early stage up to date. So far only whole ovules of special plant species[1,2] have been used as an experimental system. Our work is a new attempt to use the isolated embryo sacs for such a study. Recently, living embryo sacs and embryo sac protoplasts could be prepared by enzymatic technique[3,4,9,10]. Living embryo sac constituents were dissected out manually by micromanipulater as well[11]. In sunflower, the isolated embryo sacs, either before or after fertilization, have been tried in culture study. A few of them could keep viable up to 10 days of culture according to fluorochromatic reaction of fluorescein diacetate, though visible changes in their development have not been observed yet[4].

Another prerequisite for observing fertilization events in living state is to find out an appropriate method of vital staining. Conventional staining methods would kill the cells. Phase- or interference-contrast microscopy are incapable of distinguishing the male gametes from various particles contained in the embryo sac and discharged by the pollen tube. The DNA-binding fluorescent dye H33258 is known to be non-toxic and has been used for protoplast staining[12], heterokaryon identification[13] and also for observation on living embryo sacs[3]. The present study widened the usage of this vital stain to identify the male nuclei during double fertilization, which facilitates the work to a great extent.

At present we only carried out a static study. Further attempts will be made to find out a suitable *in vitro* condition for following the dynamic process of fertilization in the observed embryo sacs and to study the direct influence of various factors on this process.

Explanation of plates

Figs. 1-12. Fig. 1. A young organized embryo sac, showing fluorescence of egg nucleus and secondary nucleus, ×450. **Fig. 2.** A mature embryo sac after pollination, showing fluorescence of two sperms in a synergid and no fluorescence in female nuclei, ×450. **Fig. 3.** In another embryo sac, a pair of sperms overlapped with each other, showing very strong fluorescence, ×450. **Fig. 4.** The same embryo sac as in Fig. 3 at another optical section, ×450. **Fig. 5.** Fusion of sperms with egg and secondary nucleus. Fluorescence of the dispersing sperm chromatin was weakened, ×450. **Fig. 6.** Another embryo sac similar to Fig. 5, ×450. **Fig. 7.** A pair of additional sperms in an embryo sac where double fertilization was taking place, ×450. **Fig. 8.** Slightly strengthened fluorescence of zygote nucleus after fertilization and strong fluorescence of 2 dividing endosperm nuclei, ×450. **Fig. 9.** A zygote and 4 endosperm nuclei at interphase, ×450. **Fig. 10.** A dividing zygote (anaphase) and 4 endosperm nuclei in different optical plane, ×450. **Fig. 11.** Very strong fluorescence of dividing endosperm nuclei, ×350. **Fig. 12.** A multicellular proembryo, ×220. Abbreviations: e, egg nucleus; en, endosperm nucleus; m, male nucleus; s, synergid nucleus; se, secondary nucleus; z, zygote nucleus.

(Author: Zhou C. Published in: Plant Science, 1987, 52:147-151.
with 12 figures and 13 references. All figures are retained here.)

第二章 雌性器官与细胞的实验研究

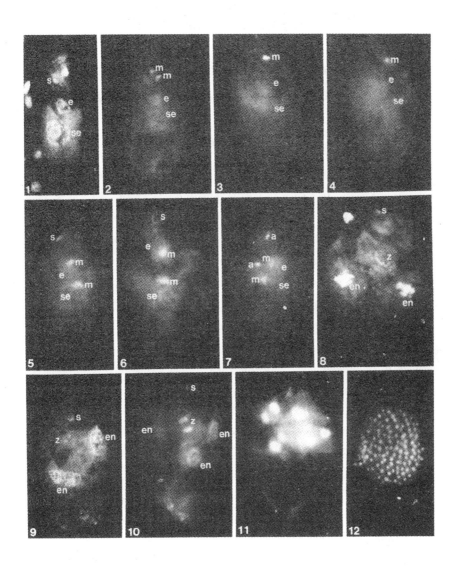

7 金鱼草珠被绒毡层壁囊的分离与鉴定

Isolation and identification of integumentary tapetal wall sac in *Antirrhinum majus* L.

Abstract

Sac-shaped wall structures were isolated from post-anthesis ovules of *Antirrhinum majus* treated with acetic anhydride-H_2SO_4 (9∶1), 50% cold chromic acid, 80% H_2SO_4-30% H_2O_2 (5∶1) or cellulase-pectinase solution. This means that the wall sac is acetolysis-resistant as well as enzyme-resistant and has some resistance to strong oxidizers. Observations on paraffin and ultrathin sections confirmed that the isolated wall sac is just the inner wall of integumentary tapetum, which encloses the embryo sac from all side but leaves an aperture at each pole. This wall showed positive reaction to Sudan IV staining and auramine O fluorescence staining but negatively reacted to phloroglucinol, indicating that it was likely to be composed of highly stable lipid substances, preferably cutin or sporopollenin. Based on all these results, it is supposed that the wall sac functions as a dam for preventing damage of young embryo and endosperm from the hydrolytic enzymes secreted by integumentary tapetum and also for chanalizing nutrient flow through the chalazal and micropylar apertures into embryo sac.

在酶法分离金鱼草胚囊的实验中,我们曾观察到分离胚囊的外周有完整的壁,它不被纤维素酶与果胶酶降解,而被金胺 O 或樱草黄染色呈现荧光。当时我们认为它可能是胚囊的角质层,而某些情况下其外表的明显网格则可能是与外周胚珠细胞壁连接处的残迹[1,2]。后来的研究发现实际情况要比原来设想的更为复杂,不仅涉及胚囊壁,而且更多地涉及珠被绒毡层壁。本文报道用实验方法分离出这种细胞壁囊状结构(后文简称壁囊),并探讨了其性质与可能具有的功能。

材 料 与 方 法

金鱼草(*Antirrhinum majus* L.)开花后 1～10d 期间,每天取胚珠固定于 Carnoy 液(3∶1)中,冷藏于70%乙醇中。

一部分胚珠用下列各项方法分离壁囊:(1)乙酰解。用醋酸酐-浓硫酸(9∶1)沸水浴 7～45min[12]。(2)50%铬酸处理[9]。(3)80%浓硫酸和30%过氧化氢混合液(5∶1)处理[11]。(4)酶解。用2%纤维素酶与2%果胶酶混合液在28℃下保温振荡4h以上[2]。(5)二乙醇胺处理。将酶解分离的壁囊置二乙醇胺(diethanolamine)中沸水浴1h[12]。

一部分胚珠按常规方法制作石蜡切片,进行胚胎学与组织化学观察:(1)爱氏苏木精整体染色。因染色偏浅,故用干涉差显微观察。(2)苏丹Ⅳ。用溶于95%乙醇和甘油等量混合液的苏丹Ⅳ饱和溶液染色 30min,分别进行明视野与荧光显微观察[4]。(3)金胺O。切片先用 pH7.2 的 Tris-HCl 缓冲液浸泡数小时,再以同一缓冲液配制的0.01%金胺O(auramine O)染色数分钟,荧光显微观察[7,8]。酶法分离的壁囊染色方法相同。(4)间苯三酚。用溶于17%盐酸的2%间苯三酚(phloroglucinol)染色数分钟,分别进行明视野与荧光显微观察。上述所有观察均用 Olympus NEW VANOX AHBS-514 型多功能显微镜。

另取部分胚珠,戊二醛、锇酸固定,乙醇、环氧丙烷脱水、Epon812包埋,在 Sorvall MT-6000 型超薄切片机上制作超薄切片,经醋酸铀、柠檬酸铅染色后,置 JEM 100 CX/Ⅱ 型电子显微镜下观察。

结 果

一、分离实验

采用下列四种方法均能溶解胚珠组织而分离出对试剂有抗性的

壁囊。

乙酰解 开花后 1~2d 的胚珠经 7min 乙酰解后所有组织均被溶解,分离不出壁囊。但从开花后 3d 起开始分离出抗乙酰解的壁囊,且随胚珠发育推迟其抗乙酰解性能增强;开花后 5~10d 的胚珠经 30min 处理后,壁囊仍然保持完整。图版 Ⅰ,1 示一个开花后 8d 的壁囊,其两端尖削而中部宽阔,呈橄榄形。囊的各部是完整的,惟两端各有一狭窄的开口。如与后文描述的切片原位观察对照,可以判断哪一端是合点或珠孔端。囊的表面呈网络状,从光学切面上可以看出这种网络是由壁囊外方短的增厚带所组成的(图版 Ⅰ,2)。增厚带是不整齐的,由此可以推测它是与壁囊相连的外周细胞壁的抗乙酰解残迹。

铬酸处理 铬酸是一种强氧化剂。50% 冷铬酸可用于溶解细胞壁中的多数成分而保留角质、栓质等[9]。金鱼草胚珠在 10℃ 下经其处理 25h 后组织溶解而分离出壁囊。壁囊的形态和乙酰解法分离者相似。在 18~20℃ 下则壁囊亦被溶解。

80% 硫酸与 30% 过氧化氢处理 本试剂氧化作用极强,可用于溶解细胞壁中纤维素、木质等成分而保留角质[11]。金鱼草胚珠在 10℃ 下以其处理 6h 后即开始分离出壁囊;24h 后则壁囊亦被溶解。

酶解 开花后 1d、3d、5d、8d 的胚珠经纤维素酶与果胶酶处理后均分离出壁囊。酶解适度时壁囊内的细胞结构可以保存;超过限度则细胞结构破坏而壁囊完好无损。图版 Ⅰ,3 示一个开花后 3d 分离的壁囊,金胺 O 染色呈明亮的荧光。其表面为网格状,合点端尤其显著,和乙酰解分离者相似。壁囊的合点端连着一团尚未被酶解的尾状物,从后文切片观察可知这是由珠柄通向合点的维管束。图版 Ⅰ,4 示过度酶解后所分离的空的壁囊,可清楚看出其两端的开口。将酶解分离的壁囊再用二乙醇胺处理以观察其反应。二乙醇胺是一种能溶解孢粉素的有机溶剂,性质和乙醇胺相近,后者曾用于溶去花粉的外壁[8,12]。由于它对纤维素壁是无效的,所以如用它直接处理完整的胚珠不能观察其对壁囊的作用。因此我们先用酶法分离出壁囊再用二乙醇胺处理。金鱼草开花后 1d、3d、5d、8d 分离的壁囊经此剂处理 1h 后,壁囊大多溶解,少数正在溶解过程中。这表明壁囊是不抗二乙醇胺的,可能含孢粉素成分。

以上各项试验结果有两方面的意义:一是找到了几种分离壁囊的方法,从而可以观察它的整体三维图像,并且为今后进一步分析它的成分提供了大量纯化的材料。二是通过分离实验可以鉴别壁囊对各种试剂的反应,因而这也是认识其化学性质的一种手段。本试验结果表明:壁囊含有抗乙酰解、抗酶解和对氧化剂有一定程度抗性的物质,但不抗二乙醇胺,性质上和孢粉素或角质相似。

二、胚胎学观察

胚胎学观察的目的是了解壁囊所处的胚珠发育时期,作为探讨其功能的依据;另一目的是为后文的组织学鉴定提供形态学对照。金鱼草开花后 1~10d 正处于由合子到胚分化的阶段,具体描述如下:开花后 1~3d(图版Ⅱ,10),合子尚未分裂,胚乳细胞已经多数。胚囊被一层珠被绒毡层细胞包围,惟两端未被封闭:合点端有一含双核的胚乳吸器通过珠被绒毡层的开口向外突出,伸入承珠盘(图版Ⅱ,11),并与由珠柄通向该处的维管束相对应(图版Ⅱ,10)。珠孔端的开口则正对着合子基部,以后形成的胚柄吸器由此处伸出(图版Ⅱ,12、13)。开花后 4~5d,合子分裂成原胚;胚乳细胞继续增多(图版Ⅱ,12)。开花后 8d 为球形胚时期。开花后 10d,胚开始分化,出现心形和鱼雷形胚;胚乳中积累大量贮藏物质(图版Ⅱ,13、14)。此时珠被绒毡层外方的组织呈现贮藏物质被消化的一圈区带,和胚周围的胚乳组织被消化的状态很相似(图版Ⅱ,15)。总的看来,开花 10d 期间的胚珠正处于原胚形成和胚分化的阶段,胚和胚乳的发育需要大量的营养供应;此时的珠被绒毡层具有向外分泌、消化周围组织中的贮藏养料的功能。对乙酰解等有抗性的壁囊正是由这一时期的胚珠中分离出来的,这对我们认识壁囊的功能很重要,后文将详细讨论。

三、组织化学鉴定

采用下列几种方法对壁囊作了原位的组织化学鉴定,以便了解其所处位置和化学性质。

苏丹 IV 反应 苏丹 IV 染色作为常规的明视野观察时可以鉴别脂类,而作为荧光显微观察时则可鉴别栓质(有荧光)与木质(无荧光)[4]。金鱼草开花后 1~10d 的胚珠切片经此法染色后,明视野中均可看到在珠被绒毡层与胚囊之间有红色反应的环带;有些切片上可看出反应的部位是珠被绒毡层的内切向壁和与之相连的径向壁内端(图版 II,16、17)。在合点端开口处反应特别明显,形成一条由它构成的孔道,胚乳吸器由此伸出(图版 II,18)。这些特点和分离壁囊的特点是完全吻合的。在荧光显微观察时,相应于壁囊的部位没有荧光反应。上述结果表明该壁含有脂类成分,但不是栓质。

金胺 O 反应 此法也是显示脂类的,曾被用于观察柱头表面的角质[7]和花粉外壁的孢粉素[8]。金鱼草开花后 1~10d 的胚珠切片用此法观察,均看到珠被绒毡层与胚囊之间有明亮的荧光环带,构成两端开口的囊状(图版 I,5)。高倍观察时可看出它紧贴珠被绒毡层内侧,与其径向壁交界处有较突出的亮点(图版 I,6)。会点端开口处两侧荧光很强,构成胚乳吸器伸出的孔道(图版 I,7)。这些情况都和苏丹 IV 反应一致。

间苯三酚反应 间苯三酚染色作明视野观察时可显示木质,而作荧光显微观察时可区分栓质(有荧光)和木质(无荧光)[4]。金鱼草开花后 8~10d 的胚珠切片经此法染色后,明视野中除珠被外表皮的增厚壁呈红色反应外,其它各种(包括与壁囊相当的部位)均为无色。在荧光显微观察时,与壁囊相当的部位亦无荧光。上述结果表明该壁既不含木质亦不含栓质。

综上所述,可以得出如下看法:第一,分离的壁囊是由珠被绒毡层的内壁构成的;第二,它由脂类组成,但不含栓质与木质,可能是角质或孢粉素。但由于这几类物质性质比较接近,组织化学只能作出大致的鉴定;准确的结论尚有待于今后的化学分析。

四、超微结构观察

为了更准确地判断壁囊的部位及其细胞学特征,作了有关的超微结构观察(图版 I,8、9)。初步结果表明:在珠被绒毡层与胚囊之间有三个壁层:一层为珠被绒毡层的内切向壁,其特征是:较厚,在浅色的壁层中含电子密度浓厚的物质。与径向壁相连处特别厚,径向壁由内向外逐渐变薄。电子浓密物质由内切向壁连通到径向壁内端,向外则逐渐消失。这种电子浓密物质可能即是正在积累中的抗乙酰解成分。一层为电子浓密而均匀的薄层,可能是角质层。一层为透明而含微纤丝结构的胚囊壁,其主要成分大概是纤维素。将超微结构与前述实验和观察结果结合起来,可以推断:珠被绒毡层内壁与角质层两层均对苏丹Ⅳ和金胺O有反应,均对纤维素酶与果胶酶有抗性。但珠被绒毡层内壁是从开花后3d起才积累抗乙酰解成分的,所以只是在这以后才能分离出抗乙酰解的壁囊。

讨 论

在不少植物的花药绒毡层中存在着绒毡层膜和/或外绒毡层膜,这是一种分布于绒毡层内表面或外表面的结构,含抗乙酰解成分,可用乙酰解处理分离出来。它们呈囊状包围内部的花粉。绒毡层膜的功能还不很清楚,据推测可能作为花粉的"培养囊"而起作用[3]。珠被绒毡层在结构与功能上和花药绒毡层有一些类似的特点,但因为它在各种植物中变化很大,亦较难操作,所以对其研究较少。Kapil等(1978)曾对此作了一个全面的综述[10],指出珠被绒毡层在胚珠发育的不同阶段具有不同的特点:在幼嫩的胚珠中,它具有分生组织的特征,通过其分裂活动维持胚珠中各个部分的协调生长。受精前后,它分化成类似分泌组织,能分泌酶类等分解外围组织,为胚囊内的细胞提供营养。在种子成熟过程中,它或者退化,或者转变为厚壁的保护组织。在珠被绒毡层和胚囊之间常有角质层间隔;珠被绒毡层的内切向壁异常增厚,其中可能含有抗乙酰解成分[5]。但是关于它的壁的性质与功能却缺乏详细深入的研究。

目前还缺乏足够的资料来对珠被绒毡层的内壁和花药绒毡层膜作出中肯的比较,更不能说二者是对等的关系。但通过本文的研究毕竟可以看出它们有若干共同的特征:第一,二者均是分布在特化的分泌组织表面;第二,二者均由抗乙酰解成分构成,均可用乙酰解处理分离;第三,二者均为囊状构造,包围着内部正在发育的细胞。这些有趣的共同特点暗示它们可能负有类似的功能,因而是值得深入研究的。

就金鱼草珠被绒毡层壁囊而言,我们推测它有双重的生物学意义:首先,它和珠被绒毡层的分泌机制有关。珠被绒毡层能向外分泌水解酶类,消化周围的组织。但如果这些酶也向内分泌,势必伤害胚囊内的细胞。壁囊的作用就像一道堤防,只允许外向的分泌而防止内向的分泌,从而保护胚和胚乳免受水解酶的侵蚀。在 Kapil 等(1978)的综述中已经讨论了这一点[10]。其次,它和养料导入胚囊的途径有关。壁囊的存在固然防止水解酶进入胚囊,同时也就阻碍了养料通过它摄入胚囊。只有壁囊两端的开口是养料摄入的渠道。而合点端的胚乳吸器,可能还有珠孔端的胚柄,便是适应这种状况的摄取养料的机构。由维管束输送的养料必须通过合点端的开口吸进胚囊。珠被绒毡层周围的消化区中的养料,也须经由合点或珠孔的开口才能被胚囊利用。以上两个方面看来是有内在联系的:前一方面是因,后一方面是果;壁囊的存在使两者巧妙地统一了起来。

应该指出,金鱼草仅仅代表变化多端的珠被绒毡层的一种类型。在其它类型中会有不同的形式来保证以上两个方面的功能。例如 Erdelska(1975)[6]研究的 *Jasione montana*(桔梗科)的珠被绒毡层便有与此不同的结构特点:在珠被绒毡层和胚囊之间有两层角质层包围着胚囊,仅在近珠孔的卵器附近角质层中断。而正是在这一部位的胚囊壁上具有发达的内突,成为养料输入的通道[6]。因此,研究各类珠被绒毡层的结构特点将有助于最终阐明它的功能。同时,对珠被绒毡层壁囊的深入研究,也许对进一步认识花药绒毡层膜的功能也会有所启发。

图 版 说 明

c.角质层 ch.合点 em.胚 en.胚乳 h.吸器 i.珠被绒毡层 m.珠孔 s.胚柄 t.内切向壁 v.维管束,w.胚囊壁,z.合子

第二章 雌性器官与细胞的实验研究

图版 I
plate I

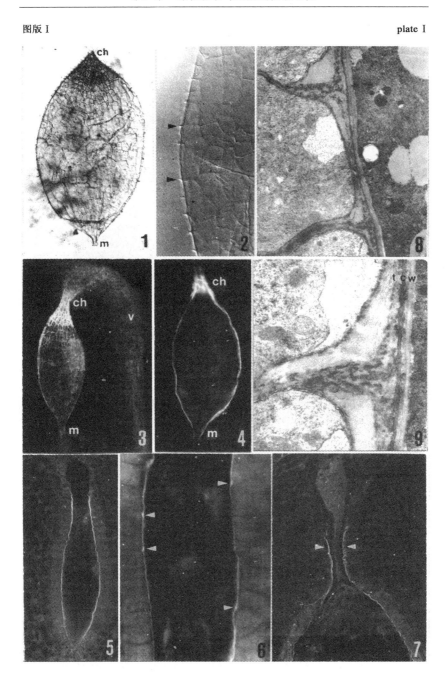

图版 II

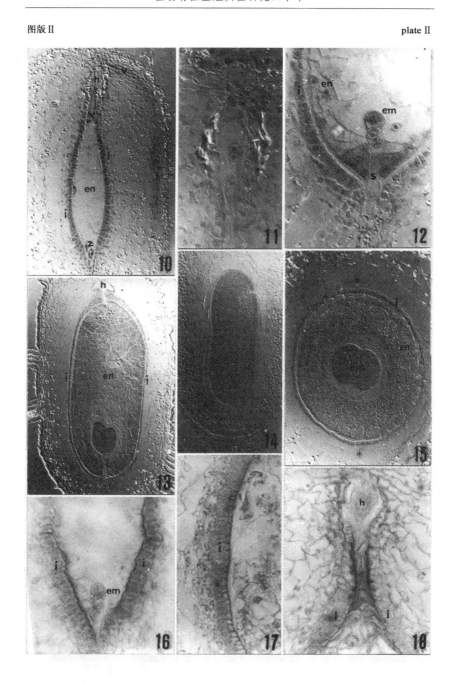

plate II

第二章 雌性器官与细胞的实验研究

图 版 I

1、2. 干涉差显微摄影。**3~7**. 金胺O染色,荧光显微摄影。**8、9**. 超薄切片电镜摄影。**1**. 开花后8d乙酰解分离的珠被绒毡层壁囊。×150 **2**. 壁囊侧面观。箭头示其外侧的增厚带。×550 **3**. 开花后3d酶法分离的壁囊表面观。×100 **4**. 开花后5d酶法分离的壁囊光学切面观。×150 **5**. 开花后1d,石蜡切片,示金胺O荧光染色的珠被绒毡层内壁。×200 **6**. 开花后3d,箭头示珠被绒毡层内切向壁与径向壁交界处的增厚点。×400 **7**. 开花后5d,箭头示合点端的珠被绒毡层孔道,胚乳吸器由此处伸出。×300 **8、9**. 开花后5d,示珠被绒毡层内切向壁、角质层与胚囊壁。**8**. ×10000 **9**. ×18000

图 版 II

10~15. 石蜡切片,干涉差显微摄影。**16~18**. 石蜡切片,苏丹IV染色,明视野摄影。**10**. 开花后2d,珠被绒毡层包围胚乳与合子,胚乳吸器由合点端开口伸出并指向维管束。×150 **11**. 开花后4d,胚乳吸器,含双核。×450 **12**. 开花后4d,原胚期,含正在分裂的胚细胞以及胚柄。×450 **13**. 开花后10d,心形胚期。×100 **14**. 开花后10d,鱼雷形胚期。×200 **15**. 开花后10d胚珠横切面。星号示珠被绒毡层周围的消化区。×200 **16**. 开花后4d的胚珠珠孔端,示苏丹IV染色的珠被绒毡层内壁。×250 **17**. 开花后3d,示珠被绒毡层内切向壁与径向壁内端。×250 **18**. 开花后5d,示珠被绒毡层的合点端孔道。×350

Explanation of plates

c. cuticle   ch. chalaza   em. embryo   en. endosperm   h. haustorium   i. integumentary tapetum   m. micropyle,   s. suspensor   t. inner tangential wall.   v. vascular bundle   w. embryo sac wall   z. zygote

Plate I

Figs. 1 and **2**. Nomarski inteference contrast microscopy. **Figs. 3-7.** Auramine O staining and fluorescence microscopy. **Figs. 8** and **9**. Ultratomy and electron microscopy. **Fig. 1.** A wall sac isolated by acetolysis at 8 days after anthesis. ×150 **Fig. 2.** Side view of the isolated wall sac. Arrowheads show its outside thickenings. ×550 **Fig. 3.** Surface view of an enzymatically isolated wall sac at 3 days after anthesis. ×100 **Fig. 4.** Optical section of an enzymatically isolated wall sac at 5 days after anthesis. ×

150　**Fig. 5.** Paraffin section of an ovule 1 day after anthesis, showing inner wall of the integumentary tapetum stained with auramine O fluorescence. ×200　**Fig. 6.** An ovule 3 days after anthesis. Arrowheads show thickenings between the inner tangential wall and the radial wall of integumentary tapetum. ×400　**Fig. 7.** An ovule 5 days after anthesis. Arrowheads show chalazal channel of the wall sac through which the endosperm haustorium stretchs out. ×300　**Fig. 8** and **9.** Ultrathin sections of ovules 5 days after anthesis, showing the inner wall of integumentary tapetum, the cuticle, and the embryo sac wall. **Fig. 8.** ×10000　**Fig. 9.** ×18000

Plate II

Figs. 10-15. Paraffin sections with Nomarski interference contrast microscopy. **Figs. 16-18.** Paraffin sections stained with Sudan IV. **Fig. 10.** 2 days after anthesis. The integumentary tapetum enclosing the endoperm cells and the zygote. Note the endosperm haustorium stretching out from the tapetum and pointing to the vascular bundle. ×150　**Fig. 11.** 4 days after anthesis. An endosperm haustorium with two nuclei. ×450　**Fig. 12.** 4 days after anthesis. A proembryo with dividing embryo cells and a suspensor. ×450　**Fig. 13.** 10 days after anthesis. Heart-shaped embryo. ×100　**Fig. 14.** 10 days after anthesis. Topedo-shaped embryo. ×200　**Fig. 15.** 10 days after anthesis. Transection of an ovule showing the digested region (asterisks) around the integumentary tapetum. ×200　**Fig. 16.** 4 days after anthesis. Micropylar part of an ovule showing the inner wall of integumentary tapetum stained with Sudan IV. ×250　**Fig. 17.** 3 days after anthesis. The inner tangential wall and its connecting sites with the radial wall of the integumentary tapetum stained with Sudan IV. ×250　**Fig. 18.** 5 days after anthesis. Densely stained chalazal channel of the tapetal inner wall. ×350

(作者:杨弘远。原载:植物学报,1989,31(11):817~823。图版2幅,参考文献12篇。仅选录图版Ⅰ、Ⅱ。参看书末彩色图版Ⅹ)

8 分离烟草胚囊的新方法及诱导卵细胞与助细胞原生质体的原位融合

A new method for embryo sac isolation and *in situ* fusion of egg and synergid protoplasts in *Nicotiana tabacum*

Abstract

A new method combining enzymatic maceration with osmotic shock was developed for isolation of living embryo sac and its protoplasts in *Nicotiana tabacum* L. The principle of this method was that the ovules submitted to enzymatic treatment and osmotic shock could release embryo sacs along with some internal ovular cells through either the funicle cut end or the micropyle. Factors affecting embryo sac isolation were investigated, including concentration of mannitol as a shock osmoticum and in enzyme solution, duration of enzymatic maceration, and duration of osmotic shock. As a result a procedure was established: Ovules at mature embryo sac stage were macerated for 2.5 h in 1%-1.5% cellulase R-10 and 0.5% macerozyme R-10 (or 1% Pectinase, Serva) dissolved in 13% mannitol solution using microshaker, followed by osmotic shock for 15-30 min with enzyme-free 8% mannitol solution and gentle agitation using a pipette. Using a capillary, 50-70 embryo sacs could be collected manually in one hour. The embryo sacs thus isolated could be kept viable from which protoplasts of egg cell and other component cells could be further isolated. An additional interesting phenomenon was that osmotic shock often caused *in situ* fusion the protoplasts of egg cell and synergids. The rate of fusion ranging 9%-71.9% could be controlled by modification of the procedure. This phenomenon merits further attention both from basic and practical point of view. The present method gives the advantages of faciliting isolation and promoting good harvest of viable embryo sacs/female protoplasts within a relative short time.

近年先后在多种植物中分离出生活胚囊、雌性细胞或其原生质体[10]。其所采取的技术路线集中体现在以烟草为材料的一系列研究中,概括起来主要有两类:一类是较长时间酶解辅以一定的机械处理[1,3,11~13,15];另一类是较短时间酶解辅以显微解剖[5,6]。前者可得大量胚囊但需时较长,用酶量较高,且胚囊及其成员细胞混杂于大量解离的体细胞中难以纯化;后者可得生活力较强的胚囊及其成员细胞,但解剖速度较慢,且需熟练的手工技巧。随着研究工作的不断深入[4,8,9],对雌性细胞离体操作技术提出了更高的要求,而目前的分离技术尚未尽如人意。我们以酶解与渗透压冲击相结合的方法,成功地在短时、低酶条件下分离出较多的生活胚囊及其成员细胞原生质体,并发现低渗作用可诱导卵器中各个细胞原生质体的原位融合。

材 料 与 方 法

实验材料为烟草(*Nicotiana tabacum* L.)品种 G-80,种植于湖北省农业科学院温室内。取成熟胚囊期的胚珠,以下述两种方法分离胚囊:(1)二步酶解——渗透压冲击(简称二步法):先将胚珠在13%甘露醇配制的酶液中处理2~2.5h,再换入8%无酶甘露醇溶液冲击,辅以用滴管轻缓吸打;(2)一步酶解——渗透压冲击(简称一步法):以8%~11%甘露醇配制酶液,胚珠直接在该液中处理。上述两种方法所用酶配比均为1%~1.5% cellulase R-10 和 0.5% Macerozyme R-10(或1% pectinase, Serva)。每毫升酶液处理3个子房的胚珠。于直径3cm平皿中,在28~30℃下保温振荡酶解[1]。本文图表所列数字均为从3个子房的胚珠获得的胚囊数。以自制吸管手工收集胚囊,转移到13%甘露醇溶液短期保存;或转移到8%~10%甘露醇溶液中诱导卵细胞与助细胞原生质体原位融合；或于含 10.5%～11.5%甘露醇的稀酶溶液中分离卵细胞原生质体。细胞生活力以荧光素二醋酸酯(FDA)测定。用 Olympus IMT-2 倒置显微镜进行荧光与干涉差观察并摄影。

结　　果

一、胚囊分离的方式

二步法和一步法均可获得较多的胚囊，其形态完整、结构清晰、生活力强。且成员细胞多已形成原生质体(图版Ⅰ,4、5)。

二步法中，胚珠细胞在 13%甘露醇配制的酶液中基本上维持渗透压平衡状态，酶解 2～2.5h 尚不足使胚囊外的体细胞组织离散。这是由于胚珠表皮的角质层妨碍酶液渗入。换入渗透压冲击液后，可见不少内部的体细胞原生质体从珠柄切口或珠孔处外逸。说明酶液首先是沿珠柄维管束及珠孔两条渠道向内渗透的，因而这两个部位也是最易酶解之处。一步法中可见相似的情况，只是体细胞的逸出比较平缓。

在酶解与渗透压冲击双重作用下，胚囊随部分胚珠体细胞而逸出。胚囊逸出的方式有三类：其一是全部或大多数内部体细胞裹挟胚囊而出，胚珠只剩下一个仍然完整的角质层空壳(图版Ⅰ,1)。这种情况多是由于酶解程度较大、低渗冲击较剧烈所致，不是分离胚囊的最佳方式；其二是胚囊从珠柄切口处逸出。胚囊首先被膨胀的内部体细胞挤向珠柄通道，进而随之逸出(图版Ⅰ,2)。这种情况在分离成熟胚囊时较为常见；其三是胚囊从珠孔处逸出(图版Ⅰ,3)。这种情况多见于较幼嫩的胚珠中。在后两种情况下，由于胚珠中只有一小部分内部体细胞逸出，因而在冲击产物中离解的体细胞数量比常规酶解法中少得多。这一特点显然有利于辨识与挑选胚囊。

二、二步法分离胚囊的条件

酶解时间　为了尽可能缩短酶解时间而又获得较多的胚囊，比较了二步法中不同酶解时间对低渗冲击效果的影响(表1)。表1中(以后各表相同),B栏数字为冲击后尚未用吸管吸打时的,包括已裸露和尚被部分体细胞包裹的胚囊数。A栏是吸打后最终获得的胚囊数。可以看出,虽然3次重复实验的数据参差不齐,但总的趋势是一致的:酶解1.5h,低渗冲击效果不明显;2~3h,冲击效果较好。但3.5h后则由于部分胚囊破裂,胚囊数反而减少。因此,以后实验均采用2.5h。

渗透压调节剂浓度　表2比较了不同浓度甘露醇溶液的冲击效果。6%以下甘露醇溶液很易使体细胞逸出,但分离的胚囊因吸涨太大而难以持久存活。易于破裂,或稍经吸打即破。一般以8%甘露醇溶液冲击效果最好。较幼嫩材料或酶解程度较大时可提高到10%。对照组(13%)只逸出未裸露的胚囊,需经较强的吸打方可使胚囊裸露。

渗透压冲击时间　由表3可见,冲击15~30min效果较好,时间更长反而不利,主要问题是已逸出的胚囊容易被吸打破坏。此外,胚珠充分吸涨后若仍未能释放胚囊,继续延长冲击时间亦难奏效。

三、一步法分离胚囊的条件

以一步法在低渗条件下直接酶解,亦可获得一定数量的胚囊。由于处在酶解和低渗双重作用下,处理时间不宜过长,渗透压不宜过低。由表4可见,一步法分离胚囊,其酶液中甘露醇浓度以10%较好,不宜低于8%。幼嫩材料可提高至11%。在实际操作中,为了获得更多生活胚囊,可先挑选收集已分离出的胚囊,然后吸打剩余的悬浮液,再收集。如此反复几次,可提高胚囊获得率。

四、低渗条件下卵细胞与助细胞原生质体的原位融合

在用此法分离胚囊的过程中,发现部分胚囊内发生卵器中各个细胞原生质体互相融合的现象。多数是3个细胞的原生质体融合(图版Ⅰ,6;图版Ⅱ,12),也有两个细胞的原生质体融合。两两融合的组合是随机的,可以是卵细胞和一个助细胞融合,也可以是两个助细胞融合(图版Ⅰ,7;图版Ⅱ,12)。中央细胞虽亦原生质体化,但通常不参与卵器原生质体的融合。其融合过程是:在低渗溶液中卵细胞与助细胞原生质体由于吸涨而彼此紧贴,进而融为一体。融合体中,各个细胞原来

处于不同位置的细胞质互相靠拢,被新汇集而成的大液泡挤向一侧。初时尚可见到各个细胞的细胞质界限,以后细胞质不易分辨,但细胞核仍是分开的(图版Ⅰ,6)。FDA 测定显示融合体荧光明亮且经久不衰。甚至当未融合的胚囊细胞原生质体已经丧失荧光以后,融合体仍然保持荧光。从表5可见,在近等渗的13%甘露醇溶液中酶解,融合率很低,即使4.5h后仍仅9.4%;采用一步法,在10.5%甘露醇溶液中酶解2.5h,融合率即达15.7%;采用二步法,在8%甘露醇溶液中冲击25min,融合率为35.3%。4h后,融合率高达71.9%。以上结果表明:第一,低渗条件对融合有明显诱导作用。第二,在低渗溶液中时间越长,融合率越高。因此可以通过人为改变溶液渗透压,增加或减少融合率。

五、胚囊及卵细胞原生质体的富集

与用其它方法分离胚囊的情况不同,用本法分离的产物中,体细胞杂质相对不多。以自制的细口径吸管在倒置显微镜下吸取胚囊,每小时可收集50~70个,形成富集的群体(图版Ⅱ,10)。收集的胚囊可暂时保存于13%甘露醇溶液中,放置3~4h仍有生活力(图版Ⅱ,11)。

卵细胞原生质体的分离可有两种方法:一是机械分离法。用解剖针刺破胚囊角质壁及中央细胞,释放出卵器。另一是将收集的胚囊在含10.5%~11.5%甘露醇配制的稀酶溶液中稍加振荡后静置,两小时内可见有胚囊原生质体游离(图版Ⅰ,8)。往往反足细胞最先脱离,然后是中央细胞,最后是卵器。在一定时间内,卵器的3个原生质体常连在一起(图版Ⅰ,9)。由于卵细胞原生质体通常略小于助细胞原生质体,这种3个原生质体暂时相连的状态无疑有利于从中识别卵细胞而加以挑选。

讨 论

本方法是根据胚珠结构上的特点而试验成功的。胚珠的全部表面均被角质层覆盖,酶液渗入胚珠的最佳途径是通过珠柄切口和珠孔两个渠道。在酶解和低渗双重作用下,初步酶解和吸涨的胚珠内部体细胞组织裹挟胚囊从这两处冲出,因而只需经过短期酶解就能分离胚囊。

与已有的胚囊分离方法相比,此法的优点在于:第一,用酶浓度低。以1%~1.5%纤维素酶和0.5%果胶酶即可。前人用酶浓度一般较高且常采用作用强烈的酶类。如:2%的纤维素酶与果胶酶[1];2%崩溃酶[3];2% cellalysin[13]。第二,时间短。一般一个上午即可获得大量胚囊。前人则需较长时间酶解和后续步骤[1,3,5,6,13]。以上两点均有利于保存胚囊的生活力,便于进一步操作。第三,此法大大减少了酶解产物中体细胞杂质的数量,故易于识别与收集胚囊并由其分离卵细胞原生质体。在雄性细胞(生殖细胞与精细胞)的分离技术中,渗透压冲击是一种有效的方法[2,10,14]。现在这一原理又首次用于分离雌性细胞取得成功,这在方法学上是有革新意义的,值得今后在其它材料中加以试验。

关于卵细胞与助细胞原生质体原位融合的现象,对雌配子操作是一种不利因素。但这一现象也有其研究意义与利用价值。单倍性的卵器细胞通过彼此融合成为二倍体或三倍体,可以作为一种新的细胞资源。由于不涉及化学药品及电刺激,其融合体生活力强而持久,可能是培养与操作的良好起点。也可以借此探讨生殖生物学的一些理论问题,如:卵细胞与助细胞融合体能否启动发育,"卵-助"融合与"助-助"融合两种融合体的发育前途有何异同,等等。这类研究将有助于增进对卵细胞、助细胞的功能及受精机制的认识。

在体细胞原生质体融合的文献中,前人已注意到低渗条件有利于融合,甚至起关键作用[7]。本实验同样证明低渗条件显然比接近等渗条件更有利于诱导卵细胞与助细胞原生质体的融合。不同的是,由于卵细胞与助细胞原生质体在原位自然聚集,仅给予低渗条件即可诱导其融合。因此,通过控制渗透压调节剂浓度及处理时间,可以按照不同的研究目的有意提高或降低融合频率。

图 版 说 明

除图5和11为荧光显微摄影外,其余均为干涉差显微摄影。A. 反足细胞

第二章 雌性器官与细胞的实验研究

图版 I   plate I

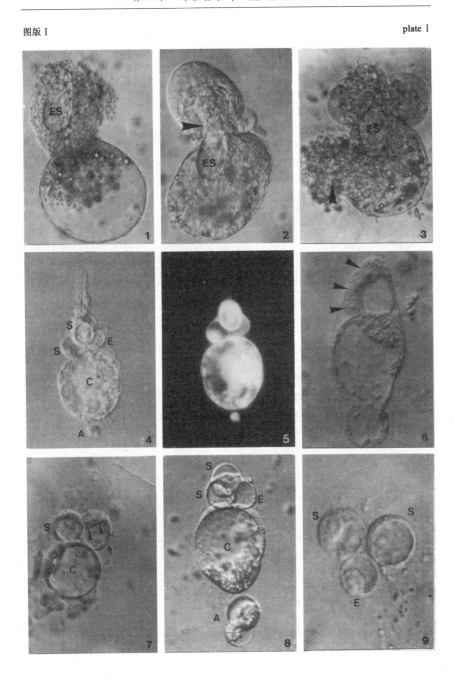

图版 II plate II

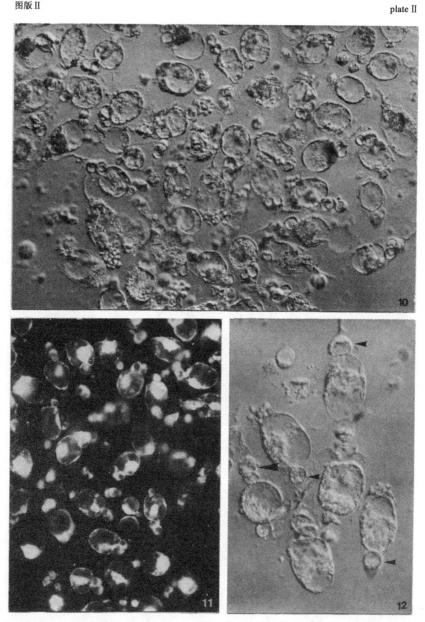

C. 中央细胞　E. 卵细胞　ES. 胚囊　S. 助细胞

图　版　Ⅰ

1~3. 胚囊逸出的不同形式。×200　1. 胚珠内部体细胞组织连同胚囊全部逸出。2. 胚囊从珠柄切口(箭头)逸出。3. 胚囊从珠孔逸出(箭头示珠柄处)。4. 分离的胚囊。×300　5. 与 4 同一胚囊,FDA 测定示胚囊细胞生活力。6. 卵器中 3 个细胞原生质体融合(箭头分别示 3 个核的位置)。×350　7. 卵细胞与一个助细胞原生质体融合(箭头分别示两个核的位置)。×366　8. 分离的卵细胞、中央细胞及反足细胞原生质体。×350　9. 分离的卵细胞与助细胞原生质体。×370

图　版　Ⅱ

10. 收集的胚囊群体。×180　11. 胚囊群体,FDA 测定示胚囊生活力。×140
12. 低渗诱导后的胚囊群体(大箭头示卵器中两个细胞融合;小箭头示三个细胞融合)。×235

Explanation of plates

All are Nomarski interference contrast micrographs except Fig. 5 and Fig. 11 which are fluorescence micrographs. A. Antipodal cell　C. Central cell　E. Egg cell　ES. Embryo sacs　S. Synergid

Plate Ⅰ

Figs. 1-3. Different modes of embryo sac (ES) release. ×200　**Fig. 1.** Release of all ovular cells together with ES.　**Fig. 2.** Release of ES through funicle (arrowhead).
Fig. 3. Release of ES through micropyle (arrowhead indicates the site of funicle).
Fig. 4. Isolated living ES. ×300　**Fig. 5.** The same ES as in Fig. 4 showing bright FCR of its component cells. ×300　**Fig. 6.** Fusion of three egg apparatus cells (arrowheads show position of three nuclei). ×350　**Fig. 7.** Fusion between the egg cell and one synergid (arrowheads show position of two nuclei). ×366　**Fig. 8.** Isolated protoplasts of egg apparatus, central cell and antipodal cells. ×350　**Fig. 9.** Isolated protoplasts of egg cell and two synergids. ×370

Plate Ⅱ

Fig. 10. A population of isolated ESs collected by a hand-made pipette. ×180

Fig. 11. A population of isolated ESs showing their FCR. ×140　**Fig. 12.** ESs with fused egg apparatus cells (big arrowhead indicates two-cell fusion; small arrowheads indicate three-cell fusion). ×235

(作者:孙蒙祥、杨弘远、周嫦。原载:植物学报,1993,35(12):893~900。图版2幅,表5幅,参考文献15篇。选录图版Ⅰ、Ⅱ。)

其它论文目录

1. 周嫦,杨弘远.诱导水稻胚囊产生愈伤组织的离体实验.遗传,1981.3(5):10~12.
2. Zhou C. Yang H Y. Yan H. Cai S. 1983. Factors affecting callus formation in unpollinated ovary culture of rice. In: Cell and Tissue Culture Techniques for Cereal Crop Inprovement. Science Press, Beijing, Gordon and Breach Science Publishers, New York, 81-94.
3. Zhou C. Yang H Y. Tian H Q. Liu Z L. Yan H. 1986. *In vitro* culture of unpollinated ovaries in *Oryza sativa* L In: Haploids of Higher Plants *in Vitro* (ed. by Hu H and Yang HY), China Academic Publishers, Beijing, Springer-Verlag, Berlin Heidelberg New York Tokyo, 165-181.
4. Yang H Y. Zhou C. Cai D T. Yan H. Wu Y. Chen X M. 1986. *In vitro* culture of unfertilized ovules in *Helianthus* annuus L. In: Haploids of Higher Plants in Vitro (ed. by Hu H and Yang H Y). China Academic Publishers, Beijing, Springer-verlag, Berlin Heidelberg New York Tokyo, 182-191.
5. 何才平,杨弘远.水稻子房培养中助细胞无配子生殖的稳定性及其发育条件的探讨.武汉植物学研究,1988.6(4):203~208.
6. 阎华,周嫦,杨弘远.向日葵未受精胚珠培养中各种影响因素的实验研究.武汉植物学研究,1988.6(4):319~326.
7. Yang H Y. Zhou C. Tian H Q. , Liu Z L. , Huang Q F. Cai D T. Yan H. 1988. *In vitro* gynogenesis in unfertilized ovaries and ovules. In: Genetic Manipulation in Crops. International Rice Research Institute and Academia Sinica, Cassell Tycooly, 58-59.

8. 阎华,杨弘远. 向日葵离体孤雌生殖的超微结构研究. 植物学报, 1989.31(1):1~6.

9. 华琳,杨弘远. 向日葵未受精胚珠培养再生植株. 武汉植物学研究, 1991.9(3)299~300.

10. 田惠桥,杨弘远. 韭菜胚囊发育与胚胎发生. 武汉植物学研究,1991. 9(1):5~10.

11. 田惠桥,杨弘远. 韭菜幼小子房培养时胚囊的发育和一些异常现象. 实验生物学报,1991.24(2):169~173.

12. 周嫦. 几种植物生活胚囊的分离与鉴定. 植物学报,1985.27(3): 258~262.

13. Zhou C. Yang H Y. 1985. Isolation of embryo sac by enzymatic maceration and its potential in haploid study. In: Genetic Manipulation in Crops Newsletter (ed. by Swaminathan MS, Hu H and Shao Q Q), Chinese Agricultural Science and Technology Press, Beijing, 68-69.

14. Zhou C. Yang H Y. 1986. Isolation of embryo sacs by enzymatic maceration and its potential in haploid study. In: Haploids of Higher Plants in Vitro (ed. by Hu H and Yang H Y), China Academic Publishers, Beijing, Springer-Verlag, Berlin Heidelberg New York Tokyo, 192-203.

15. 吴燕,周嫦. 杏叶沙参胚囊的酶法分离. 受精、合子生长与胚乳发育. 植物学报,1988.30(2):210~211.

16. 吴燕,周嫦. 桔梗科四种植物分离胚囊的比较研究. 武汉植物学研究,1988.6(4).315~318.

17. 周嫦. 用酶解方法观察卷丹的贝母型胚囊发育过程. 武汉植物学研究,1989.7(3):297~299.

18. Wu Y. Zhou C. Koop H U. 1993. Enzymatic isolation of viable nucelli at the megaspore mother cell stage and in developing embryo sacs in *Nicotiana tabacum*. Sex Plant Reprod. 6:171-175.

第三章 授粉、受精与胚胎发育的实验研究

提要

本章无论在内容上还是在我们研究所经历的时间上跨度都很大。就内容而言，既包括开花习性、人工授粉技术和受精及胚胎发育的研究，又包括雌、雄配子离体融合与合子、中央细胞及幼胚离体培养的研究。就研究历程而言，从20世纪50年代末到90年代末整整经历了40年，其间包括60年代和90年代两个高峰。实际上，我们关于受精与胚胎发育的研究并不限于本章，在其它章节中也有所涉及，如第一章第二节中关于花粉萌发与花粉管生长、第四节中关于脱外壁花粉的离体授粉、第六节中关于精细胞操作、第二章第二节中关于胚囊操作、还有第四章中有关的超微结构与细胞化学研究，都与这一主题有直接或间接的关系，只是由于考虑到本书结构的合理性，将它们纳入其它章节更为合适。

第一节论述的是几十年前的早期研究工作，以现代眼光来看显然是陈旧的。之所以收录它们是出于下面的理由：第一，这些工作中采用的方法不限于当时流行的描述胚胎学方法，而是应用了实验的方法，打下了我们此后数十年从事实验生殖生物学研究的方法学基础。尽管技术手段多属过时，但研究思路和方法与后来的工作是一脉相承的。第二，这类工作也取得了若干有一定创新意义的结果，当时所提出的有些问题至今也未见完满的解答，积累的某些基础性资料也还有一定参考价值。

第二节是一个比较集中的专题，包括20世纪90年代中期以来在烟草上所开展的离体受精研究。在国际上迄今只有德国和法国两个研究单位能够进行这方面的研究。我们应用自行研制的简易方法步入了这一前沿课题的行列。本节所反映的仅仅是这项研究的开始，以后还会继续深入下去。

第三节的工作是围绕一个总的目标，即建立合子与中央细胞的离体培养实验系统，以便从植物个体发育的起始就能进行实验研究。在烟草与水稻中我们已经创建了这样的实验系统，开展了初步的细胞生物学、分子生物学与遗传转化研究。这个方向的研究也还将在21世纪继续深入下去。

第一节 授粉与受精

提要

1959年我们发表了自己第一篇研究论文《水稻去雄方法的初步研究》,其中报道了"温汽去雄"和"二甲苯去雄"两种方法。"温汽去雄"是根据俄文文摘杂志中一项二手信息,报道有人利用一种"真空容器"中的湿热气体进行水稻去雄。我们不知"真空容器"为何物,于是试用保留热气的热水瓶倒扣稻穗,去雄效果很好。此法后被湖北省农科所所采用,代替沿用的"温汤去雄法"推广到水稻杂交育种工作中。"二甲苯去雄"则纯粹基于我们自己一次偶然的发现:当用彩色油漆标记稻花以观察开花顺序时,我们注意到被标记的稻花有提前开颖的奇特现象。我们推测可能是油漆中的有机溶剂起作用,因而试验了多种有机溶剂,以它们的气体刺激稻花,居然有不同程度的催花效果。其中二甲苯不仅催花,而且延迟花药开裂,且对雌蕊结实损害较小,比较符合去雄方法的要求。后来在"文革"期间又重复试验,证明至少在籼稻中许多有机溶剂有明显的催花效果(未发表资料),但就去雄而言,尚难达到推广的程度。这项研究从此湮没无闻。40年后的今天重新审视,觉得换一个角度来看更有意义:促进稻花开颖的因素,已知主要有温度或光照的变化,属于物理因素,而化学试剂的气态分子能够催花则是前人未曾发现的新现象,似乎至今也未见到这方面的研究报道,更不明了内中机理,值得继续探索。

关于小麦受精的胚胎学研究,中外都做了不少工作。我们在1964年发表的论文中,通过对人工授粉后系列切片的观察统计,研究了受精过程各个时段的动态变化、几种不同授粉方式下受精速率的差别以及大量授粉与限量授粉时"附加花粉管"的频率。与一般描述性研究不同的是,该文以实验处理与计量方法相结合研究受精的特点,这也是我们迈入实验胚胎学的开始。

20世纪60年代我们的重点研究材料是芝麻。围绕它的开花授粉习性、人工去雄方法、花粉萌发与花粉管生长,尤其是限量授粉对受精及其后果的影响等问题,开展了一系列研究。我们提出的芝麻人工杂交与人工自交改良技术,后被中科院油料所采纳应用于他们的育种工作中。芝麻雌蕊"年龄"与花粉生活力的研究,为有性杂交添增了一些新的基础资料。我们报道了芝麻花柱属于具引导组织的闭合型;测量了花粉管在花柱中的生长速度和进入子房的时间;发现了花粉管在花柱中大量破裂的现象,并通过实验研究提出这是一种自然生理现象的观点。

关于花粉数量对芝麻受精结实、胚胎发育和后代的影响的研究,历时三年,完成于"文革"前,但直到10年后的1975年方才发表。这项研究的初衷是企图在芝麻上重复当时有些苏联研究者的实验,印证他们提出的限量授粉可以作为一种诱变手段的理论。我们以花粉粒较大、花粉生活力较强的芝麻为材料,采取比他们更精细、规模更大的实验方法,分别在柱头上进行1粒、4粒、10粒花粉等不同数量级别的人工授粉。结果发现:无论授予几粒花粉均可结实,并且原则上每一粒花粉可以完成受精,结成一粒种子;花粉数量减少导致胚胎发育早期进度迟缓和不同步,但后期接近甚至赶上大量授粉对照;限量授粉的F_1与F_2后代植株在生活力与遗传表现上也和对照没有区别。实验得出了与预期相反的结果,也改变了我们对当时流行的"受精多重性"的看法。这篇论文记录了作者青年时代的求索精神、求实态度和严谨方法,是我们科学道路上的一块奠基石。

1 水稻去雄方法的初步研究

A preliminary study on the emasculation method in rice

在进行水稻的选种工作和遗传学研究时,经常会碰到一个具体问题,就是杂交技术的问题。由于目前水稻的杂交育种工作已经日益提上重要的地位,特别是近年来群众性的杂交工作纷纷展开,因此,这个问题的解决,就有更加迫切的意义。

水稻的杂交技术中,去雄是比较困难的一环。50多年来,已经提出了六七种不同的去雄方法,但是都还不够理想。以历来在我国应用较广的温汤去雄法为例,虽然优点很多,但也有操作不便、易损稻穗、不易择定适宜的处理温度和时间、难保杀雄作用完全等缺点。因此,不断改善水稻的去雄方法,仍然十分必要。

近年来,万安良与蒋少芳曾经各自设制出一种结构相同的夹层温水筒,以筒中的温气处理稻穗,试验结果较温汤去雄为好[3,4]。此外,据报道,约当(Jordan H. D.)在非洲西部曾经利用一种真空容器(сосуд-вакуум)中的湿热空气影响稻穗,达到开花杀雄的目的[8]。

我们对水稻的去雄方法也做了一些新的试验。试验主要沿着两个方向进行:一方面,为了改进现有的方法,我们根据高温促进水稻开花和杀死花粉、高湿阻碍花药破裂的原理,设想用普通热水瓶中保存的温热水蒸气来处理稻穗,达到催花杀雄的目的。另一方面,也着手探求新的去雄途径,即利用一定的药剂来处理稻穗,为此,我们试验了几种挥发性的药剂:特别重点地试验了二甲苯(xylol)的效能。我们希望,通过继续的研究,能够找到一种有效而方便的水稻去雄方法,符合于既能彻底去雄、又能提高雌蕊结实率的要求。

材 料 与 方 法

试验于1959年春秋两季在本系实习农场进行。试验材料以晚粳老来青为主,早粳青森五号、中籼马尾粘、晚糯荣怕黑等为辅。全部试验植株除马尾粘外一律盆栽,栽培管理方法般。

温汽去雄法于水稻开花期中每日上午进行。所用器具有:五磅装大热水瓶、一磅装小热水瓶、温度计、铁支架、剪刀、镊子等。工作时,首先将大热水瓶中的水温调节至一定高度(水温高于所需要的汽温 2~4℃,视当时外界气温高低而定)。将此温水盛入小热水瓶中片刻,复又倾回原瓶中。此时,小瓶中即保存一定温度的水汽(根据我们的观察,瓶内汽温在 10min 以内极少下降)。然后,将小瓶倒置于事先选定的稻穗上,使后者全体处于温汽的包围之中。为了节省人力,小热水瓶可以安放在普通实验室用的铁支架的铁环上。这样,经过一定时间之后,揭开小瓶,便可发现稻花开放。采用的汽温有 43~44℃、45~46℃、47~48℃ 三种,处理时间一律为 10min。包括下列三个试验处理:

处理1:不去雄不授粉:温汽促进开花后,将穗上所有未开放的花全部剪去,随即套以隔离纸袋。

处理2:去雄不授粉:促进开花后,用镊子自上而下细心地除去穗上所有开放的花中的花药,并剪去所有未开放的花,然后套袋。

处理3:去雄授粉:人工去雄后,当日上午以同一品种的花粉授粉。取粉与授粉方式是:用黑布遮光促使稻花提前开放,或稻花自然开放时,镊取花药置入去雄后的小花中。

以上处理的目的是测验温汽对花粉和雌蕊的影响,确定适宜的汽温,并阐明既经温汽处理之后,是否仍有人工去雄以彻底防止自花授粉之必要。

二甲苯去雄法的工作时间和用具与前法类似,但不需要大热水瓶,小热水瓶也可用普通玻瓶代替。方法为用一小团棉花吸收少量的二甲苯液体,置入小热水瓶中,然后如前法将小瓶倒扣于稻穗上,并用铁支架撑持。每穗处理时间 5~10min,一般 7~8min。试验处理也包括以上三种。

根据我们的试验,两人轮流使用两个小热水瓶,温汽去雄一上午可完成20穗左右,二甲苯去雄则完成穗数可能更多。外界气温较高时,大热水瓶中的水温,一上午调节一次即可。

结　果

一、温汽去雄法

促进开花的效果　在 43～44℃、45～46℃、47～48℃ 的汽温影响下,稻花都能提前开放。同时,上述汽温的差别,对于促进开花的数目关系不大。但不同的品种,温汽处理的效果可能很不一致,大穗型的中籼马尾粘,开花的数目较粳稻青森五号与老来青高出一倍以上。同一品种,由于处理时间的迟早不同,效果亦异,一般愈接近该品种自然开花期者,开花的数目愈多(表1)。此外,根据我们的观察,同一品种的不同稻穗,对温汽的反应也不一致,壮大的稻穗一般较瘦小的容易开花,开花数目也较多;极个别的稻穗,在处理时间较早的情况下,甚至当时一朵花也不开放。以上种种情况,说明温汽处理和温汤处理的作用有相似的规律[1];但是,根据我们的统计,温汽促进开花的效果,并不直接依赖于处理当日的外界天气状况(表2),这一点,似又有别于温汤处理[1],而具有万安良等所提出的温汽处理的优点[3]。

去雄的效果　温汽处理对于花粉的杀伤力,各个品种程度有所不同。例如43～44℃,处理1(不去雄不授粉),籼稻马尾粘结实率为0,粳稻青森五号为2.1%,粳稻老来青则为20.6%。同一品种,汽温愈高,对花粉的杀伤力愈大。例如老来青处理1,43～44℃,结实率为20.6%,45～46℃,降为11.0%,47～48℃,更降为5.8%(表3)。由此可见,单纯施用温汽影响,虽能杀死大部分花粉,但也如万安良试验温汽杀雄的结果相仿,对于某些粳稻作用很不完全[3]。然而,根据我们对温汽处理后开放的花朵的显微镜检查,发现由于在高温高湿条件下花药的开裂延迟,如果及时地镊去花药,柱头上一般没有花粉粒。事实上,如果在温汽处理后,随即进行人工去雄,则可获得良好效果。例如老来青处理2之结实率,43～44℃时可低至0.85%(表4)。仔细地分析了这些个别结实花朵的情况之后,我们发现,它们大部分出现在开花较多(21~24朵)的穗上,而且全都位于穗的最基部。这可能由于穗基部靠近瓶口的花蒙受的汽温过低,花粉未被杀死,加上穗上花数太多,以致当人工去雄自上而下进行时,基部花朵去雄不够及时。因此,如果今后整穗时保留花数不必太多,或者改为自下而上去雄,估计可以达到去雄更加彻底的效果。至于青森五号虽经人工去雄仍有4.5%结实,可能还和工作时间太迟(已达自然开花时间),个别花朵在处理前即已裂药有关。

对雌蕊结实率的影响 一定的温度,既能杀死花粉,又不伤害雌蕊,在温汽处理的试验中也得到证实。青森五号处理3,43~44℃,结实率为57.8%,老来青处理3,46℃以下的汽温,结实率也有42.9%(表5)。这说明,在对雌蕊授粉结实率的影响方面,温汽去雄和一般的温汤去雄效果不相上下[7]。

二、二甲苯去雄法

促进开花的效果 老来青经5~10min(一般7~8min)二甲苯气体影响后,每穗平均开花10.9朵,较温汽处理结果(9.2朵)良好。处理时愈接近自然开花,促进开花数也愈多,但处理较早时,却能较温汽去雄促使更多的花朵开放(表6)。此外,我们在青森五号上还观察到一个有趣的现象,即在下午自然开花终止之后进行二甲苯处理,甚至也能促使少数花朵开放。由此可见,二甲苯处理较之温汽处理,在促进开花方面有以下优点:第一,较短的作用时间可以取得较大的效果;第二,早晨开始工作的时间可以较为提前。

去雄的效果 老来青二甲苯法处理1,结实率为21.6%,类似于43~44℃的温汽处理结果。然而,处理后施行人工去雄,则结实率可降为0(表7)。据我们的观察,在二甲苯刺激下开放的稻花,花药大多长期不裂,甚至次日仍有不少花药包含着黄色的花粉。此外,仔细地分析了处理1的结实情况后发现,在二甲苯影响下当时开花的稻穗,结实率仅为3.4%;延迟一段时间后开花的稻穗,结实率提高为29.6%;而个别在处理同时即已裂药的稻穗,结实竟达72%之多。由此推测,二甲苯法之所以能够去雄彻底,与其说由于对花粉杀伤力大,不如说由于严重地阻碍了花药破裂,以至经过人工去雄以后,可以完全杜绝授粉的机会。当然,这一推测还有待于今后的证实。

对雌蕊结实率的影响　老来青二甲苯法处理3,结实率达68.4%(表7),远较温汽法为高。此外,根据表面观察,对籽粒的充实也无不良影响(以上情况,似乎也可作为二甲苯作用主要不在杀雄而在阻碍裂药的旁证之一)。因此,无论就去雄效果或就授粉效果而言,二甲苯法皆有很大的优越性,根据我们初步掌握的资料,可能成为一种很好的去雄方法。当然,为了证明这一点,还需要做广泛的重复试验,并且确保用此法处理后,种子后代并无异常表现。

三、其它挥发性药剂处理稻穗的效果

我们在用二甲苯法获得了良好结果之后,又在晚糯品种荣怕黑上粗放地观察了其它几种挥发性药剂的作用。由于时间不够,我们做的穗数不多(总共41穗),而且除福尔马林外,都只包括一个处理(不去雄不授粉)。根据初步观察,福尔马林气体对促进开花也有相当大的作用(平均每穗开花9.6朵),去雄效果也很好(处理2结实率为1.4%),然而对雌蕊毒害较大(处理3结实率为22.2%),甚至有使整个稻穗都变干枯的现象。其它药剂如酒精、乙醚、丙酮等的气体,也能促进少量开花(每穗平均3.2~4.2朵)。但是,因为我们工作的时间已是九十月之交,外界气温颇低,而且荣怕黑为新引来之外地品种,适应力较差,因此,这个数字只能作为进一步研究的参考。

结　　论

1. 利用热水瓶中一定温度的水蒸气处理稻穗,并结合进行人工去雄的方法,无论在促进提前开花、去雄及授粉结实方面,都有良好的效果。由于该法具有操作便利、不损稻穗、不受天气限制等特点,经过进一步改良之后,可能成为一个有效的水稻去雄方法。

2. 利用二甲苯气体处理稻穗,能在较短时间内促进较多的花朵开放。如果结合进行人工去雄,去雄效果可以彻底。另一方面,该法又能

保证雌蕊有很高的结实率。因而,就目前情况看来,是一种有前途的水稻去雄方法。除此之外,许多挥发性药剂,都能或多或少促进水稻开花。因此,沿着这一方向作广泛的试验研究,是有意义的。

(作者:周嬪、杨弘远。原载:武汉大学自然科学学报,1959,(1):24~29。图1幅,表7幅,参考文献8篇,均删去。)

2 小麦受精过程中若干问题的胚胎学研究

Embryological studies on some problems in wheat fertilization process

在被子植物受精过程的研究工作中,小麦是历来较为常用的材料之一。特别是20世纪50年代以来,工作更为广泛和深入。研究内容以受精和早期胚胎发育的形态学描述最多[32,10,21,31,38,6,7,8,11,12,36,1,2,3]。有些工作专门研究了气候与施肥等外界条件的变化对于受精和胚胎发育的影响[33,39,40,8]。近来,开始用组织化学方法追溯雌雄性细胞受精前后的生理生化变化,首先是DNA的变化[25,26,36,1,35,4,3]。此外,有些研究者还注意到多数花粉和"附加"花粉管在受精过程中的行为与作用[23,24,28,13,16,17]。总之,由于上述各方面研究的结果,许多旧的问题得到了澄清,同时也提出了一些新的问题,有待于进一步解决。

我们的研究,是想在前人工作的基础上进一步认识小麦受精过程的形态学特点,掌握在本地区条件下小麦受精过程进行的速度,同时初步探索在不同授粉方式下受精过程的差别,以便为进一步开展小麦受精和遗传的研究打下基础。

材 料 与 方 法

试验于1962—1963年在武汉大学生物系进行。总共对三批材料

1000 多张镜片进行了观察。供试品种为普通小麦(*Triticum vulgare* Host.)白玉皮和阿夫两个品种。

第一批材料:白玉皮品种内交配。1962 年 4 月 8、9 两日整穗去雄,每穗取中部发育良好的 20 朵小花于去雄后 5 日人工授以大量新鲜花粉。授粉后 0.5h、1h、2h、3h、4h、5h、6h、8h、10h、12h、24h 用 Randolph 改良的 Навашин 液和 Carnoy 液(3∶1)固定。前者石蜡切片后以海氏苏木精染色,部分切片用曙红复染。后者以整个雌蕊进行富根氏反应,然后埋蜡切片,部分切片用亮绿或固绿复染。另以一部分雌蕊不经盐酸水解进行富根氏反应做成对照切片。

第二批材料:白玉皮×阿夫品种间杂交。1962 年 4 月 18 日进行人工授粉,分大量授粉与限量授粉(20～30 粒花粉)两个处理。授粉后 9h、26h 及 51h 用 Randolph 液固定,海氏苏木精染色。

第三批材料:白玉皮×阿夫品种间杂交。1963 年 4 月 25 日授粉。分大量授粉、限量授粉(8～15 粒花粉)和衰老花粉授粉三个处理。衰老花粉是在田间阴暗条件下保存 3h 的花粉。授粉后 1h、3h、8h、24h、30h 用 FAA 固定。以整个雌蕊在爱氏苏木精中染色,然后埋蜡切片。

各批材料均于同时固定,以去雄后未授粉的小花作对照。切片厚度为 13μm、15μm 或 22μm。绘图借助于国产的显微描绘器。

结 果 与 讨 论

一、受精过程的形态学及其进行的速度

受精前的小麦胚囊,珠孔端有一个卵细胞、两个助细胞。一对很大的极核互相紧贴,并与卵器十分接近,其间有浓厚的细胞质联系。卵核和每一个极核中,各含一个明显的核仁。在胚囊的合点端是由十几个巨大细胞组成的反足器。反足细胞的核很大,核仁染色很深,且常呈种种不规则形态(图版 IV,24～31)。

授粉后半小时,在半数的胚囊内已看到花粉管进入珠孔,在助细胞内破裂,将一对精子和其它内含物质释入胚囊内卵器和极核之间。此

时两个精子互相靠近,可看到染色很深而呈螺旋状的精核(图版Ⅰ,1)。在有些镜片上,精核周围有一明亮的晕环(图1),以曙红复染,有时呈浅红色的一圈,疑是精子的细胞质。关于被子植物进入胚囊的精子是裸核还是细胞的问题,历来有所争论。在小麦方面,星川清亲(1959)[38]曾指出在胚囊内可稍稍看到精子的细胞质。胡适宜(1962)[1]亦看到精核外包有细胞质一层。其他研究者在桃、大麦、赛菊芋(*Heliopsis helianthoides*)等植物的胚囊内也看到精核外有一明亮的晕环[22,31,27],有人推测这实际上就是精子的细胞质。我们推测很可能是精子细胞质的成分比较特殊,因而在制片过程中不易保存,这一问题还有待于继续研究澄清。

授粉后1h,两个精子已经分开,其中之一进入卵细胞质,缓缓向核移动;另一精子则沿胚囊的细胞质向极核移动(图版Ⅰ,2)。授粉后3~4h,精子进入卵核,原来紧密的结构逐渐疏松开来,开始在卵核中缓缓地溶化。在富根氏反应的镜片上,受精前的卵核呈富根氏负反应或极微弱的正反应,精子呈明显的正反应,精子的入卵及在卵核中逐渐溶化的过程十分清楚。从授粉后4小时起,精子溶化处开始出现1~2个小核仁,其体积逐渐增大,形成可以明显辨认的雄核仁。精子染色物质的消散和雄核仁的产生,看来是相伴进行的,一直持续到授粉12h以后(图版Ⅱ,5~8)。Герасимова-Навашина[18]认为这表示原来处于有丝分裂末期的精核在静养状态的卵核的影响下逐渐转变为静养的状态。

精子进入极核较进入卵核略早,是在授粉后2~3h。此时精子通常进入两个极核中的一个,并与之融合。精子在极核中溶化和出现雄核仁的过程和精卵融合大体一样,但进行的速度则显著地快些,授粉后6h,100%的镜片上都看到极核中雄核仁的出现。此外,在极核中虽然一般也出现1~2个雄核仁,但在极核内出现两个雄核仁的情况,比在卵核内要多。有时极核内可出现等大的3个雄核仁,或者在1~2个雄核仁外尚有几个更小的核仁。

受精前后极核在胚囊中的位置发生很大的变动(表1;图版Ⅰ、Ⅱ)。授粉后当精子尚未接触极核时,即有部分材料的极核开始向反足器的

第三章 授粉、受精与胚胎发育的实验研究

方向移动。精子进入极核时,后者均已移动到卵器与反足器之间的中央位置。授粉后4h,受精的极核移动到反足器旁边,而以细胞质和后者紧密相接,此后极核与卵器之间的细胞质联系逐渐中断。授粉后8h受精的极核在反足器附近分裂,授粉后10～12h形成两个胚乳游离核(图版Ⅱ,7～8)。授粉后24h,四个胚乳游离核已经产生,并有一部分开始第三次分裂。这时,胚乳游离核又沿胚囊两侧向珠孔端移动,并且以细胞质带和卵器重新联系起来。

　　为了更清楚地说明受精过程进行的速度,特别是为了便于比较精卵融合与三核融合的速度,我们将第一批材料(海氏苏木精染色的镜片)的观察结果列于表2。由表中数字可以看出,当一个精子进入卵细胞质时,另一精子尚在极核之外。但随后精子进入极核却比进入卵核略早一些。至于以后的融合过程,则在极核内显然进行得非常迅速。授粉后24h,胚囊中已具备四个胚乳游离核,而合子尚处于静养的状态。以上观察结果与星川清亲(1959)[38]和 Батыгина(1962)[12]的资料基本一致。此外,根据我们对各批材料的观察,受精过程的节律基本上是稳定的,惟具体的时间变动较大,甚至相邻的前后两天授粉的材料也表现出一定的差别。至于不同气候条件下和不同年代所得到的结果,差别就更大(参看表4)。需要指出,由于我们所用的第三批材料是田间最后一批开花的植株,气温也较高,因而受精的速度偏快,受精过程的进行不够整齐。星川清亲(1960a,b)[39,40]曾用专门的实验证明气温和施肥条件的变化对花粉萌发、花粉管生长和受精过程的速度有很大影响。过高过低的气温、过多过少的氮肥均延缓受精的进行,而当两种不利因素共同作用时,延缓的效应是累加的。可以肯定,受精过程也和有机体内任何生理过程一样能够受到外界条件的影响。Модилевский 等(1958)[31]、Оксиюк 和 Худяк(1961)[33]一再指出小麦受精的速度不随气候条件而变化,对此我们不能同意。

515

表1　受精前后极核在胚囊中位置的变动(白玉皮,1962)

胚囊数　极核位置　时间	靠近卵器	开始向反足器移动	卵器和反足器的中间	靠近反足器	观察的胚囊总数
未授粉(对照)	10				10
授粉后 0.5h	7	4	1		12
授粉后 1h	3	7	3		13
授粉后 2h		1	11	2	14
授粉后 3h			5	3	8
授粉后 4h				15	15

二、不同授粉方式下受精过程的初步比较

和大量新鲜花粉授粉不同,限量授粉(8~15粒花粉)和衰老花粉授粉(在田间贮存3h的花粉)使受精率大为降低,特别是衰老花粉授粉的受精率更低(表3)。

比较三种授粉处理受精过程的形态学变化,没有看出显著的不同。但是受精的速度则有差别。虽然在不正常授粉的处理中观察的材料较少,由表4仍然可以见到,在不正常的授粉方式下,有一部分胚囊的受精速度和正常授粉(对照)下完全相同,而另一部分胚囊的受精则较对照显著延迟。这种情况在衰老花粉授粉的处理中似乎表现得较为明显。例如,在授粉后8h,对照处理的卵核内精子均已溶化,出现了雄核仁;极核受精后已分裂成2~4个胚乳游离核;而在衰老花粉授粉处理中有一张镜片,精子尚在卵核外,极核受精而未分裂。又如在授粉后24h的衰老花粉授粉处理中看到两张受精特别延迟的镜片(图版Ⅲ,10),此时精子刚刚进入胚囊,而在对照处理中合子已接近分裂,胚乳游离核达8~16个。

第三章　授粉、受精与胚胎发育的实验研究

表2　受精过程进行的速度（白玉皮，1962）

授粉后时间(h)	观察的胚囊总数	受精各时期占胚囊总数的百分率(%)											
		精卵融合					三核融合与胚乳发生						
		精子位于				雄核仁出现*	精子位于			雄核仁出现*	初生胚乳核分裂	胚乳2核	胚乳4核
		胚囊内	卵质内	卵核边	卵核内		极核外	极核边	极核内				
0.5	12	50					50						
1	17		82.4	17.6			100						
2	15		8.3	75.0	16.7		6.7	40.0	53.3				
3	15			53.8	46.2			13.3	86.7				
4	13			7.7	92.3	15.4			100	54.5			
5	18				100	5.5			100	86.7			
6	10				100	33.3			100	100			
8	18				100	28.6			33.3	100	66.7		
10	23				100	27.8						100	
12	21				100	75.0						100	
24	17				100	89.0							100

＊此项百分比数系单项计算。

表4 不同授粉方式下受精过程进行的速度（白玉皮×阿夫，1963）

授粉后时间(h)	处理	观察的胚囊总数	精卵融合与胚胎发生							观察的胚囊总数	精子位于精卵融合与胚乳发生				胚乳发生			
			精子位于				合子	二胞胚		胚囊内	极核外	极核边	极核内	2个	2个分裂中	4个	8~16个	多个
			胚囊内	卵质内	卵核边	卵核内												
1	对照	20		14	6				18		9	5	4					
	限量花粉	1		1					1			1						
	衰老花粉	1		1					1			1						
3	对照	21			5	16			19				19					
	限量花粉	3				3			2				2					
	衰老花粉	1				1			1				1					
8	对照	25				25			22					12	6			
	限量花粉	14				14			14	1		1	1	1	11	1		
	衰老花粉	3	1			2			3									
24	对照	27					27		27									
	限量花粉	6	1				5		6	1						1		
	衰老花粉	2	2						2	2								
30	对照	21						21	21								27	21
	限量花粉	5						5	6				1			1	4	5
	衰老花粉	1						1	1									1

在文献资料中,有一些关于限量授粉[15,37,34]和衰老花粉授粉[5,30]引起植物花粉管生长和受精发生延缓的报道。很可惜,这些资料多数只进行了某些现象的描述,而未提出可供分析的统计数字,因而对于延缓的程度及其普遍性很难有一全面的了解。至于小麦方面,我们还没有看到类似的胚胎学研究。通过我们自己的初步观察,发现在不正常授粉条件下情况相当复杂:大部分材料不能受精;而在能够看到受精现象的材料中,有些与正常授粉一致,有些则出现受精显著延迟的现象。尽管目前我们还不能明确指出造成同一处理内发生如此巨大差异的原因,但至少可以启发我们今后作类似的胚胎学研究时应该重视数字的统计与分析,力求避免把个别现象与一般现象混淆起来。

三、几个问题的商榷

雄核仁的数目能否代表进入雌核的精子数目? 没有受精的卵核一般只含一个核仁(雌核仁),精子进入卵核以后才可产生新的核仁(雄核仁)。过去一般即根据雄核仁的出现来判断受精是否在进行。并且有人认为,雌核内出现的雄核仁数目即等于进入雌核的精子数目,当看到两个以上的雄核仁时,便作出多精入卵的推测。近来有不少研究者[4,12,26,31]倾向于否定这种看法。我们的观察也证明,无论是卵核内或极核内,雄核仁的数目并不经常是一个。在第一批材料中,有26.8%的受精卵核内出现两个雄核仁;在第三批材料中这一数值更高(42.6%)。从时间上来看,两个雄核仁几乎是同时产生的(表5);从空间上来看,它们通常出现于一团正在溶化的精子物质附近。因此,没有任何理由认为这是两个精子入卵的结果。何况在我们所观察的材料中,超过一对精子进入胚囊的情形是比较少见的(总共只发现两个例子),更没有理由认为这是多精入卵的标志。除此以外,在没有授粉和授粉后精子尚未进入卵核的情况下,有时还看到卵核中除了一个大型的雌核仁外,尚有个别较小的核仁(例如图版Ⅰ,2)。在极核中也有这种附加小核仁的出现。这种小核仁显然不是由精子进入雌核所引起的,但其大小和染色程度和刚产生的雄核仁并没有严格的区别。这一点,在判断雌核是否受精时尤其值得注意。

图版 I

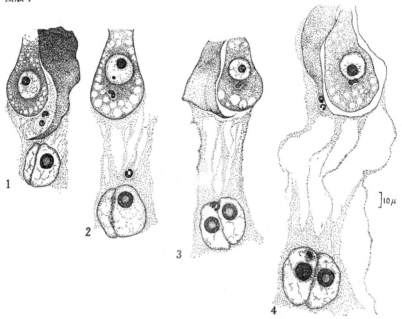

白玉皮品种内交配,大量新鲜花粉授粉,海氏苏木精染色。
1. 授粉后 0.1h,花粉管通过助细胞将一对精子释放到卵细胞和极核之间。
2. 授粉后 1h,一个精子在卵细胞质中,另一个精子趋近极核。
3. 授粉后 2h,一个精子靠近卵核,另一个精子紧贴极核。
4. 授粉后 3h,一个精子紧贴卵核,一个精子进入极核并开始溶化;另有一对弯曲的附加精子进入胚囊。

合子内雌雄核仁是否融合?(从略)

胚囊细胞质在受精中的作用 小麦成熟胚囊中卵器、极核和反足器三者之间,有细胞质联系的事实很多研究者都看到了。Александров 与 Александрова(1952)[9] 用活体观察方法发现这一细胞质联系实际上是一条运动着的细胞质流,推测它担任着将养料由反足器转运到珠孔端的功能。我们同意这一看法。受精后极核移到反足器旁边,胚囊中的细胞质也集中到这一区域,在极核和反足细胞之间形成浓厚的细

图版Ⅱ

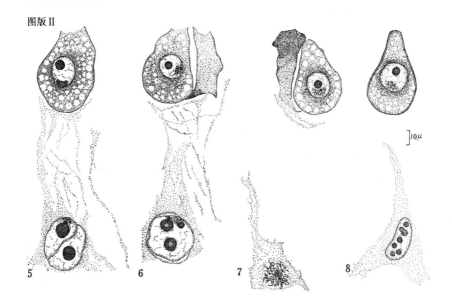

白玉皮品种内交配,大量新鲜花粉授粉,海氏苏木精染色。

5. 授粉后4h,一个精子进入卵核并开始溶化,另一个精子在极核中溶化,出现一个雄核仁。
6. 授粉后6h,一个精子在卵核中溶化,出现一个雄核仁;另一个精子在极核中溶化,出现两个雄核仁。
7. 授粉后8h,精子在卵核中溶化,出现一个雄核仁;初生胚乳核正在分裂。
8. 授粉后10h,精子在卵核中溶化,出现一个雄核仁;胚乳游离核两个(图中只有一个)。

胞质联系,而在极核和卵器之间的细胞质联系暂时中断。接着,受精的极核很快开始分裂,而卵核中精子的溶化却进行很慢。我们认为三核融合和胚乳发生速度较快的原因正是由于周围的细胞质将养料由反足器集中供应给它,而精卵融合速度较慢、合子"静养期"长的原因之一也正是由于在卵和反足器之间暂时失去细胞质的联系以致养料供应不足。当胚乳核进行了两次分裂以后,它们又逐渐向珠孔端移动,在合子和反足器之间又恢复了细胞质联系。星川清亲(1961a)[6]明确地指出,这种状况为合子的营养提供了条件,从而导致它的分裂。我们认为,进一步研究胚囊细胞质在受精过程中的动态是有必要的。

不少研究者都看到极核受精后向反足器移动的现象,但是我们注意到这一移动实际上发生在精子与极核接触之前(表1),并且在授粉而未发生受精的一部分胚囊中也看到这种移动。这促使我们推测,引起这种反应的刺激是由胚囊细胞质所传达的。胡适宜(1962)[1]发现小麦成熟胚囊各种分子的细胞质中均有线粒体存在,卵核和极核附近特别密集,反足细胞中也十分丰富。众所周知,线粒体是能量代谢的中心。很难设想在受精过程中,生理上如此活跃的细胞质不起重要的作用。

附加花粉管的行为 在大量授粉处理中,受精后相当长时期内,有一些胚囊的珠孔中尚可看到一根花粉管(图版Ⅲ,9)。在其它植物上也有类似的报道,但 Магешвари[29]认为这乃是第一个花粉管所留下的痕迹。为了判断这是否为参加双受精的第一个花粉管,我们比较观察了大量授粉和限量授粉两种处理中"多余"花粉管的情况(表6)。结果表明,在限量授粉情况下,不仅在子房中生长的"附加"花粉管很少,而且在珠孔中一般看不到有花粉管的痕迹。这使我们相信,在大量授粉时受精完成后珠孔中所出现的花粉管,在很多情况下不是最初那个花粉管的痕迹,而是以后进来的"附加"花粉管。

不仅如此,在个别情况下,我们还看到受精后有附加精子进入胚囊。例如在第一批材料授粉后3h的一个胚囊内,极核和卵细胞中已有一对精子,而在胚囊中又出现一对刚刚进来的卷曲状精子(图版Ⅰ,4)。在第三批材料授粉后8h的一个胚囊内也看到类似的情况。Модилевский 等(1958)[31]曾强调指出,他们观察过上百次小麦受精的情形,从未发现"附加"花粉管进入胚囊。但是,Батыгина(1962)[12]、胡适宜(1963)[2]都承认附加花粉管可能进入小麦胚囊。Ключарева(1957a,1962)[23,26]指出"附加"花粉管常在双受精完成后进入胚囊。她还对附加精子在胚囊中的行为进行了描述。所有这一切,使我们相信在大量授粉条件下"附加"花粉管进入胚囊是可能的。但是为什么

图版 Ⅲ

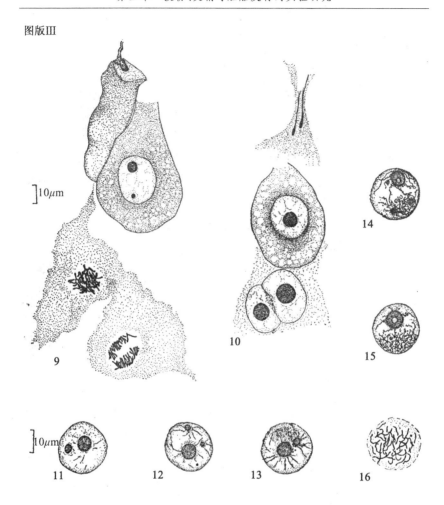

白玉皮品种内交配,大量新鲜花粉授粉,海氏苏木精染色。图10~16 白玉皮×阿夫,爱氏苏木精整体染色。

9. 授粉后24h,合子中有一个雌核仁和一个雄核仁;四个胚乳游离核正在分裂(图中只有两个胚乳核分裂);有一根附加花粉管进入胚囊。

10. 储藏3h的衰老花粉授粉后24h,卵和极核尚未受精;一对细长的精子刚进入胚囊。

11~16. 大量新鲜花粉授粉后24h,合子核由静养状态走向分裂的不同情况。

表6 "附加"花粉管在雌蕊中的生长情况（白玉皮×阿夫）

年代	授粉后时间(h)	授粉的花粉数量	观察的胚囊总数	各种状况的胚囊数			
				花粉管在子房中			花粉管在珠孔中
				多	少	无	
1962	9	大 量	20	13	7	0	7
		限 量	6	0	0	6	0
	26	大 量	11	10	1	0	2
		限 量	16	0	4	12	0
1963	8	大 量	17	—	—	—	9
		限 量	10	—	—	—	2
	24	大 量	18	—	—	—	12
		限 量	4	—	—	—	0

在我们的材料中当"附加"花粉管进入胚囊时很少同时伴随着附加精子的出现，则仍然是一个疑问。

（作者：周嫦、杨弘远。原载：遗传学集刊，1964（4）：39～47。图版4幅，表6幅参考文献40篇。仅选录图版Ⅰ、Ⅱ、Ⅲ，表1、2、4、6。）

3 芝麻授粉方法和性因素年龄的初步研究

A preliminary study on the pollination method and the viability of pistil and pollen in sesame

摘 要

芝麻是我国重要的油料作物之一。过去对它的杂交育种工作很少开展，近年来已逐渐提上日程[1]。但是，作为杂交育种工作基础之一的受精生物学研究，迄今无论国内外均进行得不多或不够全面，只是在开花结实的习性方面尚有比较详细的资料[3,4,6]。因此,许多基本问题,还有待进行深入的研究。

芝麻是自花授粉植物,材料一般较纯,但也难免因昆虫传粉而有一定的天然杂交率。据研究报道[4~6],芝麻的天然杂交率,一般为4%~5%,也有更高些的;因而,在遗传育种研究中,有必要对材料进行人工纯化。

芝麻的人工交配方法中,去雄与授粉两个环节均很方便,唯每朵花套一隔离纸袋,费时较多。过去有些研究单位采用麦秆套袋法,亦不甚适用,为此,我们在本试验中,采用去雄与授粉后不套袋的方法,效果良好。

关于芝麻雌蕊和花粉生活力的变化规律,目前我们还没有看到任何文献报道,本试验在这方面提供了初步的实验资料。

根据两年来在五个芝麻品种上的试验,得到如下初步结果:

1. 在严密的遗传研究工作中,应用扯掉花冠的去雄方法,结合套小形隔离纸袋及授以新鲜花粉,可保证没有本朵花粉和外来花粉的污染,同时授粉结果甚佳。在杂交育种工作中,则可以不套袋隔离,以便提高工作效率;但最好将同一茎枝上相邻的、正在开放的花朵去掉,并且尽量在较为稀植的条件下,应用这一方法。

2. 人工自交可应用以下方法:傍晚用短线结扎花冠顶部,翌日上午扯下花冠,将散落于冠筒内表面的花粉补授于同朵花的柱头,套以小纸袋,自交结实率可达100%,种子正常。只扎线而不辅以人工授粉,亦可获得自交种子,惟效果不及前法稳定。

3. 芝麻雌蕊能够受精的时期,只有三天左右,即自开花前一天至开花后一天。开花前半天雌蕊受精能力已达最高峰,因而杂交时可在开花当天上午去雄后即行人工授粉,而无需分段进行。

4. 芝麻花粉生活力的变化,依贮藏条件为转移。在盛氯化钙的干燥器中,贮于夏季室内条件下,花粉在三天内,受精能力无显著变化,7至11天期间,受精能力急速减退。从试验结果推测,芝麻花粉最适宜的贮存湿度是30%左右的相对湿度。

(作者:杨弘远、周嫦。原载:湖北农业科学,1964,(4):44~49。表5个,参考文献6篇,均删去。)

4 芝麻花粉在雌蕊上萌发与生长的研究

A study on pollen germination and pollen tube growth in the pistil of sesame

关于芝麻的受精过程,前人曾有若干研究(Nohara 1934; Hanawa 1953; Joshi 1961),但是对于花粉萌发和花粉管在花柱组织内生长期间的特点则还了解不够具体,有进一步阐明的必要。本文报道两年来在这方面研究的初步结果。

材 料 与 方 法

研究材料是引自中国农业科学院油料作物研究所的五个品种:安陆黑芝麻、332、荆芝 7 号、紫花叶二三、江西佛座,均按单株种植。栽培与人工传粉方法按常规进行,不赘述。

观察花柱的构造是收集未传粉及传粉后的花柱固定于 FAA 中,按常法进行石蜡切片。切片厚度为 10μm,用 Ehrlich、Haidenhain 苏木精等染色剂染色。

研究花粉管的生长,一部分依靠花柱切片,主要是应用花柱的整体解剖制片法:将花柱纵分为两半,染以醋酸洋红,经酒精脱水、二甲苯透明后,封于加拿大树胶。结果花粉管呈鲜红色,在淡红色花柱组织的背景上衬托得相当明显(图1)。

在新鲜或固定的花柱上,进行了下列物质的组织化学考察:淀粉用碘-碘化钾法;多糖用 PAS 法;脂肪用苏丹Ⅲ与苏丹黑法;RNA 用甲绿-派洛宁法;还原态抗坏血酸用硝酸银法;过氧化物酶用联苯胺-过氧化氢法;细胞色素氧化酶用氯化氢氨基二甲基苯胺-α萘酚法。以上均有重复实验。

花粉贮藏是收集新鲜花药存于盛氯化钙的干燥器中。气温 28～34℃;空气湿度 35%～55% 相对湿度。

花粉培养用 30%～40% 蔗糖、0.8% 洋菜与 50～100ppm 硼酸配制的培养基。活体观察后,用醋酸洋红固定染色,甘油胶封片。

全部工作于 1962—1963 年在武汉大学生物系实验农场及植物遗传教研室细胞胚胎实验室完成。

结　果

一、花柱的构造

芝麻有二心皮芝麻(*Sesamum indicum* Subsp. *bicarpellatum*)与四心皮芝麻(*S. indicum* Subsp. *quadricarpellatum*)两大类型(詹英贤 1960)。本工作以安陆黑芝麻与江西佛座两个品种分别代表以上两类,观察其花柱构造。

如图版Ⅰ,2 所示,花柱最外一层细胞为表皮,其内是包括数层薄壁细胞的皮层。中央为发达的引导组织,其细胞壁薄而富含细胞质,在纵切面上呈细长形。花粉在柱头上萌发后,花粉管向下生长,穿行于花柱引导组织内(图版Ⅰ,3)。皮层内方、引导组织外侧有维管束,每组维管束代表一片心皮的中脉。安陆黑芝麻有二组维管束,江西佛座有四组(稀为五组)维管束。综上所述,可以断定芝麻花柱的构造是属于具引导组织的"封闭型"。

二、花粉萌发与花粉管生长的速度

观察人工传粉后定期固定的花柱,在武汉一般夏季气候条件下,传粉后 5min 即有少数花粉粒开始萌发。15min 时,较大量的萌发开始,其中生长最快的花粉管接近长入花柱。由传粉后 30min 起,到 2h 止,用测微尺测量了生长最快的一根花粉管在花柱中的长度,求得其平均数及标准机误,并据此算出在不同时间内花粉管到达花柱中的部位及其生长速度。结果见表 1。两年的资料颇一致,332 与安陆黑芝麻生长最快的花粉管于传粉后 2h 均达到花柱全长约 4/5 处,它们的平均生长速度均为每小时 4.7mm 左右。

表1　　　　　　　　花粉管生长的速度

年代	品　　种	传粉后的时间	观察的半边花柱数	花粉管长度（毫米）$\bar{x}\pm S.E.\bar{x}$	花粉管占花柱全长的百分比（%）	花粉管生长速度（mm/h）
1962	安陆黑芝麻	2h	11	8.23±0.25	—	4.70
	332	2h	31	8.15±0.17	80.7	4.66
1963	332	15min	28	0	0	—
		30min	29	0.97±0.03	9.6	3.88
		1h	30	3.59±0.06	35.5	4.79
		2h	29	8.28±0.05	81.9	4.73

注:花粉管长度和速度是以生长最快的一根花粉管计算的。

整个花粉"群体"的萌发与生长是参差不齐的。传粉后 2h、4h、6h、8h、9.5h、11.5h 及 24h 的各批材料中,均可见到大批花粉在萌发与生长。图版Ⅰ,1 示传粉后 24h 的花柱,其中花粉管旺盛生长的情况和以前各批材料颇相仿佛。

人工传粉后定期将一部分花朵的花柱自基部切去,保留子房以观察结实情况,由结实率与种子数的多少推断在不同时期内花粉管进入子房的相对数量。实验结果和上述显微镜考察结果是一致的(表2)。传粉后 2h 切去花柱,子房不结实,表明此时尚无花粉管通过切面进入子房。3h 切去花柱者,虽不结实,但形成了若干无籽果实。这可能由于少数花粉管进入子房后,其后段又遭到切断,以致不能继续生长,而仅以其内含物刺激了子房壁的膨大。在这一点上,我们的实验与 Yasuda 等(1930)在茄子和黄瓜上的类似实验得到了一致的结果。从传粉后 4h 起,切去了花柱的子房已能正常结实,且随切除时间的延迟,每蒴果中种子数显著增加,表明在 4～6h 期间有较多花粉管进入子房,一个子房内大多数胚珠的受精是由它们来完成的。然而,甚至迟至 8h 切除花柱的子房,种子数仍未完全达到对照的水平,似又表明此后还会有花粉管继续进入子房以完成剩余胚珠的受精。

总之,我们的研究结果证实了前人(Nohara, 1934; Hanawa, 1953; Johri, 1961)的工作,即芝麻花粉管约于传粉后4h进入子房,6h大多数胚珠已接受了花粉管。但是,在前人文献中没有提到花粉"群体"萌发与生长持续时间很长的现象。我们认为,这一现象的原因和意义是值得进一步探索的问题。

表2　　　　　人工传粉后定期切除花柱的实验(1962)

品种	处　理	花数	结实率(%)	每蒴果种子数 $\bar{x} \pm S.E.\bar{x}$	无子结实率(%)
332	传粉后2h切去	16	0	0	4.5
	传粉后4h切去	17	47.0	9.5±1.95	0
	传粉后6h切去	17	94.1	52.7±3.65	0
	传粉后8h切去	17	100	63.3±3.69	0
	不切除(对照)	18	100	72.8±1.86	0
安陆黑芝麻	传粉后2h切去	13	0	0	7.7
	传粉后3h切去	19	0	0	57.9
	传粉后4h切去	34	70.7	9.7±2.96	2.9
	传粉后5h切去	20	95.0	29.0±1.85	0
	传粉后6h切去	14	92.8	44.7±1.53	0
	传粉后8h切去	14	98.6	43.9±5.43	0
	不切除(对照)	14	92.8	53.9±4.39	0

三、花粉管的生长行为

贯穿于花柱引导组织中的花粉管,一部分是生长正常的,管端钝圆或稍尖削;另一部分则表现异常特征,即管端膨大继而破裂,将内含物泄入引导组织之中。泄出的物质对于苏木精或醋酸洋红的着色情况和花粉管本身完全一致(图版Ⅰ,1、3、4)。有些较大的团块显然是由数根花粉管流出的内含物汇成的。在所研究的五个品种自然或人工传粉后不同时期的花柱中,毫无例外地观察到这种花粉管破裂的现象。

组织化学反应显示,花粉管及其泄出的内含物均含相当多的淀粉粒和其它多糖物质、脂肪、RNA、过氧化物酶与细胞色素氧化酶,但不含还原态抗坏血酸。这证明,这些团块确系由花粉管中流出,而不是由于花粉管刺激花柱组织所产生或其它来源。此外,在新鲜材料的组织化学考察中,同样看到花粉管破裂的现象,证明它不是在固定或制片过程中产生的人工赝象。

芝麻开花正当盛夏,气温高达30℃以上。为了查明花粉管破裂是否因高温所引起,1963年有意于气温较低(21℃)时固定安陆黑芝麻与江西佛座两个品种自然开花后的花柱。发现在这样的花柱中,只要有花粉管,就同样伴有破裂现象。看来,高温并非破裂的主要原因。

用贮藏的衰老花粉传粉,结果迥然不同。表3说明,花粉的衰老不仅降低了花粉管的生长速度,而且使破裂现象大为减少。花粉管本身也较细瘦些。

在人工培养的条件下直接观察了花粉管破裂的情景。在我们采用的培养基上花粉萌发率不高,但萌发很快,数分钟内便有少数花粉粒产生突起,迅速延伸成花粉管。有些花粉管在生长途中还形成很长的分支。经一定时间后,大部分花粉管尖端突然破裂,大量物质从裂口急剧涌出,在培养基表面扩展成一片。用醋酸洋红染色看到流出的物质中常含生殖核与营养核(图版Ⅰ,5)。

表3 花粉生活力对于花粉管生长速度和破裂行为的影响(1963)

处理	结实率(%)	每蒴果种子数 $\bar{x}\pm S.E.\bar{x}$	花粉管长度(毫米) $\bar{x}\pm S.E.\bar{x}$	花粉管生长速度(%)	平均每半边花柱中破裂的花粉管数
新鲜花粉传粉	100	79.7±0.81	7.28±0.078	100	极多
贮藏9.5d花粉传粉	90.0	52.4±3.74	3.93±0.076	54.0	3.30
贮藏11.5d花粉传粉	83.3	18.2±2.56	1.59±0.052	21.8	0.11

注:品种——安陆黑芝麻。

花粉管长度与速度以生长最快的一根花粉管计算。

统计结实率与种子数每处理30朵花;统计花粉管长度每处理27~30个半边花柱。

第三章 授粉、受精与胚胎发育的实验研究

图版 I
plate I

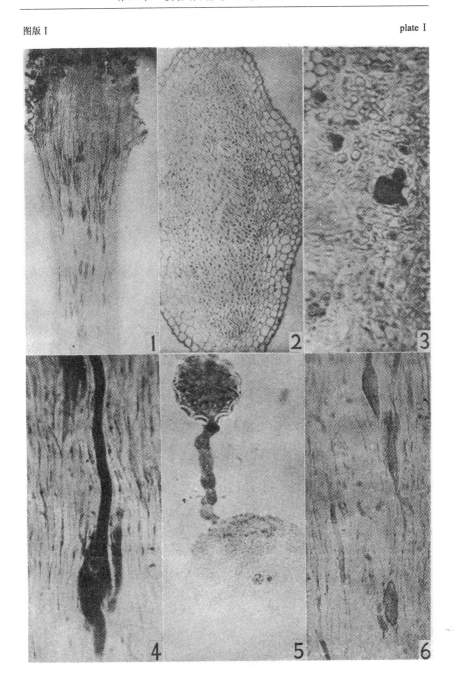

图 版 说 明

图 1. 332 芝麻人工传粉后 24h 花柱整体解剖,醋酸洋红染色。示花粉管生长情况,其中不少花粉管破裂。×40

图 2. 安陆黑芝麻未传粉的花柱横切面,Ehrlich 苏木精染色。示表皮、皮层、维管束与引导组织。×120

图 3. 安陆黑芝麻自然传粉后的花柱横切面,Ehrlich 苏木精染色。示引导组织中的花粉管及其破裂后流出的内含物。×350

图 4. 安陆黑芝麻自然传粉后的花柱纵切面,Haidenhain 苏木精染色。示花粉管在引导组织中破裂流出内含物。×300

图 5. 安陆黑芝麻花粉人工培养,醋酸洋红染色。示花粉管破裂,流出的内含物中含一个生殖核(小的)和一个营养核(大的)。×350

图 6. 安陆黑芝麻自然传粉后的花柱纵切,甲绿-派洛宁染色。花粉管及其流出的内含物中含 RNA(被派洛宁染成红色)。×200

讨 论

关于花粉管在花柱中破裂的现象,过去大多作为一种不正常行为加以报道。例如温度过高时,许多植物的花粉管尖端膨胀,然后破裂(Johri,1961)。Buchholz 等(1930)用镭射线处理传粉后的曼陀罗(*Datura*)花柱,结果后代因花粉管在花柱中大量破裂而致高度不孕。据 Lewis(1954)报道,花柱异长的大花亚麻(*Linum grandiflorum*)当与同型花交配(短柱花×短柱花)时,花粉管进入花柱后遇到不适合的渗透压很快破裂,因而不能受精。在远缘杂交情况下,发生花粉管破裂更是常有的现象。例如曼陀罗或荞麦的二倍体与四倍体之间的交配(Buchholz 等,1929;Жебрак,1963)、小麦属和还阳参属(*Crepis*)中的种间杂交、玉米的亚种间杂交(Батыгина 等,1961)以及用各种远缘花粉对玉米传粉时(Колесников,1963)均有这类描述。自然,在所有这些情况下,花粉管的破裂是一种生物学上不适应的表现,其后果是导致部分或完全的不孕。

但是在文献中也有一些资料,表明在正常的情况下也会出现花粉管的破裂。例如在向日葵(Устинова,1951)、大麦(山本正,1957)和小麦(Гаврилова,1961、1962;Батыгина,1962;Ключарева,1962;周嫦等,1964)等植物正常传粉受精的情况下,都有这样的现象。

芝麻花粉管的破裂,显然属于上述第二类情形,而且其数量和普遍程度更加引人注目。一部分花粉管的破裂并不妨碍受精的完成,并已成为芝麻传粉—受精过程中一种自然的生理现象。

这一现象有没有积极的生物学意义？有些作者推测花粉管泄出的内含物对雌蕊组织和正在发育的幼胚有深刻影响(Устинова,1951;Гаврилова,1961、1962;Ключарева,1962;Чеботарь,1963)。有些作者则指出它们有促进相邻花粉萌发和生长的作用(刘达-цзюнь,1959)。考虑到花粉内含物是一个丰富复杂的生物化学系统(Бритиков,1955;Johri等,1961;Поддубная-Арнолвди等,1961),我们认为以上设想是有一定根据的。本文的初步研究发现,在芝麻花柱中生长的花粉管及其流出的内含物中,含有颇多的营养物质和生理活性物质。这些物质是积极参与雌蕊的代谢从而影响受精与胚胎发育,抑或仅是无谓的消耗,尚需进一步用实验来解决。

(作者:杨弘远、周嫦。原载:植物学报,1964,12(3):211~216。图版1幅,表3幅,参考文献24篇。保留全部图版、表。)

5 花粉数量对芝麻受精结实、胚胎发育和后代的作用

Effect of pollen-grain number on fertilization, embryo development and progeny's characteristics in *Sesamum indicum* L.

Abstract

The effects of pollen-grain number in intra- and inter-varietal pollinations among five sesame varieties were studied. 1,4,10,20 and 40 limited pollen grains, as well as a large quantity of pollen grains which were used as a control test, were artificially pollinated onto stigma, respectively. The main results obtained are summarized as follows.

1. In spite of the somewhat lowered capsule-set percentage after pollinations with few pollen grains, the data obtained in each case indicated that one pollen-grain is basically capable to fertilize one ovule and to produce viable seed.

2. The germination and the pollen tube growth from the limited pollination showed quite normal as compared with those in the unlimited or large quantity pollination. The pollen germination observed at 30 minutes after pollination showed considerable high percentage, about 70% ~ 80%, in each case. Two hours after pollination, the growth of most pollen tubes elongated normally in style, but few tubes showed their broken tips as in natural pollination sooner or later.

3. Paraffin sections of 5 days embryos and of 9 and 15 days dissected embryos produced from both the limited and unlimited pollinations were observed with microscope. At early stage of embryo development, owing to the effects of decrease in number of pollen grains in the limited pollinations, the rate of embryo development and seed growth were somewhat retarded and asynchronous. At later stage, however, the above mentioned effects could no longer be detected. There were no or little significant differences found among the mature embryos or seeds in all cases performed.

4. The F_1 and F_2 plants produced from all the limited pollinations showed obviously normal. The dorminance and segregation ratios were not changed and no any other variations could be found among them.

5. The following problems were discussed: (a) The availability of number of pollen grains in fertilization; (b) The role played by the "superfluous" pollen grains in fertilization and their effects on capsule-setting, embryo development and, as has been reported, on the so-called variability

in progeny; (c) The so-called reliability of applying limited pollination as an effective means of plant breeding.

植物授粉的花粉数量对受精及遗传的作用,历来是遗传学中一个有争议的问题,涉及对受精作用的本质认识上的分歧。20世纪40—60年代,某些研究者的试验表明花粉数量对授粉当代及后代均有重要影响,并以此作为受精多重性理论的主要内容之一。有人还提出限量授粉是一种育种手段。1963—1966年期间,我们在芝麻上开展了本试验,得出一些不同的结果。现将主要试验结果与我们的初步看法报道如下。

材 料 与 方 法

试验在武汉大学生物系实验农场和植物细胞胚胎实验室进行。共用五个芝麻品种进行了品种内与品种间不同数量花粉授粉的比较研究。主要供试品种事先经过两代扣线自交[3]纯化。试验处理包括1粒、4粒、10粒、20粒、40粒花粉与大量花粉等级别。各处理每次实验授粉的花数,一般为20~40朵,授1粒花粉者为70~170朵。两年内共授粉3000余朵花。限量授粉方法是用黑色猪鬃蘸取一定数目的花粉,经放大镜检查无误后,授于柱头上。授粉当代结实情况的考察是在成熟时进行结实率、每蒴种子数、种子重与蒴果长度等项指标的统计。花粉在柱头上萌发和花粉管在花柱中生长情况的观察是用花柱整体解剖醋酸洋红染色法[4]。胚胎发育的观察是用石蜡切片与整体解剖法。后代按组合与处理单株种植,以观察其表现。

结 果

一、授粉当代的结实情况

表1列举了1964年几次主要实验的资料,说明如下:

表1　不同数量花粉授粉时当代结实情况的比较

处理（花粉数量）\组合	安×安①	安×332	安×紫	332×332	荆×紫	江×荆
结　实　率　(%)						
1 粒	61.4	63.5	51.6	54.8	2.4	4.0
4 粒	89.7	90.0	88.9	70.4	17.5	6.7
10 粒	100	100	100	93.8	75.9	35.7
20 粒		100		90.6	79.3	
40 粒		100		100	95.0	
大 量	100	100	100	100	86.7	100
无　子　结　实　率　(%)						
1 粒	13.3	33.3	22.6	29.0	0	0
4 粒	0	2.5	0	15.9	0	0
10 粒	0	0	0	0	0	0
20 粒		0		3.1	0	
40 粒		0		0	0	
大 量	0	0	0	0	0	0
每　蒴　种　子　数 ($\bar{x} \pm S.E.\bar{x}$)						
1 粒	1.0	1.0	1.0	1.0	1.0	1.0
4 粒	3.3±0.19	3.0±0.18	3.2±0.16	2.6±0.20	2.7±0.28	2.0
10 粒	9.3±0.26	6.1±0.39	7.3±0.37	6.0±0.37	6.7±0.36	5.2
20 粒		15.3±0.65		12.0±0.85	12.0±0.67	
40 粒		30.5±0.99		24.4±1.80	23.4±0.82	
大 量	76.4±5.23	79.8±1.64	85.3±0.61	71.2±5.20	52.5±4.59	90.5
平均结一粒种子所需要的花粉粒数						
1 粒	1.00	1.00	1.00	1.00	1.00	1.00
4 粒	1.21	1.35	1.24	1.55	1.48	2.00
10 粒	1.08	1.65	1.37	1.67	1.49	1.92
20 粒		1.31		1.66	1.67	
40 粒		1.31		1.64	1.71	
大 量						
平均一粒种子重 (mg)						
1 粒	2.92	3.43	3.33	2.69	—	—
4 粒	3.42	3.15	3.81	2.96	3.73	—
10 粒	3.55	3.63	3.85	3.30	3.45	—
20 粒		3.58		3.47	3.86	
40 粒		3.50		3.47	3.62	
大 量	3.18	3.08	3.35	3.22	3.59	—

续表

处理 (花粉数量) \ 组合	安×安①	安×332	安×紫	332×332	荆×紫	江×荆
	蒴 果 长 度 ($\bar{x}±S.E.\bar{x}$mm)					
1 粒	13.6±0.13	14.0±0.19	13.3±0.23	12.1±0.16	16.0±2.01	—
4 粒	15.7±0.37	14.7±0.22	15.1±0.25	13.3±0.24	20.0±0.43	—
10 粒	18.4±0.26	17.4±0.22	17.3±0.15	15.8±0.29	25.2±0.46	—
20 粒		20.0±0.24		18.9±0.44	30.7±0.64	
40 粒		23.9±0.25		22.1±0.68	35.3±0.76	
大 量	28.3±13.8	30.1±0.20	28.5±1.24	30.4±0.49	46.5±1.41	—

①品种名称:安:安陆黑芝麻;332:332 芝麻;紫:紫花叶二三;荆:荆芝 7 号;江:江西佛座。(以下各表均同)

结实率 在各授粉组合中,结实率均随花粉数量减少而降低,表明花粉数量不足时不利于结实。但用不同品种作母本有很大差异,如以安陆黑芝麻或 332 作母本时,不论品种内或品种间授粉,授 1 粒花粉可得到 50%～60% 以上的结实率;有一次,结实率甚至高达 80% 左右。而以荆芝 7 号或江西佛座作母本,则授 1 粒花粉时结实率很低(2%～4%)。我们分析,这种母本影响可能是由于各类品种子房的脱落性程度不同所致。安陆黑芝麻与 332 授 1 粒花粉未能结子的情况下,常产生大量无子蒴果(表1;图版Ⅰ,1)。这一现象可作为这类品种较易坐果的佐证。荆芝 7 号与江西佛座则授 1 粒花粉后子房大量脱落,不形成无子蒴果。我们曾对荆芝 7 号授 1 粒花粉的 118 朵花于授粉后 5d 与 8d 收集脱落与未脱落的子房进行解剖观察,发现其中有 45% 的子房含有一粒已膨大的胚珠;通过对一部分这类胚珠的切片检查,证明它们多已受精发育。因此,我们认为这类品种限量授粉时结实率大大降低只是表面现象,实际上受精率要高得多,不过由于花粉太少以致无法坐果而已。在棉花上,也曾报道过花粉数量不足导致大量落铃的情况[16]。所以,限量授粉时结实率的高低并不能完全反映受精率的高低。

每蒴种子数 如前所述,既然结实率受各种因素的干扰甚大,那么,把不结实的情况除外来看已结实蒴果内的种子数,似乎应是衡量授几粒花粉能结几粒种子的主要指标。在本试验中,授 1 粒花粉所结蒴果均含 1 粒种子;授 4 粒花粉者一般含 3 粒种子左右,最多可含 4 粒;授 10 粒花粉者一般含 6~8 粒种子,最多可含 10 粒;授 20 粒花粉者一般含 12~15 粒种子,最多可含 20 粒;授 40 粒花粉者一般含 23~30 粒种子,最多可含 37 粒(平均数见表 1)。由此计算出平均结 1 粒种子所需要的花粉粒数,各级限量授粉均很相近,为 1 粒或稍多于 1 粒,不超过 2 粒(表 1)。当然,授 1 粒花粉情况比较特殊,因为它要就不结实,要就结 1 粒种子(无子结实除外),所以得到的数字一律都是 1 粒,比其它处理偏低。如果考虑到多数组合授 1 粒花粉时的结实率在 50%~60% 以上,则结 1 粒种子所需的花粉数大约也是稍多于 1 粒,不超过 2 粒。这说明,在所有限量授粉处理中,落在柱头上的每一粒花粉理论上都能使一粒胚珠受精结成种子。至于实际上每蒴种子数稍低于授粉花粉数的原因,大概是因为少数花粉粒本来不健全或在授粉过程中丧失受精能力的缘故。

种子重　限量授粉所结种子和对照种子相比,不但不轻,反而常常较重。这可能是由于每一蒴果内种子数的减少相对地改善了每一粒种子的营养条件所致。Тер-Аванесян(1957)也曾持类似的看法[17]。但授 1 粒花粉所结种子则属特殊,常较对照种子小而轻,其原因将在后面胚胎学观察部分加以讨论。

蒴果大小　授 1 粒花粉能使子房膨大成一个小形蒴果;随着花粉数量增加,蒴果也相应增大(表 1;图版 I,1、2)。显然,蒴果大小是依其所含种子数的多少为转移的。在限量授粉场合,仅位于子房顶部的少数胚珠有机会受精,其所结种子恒位于蒴果上部,致使蒴果外形一律上宽下窄,呈倒锥形。

二、胚胎学观察

花粉萌发与花粉管的生长　为了了解限量授粉时花粉萌发与生长的情况,以安陆黑芝麻为材料分别授以 1 粒、4 粒、10 粒及大量花粉,于授粉后 30min 与 2h 分别切下花柱进行显微镜观察。由表 2 与图版 II,

5~8可以看出,各级限量授粉处理的花粉萌发率均颇高,达70%~80%以上。授粉后2h,花粉管大多在花柱中生长,其中仅一部分尖端破裂。和大量授粉的花粉管相比,限量授粉的花粉管看来较为细瘦。总之,本实验表明,限量授粉时多数花粉能顺利通过萌发与生长两个环节,为以后完成受精作用准备条件。至于少部分花粉不萌发或延迟萌发,或在生长中途发生破裂或其它异常情况,在大量授粉时同样可能发生,并不一定是限量授粉特有的效应。

表2　　　　　限量授粉时花粉萌发与花粉管生长情况①

观察时间	处理(花粉数量)	观察花粉粒数	萌发花粉粒数	花粉萌发率(%)	破裂花粉管(%)	未破裂花粉管(%)
授粉后30min	1粒	47	37	78.7		
	4粒	60	50	83.3		
	10粒	150	108	72.0		
授粉后2h	1粒	87	65	74.7	34.2	65.8
	4粒	116	102	87.9	22.9	77.1
	10粒	270	215	79.6	19.6	80.4

①观察的花柱数目,授1粒花粉处理为50~90个;授4粒与10粒花粉处理为15~30个。

胚的发育　关于芝麻的胚胎发育,仅知道少数研究报告[6,7]。为了比较不同数量花粉授粉时胚的发育,我们首先对自然传粉条件下芝麻的胚胎发育全过程作了系统的观察。现将主要观察结果在本文中一并作一简述(图版Ⅲ)。

芝麻的成熟胚囊含卵细胞、两个助细胞、一对极核,反足细胞早期退化。传粉后12h,花粉管通过珠孔进入胚囊,倾注其内含物(图版Ⅲ,16)。花粉管内含物十分浓厚,染色很深,经常掩盖了雌雄配子,使双受精现象难以观察清楚(Joshi也提到同样的情况[7])。一般只有一根花粉管进入胚囊,仅看到个别胚囊中有两根花粉管进入一个胚囊的情况(图版Ⅲ,17)。

极核受精后第一次分裂即形成两个胚乳细胞(图版Ⅲ,18),以后继续细胞分裂,不经过游离核时期,故属于细胞型胚乳。传粉后1天,胚囊中已有4~8个胚乳细胞。以后胚乳细胞不断增多,包围着正在发育的胚。传粉后约9d,胚的子叶开始形成,其周围相邻的胚乳细胞开始被分解吸收。种子成熟时,胚乳仅余周边的约三层细胞(图版Ⅲ,31)。

卵细胞受精后,合子有一段休眠期,此时合子显著纵向延长(图版Ⅲ,19)。传粉后3d,合子已进行过一二次横分裂,由于胚柄的延伸,将顶细胞推向胚囊中部(图版Ⅲ,20)。传粉后4d,顶细胞纵裂一或二次,形成2或4个胚细胞(图版Ⅲ,21)。第5d,形成8个胚细胞(图版Ⅲ,22)。第6d,胚细胞已行平周分裂,形成16个细胞(图版Ⅲ,23)。第7d,发育成球形胚(图版Ⅲ,24)。第8d,原胚开始分化,出现心形胚,胚柄开始退化(图版Ⅲ,25)。第9d以后,由于子叶与胚轴的纵向延长(图版Ⅲ,26),胚逐渐成为鱼雷形,胚的各部及维管束系统均逐渐分化成形(图版Ⅲ,27)。此后,胚的外形基本不变,仅体积继续增大与内含成分发生变化。随着胚的成熟,细胞中出现一种被苏木精染色很深的颗粒,并渐次增多与增大,在胚的各部分(除生长点外)和胚乳细胞中均很密集(图版Ⅲ,28、30、31)。根据PAS反应(图版Ⅲ,29)及在石蜡切片上作的苏丹Ⅲ染色反应,证明这种颗粒不是淀粉粒与油脂,因而很可能是蛋白质粒。

上述观察结果与Hanawa(1953)[6]和Joshi(1961)[7]的报道基本一致,在发育时间进度上大约比他们看到的迟一天,这大概是因为发育时所处的气温条件不同所致。根据我们自己在不同日期先后固定的三批材料,发育进度也有所不同,比上述提前或推迟一天左右。

不同数量花粉授粉时胚胎发育进度的比较研究,是以安陆黑芝麻为材料,于人工授粉后5d、9d、15d分批固定子房。5d的材料对胚珠进行石蜡切片;9d与15d的材料用整体解剖法进行观察测量。从表3、表4可以看出:在发育早期(5d和9d),花粉数量愈少,胚的发育进度愈慢,限量授粉各处理的胚长度均落后于对照胚(t测验表明差异达极显

第三章　授粉、受精与胚胎发育的实验研究

表3　不同数量花粉授粉胚胎发育进度的比较①

授粉后天数	处理（花粉数量）	观察胚数	只看见胚未见乳胚	合子未分裂尚	顶端细胞未分裂	2个胚细胞	4个胚细胞	8个胚细胞	16个胚细胞	球形胚 30μ以下	球形胚 31~40μ	球形胚 41~50μ	球形胚 51μ以上	心形胚	鱼雷形胚 500μ以下	鱼雷形胚 501~1000μ	鱼雷形胚 1001~1500μ	鱼雷形胚 1501~2000μ	鱼雷形胚 2001~2500μ	鱼雷形胚 2501μ以上
5d	1粒	42	9	1	12	5	3	2	7	1	2									
5d	4粒	32						1	2	6	21	2								
5d	10粒	32						1			16	15								
5d	大量	30						1			3	22	4							
9d	1粒	40									3		6	22	9					
9d	4粒	40													12	28	4			
9d	10粒	40														36	4			
9d	大量	40														1	34	5		
15d	1粒	40															1	6	28	5
15d	4粒	40															1		14	25
15d	10粒	40																	5	35
15d	大量	40														1		1	9	29

① 表中数字为各类胚的数目；胚的长度系指纵径。

表4　不同数量花粉授粉胚和种子长度的比较①

授粉后天数	处理（花粉数量）	观察数目	胚 长 度 $\bar{x}\pm S.E.\bar{x}$（μ）	为对照%	C.V.(%)	种 子 长 度 $\bar{x}\pm S.E.\bar{x}$（μ）	为对照%	C.V.(%)
5 d	1 粒	42	21±1.1**	47	27.9	1.39±0.06**	54	29.3
	4 粒	32	33±0.8**	73	14.3	2.38±0.04**	93	10.6
	10 粒	32	38±0.8**	84	11.7	2.50±0.03	97	6.6
	（大量）对照	30	45±1.0	100	12.1	2.56±0.02	100	5.3
9 d	1 粒	40	216±18.4**	16	55.2	2.94±0.06**	85	15.9
	4 粒	40	636±28.0**	46	27.9	3.27±0.03**	95	5.9
	10 粒	40	872±21.2**	62	15.5	3.33±0.03	97	6.1
	（大量）对照	40	1380±18.0	100	8.2	3.44±0.02	100	3.9
15 d	1 粒	40	2276±42.8**	87	11.9	3.27±0.03**	93	5.5
	4 粒	40	2504±35.2	96	8.9	3.39±0.04*	97	6.7
	10 粒	40	2608±16.4	100.8	3.9	3.44±0.07	99	13.9
	（大量）对照	40	2588±51.6	100	12.6	3.49±0.02	100	3.7

① 将各级限量授粉处理与对照比较，进行 t 测验。**表示差异极显著；*表示差异显著；未标星号表示差异不显著。

著)。此外,从发育的整齐度看,也是花粉数量愈少,发育愈不同步,这由胚长度的变异系数(C.V.)也可以看出。例如特别明显的是,授粉后5d,对照胚大多已达到球形阶段,而授1粒花粉的胚则包括由合子到球形胚各种发育状态。(图版Ⅱ,9~15)示授粉后5d各处理有代表性的胚状态。

但是随着发育的推进,各处理在胚发育进度和整齐度方面的差异逐渐减小。授粉后15d,授4粒与10粒花粉的胚已完全赶上对照胚。授1粒花粉的胚虽仍赶不上对照胚,但差距也大为缩小。关于花粉数量减少导致胚发育延缓的现象,前人也有类似报道[5,12,18],但他们没有指出发育后期差异是否减小以至消失这一重要情况。

种子的生长 由表4种子长度一栏可以看出,种子的生长和胚的发育表现相似的规律:早期显示明显的花粉数量效应,而至后期则这种效应有逐渐消失的趋势。联系本文第一部分所述种子重量的资料,即限量授粉的种子(授1粒花粉者除外)常较大量授粉的种子为重,更可以看出上述趋势的延续。我们推测,这里似乎有两种互相矛盾的因素影响着种子的生长:一种是花粉数量因素——花粉数量减少使种子(与胚)早期生长受抑;一种是种子数量因素——蒴果内种子数目的减少导致每一种子营养条件的相对加强。随着发育的推进,前一种因素的作用逐渐让位于后一种因素的作用。至于授1粒花粉的种子至终赶不上对照种子,看来主要由于早期差异太大的缘故。

三、后代的表现

关于芝麻的遗传,国内外有一些研究[1,7]。本试验在前人研究的基础上选取具有明显相对性状的品种作亲本进行不同数量花粉授粉的实验,以观察后代表现。

子一代的表现 如表5所示,根据两个品种内授粉组合和三个品种间授粉组合的F_1植株的观察,各级限量授粉和大量授粉没有差别:在品种内授粉情况下,各处理的全部单株均表现该品种典型的性状;在品种间杂交情况下,各处理的全部单株均表现该组合典型的显性性状,未发现限量授粉导致亲本一方遗传传递力量的削弱或加强,也没有出现特殊的新变异。

对两个组合 F_1 植株的产量构成因素进行了考察(表6)。单株结蒴数的统计分析表明,各级限量授粉处理与大量授粉之间看不出显著差异。平均单蒴种子重也没有明显差异。逐株的观察也未发现限量授粉后代出现畸形、不育、迟熟等特殊变异株,或特殊优良的变异株。图版Ⅰ,3、4 示这两个组合子一代成熟植株的状态。

子二代的表现　1963年我们曾对安陆黑芝麻×紫花叶二三组合进行了不同数量花粉授粉试验,当时采用的花粉数量为1粒、4粒、8～10粒、16～20粒、40～50粒花粉与大量授粉,当代结实情况与子一代表现与1964年所做授粉试验的结果一致,故不赘述。1965年对其子二代共1 000多个单株进行了观察(表7)。结果表明:各级限量授粉处理的性状分离比率与对照均很接近,没有出现明显的偏离,也没有看到特殊新性状的出现。

表5　不同数量花粉授粉子一代的性状表现

观察材料		观察株数	性状					
			分枝性	每节花数	花色	蒴果棱数	蒴果长度	种子色
亲本	安陆黑芝麻	—	分枝	单花	白	四棱	中蒴	黑
	332	—	单干	三花	白	四棱	中蒴	白
	紫花叶二三	—	分枝	三花	紫	四棱	中蒴	褐
	荆芝7号	—	单干	三花	白	四棱	长蒴	白
子一代	安×安限量授粉①	98	分枝	单花	白	四棱	中蒴	黑
	安×安大量授粉	118	分枝	单花	白	四棱	中蒴	黑
	332×332 限量授粉	223	单干	三花	白	四棱	中蒴	白
	332×332 大量授粉	170	单干	三花	白	四棱	中蒴	白
	安×紫限量授粉	230	分枝	单花	紫(略浅)	四棱	中蒴	黑
	安×紫大量授粉	98	分枝	单花	紫(略浅)	四棱	中蒴	黑
	安×332 限量授粉	237	分枝	单花	白	四棱	中蒴	黑
	安×332 大量授粉	106	分枝	单花	白	四棱	中蒴	黑
	荆×紫限量授粉	225	分枝	三花	紫(略浅)	四棱	长蒴(略短)	褐
	荆×紫大量授粉	123	分枝	三花	紫(略浅)	四棱	长蒴(略短)	褐

①限量授粉包括1粒、4粒、10粒、20粒、40粒花粉等处理的总株数。

表6　限量授粉与大量授粉子一代植株产量构成因素的比较

组合	处理 (花粉数量)	单 株 结 荫 数①				平均单荫 种子重(克)
		$\bar{x} \pm S.E.\bar{x}$	自由度	处理与对照 相比的 t 值	5% t	
安×332	1　粒	93.2±5.75	20	0.41	2.086	0.25
	4　粒	102.0±7.08	30	0.58	2.042	0.22
	10　粒	106.1±9.05	25	0.87	2.060	0.24
	大量(对照)	96.6±6.11				0.23
332×332	1　粒	135.4±13.55	35	1.42	2.030	0.21
	4　粒	119.9±13.20	40	0.41	2.021	0.22
	10　粒	116.0±10.13	40	0.16	2.021	0.20
	20　粒	124.1±5.71	45	1.36	2.014	0.20
	40　粒	119.3±5.92	45	0.68	2.014	0.21
	大量(对照)	114.2±4.62				0.20

①各处理单株结荫数与对照相比较的 t 值均未达到5%,表示处理与对照间没有显著差异。

表7　不同数量花粉授粉子二代性状分离比率的比较①

性状 处理 (花粉数 量,粒)	每节花数		花　色			种子色		
	单花 (母本型)	三花 (父本型)	白 (母本型)	各种浅紫 (中间型)	紫 (父本型)	黑 (母本型)	深褐 (中间型)	褐 (父本型)
1	67.7	32.3	16.1	59.4	24.5	54.4	23.6	20.0
4	70.2	29.8	14.4	72.2	13.4	56.8	23.7	19.5
8～10	71.0	29.0	9.8	73.8	16.4	60.4	19.6	20.0
16～20	68.7	31.3	11.9	79.0	9.1	58.4	21.6	20.0
40～50	67.1	32.9	5.9	82.4	11.7	55.8	22.3	21.9
大　量	66.2	33.8	6.2	79.5	14.3	60.2	20.9	18.9

①表中数字为各种分离类型株数占观察总株数的百分比。观察株数,各处理为130～300株,总计1 003株。

讨 论

一、受精作用的基本形式是"一对一"还是"多对一"？

细胞学上早已确认,受精是一对精卵的融合。然而在自然状态下,落在柱头上的花粉经常为数很多,远远超过子房内的胚珠数。这些大量的花粉是否为受精作用必不可少？换句话说,为了完成一粒胚珠的受精,原则上只需要一粒花粉还是多粒花粉？这涉及受精作用的基本形式是"一对一"还是"多对一"的问题。从19世纪以来,包括达尔文、孟德尔在内的一些研究者就曾注意到这一问题,并试图通过限量授粉实验加以阐明,但早期的工作大多规模很小,说服力不够。20世纪40年代以后,不少苏联研究者又重新开展规模较广泛的研究来探讨花粉数量在受精中的作用[10~14,16~18]。他们发现:花粉数量不足时,受精过程破坏,胚胎发育延缓,结实率降低,后代发育异常,因而认为一粒花粉不能满足一个胚珠受精的需要,必须有多数花粉参加。有人还应用放射性同位素示踪法[15]和限量授粉法[14]计算出形成一粒种子所需要的花粉粒数目,一般都在几粒以上。

但是,检查这些资料,却发现情况并不完全一致。即以最常采用的研究材料紫茉莉来说,不同作者的试验结果出入甚大,同是授1粒花粉的处理,有的结实率很低,有的却相当可观。有的资料说明授粉时的植株及气候条件对限量授粉结果影响很大[11]。有人在笋瓜和西葫芦上做实验,结一粒种子所需要的花粉粒数,最多达83.5粒,一般3~6粒,但也有只需1.1粒花粉就能结一粒种子的情况[14]。这一切说明,否定"一对一"的论据是不够充足的。

根据本文在芝麻方面的试验结果,授1粒花粉的处理在良好条件下有相当高的结实率;在各级限量授粉处理中,从授粉花粉数与所结种子数的比例来看,理论上每一粒花粉都能使一个胚珠受精结成一粒种子;并且限量授粉产生的后代是完全正常的。这说明受精作用基本上可以按"一对一"的方式进行,大量花粉并不是绝对必需的。我们认为,受精过程包含许多复杂的矛盾,但其中的主要矛盾是精卵以"一对一"方式进行的融合。这本已为细胞学和遗传学中的基本事实所证明,限量授粉实验不过是在排除了"多余"花粉的条件下进一步证明了这一点。

二、"多余"花粉起什么作用？

承认受精的基本形式是"一对一",并不否认"多余"花粉的作用。大量资料证明:花粉具有高度的生理活性,能分泌植物激素、酶及其它生理活性物质,通过与雌蕊的代谢,对受精及其结果发生一定的影响。花粉对于促进果实的初始发育和防止落果起重要作用。花粉物质对种子生长也有作用。在某些情况下,"多余"花粉管可能进入胚囊,甚至多精入卵,这看来对受精和早期胚发育会有一定作用。因而,落在柱头上的大量花粉的作用决不限于保证精卵相遇的机会和提供最适合的雄配子,而且还在于创造一个有利于受精和早期胚胎发育的生理环境。在异常的受精条件下,例如在不亲和交配时,"多余"花粉更显示其重要意义。"花粉蒙导"作用已为不少实验所支持[2],在我国农民育种家的广泛杂交育种实践中,有许多事实看来也可以用花粉的蒙导作用来解释。最近,有人发现花粉壁中有一种蛋白质性质的"辨认物质"(recognition substances),由亲和花粉提取出来的这种物质有助于克服种间杂交与自交不亲和[8],这对于阐明花粉蒙导的物质基础提供了新的资料。在本文关于芝麻的试验中,花粉数量的减少虽不妨碍基本受精过程的完成,但仍可看出以下几方面的影响:(1)由于花粉数量太少,不足以提供与诱导雌蕊产生足够的生长物质来促进子房膨大与坐果,致使子房脱落,结实率降低,这在某些品种尤其严重;(2)由于花粉数目有限,其中可能有一部分花粉缺乏生活力,一部分花粉管在生长中途破裂,致使受精率有所降低;(3)花粉数量过少对胚和种子的早期发育有不利影响,表现在发育与生长的延缓和不整齐。这说明尽管受精的基本过程是"一对一"式地完成的,但"多余"的花粉仍有一定的生理作用。

图版 I

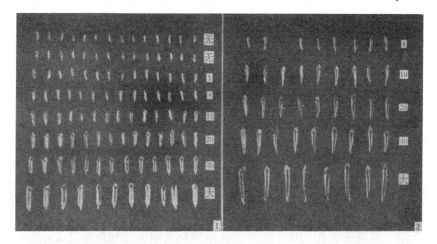

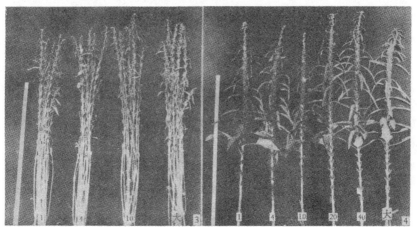

图 1~2. 不同数量花粉授粉成熟蒴果的比较。1.332×332；2. 荆芝 7 号×紫花叶二三。

图 3~4. 不同数量花粉授粉子一代成熟植株的比较(标尺 1 米)。

3. 安陆黑芝麻×332(5 株一束)。

4. 332×332(单株)。1：授 1 粒花粉； 4：授 4 粒花粉； 10：授 10 粒花粉； 20：授 20 粒花粉； 40：授 40 粒花粉；大：授大量花粉)

图版 II　　　　　　　　　　　　　　　　　　　　　　　　　　　plate II

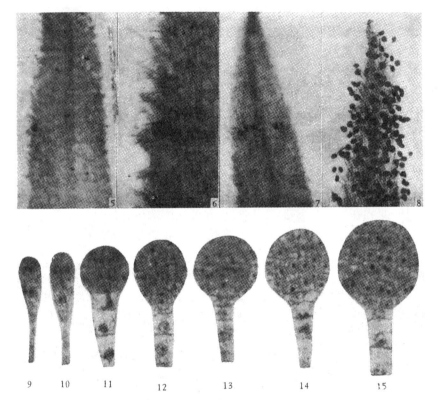

图 5~8. 不同数量花粉授粉时花粉柱头上的萌发情况（安陆黑芝麻，授粉后 30min，醋酸洋红染色） **5**. 授 1 粒花粉（×40）； **6**. 授 1 粒花粉（×120）； **7**. 授 4 粒花粉（×40）； **8**. 授大量花粉（×40）。

图 9~15. 不同数量花粉授粉时早期胚胎发育进度的比较（安陆黑芝麻，授粉后 5d，石蜡切片，铁矾苏木精染色，×700） **9**. 顶细胞尚未分裂； **10**. 两个胚细胞； **11**. 16 个细胞的胚； **12**. 球形胚； **13**. 授 4 粒花粉，球形胚； **14**. 授 10 粒花粉，球形胚； **15**. 授大量花粉，球形胚。

我们认为,对于"多余"花粉的作用,应当区分不同情况,给予恰如其分的估计。应当区分它们在正常受精条件和特殊受精条件下,在一般情况下和个别情况下,在生理方面与遗传方面所起的作用。"受精多重性"理论的提出,促使人们重视"多余"花粉的作用;然而过去有时把它强调到不适当的程度,甚至与受精的基本过程相提并论,看来混淆了矛盾的主次。

三、限量授粉是不是一种有效的育种途径?

前人的资料中,有些提到限量授粉导致后代遗传规律的反常和出现新变异。其中有的提到限量授粉后代经常出现与亲本迥然不同的新类型。在紫茉莉上,发现有多方面的、大幅度的变异,如极端迟熟、不育或常绿型的新类型[13]。Тер-Аванесян(1957)在棉花、小麦等作物上进行的试验中看到限量授粉子一代表现多样性,其中包括若干优良的变异株;变异在子二代常能稳定遗传,并可从中选出有价值的品系;因而提出限量授粉是一种有效的育种途径[17]。我们在开展本试验时,也曾希望通过这一方法引起变异,为芝麻育种探索一条新途径。但通过实践,看不到限量授粉与大量授粉后代间有明显区别,因而对这一方法的有效性提出疑问。当然,由于我们只研究了芝麻一种作物,尚不能作出概括一切的结论。这里只提出我们的看法,供大家讨论。

(作者:杨弘远、周嫦。原载:遗传学报,1975,2(4):322~331。图版3幅,表7幅,参考文献18篇。选录全部表格与图版Ⅰ、Ⅱ。)

第二节 离体受精

提要

被子植物的离体受精是20世纪90年代初由德国研究者首先突破

的。这一重大成就有三项技术前提:一是精细胞与卵细胞的分离;二是一个精细胞与一个卵细胞的成对融合;三是融合产物(离体合子)的微室饲养。其中第一项技术我们当时已经建立,第三项技术是当时预期经过努力可以达到的,唯有第二项技术难度最大。德国研究者是应用微机操纵挑选单个性细胞进行"微电融合",而我们缺乏这样的设备条件,如何实现单对雌、雄性细胞的融合是面临的主要挑战。

解决这一难题的途径是立足国内现实条件,创造性地将常用的PEG诱导大量原生质体融合技术加以改造,使之适应单对原生质体的融合。本研究组首先以烟草叶肉原生质体做试验,用微吸管挑选一对原生质体置于PEG微滴中进行融合取得成功。此法克服了一般群体融合中的盲目性,排除了一方原生质体自相融合、多个原生质体互相融合以及未融合原生质体等等混杂,保证了融合产物来自指定的一对原生质体。接着,将上述技术应用于烟草雌性细胞的融合,成功地实现了卵细胞、中央细胞、助细胞之间及与雄性细胞或体细胞之间的各种组合的融合,其中主要是雌性与雄性细胞的融合。对不同组合的融合特点与技术要求也进行了研究。此外,还建立了微滴培养与微室饲养两种培养技术,使少数原生质体得以发育成多细胞团,为今后性细胞融合产物的培养奠定了基础。

以上述自行创建的PEG诱导单对雌、雄性细胞融合技术开展国际合作,在意大利锡耶那大学继续深入进行了烟草离体受精的细胞生物学研究。应用视频增差显微术观察记录了卵细胞或中央细胞为一方、精细胞或生殖细胞为另一方的融合过程的生活动态,从雌、雄性细胞互相贴合开始到雌、雄核融合为止,包括一系列过去未曾观察到的细胞学事态。这是国际上首次对离体受精过程生活动态的连续追踪观察,也是首次以双子叶植物为对象的离体受精实验。本书仅录用了在《植物学报》上发表的简报一篇,以后该项研究又在国际刊物上发表系列论文多篇。

1 用聚乙二醇诱导选定的成对原生质体间的融合

Polyethylene glycol-induced fusion of selected pairs of single protoplasts

Abstract

A new method for polyethylene glycol (PEG)-induced fusion between single protoplasts was developed. The protoplasts were prepared from tobacco leaves. Under an inverted microscope, two defined protoplasts were selected with a hand-made micropipette and transferred into a droplet of fusion solution containing 25% PEG (M.W. 6000), 0.1 mol/L mannitol and 0.01 mol/L $CaCl_2 \cdot 2H_2O$ (pH 5.6). Slightly moving the pipette caused the protoplasts to contact and adhere to each other, the fusion pairs were then transferred into a solution containing 10% PEG, 0.35 mol/L sucrose and 0.01 mol/L $CaCl_2 \cdot 2H_2O$ (pH 5.6) for approximately 10 min, followed by subsequent washing with a solution containing 0.45 mol/L sucrose and 0.04 mol/L $CaCl_2 \cdot 2H_2O$ (pH 7-9). Compared with conventional fusion methods adopted to protoplast population, the present method can avoid either blind fusion of protoplasts belonging to one partner and fusion among multiple protoplasts, or the presence of unfused protoplasts, thus ensure the fusion to be precisely at the level of a selected pair of single protoplasts. Moreover, it is simple and convenient enough to show its potentiality for wide application in somatic hybridization and particularly in the case of small quantity of parental protoplasts such as *in vitro* intergametic fusion studies.

通过原生质体或亚原生质体间的融合进行细胞杂交或细胞重建,是植物细胞工程的重要手段。现有的融合技术中,以聚乙二醇(PEG)诱导融合和电融合二者占主要地位。其中 PEG 融合技术具有设备简单、操作方便等优点,至今仍然广泛采用。但现行 PEG 融合技术是在原生质体群体的条件下进行的,因而在融合产物中除了所欲获得的杂种细胞外,还不可避免地含有未融合的原生质体、一方亲本原生质体之间的融合体以及多个原生质体的融合体,从而增加了融合后筛选的困难。此外,在有些情况下,可供融合的亲本原生质体数量很少,现行的 PEG 法亦难适用。Koop 等创建的成对原生质体微电融合技术克服了上述缺点,使原生质体融合技术得以精确化,不仅成功地用于体细胞原生质体[1]、亚原生质体[2,3]的融合及细胞器的转移[4~6],而且还被 Kranz 等采用,成功地进行了玉米精、卵的体外融合[7],并得到了再生植株[8]。然而,这种微电融合技术需要特殊仪器,难以推广应用。如果能够使 PEG 融合技术提高到成对原生质体融合的水平,则将兼有简易与精确两方面的优点。为此,我们做了以下实验,并获得肯定的结果。

材 料 与 方 法

实验材料为烟草(*Nicotiana tabacum* L.)品种"G-80"。

一、原生质体分离

选用大田种植的幼苗 10~20cm 长的叶片或试管内种植的无菌苗 1.2~2.5cm 长的叶片,撕去下表皮,将小块叶片置于酶解液中。在 25℃下保温 4~6h,用 200 目不锈钢网过滤去残渣,离心去酶液。原生质体以 0.45mol/L 蔗糖溶液漂浮纯化并洗涤两次备用,或转入 0.48mol/L 甘露醇中沉降与保存备用。

二、原生质体融合

采用以下两种方法:(1)将一滴融合液加在载玻片上。在倒置显微镜下以手工制备的微吸管(尖端口径约200μm),由亲本原生质体群体中各吸取1个原生质体,置于融合液小滴中。通过手工操作微吸管轻轻移动原生质体,使之接触与粘连。然后将已紧密粘连的一对原生质体转入中继液中,静置片刻,再转入清洗液中除去残余的PEG。最后将其转入培养基中。(2)将约2ml融合液加入3cm直径的培养皿中,在融合液上覆盖一薄层石蜡油。用微吸管将选定的一对原生质体置入融合液中,按前述步骤进行融合(原生质体酶解、融合、中继和清洗4种溶液的成分见表1)。

表1 融合程序中所用各种溶液成分表

Table 1 Composition of solutions used at different stages of fusion procedure

成分 Component	酶解液 Enzyme solution	融合液 Fusion solution (1)	中继液 Fusion solution (2)	清洗液 Washing solution
果胶酶 Pectinase	0.5%~1%	–	–	–
纤维素酶 Cellulase R-10	1%~1.5%	–	–	–
甘露醇 Mannitol	0.48mol/L	0.1mol/L	–	–
蔗 糖 Sucrose	–	–	0.35mol/L	0.45mol/L
PEG(6000)	–	25%	10%	–
$CaCl_2 \cdot 2H_2O$	–	0.01mol/L	0.01mol/L	0.04mol/L
KH_2PO_4	0.1mg/ml	0.1mg/ml	0.1mg/ml	–
pH	5.6	5.6	5.6	7~9

三、观察与摄影

以FDA染色观察融合体活性;以DAPI染色观察融合体细胞核。在Olympus CK2倒置显微镜下操作,用Olympus IMT-2倒置显微镜摄影。

结　果

经蔗糖漂浮并洗涤后的烟草原生质体基本上是纯净和有生活力的。挑选其中状态最好的叶肉原生质体或不含叶绿体的维管薄壁细胞原生质体，或用大小不等的原生质体为材料，探索并建立原生质体成对融合的操作技术。

一、融合过程

用微吸管将选出的一对叶肉原生质体释入融合液中以后（图版Ⅰ,1），通过手工操作使其相互接触与粘连（图版Ⅰ,2），此时2个原生质体会发生猛然的相向压缩，继而每一个原生质体由原来的球形变成半球形状态，两者紧紧粘合（图版Ⅰ,3）。这一过程一般在2min内完成。显微观察显示：如果仅有原生质体间的粘连而不发生猛然的相向压缩，则不能继续进行融合，稍加触动即可重新分开；即使通过相向压缩过程也还有重新脱离的可能。这种情况多在将融合产物直接转入清洗液时出现。为了防止上述逆向过程，有必要将融合产物转入含低浓度PEG的中继液中放10min以上，以利融合的完成。在中继液中两原生质体间尚可见一条或深或浅的界线，再转入清洗液后，界线逐渐模糊甚至消失（图版Ⅰ,4,5）。经DAPI染色荧光镜检，内含2核，是为同核体（图版Ⅰ,6）。用微吸管将逐对融合的原生质体加以富集，一起转移、清洗（图版Ⅰ,9,10），以便进行后续的培养。荧光素二醋酸酯染色反应显示，融合体都是生活的（图版Ⅰ,11）。

二、操作方法

单对原生质体的融合全过程必须在倒置显微镜下进行。为了使所观察的对象在操作过程中不致从视野中消失，提高工作效率，可根据不同情况分别采取下列几种具体方法：

（1）当供试双方原生质体数量较多时，可分别将两滴原生质体悬浮液加在载玻片左侧,而在右侧加一大滴融合液,移动载物台,从左侧选取一对原生质体释入右侧的融合液中使之融合。这样可以反复进行多对原生质体的融合。(2)当一方原生质体数量很少时,只将数量多的一方原生质体悬浮液滴于载玻片左侧。操作时先用微吸管吸取数量少的一个原生质体并保存在吸管中,然后由载玻片左侧再吸入另一个原生质体,一同释放到融合液中。(3)当双方原生质体数量均很少而体积又较小时,可将一方的原生质体先转移到另一方原生质体近旁,然后将两者一同吸入微吸管,转入融合液中。在吸取和释放过程中注意使2个原生质体尽量靠近,以便于在融合液中寻找和操作。(4)在载玻片上操作比较方便,效率高,观察及摄影效果亦较好。但溶液因易蒸发而改变浓度,需及时补充新的溶液,否则难以进行多次的融合操作。为此,可改用小培养皿,在融合液上覆盖一层石蜡油以防水分蒸发。这样操作虽不如前法方便,但可连续进行多次融合而不必更换溶液。

三、关键因素

PEG 浓度 在一定温度条件下,PEG 浓度对融合速率影响很大。5%以下浓度只能使其粘连,不能完成融合;25% PEG 可使融合快速进行;15%左右的 PEG 融合速度适中,便于仔细观察融合过程及显微摄影。

渗透压调节剂 在维持原生质体活力的条件下,一般在较低渗的介质中有利于融合[9]。本实验是将原生质体由含 0.45mol/L 蔗糖或含 0.48mol/L 甘露醇的保存液中转入含 0.1mol/L 甘露醇的融合液中融合,然后在随后的中继液和清洗液中逐步恢复到 0.45mol/L 蔗糖浓度,这样有利于融合体的稳定。然而,在为防止水分蒸发而在融合液中补加低渗溶液时,渗透压的骤降反而会使已经粘连一体的原生质体(图版Ⅰ,7)重新分开(图版Ⅰ,8)。

原生质体粘底现象 原生质体在 PEG 中极易粘附于玻皿底部。在成对原生质体融合中如果发生这种现象,对融合及融合体的转移都很不利。为了证实这一认识,作者分下列 3 种处理作了比较试验,每组做 50 对原生质体:第 1 组,在融合过程中控制每对原生质体不使粘底。结果 100%都可完成融合;第 2 组,先使一方原生质体粘底,再将另一

第三章 授粉、受精与胚胎发育的实验研究

图版 I   plate I

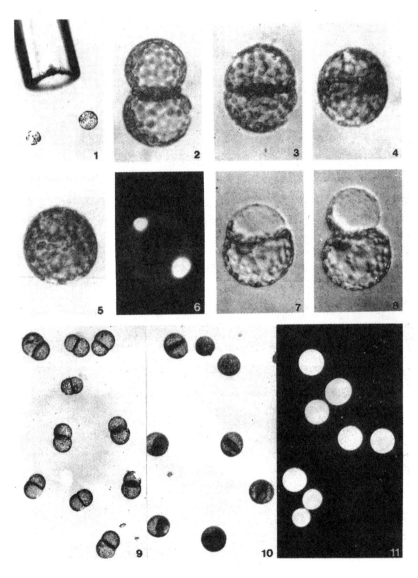

方原生质体移近与之粘连,结果仅 24% 完成融合,余则停滞于葫芦形状态等不同阶段;第 3 组,将双方原生质体两两成对靠拢,但在相互粘连前先使之粘底。结果 100% 均停滞在仅仅相贴的状态而不能通过相向压缩完成融合。此时,原生质体粘底的拉力和原生质体相向压缩的拉力相拮抗,常使原生质体变形拉长。

讨 论

传统的 PEG 诱导的原生质体融合技术是以有大量的供试原生质体为前提的。即使是其中的"小规模融合"[10],仍然是基于原生质体群体中的随机融合,不能摆脱融合结果的盲目性。本文提出的 PEG 诱导的成对原生质体融合方法,是 PEG 技术的一次重要革新。其优点在于:第一,无需考虑原生质体纯化程度、融合时双亲原生质体密度等因素,可以简化前期处理;第二,特别适用于供试原生质体数量极少的场合,例如在进行雌雄配子体外融合实验时,供试的卵细胞数量很有限,在这种情况下,群体融合是无法奏效的,唯有成对融合可行;第三,可以确保融合体均来自选定的双亲原生质体,排除未融合的原生质体、亲本原生质体自相融合、多个原生质体融合等等不希望出现的情况;第四,因此也无需采用群体融合时必经的各种筛选系统,可以直接将融合体转入培养,而融合体的筛选常是传统 PEG 融合技术的重要限制因素;第五,采用这一系统还可以准确追踪观察融合过程和融合后的发育过程,便于开展有关细胞生物学研究。总之,PEG 诱导的成对原生质体融合方法具有与成对原生质体微电融合方法异曲同工的效果,而又无需后者特殊的仪器设备,因而,可以广泛采用。当然,由于成对融合体数量较少,一般的大量培养方法显然是不行的,必须有特殊的微量培养或饲养系统,方能促使融合体分裂、发育以至再生植株。在这方面,前人已有成功的经验[8,11~14]可供借鉴。

图 版 说 明

除图 6 和图 11 为荧光显微摄影、图 7 和图 8 为 Nomarski 干涉差摄影外,其余均为明视野图像。

1. 由微吸管释出的一对叶肉原生质体。×100 2~5. 一对原生质体的融合过程。注意此过程中两原生质体间的界线逐渐消失。×600 6. DAPI 染色,示同核体中的两个核。×600 7、8. 在融合过程中由于渗透压骤然降低,使一对本已粘连的原生质体重新部分脱离。×600 9、10. 用微吸管富集的处于不同融合阶段的融合体群体。×170 11. 富集的融合体群体,FDA 染色示生活力。×170

Explanation of plate

All photographs are bright-field images except **Figs. 6** and **11** (fluorescence images) and **Figs. 7** and **8** (Nomarski interference contrast images). **Fig. 1.** A pair of protoplasts released from a micropipette. ×100 **Figs. 2-5.** Fusion process of a pair of protoplasts. Note that the boundary between them gradually disappeared. ×600 **Fig. 6.** A homokaryon stained with DAPI. ×600 **Figs. 7** and **8**. Partial detachment of two already adhered protoplasts by sudden lowering of osmotic pressure of the medium. ×600 **Figs. 9** and **10**. Populations of the fusion products enriched at different stages of fusion process. ×170 **Fig. 11.** A population of fusion products showing their viability by fluorochromatic reaction. ×170

(作者:孙蒙祥、杨弘远、周嫦。原载:植物学报,1994,36(7):489~493。

图版 1 幅,表 1 幅,参考文献 14 篇。选录全部图、表。)

2 烟草雌性细胞原生质体的融合实验

Single-pair fusion of various combinations between female gametoplasts and other protoplasts in *Nicotiana tabacum*

Abstract

We have reported a new method for fusion between single protoplasts using polyethylene glycol (PEG) as an inducing agent. This method has lead to the successful fusion of female gametoplasts with other kinds of protoplasts. The female gametoplasts described here, in a broad sense, include the egg cell (E), central cell (C) and synergid (S). One of the female gametoplasts was selected and fused with another female, male (generative cell, G) or somatic (mesophyll, M) protoplast. Various combinations were involved: E+S, E+C, E+G, E+M, C+C, C+S, C+G, C+M, S+S, S+G, S+M, etc. Briefly, the authors were able to choose any desired combination to realize single-pair fusion by the new PEG method. For the purpose of culturing such fusion products were limited in number, the authors had done some preliminary experimets using mesophyll protoplasts as feeder cells. Two methods were adopted: the microdrop culture, and the millicell culture with feeder cells. The mesophyll protoplasts were precultured for 2-3 days in large population before they were used as feeder cells. One or several protoplasts were cultured in a microdrop or a millicell and were induced to the formation of small cell clusters. This result indicated that the culture methods might also be suitable for culturing the products from fusion of female gametoplasts and other protoplasts in this plant species.

性细胞原生质体融合(包括性细胞原生质体间、性细胞与体细胞原生质体间的融合)是性细胞离体操作的一个重要方面。近年这一领域进展迅速[1]。但几乎所有工作都集中在雄性细胞的融合方面,迄今只在玉米一种植物中以雌性细胞进行过融合实验,取得了精、卵体外融合并再生植株的成功[2~5]。烟草这一模式植物在原生质体融合与性细胞分离两方面均具有较好的工作基础。最近,我们先后建立了分离烟草胚囊与雌性细胞[6],用聚乙二醇(PEG)诱导选定的成对原生质体融合[7]两项实验技术,这就有可能开展数量有限的雌性细胞原生质体的各种组合的融合实验,观察各种组合的融合行为,并探讨 PEG 诱导的成对原生质体融合技术应用于性细胞融合的特点。本文主要报道上述实验结果。此外,为下一步进行少量融合体的培养,初步建立了两种简便有效的微培养方法。

材 料 与 方 法

一、材料

烟草(*Nicotiana tabacum* L.)品种"G-80"。

二、雌性细胞原生质体分离

用前文方法[6]分离出雌性细胞原生质体后,以微吸管收集并保存于含10%～12%甘露醇、1%葡萄糖、0.3%聚乙烯吡咯啉酮(PVP)、0.25%葡聚糖硫酸钾(PDS)的溶液中备用。

三、生殖细胞原生质体分离

参照吴新莉和周嫦方法[8],将花粉萌发液的蔗糖浓度改为32%。

四、叶肉原生质体制备

按前文方法[7],甘露醇浓度为9.1%,并补充1%葡萄糖。

五、原生质体成对融合

融合液PEG浓度为15%～25%,甘露醇浓度为5.5%,其余成分与前法相同[7]。清洗液成分同雌性细胞保存液。

六、原生质体大量培养

将纯化的叶肉原生质体以KM8p培养基[9](附加1%葡萄糖、0.2%甘露醇、0.2%山梨醇、8.6%蔗糖、0.06% 2-N-吗啉乙磺酸(MES)、1mg/L 2,4-D、1mg/L BA,不加维生素A、D_3、水解酪蛋白、椰子汁,pH5.7)进行液体浅层培养,密度调至10^4～10^5个/ml。

七、微滴培养

参考 Koop 和 Schweiger 方法[10]并作如下改变:在 3cm 培养皿中平铺一薄层石蜡油,以手工操作吸取预先大量培养 1d 的原生质体,以 100~1200nl 不同体积的小滴注入油层,并使之粘附于皿底,每滴含 1~100 个原生质体。每次实验设 1~2 皿大量培养对照。

八、微室饲养培养

以大量培养的叶肉原生质体作饲养细胞。将盛有 5~10 个原生质体的微室(Millicell-CM PICM 01250 MILLIPORE)置于 3cm 培养皿中央(图版Ⅱ,27),皿中盛饲养细胞,静置或低速振荡培养。每次实验设 1~2 皿大量培养对照。

九、观察与摄影

在 Olympus CK2 倒置显微镜下操作,以 Olympus IMT-2 倒置显微镜进行荧光与干涉差观察及摄影。5~10μg/ml FDA 染色观察生活力。

结　　果

一、雌性细胞间的融合

用于融合的雌性细胞包括卵细胞、助细胞及中央细胞。进行了下列组合的融合实验:卵细胞+助细胞(图版Ⅰ,1~3)、卵细胞+中央细胞(图版Ⅰ,4~6)、助细胞+助细胞(图版Ⅰ,7~10)、助细胞+中央细胞、中央细胞+中央细胞(图版Ⅰ,11~13)。雌性细胞在 PEG 中粘连,相向压缩及完成融合的一般规律与体细胞融合相似[7]。上述组合可分为 3 类情况:一是融合双方体积均较大,如中央细胞间的融合,在融合液中易于观察与操作,很快可使两者粘连融合(图版Ⅰ,11~13)。二是融合双方体积均较小,如卵细胞与助细胞间的融合(图版Ⅰ,1~3),观察与操作难度较大。三是双方大小悬殊,如中央细胞与卵细胞的融合

(图版Ⅰ,4~6),由于两者体积及比重差别很大,在融合液中沉降及移动速度不同,操作更为困难。尽管上述三种类型难易不同,需要不同的操作技巧,但一旦互相粘连,即可迅速融合。烟草各类雌性细胞均有明显的大液泡,故较易观察融合过程中的形态变化。曾经观察到一对具几个液泡的助细胞的融合过程:首先质膜融合;然后双方相邻的小液泡融合贯通;再后,融合体中的数个小泡逐渐合并为一个大液泡;最后,细胞质在大液泡挤压下汇集于同一侧,但二核仍然分开,至此两细胞在形态上成为一个整体(图版Ⅰ,7~10)。这种先质膜、再液泡、最后细胞质融合的顺序,与在分离胚囊时诱导的卵器细胞原位融合的情形相似[6]。

二、雌、雄性细胞间的融合

雌、雄性细胞间的融合,包括卵细胞+生殖细胞(图版Ⅰ,14~16)、助细胞+生殖细胞、中央细胞+生殖细胞(图版Ⅱ,17~19)等组合。分离的生殖细胞初时呈梭形,两端有尾状突出,有时可见两尾端向一侧弯曲并连成环状,与 Theunis 等以压片法分离烟草生殖细胞时的情况相似[11]。一段时间后,一部分生殖细胞变为圆球形,估计已形成原生质体,另一部分仍呈梭形。取球形生殖细胞用于融合。

雌性细胞与生殖细胞的融合较难操作,主要由于后者体积很小,易于丢失。故常需以数个生殖细胞对一个雌性细胞的比例释入融合液,方能保证一对一融合的完成。生殖细胞的胞质很少,一经融合很快消失在卵细胞或中央细胞胞质中。其融合过程在光镜观察下犹如一个小细胞直接进入一个大细胞,对后者的形态结构影响很小,看不到液泡合并、胞质汇流等现象。这是雌、雄性细胞融合不同于其它融合组合的明显特点。FDA 染色显示卵细胞与生殖细胞的融合体有生活力(图版Ⅰ,16)。

三、雌性细胞与体细胞的融合

以叶肉原生质体作为体细胞一方与雌性细胞融合,包括卵细胞+体细胞(图版Ⅱ,20~22)、助细胞+体细胞、中央细胞+体细胞(图版Ⅱ,

23、24)、卵细胞与助细胞的原位融合体+体细胞等。不同组合在融合过程中的特点有很大差别。大体上有两种情况:一种如中央细胞与体细胞融合,融合难度较性细胞之间的融合要大,表现在两细胞相互压缩速度较慢,细胞间的界限不易消失。即使在质膜已融合、叶肉原生质体中的少数叶绿体已流向中央细胞的情况下,两者胞质仍不汇合(图版Ⅱ,24),因而也就不能很快形成稳定的融合体。另一种情况是较小的性细胞,如卵细胞与体细胞融合(图版Ⅱ,20~22),两者一经粘连即迅速完成融合。看不到体细胞有任何形态上的变化。由于大量叶绿体的遮掩,融合过程中卵细胞核、质及液泡的动态均难以分辨。

四、原生质体微培养

微滴培养 试验了各种体积的微滴及每滴接种不同数目的原生质体,各种组合均能诱导细胞分裂。从操作难易、细胞分裂频率高低、观察效果等方面综合评价,以每滴300~500nl,含5~10个原生质体最好(图版Ⅱ,25)。以手工操作较易完成。平均分裂频率为10%,最佳情况下,一滴8个细胞中有5个启动分裂。并且在此密度下,原生质体互不重叠,观察效果很好。(图版Ⅱ,26)示微滴中原生质体完成一次分裂。

微室饲养培养 以经过2~3d大量培养的原生质体作饲养细胞,用于饲养微室中的5~10个原生质体。接种时饲养细胞本身尚未分裂,培养6d后出现第1次分裂。被饲养的细胞晚1~2d出现第1次分裂(图版Ⅱ,28)。培养10d后,统计分裂频率可达40%。第1次分裂后4d有细胞团生成(图版Ⅱ,29)。

讨 论

性细胞的体外融合是研究性细胞特点、探讨受精机制的有力手段,也是性细胞工程研究的重要方面。雌性细胞的融合实验,迄今只在玉米中有过报道[2~5,12]。其主要技术关键:一是应用了微电融合方法[13]进行了单对原生质体的融合;二是采用微室饲养方法培养融合体。我

第三章 授粉、受精与胚胎发育的实验研究

图版 I   plate I

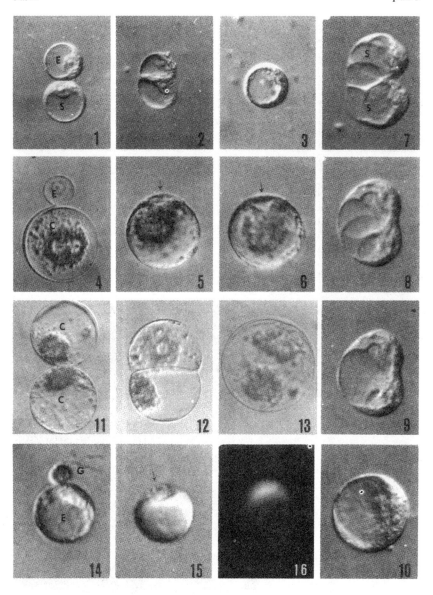

565

图版 II

plate II

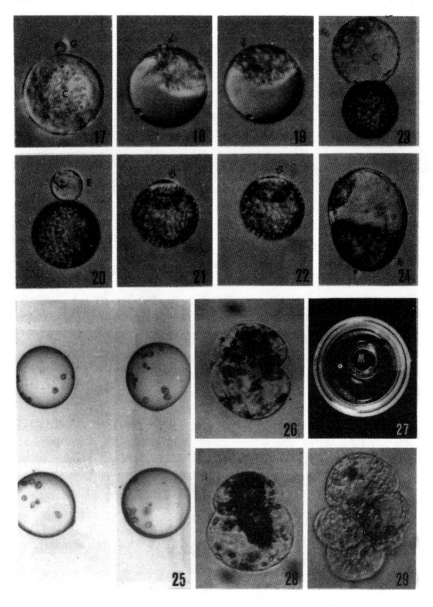

们以 PEG 诱导的单对细胞融合技术,在相对简便的条件下应用于雌性细胞融合的广泛组合取得成功,表明在缺乏特制微电融合设备的条件下,上述方法用于开展配子体外融合等精密实验同样是有效的,并且由于其简便易行,更有推广价值[7]。根据我们的体验,性细胞之间的融合较之体细胞之间的融合在同等条件下似乎更易完成。这是否暗示性细胞膜具有某种不同于体细胞膜的特性,值得进一步探讨。

性细胞单对融合产物数量不多,难以大量培养。唯有采用微培养才是可行的途径。根据 Koop 等创建的微机自动控制的微培养思路[10],本实验以简便的手工操作方法试验少数叶肉原生质体的微滴培养和微室饲养两种方法,取得初步结果。一般建立饲养系统,首先要求建立适宜的胚性细胞系作为饲养细胞[3,5,14,15],而本实验中以几乎同步培养的叶肉原生质体作饲养细胞,更为简便省时。两种微培养方法中,微滴培养的特点是便于追踪观察早期分裂动态,微室饲养培养则较易诱导细胞团形成,便于更换培养基,有利于持续培养。在此基础上,下一步进行烟草雌性细胞融合产物的培养将是有希望的。

图 版 说 明

除图 16 为荧光显微摄影、图 25 和图 27 为明视野显微摄影外,其余均为干涉差显微摄影。

图 版 Ⅰ

1~3. 卵细胞(E)与助细胞(S)融合过程。×490　4~6. 卵细胞(E)与中央细胞(C)融合过程(箭头示融合位点)。×300　7~10. 一对助细胞(S)融合过程,注意其液泡合并过程。×600　11~13. 一对中央细胞(C)融合过程。×210　14、15. 生殖细胞(G)与卵细胞(E)融合过程,箭头示融合位点。×700　16. FDA 荧光染色示生殖细胞与卵细胞融合体活性。×700

图 版 Ⅱ

17~19. 生殖细胞(G)与中央细胞(C)融合过程,箭头示融合位点。×450　20~22. 卵细胞(E)与叶肉原生质体融合过程,箭头示融合位点。×350　23、24. 中央细胞(C)与叶肉原生质体融合过程,箭头示已移入中央细胞的叶绿体。×230

25. 含 5~10 个原生质体微滴。×35 **26**. 微滴内细胞分裂 1 次。×350 **27**. 微室（M）饲养培养装置。×1 **28**. 微室内细胞分裂 1 次。×400 **29**. 微室内形成的细胞团。×300

Explanation of plates

All are Nomarski interference contrast micrographs except Fig. 16 which is fluorescence micrograph, and Fig. 25 and 27 which are bright field photos.

Plate I

Figs. 1-3. Fusion of an egg cell (E) and a synergid (S). ×490 **Figs. 4-6.** Fusion of an egg cell (E) and a central cell (C). Arrows indicate the fusion sites. ×300 **Figs. 7-10.** Fusion of two synergids (S). Note the small vacuoles fusing into a large one. ×600 **Figs. 11-13.** Fusion of two central cells (C). ×210 **Figs. 14, 15.** Fusion of a generative cell (G) and an egg cell (E). Arrow indicates the fusion site. ×700 **Fig. 16.** An egg cell fused with a generative cell showing FCR. ×700

Plate II

Figs. 17-19. Fusion of a generative cell (G) and a central cell (C). Arrows indicate the fusion site. ×450 **Figs. 20-22.** Fusion of an egg cell (E) and a mesophyll cell. Arrows indicate the fusion sites. ×350 **Figs. 23, 24.** Fusion of a central cell (C) and a mesophyll cell. Arrow indicates some chloroplasts moving into the central cell. ×230 **Fig. 25.** Microdroplets with 5-10 protplasts inside. ×35 **Fig. 26.** The first division of a cell within a microdroplet. ×350 **Fig. 27.** A millicell (M) feeder device. ×1 **Fig. 28.** The first division of a cell within a millicell. ×400 **Fig. 29.** A cell cluster formed in a millicell. ×300

（作者：孙蒙祥、杨弘远、周嫦。原载：植物学报,1995,37(1)：1~6。图版 2 幅,参考文献 15 篇,选录图版 I、II。）

3 烟草中央细胞离体受精过程中雌雄核融合生活动态的记录

The first record of dynamics of male-female nuclear fusion in viable tobacco central cell during *in vitro* fertilization

Abstract

In vitro double fertilization in tobacco (*Nicotiana tabacum* L.) was carried out and one of its significant events, the dynamics of bisexual nuclear fusion in the viable central cell, was observed by video-enhanced microscopy for the first time. The observation revealed that after *in vitro* fertilization the male nucleus was first fused with one of the polar nuclei, then the other. The whole fusion process could be finished within two seconds. Morphologically the fusion was very similar to common protoplast fusion mediated by polyethylene glycol. It also went through a series of processes, namely touch, adherence, membrane fusion and content mixture. The male nucleolus moved closely towards the female one but no further fusion was recorded although eventually a big nucleolus was observed in the primary endosperm nucleus. The technique for *in vitro* fertilization and the observation of the nuclear fusion process may enable us to peep at the mechanism of male and female gamete fusion.

受精作用一直是植物生殖发育生物学研究的热点课题。近年发展尤为迅速。特别是诸如偏向受精等新概念的提出更进一步推动了对双受精作用的寻微探秘,日益显现出这一过程的精巧与复杂。但限于体内研究的局限性,对其中一些关键环节,如雌雄配子间的识别、配子融合过程中的相互作用、雄核在雌性细胞内迁移的动力学及雌雄核融合的时间进程与机制等仍知之甚少。离体受精操作及相关技术的建立[1~3]为探讨上述问题提供了新途径。我们在过去工作的基础上以烟草为材料进行了离体双受精研究,以视频增差显微观察系统首次记录到在生活状态下精核进入中央细胞后与雌核融合的动态过程。

材料与方法

所用烟草(*Nicotiana tabacum* L.)种植于意大利锡耶那大学环境生物学系植物园温室内,25~28℃,16h/8h 昼夜交替光照。开花前 1d 去雄。去雄后两天取子房按孙蒙祥等[4]的方法分离中央细胞;参照莫永胜等[5]的活体-离体法分离精细胞;授粉后 36h 切下花柱。培养基成分为 20% Sucrose, 0.01% H_3BO_3 和 0.01% $CaCl_2$, pH5.4。冲击液成分为 9.6% Mannitol;依孙蒙祥等[6]的方法融合中央细胞与精细胞。融合液成分为 17.5% PEG, 5mmol/L $CaCl_2$;依孙蒙祥等[3]的方法将受精的中央细胞转入 MS 无激素培养基(附加 3.0% glucose, 5.4% mannitol, pH5.7)于室温(22~25℃)下作微滴培养;细胞及核融合在倒置显微镜下操作并观察。通过视频增差系统录像并以与之相连的激光共聚焦扫描显微镜(BIORED)监视器存取图像。

结果和讨论

分离后的中央细胞内各种内含物以多种形式快速运动,显示出旺盛的生活力。在显微镜下连续观察 30min 仍可见活跃的胞质运动。在分离的所有中央细胞中未见到受精前极核融合的现象。两个极核一般被大量储藏性物质的颗粒包裹(图版Ⅰ,1)。以 K-K1 溶液作细胞的整体染色时这些颗粒很难着色,将其分离后染色鉴定表明确为淀粉粒,但较之典型的贮藏性淀粉着色为浅(图版Ⅰ,2),分离后的中央细胞已完全无壁,在任一方向与精子融合都易完成(图版Ⅰ,3~5)。将融合后的中央细胞转入培养基微滴中培养,并追踪观察。当精核靠近极核时首先被挡在淀粉粒包被层以外(图版Ⅰ,6)。核的运动在此停滞一段时间,然后进入该层与两极核靠近并被共同的淀粉粒层所包被,进而与其中一个极核紧密接触。在此又有一明显停滞时期(图版Ⅰ,7)。以后雄核与两极核中的一个开始粘连。一旦粘连开始,两核彼此相向压缩的速度非常快(图版Ⅰ,8),直至两核间核膜接触的面积达到最大值后才停止进一步压缩,此时两核在整体上组成一个球体。很快,两核间的膜消失,雄核仁向雌核仁靠近,表明有核质汇流(图版Ⅰ,9)。从开始粘连到融合结束的整个过程可在2s内完成,而从两核明显开始相向压缩到两核间界限消失只需不足1s。在融合的整个过程中,雌、雄核仁始终可见。融合完成后两核仁彼此相接(图版Ⅰ,10),但没有观察到两性核仁进一步融合为一体的过程。最后三核融合为一体,可见一特大核仁(图版Ⅰ,11)。

我们的观察表明:(1)烟草中央细胞离体受精过程中精核首先与一个极核融合,然后再与另外一个极核融合。在玉米中曾报道过中央细胞的受精有两种核融合的形式,即精核先与一极核融合,然后再与另一极核融合,或精核与中央细胞次生核融合[7]。但在本实验中未见受精前有次生核形成。(2)核融合的速度非常快,仅以秒计。通常温度对细胞融合速度是有影响的,可能在自然条件下植物体内的温度由于季节和生长条件的关系较室温(22℃)为高,推测体内实际融合速度会更快。(3)整个融合过程与一般细胞的PEG诱导融合过程非常相似,经接触、粘连、压缩、界膜消失和内含物汇流等过程,尤其与我们曾观察到的卵器细胞的原位自发融合相似[4]。其共同特点为当压缩至最大限度时界膜随即消失,而PEG诱导的细胞融合其界膜常需较长时间,甚至清洗过后才消失。(4)在融合过程中始终可见两性核仁,证明烟草中央细胞的受精属于有丝分裂前型。以往对受精类型的划分主要集中于卵细胞的受精方面,较少注意中央细胞。由于中央细胞核数目多,分裂较早及三倍染色体等特点,其受精过程较之卵细胞可能更加复杂,值得进一步探讨。(5)人工培养条件虽与胚囊内自然条件有较大差异,但从核融合及胞质运动的情况看其细胞内微环境仍相对稳定。融合后的中央细胞仍可进行核的分裂(图版Ⅰ,12),表明其自然发育规律至少在早期阶段并未因离体操作而有较大改变。

离体受精技术使在生活状态下观察雌雄核融合成为现实。但我们从烟草的工作中发现仍有许多尚待克服的困难:(1)融合过程对光照非常敏感。长时间暴露在显微镜光照下,融合过程可停滞在任一阶段,甚至丧失细胞活性,故无法保持持续观察记录。而融合本身又是一极快的过程,如何准确地捕捉发生融合的 2s 进行观察记录,同时又能最大限度地减少观察次数以保持细胞活性使之更接近体内自然状态,殊非易事;(2)烟草中央细胞含大量淀粉粒,人工受精后的培养过程中只在少数中央细胞内淀粉粒明显减少而将核界限暴露出来。多数因淀粉粒干扰而难以做清晰的细胞学观察;(3)中央细胞是生活的,其胞质仍在不断运动,细胞质索常常掩盖住观察目标,只有在核靠近细胞边缘、在细胞的最上缘或最下缘时,观察才较为清晰。基于上述困难,欲在统计量的基础上对融合过程作出较准确的时间进程表尚待进一步努力。

(作者:孙蒙祥、Moscatelli A、杨弘远、Crest M。原载:植物学报,1999,41(8):906~908。图版1幅,参考文献7篇,均删去。)

第三节 合子、中央细胞与幼胚的操作

提要

20 世纪 90 年代以来,由于卵细胞、合子及受精前后中央细胞的分离和离体受精的成功,合子与中央细胞的离体培养被提上研究日程。德国研究者率先完成玉米离体合子培养再生植株。随后,大麦、小麦、水稻的自然合子培养也相继再生了植株。玉米离体受精的中央细胞也

发育成类似胚乳的构造。这一连串研究进展预示了以离体实验系统对植物胚胎发生及胚乳发生的最早阶段开展基因表达与遗传操作研究的广阔前景。

我们也积极投入了这一研究领域。在起步较迟的情况下,我们选择了国际上尚无先例的双子叶植物烟草及当时尚未见报道的重要禾谷类作物水稻作为重点研究对象。从1996年起,用了5年时间建立了这两种植物的合子、中央细胞、初生胚乳细胞的分离与早期培养体系,进行了较为细致的细胞生物学研究。

在烟草中,首先要解决受精后胚囊及其中所含合子的分离方法。受精后,胚珠组织发生了变化,不能采取分离受精前胚囊的方法。我们摸索了较适宜的"酶解-研磨法",解决了这一问题。应用微室饲养技术,分离的合子及二细胞原胚可以离体发育至多细胞阶段;合子的第一次分裂频率可达60%左右。未受精中央细胞及稍后的初生胚乳细胞培养亦可发育至多细胞阶段。

在水稻中,我们于1998年在国际上首次报道了卵细胞与合子的分离,进行了相应的荧光显微观察。为了开展难度较大的合子培养,先研究了水稻幼胚培养技术,使授粉后2~3d的幼胚再生了植株,比前人幼胚培养的起点提早了2d。接着进行合子培养,使合子第一次分裂频率达60%以上,并发育成多细胞团。受精前、后的中央细胞培养则诱导了游离核分裂。

在初步建成上述各种离体培养系统之后,我们开始着手利用它们进行下列三方面的探索:一是受精与早期胚胎发生过程中的基因表达。我们已在烟草、水稻中由原胚与刚分化的幼胚构建了cDNA文库,着手筛选原胚分化期间的优势表达基因。二是幼胚与合子的遗传转化。用电激法将报告基因导入烟草原胚获得瞬间表达之后,又进一步利用自行设计的装置,对少量烟草合子进行电激处理,电激后的合子表达了GFP基因,并存在发育的潜能。三是有关细胞生物学研究。不仅对分离的小麦与水稻合子进行了钙和钙调素的荧光显微定性观察,而且今后拟利用新装备的共聚焦激光扫描显微镜与Cooled CCD等设备开展定量的检测。以上三方面研究目前仅仅开始,以后将继续深入下去。

1 烟草受精后胚囊和合子的分离及合子的离体分裂

Isolation of fertilized embryo sacs and zygotes and triggering of zygote division in vitro in *Nicotiana tabacum*

Abstract

Three methods were established to isolate fertilized embryo sacs in *Nicotiana tabacum*, i. e. enzymatic maceration combined either with shaking, microdissection or grinding, respectively. Living fertilized embryo sacs of various developmental stages after fertilization could be isolated successfully by these methods. Each method had its own adoptation to the materials of different developmental stages. Among them the method of enzymatic maceration combined with grinding was the best: Ovules were first treated in enzymatic mixture (1% Cellulase R-10, 0.5% Macerozyme R-10, 12% mannitol, pH5.7) for about 30 min. Then droplets of the ovule suspension were gently grinded by a flat-headed glass rod. After grinding several droplets of mannitol solution (8%-10%) were added for releasing and washing embryo sacs. Compared with the other two methods this method was more convenent and had higher isolation efficiency. Isolation of fertilized embryo sacs offered a good means for microscopic observation on the postfertilization development including synergid degeneration, endosperm formation and zygote changes without interference by the surrounding sporophytic tissue. Living zygotes and endosperm cells could be further isolated by a second enzymatic maceration procedure followed by brief micromanipulation. Several characters had been found to distinguish the protoplasts of free zygotes from those of other cell sources. Isolated zygotes were cultured in microchambers (Millicell-CM) fed with macrocultured mesophyll protoplasts. The first division of zygotes was induced, resulting in proembryos consisting of two cells.

未受精胚囊及卵细胞的分离已在多种植物中取得成功[1~3]。10 年来先后建立了各具特色的多种分离技术[4~8],推动了性细胞离体操作的飞速发展。但另一方面,受精后胚囊及自然合子的分离因难度很大,进展甚微。本文研究了以不同方法分离烟草受精后生活胚囊及合子的效果,并在离体培养条件下启动了合子的第一次分裂。

材 料 与 方 法

一、试验材料

供试材料为烟草(*Nicotiana tabacum* L.)品种"G-80"。大田材料 4 月上旬播种,5~10 月采花;温室材料 9 月底播种,次年 2~5 月采花。

二、受精后胚囊分离

取受精后子房剥取胚珠,以下列 3 种方法分离:(1)酶解-振荡法:将胚珠投入含 10% 纤维素酶、0.5%~0.8% 离析酶、0.5% 半纤维素酶、10% 甘露醇、pH5.7 的酶液。每毫升酶液处理 1~2 个子房的胚珠(下同),在 30℃下保温振荡 0.5h,再静置 0.5h。如此反复两次,共处理 2~2.5h。用滴管轻轻吸打以释放胚囊;(2)酶解-解剖法:将胚珠投入含 1% 纤维素酶、0.5% 离析酶、12% 甘露醇、pH5.7 的酶液中,在 30℃下静置 25~45min。去酶液后以 12% 甘露醇洗胚珠 2~3 次。在 Olympus CK2 倒置显微镜下以自制玻璃解剖针解剖出胚囊;(3)酶解-研磨法:酶解及洗涤同(2)法。以滴管吸取 1 滴胚珠悬浮液置于小培养皿底部,用平头玻棒轻轻挤压研磨数次后,加入 8%~10% 甘露醇溶液。以自制微吸管将上述 3 种方法分离的胚囊收集于洗液中,洗液成分为:10% ~ 12% 甘露醇,0.25% 聚乙烯吡咯啉酮,0.25% 葡聚糖硫酸钾。

三、合子及胚乳细胞的分离

将以酶解-研磨法分离的胚囊在含 1% 纤维素酶、0.5% 离析酶和 10% 甘露醇的酶液中室温下酶解 1 ~ 2h,然后在倒置显微镜下解剖出合子与胚乳细胞及其原生质体。

四、合子的培养

将分离到的合子原生质体以洗液(配方同上)洗 2 ~ 3 次,并去除可能裹杂的其它细胞原生质体。用微吸管将合子转入底部覆盖有 KM8p 培养基[9]的微室(Millicell-CM PICM 01250, MILLIPORE)中,在 25℃下进行饲养培养[10]。

五、观察与摄影

胚囊及合子生活力以荧光素二醋酸酯(FDA)染色测定。细胞壁以荧光增白剂染色鉴定。用 Olympus IMT-2 倒置显微镜进行荧光与干涉差观察并摄影。

结　　果

一、子房长度与内部发育时期的对应关系

为取材方便,观察、比较了花外部形态变化与内部发育时期的关系。开花当天花药开裂,开花后 1 ~ 2d,花冠由粉红变为淡紫,此时已受精。开花后 3 ~ 4d,花冠褐色并枯萎,易脱落。其冠长可随栽培条件而异,花色则因季节气候而变,惟子房长度较为稳定,宜作为取材标准。表 1 列出子房长度与内部发育时期的对应关系。

表1　　　　　　子房长度与内部发育时期的对应关系

Table 1　Correlation of ovary length with internal developmental stage

开花天数 Days after flowering	子房长度 Ovary length (mm)	内部发育时期 Internal developmental stages
0	5.0~7.0	成熟胚囊。卵细胞有大液泡,核明显,两个极核。 Mature embryo sac. Large vacuole and clearly observed nucleus in egg cell. Two polar nuclei.
1~2	6.5~8.0	刚受精。合子有大液泡,核不明显,1个助细胞退化,初生胚乳核形成。 Just fertilized. Zygote with large vacuole and undistinguishable nucleus. One synergid obviously degenerated. Primary endosperm nucleus.
3	8.0~10.0	合子大液泡消失,核不明显,宿存助细胞液泡变小,初生胚乳细胞分裂1次。 Nucleus undistinguishable and large vacuole disappeared in zygote. Vacuole getting smaller in persistant synergid. Two endosperm cells.
4~5	9.0~11.0	合子体积增大,两个助细胞均退化。胚乳细胞3个以上。 Zygote enlarged in volume. Both synergids degenerated. Three or more endosperm cells.
6~7	10.0~14.0	合子伸长,核显著,或已分裂成二细胞原胚,胚乳细胞多个。 Zygote elongated with clearly observed nucleus or two-celled proembryo. Multiple endosperm cells.

二、三种分离方法的比较

酶解-振荡法、酶解-解剖法和酶解-研磨法均可分离出受精后各发育时期完整的生活胚囊。3种方法在操作与效果上各有其特点及局限性。

振荡法操作简便,适于大量材料的处理。可直接得到原生质体化的胚囊,但酶解时间较长,且难以分离初生胚乳细胞分裂 1 次后各时期的胚囊;解剖法酶解时间短,有利于保持胚囊的生活力,其胚囊成员细胞较接近活体内自然状态,但需熟练的操作技巧和较长的操作时间,适于分离胚珠体积较大,含多个胚乳细胞时期的胚囊;研磨法酶解时间亦短,可同时得到合子原生质体化或接近体内自然状态的胚囊,分离效率明显为高,但分离受精后早期胚囊较为困难。综合比较上述 3 种方法,以酶解-研磨法较为实用,故本文进一步实验均以此法分离胚囊与合子。

三、受精后胚囊的离体显微观察

分离的胚囊可清楚地观察受精后的发育变化(图版Ⅰ,1~5)。刚受精的胚囊在形态上与未受精胚囊相似,但可根据如下特征区别两者:(1)受精后胚囊在珠孔端常附有较长的尾状结构,由助细胞丝状器、花粉管及珠孔周围细胞组成(图版Ⅰ,1);(2)受精后 1 个助细胞明显退化,其胞质及液泡界线已消失(图版Ⅰ,6);(3)已形成一大的初生胚乳核,核内可见一大核仁,核的位置已由卵器附近向合点端移动(图版Ⅰ,6)。

受精后,初生胚乳细胞很快分裂,两子细胞形态大小接近(图版Ⅰ,2),再次横分裂,形成 3~4 个大小不等沿胚囊长轴排列的胚乳细胞(图版Ⅰ,3),此后各子细胞先后纵裂,形成 8 个细胞,胚囊体积明显增大(图版Ⅰ,4)。进一步分裂,胚乳细胞排列无序,数目难计(图版Ⅰ,5)。

刚受精时 1 个助细胞已呈退化状态,宿存助细胞自初生胚乳细胞第一次分裂时起液泡逐渐变小(图版Ⅰ,7),至形成 4 个胚乳细胞时明显退化(图版Ⅰ,8),并在 8~16 个胚乳细胞时消失。

受精后合子的大液泡很快消失(图版Ⅰ,7),以后合子体积增大,形状变长,因酶解程度与操作方法不同而呈各种形态(图版Ⅰ,9;图版Ⅱ,10~12),但易与其它细胞区别。

四、合子及胚乳细胞原生质体的分离

将分离到的胚囊进一步适度酶解,辅以显微解剖,可得到生活的合子原生质体(图版Ⅱ,13、14)及胚乳细胞原生质体。FDA 染色显示了合子较强的生活力(图版Ⅱ,15),荧光增白剂染色无荧光反应。

不同时期的胚囊,分离合子的难易各异。早期胚乳细胞少,合子易从胚囊中逸出。形成多个胚乳细胞后,合子与胚乳细胞联系紧密,不易分离,分离的合子上常粘附破损胚乳细胞的残留物,难以除去。

游离的合子原生质体有时易与胚囊周围的体细胞原生质体混淆,但体细胞内常有折光性强的颗粒物质,核位置难辨;合子则无折光性强的颗粒,胞质浓厚,核位置可辨,且核占细胞较大比例。据此可区别两者。

五、合子的培养

曾试验比较了多种培养方法,以微室饲养培养效果最好,初步结果表明,以预培养 6~8d,已分裂 1~2 次的烟草叶肉细胞原生质体作饲养细胞较为适宜。刚接种时合子原生质体呈圆形,胞质均一。培养 3d 后,略呈椭圆,胞质中出现颗粒状物质。培养 6d,合子启动分裂(图版Ⅱ,16),有时分裂面不甚规则。形成的两子细胞只有大小的区别,无形态上的明显差异。与活体内形成的二胞原胚显著不同(图版Ⅱ,17)。

讨 论

一、受精后胚囊及合子的分离

10 年前,Allington[11] 就曾尝试过用显微解剖法分离禾本科植物受精后胚囊及其成员细胞,但未能成功地分离完整的生活细胞。此后,Compenot 等[12] 和 Möl 等[13] 先后从事玉米受精后胚囊的培养,因未能分离出完整的胚囊,而将胚囊连同周围大块珠心组织一同培养得到了

完整植株。Huang 和 Russell[14]、Huang 等[15]曾以短时酶解后手工解剖的方法在有菌条件下获得过少量烟草受精后胚囊。最近,Holm 等[16]绕过胚囊分离的难关,直接解剖未经酶解的大麦子房,获得了生活的合子。其关键在于事先对大麦子房进行精细的切割,使胚囊珠孔端近于裸露,然后施以显微解剖。这要求精细的工具且对较小的胚珠很难适用。总之,与未受精胚囊及卵细胞的分离相比,受精后胚囊及合子的分离进展较慢。迄今尚未见能大量分离合子的报道。上述大麦合子的分离是目前唯一成功的一例。

分离受精后胚囊的主要困难在于胚珠受精后在结构及成分上发生了种种变化:花粉管的插入使珠孔通道被堵塞,酶液难以渗入;丝状器与珠孔端组织结合紧密,致使胚囊固着于原位;此外,受精后胚珠合点端组织的结构与成分难酶解,阻碍了胚囊由珠柄维管束切口处逸出。这是包括酶解-渗透压冲击法[8]在内的现有胚囊分离方法不能奏效的原因。实验表明,仅靠酶解作用尚不足以分离受精后胚囊,必须辅以一定的机械外力。本文报道的酶解-研磨法有酶解时间短和操作简便的优点,又避免了显微解剖操作上的困难,明显提高了效率。尤其对如烟草等胚珠体积较小的材料更有实用意义。由分离的胚囊进一步分离合子,在操作及识别上都更简易而准确。

二、合子的离体培养

合子培养中首先要解决的是密度问题,饲养培养看来是目前解决这一问题的较好方法。除了分离合子本身的状态外,选择何种培养物作为饲养细胞是培养成功的关键因素。Holm 等[16]在大麦合子培养中强调了以大量培养的小孢子作为饲养物的重要性。认为小孢子同步的、旺盛的胚胎发生能力对合子的离体持续发育是至关重要的,而小孢子的不同发育时期又明显影响合子离体发育的式样。以此法培养自然合子首次再生了植株。Kranz 等[17]则以悬浮细胞作饲养物,诱导玉米人工合子的高频分裂及植株再生。表明饲养细胞本身具有胚胎发生能力并非必要条件。本文以经预培养的叶肉细胞原生质体作饲养细胞也启动了合子分裂。对持续发育的影响有待观察。

在离体条件下合子的发育模式是否与体内一致,特别是胚胎发生早期式样是否受外界条件控制是很有意义的理论问题。目前国际上对人工合子及自然合子的培养虽各有成功的一例,但尚缺乏详细的胚胎发生过程的报道,这是值得进一步研究的。合子的分离与培养使胚胎培养的研究由以往的原胚培养进一步提早到单细胞的起点,这对于建立在离体条件下研究胚胎发生全过程的实验体系,并依仗这一体系深入开展有关发育生物学研究,以及开拓细胞工程新技术展现了重大的前景。

(作者:傅春梅、孙蒙祥、周嫦、杨弘远。原载:植物学报,1996,38(4):262~267。图版2幅,表1幅,参考文献17篇。仅选录表1。)

2 烟草合子与二胞原胚在离体培养中的发育

In vitro development of zygotes and two-celled proembryos in *Nicotiana*

Abstract

Fertilized embryo sacs of *Nicotiana tabacum* cultivar G-80 and *N. tabacum* var. *macrophylla* were isolated by enzymatic maceration and grinding, then the zygotes or two-celled proembryos were isolated by microdissection. Three to five zygotes or two-celled proembryos were embedded in agarose in microchambers (Millicell-CM, MILLIPORE) fed with 3-or 4-day-old mesophyll protoplasts of *Nicotiana tabacum*, *N. tabacum* var. *macrophylla* or *N. rustica* and cultured in KM8p basic medium supplemented with various additions at 25℃ in darkness. The zygotes underwent first division after 3d culture and the division frequency was approximately 60%. Microscopical observations showed that the first cell division was unequal in most of the zygotes. Few proembryos and multicellular structures were acquired after 12d culture. The two-celled proembryos also were induced to form multicellular proembryos. We also studied the effects of the isolation method, zygote stage and the genotype and age of feeder cells on *in vitro* development of zygotes.

被子植物的合子由于被层层孢子体与配子体组织包裹,而本身又相当柔嫩,因此其分离和培养难度均较大。虽然近年取得了重大的突破,在三种禾本科植物(玉米、大麦、小麦)中合子培养再生植株成功[1~3],但在双子叶植物中仅本实验室在烟草中作了尝试,少数合子培养启动了一次分裂[4]。本文进一步研究了烟草合子和二胞原胚的培养,取得了较高的分裂频率,并发育到幼小原胚或细胞团,还初步研究了不同因素对合子早期离体发育的影响。

材料与方法

一、材料

试验材料:烟草(*Nicotiana tabacum* L.)品种"G-80"、大叶烟草(*Nicotiana tabacum* var. *macrophylla*)和黄花烟草(*Nicotiana rustica* L.),取三者幼苗叶片分离叶肉原生质体,取前二者盛开的花分离胚囊。

二、合子和二胞原胚的分离

先按傅春梅等的酶解-研磨法[4]分离受精后胚囊,然后采用2种方法分离合子和二胞原胚:(1)酶解法:将胚囊在含0.5%纤维素酶(Cellulase R-10)、0.3%离析酶(Macerozyme R-10)和10%甘露醇的溶液中25℃下酶解0.5~1h,然后用微吸管轻轻吸打数次,使合子或二胞原胚分离;(2)解剖法:将分离的胚囊置于10%甘露醇中,用自制玻璃针直接解剖胚囊。

三、合子和二胞原胚的培养

分离的合子和二胞原胚以下列两种方法培养:方法一是将微室(Millicell-CM, MILLIPORE)置于盛大量培养的叶肉原生质体饲养物的培养皿中,微室中预先加入约 100μl 培养基,然后将 3~5 个合子或二胞原胚转入微室之中;方法二是用微吸管吸取 0.5% 的低熔点琼脂糖(Agarose TypeI, Low EEO, Sigma)在微室内膜上做数个微滴(每滴约 10μl),将合子或二胞原胚转入微滴中,在 4℃ 下待琼脂糖凝固后,将微室置于叶肉原生质体中饲养。以上两种方法均能诱导合子和二胞原胚分裂,而以固体培养较易定点观察。培养基为 KM8p[5],附加 0.05mol/L 葡萄糖,0.25mol/L 蔗糖,0.15mol/L 甘露醇,0.1mol/L 山梨醇,0.5mg/L 6-BA,0.5mg/L NAA,pH5.7。在 25℃ 与黑暗条件下静置培养。

四、观察与摄影

在 Olympus CK2 倒置显微镜下操作,以 Olympus IMT-2 倒置显微镜进行观察与摄影。由于培养材料是透过培养皿和其中的微室进行活体摄影的,因此虽以 Nomarski 干涉差方式摄影,分辨率仍然受到影响。

结 果 与 讨 论

一、合子在离体培养中的发育

刚分离的胚囊绝大部分由不同程度原生质体化的胚乳细胞构成,合子仅在珠孔端占很小的位置(图版Ⅰ,1、2)。刚受精的早期合子分离后常呈圆球形(图版Ⅰ,3),而晚期合子分离后多呈椭圆形或卵形(图版Ⅰ,2)。先后作了 64 个微室培养,共计 242 个合子。培养 2d 后,合子伸长(图版Ⅰ,4)。3~4d 时,多数合子已经以不等或均等分裂方式形成两个大小不等(图版Ⅰ,5、6)或相等(图版Ⅰ,7)的子细胞。培养 5~6d 后,由不等分裂所产生的二细胞原胚中,相当于体内顶细胞的较小细胞进行纵分裂(图版Ⅰ,8、9)。培养12d后,有的合子继续分裂形成原胚状结构(图版Ⅰ,10)或不定形的细胞团(图版Ⅰ,11)。

二、影响合子离体发育的因素

合子分离方法的影响 在所采用的两种分离方法中,酶解法较为简便,而解剖法较为费时且需熟练的技巧。但从培养结果看,解剖法分离的合子启动第一次分裂时间较早,分裂频率远高于酶解法,并且至今所获多细胞结构均来自解剖法(表1)。总的看来,解剖法显然优于酶解法。Leduc等的专门实验指出:酶处理不利于玉米合子的持续分裂和产生正常胚状结构[6]。因此,尽量减少分离过程中酶处理的有害影响是合子培养中一个值得注意的因素。

表1 分离方法对烟草合子培养的影响*

Table 1　Effect of isolation methods on zygote culture in *Nicotiana*

分离方法 Isolation method	培养合子数 No. of Zygotes cultured	合子第一次分裂 Zygote first division			形成细胞团数 No. of multicellular structures
		时间 Time (d)	分裂数 No. of zygotes divided	频率 Frequency (%)	
解剖法 Dissection	84	3.5	51	60.7	6
酶解法 Enzymatic maceration	13	5.0	5	38.5	0

* 饲养细胞为培养3~4d分裂1~2次的叶肉原生质体。

　　Feeder cells were mesophyll protoplasts cultured for 3-4 days and divided once or twice.

合子发育时期的影响 不同发育时期的合子均能在离体条件下发育,但表现有所差别:与受精后不久的早期合子相比,晚期合子一般启动第一次分裂早1~2d,并且全部为不等分裂方式(早期合子约1/3为均等分裂);此外,晚期合子通常发育为原胚状结构,而早期合子则除发育为原胚外,也可形成无规则的细胞团。晚期合子之所以离体发育较好,可能是由于在体内已经通过了由 G_1 期 S 期向 G_2 期的转变,更接近 M 期之故,需要进一步应用细胞光度测定等方法加以验证。

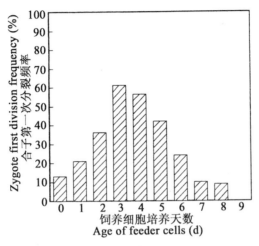

图 1 饲养细胞培养天数对合子培养的影响

Fig. 1 Effect of the age of feeder cells on zygote culture

饲养细胞种类与培养天数的影响 在没有饲养细胞的条件下,合子与二胞原胚均不能诱导分裂。采用烟草,大叶烟草,黄花烟草三种叶肉原生质体分别饲养烟草与大叶烟草合子,结果均能诱导合子分裂,频率为 45.4% 至 62.5% 不等,差异不明显。Holm 等以大麦小孢子饲养大麦和小麦的合子均再生植株[2]。Leduc 等以大麦小孢子,玉米悬浮细胞,分别饲养玉米合子亦均获成功[3]。表明在一定范围内采用远缘饲养物也是有效的。以不同培养天数的叶肉原生质体饲养合子,结果是有规律的(图1)。刚分离的叶肉原生质体饲养效果较差,随着叶肉原生质体培养天数的增加,其饲养效果上升,以培养 3~4d 的饲养物诱导合子分裂频率最高,达 56.5%~61.1%,以后又逐渐降低,培养 9d 以后的饲养物无诱导能力。在大麦合子培养中,饲养小孢子的发育时期也有重要作用,以培养 14d 的小孢子效果最好[2]。

三、二胞原胚的培养

分离的二胞原胚由顶细胞与基细胞组成,前者较小而呈半球形,后者较大呈长形(图版Ⅰ,12)。一共培养 81 个二胞原胚,启动 35 个分裂发育。培养 3d 时顶细胞第一次分裂,5d 时第二次分裂(图版Ⅰ,13)。以后,顶细胞与基细胞各自继续分裂而形成多细胞原胚(图Ⅰ,14)。若分离二胞原胚时酶处理较长,易使顶细胞与基细胞脱离(图版Ⅰ,15)。在此种情况下,二者均可在培养过程中分裂而形成细胞团(图版Ⅰ,16)。

胚胎培养是植物细胞工程的重要手段,也是研究植物胚胎发育的重要实用方法。幼胚特别是原胚培养的技术难度远较近成熟胚培养为大,至今成功的物种不多,且多为多细胞原胚的培养[7]。至于合子培养,则是近几年方才开始[1~3]。胚胎培养的起点由成熟胚向幼胚再向合子时期提早的趋势,标志着培养技术的不断进步。从而有可能在人工控制的离体条件下研究植物个体最早阶段的发育规律。合子培养有两项主要的技术前提:一是合子分离技术;二是微培养技术,而微室饲养[8]是至今唯一用于合子培养取得成功的方法。本实验在改进了原有烟草合子分离方法[4]的基础上,进一步试探了合子与二胞原胚的微室饲养,取得了一定的结果,合子第一次分裂频率较高,少数可发育至多细胞原胚或细胞团。进一步的离体发育有待继续研究。

图 版 说 明

图 版 Ⅰ

图 1. 刚分离的受精后胚囊,箭头示合子。×220

图 2. 图 1 胚囊内合子的放大。×670

图 3. 分离的合子。×670

图 4. 培养 2d,合子引长。×670

图 5. 培养 4d,合子第一次分裂。×670

图 6. 图 5 的 DAPI 荧光染色,示细胞核。×670

第三章 授粉、受精与胚胎发育的实验研究

图版 I

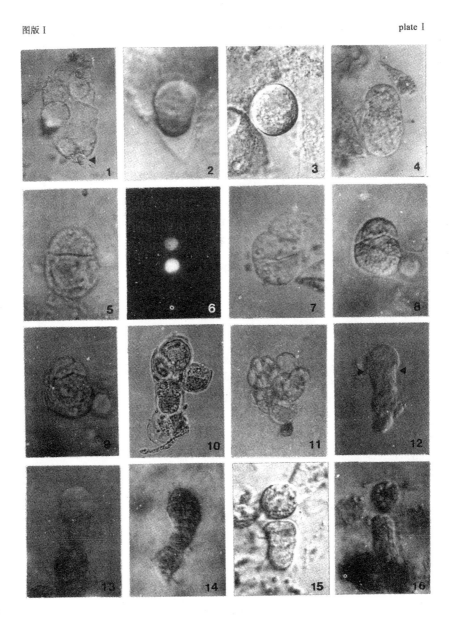

plate I

图 7. 合子第一次均等分裂。×670
图 8. 合子第一次不等分裂。×670
图 9. 图 8 中合子第二次分裂。×670
图 10. 培养 10d,形成原胚状结构(摄影时已开始退化)。×400
图 11. 培养 13d,形成细胞团。×240
图 12. 分离的二胞原胚,箭头示顶细胞与基细胞分界。×400
图 13. 二胞原胚顶细胞分裂二次。×400
图 14. 二胞原胚分裂多次。×400
图 15. 分离的二胞原胚,顶细胞与基细胞已分开。×400
图 16. 图 15 中顶细胞与基细胞分别分裂多次。×400

Explanation of plate

Plate I

Fig. 1. Freshly isolated fertilized embryo sac. Arrow head indicates the zygote. ×220

Fig. 2. Magnification of the zygote in Fig. 1. ×670

Fig. 3. A single isolated zygote. ×670

Fig. 4. Zygote elongation after 2d of culture. ×670

Fig. 5. Zygote first division after 4d of culture. ×670

Fig. 6. DAPI fluorescence of Fig. 5 showing two nuclei. ×670

Fig. 7. Zygote first equal division. ×670

Fig. 8. Zygote first unequal division. ×670

Fig. 9. Second division of the same zygote as in Fig. 8. ×670

Fig. 10. A proembryo-like structure derived from zygote after 10d of culture. (already showed sign of degeneration). ×400

Fig. 11. A multicellular cluster derived from zygote after 13d of culture. ×240

Fig. 12. An isolated two-celled proembryo. Arrow heads indicate the boundary line between apical cell and basal cell. ×400

Fig. 13. The apical cell of two-celled proembryo had undergone two divisions. ×400

Fig. 14. A multicellular proembryo derived from two-celled proembryo. ×400

Fig. 15. An isolated two-celled proembryo. Its apical cell and basal cell had separated from each other. ×400

Fig. 16. The apical cell and basal cell in Fig. 15 divided many times individually. × 400

(作者:李师弟、傅缨、孙蒙祥、杨弘远、周嫦。原载:实验生物学报,1998,31(3):309~315。图1幅,图版1幅,表1幅,参考文献8篇。选录全部图、表。)

3 烟草未受精中央细胞及其它胚囊细胞的离体分裂

In vitro divisions of unfertilized central cells and other embryo sac cells in *Nicotiana tabacum* var. *Macrophylla*

Abstract

The embryo sacs and female cells could be isolated from the unfertilized ovules of *Nicotiana tabacum* L. var. *macrophylla* which were treated in a solution containing 1.5% cellulase R-10, 1% Macerozyme R-10, 10% mannitol, 10 mmol/L $CaCl_2$, pH5.8 for 3h followed by given slight pressure with a micropipette. The central cells could be kept viable for 10h and the egg cells for 3h in 10% mannital. Sometimes, the *in situ* fusion products of egg cell and synergid protoplasts could be obtained and kept viable for at least 5h. The high concentration (20 mg/L) of 2,4-D was used in enzyme solution to induce the division of the unfertilized central cells and other megagametophytic cells in subsequent culture. Treatment of 2,4-D together with enzymatic maceration of ovules was proved to be better than its direct treatment of isolated embryo sac or its component cells. Isolated embryo sacs were cultured in microchambers (Millicell-CM PICM 01250 MILLIPORE) fed with divided mesophyll protoplasts of *Nicotiana* rustica L. The medium was KM8p medium supplemented with 1% glucose, 0.1 mol/L mannitol, 0.1 mol/L sorbitol, 0.25 mol/L sucrose, 1 mg/L BA, 6% to 10% coconut water, and 0.15% low gelling agarose. Division of central cells, antipodal cells and the *in situ* fusion products of egg cell and synergid protoplasts were induced. The unfertilized central cell was for the first time to be induced *in vitro* to develop into small cell clusters.

自20世纪70年代中期以来,未传粉子房和胚珠的离体培养已在多种植物中取得成功,得到的单倍体植株来源于胚囊中的卵细胞、助细胞以及反足细胞[1]。而分离的未受精胚囊及其成员细胞的离体培养虽屡经尝试,迄今只有Kranz等[2]诱导了玉米未受精卵细胞分裂形成小愈伤组织,至于中央细胞与其它雌配子体细胞则无离体分裂的报道。本文报道大叶烟草未受精中央细胞首次培养成细胞团及其它胚囊细胞启动离体分裂的实验结果。

材 料 与 方 法

一、未受精胚囊及其成员细胞的分离

取大叶烟草(*Nicotiana tabacum* L. var. *macrophylla*)未开放花中的胚珠,参考孙蒙祥等[3]的一步酶解-渗透压冲击法,以1.5% cellulase R-10、1% macerozyme R-10、10%甘露醇、10 mmol/L CaCl₂保温振荡3h后,以酶基液(10%甘露醇、10mmol/L CaCl₂)换去酶液,用自制微吸管轻压胚珠,使胚囊及其成员细胞从珠孔处逸出。

二、2,4-D 处理

(1)参考Kranz等[2]的方法,将分离的胚囊及其成员细胞分别置于含10mg/L、20mg/L、40mg/L 2,4-D的酶基液中处理1h,然后培养。(2)在酶解液中添加同样浓度的2,4-D于酶解同时处理胚珠。

三、未受精胚囊及成员细胞的培养

采用 KM8p 培养基[4],附加 1% 葡萄糖、0.1mol/L 甘露醇、0.1mol/L 山梨醇、0.25mol/L 蔗糖、1mg/L BA、6% ~ 10% 椰乳(Sigma), pH5.8,根据实验需要添加不同浓度超低熔点琼脂糖(Sigma),分别以微滴培养法[5](每体积为 500nl 的小滴中含 5 ~ 8 个细胞)、琼脂糖珠饲养法[6](琼脂糖浓度为 0.6%)和微室饲养法[5]三种方法培养。饲养细胞采用大量培养正在分裂的黄花烟草(*Nicotiana rustica* L.)叶肉原生质体。

四、观察与摄影

以荧光素二醋酸酯(FDA)染色观察生活力;以 DAPI 染色观察细胞核;以荧光增白剂(Calcofluor white ST)染色观察细胞壁。在 Olympus CK2 倒置显微镜下操作,用 Olympus IMT-2 倒置显微镜和 Olympus BHS 顺置显微镜观察摄影。

结 果 与 讨 论

一、胚囊分离技术的改进

用孙蒙祥等[3]的酶解-渗击二步法和一步法必须经吸打后才可获得较大数量的胚囊及其成员细胞,而本文所用改良方法只用微吸管在酶解后的胚珠上略微施加压力,就可使胚囊从胚珠中逸出,并可立即吸出培养。FDA 染色表明,用微吸管挤出的胚囊细胞生活力明显强于吸打后分离的细胞,而且一经逸出即转入培养基的胚囊细胞也比浸于酶基液里等待收集的细胞生活力强。生活力的强弱是决定胚囊细胞培养前途的关键因素之一,因此虽然本法分离速度较慢且需要一定的操作技巧,但更适合于此后的离体培养。

因分离过程中酶液渗入胚珠的程度不同,分离出的产物主要分以下 3 种:(1)成员细胞接近体内自然状态的胚囊。(2)成员细胞原生质体化的胚囊(图版Ⅰ,1),其中一部分的卵器细胞发生如孙蒙祥等[3]所报道的原位融合(图版Ⅰ,2~5)。(3)游离的胚囊细胞原生质体(图版Ⅰ,6)。刚分离出的中央细胞、两个极核都尚未融合成次生核(图版Ⅰ,7)。FDA 测定表明,中央细胞一般生活力较强,在酶基液中保存 10h 后仍有较强荧光;而单独的卵器成员细胞在保存 3~4h 后多数死亡。不过卵器细胞的原位融合体生活力较强,保存时间可达 5h 以上。在分离过程中,反足细胞易丢失。

二、2,4-D 处理方法的选择

未受精的胚囊细胞在自然情况下一般不分裂,要诱导其在离体条件下生长发育需有一定的外界刺激。Kranz 等[2]用 25~40mg/L 的 2,4-D 处理玉米未受精卵细胞能诱导其启动分裂。但若沿用其方法将分离产物直接浸于高浓度 2,4-D 溶液中对其生活力影响极大,培养后无发育前途,特别是用 40mg/L 2,4-D 处理后,培养 20h 就已褐化死亡。这大概是由于 2,4-D 作为双子叶植物除草剂,烟草和玉米对其敏感程度不同所致。若在酶解液中添加 2,4-D 于酶解同时处理胚珠,由于胚囊为珠被、珠心组织包裹,经 40mg/L 2,4-D 处理后,培养 48h 仍能保持较好的生活状态,但最终褐化死亡;而用 20 mg/L 2,4-D 处理过的材料,中央细胞、卵器细胞原位融合体和反足细胞均能启动分裂(图版Ⅰ,8~10);10 mg/L 2,4-D 处理过的材料在培养基中能存活数天,但不分裂。因此,我们最终选用酶解与 20 mg/L 2,4-D 同步处理的方法。

三、不同培养方式的比较

所用 3 种方法的结果大不相同。没有饲养细胞的微滴培养虽然密度达到了 10^4 个/ml,但被培养的细胞仍在 1~2d 内褐化死亡;琼脂糖珠饲养法能使被培养物存活 3~5d,但不能启动分裂,最终死亡;而用微室饲养法却能使未受精的胚囊各种成员细胞离体分裂,用荧光增白剂染色,能检测到分裂后子细胞的间壁(图版Ⅰ,10、12),中央细胞第一次分裂时体积变化不大,多为均等分裂,分裂面一般垂直于胚囊长轴。在微室内的琼脂糖培养基中,中央细胞能分裂几次形成多细胞团

(图版Ⅰ,11、12)。卵器细胞原位融合体与反足细胞只观察到一次分裂(图版Ⅰ,9、10)。至于未受精卵细胞则未能诱导分裂。

四、琼脂糖浓度的影响

用微室饲养法,共培养了184个中央细胞及其它胚囊细胞,调整微室内培养基中琼脂糖的浓度能大大提高分裂率。在液体培养基里,目前仅有中央细胞能进行一次分裂,频率为4.5%,而0.15%琼脂糖的半固体介质中,14.1%的中央细胞和12.5%的卵器细胞原位融合体均能启动分裂,但当琼脂糖浓度达到0.3%以上,培养基呈固体状时,培养物却无一分裂。合适浓度琼脂糖的良好作用可能是因其所营造的半固体稳定环境同胚囊细胞在体内受一定束缚的状况有所类似。Mól等[7]认为,将玉米受精后胚囊埋入琼脂糖培养基,就不会在培养皿移动时改变已建立的营养梯度和生长因子梯度,并且指出这与Van Lammeren(博士论文)的观点相似,即营养供应途径与合子和胚的极性有关。此外,在移动培养皿时,液体培养基中的细胞会漂移甚至粘结,而半固体培养基则能避免这种情况,便于跟踪观察。琼脂糖的固定还可为将来进行显微注射等操作带来便利。

中央细胞是双受精过程中精子的受体之一,是胚乳的前身,生长旺盛期或成熟期的胚乳培养已在多种植物中再生了植株[8],Kranz和Dresselhaus[9]在其综述中也简单提到中央细胞与精子的人工融合产物能在离体条件下生长(至于生长至何种程度却未说明),但至今未有未受精的中央细胞启动分裂的报道,这说明诱导未受精中央细胞分裂存在较大难度。而在本实验中,选用与酶解同步的2,4-D预处理方法和成分复杂的半固体培养基,分离的中央细胞比其它离体胚囊细胞更易启动并能继续分裂,这与分离方法、预处理诱导方法的改进和合适培养基的选择以及中央细胞生活力较强有一定关系。此外,中央细胞的二倍体性质对启动分裂的作用及细胞分裂时极核的动态还有待进一步研究。

图版 I   plate I

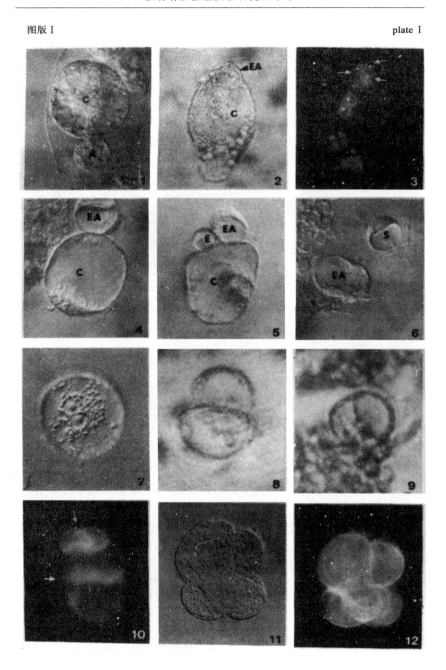

第三章 授粉、受精与胚胎发育的实验研究

图 版 说 明

A. 反足细胞 C. 中央细胞 E. 卵细胞 EA. 卵器细胞原位融合体 S. 助细胞

图 版 Ⅰ

除 3、10 和 12 为荧光显微摄影外,其余均为 Nomarski 干涉差显微摄影;除 3、10、11 和 12 为顺置显微镜摄影外,其余均为倒置显微镜摄影。**1.** 分离胚囊中原生质体化的中央细胞与反足细胞。×320 **2.** 分离胚囊中卵细胞与两个助细胞原生质体原位融合。×350 **3.** 与图 2 同一胚囊,DAPI 显示细胞核(箭头示融合体三核)。×340 **4.** 分离的中央细胞与卵器 3 个细胞原位融合体。×460 **5.** 2 个助细胞原位融合体、卵细胞原生质体及中央细胞。×370 **6.** 卵细胞与 1 个助细胞原位融合体及 1 个单独的助细胞原生质体。×460 **7.** 中央细胞原生质体。×380 **8.** 中央细胞分裂一次。×430 **9.** 反足细胞分裂一次。×780 **10.** 荧光增白剂示细胞壁(箭头示中央细胞和卵细胞原位融合体分别分裂一次。×380 **11.** 中央细胞分裂形成小细胞团。×330 **12.** 与 11 同一细胞团,荧光增白剂示细胞壁。×330

Explanation of plate

A. Antipodal cell, C. Central cell, E. Egg cell, EA. Fusion product of egg apparatus S. Synergid

Plate Ⅰ

All are Nomarski interference contrast micrographs except Figs. 3, 10 and 12 which are fluorescence micrographs; all are micrographs under inverted research microscope except Figs. 3, 10, 11 and 12. **Fig. 1.** Central cell and antipodal cell protoplasts in an isolated embryo sac (ES). ×320 **Fig. 2.** Fusion product of the egg cell and two synergids in an isolated ES. ×350 **Fig. 3.** The same ES as in Fig. 2. Arrowheads show three nuclei of three egg apparatus cells stained with DAPI. ×340 **Fig. 4.** An isolated central cell protoplast and fusion product of three egg apparatus cells. ×460 **Fig. 5.** A fusion product of two synergid protoplasts, an egg cell protoplast and a central cell. ×370 **Fig. 6.** Fusion product of the protoplasts of an egg cell and one synergid and a separated synergid protoplast. ×460 **Fig. 7.** A central cell protoplast. ×380 **Fig. 8.** Two cells divided from a central cell. ×430 **Fig. 9.** Two cells divided from an antipodal cell.

×780　**Fig. 10.**　Four cells divided from a central cell and a EA respectively; stained with Calcofluor white ST. ×380　**Fig. 11.**　A cell cluster divided from a central cell. ×330　**Fig. 12.**　The same cell cluster as in Fig. 11, showing cell wall stained with Calcofluor white ST. ×330

(作者:傅缨、孙蒙祥、杨弘远、周嫦。原载:植物学报,1997,39(8): 778~781。图版1幅,参考文献9篇。选录图版Ⅰ。)

4　植物幼小原胚的电激转化

Transformation of tobacco young proembryos by electroporation

电激法是植物遗传转化的一种有效方法,已在悬浮培养细胞[1]、原生质体[2,3]、组织[4]、花粉[5,6]等材料上导入外源基因获得瞬间或稳定表达,但是由于这些材料有培养时间较长、易发生突变、再生困难等缺点,转基因植株常表现不正常和育性较低,而植物胚胎以其离体发育和再生能力强、体外培养发育较为正常等特点,近年来被视为植物转化的理想材料,通过电激方法,Topfer等人[7]首先将外源基因导入小麦近成熟胚获得了外源基因的瞬间表达。随后,在水稻[8]、玉米[9~11]、小麦[12]等多种植物的成熟胚和近成熟胚中电激转化获得了外源基因的瞬间表达或稳定表达。Ke等人[11]用电激法转化玉米近成熟胚获得转基因植株。迄今植物胚胎转化主要限于成熟胚和近成熟胚幼胚,它们含有成千个细胞,而电激后仅少数细胞得到转化,大多出现遗传嵌合现象,需要进行大量的筛选和鉴定工作。植物的幼小原胚(包括尚未开始分化的胚与胚柄)只有几个到几十个细胞,应用电激转化理应较成熟胚和近成熟胚优越,但是由于原胚的收集与操作较为困难,迄今关于这方面的工作还未见报道。我们在已建立的烟草原胚大量分离方法基础上[13],试探了原胚的电激转化,获得了 GUS(β-glucuronidase)和 GFP (green fluorescent protein)两种报告基因的瞬间表达,并比较了幼小原胚和较大的原胚(球形胚)电激导入的效果,证明了植物幼小原胚电激转化的可行性和优越性。

材料与方法

一、材料

普通烟草(*Nicotiana tabacum* L.)品种 W38,取开花后 4~6d 的原胚。电激仪为 LN-101 基因脉冲导入仪(天津飞龙电子仪器厂)。质粒:pBI221 和 pGFP221。pBI221 购自 Clonetech 公司,含有 GUS 编码区,pGFP221 是将 pBI221 中的 GUS 基因替换成 GFP 基因。其中 GFP 基因来自 mGFP$_4$ 质粒。KM8p[14]基本培养基附加 0.05mol/L 葡萄糖、0.15mol/L 蔗糖、0.15mol/L 山梨醇、0.5mg/L 6-BA、0.5mg/L NAA。

二、方法

(1)原胚的分离。主要按照文献[13]的两种方法:一种是将胚珠置于含 1% 纤维素酶(Cellulase R-10)、0.8% 离析酶(Macerozyme R-10)和 0.5mol/L 甘露醇的酶解液中,室温静置 0.5h 后,用吸管吸打分离出原胚;另一种是直接用玻璃微针解剖胚珠分离原胚。(2)电激处理方法。用微吸管将分离的原胚每 50~100 个为一组放入电激小槽中,加入 0.2ml 含 pBI221 或 pGFP221 质粒 50μg/ml、小牛胸腺 DNA(C. T. DNA, SABC)100μg/ml 的 MES(2-N-morpholinoethane acid)缓冲液[15](35mmol/L 天冬氨酸钾盐,35mmol/L 谷氨酸钾盐,5mmol/L 葡聚糖酸钙,5mmol/L MES 和 0.45mol/L 甘露醇,pH = 6.0)。调节各种电激参数进行电激处理,对照组原胚在无质粒 DNA 的相同缓冲液中电激处理。(3)原胚的培养。用微吸管吸取电激处理后的原胚,置冰上 10min 后,用 KM8p 培养基洗 3 次,置入 24 孔培养板(Nunc)中 25℃培养 1~2d。(4)原胚生活力的检测。将培养后的原胚置于含 0.05mg/ml 荧光素二醋酸酯(fluorescein diacetate, FDA)和 0.3mol/L 蔗糖的检测液中。

5~10min后,在倒置荧光显微镜(Olympus IMT-2)下 B 激发,有生活力的原胚呈绿色荧光。(5) GUS 基因和 GFP 基因瞬间表达的检测。用 X-gluc (5-bromo- 4-chloro-3-indolyl-β-D-glucuronic acid, Molecular Probes)作为 GUS 组织化学反应底物。染液组成按文献[16]的配方:1 mg/ml X-gluc,0.5mmol/L $K_3[Fe(CN)_6]$,0.5mmol/L $K_4[Fe(CN)_6]$,0.03% Triton X-100,0.1 mol/L 磷酸-柠檬酸缓冲液,pH=5.0。37℃下染色过夜,统计有蓝色反应的原胚频率。GFP 检测在倒置荧光显微镜下 B 激发,直接统计发绿色荧光原胚的频率。(6)原胚细胞数目的统计,幼小原胚细胞数目可直接在显微镜下计数。较大的球形胚在 4% 多聚甲醛中室温固定 5h,PBS 洗 3 次,置于酶液(成分同上)中 2h 后,轻轻挤压胚使细胞分离,用吸管吸于血球计数板上,在显微镜下计数。

结　果

一、报告基因在电激原胚中的瞬间表达

实验共分离出 1 669 个幼小原胚和 2 304 个球形胚,分多种组合进行电激。首先检测了烟草胚胎发生过程中的内源 GUS 活性,经过底物反应后原胚不显蓝色(图版Ⅰ,7,附本刊后,下同),表明在实验条件下无内源 GUS 活性干扰。经过适当条件的电激后,部分球形原胚和幼小原胚可检测到 GUS 基因的瞬间表达。在幼小原胚中的表达呈现多种形式:(1)在原胚的所有或部分胚细胞中表达(图版Ⅰ,1、2);(2)在原胚的胚柄,尤其是其珠孔端吸器细胞中表达(图版Ⅰ,3);(3)在原胚的胚体和胚柄中都表达(图版Ⅰ,4)。球形胚中可观察到多个 GUS 阳性反应细胞(图版Ⅰ,6)。未加质粒的对照组中原胚均无蓝色反应(图版Ⅰ,5)。同时也进行了 GFP 基因的转化,电激后的原胚呈强烈的绿色荧光(图版Ⅰ,8),对照组无荧光。

二、电场强度对原胚和球形胚生活力的影响

一般来说,电激产生的高场强脉冲对材料有一定的损伤,因此大多

采用较低场强和较长脉冲时间进行电激转化。以 FDA 来检测电激后培养 2d 的原胚的活性(表1),当脉冲时间常数约为 13ms 时,随着电场强度的增加,胚的活性逐渐降低。幼小原胚的半致死电场强度约为 750V/cm,球形胚由于细胞数目多、结构致密,对电场强度的耐受性较大,其半致死电场强度约为 1 250V/cm。

表1　电场强度对烟草原胚和球形胚生活力的影响

电场强度 ($V \cdot cm^{-1}$)		400	500	750	1 000	1 250	1 500	2 000
生活力(%)	球形胚	96	87	77	65	54	37	11
	幼小原胚	71	65	48	31	6	0	0

三、不同发育程度原胚电激转化效果的比较

如图 1 所示,幼小原胚和球形胚的电激转化最佳条件有较大的差异。以整个原胚中凡有蓝色反应部位作为 GUS 阳性结果作为"整体转化频率",幼小原胚的整体转化频率在电场强度为 750V/cm、电容 21μF、脉冲时间 13ms 时达到最高(2.2%);球形胚在电场强度 1 250V/cm、电容 21μF、脉冲时间 11ms 时,整体转化频率达到最高

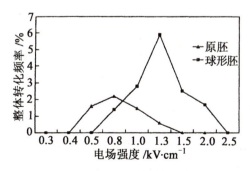

图 1　原胚和球形胚整体转化频率的比较

(5.9%)。若以蓝色反应细胞占整个原胚细胞总数的比例作为"有效转化频率",则幼小原胚的有效转化频率远较球形原胚为高。幼小原胚的细胞总数为 8~32 个,球形胚细胞总数为 250~400 个。如图 2 所示,幼小原胚的有效转化频率最高为 7.3×10^{-3},而球形胚的最高转化频率仅为 1.1×10^{-3},前者约为后者的 7 倍。

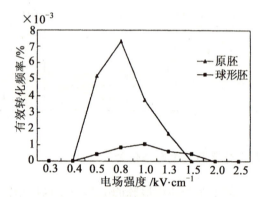

图 2　原胚和球形胚的有效转化频率的比较

四、原胚分离方法对电激转化效果的影响

将酶解和解剖两种方法分离的幼小原胚分为 2 组,以 GUS 作报告基因,在电场强度 750V/cm、电容 21μF 的条件下电激,两者的整体转化频率有较大差异:酶解分离的 234 个原胚中 5 个呈 GUS 阳性反应,转化频率为 2.1%;解剖分离的 159 个原胚中 2 个呈 GUS 阳性反应,转化频率为 1.3%。可以看出,酶解分离原胚的电激转化频率约为解剖分离原胚的两倍。

五、原胚的 GFP 基因电激转化

对幼小原胚进行了 GFP 基因的转化,当电场强度为 500~1250V/cm 时,可检测到绿色荧光,其中电场强度 750V/cm、电容 21μF,其整体转化频率最高(1.9%)。

讨 论

植物多细胞组织的电激转化,由于依赖细胞壁和细胞膜在短时、高脉冲下产生的通透性,转化的细胞大多为组织表层细胞,从而导致转化组织表现出嵌合性,植物胚的转化也不例外。Songstad 等人[10]观察到电激转化后外源基因在玉米的未成熟胚表层部分细胞中表达;Klöti 等人[12]也发现外源基因在小麦成熟胚中表达呈嵌合现象。我们在电激后的烟草原胚中也观察到这一现象。不过,从幼小原胚和球形胚的有效转化频率可以看出,以幼小原胚为电激转化材料远为优越,这是因为其细胞数目仅 8～32 个,且多数细胞都暴露在表面,增加了外源基因导入的几率,幼小原胚的最高有效转化频率为 7.3×10^{-3},通常转化的细胞为 1～5 个左右,转化的细胞占整个原胚细胞总数的比例为 30% 左右,这样将有可能大大减少再生植株的嵌合性。

关于植物胚的预处理对电激的影响,前人有不同的实验结果,李宝健等人[8]预先用人工方法对水稻成熟胚制造伤口,然后再电激转化,获得了转基因水稻植株。D'Halluin 等人[9]研究了 NPT Ⅱ 基因在玉米近成熟胚中的时序表达,发现不经酶解或机械损伤预处理的胚检测不到 NPT Ⅱ 活性,认为机械损伤或酶解预处理是电激转化胚的不可缺少的条件。Songstad 等人[10]将 GUS 基因和花青素基因转入经热激预处理的玉米近成熟胚也获得瞬时表达。但 Klöti 等人[12]报道,不经任何预处理的小麦成熟胚电激后,在盾片中获得 GUS 基因和花青素基因的瞬间表达。我们初步比较了酶解处理和直接解剖两种方法的效果,发现两者皆可获得外源基因的表达,但前者较后者转化频率约高 1 倍,表明酶解损伤细胞壁后使质粒更易穿透。

(作者:李师翁、王胜华、杨弘远。原载:科学通报,2000,45(6):598～601。图版 1 幅,图 2 幅,表 2 幅,参考文献 16 篇。选录图 1、2 与表 1。)

5 Gene transfer into isolated and cultured tobacco zygotes by a specially designed device for electroporation

The zygote of flowering plants is deeply embedded in the embryo sac and ovule. In some species, however, the inaccessibility of the zygotes can be overcome through mechanical isolation (Kranz and Lörz 1993) or enzymatic maceration (Fu et al. 1996), which makes the culture of the zygotes and direct gene transfer into them possible. Plant regeneration from *in vitro*-generated zygotes was first reported in maize (Kranz and Lörz 1993). Subsequently, isolated natural zygotes of four cereal crops (maize, barley, wheat and rice) have been regenerated into fertile plants (Holm et al. 1994; Leduc et al. 1996; Kumlehn et al. 1998; Zhang et al. 1999). There have, however, been few attempts to culture the zygotes of dicotyledonous plants. Until now, among dicots only tobacco zygotes have been cultured to multicellular structures (Li et al. 1998). As for genetic manipulation of isolated zygotes, it could help to investigate gene expression at the initial stage of ontogenesis. Leduc et al. (1996) first reported that maize zygotes in culture could express two microinjected reporter genes. In this paper, we report our methods of gene transfer into isolated and cultured zygotes of tobacco (*Nicotiana tabacum* L.) using a specially designed simple device for electroporation. Transient expression of the GFP gene in the electroporated zygotes confirmed the utility of this method.

Materials and methods

Plant materials

Plants of *Nicotiana tabacum* L. cv. W38, *N. tabacum* var *macrophyl-*

la and *N. rustica* were grown in the greenhouse at 25℃ during the day and 20℃ at night. Flowers were harvested 40-55 h after pollination to collect the ovules containing the zygotes.

Zygote isolation

The protocol used for isolating zygotes was similar to that established previously (Li et al. 1998). Ovules dissected from three ovaries were placed into 1.5 ml of enzymatic maceration mixture containing 1.5% cellulase R-10, 1.0% macerozyme R-10 and an isolation solution (3 mM MES, 3mM PVP and 0.5M mannitol, pH5.7). After 0.5h of incubation at 25℃ in a microshaker, the ovules were washed three to four times with the isolation solution, then collected and distributed into several drops on a slide. The ovules were then pressed gently with a small glass pestle, thereby liberating the embryo sacs, which were collected with a micropipette under an inverted microscope. After three rinses with the isolation solution to remove ovular tissue debris, the embryo sacs were transferred to a fresh drop of isolation solution and the zygotes were isolated. Two protocols for isolation were adopted. One consisted of incubating the embryo sacs in 200μl of enzymatic maceration mixture containing 0.5% cellulase R-10, 0.3% pectinase and isolation solution. The second method consisted of microdissecting the embryo sacs directly to isolate the zygotes in an enzyme-free condition. The isolated zygotes were transferred into 12-mm-diameter Millicell-CM microchambers (Millipore) and embedded in about 0.5μl of medium containing 1% agarose (Agarose Type IX, Sigma), which was then solidified for 6 min at 4℃. Generally, 5-6 zygotes were embedded in separate agarose drops in a millicell.

Zygote culture

For zygote culture, we used KM8p medium (Kao and Michayluk 1975) supplemented with 0.05 mol/L glucose, 0.25mol/L sucrose,

0.15mol/L mannitol, 0.1mol/L sorbitol, 0.5mg/ml NAA, 0.5mg/L 6-BA and 10% coconut water, pH5.7. The millicell was placed in a Petri dish containing dividing mesophyll protoplasts as feeder cells, which were maintained in KM8p medium. The zygotes were cultured in the dark at 25℃.

Zygote electroporation

Plasmid pBG221 was used as a reporter gene. This plasmid is derived from plasmid pBI221 (Clontech) and contains the GFP (green fluorescent protein) gene instead of the GUS gene that is present in pBI221.

A Gene Pulser (Feilong Co, Tianjin, China) with a capacitance extender was used for electroporation. A special reaction chamber was constructed for this purpose in which several zygotes could be easily treated (Fig.1). The device contained a chamber made from organic glass, two aluminium electrodes, a 12-mm-diameter Millicell-CM microchamber (Millipore) and a plastic holder. The electric field was applied between the two electrodes, which were 0.5 cm apart. The millicell was cut short to enable it to be inserted into the electroporation chamber, thereby providing a position to localize the zygotes. A hole was made in the middle of the plastic holder, and the size of the hole was adjusted to just fit the millicell. The holder facilitated the insertion and removal of the millicell.

In preparation for electroporation, the millicell containing the embedded zygotes was put into the holder, which was then inserted into the electroporation chamber. The electroporation chamber was filled with 0.5ml chilled electroporation buffer containing 35 mM aspartic acid monopotassium, 35mM glutamatic acid monopotassium, 5mM calcium gluconate, 5mM 2-(N-morpholino)-ethanesulfonic acid (MES) and 0.5M mannitol, pH5.8 (Tada et al. 1990). After plasmid DNA (100μg/ml) was added, a single electric pulse was discharged at a field strength of 900 V/cm with a capacitance of 21 μF and pulse time 13 ms. Immediately after delivery of the pulse, the millicell was taken out by the holder and washed three times with chilled KM8p medium. Electroporated zygotes were then cultured as described above.

Transient expression of the GFP gene was detected 24-30h after electroporation. Expression of the GFP gene was observed using a Leica TCS confocal microscope equipped with a Kr-Ar laser. Observations and recordings were made using single channel image collection. The 488-nm wavelength excitation line of the laser was used for GFP fluorescence.

Results and discussion

In vitro development of freshly isolated zygotes

Zygotes could be isolated from fertilized embryo sacs by microdissection or by enzymatic maceration. In our experiments, 84 millicells were used to culture 406 zygotes. Freshly isolated zygotes were initially spherical (Fig. 2A), elongating after 1 day in coculture with mesophyll protoplasts. The zygotes underwent first division after 3 days in culture. Most of the zygotes divided unequally to form a large basal cell and a small apical cell (Fig. 2B). However, a pattern of equal division was also observed in a few zygotes. After 4 days of culture, the zygotes underwent a second division with a division plane perpendicular to the first division plane (Fig. 2C). Sustained development (Fig. 2D) could lead to the formation of globular embryos (Fig. 2E) after 12 days in culture.

With respect to culturing the zygotes, the two isolation methods were compared. The enzymatic maceration technique was easier to perform than the microdissection technique, which needed more time and skill. However, the culture results indicated that the first division of zygotes isolated using the microdissection method was 1-2 days earlier than that of zygotes obtained from enzymatic maceration. Moreover, the frequency of first division of the former (61.2%) was much higher than the frequency of division of the latter (30.5%). No multicellular proembryos were obtained using enzymatic maceration. Leduc et al. (1995) reported that minimal enzymatic treatments had a deleterious effect on the development of isolated maize embryo sacs in culture. This was also observed in our experiments; the enzyme treatment showed a harmful effect on the delicate zygotes.

Due to the limited number of cultured zygotes, feeder cells were indispensable. In our experiments, we used three kinds of mesophyll protoplasts as feeder cells: *Nicotiana tabacum*, *N. tabacum* var. *macrophylla* and *N. rustica*. All of these could induce the zygotes to divide, and the division frequencies were similar, ranging from 45.4% to 62.5%. Holm et al. (1994) regenerated fertile plants from barley and wheat zygotes using barley microspores as nurse cells. Leduc et al. (1996) reported plant regeneration from natural zygotes of maize by coculture using barley microspores, moss suspension cells or maize suspension cells. It seems that the species of the feeder cells is not specific to success in zygote culture.

In vitro development of electroporated zygotes

The immobilization of the isolated zygotes in agarose and the use of the specially designed electroporation device provided an efficient procedure to reduce the loss of viable zygotes during the electroporation process. Most of the electroporated zygotes could be easily collected and transferred for further treatments using this technique. In our electroporation experiments, as shown in Table 1, the electroporated zygotes could be cultured to form proembryos. However, the first division frequency of electroporated zygotes (54.6%) was slightly lower than that of freshly isolated, unelectroporated zygotes (61.2%). The electroporated zygotes could develop into globular proembryos, but the frequency at which this could occur (3.1%) was much lower than that shown by the unelectroporated zygotes (8.7%). Zygotes undergoing electroporation treatment in general still maintained the capability to develop under *in vitro* culture conditions.

Transient expression of the GFP *gene in electroporated zygotes*

Expression of the GFP gene was obtained in electroporated zygotes after 2 days in culture and produced a green fluorescence (Fig. 2F). These phenotypes were not observed in the control zygotes electroporated in the absence of the plasmid. Regarded as a nondestructive reporter gene, the GFP gene can be directly detected using fluorescence microscopy. In the millicell, however, the detection of GFP fluorescence of the transgenic zygotes was complicated by the strong autofluorescence of the feeder cells. For this reason, we had to remove the zygotes from the millicell for GFP detection. Of the 303 electroporated zygotes examined for transgene expression 2 days after culture, 8 showed fluorescence – a frequency of 2.6%. The transformation frequency of microinjected maize zygotes was 3.5% (Leduc et al. 1996). Our results demonstrated once again that isolated zygotes can be transformed.

Table 1 Comparison of *in vitro* development between electroporated and unelectroporated zygotes

	No. of zygotes cultured	Zygote first division		Globular embryos formation	
		No.	Frequency(%)	No.	Frequency(%)
Electroporated zygote	97	53	54.6	3	3.1
Unelectroporated zygote	206	126	61.2	18	8.7

As a direct method for gene transfer, electroporation has been applied to plant embryos. Using this technique, it has been possible to show exogenous gene expression in mature and immature embryos of maize (Songstad et al. 1993), wheat (Klöti et al. 1993) and rice (Dekeyser et al. 1990). Recently, we also observed GUS and GFP genes with transient expression in tobacco proembryos (Li et al. 2000). In this research, we refined electroporation techniques to transfer foreign gene into the zygotes, thus establishing an alternative to microinjection for investigating gene expression during *in vitro* development of zygotes and early embryogenesis.

(Authors: Li ST, Yang HY. Published in: Plant Cell Reports,
2000, 19: 1184-1187, with 2 Figures, 1 table and
14 references. Only retained Tab. 1 here.)

6 小麦分离合子与幼胚中膜钙和钙调素的分布

Distribution of membrane-bound calcium and activated calmodulin in isolated zygotes and young embryos of *Triticum aestivum*

摘 要

用非酶手工显微解剖法分离小麦(*Triticum aestivum* L.)合子与幼胚。应用金霉素(CTC)和氟奋乃静(FPZ)荧光探针,观察了合子与胚胎发育过程中膜钙和活化钙调素的分布,合子原生质体 CTC 荧光呈极性分布,二细胞至十几个细胞原胚时期各细胞间的 CTC 和 FPZ 荧光均匀分布,幼小梨形胚胚体 CTC 和 FPZ 荧光稍强于胚柄,接近分化的梨形胚 CTC 荧光仅位于胚体与胚柄之间的几层细胞和胚体腹侧中部的几个细胞中,胚芽刚分化时,胚基部 CTC 荧光较强,FPZ 荧光则仅出现在胚基部。讨论了合子分离技术和膜钙与钙调素(CaM)在胚胎发育中的可能作用。

Abstract

The method of non-enzymatic, manual microdissection was established to isolate zygotes and young embryos in *Triticum aestivum* L. The distribution of membrane-bound calcium and activated calmodulin in the isolated zygotes and young embryos was visualized by chlorotetracycline (CTC) and fluphenanize (FPZ) fluorescence probe respectively. The CTC fluorescence was polarly distributed in the zygote protoplast. The distribution of the CTC and FPZ fluorescence from two-celled embryos to multicellular embryos was observed. In the young pear-shaped embryos the CTC and FPZ fluorescence of the embryos was slightly higher than that of the suspensor. In a pear-shaped embryo beginning with differentiation the CTC fluorescence was restricted to several-layer of cells between embryo and suspensor and several ventral cells of the embryo. In the embryos with newly differentiated plumule the basal part of the embryo possessed a higher CTC fluorescence, while the FPZ fluorescence was only distributed in the basal part. It indicated that the distribution of CTC and FPZ fluorescence was in coincidence with the sites that plumule and radicle were beginning to differentiate. The technique of isolated zygotes and the possible function of calcium and calmodulin during embryo development are discussed.

(作者:赵洁、周嫦、杨弘远。原载植物学报,1998,40(1):28~32。图版2幅,参考文献24篇,均删去。)

7 水稻卵细胞与合子的分离

Isolation of egg cells and zygotes in *Oryza sativa*

摘 要

采用两种方法分离水稻卵细胞与合子。其中横断胚珠法易使卵细胞与合子保持原位,利于追踪观察和分离,效果较好。一般 2h 内可剥取 20 个子房,从中分离出 5~8 个合子。刺压法是施加外力改变中央细胞的膨压,致使合子逸出。由于难以控制挤压力,易使合子破裂或随珠心残渣裹挟而出,不易识别。纵剖胚珠常导致合子的移位而被致密的珠心组织包裹,难以辨别和操作。

成熟卵细胞与早期合子(开花后 2~6h)的分离物有两种形状,一种呈梨形,与体内的自然状态相似,但在短时间内(3~6min)即变为圆球形。另一种呈圆球形(图版Ⅰ,1~5),这是由于在分离操作时中央细胞受损引起膨压改变,使卵与合子失去与珠心组织和自身不完整细胞壁的联系,从而变为球形。开花后 8~10h 分离的晚期合子大多仍保持梨形(图版Ⅰ,6)。经 CW 染色,鉴定球形合子无壁成分,表明为原生质体。分离的卵细胞或合子常与助细胞脱离,有时也可连同助细胞被分离出来(图版Ⅰ,7)。FDA 染色表明分离的卵细胞与合子具有较强的生活力,在分离介质中,至少可存活 2h。

成熟卵细胞和早期合子的核位于细胞近中央处,细胞质密集于核周围,细胞外周有许多小液泡,在合点端分布有较大的液泡。在分离后 5min 左右,细胞质向外围有所扩散,液泡的极性化变得不甚明显。晚期合子细胞质浓密,周边散布有小液泡。成熟卵细胞的直径为 33~36 μm,球形合子的直径为 37~41 μm。3 个品种的子房长度差异较大,但胚珠大小相近,卵细胞与合子的大小与分离频率也类似。

水稻离体合子原生质体 CTC 荧光在合点端一侧稍弱于珠孔端(图版Ⅰ,4),FPZ 荧光分布较均匀(图版Ⅰ,5)。表明合子中含有丰富的膜结合钙与活化钙调素。

助细胞呈牛角形,退化助细胞色泽深厚。偶尔可分离出助细胞原生质体。反足细胞数目有很大差异,有 6~20 个或更多(图版Ⅰ,12),较易分离。

开花后 13~18h,合子已分裂形成 2-细胞原胚(图版Ⅰ,8、9);开花后 1d,形成 4~8 个细胞原胚(图版Ⅰ,10、11)。此时细胞个数增加,但体积无明显增大。2-细胞原胚至多细胞原胚的发育时间及形态特征与前人切片原位观察结果一致。

玉米离体受精程序中成功采用短时酶解胚珠,然后用微针解剖出卵细胞的方法[1]。大麦与小麦自然合子培养中采用非酶处理的解剖技术,以镊尖由胚珠刺入中央细胞,引起后者膨压剧变致使卵细胞逸出[2]。双子叶植物烟草采用酶解-研磨法从烟草受精后胚珠中分离出胚囊,再由胚囊分离合子的方法[3]。

本实验不经酶处理,用自制的玻璃微针成功分离出水稻的卵细胞与合子。操作方法简便,不受品种限制,为进一步的培养工作与发育分子生物学研究奠定了技术基础。

Abstract

A method of non-enzymatic, manual microdissection was established to isolate egg cells and zygotes in *Oryza sativa* L. Generally 5 to 8 protoplasts of egg cells or zygotes could be isolated from 20 ovaries in 2h. Fluorochromatic reaction proved that the isolated cells were viable and could survive in the isolation medium consisting of 0.5 mol/L mannitol and 5 mmol/L $CaCl_2$ for at least 2h. The egg cells and early zygotes usually turned to spherical shape during the course of isolation, whereas the late zygotes often maintained pear-shaped. The rounded zygotes were proved to be true protoplasts by Calcoflour White fluorescent staining. Chlorotetracycline and fluphenanize fluorescence treatment demonstrated that the isolated zygotes were rich in membrane-bound calcium as well as calmodulin.

(作者:韩红梅、赵洁、施华中、杨弘远、周嫦。原载:植物学报,1998,40(2):186~188。图版1幅,参考文献4篇,均删去。)

8 In vitro development of early proembryos and plant regeneration via microculture in *Oryza sativa*

More than fifty years have passed since the pioneer work of young embryo culture in *Datura* (Van Overbeek et al., 1941). Embryo culture techniques can be applied in plant breeding as well as basic studies on the physiological and molecular aspects of embryo development. With the refinement of culture techniques, proembryos in more and more plant species could be cultured successfully. However, the culture of proembryos is not routine for many plants.

Rice is the most important cereal crop in Asia. There have been many reports about embryo culture in rice, including increased frequency of plant regeneration (Lai, 1983; Rance et al., 1994), analysis of hereditary characters, variation in regenerated plants (De Guzman, 1983; Lai and Liu, 1986; Peng and Hodges, 1989) and genetic transformation (Rainieri et al., 1990; Li et al., 1991). Nevertheless, embryos have been cultured in a limited number of rice cultivars and their regeneration frequencies still were low. The earliest proembryos successfully cultured were 4-day-old embryos; culture of younger proembryos has not yet been reported. In this communication, we report the results on the successful culture of 2-to 4-day-old proembryos, as well as describe key factors involved in this process in four rice cultivars including both *japonicas* and *indicas*.

Materials and methods

Four rice (*Oryza sativa* L.) cultivars, Xiao-gan 01 (early *indica*), Zao-xian 921 (early *indica*), Chun-jiang 05 (early *japonica*) and Chun-jiang 06 (late *japonica*), were used as materials. Ovaries 2 to 4 days after anthesis were sterilized with 70% ethanol for 0.5-1 min and 0.05% $HgCl_2$ solution for 8-12 min, washed with sterile distilled water 3-4 times, and then soaked in an isolation solution (10 mmol/L $CaCl_2$, 0.7 mmol/L KH_2PO_4, 1.6 mmol/L $MgSO_4$ and 0.26mol/L maltose). The proembryos were isolated with glass microneedles under an Olympus CK2 inverted microscope and transferred into the medium with a micropipette. After incubated at 28±2℃ in darkness for 10-12 days the proembryos were cultured in a 16h photoperiod.

第三章 授粉、受精与胚胎发育的实验研究

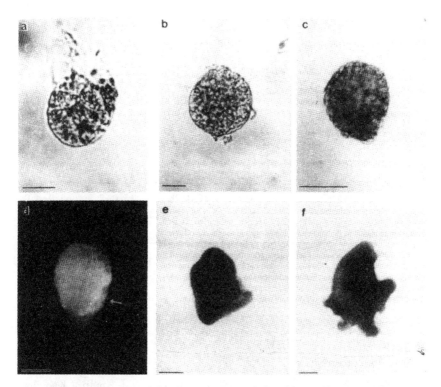

Figure 1. Growth and development of 2 to 3-day-old proembryos in culture. **a.** A 2-day-old proembryo at inoculation. **b.** A 2-day-old proembryo after 2 days of culture. The bar represents 30 μm in a-b. **c.** A 3-day-old proembryo after one day of culture. **d.** Formation of a small hollow area at the lateral central part of a 3-day-old proembryo after 2-4 days of culture. Arrow indicates horrow area. The bar represents 100 μm in (c-d). **e.** Formation of coleoptile primodium from the peripheral basal cells of a 3-day-old proembryo. **f.** Formation of radical primodium in opposite part to the plumule of a 3-day-old proembryo. The bar represents 200μm in (e-f).

N_6 medium (Zhu et al., 1975) supplemented with 9.5% maltose, 300 mg/L CH, 10% CM (Sigma), 610 mg/L 14 kinds of mixed amino acids (phenylalanine, glycine, alanine, threonine, serine, tyrosine, asparatic acid, glutamic acid, glutamine, arginine, lysine, histidine, proline,

Table 1 Difference of growth and development among rice cultivars during proembryo culture*

Cultivars	Proembryo age (day)	Inoculation number	Growth rate[1]	Induction frequency (%)[2]	
				Embryogenesis	Callus
Xiao-gan 01	3	36	1.13±0.27	16.7±4.9	41.7±2.3
Zao-xian 921		33	0.98±0.15	0	63.6±5.1
Chun-jiang 05		43	2.10±0.28	41.9±3.7	53.5±3.3
Chun-jiang 06		42	3.48±0.24	38.1±2.2	50.0±2.8
Xiao-gan 01	4	36	2.03±0.08	33.3±5.0	66.7±5.0
Zao-xian 921		40	1.51±0.15	15.0±2.3	65.0±4.0
Chun-jiang 05		38	4.56±0.17	44.7±3.5	55.3±3.5
Chun-jiang 06		36	5.79±0.20	41.7±2.6	50.0±2.7

* N_6 medium supplemented with 10% CM and 14 amino acids.
1. Data represent the mean values ±S.D.
2. Data represent the mean values ±S.D from three-five independent repeats.

valine), 2 mg/L BA and IAA, pH5.8, was used for growth and development of the proembryos. Several other basic media, AA (Thompson er al., 1986), MS (Murashige and Skoog, 1962) and Km8p (Kao and Michayluk, 1975), were also used as comparisons. All media were filter-sterilized by 0.45 μm filter membrane. N_6 medium supplemented with 0.1-0.5 mg/L BA and IAA and 3% maltose was used to induce shoots and roots for plantlet regeneration. When grown to 8-10 cm height, the plantlets were transplanted into pots.

A liquid droplet microculture technique was used to incubate 2-day-old proembryos. A 3 cm diameter Petri dish with four droplets of 5 μl medium was placed in a 6 cm diameter Petri dish with sterile water and sealed with parafilm. Six proembryos (70-100μm in length) were inoculated in a droplet. During the first 8 days, the same volume of fresh medium was added into each droplet at 2 days intervals. After 8 days, 150-200μl medium was added into each droplet. Another microculture system, a shallow-layered liquid culture technique, was adapted for larger proembryos: 250-300μl medium in a 1 cm diameter microchamber was used to incubate 10-12 3-day-old proembryos (130-160 μm in length) or 8-10 4-day-old transition-staged proembryos (220-250 μm in length). At 2-day intervals, 100μl fresh medium was added into the microchamber.

Each kind of medium was tested with 3-5 replications with 30-50 proembryos in total. About 1400 proembryos were cultured. After the proembryos had grown for 8 days, their length was measured with a micrometer to calculate the average growth rate (length of proembryos at 8 days length of proembryos at inoculation/length of proembryos at inoculation). After 14 days of culture, the percentages of embryogenesis, callus formation and total induction were calculated.

Results

Development of cultured proembryos and plant regeneration

Two-day-old proembryos were pear shaped and undifferentiated, about 80-100 μm in length containing 32-100 cells (Figure 1a). After 2 days of culture, the proembryos started to grow (Figure 1b). Some embryos continued to grow into a spherical shape; the others became elliptic. After 10-15 days of culture, calli visible to the naked eye were produced from the proembryos. After 25-30 days of culture, plantlets were regenerated through organogenesis when the calli had been transferred to a differentiation medium. Plant regeneration via *in vitro* embryogenesis was not observed.

Three-day-old proembryos also had an undifferentiated pear shape. Different morphological changes and developmental pathways were observed in culture. One pathway was similar to *in vivo* embryogenesis. At 0-2 days of culture, the proembryos grew evenly at all parts (Figure 1c). 2-4 days later, at the central of their lateral side, a small hollow area appeared from which primodia of plumule (Figure 1d) and coleoptile (Figure 1e) differentiated. The radicle was then formed in the opposite pole to the plumule (Figure 1f). When embryos matured, they germinated into normal plantlets after transfer onto solid medium supplemented with 0.1-0.5 mg/L IAA (Figure 2a-e). The frequency of regenerated plantlets was up to 60%-80%. The plantlets were transplanted into pots when they reached a height of 8-10 cm. After 3 months, mature fertile plants were obtained (Figure 2f). The other pathway was callus formation and plantlet regeneration via organogenesis. Such proembryos grew evenly or unevenly, giving rise to cylindrical (Figure 3a, b), fan-shaped or irregular calli (Figure 3c), which were able to proliferate significantly when transferred onto the medium supplemented with 1.0-2.0 mg/L 2,4-D and 0.5 mg/L kinetin (Figure 3d). Clumps of shoots were produced from the calli on differentiation medium (Figure 3e). In culture, some deformed embryos with only plumule germination were also observed (Figure 3f).

Four-day-old proembryos were pear shaped with the first sign of differentiation. Their developmental pattern was similar to that of 3-day-old proembryos, but their growth rate and the frequency of morphogenesis were higher.

Factors affecting proembryo development in vitro

Difference among cultivars Among four cultivars tested, the frequency of embryogenesis was highest in Chun-jiang 05, 41.9% for 3-day-old proembryos and 44.70% for 4-day-old proembryos. The growth rate was highest in Chun-jiang 06 up to 3.48 for 3-day-old proembryos and 5.79 for 4-day-old proembryos. In general, the *japonica* cultivars were easier to culture than the *indica* cultivars (Table 1).

Effect of media In our experiments, four kinds of media were included: i.e., AA, MS, N_6 and Km8p (Table 2). Two-day-old proembryos could survive and grow in N_6 and Km8p, but not in AA and MS. The fre-

Figure 2. Development and plant regeneration of 3-day-old proembryos in culture. **a.** Embryo developed from proembryos. **b-c.** Germination of mature embryos. The bar represents 1000 μm in (a-c). **d-e.** Regenerated plantlets from the embryos. The bar represents 2000 μm in (d) **f.** Mature plants transplanted in the pot after 3 months.

quency of their callus formation was up to 66% in Km8p. In all four kinds of media, 3-4-day-old proembryos could develop and regenerate plantlets. Among them, N_6 was also the most efficient taking account of the frequency of embryogenesis. Embryogenesis failed to occur in MS for all proembryos.

Effect of coconut milk Among four sets of N_6 media supplemented with 0%, 5%, 10% or 20% CM, respectively, the growth rate and especially the frequency of embryogenesis increased with the increase of CM concentration (Table 3). In the medium without CM, proembryos could grow and regenerate plantlets only by organogenesis.

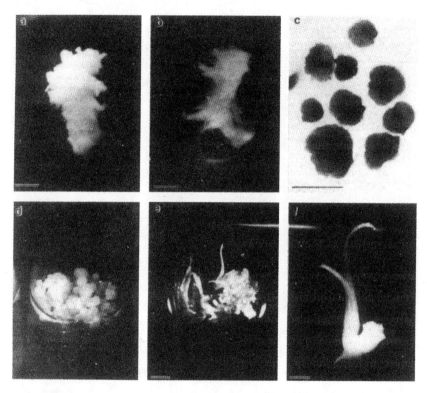

Figure 3. Growth and development of 3-day-old proembryos in culture. **a-b.** A Cylinder-shaped callus derived from 3-day-old proembryos. **c.** Fan-shaped and irregular calli derived from 3-day-old proembryo. The bar represents 1000μm on (a-c). **d.** Proliferation of calli from 3-day-old proembryos. **e.** Clumps of buds regenerated from proembryos-derived calli. The bar represents 4000μm. **f.** A deformed embryos with only plumule germination. The bar represents 2000μm.

Effect of amino acids Taking N_6 medium without amino acids as a control, we tested the effect of supplement with 4 or 14 kinds of amino acids (Table 4). There was no embryogenesis whether in the control medium or in the medium with 4 kinds of amino acids, even if CM existed. When cultured in the medium with 14 kinds of mixed amino acids, proembryos underwent sustained development according to embryogenesis pattern, with the frequency of 38%-42% for 3-day-old proembryos and 42%-45% for 4-day-old proembryos. After 14 days of culture, the total induction frequency reached 95%-100% in Chun jiang 05 and 88%-91% in Chun-jiang 06.

Table 2 Effect of media on growth and development of cultured proembryos*

Proembryo age (day)	Medium	Inoculation number	Growth rate[1]	Induction frequency(%)[2]	
				Embryogenesis	Callus
2	AA	42	0	0	0
	MS	30	0	0	0
	N_6	48	3.21±0.39	0	58.3±5.7
	KM8p	60	4.71±0.41	0	66.7±3.2
3	AA	36	4.39±0.27	22.2±3.4	61.1±2.9
	MS	35	2.38±0.42	0	28.6±6.3
	N_6	42	3.48±0.16	38.1±5.2	50.0±3.0
	KM8p	30	3.86±0.23	16.7±4.1	40.0±2.5
4	AA	32	5.92±0.55	28.1±2.3	65.6±4.7
	MS	30	3.06±0.20	0	30.0±7.4
	N_6	36	5.79±0.33	41.5±4.2	50.0±5.0
	KM8p	40	5.46±0.24	25.0±3.0	50.0±3.3

* Medium supplemented with 10% CM and 14 amino acids.
Cultivar Chun-jiang 06.
1 and 2: See Table 1.

Table 3 Effect of coconut milk on growth and development of cultured proembryo*

Proembryo age (day)	Coconut milk (%)	Inoculation number	Growth rate[1]	Induction frequency (%)[2]	
				Embryogenesis	Callus
3	0	28	0.92±0.18	0	53.6±5.2
	5	30	1.06±0.21	10.0±2.2	46.7±2.3
	10	36	1.14±0.11	16.7±2.7	41.7±3.5
	20	35	1.74±0.10	28.6±3.2	31.4±1.6
4	0	29	1.24±0.16	0	89.7±6.8
	5	33	1.80±0.12	27.3±1.3	69.7±3.3
	10	36	2.03±0.13	33.3±3.6	66.7±3.6
	20	30	3.15±0.17	36.7±2.8	63.3±2.8

* N_6 medium supplemented with 14 amino acids.
Rice cultivar was Xiao-gan 01.
1 and 2: See Table 1.

Discussion

It was usually necessary to develop sophisticated culture methods for *in vitro* growth and development of proembryos. Monnier (1978) devised a culture system in which two media with different compositions and osmotic pressure were poured in juxtaposition into the periphery and center of a Petri dish. This culture system ensured a continual variation of osmotic pressure and constant supply of nutrient substances for culturing proembryos only 50μm long (about 100 cells) in *Capsella*. Liu et al. (1993) developed a culture system of two agar layers with the top layer containing higher osmolarity than the bottom layer. In this system, proembryos as small as 35 μm (8-36 cells) from *Brassica juncea* could grow and develop into normal, mature embryos with an efficiency of at least 75%. The double layer system with different media was also used to culture globular embryos of wheat, and the frequency of direct embryogenesis was up to 47% (Fischer and Neuhaus, 1995). Pullman and Gupta (1994) invented a multistage culture method for coniferous proembryos, including three culture steps from proembryo to mature embryo stages. In our experiments, we adopted a liquid droplet micro-culture technique for 2-day-old proembryos in rice (80-100 μm in length). In order to maintain nutrient supply and gradually reduce osmotic pressure, fresh medium was added into the droplet at 2 day intervals. In order to maintain sufficient density of the cultured proembryos, the droplet was reduced to a volume of only 5 μl. Proembryos could form calli with an efficiency up to 66% and regenerate plants. 3 to 4-day-old proembryos were too big to incubate in such small droplets and were reared by another shallow-layered microculture technique. Nearly all the proembryos could develop and quite a proportion went along normal embryogenic pathway. Both these two methods were easy to manipulate and showed quite high efficiency.

Table 4 Effect of amino acids on growth and development of cultured proembryos*

Cultivars	Proembryo age(day)	Amino acids number	Inoculation number	Growth rate[1]	Induction frequency(%)[2]	
					Embryogenesis	Callus
Chun-jiang 05	3	0	32	1.14±0.13	0	75.0±5.8
		4	33	1.27±0.24	0	90.9±4.7
		14	43	2.10±0.11	41.9±3.7	53.5±3.2
	4	0	33	2.22±0.12	0	97.0±7.2
		4	30	2.76±0.30	0	100
		14	38	4.56±0.26	44.7±2.8	55.3±2.8
Chun-jiang 06	3	0	38	2.38±0.28	0	50.0±3.2
		4	32	2.30±0.17	0	75.0±2.5
		14	42	3.48±0.12	38.1±3.9	50.0±1.6
	4	0	30	3.39±0.13	0	60.0±4.2
		4	31	3.95±0.20	0	74.2±2.1
		14	36	5.79±0.32	41.7±3.6	50.0±2.7

* N_6 medium supplemented with 10% CM.

1 and 2: See Table 1.

Supplement of CM and amino acids into the medium was important for proembryo development (Raghavan, 1980; Collins and Grosser, 1984). CM was first used by Van Overbeek et al. (1941) for culturing proembryos of *Datura stramonium* and has extensively been utilized later in many other plants. Liu et al., (1993) demonstrated that 30% CM had a clear promoting effect on *Brassica juncea* embryo development. Norstog (1961) pointed out that barley proembryos (about 100 cells, 60 μm in length) could not survive without CM, but could grow in a medium supplemented with CM and amino acids. The addition of 1%-15% CM to medium enhanced callus formation, normal development and germination of barley immature embryos in length of 0.35–0.55 mm (Cameron-Mills and Duffus, 1977). We confirmed that addition of 14 different amino acids and 10%-20% CM were indispensable factors for *in vitro* embryogenesis in rice.

As compared with 3 to 4-day-old proembryos, it was much more difficult to culture 2-day-old embryos which followed the pathway of organogenesis but not embryogenesis. This observation indicated that there may be some key factors indispensable to 2-day-old prembryo development that need further investigation. For instance, an appropriate feeder cell system combined with milli-cell culture might fit in with the proembryos at 2 days or even earlier stages, as was recently used for culture of either *in vitro* zygotes in maize (Kranz and Lörz, 1993) or natural zygotes in barley and wheat (Holm et al., 1994; Kumlehn et al., 1998), although to find and maintain an efficient feeder system will not be as convenient as the methods presented here.

(Authors: Zhao J, Zhou C, Yang HY. Published in: Plant Cell, Tissue and Organ Culture, 1999, 55: 167-174, with 3 figures, 4 tables and 24 references. All figures and tables are retained here.)

9 Isolation and *in vitro* culture of zygotes and central cells of *Oryza sativa* L.

In angiosperms, double fertilization and embryogenesis takes place in the female gametophyte, the embryo sac, which is in turn embedded in the ovule and ovary. The successful isolation of viable embryo sacs has provided possibilities for further manipulation of the female gametes and zygotes under *in vitro* conditions (Zhou and Yang 1985; Theunis et al. 1991). Since the beginning of the 1990s, great progress has been made in *in vitro* fertilization and culture of the *in vitro* produced zygotes in maize (Kranz et al. 1991; Kranz and Lörz 1993). Culture of isolated natural zygotes has also led to plant regeneration in barley and wheat (Holm et al. 1994; Kumlehn et al. 1998). In other attempts along this direction, multicellular structures have been obtained from *in vitro*-produced zygotes of wheat (Kovács et al. 1995), *in vitro* hybrid zygotes between maize and other gramineceous species (Kranz and Dresselhaus 1996), natural zygotes of tobacco (Li et al. 1998), unfertilized egg cells of tobacco (Tian and Russell 1997), unfertilized central cells of tobacco (Fu et al. 1997) and *in vitro*-fertilized central cells of maize (Kranz et al. 1998). However, in rice, one of the most important cereal crops, a comparable investigation has not been reported up to now.

We have isolated unfertilized and fertilized egg cells and early proembryos in rice by means of non-enzymatic, manual microdissection (Han et al. 1998). Here we report results from subsequent investigations concerning the efficiency of two isolation techniques, *in vitro* culture of unfertilized and fertilized egg cells and central cells, cellular characteristics of the isolated and cultured cells and some key factors involved in the culture system.

Materials and methods

Two rice (*Oryza sativa* L.) cultivars, 'Chun-jiang 05' (early *japonica*) and 'Xiao-gan 01' (early *indica*), were used for isolation of the embryo sac cells. The former was also used for the establishment of the feeder cell line. Rice plants were grown in pots under natural conditions. Unpollinated and pollinated ovaries were sterilized with 70% ethanol for 0.5 min and 0.05% $HgCl_2$ solution for 10-12 min, washed with sterile distilled water three to four times and soaked in an isolation solution (10 mM $CaCl_2$, 0.7mM KH_2PO_4, 1.6 mM $MgSO_4$ and mannitol). The concentration of mannitol was adjusted to 0.55M for isolating egg cells/zygotes and to 0.6-0.65 M for isolating central cells. Two methods were adopted for isolating embryo sac cells. One was non-enzymatic, manual microdissection (Han et al. 1998). Ovules were dissected out from the ovaries and soaked in the isolation solution. Embryo sac cells, except the central cell, could be isolated with glass microneedles under an Olympus CK2 inverted microscope. The other method consisted of a short pulse of enzymatic treatment followed by manual microdissection. Forty to fifty ovules were collected per sample and incubated at 28-30℃ in 0.5 ml of an enzyme mixture solution (0.8% pectinase, 0.2% pectolyase Y23, 0.5% cellulase Onozuka R-10, 0.5% hemicellulase and mannitol, pH5.8) for 10-15 min for egg cells/zygotes and 20-30 min for central cells. After incubation, the ovules were washed with the isolation solution three to four times for 4-5 min each to remove the enzymes. For isolation of egg cells and zygotes the micropylar portion of the treated ovules were dissected carefully with two glass microneedles under an inverted microscope. For isolation of central cells, it was necessary to remove some nucellar tissues from the micropylar or the chalazal pole of the ovules after which the central cells could be pushed out from the opposite pole with the needle.

N_6 or KM8p basic medium supplemented with 4% mannitol, 6% sucrose, 10% CM, 0.02% LH, 0-10 mg/L 2,4-D and 0.5 mg/L 6-BA, pH 5.8, was used for culture of the isolated cells. A microchamber (Millicell-CM PICM 012 50 Millipore) containing 0.1-0.15 ml medium was placed into a 3.0-cm-diameter Petri dish with 1.0 ml medium. Three to eight isolated cells were inoculated into each microchamber. Vigorously dividing rice suspension cells derived from young embryos were added into the Petri dishes as a feeder. The dishes were incubated at 25-26℃ in darkness.

Both newly isolated and incubated cells were stained with 1 mg/ml CW to detect the cell wall. The cells were observed and photographed through the millicell microchambers under an Olympus IMT-2 inverted microscope with bright field, Nomarski interference contrast and fluorescence equipment. The diameter of the isolated cells was measured with a micrometer.

Results

Efficiency of isolation and characteristics of the isolated cells

All component cells of the embryo sac were successfully isolated by both the non-enzymatic and enzymatic methods, but central cells could only be isolated by the enzymatic method. The isolation frequency was higher using the enzymatic method. In general, from approximately 40 ovules, 6-12 egg cells could be obtained by the non-enzymatic method and 10-16 egg cells by the enzymatic method. However, excessive enzyme treatment was harmful to cell viability and further *in vitro* culture, as mentioned below.

The unfertilized egg cells attained a spherical shape soon after isolation. CW-staining confirmed that they were naked protoplasts. The immature egg protoplasts isolated 1-2 days before anthesis showed polarity in that the nucleus with the cytoplasm mainly concentrated at one pole and the big vacuole at the other (Fig. 1a). At stages closer to anthesis, the polarity of the egg protoplasts became less evident and finally disappeared. Its dense cytoplasm was evenly distributed around the centrally located nucleus, and a lot of small vacuoles were scattered in the periphery of the cell. The diameter of the unfertilized egg cell was about 40-45 μm.

Isolated zygotes were also spherical protoplasts. It was difficult to identify the nucleus as it was often hidden in the abundant cytoplasm. A zygote was morphologically so different from an unfertilized egg cell that they could be easily distinguished (Fig. 1b). The diameter of the zygote was about 43-47 μm, slightly larger than that of the unfertilized egg cell.

Synergids could be confused with the egg cell due to their close proximity with the latter and their similarity in size, especially when isolated. However, they could be distinguished by careful examination. The two synergids isolated 1-2 days before anthesis were also spherical protoplasts and similar in size. It was difficult to see the nucleus, which was hidden in the surrounding cytoplasm. The cytoplasm was evenly distributed throughout the cell, and a large vacuole was not apparent (Fig. 1c). Close to anthesis, one of the synergids degenerated and shrunk, attaining an irregular shape with a dark colour. The other persistent synergid retained its original shape *in situ* (Fig. 1d), but became spherical when isolated (Fig. 2i). Isolated synergids were usually separated from the egg cell, but in a few cases they adhered to each other (Fig. 2i). The persistent synergid was slightly smaller than the egg cell, with a diameter of 35-38 μm.

The central cell containing two polar nuclei is the largest cell of the embryo sac and is pear-shaped *in situ*. Most of the central cells usually became spherical when isolated (Fig. 1e), occasionally they became elliptic or retained their shape *in situ* (Fig. 1f). The cells showed polarity, with the nucleus and cytoplasm located at one pole and a large vacuole at the other. In most cases, two polar nuclei or, after fertilization, one primary endosperm nucleus were seen. The cytoplasm was denser in the fertilized central cell than in the unfertilized one. CW-stained cell-wall material remained after isolation (Fig. 1g). In some cells, the wall became incomplete during the course of isolation and collection. Due to its large volume, the central cell often separated into two or three individual subunits. The sizes of the spherical central cells when isolated were similar before and after fertilization, with a diameter of 180-195 μm.

The antipodal cells were located at the chalazal pole of the embryo sac, opposite to the egg cell and synergids. When the non-enzymatic method was used, the isolated antipodal cell cluster contained more than ten cells, all retaining their original size and shape (Fig. 1h). After enzymatic treatment, individual antipodal cells were obtained and became spherical, respective of the presence or absence of a cell wall (Fig. 1i). Their cytoplasm was dense and evenly distributed with many of granules, possibly starch grains. Sometimes, several large vacuoles in a cell were visible. During the isolation procedure, several antipodal cells often fused into one large protoplast.

Division of embryo sac cells during in vitro microculture

After 2-10 h of culture, a newly formed cell wall was observed around the surface of the zygote protoplasts (Fig. 2a). After 1-2 days, the zygotes were smaller in size than newly isolated ones and had started to divide into two daughter cells (Fig. 2b). After 3-6 days of culture, multicellular proembryo-like structures were formed, which usually exhibited a compact globular shape (Fig. 2c). Cell division of the unfertilized egg cells was also observed after 2-3 days of culture in the medium supplemented with 10 mg/l 2,4-D, followed by subsequent culture in the medium with 2 mg/l 2,4-D. However, additional divisions have so far not been observed.

After 1-3 days of culture, the amount of cytoplasm and number of small vacuoles in the isolated fertilized central cells increased becoming more prominent, while the big central vacuole became smaller (Fig. 2d, e). The nucleus divided into several free nuclei located at the periphery of the cell (Fig. 2f). After 3-4 days, some central cells divided into two cells in which the cytoplasm and nuclei were distributed close to the newly formed cell plate (Fig. 2g); some stretched and formed irregular protrusions (Fig. 2h). The unfertilized central cells could also be induced to nuclear division after 2 days, but further development was not observed.

In synergids isolated between 1 day before pollination to 5-6 h after pollination, the nucleus and cytoplasm of the degenerated synergids all disappeared after 1 day of culture, resulting in a hollow cell with cell wall only; the persistent synergids were filled with dense cytoplasm (Fig. 2i). However, division of this cell was not observed.

Factors affecting in vitro culture of zygotes and central cells

Isolation method The presence of enzymes during the isolation procedure had a negative influence on the efficiency of *in vitro* culture of zygotes. The frequency of the first zygotic division as well as the frequency of multicellular structures was higher using non-enzymatic microdissection than using the enzymatic method (Table 1).

2,4-D concentration Zygotes could be induced to divide when they were cultured in Km8p medium with or without 2 or 10 mg/L 2,4-D. Multicellular clusters were induced only when the zygotes were subsequently incubated in the medium with 2 mg/L 2,4-D; no further development was seen in medium with or without 10 mg/L 2,4-D. In the presence of 2,4-D (2 or 10 mg/L), central cells showed cellular and nuclear divisions; medium without 2,4-D did not support these divisions.

Feeder system In microchamber culture, division of the embryosac cells was induced by co-cultivation with vigorously dividing rice suspension cells. Without feeder cells, they could not divide at all. The data for division of zygotes and central cells in microculture are shown in Table 2. It is evident that the zygotes divided more readily than the central cells and could develop into proembryo-like structures, whereas the central cells could not.

第三章 授粉、受精与胚胎发育的实验研究

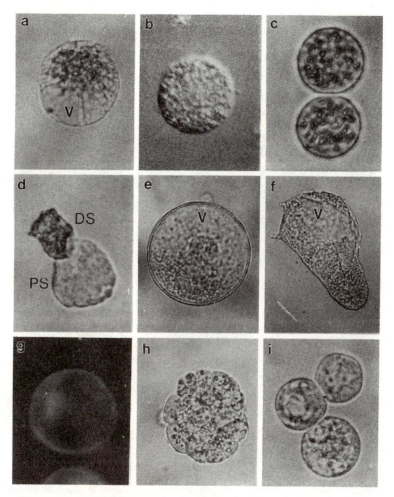

Fig. 1 a-i Cellular features of various isolated embryosac cells in rice. **a.** Egg protoplasts 1 day before anthesis, showing polarity. V: Vacuole. ×580. **b.** Zygote protoplasts with abundant cytoplasm 9 h after fertilization. ×600. **c.** Two synergid protoplasts 2 days before anthesis with evenly distributed cytoplasm and similar size. ×710. **d.** A pair of isolated synergids close on anthesis, showing the degenerated synergid (DS) and the persistent synergid (PD). ×750. **e.** An isolated central cell 1 h after pollination, showing a polarized distribution of the cytoplasm and vacuole (V). ×260. **f.** An isolated central cell with original *in situ* shape. ×200. **g.** An isolated central cell with its whole wall stained by CW. ×300. **h.** An isolated antipodal cell cluster with its original size and shape. ×330. **i.** Individual antipodal cells after enzymatic treatment that causes the cells to attain a spherical shape. ×420

629

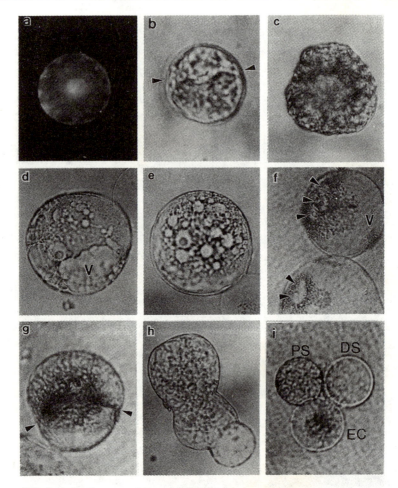

Fig. 2 a-i Zygotes and central cells after *in vitro* culture. **a.** A zygote forming a new cell wall 2 h after culture (CW staining). ×780. **b.** Bicellular embryo. Arrowheads indicate the position of the separating cell wall. ×910. **c.** Subsequent divisions of zygotes 6 days after culture leading to proembryo-like structures. ×450. **d、e.** A fertilized central cell photographed 2-3 days after culture, showing increase in its cytoplasm and reduction in vacuolar (V) size. ×240. **f.** Central cells with a number of nuclei (arrowheads) 2 days after culture. ×300. **g.** Bicellular structure derived from the central cell 4 days after culture. Arrowheads indicate cell wall. ×360. **h.** Irregular growth of fertilized central cell 3 days after culture. ×240. **i.** Two synergids adhering to an egg cell (EC) 1 day after culture, showing that the degenerated synergid (DS) is a hollow cell without cytoplasm and nucleus while the persistent synergid (PS) is filled with dense cytoplasm. ×350

第三章 授粉、受精与胚胎发育的实验研究

Table 1 Effect of isolation method on division of zygotes in rice *

Isolation method	Number of zygotes cultrured	First division		Multicellular structure	
		Number	Frequency (%)	Number	Frequency (%)
Non-enzymatic dissection	39	25	64.1	12	30.8
Enzymatic maceration-dissection	11	5	45.5	2	18.2

* The medium was Km8p supplemented with 2 mg/L 2,4-D

Table 2 Division of zygotes and central cells in rice during microculture *

Cell type	Number of cells cultured	First Division		Multicellular structure		Free-nuclei division	
		Number	Frequency (%)	Number	Frequency (%)	Number	Frequency (%)
Fertilized egg cell	56	31	55.4	14	25.0	0	0
Fertilized central cell	25	2	8.0	0	0	3	12.0
Unfertilized central cell	39	1	2.6	0	0	3	7.7

* The zygotes were isolated by non-enzymatic and enzymatic methods; the central cells were isolated by the enzymatic method. Both were total numbers cultured in Km8p and N_6 media supplemented with 2 mg/L 2,4-D

Discussion

The manipulation of sexual cells in higher plants is relevant for both fundamental research and genetic engineering. It allows for techniques such as *in vitro* fertilization and zygote culture, which can complement the *in situ* study on fertilization and embryogenesis. Such systems render the initial programme of plant ontogeny controllable under *in vitro* conditions, and also make it possible to transform plants from the earliest stage of plant development. In this paper, we report, for the first time, the division of zygotes up to proembryo-like structures in rice, which is one of the most important cereal crops. Here we discuss two key factors involved in this system.

Isolation technique

So far, two methods, enzymatic and the non-enzymatic have been used to isolate embryo sac cells. A typical example of the former is the one used to isolate egg cells and central cells in maize. A short pulse (30 min) of enzymatic maceration of the ovules did not show any harmful effect on the isolated female cells, as judged by subsequent *in vitro* intergametic fusion and culture of the fused products that resulted in a final regeneration of fertile plants (Kranz and Lörz; 1993). The enzymatic method also supported final formation of endosperm-like structures from the fertilized central cells (Kranz et al. 1998). Nevertheless, there is contradictory evidence that even a very short enzymatic treatment (2 min) has a harmful effect on embryo sac culture in maize (Leduc et al. 1995). For triggering of the first zygotic division and further divisions in tobacco zygote culture, the non-enzymatic method appears to be superior to the enzymatic method (Li et al. 1998). It is apparent that gametes isolated without enzymes are more viable.

Exogenous hormone

The exogenous hormones, 2-methyl-4-chlorophenoxy acetic acid (MCPA) and 4-amino-3,5,6-trichloropicolinic acid (Picloram), are indispensable factors for inducing haploid plants in unpollinated rice ovary culture (Zhou and Yang 1991). A similar situation is apparent in microculture of isolated female cells. Maize unfertilized egg cells could be induced to divide and form multicellular structures *in vitro* at a frequency of 6% by a very high dose of 2,4-D(25-40 mg/L) (Kranz et al. 1995). Unfertilized tobacco egg cells divided only once in medium supplemented with 1 mg/L 2,4-D (Tian and Russell 1997). For zygote culture in barley, the frequency of embryo-like structures and regenerated plants was higher in the presence of 1 mg/L 2,4-D than without 2,4-D (Holm et al. 1994). Similarly, only in the presence of a high concentration of 2,4-D (20 mg/L) could unfertilized tobacco central cells be induced to divide (Fu et al. 1997). In our experiments, rice zygotes and central cells were induced to divide by 2 mg/L 2,4-D. These experiments demonstrated that exogenous hormone has a definite effect on the *in vitro* development of female cells in various species.

(Authors: Zhao J, Zhou C, Yang HY. Published in: Plant cell Reports, 2000, 19: 321-326, with 2 figures, 2 tables and 18 references. All figures and tables are retained here.)

10 水稻原胚和刚启动分化的幼胚 cDNA 文库的构建与分析

Construction and analysis of cDNA libraries from proembryos and just differentiating young embryos in rice

Abstract

Two cDNA libraries were constructed from microdissected 214 rice proembryos (2-3 d after pollination) and 121 just differentiating young embryos (3-5d after pollination) respectively through RT-PCR technique. The primary libraries had a total of 3.7×10^6 phages for the proembryos and a total of 2.5×10^6 phages for the just differentiating young embryos, in which 96% of the phages were recombinants. Insert sizes ranging from 400 bp to 3500 bp were obtained. All of the above mentioned accorded with the general requirements of cDNA library construction.

由原胚向开始形成胚器官的分化胚过渡,必定有一系列特异表达基因的调控。对这些调控胚分化的基因的结构与功能研究是了解胚胎分化机制的关键。构建 cDNA 文库进行差异筛选是一种广泛应用于分离植物有性生殖发育调控基因的方法。胚胎发生早期的原胚和分化胚因体积小而很难获得大量的 mRNA。通过反转录-多聚酶链反应(RT-PCR)技术,可以利用微量的 mRNA 构建 cDNA 文库,从而分离出特异表达基因。Dresselhaus 等[1]用 RT-PCR 技术,从 128 个玉米卵细胞和 104 个体外融合的合子提取 mRNA,分别构建了其 PCR-cDNA 文库。他们再用合子 cDNA 文库的 4000 个克隆分别与合子 cDNA 文库和卵细胞 cDNA 文库进行杂交,分离出全长编码钙网蛋白(calreticulin)的 cDNA[2]。Breton 等[3]用 100 个玉米转型期胚构建了 cDNA 文库,并用矮牵牛异源探针分离出早期胚胎发生过程中的钙调素基因。Chen 等[4]用水稻传粉 10d 后的已分化胚构建 cDNA 文库,再用幼叶 poly(A)RNA 作探针分离出 37 个胚胎特异表达 cDNA 克隆。但是,对于更为难以获得的水稻早期原胚和分化胚,其文库构建至今尚未见报道。我们为了探索胚胎分化这一关键事件的基因表达调控机制,利用显微分离技术获得传粉后 2~3d 的原胚和传粉后 3~5d 的刚启动分化的幼胚,通过 RT-PCR 技术,成功地构建了水稻原胚和刚分化幼胚的 cDNA 文库,以求在今后的进一步研究中分离调控胚胎分化的特异表达基因。

材 料 与 方 法

实验材料为水稻(*Oryza sativa* L.)品种"春江 5 号"。取传粉后 2~5d 的子房置于滴有焦碳酸二乙酯处理过的蒸馏水(DEPC-H_2O)的载玻片上。在体视显微镜或倒置显微镜下剖开子房和胚珠。用吸管将分离的倒卵形原胚和梨形分化胚(图 I,a、b)分别移至装有 CPW(10mmol/L $CaCl_2 \cdot 2H_2O$,0.7mmol/L KH_2PO_4,1.6mmol/L $MgSO_4$,0.5mmol/L 甘露醇,pH5.8)的离心管中。于液氮中预冷后置于 -80℃ 冰箱保存。将材料分别置 CPW 和 DEPC-H_2O 中保存 6d 后用荧光素二醋酸(FDA)检测活性,在显微镜检时均测到较强的荧光。说明胚胎保持活性,可以作为构建 cDNA 的起始材料(图Ⅰ)。

图Ⅰ

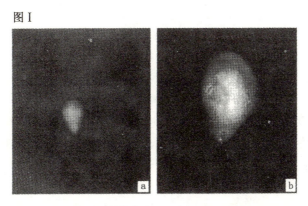

图Ⅰ. a.原胚;b.刚分化胚。
Fig. 1. a. Proembryo; b. Just differentiating young embryo.

用购自 QIAGEN 公司的试剂盒 Oligotex™ Direct mRNA Kit 直接从 214 个原胚和 121 个刚分化幼胚分别提取 mRNA。用 5′RACE System(Giboc 公司)将原胚 mRNA 和刚分化幼胚 mRNA 分别进行 RT-PCR。引物为:AAP:5′-GGC CAC GCG TCG ACT AGT ACG GGI IGG GGG IIG-3′; AUAP:5′-GGC CAC GCG TCG ACT AGT AC-3′;GSP1:5′-GGA AGA ATG CGG CGG CTT TTT TTT TTT TTT-3′;GSP2:5′-GGA AGA ATG CGG CGG CTT TT-3′。以 GSP1 为引物,将 mRNA 反转录成单链 cDNA。经纯化后,进行 poly C 加尾。以 AAP 和 GSP2 为引物,用 Vent DNA polymerase(Biolab 公司)在 Gene Cycler™(Bio-Rad 公司)上对加尾的 cDNA 进行 PCR 扩增;94℃ 1 min; 55℃ 1 min; 72℃ 5 min,30 个循环;72℃保温 10min。取 10% 扩增产物进行第二次 PCR 扩增:94℃ 1 min; 55℃ 1 min; 72℃ 5 min,25 个循环;72℃保温 10min。

扩增产物经 Advantage™ PCR-Pure Kit(Clontech 公司)纯化后,用 Not Ⅰ 和 Sal Ⅰ 限制性内切酶(B. M. 公司)双酶切,再与 λZIPLOX™, Not Ⅰ-Sal Ⅰ Arms(Gibco 公司)连接。刚分化幼胚连接产物用 Ready. To. Go Lambda Packing Kit(Pharmacia 公司),原胚连接产物用 Packegene

Lambda DNA Packaging System(Promega 公司)分别包装,再转染于大肠杆菌(E. coli)Y1090(ZL)。用 SM 缓冲液(NaCl 5.8 g/L·MgSO$_4$7H$_2$O 2g/L·Tris 50 mmol/L, pH7.5)回收噬菌体。取一部分加入 3% 氯仿,置 4℃ 备用,另一部分加入 3% 氯仿和 7% 二甲基亚砜(DMSO), -70℃ 保存。取 10μl 噬菌体液,用 E. coli DH10B 亚克隆。

取经稀释的噬菌体液转染 E. coli DH10B 在含 2 mmol/L 异丙基硫代-β-D-半乳糖苷(IPTG), 0.01% 5-溴-4-氯-3-吲哚-β-D-半乳糖苷(X-gal)和 100μg/ml 氨苄青霉素的 LB 培养基上进行连接效率测定。在 E. coli DH10B 亚克隆平板上,随机挑取单菌落进行液体培养,提取质粒 DNA 后,用 Not Ⅰ 和 Sal Ⅰ 双酶切,用 1% 琼脂糖凝胶电泳,分析 cDNA 插入片段的大小。

结 果 和 讨 论

实验结果证明,原胚 cDNA 连接反应物用 Packegene Lambda DNA Packaging System 包装,得包装反应物 500μl,转染 E. coli Y1090(ZL)后,得噬菌斑 3.7×10^6 个。刚分化幼胚cDNA连接反应物用 Ready. To. Go Lambda Packing Kit 包装,得包装反应物 525μl,转染 E. coli Y1090(ZL)后,得噬菌斑 2.5×10^5 个。

原胚和刚分化幼胚包装反应物转染 E. coli Y1090(ZL)获得的噬菌斑分别为 10^8 个和 10^5 个以上,均达到初始文库 cDNA 包容量的要求。Dresselhaus 等[1]用 128 个玉米卵细胞构建的初始文库得噬菌斑 6.8×10^6 个,Breton 等[3]用 100 个转型胚构建的初始文库得噬菌斑 2.5×10^5 个。他们用的是 Stratagene 公司的 Gigapack Gold System,这是目前已知的最好的包装系统。我们分别用的是 Pharmacia 公司的 Ready. To. Go Lambda Packaging Kit 和 Promega 公司的 Packagene Lambda DNA Packaging System,前者是固体而后者是液体。固体系统易于运输和保存,-20℃ 条件下就能储存 6 个月左右。液体系统则需要干冰运输,-70℃ 储存。但是,我们经多次实验证明液体系统包装效率远较固体系统为高(10 倍以上)。因此,要构建高质量的 cDNA 文库,应尽量使用液体包装系统。

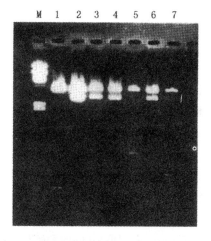

图 Ⅱ 原胚文库插入片段检测。

Fig. Ⅱ. Examination of insert sizes in the clones of proembryo library. (M) Molecular weight marker: λDNA/*Hind*Ⅲ.

用 SM 收集原胚文库噬菌体 50ml,取 10μl 稀释后用于文库噬菌斑滴度测定,滴度为 1.04×10^{11} pfu/ml;用 SM 收集分化胚文库噬菌体 50ml,取 10μl 稀释后用于文库噬菌斑滴度测定,滴度为 1.89×10^{11} puf/ml。文库噬菌体的滴度符合文库保存和筛选实验的要求[3,4]。对在含 X-gal/IPTG 的平板上培养的重组菌落(无色)和非重组菌落(蓝色)计数,得原胚的连接效率为 96.4%;刚分化幼胚的连接效率为 95.6%,连接效率符合文库构建要求[3]。从原胚和刚分化幼胚文库分别随机挑取的 7 个菌落提取质粒,经酶切、电泳检查,均含有插入片段,其长度在 0.4~3.5 kb 之间,符合文库构建要求(图Ⅱ)。

我们用材料难以获得的少量水稻原胚和分化胚成功地构建了 cDNA 文库,为分离胚胎分化的调控基因奠定了基础,也为进一步构建卵细胞和合子 cDNA 文库积累了资料。

(作者:陈绍荣、李师弟、吕应堂、杨弘远。原载:植物学报,2000,42(2):214~216。图 2 幅,参考文献 4 篇,选录图Ⅰ、Ⅱ。)

11 一种适于植物幼胚 mRNA 整体原位杂交的方法

An approach to mRNA whole-mount *in situ* hybridization for plant young embryos

Abstract

A mRNA whole-mount *in situ* hybridization method is reported here for quick, direct analysis of the spatial and temporal mRNA expression patterns in plant young embryos. A cDNA clone *THE*3 (tobacco heart-shaped embryo 3) was isolated by differential screening from tobacco (*Nicotiana tabacum* L.) heart-shaped embryo cDNA library as compared with the globular embryo cDNA library. The distribution of *THE*3 mRNA in tobacco heart-shaped embryos and globular embryos was investigated by a whole-mount *in situ* hybridization technique, showing that *THE*3 is preferentially expressed in heart-shaped embryos.

mRNA 组织原位杂交已广泛应用于植物组织、细胞的基因时空表达研究[1]。目前主要有 3 种方法：一是常规组织切片 RNA 原位杂交[1]，广泛适用于各种组织；二是大量细胞的整体 RNA 原位杂交，已应用于花粉管[2]和藻类合子[3]；三是微量细胞的整体 RNA 原位杂交，用于分离胚囊[4]。整体原位杂交具有易操作、周期短、整体效果明显等特点，是一种较新的 RNA 组织原位杂交方法。胚胎是个体发育的最初阶段，整体原位杂交作为检测胚胎发育过程中基因表达整体效果的快速、有效手段，已在果蝇[5]、*Dictyostelium*[6]等少数几种动物中应用，但在植物胚胎发育方面迄今未见报道。我们以本实验室从烟草心形胚 cDNA 文库中分离出的 *THE*3 克隆作为探针，进行了烟草球形胚和心形胚的 mRNA 整体原位杂交，初步证明该 cDNA 在心形胚中优势表达。

材料与方法

供试材料为普通烟草(Nicotiana tabacum L.)品种"W38",取开花后 8~14d 的球形胚和心形胚。

从 200 个球形胚和 150 个心形胚分别构建球形胚和心形胚 cDNA 文库并进行差异筛选。在心形胚文库中筛选出 cDNA THE3(tobacco heart-shaped embryo 3),长度约为 1.2 kb,克隆于 pZL1 载体中,克隆位点为 Sal I/Not I,两端分别有 SP6 和 T7 RNA 聚合酶启动子。正义和反义 RNA 的标记按 Boehringer-Mannheim 说明书的方法。胚的分离主要按照本实验室已建立的方法[7]:将胚珠置于含 1.0% 纤维素酶(Cellulase R-10)、0.8% 离析酶(Macrozyme R-10)、10% 甘露醇的酶解液中,室温静置 0.5 h 后,用吸管吸打分离出胚。将分离的胚收集在 1.5ml eppendorf 管中,此后操作均在 eppendorf 管中进行,加入含 4% 多聚甲醛的 PBS 缓冲液(133mmol/L NaCl、2.7mmol/L KCl、10mmol/L Na_2HPO_4、1.8mmol/L KH_2PO_4,pH7.4),4℃固定过夜。PBS 洗 3 次(每次 5min),置于含 20μg/ml 蛋白酶 K(Merck)的 PBS 中 30~50min。PBS 洗 3 次(每次 5min)后,进行预杂交,预杂交液为 4×SSC(1×SSC:150mmol/L NaCl、15mmol/L 柠檬酸钠,pH7.0),1×Denhardt 试剂,0.5g/L 鲑鱼精 DNA(Gibco),0.5g/L 酵母 RNA,60% 去离子甲酰胺(Sigma)。于 42℃摇床以 150r/min 低速振荡预杂交 3h 后,弃掉预杂交液,加入杂交液于 42℃继续低速振荡杂交 20h。杂交液是预杂交液成分中附加 50~100ng/ml Digoxigenin 标记的探针。杂交后用 2×SSC、1×SSC、0.5×SSC、0.2×SSC 逐级洗脱剩余探针,每级 42℃ 低速振荡 20min。然后将胚置于 PBT 缓冲液(PBS 附加 0.05% Tween 20 和 0.2% blocking reagent(Boehringer-Mannheim))中室温封阻 1h 后,弃去封阻液,加入含 1/500 稀释 anti-DIG-AP 的 PBT 免疫反应液,4℃过夜。用 PBT 洗 3~4 次(每次 15min),将胚吸于另一新的 1.5ml eppendorf 管,加入 detection buffer(100mmol/L Tris、100mmol/L NaCl、50mmol/L $MgCl_2$,pH9.5),室温平衡 10min 后,将胚置于含 4.5μl/ml NBT(nitroblue tertazolium)、3.5μl/ml BCIP (5-bromo-4-chloro-3-indolyl phosphate)的 detection buffer 中显色 2～5h,中途可吸取少量样品镜检蓝色反应程度。用 TE(8.0)洗 2 次终止反应,加入 80% 甘油后显微镜观察与摄影。

结 果 与 讨 论

烟草胚胎发育过程中球形胚向心形胚的转变是胚胎分化的重要阶段,其基因的表达和调控必然发生深刻的变化。本实验室从烟草球形胚和心形胚分别构建 cDNA 文库,通过差异筛选由心形胚文库中分离出 THE3 克隆(另文发表),以之作为探针进行了烟草球形胚和心形胚的 mRNA 整体原位杂交。实验共重复 3 次,每次球形胚和心形胚各 50 个,置于同一 eppendorf 管中以保证两者杂交条件相同。用 SP6 RNA 聚合酶体外转录的 THE3 反义 RNA 探针与烟草胚进行整体原位杂交,心形胚呈很深的蓝紫色(图 I,a),而球形胚仅显示较淡的颜色(图 I,

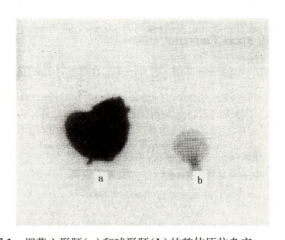

图 1 烟草心形胚(a)和球形胚(b)的整体原位杂交
Fig. 1. a, detection of *THE*3 mRNA in heart-shaped embryo, showing strong positive reaction, ×300; b, detection of *THE*3 mRNA in globular embryo, showing weak reaction, ×300.

b)。用 T7 RNA 聚合酶体外转录的 *THE3* 正义 RNA 探针与胚整体原位杂交,除褐色背景外无蓝色阳性反应。初步证明 *THE3* 基因在烟草心形胚中优势表达。

本实验结果表明,mRNA 整体原位杂交应用于幼胚具有以下优点:(1)实验时间短。整个流程只需 4~5d,而常规切片 mRNA 原位杂交需要 2~3 个星期;(2)杂交信号均一。杂交信号不均一是切片原位杂交常见的现象,而整体原位杂交全部操作都在 eppendorf 管中进行,且置于摇床低速振荡,确保了杂交的均一性。(3)获得胚的整体杂交信息。但整体原位杂交也有一定局限性,较难应用于较大的胚内部细胞的 mRNA 定位。因此,所用的胚愈幼小,应该愈能反映其内部组成细胞的 mRNA 定位情况。

(作者:李师翁、吕应堂、杨弘远。原载:植物学报,2000,42(1):105~106。图 1 幅,参考文献 7 篇。选录图 1。)

其它论文目录

1. 杨弘远、周嫦. 关于水稻开花与结实习性的一些问题. 武汉大学自然科学学报, 1960. (3):64~78.
2. Sun M X. Moscatelli A. Yang H Y. Cresti M. 2000. In vitro fertilization in *Nicotiana tabacum* L. : Fusion behavior and gamete interaction traced by video-enhanced microscopy. Sexual Plant Reproduction, 12: 267-275.
3. Sun MX. Moscatelli A. Yang HY. Cresti M. 2000. In vitro double fertilization in *Nicotiana tabacum* (L.): polygamy compared with selected single pair somatic protoplast and chloroplast fusions. Sexual Plant Reproduction, 13: 113-117.
4. 夏惠君、周嫦. 烟草原生质体微滴培养及细胞早期分裂的定点观察. 实验生物学报, 1989. 4(22):477~481.
5. 李仕琼、杨弘远. 甘蓝型油菜下胚轴原生质体培养及植株再生. 武汉大学学报(自然科学版), 1992. (3):97~101.
6. Wu Y. Haberland G. Zhou C. Koop, H. U. 1992. Somatic embryogenesis, formation of morphogenetic callus and normal dovelopment in zygotic embryos of *Arabidopsis thaliana* in vitro. Protoplasma, 169: 89-96.
7. 何龙飞、周嫦. 玉帘的组织和原生质体培养. 武汉大学学报(自然科学版), 1995. 41(2):213~217.
8. 傅缨、孙蒙祥、杨弘远、周嫦. 烟草中央细胞体内与离体发育中的细胞化学变化. 植物研究, 1998. 18(3):346~351.
9. Li S T. Fang, K F. Yang H. Y. 2000. *In vitro* development of tobacco

primary endosperm cells by microculture. Acta Botanica Sinica, 42(5): 542-544.
10. Fu. Y. Yuan M. Huang B Q, Yang H Y. Zee S Y. O'Brien, T. P. 2000. Changes in actin organization in the living egg apparatus of *Torenia fournieri* during fertilization. Sexual Plant Reproduction, 12: 315-322.

第四章 有性生殖的超微结构与细胞化学研究

提要

前面三章所介绍的实验研究,是我们研究工作的主体。这一章则是我们从事离体实验研究的同时所进行的体内的描述性研究。这一部分工作主要是在近10年期间完成的,在超微水平上由单纯形态结构的观察发展到探索结构与功能的关系。研究手段主要是超微细胞化学、免疫细胞化学以及分子原位杂交。研究对象主要是雌蕊、胚珠、胚囊及其中发生的受精和胚胎发生过程。

水稻与向日葵是我们研究离体雌核发育的重点材料(见第二章第一节)。为了配合离体雌核发育的研究,对这两种植物的胚囊中受精前后的发育过程进行了超微结构观察。在向日葵的研究中还派生出关于珠孔结构的研究。珠孔是花粉管进入胚珠的门户,在文献中基本上缺乏这方面的超微结构研究资料,我们认为有一定特色,以后就由向日葵珠孔扩大到棉花与油菜珠孔的研究。

钙在植物受精过程中具有重要作用(参看第五章有关综述)。钙信号通过一系列转导系统影响代谢过程,其中钙调素是钙信号系统中最为广泛的钙靶蛋白。在我们从事的细胞化学研究中,钙与钙调素是一个重点。本章主要介绍雌性系统中有关这方面的研究。有关花粉管与合子离体实验中钙与钙调素的工作则分别见于第一章第二节和第三章第三节。

此外,在ATP酶和植物激素的细胞化学定位方面也做了一些工作。ATP酶的超微细胞化学定位是在20世纪90年代初所作的,通过金鱼草与向日葵胚珠与胚囊中ATP酶的分布特点,提出了对胚囊营养途径的看法。植物激素的免疫细胞化学定位则是90年代末期刚刚开始的工作,以烟草为材料对几种植物激素在受精前后的分布进行了"扫描式"的初步研究。

第四章 有性生殖的超微结构与细胞化学研究

第一节 胚囊和珠孔的超微结构研究

提要

截至20世纪80年代末,国际上关于水稻胚囊的研究还只停留在光镜观察水平上,这种状况和水稻本身的重要地位极不相称。我们在进行水稻离体雌核发育的胚胎学研究时,认为有必要进一步了解它的胚囊在受精前后的超微结构特点。这也是我们整个研究工作由光镜水平进入电镜水平的开端。研究结果就水稻卵细胞是否存在极性、助细胞退化的特点、反足细胞中的无丝分裂与不完全壁等问题提出了见解。

向日葵是我们研究离体雌核发育的另一主要对象。在进行向日葵离体孤雌生殖超微结构研究的同时(参看第二章第一节),通过与美国研究者的合作对其活体内胚囊受精前后的超微结构变化也作了详细的研究。虽然这是国际上有关向日葵胚囊超微结构的第二篇报道,但在若干方面对前人观察结果作了重要补充,在另一些问题上则提出了修正意见。特别需要指出的是:当时国际上开始提出"雌性生殖单位"的新概念,但尚无实际的资料支持这一观点。我们以向日葵受精前后卵细胞、助细胞与中央细胞的结构变化为实例,详细论证了"雌性生殖单位"的实际表现形式,从而上升到理论研究的层次。

在从事向日葵超微结构研究的过程中注意到一个情况,即文献资料中几乎没有关于植物珠孔超微结构的报道。在由传粉至受精的花粉管生长轨道中,珠孔是一个重要而却唯一缺乏应有重视的环节。原因之一可能是由于在一般胚珠纵切面上难以觉察珠孔的特征。我们对向日葵较长的珠孔进行一系列横向的半薄切片与超薄切片,发现了相当新鲜的现象:首先,所谓"珠孔",在向日葵中实际上并非一个孔道,而是由珠被合拢而成的闭合结构,从而澄清了以往关于珠孔的含糊概念。其次,向日葵珠孔系由类似花柱引导组织的细胞构成,有发达的胞间基质,花粉管在胞间基质中通行。第三,向日葵珠孔是不对称结构,花粉管主要在偏向珠柄一侧的引导组织中生长,由此推论这和该侧的助细胞退化有因果联系。

既然珠孔在向日葵中是闭合型,那么在其它植物中的表现如何呢? 是否也像花柱结构那样存在闭合型与开放型呢? 为此,我们以后又在棉花与油菜中对此作了研究。结果表明:棉花的外珠孔基本上是闭合的,内珠孔完全是闭合的;而油菜外珠孔基本上是开放的,内珠孔则是闭合的。总的看来,以上三种植物的珠孔基本上均属闭合型,与它们的花柱结构类型相吻合。我们原拟继续对具开放型花柱的植物(如百合)进行珠孔结构的研究,以验证是否确实存在这种相关性,可惜时过境迁未能如愿。

1 水稻胚囊超微结构的研究

An ultrastructural study of embryo sac in *Oryza sativa* L.

Abstract

Rice embryo sac consists of an egg cell, two synergids, a central cell with two polar nuclei, and a multicellular antipodal mass. The egg cell bears no wall at its chalazal end. Its nucleus is centrally located. The greater part of its cytoplasm surrounds the nucleus, while vacuoles of various sizes are distributed at the peripheral region. The mature egg cell appears to be an inactive cell in the light of its organelle state. The two synergids show no difference with each other one day before anthesis, but by the time just prior to anthesis, one of them has degenerated. The synergids bear no wall at the chalazal end and have a thick wall at the micropylar end, where a filiform apparatus is formed. The nucleus occupies a mid-lateral position with the majority of organelles distributed at the micropylar part and many vacuoles at the chalazal part. The central cell is occupied by a large vacuole which presses the cytoplasm to a thin peripheral layer. The polar nuclei, partially fused with each other, are located in proximity to the egg apparatus and surrounded by organelles. Long, parallel endoplasmic reticulum are seen near the plasma membrane bordering the egg apparatus. A thick, multilayered embryo sac wall which bears numerous finger-like ingrowths seperate the central cell from the nucellus. The antipodal cells keep dividing before and even after fertilization, proliferating into a multicellular mass. Amitosis accompanied with freely growing walls is a usual phenomenon in this proliferation process. The cells contain abundant, active organelles and have many ingrowths at the wall bordering the nucellus. Six hours after fertilization, double fertilization is finished, resulting in a zygote and several endosperm nuclei. The zygote has now formed a new wall at its chalazal end. Its cytoplasm shows no significant changes except that more polysomes are seen and the vacuoles become smaller in size. In contrast, the cytoplasm of endosperm becomes more abundant and active than before. The degenerated synergid elongates and embraces the chalazal part of the zygote. Besides formerly existing electron-dense materials, many starch grains and spherical bodies are seen in the degenerated synergid as a result of pollen tube discharge. After fertilization, both synergids are characterized by newly formed walls at their chalazal part. To our knowledge, the present paper is the first report on ultrastructure of rice embryo sac. Based on the results obtained by this paper and the reports on other graminaceous species, a discussion is made around the following problems as polarity of the egg cell, degeneration of the synergid, changes of the boundary wall between egg apparatus and central cell, and division of antipodal cells.

胚囊的结构与功能是植物生殖生物学研究的一个重要方面。近

20年来已有几十种被子植物作过胚囊超微结构的研究,获得了较以往光学显微镜研究深入得多的知识[13,23]。在有重要经济价值的禾本科植物中,被研究过的有玉米[8,9]、大麦[4~7]、小麦[25]和小黑麦[11,12]。但关于水稻胚囊还未见超微结构方面的报道,因此我们对于水稻胚囊结构与功能的认识还停留在光镜观察的水平上。这种状况和水稻本身的重要地位是不相称的。本文第一次从超微结构的角度研究了水稻胚囊中各个组成细胞受精前后的特点。

材 料 与 方 法

供试材料为水稻(Oryza sativa L.)品种"景洪2号"和"鄂晚3号",均属粳稻类型。取材分为三期:开花前1d(近成熟胚囊)、即将开花(成熟胚囊)和开花后6h(受精后胚囊)。将子房切下,用0.2mol/L二甲胂酸钠缓冲液(pH7.4)配制的4%戊二醛与1%锇酸先后固定,每次均在4℃冰箱中过夜。此后,经乙醇系列脱水、环氧丙烷过渡,用Epon812包埋。在Sorvall MT-6000型超薄切片机上用玻璃刀切片。根据半薄切片的检查,选择适当的部位做超薄切片。后者经醋酸双氧铀与柠檬酸铅先后染色,置于JEM-100CX/Ⅱ型透射电子显微镜下观察和摄影。

结 果

开花前1d,胚囊接近成熟,已形成一个卵细胞、两个助细胞、一个中央细胞(含二极核)和多个反足细胞。即将开花时,胚囊完全成熟,一个助细胞退化。开花后6h,卵细胞已受精形成合子,退化助细胞与宿存助细胞仍在,极核受精后已分裂成数个胚乳游离核,反足细胞仍在继续增殖。以下分别描述胚囊中各个成员细胞受精前后的超微结构特征。

第四章 有性生殖的超微结构与细胞化学研究

一、卵细胞

图版Ⅰ,1示一个成熟卵细胞的完整切片,其细胞呈梨形,珠孔端狭窄而合点端宽阔。核居细胞中部(图中核偏珠孔端是因切片稍倾斜所致),核质均匀,核仁较小而致密。细胞质大多聚集在核周围,而液泡则分散在外围。合点端的液泡比珠孔端更为发达。细胞的珠孔端有壁,愈向合点端壁愈薄,合点端完全无壁而仅以质膜与中央细胞为界(图版Ⅰ,4)。卵细胞与中央细胞及助细胞之间的细胞壁上未见胞间连丝。

卵细胞的胞质中含线粒体、质体、内质网、高尔基体、核糖体、小泡、脂滴等细胞器。在由开花前1d到即将开花的成熟过程中,细胞器的状态有如下一些变化:线粒体起初电子密度较高,以后多数电子密度降低,嵴变为不发达状态。质体原来不含淀粉,到成熟卵时期部分质体积累淀粉而形成造粉体。内质网不多,常单个存在于细胞边缘,起初为粗糙型,到开花前不久其上所附核糖体减少。游离核糖体很丰富。高尔基体由几个扁平的潴泡组成,其末端稍膨大,呈不活跃状态。近成熟卵中未见脂滴,成熟卵中在细胞核周围有一些脂滴。总体来看,卵细胞在成熟过程中细胞器变为不活跃状态,显示其合成与代谢的低落,这和其它植物的观察结果是一致的[13,23]。图版Ⅰ,2示成熟卵细胞中的细胞器。

开花后6h,卵细胞已受精形成合子(图版Ⅰ,3;图版Ⅱ,10)。此时,其合点端形成新壁(图版Ⅰ,5)。合子的核、质和液泡分布状态与受精前相比变化不大,但游离核糖体有形成多聚核糖体的趋势,出现许多短小而膨胀的内质网和小泡,液泡体积比受精前减小。

二、助细胞

图版Ⅱ,6示开花前1d卵器的一部分,可以看出此时两个助细胞在超微结构上没有差别。它们的共同特征是:珠孔端壁很厚,但尚未形成丝状器。愈向合点端壁愈薄,合点端无壁。细胞核位于中部而贴近侧壁,核质均匀,核仁致密,有小核仁。细胞质主要分布在珠孔端,液泡主要分布在合点端。

即将开花的成熟胚囊中,两个助细胞表现明显的差异(图版Ⅱ,7)。其中一个助细胞呈退化状态,充满电子致密物质,核和细胞器难以分辨。电子致密物质还分布到另一助细胞的合点端及其与卵细胞的交界处。另一个助细胞(即宿存助细胞)保持原有的极性状态,珠孔端的细胞质区域和合点端的液泡区域更为明显。细胞质中含大量质体、线粒体、高尔基体、粗糙内质网、脂滴等。此时,丝状器已经形成,在切片上可以看到丝状器的许多内突(图版Ⅱ,8)。

开花后6h,退化助细胞外形呈牛角状,其合点端延伸成狭窄的钩状环抱着合子(图版Ⅱ,10;图版Ⅰ,3)。此时,其合点端已形成完整的壁。细胞内含物中除原有的电子致密物质外,还混杂着许多电子透明的椭圆形物体和中等电子密度的小球。它们和Mogensen(1982)在大麦受精过程中所描述的花粉管释放的淀粉粒与多糖分泌小泡可能是类似的[18]。由此可以推测在水稻中也是花粉管进入退化助细胞并在其中释放内含物。宿存助细胞中各种细胞器的状态看不出明显变化,只是液泡化程度更高(图版Ⅰ,3;图版Ⅱ,10)。它的合点端也形成新壁,壁的厚薄不匀(图版Ⅱ,9)。

三、中央细胞

中央细胞大部分空间被一个大液泡占据,细胞质分布在周边区域。在靠近卵器处有两个极核,其周围细胞质较密集。极核核质均匀,核仁大而致密,其中有一些小的核仁液泡。从开花前1d起,两个极核就互相紧贴,在接触处形成融合桥(图版Ⅲ,11)。极核的核膜上可以看到众多核膜孔(图版Ⅰ,4;图版Ⅲ,12)。细胞质中含质体、线粒体、高尔基体、内质网、游离核糖体、脂滴、小泡等。内质网一般为粗糙型,很长。在靠近卵器合点端的区域,长形的内质网与中央细胞的质膜常平行排列,有时与质膜紧贴(图版Ⅰ,4、5;图版Ⅱ,9)。中央细胞的核糖体密度较卵细胞为低。

中央细胞与珠心相接处有很厚的胚囊壁(图版Ⅲ,13)。此壁由多层组成,每层均有密集的微纤丝,显然由纤维素构成,层间有电子致密

第四章　有性生殖的超微结构与细胞化学研究

带。壁的最外方有一层很厚的电子致密物质,可能是角质。胚囊壁上到处分布着指状内突,珠孔端尤为密集。

开花后 6h,已形成数个胚乳游离核。细胞质中核糖体密度增高并形成多聚核糖体。线粒体数量很多,具发达的嵴,从(图版Ⅲ,14)的照片上可以看到有些线粒体正在分裂。质体中出现淀粉。内质网短而膨胀呈小泡状。小泡很多。高尔基体很少。与受精前的中央细胞质相比,此时细胞器的合成与代谢显得很旺盛。

四、反足细胞

在胚囊成熟过程中,反足细胞不断增殖,形成多细胞的群体(图版Ⅳ,15)。根据对多个胚囊的观察,反足细胞的增殖主要是通过无丝分裂。常常是核仁先分裂,在一个核中形成多核仁,然后核也分裂成几个。图版Ⅳ,16 示一个核通过无丝分裂形成数个新核的状态,每一子核有一个核仁,子核间的联系尚未完全断绝。与核无丝分裂相伴发生的是,胞质分裂多采取壁的自由生长方式,即由侧壁上产生突起作向心生长,逐渐将细胞隔开(图版Ⅳ,15)。壁的生长可以是单向的,也可以是双向的,即由两个相对生长的壁尖端逐渐汇合(图版Ⅳ,17)。如果壁的生长不完全,在一个细胞中就可以包括几个细胞核而形成多核体。到开花后 6h,核无丝分裂和壁自由生长的过程仍在继续进行。

反足细胞含有丰富的细胞器,电子密度普遍很高(图版Ⅳ,17、19)。线粒体很多,有些电子致密,具发达的嵴;有些比较透明,嵴不发达。在开花前 1d 的反足细胞中含有造粉体,而成熟时只有电子致密的质体,没有造粉体。内质网发达,多为粗糙型,有的呈同心圆状(图版Ⅳ,18)。游离核糖体密度很高。高尔基体和小泡也很多。整个反足细胞呈现代谢活跃的状态。

反足细胞群的外围有细胞壁包围。其与珠心相邻的一面有发达的壁内突(图版Ⅳ,19)。群体内部各个细胞之间的界壁厚薄不匀,其间有胞质管道相通。

653

讨 论

一、关于卵细胞的极性

大多数被子植物的卵细胞具有明显的极性,即核和细胞质偏合点端,珠孔端有一个大液泡[13,23]。禾本科植物的卵细胞则表现不同的特点,它们基本上都是核居细胞中央,细胞质围绕着核周围,液泡多数,且分布于周边区域。只是在细节上有些差别,如玉米卵细胞核略偏合点端,液泡主要分布在近珠孔端[9];*Stipa elmeri* 的卵细胞呈现多样性,有些以合点端液泡较发达,有些以珠孔端液泡较发达[15];小麦[25]、小黑麦[11]和本文所描述的水稻相似,液泡主要分布在合点端。总之,在禾本科植物的卵细胞中看不出其它被子植物卵细胞中典型的极性现象。甚至在水稻开花后 6h 的合子中也看不到极性。那么,以后原胚的极性是如何建立起来的呢?Maze 与 Lin 曾推测:高度极性化的宿存助细胞可能有助于卵细胞以后极性的建立[15]。这是一个需要继续探讨的问题。

二、关于助细胞的退化

许多被子植物的助细胞之一在受精前退化,多数是在传粉后花粉管尚未到达胚囊前退化;有些则是在花粉管进入助细胞后引起它退化[13,22]。禾本科植物由于传粉到受精时间较短,助细胞退化的时间并非都有明确的判断。玉米[8]、大麦[4]是在传粉后花粉管到达胚囊前一个助细胞退化。针茅[15]、小黑麦[11,12]在胚囊成熟时两个助细胞具有大致相同的超微结构,以后花粉管进入一个退化的助细胞,但它何时退化并不清楚。小麦的情况很特殊,它在传粉前两个助细胞均表现相似的退化迹象[25]。在水稻方面,戴伦焰等(1964)认为一个助细胞较早消失;另一个助细胞紧贴卵细胞一侧而生长并环抱后者,一直保持到受精后[3]。吴素萱等(1965)也提到:助细胞之一退化较早,因而同时看到两个助细胞的机会不多;另一助细胞环抱卵细胞,一般也呈逐渐消失现象,有时也可保存较长时间[2]。根据本文的观察,水稻即将开花的胚囊中仍有两个助细胞,其中一个虽然退化,但并未消失。开花后 6h 两个助细胞依然存在,其中退化助细胞中有花粉管释放物,也正是这个退化助细胞环抱卵细胞的合点端。看来,水稻开花前只存在一个助细胞的结论是需要重新商榷的。

三、关于卵器与中央细胞之间壁的变化

被子植物的卵细胞和助细胞成熟时合点端没有细胞壁或不具完整的壁,受精后合子形成完整的壁[13,23]。但只有少数文献中提到受精后助细胞壁的变化。Mogensen 曾在角胡麻属的 *Proboscidea louisianica*[16,17]和烟草[19]上报道,受精后无论退化的或宿存的助细胞均被完全壁包围,并推测花粉管释放的多糖分泌小泡可能参与助细胞新壁的形成。本文在水稻中也观察到受精后两个助细胞合点端均形成新壁。如果这一现象在更多的植物中得到证实,则受精前后助细胞和卵细胞经历相同的壁的变化。进一步说,由于这部分壁也是中央细胞共有的,我们不妨认为实际上是涉及卵器和中央细胞之间一部分界壁的变化,而不限于卵细胞或助细胞本身;它的消失与再生也就可能涉及三者的相互作用。一个值得注意的现象是:在玉米[8]、白花丹[6]、菠菜[24]和本文在水稻中,都观察到中央细胞靠近卵器处有平行排列的内质网系统。Fougère-Rifot(1981)认为此处的内质网是一种一面有核糖体、另一面没有核糖体的非对称形态,它可能与卵细胞合点壁的变薄有关[10]。那么,它是否也可能与受精后卵器与中央细胞之间壁的再生有关呢? 特别是,退化助细胞中壁的再生很难从它本身的活动来解释,倒是解释为中央细胞的活动似更合理。因此,我们认为应该从三种细胞(实际上是四个细胞)之间的相互作用来进一步研究这个问题。

四、关于反足细胞分裂的特点

禾本科植物一般具有多个反足细胞。但文献中很少描述反足细胞增殖的方式。只有水稻,光镜观察一再提到反足细胞的增殖以无丝分裂为主要方式[1~3,14]。其实在大麦[5]和小麦[25]曾描述过反足细胞的

核与核仁呈不规则形态;玉米的反足细胞呈多核体结构[9]。它们不是没有无丝分裂的可能。本文进一步从超微结构观察证明了水稻反足细胞的无丝分裂,同时还观察到与此相伴的壁自由生长现象。在玉米[9]、荠菜[21]、向日葵[20]的反足细胞中都描述过壁的自由生长和不完全壁现象,但没有把它和无丝分裂联系起来。我们认为,壁的自由生长虽不一定以无丝分裂为前提,但无丝分裂之后如有胞质分裂,则必然要通过壁的自由生长方式。因此二者之间的关系还值得进一步研究。

图 版 说 明

A. 反足细胞　CC. 中央细胞　DC. 退化助细胞　E. 卵细胞　En. 胚乳　Nu. 珠心　PN. 极核　PS. 宿存助细胞　S. 助细胞　Z. 合子　a. 造粉体　d. 高尔基体　er. 内质网　ta. 丝状器　gw. 自由生长壁　lb. 脂滴　m. 线粒体　n. 细胞核　p. 质体　st. 淀粉粒　v. 液泡　ve. 小泡　w. 细胞壁　wi. 壁内突

图 版 Ⅰ

卵细胞: **1.** 成熟卵细胞整体观。×2000　**2.** 成熟卵细胞质。×20000　**3.** 合子、宿存助细胞和退化助细胞。箭头示退化助细胞合点端的一小部分。×1500　**4.** 卵细胞合点端与中央细胞交界处。箭头示两个细胞以质膜为界。三角号示极核的核膜孔。×23000　**5.** 合子合点端的新壁形成。×20000

图 版 Ⅱ

助细胞: **6.** 开花前1d卵器的一部分,示两个助细胞在超微结构上尚无差别。×2800　**7.** 即将开花时的卵器,一个助细胞退化,另一个宿存。×2800　**8.** 宿存助细胞的珠孔端,示丝状器及各种细胞器。×10000　**9.** 受精后宿存助细胞合点端形成的新壁。×20000　**10.** 受精后退化助细胞中有花粉管释放的大量淀粉粒。其合点端环抱合子(箭头)。×2800

图 版 Ⅲ

中央细胞: **11.** 近成熟胚囊中的极核。箭头示融合桥。×4000　**12.** 成熟胚囊中极核附近的细胞质,三角号示极核的核膜孔。×12000　**13.** 胚囊壁及其内突。×22000　**14.** 开花后6h的胚乳细胞质。×22000

第四章 有性生殖的超微结构与细胞化学研究

图版 I plate I

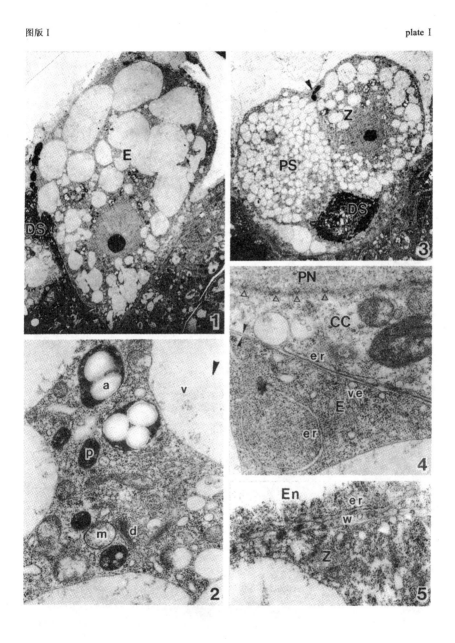

图版 II plate II

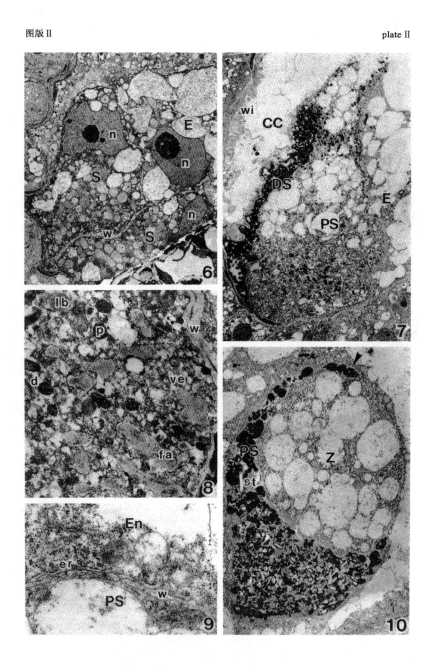

第四章 有性生殖的超微结构与细胞化学研究

图版Ⅲ plate Ⅲ

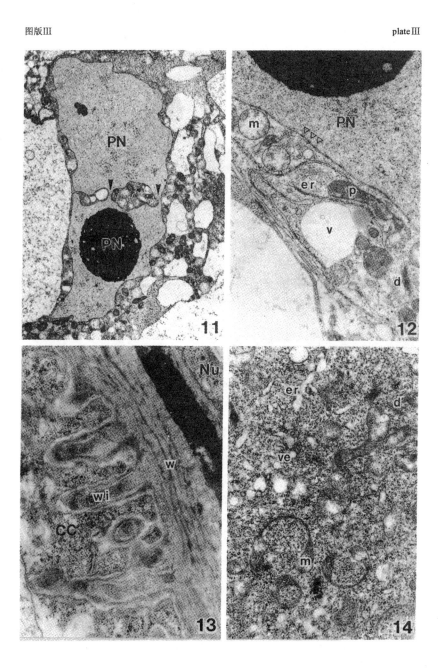

图版Ⅳ plate Ⅳ

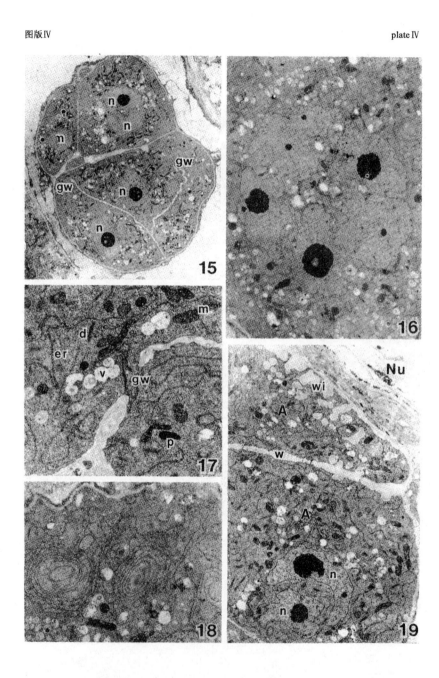

第四章 有性生殖的超微结构与细胞化学研究

图 版 IV

反足细胞: **15.** 增殖中的反足细胞群。×2000 **16.** 反足细胞核的无丝分裂。×4000 **17.** 反足细胞中的细胞器和自由生长的壁。×10000 **18.** 反足细胞中的同心圆状内质网。×8000 **19.** 反足细胞,示与珠心相接处的壁内突。×6000

Explanation of plates

A. Antipodal cell, CC. Central cell, DC. Degenerated synergid, E. Egg cell, En. Endosperm, Nu. Nucellus, PN. Polar nucleus, PS. Persistent synergid, S. Synergid, Z. Zygote, a. Amyloplast, d. Dictyosome, er. Endoplasmic reticulum, fa. Filiform apparatus, gw. Growing wall, lb. Lipid body, m. Mitochondrium, n. Nucleus, p. Plastid, st. Starch grain. v. Vacuole, ve. Vesicle, w. Wall, wi. Wall ingrowth.

Plate I

Egg cell: **Fig. 1.** Whole view of a mature egg cell. ×2000 **Fig. 2.** Cytoplasm of a mature egg cell. ×20000 **Fig. 3.** Zygote, Persistent synergid, and degenerated synergid. Arrowhead shows the chalazal tip of the degenerated synergid. ×1500 **Fig. 4.** Boundary region between the egg cell and the central cell. Arrowheads show plasm membranes of both cells. Triangles show nuclear pores of the polar nucleus. ×23000 **Fig 5.** New wall formation at the chalazal end of a zygote. ×2000

Plate II

Synergids: **Fig. 6.** A portion of egg apparatus one day before anthesis. No differences are seen between the two synergids. ×2800 **Fig. 7.** Egg apparatus just prior to anthesis showing a persistent synergid and a portion of the degenerated synergid. ×2800

Fig. 8. Micropylar end of a persistent synergid showing the filiform apparatus and various organelles. ×10000 **Fig. 9.** Newly formed chalazal wall of a persistent synergid 6 hours after anthesis. ×20000 **Fig. 10.** A degenerated synergid containing numerous starch grains discharged by pollen tube. Arrowhead shows the tip of the synergid embracing the chalazal end of the zygote. ×2800

Plate III

Central cell: **Fig. 11.** Polar nuclei in a nearly matured embryo sac. Arrowheads show fusion bridge. ×4000

Fig. 12. Cytoplasm adjacent to polar nuclei in a mature embryo sac. Triangles show nuclear pores of the polar nucleus. ×12000　**Fig. 13.** Embryo sac wall and its ingrowths. ×22000　**Fig. 14.** Cytoplasm of the endosperm 6 hours after anthesis. ×22000

Plate Ⅳ

Antipodal cells：**Fig. 15.** A proliferating antipodal population. ×2000　**Fig. 16.** Amitosis of an antipodal nucleus. ×4000　**Fig. 17.** Freely growing walls in antipodal cell. ×10000　**Fig. 18.** Concentric endoplasmic reticulum in antipodal cell. ×8000　**Fig. 19.** Antipodal cells, showing the ingrowths at the wall bordering nucellus. ×6000

（作者：董健、杨弘远。原载：植物学报，1989，31（2）：81～88。图版4幅，参考文献25篇。选录全部图版。）

2　向日葵胚囊的超微结构和"雌性生殖单位"问题

Ultrastructure of sunflower embryo sac in respect to the concept of female germ unit

Abstract

This paper deals with the ultrastructural changes in sunflower (*Helianthus annuus* L.) embryo sacs before and after fertilization with particular emphasis on the wall change events. At early stage, the embryo sac component cells are separated with each other by complete walls. During maturation, a common wall zone bordering the egg cell, synergids and the central cell becomes thinner and eventually disappears, leaving only two layers of plasma membrane for the future target zone of double fertilization. The egg nucleus and the secondary nucleus move toward each other near to the plasma membrane. After pollination one of the synergids deposits electron dense materials on its tonoplast, which is the first sign of degeneration; the other synergid has no such changes. The pollen tube penetrates the degenerated synergid and the discharged contents, including two sperms, move along a gap between the two plasma membranes. Each sperm fuses with the egg or the central cell, respectively. After fertilization, the fertilized egg and secondary nuclei move away from each other. A new wall regenerates along the previously naked zone, separating the zygote, primary endosperm cell and persistent synergid. The zygote soon enters into mitosis, even while wall formation continues. Multilayered endoplasmic reticulum and dictyosomes, which are located in the central cell and along the new wall, secrete numerous vesicles into the forming wall. Based on present observations and previously reported facts in other plant materials, it is concluded that the egg, synergids and the central cell make up a female germ unit. This unit is formed prior to fertilization by degradating a portion of the common wall, which, together with other important features, facilitates gametic fusion. Once fertilization is finished, the unit is dismissed and the zygote and persistent synergid form a wall over their entire surface. The antipodal cells, a member of embryo sac also, are not a participant of the female germ unit.

关于向日葵胚囊的超微结构,迄今只有 Newcomb 在 20 世纪 70 年代进行过系统的研究[15~17]。最近几年,我们在研究向日葵离体诱导的孤雌生殖过程的超微结构变化时[2],同时也观察了其体内胚囊发育的超微结构。结果除了在一些具体方面填补了 Newcomb 研究的空白或提出修改意见外,还促使我们从总体角度思考当前植物生殖生物学中的一个新概念——"雌性生殖单位"(female germ unit)。这一概念是和"雄性生殖单位"(male germ unit)同时提出来的[8],但却缺少像后者那样具体的研究和深入的阐述。本文试图提出我们的初步看法。

材料与方法

剥取向日葵(*Helianthus annuus* L.)品种"当阳"开花前1d、2d、3d和传粉后2h、4h、8h及1d、2d的胚珠,切除其合点端大部分,保留珠孔端1~2mm长的含胚囊的一小段。用3%戊二醛与1%锇酸(均用pH7.2的磷酸盐缓冲液配制)先后固定。经各级乙醇与环氧丙烷脱水后,用Epon 812或Poly/Bed 812树脂渗透与包埋。在Sorvall MT 6000或Reichert超薄切片机上用玻璃刀或钻石刀切片。醋酸双氧铀与柠檬酸铅染色。在JEM-100 CX/Ⅱ或Zeiss-EM10透射电镜下观察与摄影。本实验的部分工作在武汉大学生物系植物细胞胚胎学研究室与本校分析测试中心电镜室完成;部分工作在美国俄亥俄州立大学生物科学院完成。

结　果

向日葵胚囊由一个卵细胞、两个助细胞、一个中央细胞(含次生核)和两个反足细胞(含多核)组成。胚囊外有珠被绒毡层包围。图1为胚珠半薄切片甲苯胺蓝染色,示胚囊成熟过程不同时期的图像。关于卵细胞、助细胞与中央细胞的超微结构变化分别描述如下。

一、卵细胞

从开花前3d到传粉前是卵细胞由幼嫩发育到成熟的阶段,除体形增大外,内部细微结构亦发生一系列变化。开花前3d的卵细胞,外周以完全壁与助细胞、中央细胞相分隔,壁上有胞间连丝保持与相邻细胞的沟通(图版Ⅰ,1、2)。卵核位于细胞合点端;珠孔端为大液泡所占据;细胞质中含丰富的细胞器。开花前2d,卵细胞合点壁开始变薄,壁的外侧中央细胞质中有多层内质网(图版Ⅰ,3)。开花前1d,合点端约占整个卵细胞三分之一部位的细胞壁完全消失,只有原生质膜与中央细胞的质膜相贴近。接近开花时,这两层质膜互相脱离而形成一个间隙区,其中包含电子浓密物质和小泡。关于近成熟和成熟卵细胞合点端界面的特点,我们在前一篇论文中已经描述,并附有相应的照片[2],因此在本文图版中不再重复。Newcomb(1973)曾报道向日葵成熟卵细胞合点端具不完全壁[15]。根据本文的观察,我们认为可能是由于他取材偏早,当时该部位的壁尚未完全消失,成熟时应是完全无壁的。除了壁的显著变化外,卵细胞成熟过程中还有其它一些明显的变化,如细胞核更贴近合点端;珠孔端大液泡变为分散的小液泡;细胞质中油滴增多等。

受精后有一个短暂的合子间期,仅数小时。此时内部变化十分迅速:核由贴近合点端向珠孔方向略移,回复到较幼嫩卵细胞中卵核的位置(图版Ⅰ,4)。细胞质中内质网和高尔基体增多并呈活跃状态,质体与线粒体分裂,多聚核糖体增多(图版Ⅱ,5)。珠孔端有许多小液泡和一些自体吞噬泡。合点端开始形成新壁。新壁是不均匀的,有些区段很厚,有些区段很薄。在新壁的外侧,初生胚乳细胞中有丰富的高尔基体与内质网。可以看到许多小泡由初生胚乳细胞一方掺入到正在形成的新壁中(图版Ⅱ,5)。由于合子间期极短,壁的合成速率可能赶不上细胞周期的进度,因此直到合子分裂时新壁的形成尚未完成,而有一段二者同步进行的时期,这是和其它许多植物不同的特点;在其它植物中,合子完全壁的形成是在分裂之前结束的[14]。

合子经过几次分裂,形成具胚和胚柄的原胚。原胚外周壁稍厚,没有胞间连丝和相邻细胞相通。内部隔壁很薄,有胞间连丝。胞质中有密集的核糖体和活跃的细胞器。油滴消耗殆尽。胚柄细胞中有较多的自体吞噬泡(图版Ⅱ,6)。

二、助细胞

开花前3d,两个助细胞在超微结构上没有差别。它们均被完全壁包围。核位于细胞中部。液泡分散而以合点端者较大。珠孔端尚未形成钩状,丝状器刚刚出现(图版Ⅱ,7)。随着胚囊的逐渐成熟,助细胞极性加强。合点端出现大液泡,核被挤到细胞侧面。合点端细胞壁消失(图版Ⅱ,8)。珠孔端出现钩状,丝状器形成。开花当天,助细胞壁上沉积电子浓密物质,尤以珠孔端的壁上为甚。在邻近的珠被绒毡层壁上也有类似的性质不明的物质,但不如在助细胞壁上者密集(图版Ⅲ,9)。

传粉后但花粉管尚未到达胚囊之前,一个助细胞的液泡膜上沉积电子浓密物质,液泡内部也有类似的絮状物;但另一个助细胞中则看不到这种变化(图版Ⅲ,10)。这种变化是助细胞开始退化的征兆。由此可以认为,向日葵助细胞的退化与多数植物中相似,也是发生在花粉管到达之前。Newcomb(1973)当初没有弄清这一情况[16],大概是因为向日葵由传粉到受精的间隔时间很短,没有抓住助细胞开始出现退化迹象的时机。

两个助细胞中,哪个退化、哪个宿存,不是随机发生的,而是和其所处位置有关。在我们所观察的64个这一时期的胚珠中,52个胚珠(占81%)的退化助细胞偏向珠柄一侧;远离珠柄一侧的多为宿存助细胞。Mogensen(1984)首次揭示大麦助细胞退化与其在胚珠中所处位置的关系,指出退化助细胞多位于近胎座的一侧[12]。现在在向日葵中也观察到类似现象,这是很有趣的。循此追溯,将阐明助细胞选择性退化的机理,使我们对受精生物学中这一环节有更深入的认识。

受精后,宿存助细胞合点端形成新壁。新壁的外方初生胚乳细胞中有多层平行排列的内质网,和合子的合点端外侧的情况相似。Newcomb(1973)也曾指出宿存助细胞形成完整壁的现象[16]。

三、中央细胞

中央细胞幼期呈狭长形。次生核球形,和卵细胞之间有一定距离。细胞质中有无数分散的小液泡(图版Ⅰ,2)。在成熟过程中,液泡化加剧,伴随着中央细胞的膨大。受精前,次生核向卵细胞贴近,并变成近半球形,一面略向内曲而与卵的合点端呈平行弧状(图1D;图版Ⅰ,3)。细胞质多分布于珠孔端,液泡多分布于合点端,从而使中央细胞的极性加强,油滴显著增多。与此同时,中央细胞由幼期具完全壁演变为其珠孔端与卵器之间的界壁消失,而其合点端与反足细胞的界壁和其侧面的周界壁(即胚囊壁)则变得更厚(图版Ⅲ,11、12)。上述各种变化,为中央细胞的受精作了准备。

受精后,初生胚乳核离开合子向合点方向移动并很快分裂。经数次游离核分裂后转入胚乳细胞形成。细胞形成的方式是由珠孔端开始,以壁的自由生长方式形成新壁,并逐渐向合点方向推移,使整个胚囊中央细胞分隔成多个胚乳细胞(图版Ⅲ,13)。

讨 论

Dumas(1984)等提出:被子植物的雌性生殖单位应由卵细胞、助细胞、中央细胞所组成,其主要特征是在受精的靶区缺少细胞壁,从而使雌、雄性细胞得以原生质体状态互相融合[8]。雌性生殖单位是一个新概念。它的提出是否有充足的理论与事实依据?是否对我们以新的眼光认识胚囊的结构与功能及受精生物学有所启迪?这是需要植物生殖生物学界共同关心的问题。因此,尽管我们的看法是不成熟的,仍愿提出来供大家讨论。

首先,雌性生殖单位概念的成立应该具备两个基本条件:第一,它应该是由几个细胞组成的一个结构单位;第二,它应该是一个和生殖直接相关的功能单位。在胚囊的四种成分中,卵细胞与中央细胞是精子的受体,助细胞是精子释放和通向前二者的场所。三者以无壁的特定质膜区相接触,共同组成一个紧密的实体以完成受精作用,因此在结构与功能两方面均符合上述两个基本条件。反足细胞虽然亦属胚囊成员,但和其它成员细胞之间始终有壁相隔,并且其功能偏于营养方面而与生殖无直接关系,所以不应包括在雌性生殖单位之内。

其次,雌性生殖单位是当胚囊发育到一定临界状态、即准备受精之前才组成的一个暂时性的单位(雄性生殖单位同样如此)。Cass(1985,1986)等在大麦[6,7]、本文在向日葵中均发现,幼期的卵细胞、助细胞、中央细胞之间是具有完全的隔壁的,这是当胚囊由游离核转变为细胞状态后,为保证其各成员细胞的分化所必须具备的条件。直到发育充分成熟,三者之间的界壁方才局部地消失。更有甚者,陈一明与王伏雄在扁豆(1988)[1]、我们在向日葵中[2]均观察到：临近受精时卵器和中央细胞之间的双层质膜互相分离，形成充满某种物质的间隙区，似乎为精子到达受精的靶区准备了通道。此外，助细胞之一的预先退化、卵核与次生核的相向贴近，以及其它各种内部变化，也是为受精作准备。雌性生殖单位就是以这种准备受精的状态出现的。一旦受精完成，在原来壁消失的部位再生新壁，合子重新被完全壁所包围，是一般植物的共同特点[9,18]。在某些情况下，合子甚至被胼胝质[10]、角质[4]或栓质[5]的壁包围，以加强其与周围细胞的隔离机制。助细胞在受精之后同样再生新壁的现象，在其它材料[3,11,13]和向日葵中均有证明。此外，受精后合子核与初生胚乳核向相反方向移动。这些事态说明，雌性生殖单位在其功能结束后即行解散，以保证胚和胚乳各自的发育。总之，我们应将雌性生殖单位的组成和解散看成是植物有性生殖过程中一系列有序变化的重要一环。

再者，以往偏重从一个细胞的角度孤立地看待卵细胞受精前后壁的变化。而如果把雌性生殖单位作为一个整体来考察，则卵细胞的合点壁并非仅属于卵细胞，而实际上是卵细胞、助细胞、中央细胞三者共同的界壁；它的消失与重建当然不仅是卵细胞本身的活动，而可能涉及三者之间的相互作用。董健与杨弘远(1989)论证了这一观点，并援引前人报道的中央细胞贴近卵器处有平行排列的内质网现象，推测中央细胞可能参与这一共同界壁的降解与重建的活动[3]。我们在向日葵离体孤雌生殖[2]和自然有性生殖过程中均观察到这一现象，并发现初生胚乳细胞分泌小泡参与新壁的建造。这些初步的观察启示我们，今后应从雌性生殖单位的整体角度进一步研究其组成细胞之间的相互作用。

图 版 说 明

A. 反足细胞　C. 中央细胞　DS. 退化助细胞　E. 卵细胞　EN. 胚乳　P. 原胚　PS. 宿存助细胞　S. 助细胞　SN. 次生核　T. 珠被绒毡层　Z. 合子　av. 自体吞噬泡　c. 角质层　er. 内质网　f. 丝状器　w. 细胞壁

第四章　有性生殖的超微结构与细胞化学研究

图　版　Ⅰ

1. 开花前 3d 的卵细胞，示其合点端有完全壁。×8000　**2.** 开花前 3d 的胚囊珠孔端部分，示卵细胞、助细胞与中央细胞一部分。　×1800　**3.** 开花前 2d 卵细胞合点端的壁变薄，其中央细胞一侧有内质网沿壁排列。　×13000　**4.** 传粉后 3～6h 的合子与宿存助细胞，其合点端开始形成新壁。×1600

图　版　Ⅱ

5. 合子合点端新形成的不完全壁。箭头示初生胚乳细胞一侧有众多小泡参与壁的形成。×14000　**6.** 原胚与相邻的胚乳细胞。×1600　**7.** 开花前 3d 的助细胞，合点端具完全壁。×3200　**8.** 一对成熟的助细胞。×3200

图　版　Ⅲ

9. 助细胞珠孔端，示壁中有电子浓密物质。×3200　**10.** 传粉后花粉管进入以前，一个助细胞的液泡膜上与液泡内出现浓密物质（箭头）；另一助细胞则无。×3200　**11.** 中央细胞与反足细胞之间的隔壁具胞间连丝。×6400　**12.** 中央细胞的侧壁（即胚囊壁）上无胞间连丝。×6400　**13.** 胚乳中的自由生长壁（箭头）。×5100

Explanation of plates

A. Antipodal cell　C. Central cell　DS. Degenerated synergid　E. Egg cell　EN. Endosperm　P. Proembryo　PS. Persistent synergid.　S. Synergid　SN. Secondary nucleus　T. Integumentary tapetum　Z. Zygote　av. Autophagic vacuole　c. Cuticle　er. Endoplasmic reticulum　f. Filiform apparatus　w. cell wall

Plate Ⅰ

Fig. 1. An egg cell 3 days before anthesis showing complete wall at its chalazal end. ×8000　**Fig. 2.** Micropylar part of the embryo sac 3 days before anthesis showing the egg, synergids and central cell. ×1800　**Fig. 3.** Chalazal end of an egg cell 2 days before anthesis with thinner chalazal wall. Note the endoplasmic reticulum along the wall on the central cell side. ×13000　**Fig. 4.** Zygote and persistent synergid 3—6h after pollination with new wall formation at their chalazal ends. ×1600

图版 I plate I

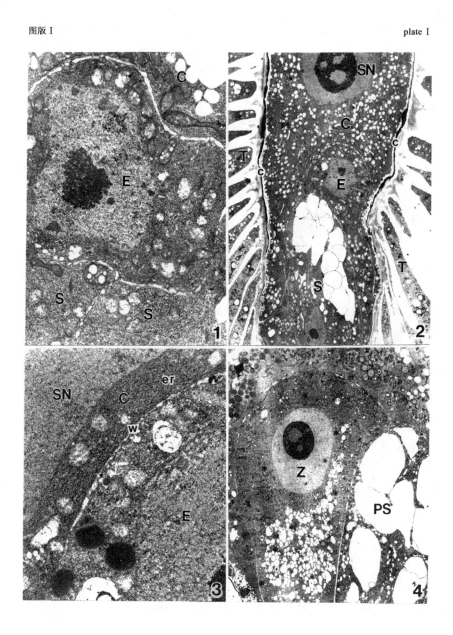

670

第四章 有性生殖的超微结构与细胞化学研究

图版 II plate II

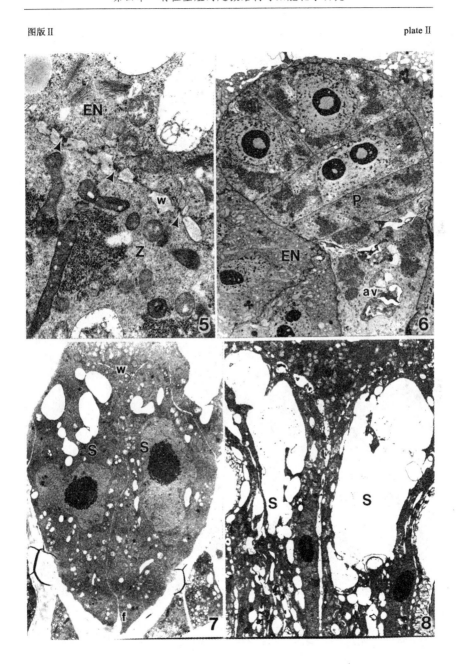

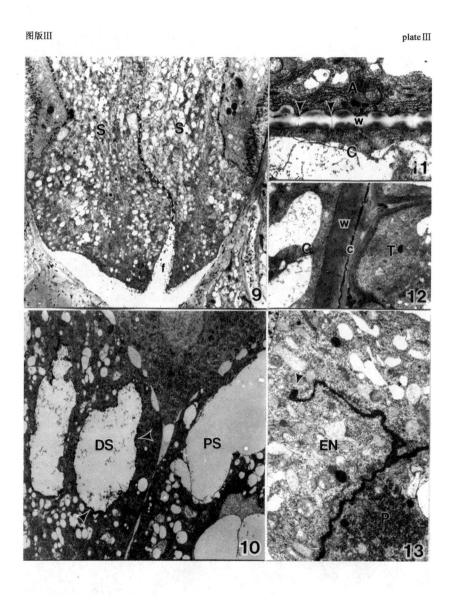

Plate II

Fig. 5. Newly formed incomplete cell wall at the chalazal end of the zygote. Arrowheads show numerous vesicles participating in wall formation from the side of the primary endosperm cell. ×1400 **Fig. 6.** A proembryo and adjacent endosperm cells. ×1600 **Fig. 7.** Synergids 3 days before anthesis with complete walls at their chalazal ends. ×3200 **Fig. 8.** A pair of mature synergids. ×3200

Plate III

Fig. 9. The micropylar part of synergids showing electron dense materials in their walls. ×3200

Fig. 10. After pollination but before entry of pollen tube, one synergid shows dense materials on the tonoplasts and inside the vacuoles (arrowheads); the other one has no such changes. ×3200 **Fig. 11.** The cell wall bordering the central cell and antipodal cells with plasmodesmata. ×6400 **Fig. 12.** Side wall of the central cell (the embryo sac wall) without plasmodesmata. ×6400 **Fig. 13.** Free growing wall in the endosperm (arrowhead). ×5100

(作者:阎华、杨弘远、Jensen. 原载:植物学报,1990,32(3): 165~171。图版3幅,图1幅,参考文献19篇。选录全部图版。)

3 Ultrastructure of the micropyle and its relationship to pollen tube growth and synergid degeneration in sunflower

Transmission electron microscope studies have been made of most female tissues along the pathway of pollen tube growth: the stigma (Bell and Nicks 1976; Jensen and Fisher 1969,1970; Wilms 1980a), the style (Crestiet al. 1978; Dashek et al. 1971; Knox 1984; Linskens 1974a, b; Vasil 1974), the nucellus (Wilms 1980b), and the synergid (Van Went and Willemse 1984). The only exception is the micropyle. Little information is available concerning its ultrastructure, although light microscopic studies have been made in a few species (Brough et al. 1933; Chao 1971; Gelin 1936; Maheshwari 1950). However, to understand the micropyle is important for understanding the interaction that occurs between the pollen tubes and the ovule prior to fertilization. Our paper presents our preliminary studies on the ultrastructure of the micropyle and its relation to pollen tube growth and synergid degeneration in sunflower.

Materials and methods

Ovules of sunflower (*Helianthus annuus* L. cv 'Dang Yang') were collected 1, 2, and 3 days before anthesis, and 6h and 1 day after pollination. The chalazal portion of the ovule was cut off and discarded. Only 1-2 mm of the micropylar portion of the ovule was used. This was fixed in 2% glutaraldehyde in phosphate buffer at pH 7.3 for 6h at room temperature or overnight at 4℃, rinsed in the same buffer, and post-fixed in buffered 1% osmium tetraoxide for 4h at room temperature or overnight at 4℃. Following thorough rinsing, the material was dehydrated in a graded ethanol series with a final passage in propylene oxide. The tissue was then infiltrated and embedded in Epon 812 or Poly/bed 812. Ultrathin sections were cut with a diamond knife on a Reichert ultramicrotome and post stained with uranyl acetate and lead citrate. Ultrastructural observations were made with a Zeiss-EM 10 electron micrsocope at 80 kv. Thick sections 1-2 μm for light microscope observation were cut with a glass knife and stained with toluidine blue.

Results

The sunflower ovule is anatropous, unitegmic, and tenuinucellate (Fig. 1). The embryo sac is enveloped by the integument, which has two branches in its micropylar region. One branch of the integument is on the side proximal to the funicle and the other, on the side distal to the funicle (Figs. 1,2). The two branches close together to form a micropyle that functions as a passage for the pollen tube as it grows to the embryo sac (Figs. 1-7).

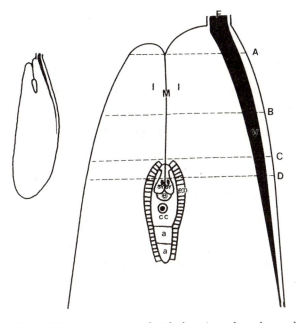

Plate explanation. All figures are oriented with the micropyle end towards the *top* of the page and the funicle and vascular bundle to the *right*. The following abbreviations are used in the figure captions: *a* antipodal, *cc* central cell, *d* dictyosome, *ds* degenerated synergid, *E* egg, *en* endothelium, *er* endoplasmic reticulum, *F* funicle, *fa* filiform apparatus, *I* integument, *iec* inner epidermal cell, *im* intercellular matrix, *l* lipid body, *M* micropyle, *m* mitochondrium, *N* nucleus, *p* pocket, *ps* persistent synergid, *pt* pollen tube, *rer* rough endoplasmic reticulum, *s* starch grain, *sy* synergid, *tt* transmitting tissue, *V* vascular bundle, *w* wall, *z* zygote

Fig. 1. Diagrammatic longitudinal section of a mature ovule of sunflower. The micropyle is formed by one integument and lies close to the funicle

The micropyle is closed (Figs. 13, 15). The integument is divided into two branches that are appressed to each other so that no micropylar canal is present (Figs. 1, 13). Where the branches meet, two layers of continuous, thin cuticle cover the epidermis of the integument (Figs. 14, 19, 20). In transverse section, the micropyle and the cuticle curve partially around the vascular bundle (Figs. 2-4, 13).

The micropyle is asymmetrical. As a boundary line, the cuticle layers divide the micropyle into two halves. The portion on the side proximal to the funicle and the vascular bundle is usually wider from the cuticle layer to the epidermal cells than the similar part on the opposite side (Figs. 13, 14, 17). The region adjacent to the micropyle on the funicle side resembles a transmitting tissue in ultrastructure. The parenchymatous cells of the transmitting tissue are thin walled and ovoid in shape (Figs. 13, 16). The intercellular spaces are large and filled with an electron-dense matrix (Figs. 13, 15, 16). The cells are usually elongated in the axial direction, forming an oblique angle with the cuticle of the micropyle (Figs. 17, 18). The epidermal cells of the micropyle are oriented toward the embryo sac (Figs. 17, 18). The cells of the transmitting tissue contain many dictyosomes, which appear to be actively forming vesicles (Figs. 12, 16). These cells also contain abundant rough endoplasmic reticulum (Fig. 16) and well-developed mitochondria. Three days before anthesis this region contains numerous amyloplasts with large starch grains (Fig. 8). By 2 days before anthesis the number of starch grains decreases, while pockets containing fibril material can be frequently observed near to or in contact with the plasma membrane (Figs. 9, 10). Some of the pockets appear to be in the processes of exocytosis (Figs. 10, 11). The electron density of the pockets is usually less than that of the cell walls and the intercellular matrix (Figs. 10-12, 14, 16), which are almost similar at this stage (Figs. 8, 10, 11). Before pollination the intercellular space becomes enlarged and the intercellular matrix is now abundant. This build up in the intercellular matrix is correlated with the decrease and disappearance of starch from the plastids. It is also correlated with the disappearance of the pockets of fibrilar material that were present in the cytoplasm of the transmitting tissue (Fig. 13). When the cells mature, the intercellular matrix appears to be more electron dense than that of the cell walls (Figs. 15, 17).

In contrast, on the side distal to the funicle and the vascular bundle, the cells adjacent to the micropyle are polygonal in cross section (Figs. 2, 4). These cells are in close contact with each other, leaving little or no intercellular space and containing less intercellular matrix (Figs. 13, 17). The area occupied by the transmitting tissue is much smaller here than it is on the opposite side. The walls of the epidermal cells are not separated by an intercellular matrix and the cells are directly connected with each other via plasmodesmata (Figs. 13, 17). In some cases, usually at the chalazal end of the micropyle, special "key junctions" (Fig. 15 arrows), i.e., ingrowth of one cell wall into the neighboring cell, provide direct plasmodesmata links between the epidermal cells. The cells are arranged parallel to the micropyle in their vertical axis (Figs. 17, 18). Rough endoplasmic reticulum and dictyosomes are abundant (Figs. 15, 18, 21), but the number of the dictyosomes is less than that on the funicle side. Lipid bodies are present 3 days before anthesis (Fig. 8) and disappear at maturity (Fig. 13). No starch grains are observed on this side.

The micropyle varies in form along its length, as does the extent of the transmitting tissue associated with it. At the micropylar end, where it is distant from the embryo sac, the micropyle is large (Figs. 2, 3, 13). The cuticle associated with the epidermal cells of the two parts of the micropyle is closely appressed and continuous (Figs. 14, 17). The amount of matrix between the cuticle and the cell wall is different on the two sides of the micropyle: the amount of matrix associated with the epidermal cells close to the funicle is much greater than the amount associated with the opposite

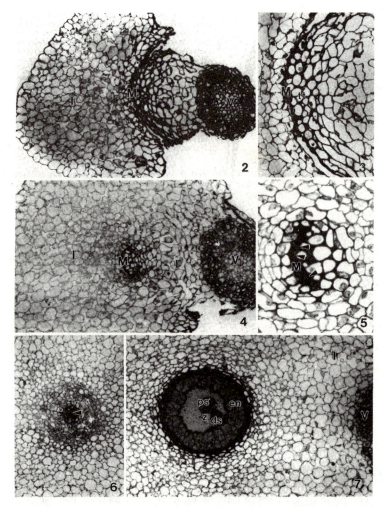

Figs. 2-7. Light micrographs of 1-2 μm cross section of a ovule. **Fig. 2.** Corresponds to *level A* in Fig. 1. Note the size of the micropyle. ×180. **Fig. 3.** Enlarged view of Fig. 2 showing the entrance to the micropyle. Note the transmitting tissue located close to the micropyle on the side proximal to the vascular bundle. ×400. **Fig. 4.** Cross section of the ovule at *level B* of Fig. 1. Note the size and shape of the micropyle. ×200. **Fig. 5.** Enlarged view of Fig. 4 showing pollen tubes (*arrowhead*) in the micropyle. ×500. **Fig. 6.** Cross section of the micropyle (*arrowhead*) at *level C* of Fig. 1. The micropyle is much smaller than in Figs. 2, 4. ×200. **Fig. 7.** Cross section of the ovule at *level D* of Fig. 1. Note that the degenerate synergid is located proximal to the vascular bundle while the persistent synergid is distal. ×200

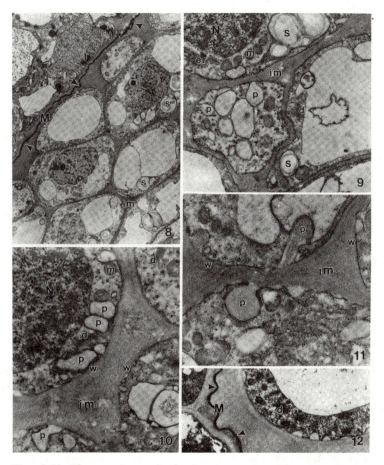

Figs. 8-12. Electron micrographs showing cross section of the micropyle at an early development stage. **Fig. 8.** A micropyle 3 days before anthesis showing abundant larger starch grains present in the cells near the micropyle on the side proximal to the funicle, and lipid bodies on the other side. *Arrowhead* points to the cuticle layer. ×4000. **Fig. 9.** The transmitting tissue near the micropyle on the side proximal to the funicle showing starch grains and pockets with fibrils. Note the ribosomes in the nucleus. ×13000. **Fig. 10.** The transmitting tissue near the micropyle on the side proximal to the funicle showing pockets with fibrils near or in contact with the plasma membrane. ×20000. **Fig. 11.** The transmitting tissue near the micropyle on the funicle side at 2 days before anthesis showing pockets with microfibril in the process of exocytosis. Note the density, orientation of microfibril in pockets, cell wall, and intercellular matrix. ×20000. **Fig. 12.** A micropyle showing abundant dictyosomes in the cells on the side proximal to the funicle. *Arrowhead* points to the cuticle. ×10000

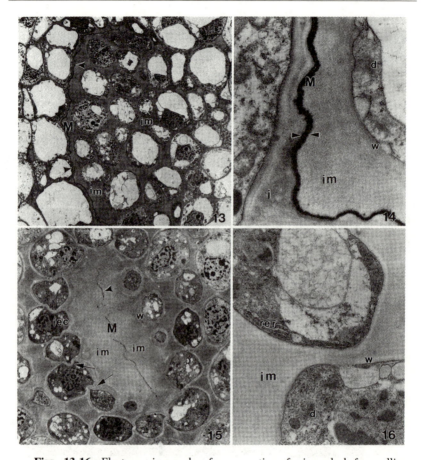

Figs. 13-16. Electron micrographs of cross section of micropyle before pollination. **Fig. 13.** Micropyle at *level B* of Fig. 1. Note the differences on the opposite sides of the micropyle and the distribution of the intercellular matrix. *Arrowhead* points to cuticle. ×1600. **Fig. 14.** Micropyle at *level B* of Fig. 1. Note that the epidermal cuticle layers (*arrowhead*) divide the micropyle into two halves that differ in width. ×25000. **Fig. 15.** Micropyle at *level C* of Fig. 1. Differences in the thickness of the micropyle on both sides are much less than that in Fig. 13. Cuticle (*arrowhead*) is discontinuous. Note "key junction" (*arrow*) in inner epidermal cells on the side distal to the funicle. ×2500. **Fig. 16.** Inner epidermal cells on the side proximal to the funicle showing active dictyosomes and rough endoplasmic reticulum. ×16000

第四章 有性生殖的超微结构与细胞化学研究

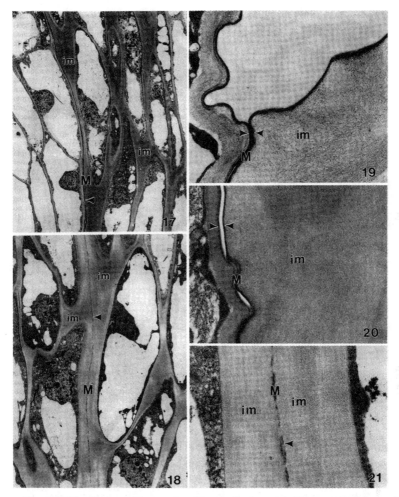

Figs. 17-21. Electron micrographs of longitudinal section of micropyle before pollination. **Fig. 17.** Micropyle at *level B* of Fig. 1. Note the distribution of transmitting tissue. *Arrowhead* points to cuticle. ×1600. **Fig. 18.** Micropyle at *level C* of Fig. 1. Compare the micropyle and cuticle (*arrowhead*) with those at *level B* of Fig. 17. × 2000. **Fig. 19.** The entrance to the micropyle at *level A* of Fig. 1. Showing inner epidermis on both sides in contact and cuticle (*arrowhead*) covering the surface of inner epidermis. ×25000. **Fig. 20.** Micropyle at *level B* of Fig. 1. Note the two continuous cuticle layers (*arrowhead*) and the different width of the micropyle on both sides. × 1600. **Fig. 21.** Micropyle at *level C* of Fig. 1. Note the discontinuous cuticle (*arrowhead*) and the almost equal width of the micropyle on each side. ×16000

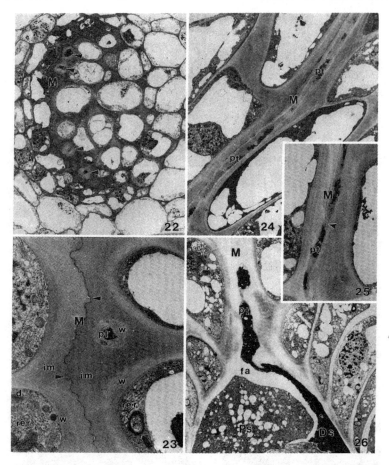

Figs. 22-26. Electron micrographs of micropyle after pollination. **Fig. 22.** Cross section of micropyle at *level B* of Fig. 1 showing four pollen tubes (*arrowhead*) passing through micropyle. Note the position of the pollen tubes. ×1600. **Fig. 23.** A pollen tube passes through the micropyle on the side proximal to the funicle. Note dictyosomes and rough endoplasmic reticulum in inner epidermal cells. *Arrowhead* points to cuticle. ×10000. **Fig. 24.** Longitudinal section of micropyle showing a pollen tube growing in the micropyle on the funicle side. ×2500. **Fig. 25.** Longitudinal section of the micropyle close to the embryo sac. Note a pollen tube penetrating the discontinuous cuticle (*arrowhead*) and growing into the micropyle on the side distal to the funicle. ×3500. **Fig. 26.** Micropyle and the filiform apparatus of the synergids. A pollen tube has grown through the micropyle and penetrated the filiform apparatus of the degenerate synergid that is located on the side proximal to the funicle. ×2500

epidermal cells (Figs. 14, 19). At this same level, the transmitting tissue is extensive on the side toward the funicle (Figs. 13, 17). As the micropyle approaches the embryo sac, it becomes less extensive in area (Figs. 4, 6, 15) and the transmitting tissue decreases in amount, but it is now present on both sides of the micropyle (Figs. 15, 18). At this level the epidermal layers partially fuse and the cuticle becomes discontinuous. The matrix is now almost equal in amount on both sides of the micropyle (Figs. 15, 18, 21). Near the embryo sac the cuticle is connected with the common synergid wall at the filiform apparatus.

After pollination, one to four pollen tubes are seen in a micropyle (Figs. 5, 22), but only a single tube enters the embryo sac, the others remaining behind in the micropyle. The pollen tubes grow through the intercellular matrix (Fig. 23). During their passage through the micropyle, most pollen tubes are restricted to the side proximal to the funicle and near to the cuticle (Figs. 22, 24). Occasionally a few pollen tubes penetrate the discontinuous cuticle close to the embryo sac and grow into the side distal to the funicle (Fig. 25).

One of the synergids begins to degenerate before the entry of the pollen tube. The pollen tube then penetrates into the filiform apparatus of this degenerating synergid (Fig. 26). There is a greater tendency (81%) for the degenerating synergid to be located toward the funicle and the vascular bundle, i. e., on the same side of pollen tube pathway (Yan et al. 1990 a b). In most cases the synergids on the side distal to the funicle are persistent (Figs. 7, 26).

Discussion

The present study reveals the following ultrastructural features of the micropyle in sunflower.

1. The micropyle is closed and not an open canal as usually described (Maheshwari 1950; Van Went and Willemse 1984). Open micropyles have been reported in some plants, such as *Greoillea robusea* (Brough 1933), *Paspalum orbiculare* (Chao 1971), and *Costus cuspidatus* (Grootjen and Bouman 1981). In *Ficus* (Condit 1932), *Fouquieria* (Khan 1943), and *Cynomorium* (Steindl 1945) the integumentary cells are closely appressed to each other so that the micropylar canal is extremely narrow and imperceptible. Perhaps micropyles, as styles, also possess two basic types: open and closed. The closed micropyle probably relates to the

unitegmic and tenuinucellate structure of the ovule. In such cases the nucellus degenerates during the development of the megagametophyte and the mature embryo sac is surrounded by the integument alone. If the micropyle was open in these cases, the embryo sac would be open to the cavity of the ovary. Another possible explanation is that the type of micropyle corresponds to the type of the style. The style of sunflower is closed (Vithanage and Knox 1977), which is characteristic of the styles of many dicots. Whether or not closed micropyles are present in other species is not known. Further investigations will be needed to discover if there is a regular pattern of distribution of closed micropyles among families of angiosperms.

2. The micropyle possesses transmitting tissue. The cells of transmitting tissue are usually elongated and fusiform in the axial direction. They are characteristically rich in endoplasmic reticulum (ER) and dictyosomes. The rough ER is related to the formation of proteins in the intercellular matrix in the style of *Lycopersicon* (Cresti et al. 1976). The transmitting cells of sunflower have a thin cell wall with a large amount of intercellular matrix. This material appears to be secreted by the transmitting cells and present in the intercellular spaces. The intercellular matrix is a wall-like structure with a higher electron density than the primary wall and a different orientation of fibrils. The intercellular matrix is assumed to be composed of polysaccharides. In stylar transmitting tissue the intercellular matrix is composed of pectin-containing substances as in *Diplotaxis* and *Petunia* (Kroh 1973; Kroh and Van Brakel 1973; Kroh and Helsper 1974). In *Lycopersicon* (Cresti et al. 1976), tobacco (Bell and Hicks 1976), *Malus* (Ciampolini et al. 1978), and *Prunus* (Cresti et al. 1978), the styles at an early developmental stage have a matrix that is composed of pectic polysaccharides. Later in development, as the style matures, proteins are added. Since sunflower pollen has only limited food reserves, and a relatively large amount of wall is synthesized during growth, it is reasonable to assume that at least part of the substrate needed for the synthesis of this wall is derived from the polysaccharides in the intercellular matrix of the long micropyle. Ultrastructural and cytochemical studies suggest that the intercellular matrix in the stigma and style serves as nutritional material for growing pollen tubes (Engels 1974; Kroh and Van Bakel 1973; Kroh et al. 1979; Loewus and Labarca 1973). In some open micropyles there are specially modified cells lining the micropylar canal, as in *Berkheya* (Gelin 1936), *Grevillea* (Brough 1933), and *Cynomorium* (Steindl 1945).

These cells become mucilaginous or glandular and seem to contribute to the nutrition of the pollen tube. In *Paspalum orbiculare* (Chao 1971) PAS-positive substances have been observed in the micropyle. In sunflower the pollen tubes grow through an intercellular matrix in the region of the micropyle, but in no other zone of the transmitting tissue. Thus, the transmitting tissue provides not only nutrition for the pollen tube, but possibly also a means for the passage of hormonal and other signals between the pollen tubes and the synergids. The nature and composition of the intercellular matrix in the micropyle remains to be clarified.

3. The micropyle is asymmetrical. On the funicle side the distance is greater from the cuticle layer to the inner epidermal cells than it is on the opposite side. The transmitting tissue is limited to the funicle side of the micropyle, except at the level close to the embryo sac. Here the transmitting tissue decreases in size and lies on both sides of the micropyle. Usually there are larger intercellular spaces and the amount of intercellular matrix is greater on the funicle side than on the opposite side. In addition, at an early stage of development large and abundant starch grains are present on the side proximal to the funicle, while lipid bodies are concentrated on the side distal to the funicle. In brief, the part of the micropyle on the side proximal to the funicle is well developed and provides optimal conditions for the growth of the pollen tubes. It is probable that the asymmetrical character of the micropyle is caused by the anatropous structure of the ovule. Further research should be undertaken to see if the asymmetry of the micropyle can be found in other plant taxa with anatropous ovules.

4. The morphology and anatomy of the micropyle vary along its length. The micropyle, being larger at the portion distal to the embryo sac and smaller closer to the embryo sac, may function as a guide to the pollen tubes, directing them toward the embryo sac. The micropyle is better developed on the side proximal to the funicle than on the opposite side, especially at the portion distal to the embryo sac. The differences between the sides decrease in magnitude from the micropylar end to the chalazal end. The cuticle layers are separated or lie side by side at the portion distal to the embryo sac, but fuse and become discontinuous at the portion close to the embryo sac.

In most species, only one pollen tube enters an ovule (Maheshwari 1950). In *Pisum sativum* two or more pollen tubes were frequently seen at the entrance to the micropyle, but only one actually entered in (Cooper

1938). A similar situation was reported in *Phaseolus vulgaris* (Weinstein 1926) and *Hordeum* (Pope 1946). In sunflower, however, two or more pollen tubes in one micropyle is a common occurrence, although only one enters the embryo sac. How the embryo sac is able to admit the one pollen tube while excluding the others is still a question.

The present data indicates a close relationship among the structure of the micropyle, pollen tube growth, and synergid degeneration. We suggest the following hypothesis to explain this relationship. (1) Asymmetry of the micropyle causes pollen tubes to grow on the more developed side, which is proximal to the funicle. Numerous authors assume that the growth of the pollen tube toward and into the micropyle is directed chemotropically by substances produced by the ovule and secreted through the micropyle. It seems logical to suggest that the growth of the pollen tubes on the funicle side of the micropyle is the result of the more abundant intercellular matrix present there. An additional factor may be the possibly greater concentration of chemotropic substances on this side. (2) The presence of the pollen tubes in the micropyle may signal the embryo sac through the release of hormones. These would diffuse ahead of the pollen tube to the embryo sac and lead to the degeneration of one of the synergids. As the pollen tube passes through the micropyle on the side proximal to the funicle, the hormone would diffuse mainly on the side that is isolated by the cuticle. This would lead to the degeneration of the synergid on the same side. Stanley and Linskens (1974) reported that pollen tubes increase the amount of gibberellic acid (GA) present in the pistils in which they are growing. In cotton the hormone triggering the degeneration of the synergid is believed to be GA (Jensen et al. 1983), although this remains to be shown in sunflower. (3) The degenerating synergid, principally the one proximal to the funicle, may release chemotropic substances that attract the pollen tube into it. Jensen et al. (1983) suggested that the vacuoles of the synergids in cotton contain large amounts of a soluble calcium salt that is released when the synergid degenerates. The pollen tube follows the calcium gradient into the degenerating synergid. Recently Chaubal and Reger (1990) reported that in wheat the synergids contain high concentrations of calcium. In some species at least pollen tubes are positively chemotropic to calcium gradients (Mascarenhas and Machlis 1962a, b, c). Studies are needed to determine if calcium is a chemotropic substance for pollen tubes in sunflower. (4) As the micropyle reaches the embryo sac, the differences between the two sides decrease. In addition, the cuticle layers in the micropyle are discontinuous close to the embryo sac. Therefore, any hormone released by the pollen tubes occasionally may diffuse from the funicle side to the opposite side and induce the synergid distal to the funicle to degenerate. The pollen tube would then penetrates the discontinuous cuticle, grow on the other side, and finally enter the synergid on the side distal to the funicle. This may explain why there are about 19% degenerate synergids located distal to the funicle.

(Authors: Yan H, Yang HY, Jensen WA. Published. in: Sexual Plant Reproduction, 1991, 4: 166-175, with 26 figures and 40 references. All the figures are retained here.)

4 陆地棉珠孔的结构及花粉管在其中的生长途径

Structure of micropyle in *Gossypium hirsutum* and the pathway of pollen tube growth

Abstract

The micropyle of *Gossypium hirsutum* consists of an exostome and an endostome. On semithin serial transections, the exostome was visualized as a branched narrow gap except its outer and inner openings, whereas the endostome only a narrow linear gap. Ultrastructurally, the micropyle gap was formed by the integumental epidermal cells coated with a cuticle. The cells lining the micropyle gap were characterized by a large nucleus and abundant organelles as mitochondria, plastids, rough endoplasmic reticulum, vesicle-secreting dictyosomes and small vacuoles. One or two pollen tubes were seen growing through the exostome and endostome gaps. Thus, the micropyle in cotton was basically a closed type as has been found in sunflower, but similar asymmetrical structural features were not observed.

在大多数被子植物中,花粉管经珠孔进入胚珠进而实现受精,但珠孔结构及花粉管在其中的生长途径是植物受精生物学研究中较薄弱的一个环节。迄今仅在向日葵中从超微结构水平上作了详细的研究[1],因而需要扩大研究的对象。陆地棉花粉管生长所经雌蕊组织中,关于柱头、花柱[2,3]、珠心[4,5]和助细胞[6]的超微结构已有较详细的研究,但涉及珠孔这一环节只有简单的描述[7]。本文报道我们在这方面的研究结果。

材料与方法

取陆地棉(*Gossypium hirsutum* L.)授粉前和授粉后22h的胚珠,切除其合点端部分,用2.5%戊二醛与2%多聚甲醛混合液、1%锇酸(用pH7.0的磷酸盐缓冲液配制)先后固定。经乙醇系列脱水,环氧丙烷过渡,Epon 812包埋,在Sorvall MT-6000型超薄切片机上切片。半薄切片经甲苯胺蓝-O染色后用Olympus BH-2型显微镜观察、摄影,超薄切片经醋酸双氧铀与柠檬酸铅先后染色,用JEM-100cx/Ⅱ型透射电子显微镜观察和摄影。

结　果

陆地棉胚珠为倒生、双珠被、厚珠心,其珠孔包括由外珠被形成的外珠孔(exostome)和由内珠被形成的内珠孔(endostome)。两层珠被的外周组织中均有不少细胞富含色素物质[1],而珠孔周围的珠被组织则不含。外珠孔结构较复杂,根据纵切面(图版Ⅰ,1)和不同部位的横切面(图版Ⅰ,2~4)的观察,除其外端和内端为开口外,中段是闭合的。闭合部分的珠被表皮细胞相互会合处的横切面上的缝隙呈分枝状(暂称珠孔隙)。超微结构观察可以看出(图版Ⅰ,6、7),会合后形成外珠孔的珠被表皮细胞,细胞壁一般较薄,但面向珠孔隙的壁比其余各面的壁厚些,而且此面的壁有一角质层覆盖。细胞核较大而异染色质不多。液泡多为分散的小液泡。细胞器丰富,含大量的线粒体和粗面内质网,质体中含有淀粉粒,高尔基体周围有众多小泡。珠孔隙中可观察到电子浓密的物质。在纵切面上,内珠孔的位置和外珠孔不在一条直线上(图版Ⅰ,1)。内珠孔的结构从外至内都是闭合的,形成内珠孔的珠被表皮细胞相互紧贴。在横切的半薄切片上,内珠孔隙略呈弧线形(图版Ⅰ,5)。内珠孔两侧的珠被表皮细胞超微结构特征与外珠孔者相似(图版Ⅰ,9)。其珠孔隙中较宽处可见絮状的电子浓密物质(图版Ⅰ,10)。在所观察的胚珠中,通常只有1~2个花粉管进入珠孔。由于外珠孔与内珠孔不在一条直线上,花粉管从外珠孔到达内珠孔,要在内、外珠被之间经历一段短距离的曲折生长[1]。在外珠孔的内、外两端的开口中,花粉管贴近珠被表皮细胞生长;在外珠孔中段的闭合部分,花粉管在珠孔隙中通过(图版Ⅰ,8)。花粉管通过内珠孔时,也是在珠孔隙中生长(图版Ⅰ,11)。观察到花粉管周围有退化的珠被细胞。在外珠孔的开口中,花粉管的横切面多呈椭圆形,而在外珠孔的闭合部分和内珠孔中,花粉管的横切面多呈三角形或不规则形状。花粉管壁为多层结构:外面的果胶层薄而不明显;中间的纤维素层很厚;最内的胼胝质层电子密度较低。管内含众多小液泡和电子浓密物质,细胞器界限不清,可能所观察的系花粉管的后段。

讨 论

在珠孔受精过程中,珠孔是花粉管进入胚珠与胚囊的必经通道,应有重要的生物学意义。然而过去对其研究比较忽视,只把它作为一个简单的孔道看待。实际上珠孔结构相当复杂。在向日葵中发现[1]:其珠孔并非开放的孔道,而是闭合的结构;并且其两侧是不对称的,近珠柄侧有引导组织,花粉管多在此侧生长;这一特点可能和向日葵的退化助细胞大多位于近珠柄侧有关。从方法学的角度看,上述研究结果的获得,主要是由于对珠孔进行了系统的横切面观察,加上纵切面的观察,才能对珠孔的整体结构与细微特征有所认识。本文从对陆地棉珠孔的观察结果得出如下几点看法:(1)外珠孔是开放与闭合并存结构,内珠孔是闭合的结构。但外珠孔的内、外开口不应被看做外珠孔本身的组织结构,因此从总体上看陆地棉的珠孔基本上属于闭合型,而非以往认为的开放的孔道[1]。在被子植物中,珠孔是否和花柱一样也包括开放与闭合两大类型,是需要继续研究的问题。(2)在横切面上,外珠孔隙呈分枝状,而内珠孔隙略呈弧线形。这表明外珠被生长过程中速度不一致,前沿形成几个凸出部分,因而在会合成外珠孔时产生几个缝隙;而内珠孔则是由两个前沿凸出部分会合成的单线的珠孔隙。(3)珠孔两侧的细胞在超微结构上显示较活跃的生理特征,且在珠孔隙中存在絮状的电子浓密物质。这些特点显示了为花粉管生长提供必要条件的功能。但在陆地棉中尚看不出如向日葵所具有的引导组织。(4)以珠柄维管束为基准,看不出其珠孔结构具有和向日葵相似的不对称性。

(作者:张劲松、何才平、杨弘远。原载:植物学报,1994,36(9): 727~729。图版1幅,参考文献7篇,均删去。)

5 甘蓝型油菜珠孔与胚囊的超微结构研究

Ultrastructural studies on the micropyle and embryo sac in *Brassica napus* L.

摘 要

甘蓝型油菜的外珠孔基本上为开放结构,内珠孔为闭合结构。胚囊成熟时反足细胞已退化,仅由二个助细胞、一个卵细胞、与一个中央细胞组成。助细胞极性不明显,二个助细胞在花粉管到达前均显示同等退化迹象。卵细胞极性明显。中央细胞中含结构独特的质体。受精前,卵与中央细胞的质膜之间含有特殊的电子致密物质。

Abstract

The micropyle of *Brassica* napus includes an exostome and an endostome. The exostome is basically open, but the endostome is a closed structure with only narrow gaps between two cuticle layers of the inner integumental epidermis. The embryo sac is composed of two synergids, an egg cell and a central cell; the antipodal cells are degenerated before the embryo sac maturation. The synergid has a micropylar filiform apparatus, but not a chalazal vacuole, and its nucleus is centrally located. Therefore it has not a typical polarity as synergids of most other plant species. Before pollination, the synergids of some ovules already show symptom of degeneration. After pollination but before entry of pollen tube, two synergids of the same embryo sac degenerate simultaneously. The egg cell is a typical polarized cell, with a big micropylar vacuole and a chalazally located nucleus. Before fertilization, the arch-shaped gap between two plasma membranes of the egg cell and central cell is filled with electron-densed granular and amorphous substances. The central cell possessed several big chalazal vacuoles; the polar nuclei and most cytoplasm are micropylarly located. The plastids in the central cell is characterized by big volume and quite prominent lamellae. The micropylar portion of the embryo sac wall bear well-developed ingrowths, which gradually smooth down towards the chalazal end.

(作者:孙芹、张劲松、杨弘远。原载:植物研究,1996,16(1): 92~95。图版2幅,参考文献11篇,均删去。)

第二节 雌蕊中钙与钙调素的细胞化学定位

提要

钙在花粉萌发与花粉管生长中的作用是数十年来长盛不衰的研究热点。我们参与这方面的研究见第一章第二节。然而,关于钙在雌蕊组织内的分布及其与花粉管体内生长的关系则研究甚少,迄今国际上比较集中研究的是钙与助细胞退化的关系。我们从20世纪90年代初即开始从事这方面的研究,涉及材料包括向日葵、棉花、油菜、水稻,研究对象覆盖从柱头到助细胞整个花粉管生长途径中钙的分布。研究工作取得了以下几方面的主要结果:第一,柱头乳突、花柱引导组织、子房壁内表皮、珠孔、助细胞等整个花粉管轨道中,钙的分布较相邻组织中更为密集,且多分布于质外体系统(细胞外基质、细胞壁)中,表明这种分布与花粉管生长有密切的关系。其中,珠孔作为花粉管进入胚珠的入口,其钙的密集尤为显著,这是前人未曾证明的。第二,关于助细胞中存在钙的超常含量,前人已在珍珠谷与小麦两种禾本科植物中作过研究。我们则对其它几种植物作了研究,包括几种双子叶植物及水稻,从中揭示出,钙在不同植物助细胞中的分布与变化具有不同的特点,并非均与珍珠谷及小麦一致。在油菜中,还首次应用图像分析技术对胚囊各成员细胞中特别是助细胞中的钙作了定量测定。以上这些研究资料论证与补充了我们在1994年发表的"受精过程中助细胞退化机理的研究进展"一文中提出的假设,并充实了1999年发表的"钙在有花植物受精过程中的作用"一文的内容(以上2篇专论见第五章)。

钙调素是重要的钙靶蛋白。我们关于钙调素的研究,在第一章第二节、第三章第三节中均有涉及。本节所收集的则是用RNA原位杂交技术及免疫细胞化学技术对钙调素mRNA与蛋白进行定位的研究。研究材料为水稻与烟草,研究范围涉及花药与花粉发育、雌蕊与胚囊、胚与胚乳发育等。在水稻中,平行地研究了钙调素mRNA与蛋白两个表达水平的特点。在烟草中,对分离的胚囊进行了整体原位杂交与整体免疫细胞化学反应,以研究钙调素mRNA与蛋白在受精前后胚囊中分布的变化。这些研究只是初步的,今后尚需继续深入。

1 Ultracytochemical localization of calcium in the embryo sac of sunflower

Calcium is a chemotropic substance inducing the directional growth of pollen tubes in *Antirrhinum majus*, *Narcissus pseudonarcissus* and *Clivia miniata* (Mascarenhas & Machlis, 1962). Pollen germination and normal pollen tube growth *in vitro* have been reported to depend on the optimal concentrations of Ca in the medium (Brewbaker & Kwack, 1963). It can be expected that *in vivo* as *in vitro*, Ca is also essential for pollen germination and tube growth. Using chlorotetracycline and X-ray microanalysis, Bednarska (1989) demonstrated that the receptive surface of the stigma of *Ruscus aculeatus* is rich in calcium. These results implicated the role of calcium in the regulation of pollen germination *in vivo*. The tissues of the gynoecium also contain considerable amounts of calcium (Mascarenhas, 1966; Glenk et al., 1971; Day et al., 1971). Jensen (1965) reported that the synergids of cotton are rich in ash by microincineration and that the main constituent in this ash is presumably calcium (Jensen et al., 1983). Recently, it was confirmed that synergids contain relatively high calcium in unpollinated wheat ovaries using energy-dispersive X-ray microanalysis (Chaubal & Reger, 1990). Ultrastructural studies have shown that in many species, one of the synergids degenerates before the pollen tube reaches the embryo sac and that the growing pollen tube will enter the degenerating synergid (Jensen et al., 1983; Yan et al., 1990). Nevertheless, the relations between calcium, synergid degeneration and the entry and discharge of the pollen tube remain unclear.

The precipitation of cations by adding potassium pyroantimonate to the fixative was first introduced by Komnick and modified by Klein et al. (1972). This method can be used to localize calcium ions selectively in ultracytochemistry (Wick & Hepler, 1982) and has been applied to plant material (Moore, 1986; Vaughan et al., 1987). The present paper reports the results of studies on embryo sacs of sunflower from before pollination to after fertilization using this method.

Materials and methods

The ovules of sunflower (*Helianthus annuus* L.) collected from the field at Wuhan University were studied at four developmental stages: 24 h prior to pollination, 2.5, 3.5 and 24 h following pollination.

The cytochemical localization procedures for calcium follow Moore's (1986) as modified by Vaughan et al. (1987). A 5% solution of potassium pyroantimonate (The Fourth Chemical Factory in Shanghai) was obtained in redistilled water, gently boiled for at least 1 h and allowed to cool. This solution was used to prepare fixatives and wash tissues. The ovules were submerged in a fixative (prepared immediately prior to use) of 2% glutaraldehyde/2.4% formaldehyde/2% potassium pyroantimonate buffered in 50 mmol/L potassium phosphate (pH 7.8). Each of the ovules was cut in the fixative and the micropylar portion containing the embryo sac was retained. After 4 h in this fixative, the tissues were washed for 2 h in 2% potassium pyroantimonate buffered in 50 mmol/L potassium phosphate (pH 7.8), postfixed in 1% osmium tetroxide/2% pyroantimonate (pH 7.8) for 2 h, dehydrated in a graded ethanol series and embedded in Poly/Bed 812 (Polysciences) via 1, 2-epoxypropane. Ultrathin sections were obtained with a Sorvall MT-6000 ultramicrotome equipped with a glass knife. After post-stained with uranyl acetate, the sections were viewed and photographed with a JEM-100 cx/Ⅱ transmission electron microscope.

For control experiments, grids with sections showing pyroantimonate precipitates were floated on a solution of 1 mmol/L EGTA (pH 8.0) at 60 ℃ for 0.5 h. This treatment is used to remove the calcium pyroantimonate precipitation from the sections.

Results

Cytochemical procedures used in this study were consistently specific for calcium and gave quite precise ultrastructural localization. The calcium precipitates are mainly granular, characte-rized by alternating layers of electron dense and electron lucent material (Pl. I :1). Sometimes, two or more of these deposits may merge (Pl. I :2). Their size and structure are similar to the "spherite" in the digestive and secretory cells of the midgut gland of agelenid spider *Coelotes terrestris* (Ludwig & Alberti, 1988). Because of this we have adopted the same term for the precipitation observed. Such spherites appear in the nucleus, cytoplasm, plasma membrane cell wall and the intercellular spaces. Some larger spherites are observed occasionally (Pl. I :3) and some appear to have collapsed (Pl. I :4). Another type of deposit fibrilliform deposit (Pls. I :5; II :9, 14) is mainly present in the vacuoles. Most nucleoli and oil drops are uniformly electron-dense. X-ray diffraction analysis has indicated that all of these precipitates are not crystals. In the control, all precipitates were removed almost completely at each sampling time. Sometimes granular lucent pores are revealed (Pls. II :13; III :16; V :24), confirming the presence of calcium in these pyroantimonate precipitates.

Before pollination

Large amounts of calcium-antimonate precipitates were observed in the filiform apparatus of synergids and contiguous micropylar channel (Pl. I : 6). Precipitates were located throughout the synergids, including their nuclei, cytoplasm, plasma membranes and walls, but much less in their vacuoles (Pl. I :7,8). The amount of precipitates increases gradually from the micropylar end to the chalazal end of the cell. No difference can be seen between the two synergids at this stage. In the egg cell and the central cell, the precipitate is also in polar distribution. However, the polarities are opposite in the two cells, namely small amounts of precipitates (mainly fibrilliform) are found in the micropylar portion of the egg and the chalazal portion of the central cell while large amounts of precipitates (mainly granular) in the chalazal portion of the egg cell and the micropylar portion of the central cell (Pl. II :9-11,14). The nuclei of the two cells contain a number of randomly dispersed deposits, but their nucleoli are uniformly electron-dense (Pl. II:10,11). The zigzag plasma membrane of micropylar end of the central cell approaches the smooth plasma membrane of chalazal end of the egg. Large amounts of precipitates are also seen in the gap between the plasma membranes where a common wall borders the egg. (Pl. II :10-13). Heavy deposits are also found in the wall ingrowths of the central cell and the contiguous apoplast outside the embryo sac in the micropylar region (Pl. III :15, 16).

After pollination and before fertilization

Fertilization does not take place until 3 h after pollination, but differences in calcium content of the two synergids can be detected (Pl. III :17, 18). One synergid remains as it was before pollination with regard to the amount and distribution of precipitates. In contrast, the other one shows a marked increase in the concentration of precipitates. This increase appears directly proportional to the previous distribution of precipitates. In other words, the distribution of the precipitates remains the same, but their amount increases dramatically. Thus, calcium deposits become massed in the chalazal portion of the cell. In both synergids, the amount of precipitates is lower in the mitochondria, plastids, Golgi bodies and vacuoles, but higher in the nucleus, particularly in the endoplasmic reticulum and tonoplasts (Pl. Ⅳ:19,20).

In sunflower, one synergid is always located closer to the funicle than the other. It is this synergid that routinely has higher amount of calcium-antimonate precipitates. Of nine ovules examined, eight showed the synergid closer to the funicle to be the one with the higher deposit content. Although the sample is small, we feel that the results are highly significant.

The egg and the central cell show a pattern of precipitates similar to that seen before pollination, with the exception of the nuclei where the amount of precipitates appears greater than before. The common wall bordeing the egg, the central cell, and the synergids disappears; but the gap between the cells, which is extensive in some areas, is still rich in precipitates (Pl. Ⅳ:21). In the embryo sac wall, the precipitates appear smaller and less than before. This is particularly true where wall ingrowths of the central cell are present (Pl. Ⅳ:22).

After fertilization

After fertilization, the greater electron density seen in the degenerated synergid may be related to its increased calcium precipitates (Pl. Ⅴ:23, 24). Many smaller granules of precipitates can be seen within the darkened mass of the degenerated synergid. This is particularly noticeable in the wall. The persistent synergid remains unchanged with regard to the content and distribution of precipitates (Pl. Ⅴ:25, 26).

One day after pollination, the precipitates are high in the nuclei and plasma membranes of the suspensor cells, but low in the cells of the embryo proper. The exceptions are some filiform precipitates that appear in the vacuoles and some spherite precipitates near the plasma membrane. Little precipitates are seen in the nuclei of the embryo cells and the endosperm, including the wall ingrowths of the embryo sac. The antipodal cells show little precipitates at any stage of development.

Discussion

The data presented here extend those obtained by Jensen (1965), Jensen et al. (1983), and more recently, by Chaubal and Reger (1990). This is the first time that calcium has been localized in the embryo sac using the pyroantimonate precipitation technique. It allows us to demonstrate a calcium gradient within the synergids—lower calcium content in the micropylar end to higher in the chalazal end. This gradient is the same in both synergids before pollination.

After pollination, but before fertilization, the calcium content of one synergid increases dramatically. Interestingly, the pattern of the distribution of the calcium remains the same but the amount increases. Yan et al. (1990) showed in sunflower that one of the two synergids began to degenerate at this stage and that this synergid was the one closer to the funicle. This was the case in 52 of the 64 examined ovules at this stage. In the present data, 8 out of 9 ovules show that the synergids with higher calcium content are always the ones closer to the funicle. Therefore, we conclude that the synergid with higher concentration of calcium is the degenerating synergid and that it is this synergid that will receive pollen tube.

Mascarenhas and Machlis (1962) verified earlier that calcium has the ability to direct pollen tube growth *in vitro* and suggested that there is a calcium gradient in the tissues of gynoecium from stigma to embryo sac for ensuring the orientation of pollen tube to embryo sac. This gradient was indeed found in *Gladiolus* (Day et al., 1971), but no similar results were obtained in several other species (Mascarenhas, 1966; Glenk et al., 1971). Therefore, in many species, this gradient, if present, is only local and temporal (Mascarenhas, 1975; Heslop-Harrison, 1987). The results reported here indicate that after pollination the calcium content in the degenerating synergid is greater than that in the persistent one. This may be why the pollen tube always grows into the degenerating synergid other than the persistent one or other places. Moreover, the density of calcium increases gradually from the micropylar to the chalazal end of the synergid. This gradient may be related to the limiting growth of the pollen tube in the degenerating synergid. As the gradients are the same in both synergids before pollination, it seems that the increase in calcium content in the synergid, which the pollen tube will enter, occurs only after pollination. Yan et al. have recently shown that in sunflower most pollen tubes growing in the micropyle are located at the side proximal to the funicle, and that this may be the cause of preferential degeneration of the synergid located at the same side (Yan et al., 1991).

In this study, the calcium deposition seems to be associated with the ER. This is in contrast to Jensen's (1965) data on cotton which suggest that the calcium is located in the vacuoles. Indeed, the ER membrane system has been identified as having an affinity for calcium ions tenfold higher than other organelles, such as mitochondria (Dieter & Marme, 1983). A family of proteins with the capacity of binding significant amounts of calcium in the millimolar range has been found inside the ER (Macer & Koch, 1988). Finally, in plant cells, the ER with its associated vesicles has been shown to have the capacity of storing and transporting calcium analogous to the sarcoplasmic reticulum (SR) in animal muscular cells, and the ER is believed to be the main regulator of cytoplasmic calcium concentration (Lee et al., 1986; Macer & Koch, 1988). Thus, the present data provide another case of the ER storing calcium for the physiological functions of synergids.

In cotton, Jensen et al. (1983) have shown that GA (gibberellic acid) can substitute for the stimulus from the pollen tube to trigger the degeneration of one of the two synergids in unpollinated ovules. They felt that the pollen tube in the style releases GA which diffuses ahead of the tip of the pollen tube causes the synergid to degenerate. GA has also been shown to stimulate calcium ion movement from apoplast to symplast (Moll &

Jones, 1981) and accumulation in the ER (Bush & Sze, 1989). However, when the concentration of calcium ions is high enough (approx. 10^5 mol/L), the ER will disassociate (Koch & Booth, 1988). Based on these observations and our own data, we suggest the following series of events. The heavy accumulation of calcium in the synergid is stimulated by GA from the pollen tube and results in the disassociation of the ER in the synergid destined to degenerate. The amount of calcium and hydrophilic enzymes released from the broken ER may be great enough to stop cytoplasmic streaming (Kohno & Shimmen, 1988; Sun et al., 1989) and result in the degeneration of the synergid. Thus, we suggest that the heavy accumulation of calcium in the ER is presumably one of the causes and not the result of synergid degeneration.

Another interesting feature is the change in the amounts of pyroantimonate precipitates in the central cell wall ingrowths and the contiguous apoplast outside the embryo sac. Before pollination copious amounts of precipitats are seen there, but after pollination the amount decreases and finally nearly disappears at the proembryo stage. Based on the cytochemical localization of ATPase in sunflower ovules, we found that the site of the wall ingrowths of the central cell is the main passage for nutrients entering the embryo sac (He & Yang, 1991). The results reported here indicate that the passage may function in the entry of messengers, including calcium, into the embryo sac.

We also show the presence of large amounts of calcium deposits in the nucleus of egg and the secondary nucleus of central cell. The gap between the plasma membranes of these two cells also contains many calcium deposits that seem to be derived from synergids. Yan et al. (1990) believed this gap is the forward position of two sperms for future double fertilization. Two sperms were found in a similar gap in the synergidless angiosperm *Plumbago zeylanica* (Russell, 1989). Taking into consideration that abundant calcium released from the ER in the egg triggers the sperm and egg activation, fusion and the development of zygote at fertilization in animals (Poenie & Epel, 1987), we are tempted to speculate that the presence of calcium there may have something to do with double fertilization.

Explanation of plates

A. Apoplast, c. Cuticle, CC. Central cell, d. Dictyosome, DS. Degenerating synergid, E. Egg, EN. Egg nucleus, En. Endosperm, ER. Endoplasmic reticulum, FA. Filiform apparatus, g. Gap, IT. Integumentary tapetum, M. Micropyle, m. Mitochondria N. Synergid nucleus, Nu. Nucleolus, o. Oil drop, PS. Persistent synergid, S. Synergid, SN. Secondary nucleus, v. Vacuole, w. Cell wall, wi. Wall ingrowths, Z. Zygote, ZN. Zygote nucleus.

Plate I

1-5. High-magnification images showing the structure of Ca-antimonate deposits.
1. Spherites. ×174000 **2.** Fusion of more spherites. ×174000 **3.** A large sperite. ×15000 **4.** A disintegrating spherite. ×10800 **5.** Fibrilliform deposit in a vacuole. ×19000 **6-8.** Localization of Ca-antimonate in the micropyle and two synergids at one day before pollination. All figures are oriented with the micropylar pole upward. Note the continuously increasing deposit gradients from the micropylar to the chalazal end. No obvious difference is visualized in Ca-antimonate content between two synergids. **6.** Micropylar portion of the synergids. ×4850 **7.** Middle portion of the synergids. Arrowheads show the wall of synergids. ×4850 **8.** Chalazal portion of the synergids, arrowheads show the wall between two synergids. ×3200.

Plate II

9-14. Localization of Ca-antimonate at one day before pollination. **9.** Transection of an embryo sac, showing similar amounts of spherites in two synergids. Note some fibrilliform deposits in vacuoles at the micropylar portion of the egg. ×2850 **10.** Transection of an egg at its chalazal portion, showing spherites concentrated in its nucleus, especially in nucleolus. Oil drops (hollow arrowhead) in the central cell are also highly electron dense. The gap (solid arrowhead) between the egg and the central cell contains many deposits. ×2800 **11.** Deposits located in egg nucleus, secondary nucleus, espe-

图版 I   plate I

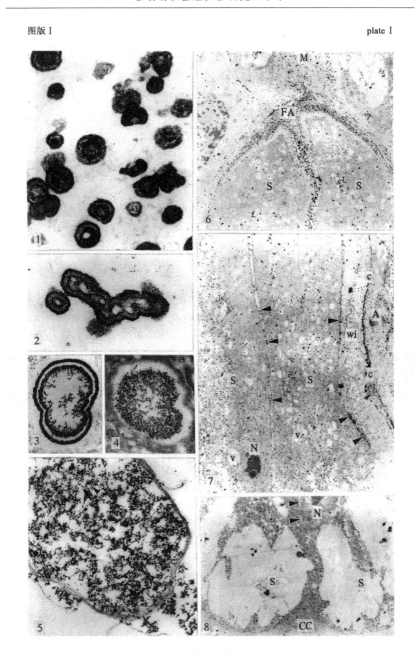

第四章 有性生殖的超微结构与细胞化学研究

图版Ⅱ   plate Ⅱ

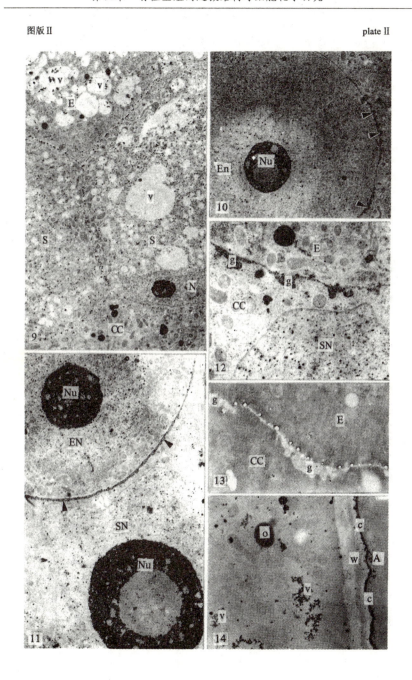

图版III plate III

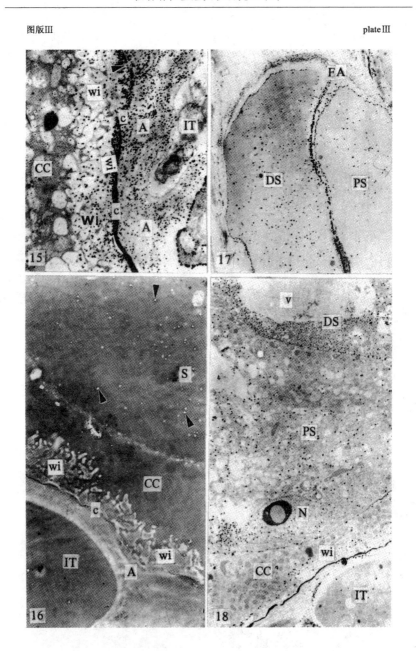

第四章 有性生殖的超微结构与细胞化学研究

图版Ⅳ   plate Ⅳ

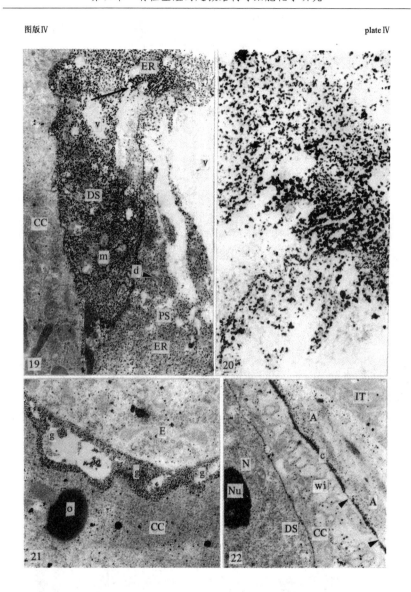

图版 V plate V

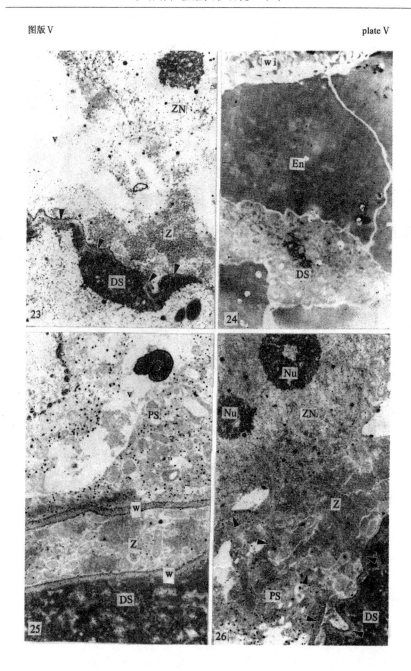

cially in their nucleoli and the gap (solid arrowheads) between the egg and central cell. ×4350 **12.** Higher magnification image of the gap between egg and central cell. ×8700

13. A control of the gap between the egg and central cell. Note most of the electron lucent pores are associated with the plasma membrane of the egg. ×10800 **14.** Chalazal portion of a central cell, showing fibrilliform deposits in its vacuoles and some spherites in its wall and the apoplast outside the wall. ×7200.

Plate III

15-16. Localization of Ca-antimonate deposits at one day before pollination. **15.** High magnification image of the wall ingrowths of the central cell. Arrowhead shows the interval of the cuticle. ×10800 **16.** A control. The wall ingrowths of the central cell and the apoplast outside the embryo sac are electron lucent. Note also some lucent pores (arrowheads) in the synergid. ×5400   17-18. About 2.5 h after pollination. **17.** Ca-antimonate deposits located in two synergids, especially in their filiform apparatus. Note more deposits in the degenerating synergid than those in the persistent one. ×4350

18. Middle portion of the two synergids, showing remarkable difference in Ca-antimonate deposit content between the degenerating and persistent synergid. ×7800

Plate IV

19-22. Localization of Ca-antimonate at about 2.5 h after pollination. **19.** The chalazal part of two synergids, showing obvious difference in Ca-antimonate-deposit density between them. ×1800 **20.** Magnification of the site showed by an arrow in the degenerating synergid in Fig. 19. Note the deposits seem to be associated with disintegrated ER. ×28000 **21.** Numerous small deposits located in the gap between egg and central cell. ×15000 **22.** Ca-antimonate deposits in the wall ingrowths of central cell and apoplast outside it become less than before pollination. Arrowheads show the intervals of the cuticle surrounding the embryo sac. ×8700.

Plate V

23-26. Localization of Ca-antimonate deposits just after fertilization. **23.** Numerous small deposits located in electron dense degenerated synergid, especially in its wall (arrowheads) bordering the zygote. ×7200 **24.** A control, showing electron lucent degenerated synergid after treatment with EGTA. ×5400 **25.** Two synergids and the micropylar part of a zygote. ×7200 **26.** The persistent synergid, degenerated synergid and zygote. Arrowheads show cell walls. ×7200

(Authors: He CP, Yang HY. Published in: Chinese Journal of Botany, 1992, 4(2): 99-106, with 5 plates and 29 references.

All the plates are retained here.)

2 向日葵柱头、花柱和珠孔中钙分布的超微细胞化学定位

Ultracytochemical localization of calcium in the stigma, style and micropyle of sunflower

Abstract

Calcium was localized ultracytochemically in the stigma, style and micropyle of sunflower (*Helianthus annuus* L.) by pyroantimonate precipitation technique. To identify the element constitution of the pyroantimonate deposits, wave-dispersive X-ray microanalysis (WDX) method was employed in addition to the energy-dispersive X-ray microanalysis (EDX) and the deposits observed were directly proved as calcium pyroantimonate. In the stigma, calcium was more abundant on the receptive surface, especially outside and inside the papillae, than on the non-receptive surface. In the style, more calcium was seen in the transmitting tissue as compared with the adjacent parenchymatous tissue, and was concentrated at the intercellular matrix and the boundary between the cell wall and the intercellular matrix. In micropyle region, the transmitting tissue on the side proximal to the funicle contained large amount of calcium, where as more calcium was localized in the intercellular matrix. As for the pollen tubes growing along the gynoecium, calcium was mainly localized at the pectin layer of the tube wall after 1.5 hours of pollination.

钙在花粉萌发和花粉管向性生长中的重要作用已为多种植物的离体花粉培养实验所证实[1]。对柱头、花柱和珠孔等花粉萌发和花粉管生长所经的雌蕊组织中钙的分布进行研究,则可以提供另一方面的证据。Bednarska[2,3]用金霉素(CTC)荧光标记法和 X 射线微区分析法证明假叶树的柱头表面富含钙,并用 ^{45}Ca 标记试验观察到柱头上的钙被花粉吸收。毛节锜等[4]用焦锑酸盐沉淀法定位了甘蓝型油菜柱头和花柱等雌蕊组织中的钙。Chaubal 和 Reger[5~7], Huang 和 Russell[8] 先后用 X 射线微区分析、焦锑酸盐沉淀或金霉素(CTC)荧光标记法观察到小麦、珍珠谷和烟草的助细胞中有较高含量的钙。何才平和杨弘远用焦锑酸盐沉淀法研究了传粉与受精前后向日葵胚囊中钙的动态变化,发现传粉后退化助细胞中钙含量明显增加[9]。何才平还初步观察了向日葵珠孔中钙分布的特点(博士论文,未发表资料)。本文对向日葵花粉萌发和花粉管生长所经柱头、花柱和珠孔中钙的分布作了比较系统的研究。在这一研究中,应用 X 射线能谱(EDX)和波谱(WDX)两种方法,对富含焦锑酸盐沉淀的微区进行了定性分析,为焦锑酸盐沉淀法的研究结果提供了直接而可靠的依据。

材 料 与 方 法

取向日葵(*Helianthus annuus* L.)授粉前和人工授粉后 1.5h 的柱头、花柱和胚珠,以焦锑酸盐沉淀法制样[9]。柱头和花柱切成 2mm 左右的小段,胚珠取其珠孔端 1~2mm,经戊二醛前固定和锇酸后固定(固定液和洗涤缓冲液中均含 2% 焦锑酸钾),乙醇系列脱水,环氧丙烷过渡,在 Epon 812 树脂中包埋。用 Sorvall MT-6000 型超薄切片机切片,超薄切片经醋酸双氧铀染色 30min 后用 JEM-100cx/Ⅱ型透射电子显微镜观察和摄影。

X射线微区定性分析是用HITACHI X-650型扫描电子显微镜上配置的X射线波谱仪(WDX)和PHILIPS PV 9100型X射线能谱仪(EDX)完成的。先将包埋块做半薄切片,暴露样品所要分析的部位。再用HUS-5GB型高真空喷镀仪镀碳后,用扫描电镜观察。根据样品表面的两次电子像和邻近的半薄切片的光镜观察,结合透射电镜下焦锑酸盐沉淀分布的特点,选定沉淀最密集的微区进行X射线能谱分析。在加速电压20kV,束流1×10^{-10}A的条件下,取得该微区的特征X射线能量谱。同时分别对样品中沉淀很少的微区和样品以外的树脂进行能谱分析作为对照。波谱分析是在加速电压25kV,束流1.2×10^{-9}A的条件下进行的,采用PET分析晶体。

结　果

一、X射线微区定性分析(从略,参看下文)

二、钙的超微细胞化学定位

柱头和花柱　向日葵柱头两裂片相对的内表面为花粉接受面,外表面为非接受面。接受面外表皮为乳突细胞,紧接乳突细胞之下是引导组织[10]。花柱属闭合型,其中央具引导组织而与柱头的引导组织相连。

通过对授粉前柱头不同部位的横切面和纵切面的观察,发现接受面分布较多Ca沉淀,非接受面一侧则很少。乳突细胞外存在密集的颗粒状Ca沉淀(图版I,1)。乳突细胞的液泡中有很多细小的絮状Ca沉淀,细胞质中也有一些很小的Ca沉淀,细胞壁中则很少(图版I,2)。在乳突细胞之下的引导组织的胞间基质中也观察到相当密集的Ca沉淀。

授粉前花柱引导组织中有大量的Ca沉淀,且主要分布在细胞之间。从纵切面(图版Ⅰ,3)和横切面(图版Ⅰ,4、5)上可以看出:细胞壁内层几乎没有Ca沉淀;细胞壁外层和胞间基质看不出明确的界限,二者相接之处是Ca沉淀最为密集的部位;胞间隙的中央也有不少细小的 Ca 沉淀颗粒,但分布不均匀;细胞内的 Ca 沉淀除零星的颗粒外,多以絮状存在于某些液泡中。引导组织周围的薄壁组织中 Ca 沉淀明显地比引导组织中的少(图版Ⅰ,6)。

授粉后 1.5h 柱头与花柱中 Ca 的分布未见明显变化,但在引导组织的横切面上看到花粉管壁最外层,即果胶质层及靠近管壁的胞间基质中,有较多的 Ca 沉淀(图版Ⅰ,5)。

珠孔 向日葵的珠孔属闭合型,珠孔区的珠被细胞特化成引导组织,在珠孔外端,以角质层为界,特化区域局限在其近珠柄的一侧[11]。观察表明,授粉前近珠柄侧的发达的胞间基质中有大量 Ca 沉淀颗粒,细胞壁中也有;远珠柄侧胞间隙不发达,Ca 沉淀也不多,细胞壁中几乎缺如(图版Ⅱ,7、8)。在珠孔内端,角质层两侧均特化为引导组织状[11],此处两侧 Ca 沉淀分布的差异不明显(图版Ⅱ,9)。

授粉后 1.5h,珠孔中 Ca 沉淀分布的变化不大,仅珠被细胞中的 Ca 沉淀多与质膜相结合。花粉管及其周围 Ca 沉淀的分布与前述花柱中的情形相似(图版Ⅱ,10)。远离珠孔的珠被组织中钙沉淀则较少(图版Ⅱ,12)。在珠孔端的珠柄维管束横切面上,很多颗粒状的钙沉淀与维管束细胞的质膜相联系(图版Ⅱ,11)。

讨 论

迄今已有的研究工作还不能完全阐明 Ca 在花粉管生长中的作用机制。Ca 是否为对花粉管定向生长起决定作用的向化性因子?是否在柱头到胚囊之间存在一个钙的梯度,引导花粉管向胚囊生长[1,12]?解答这些问题,除了依靠花粉管离体生长实验系统的研究外,还需要以雌蕊组织为对象进行深入的研究。向日葵是适合这种研究的材料之一。在向日葵中,大多数花粉管通过珠孔时在其近珠柄侧的引导组织中生长;与此相关,胚囊的两个助细胞中亦多是近珠柄侧的一个退化。由此 Yan 等[11]推测可能珠孔的非对称结构、花粉管生长与助细胞退化三者之间有某种规律性的联系。He 和 Yang[9]发现,向日葵传粉前两个助细胞中均有 Ca,而传粉后近珠柄侧的助细胞中 Ca 含量大增[9]。本文的结果显示,在柱头接受面和花柱、珠孔引导组织等花粉萌发和花粉管生长所经的雌蕊组织中 Ca 含量较高,珠孔近珠柄侧的引导组织中 Ca 含量明显高于远珠柄侧。这表明向日葵雌蕊组织中的 Ca 对花粉管生长是有重要作用的。

　　Ca 在柱头、花柱和珠孔的细胞外的分布呈现一定的规律性。除了珠孔引导组织近珠柄侧的细胞壁中有 Ca 沉淀外,所有的纤维素壁中几乎没有看到 Ca 沉淀,而富含果胶质的胞间基质尤其是胞间基质与细胞壁相接处,Ca 沉淀却相当丰富。Sanders 和 Lord[12]通过对蚕豆等几种植物的研究,认为花柱胞外基质(extracellular matrix)对于花粉管生长有重要功能。最近 Li 等[13]报道,在有 Ca 存在的离体实验条件下,对于多种具闭合型花柱的植物,其花粉管壁果胶质层中作周期性环状分布的非酯化果胶质能凝胶化以提高花粉管壁的强度。本文观察到雌蕊引导组织的富含果胶质的胞间基质以及花粉管壁的果胶质层中有较多的 Ca。由此可以推测其分布与加强细胞壁的强度有关。至于 Ca 在雌蕊组织中是否确有梯度变化,以及 Ca 是否确为花粉管生长的向化性因子,这涉及更为复杂与深刻的问题,有待进一步研究。

(作者:张劲松、杨弘远、朱绫、童华。原载:植物学报,1995,37(9):691~696。图版2幅,图2幅,参考文献13篇,均删去。)

3　陆地棉雌蕊的花粉管生长途径中钙分布的超微细胞化学定位

Ultracytochemical localization of calcium in the pollen tube track of cotton gynoecium

Abstract

Cotton (*Gossypium hirsutum* L.) stigma, style and ovule segments were fixed pre-and post-pollination in glutaraldehyde and osmium solutions containing potassium pyroantimonate, processed with conventional dehydration and infiltration procedure, and embedded in Epon 821 resin. To identify the element constitution of the pyroantimonate deposits in the samples, wave-and energy-dispersive X-ray microanalyses were employed, and it was confirmed that the deposits were truly calcium pyroantimonate. Transmission electron microscopic observations revealed that the deposits were more abundant at the whole pollen tube track as compared with the adjacent tissues. They were mainly localized in the apoplast system, i. e. the intercellular matrix of the stigmatic tissue, the outer cell wall layers of stylar transmitting tissue, the intercellular openings and gaps of micropyle, and the degenerated column of nucellus. It was reported that these apoplast spaces were just the positions where pollen tubes grew. The pollen tube tip growing in the micropyle was also rich in calcium, which was especially localized at the new wall, vesicles and mitochondria. All these observations suggested that calcium in the gynoecium tissues may play an important role in the pollen tube growth *in situ.*

作者在前文中曾报道向日葵雌蕊组织中钙分布的超微细胞化学定位[1]。Jensen 和 Fisher 早在 20 世纪 60 年代就详细研究过棉花柱头、花柱[2,3]、珠心[4,5]等组织的超微结构及花粉管在其中的生长方式。最近张劲松等[6]又对其珠孔结构作了研究。这些资料为进一步研究棉花雌蕊的花粉管生长途径(柱头毛→花柱引导组织→子房壁表面→外珠孔→内珠孔→珠心细胞柱)中钙分布的规律提供了良好的背景。本文的目的是试图揭示在棉花中是否和在向日葵中类似,钙分布也是在雌蕊的花粉管生长途径中较其它组织中更为密集,以及钙在这些组织的细胞内外分布状况,以求从体内的角度认识雌性组织中钙与花粉管生长的关系。

材 料 与 方 法

陆地棉(*Gossypium hirsutum* L.)授粉前和授粉后2、5h的柱头与花柱和授粉后11、22h的胚珠。柱头与花柱切成1~2mm小段,胚珠切取珠孔端1mm或解剖出珠心。按焦锑酸盐沉淀法制样[7]。材料先后以含2%焦锑酸钾的戊二醛与锇酸固定,乙醇脱水,环氧丙烷过渡,Epon 812树脂包埋。用Sorvall MT-6000型超薄切片机切成超薄切片,经醋酸双氧铀染色10min或不染色,用JEM-100/Ⅱ型透射电镜观察摄影。

为证明上法的可靠性,对样品中的沉淀进行了X射线微区定性分析。先将包埋块切成半薄切片,待暴露出欲观察的内部结构后,将包埋块的暴露面用HUS-5GB型高真空喷镀仪镀碳,再用Hitachi X-650型扫描电镜配置的能谱仪与波谱仪分别观测。能谱分析在加速电压20kV、束流1×10^{-10}A下进行;波谱分析在加速电压25kV、束流1.2×10^{-10}A下进行。

结　　果

一、X射线微区定性分析

在焦锑酸盐沉淀密集的微区所取得的X射线能谱图上,有3个Sb元素峰、2个Cl元素峰、1个K元素峰和1个Os元素峰(图1A)。在沉淀稀少的微区,Cl峰与前相当,Os峰较低,Sb与K峰更低(图1B)。在空的树脂微区,则仅有Cl峰(图1C)。由此推论,Cl来自包埋树脂;Os来自锇酸;K来自焦锑酸钾。由于图1A的Sb峰远高于图1B,因此Sb显然主要来自焦锑酸钙沉淀中的锑。但因能谱分析的分辨率有限,不足以区分Sb与Ca,故后者被前者掩盖而不能显示。在X射线波谱图上,则可明显看出在2个Sb峰之间有1个Ca峰(图2)。这就确证本实验中所观察的反应沉淀是焦锑酸钙。

第四章 有性生殖的超微结构与细胞化学研究

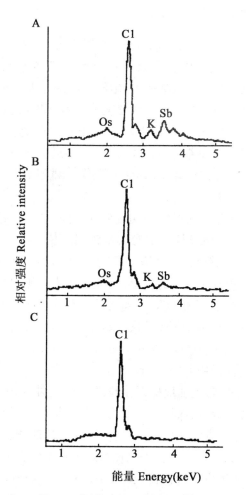

图 1 不同微区的 X 射线能谱图
A. 焦锑酸盐沉淀密集的微区;B. 焦锑酸盐沉淀很少的微区(对照1);C. 样品以外树脂上的微区(对照2)。

Fig. 1 Energy-dispersive X-ray spectra acquired from area rich in pyroantimonate deposits (A), area lacking pyroantimonate deposits (B) and area of samplefree resin (C)

二、钙的超微细胞化学定位

柱头和花柱 本文所观察到的陆地棉的柱头和花柱结构与前人报道的一致[2]：柱头表面中央有众多单细胞毛,花粉管由此进入柱头。开花当天,毛的中央为大液泡所占,胞质多已退化为电子浓密状态(图版Ⅰ,1)。柱头毛和花柱引导组织之间有多层薄壁细胞。紧接毛下的几层薄壁细胞有大的胞间隙；愈近花柱引导组织,胞间隙愈小直至缺如,同时由上至下薄壁细胞的壁渐厚,直至接近引导组织厚壁的程度。花柱属闭合型,引导组织居中,其细胞长形,侧壁厚而分为4层,胞间隙不发达(图版Ⅰ,2)。

授粉前,柱头毛与薄壁组织相接处的胞间隙中充满焦锑酸钙沉淀(图版Ⅰ,1)。花柱引导组织中有大量钙沉淀位于细胞侧壁中。从横切面(图版Ⅰ,2)和纵切面(图版Ⅰ,3)上均可看出：钙集中分布于细胞壁的外层,即花粉管所生长的部位,而壁的内层及胞质中则很少。在侧壁的胞间连丝内也有钙沉淀。不同切段的花柱引导组织,钙的分布规律一致。引导组织外的薄壁组织中则几乎没有钙沉淀(图版Ⅰ,4)。授粉后5h,在花柱引导组织中生长的花粉管,钙的分布和后文描述的珠孔中的花粉管(图版Ⅱ,9)一致。

珠孔 过去曾报道陆地棉珠孔的超微结构：外珠孔的外、内两端各有一开口,中段闭合；内珠孔完全闭合成缝隙状[6]。授粉前,外珠孔外端开口中有十分密集的钙沉淀,而构成珠孔的外珠被表皮细胞中则很少(图版Ⅰ,5)；远离外珠孔的外珠被组织中几乎没有钙沉淀。中段闭合处的角质层两侧有不少沉淀(图版Ⅰ,6)。内端开口中也充满沉淀(图版Ⅱ,7)。在内珠孔的珠孔隙中有密集的沉淀(图版Ⅱ,8示其高倍放大图像)。远离内珠孔的内珠被组织中几乎没有钙沉淀。

授粉后珠孔区组织中的钙分布未见明显变化。授粉后11h,在不同胚珠的珠孔中观察到花粉管。图版Ⅱ,9示外珠孔开口中的一根花粉管尖端细胞器区段的横切面。可以看出：除花粉管周围有密集的钙沉淀外,花粉管本身的细胞壁、小泡、线粒体中亦富含钙沉淀。这似乎表明花粉管和珠孔基质间有密切的钙离子交流。图版Ⅱ,10示另一根

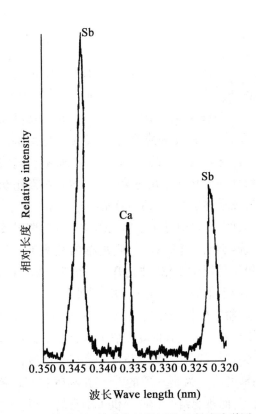

图 2 焦锑酸盐沉淀密集的微区的 X 射线波谱图

Fig. 2 Wave dispersive X-ray spectrum acquired from the area rich in pyroantimonate deposits

花粉管尖端较后区段的纵切面,可见到其胞质中有一些钙沉淀,但线粒体和花粉管壁中钙已减少。

珠心　陆地棉具厚珠心。珠心珠孔端有一特化的细胞柱由珠孔通达胚囊,其细胞在花粉管到达珠心前即已退化;花粉管到达珠心后,在退化细胞柱的胞壁中生长[4,5]。授粉前,当细胞柱尚未退化时,其细胞中钙沉淀不多(图版Ⅱ,11)。授粉后 11h 与 22h,细胞柱的退化细胞中有电子致密的退化原生质,细胞壁中密集钙沉淀(图版Ⅱ,12)。

讨 论

迄今关于雌蕊组织中钙分布的超微细胞化学定位,正式报道者为数不多。在假叶树[7,8]、甘蓝型油菜[9]、向日葵[1]中的观察表明,凡是花粉萌发与花粉管生长所经的雌蕊组织中,均有高含量的钙。本文在陆地棉中又一次证明:在柱头薄壁组织、花柱引导组织、外珠孔的内外开口、外珠孔与内珠孔的闭合缝隙、珠心退化细胞柱等花粉管生长途径中,钙的密集程度均较其它非花粉管生长途径的雌蕊组织中显著为高。另一重要点是:在上述组织中,钙多分布于质外体系统中。如向日葵的柱头乳突细胞表面、花柱引导组织胞间基质与细胞壁果胶质层、珠孔引导组织胞间基质与细胞壁等处[1],陆地棉的柱头毛与薄壁组织胞间隙、花柱引导组织外层细胞壁、珠孔开口与缝隙、珠心退化细胞柱细胞壁等处,均为钙最密集的部位,它们均属质外体系统。

关于钙对花粉管生长的作用,已有的研究结论大多依据离体实验结果,而很少依赖对体内生长的花粉管的观察。本文观察到:在珠孔中生长的花粉管尖端细胞器区,钙含量高于其后方区段,这和人工萌发花粉管中存在钙梯度的行为是一致的[10]。花粉管尖端钙梯度的维持需要由外界吸收钙,而雌蕊组织为此提供了丰富的钙源。前人已发现,在向日葵珠孔中,花粉管是在其引导组织的胞间基质中生长[11];而在陆地棉花柱中,花粉管则是在其引导组织的外层细胞壁中生长[2]。这些质外体系统中的钙无疑对花粉管生长有重要作用。

图 版 说 明

取材时期:图1~8、11. 授粉前。图9、10、12. 授粉后11h。

c. 角质层 im. 胞间基质 m. 珠孔 mt. 线粒体 pt. 花粉管 h. 柱头毛 w. 细胞壁

图 版 I

1. 柱头纵切面,示柱头毛与其下方薄壁细胞间的焦锑酸钙沉淀。×7000 **2.**

花柱引导组织横切面,示侧壁外层的钙沉淀。×9500 **3.** 花柱引导组织纵切面,示侧壁外层的钙沉淀。×10400 **4.** 花柱引导组织外围的薄壁组织横切面,其中很少钙沉淀。×4900 **5.** 外珠孔外端开口部位横切面,其中含密集的钙沉淀。×6500 **6.** 外珠孔中段闭合部位横切面,示角质层两侧与珠被细胞间质中的钙沉淀。×6400

图 版 Ⅱ

7. 外珠孔内端开口部位横切面,其中含密集钙沉淀。×14200 **8.** 内珠孔闭合缝隙中的钙沉淀高倍放大图像。×39800 **9.** 在外珠孔中生长的花粉管尖端横切面,钙沉淀在花粉管壁与线粒体中较多,质膜附近的小泡中富含钙(箭头)。×19100 **10.** 在珠孔中生长的另一根花粉管尖端较后区段纵切面(尖端未切到)。×7000 **11.** 珠心细胞柱退化前的横切面,细胞核中有钙沉淀,但细胞壁中不多。×9700 **12.** 珠心退化细胞柱横切面,钙沉淀密布于细胞壁中。**10800**

Explanation of plates

Samples collected prior to pollination (Figs. **1 ~ 8 and 11**) and 11 h after pollination (Figs. **9, 10 and 12**).

c. Cuticle layers   h. Stigma hair   im. Intercellular matrix   m. Micropyle   mt. Mitochondria   pt. Pollen tube   w. Cell wall

Plate Ⅰ

Fig. 1. Transection of stigma, showing calcium pyroantimonate deposits in the intercellular matrix between the stigma hair cell and the parenchymatous cell. ×7000 **Fig. 2.** Transection of style transmitting tissue, showing calcium deposits in the outer layers of lateral cell wall. ×9500 **Fig. 3.** Longisection of stylar transmitting tissue, showing calcium deposits in the outer layers of lateral cell wall. ×10400 **Fig. 4.** Transection of parenchy matous tissue surrounding the transmitting tissue. Not the few localized deposits. ×4900 **Fig. 5.** Transection of the outer opening of exostome, showing abundant calcium deposits. ×6500 **Fig. 6.** Transection of the closed middle part of exostome. Calcium deposits localized on both sides of the cuticle layers and intercellular space of the exostome gap. ×6400

Plate Ⅱ

Fig. 7. Transection of the inner opening of endostome, showing the abundant calci-

图版 I   plate I

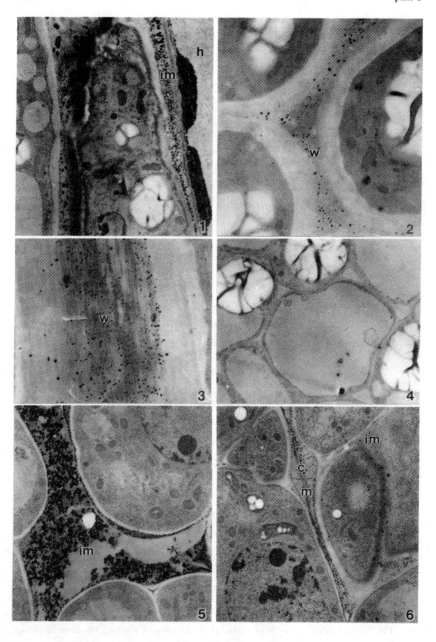

第四章 有性生殖的超微结构与细胞化学研究

图版 II   plate II

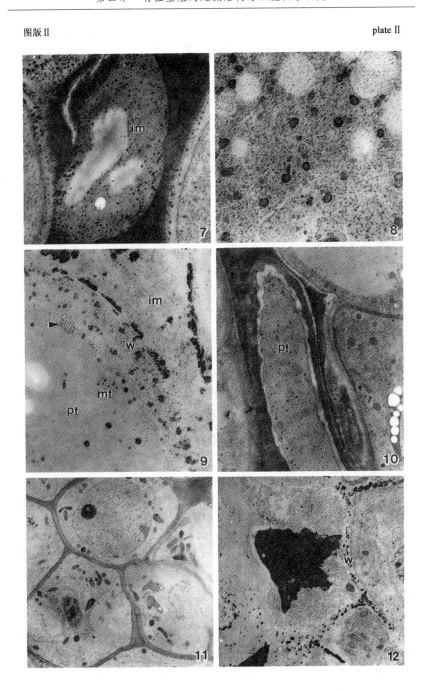

um deposits. ×14200 **Fig. 8.** High magnification showing the calcium deposits localized in the closed endostome gap. ×39800   **Fig. 9.** Transection of the pollen tube tip growing at the exostome. Calcium deposits are mostly rich in the tube wall and mitochondria. The vesicles adjacent to the plasma membrane are rich in calcium (arrow). ×19100   **Fig. 10.** Longisection of another pollen tube growing at the micropyle, showing portion of pollen tube behind the tip, which is not shown in the same section. ×7000 **Fig. 11.** Transection of the nucellar column before its degeneration. Calcium deposits are localized in the nuclei, but few in the walls. ×9700   **Fig. 12.** Transection of the degenerated nucellar column, showing abundant calcium deposits in the cell walls. ×10800

(作者:张劲松、杨弘远。原载:植物学报,1997,39(2):121~125。图版2幅,图2幅,参考文献11篇。选录图版Ⅰ、Ⅱ,图1、2)

4 甘蓝型油菜授粉前后珠孔和胚囊中钙分布的超微细胞化学定位

Ultracytochemical localization of calcium in micropyle and embryo sac of *Brassica napus* before and after pollination

Abstract

Potassium antimonate was used to localize Ca^{2+} in the micropyle and embryo sac of *Brassica napus* L. before and after pollination. To identify the nature of the pyroantimonate deposits, energy-dispersive X-ray microanalysis (EDXA) was employed and the deposits were proved to contain calcium pyroantimonate. Image processing system was employed to measure the volume density and the diameter of the deposits. Before and after pollination, calcium was more abundant in the exostome and endostome as compared with the other regions of the integuments, and was concentrated at the apoplast system, i. e. the intercellular matrix of the micropyle canal and the cell wall. Before pollination, each of the two sister synergids accumulated more calcium than the other embryo sac cells. Although the mean diameter of the deposits in the synergid was only two-thirds as that in the egg cell and central cell, the volume density of the deposits in the synergid was about 2.5 times and 1.9 times as that in the egg cell and the central cell respectively. The filiform apparatus and the nucleus had the most abundant calcium within a synergid. After pollination both sister synergids degenerated conspicuously and were characterized by much more deposited calcium (about 2.4 times more than before); and the diameter of the deposits decreased dramatically, which was less than one-third as before. The relationship between calcium distribution and synergid degeneration as well as its functions was discussed.

植物体内，钙直接或通过钙调蛋白及其受体间接调节许多代谢过程和发育事态，是植物细胞刺激-反应信号转导机制中的关键信号分子[1]。被子植物双受精包含许多复杂精巧的发育事态，其时间节奏和空间布局十分严密，研究钙在其中的时空变化对于探讨钙在植物生殖过程中的作用机制有着重要意义。钙在花粉萌发和花粉管向性生长中的重要作用已为多种植物的离体花粉培养实验[2]和对雌蕊花粉管生长途径中钙分布的超微细胞化学定位[3~5]所证实。Jensen[6]和Jensen等[7]用显微灰化法发现助细胞中富含灰分并推测其主要成分是钙。Chaubal和Reger[8~10]、Huang和Russell[11]先后用X射线微区分析、焦锑酸盐沉淀或金霉素荧光标记法观察小麦、珍珠谷和烟草几种植物胚珠和胚囊中总钙、松弛结合钙和膜钙的分布，发现助细胞中确有高含量钙。He和Yang[12]用焦锑酸盐沉淀法对传粉和受精前后向日葵中钙的动态变化的研究发现传粉后退化助细胞中钙含量明显增加。助细胞退化是大多数被子植物受精的不可缺少的前提，退化助细胞是花粉管通过珠孔进入胚囊终止其生长并释放雄配子的场所[13]。助细胞中的高钙含量可能与其上述功能密切相关。本文用焦锑酸盐沉淀结合 X 射线微区分析和细胞形态立体计量的方法,比较系统地研究了授粉前后甘蓝型油菜胚珠特别是珠孔与助细胞中钙分布的特点。

材 料 与 方 法

一、电镜制样与 X 射线微区定性分析

供试材料为甘蓝型油菜(*Brassica napus* L.)品种"4312"。取授粉前和人工辅助授粉后 22h 的胚珠,以焦锑酸盐沉淀法制样[12]。材料以含 2% 焦锑酸钾的戊二醛和锇酸先后固定,乙醇系列脱水,环氧丙烷过渡,Epon 812 树脂包埋。在 Sovall MT-6000 型超薄切片机上切片。超薄切片经醋酸双氧铀染色 30min 或不染色,用 JEM-100 CX/Ⅱ 型透射电镜观察与摄影。X 射线能谱微区定性分析按张劲松等[4]的方法进行。

二、钙沉淀颗粒的形态计量

细胞中焦锑酸钙颗粒的体密度(Vv)为单位体积内该颗粒的体积,可按公式 $Vv = \sum_{i=1}^{n} A_{xi} \Big/ \sum_{i=1}^{n} A_{ri}$ 计算[14]。A_{xi} 和 A_{ri} 分别为第 i 张照片上钙颗粒的截面积和参照系的截面积。细胞内钙颗粒的截面积的平均等效圆直径 $D = \sum_{j=1}^{n} d_j \Big/ n$,$d_j$ 为第 j 个钙颗粒截面的等效圆直径。上述 A_{xi}、A_{ri}、d_j 均用图像分析仪测得。电镜照片通过摄像机输入 KONTRON IBAS-2000 型图像处理系统进行图像分析。由于钙沉淀颗粒的电子密度较细胞其它成分显著为高,图像处理仪可通过二值图分割,留下感兴趣的钙颗粒并分析计算出钙颗粒截面面积和等效圆直径等参数。再通过计算求得体密度和平均直径,并用 t 检验求得各类各组间的参数差异程度。

结 果

一、X 射线微区定性分析

在焦锑酸盐沉淀密集的微区所取得的 X 射线能谱图上,观察到 3 个 Sb 峰、2 个 Cl 峰、1 个 K 峰和 1 个 Os 峰;在沉淀很少的微区的能谱图上,Cl 峰与前相当,Os 峰较低;Sb 与 K 的峰仅略高于本底;在空白树脂微区的能谱图上,则只有 Cl 峰(与文献[4,5]一致,故图形从略)。由此推论 Cl 来自包埋树脂,Os 来自固定剂锇酸;K 来自焦锑酸钾[4]。由于沉淀密集微区的 Sb 峰显著高于沉淀很少的微区,显然 Sb 主要来自焦锑酸钙沉淀中的锑。因为 Ca 的两个峰($Ka = 3.69$ keV,$K_{\beta 1} = 4.01$ keV)与 Sb 的两个峰($La = 3.60$ keV,$L_{\beta 2} = 4.10$ keV)十分接近而互相重叠,而能谱图中并没有发现 Na、Mg 等其它可与焦锑酸钾反应生成沉淀的元素的峰,故可推论 Ca 峰被 Sb 峰掩盖,而判断所分析的沉淀为焦锑酸钙沉淀[9](以下称钙沉淀)。

二、钙沉淀的形态计量

助细胞、卵细胞和中央细胞中的钙沉淀的体密度(Vv),平均等效圆直径与标准差见表 1。以助细胞 1 的参数值为对照组分别与其它各组数据配对进行 t 检验。表 2 显示授粉前助细胞不同部位的钙沉淀的体密度(Vv)和平均等效圆直径。以丝状器的参数值为对照分别与其它各组数据配对进行 t 检验。统计结果表明,授粉前两个助细胞差异不明显,但 Vv 值均显著高于卵细胞与中央细胞,约为前者的 2.5 倍和后者的 1.9 倍;D 值则明显小于卵细胞和中央细胞,约为其 2/3。助细胞中,丝状器和细胞核的 Vv 值明显高于细胞质,而 D 值则差异不大。授粉后退化助细胞 Vv 显著升高,约为授粉前的 2.4 倍;D 值显著减小,不足授粉前的 1/3。

三、钙的超微细胞化学定位

甘蓝型油菜胚珠为倒生型,有内、外珠被。胚囊成熟时,包围胚囊珠孔端及侧面的珠心组织均退化。胚囊含两个助细胞,1个卵细胞和1个中央细胞,反足细胞退化。

珠被和珠孔 油菜的珠孔由外珠被构成的外珠孔和内珠被构成的内珠孔两部分组成。外珠孔基本上是开放的结构,内珠孔则是闭合的结构。两侧的内珠被表皮细胞角质层常互相联系,只在局部地点分开[15]。传粉前,其外端开放处外珠被表皮角质层之间有大量密集的钙沉淀,细胞壁中也有(图版Ⅰ,1);而一般外珠被细胞中钙沉淀多与液泡膜相联系,壁和细胞基质中几乎缺如。内珠孔角质层两侧的细胞壁和胞间基质中有大量钙沉淀颗粒,紧邻内珠孔的内珠被细胞的胞质中也有钙沉淀分布,而细胞核中少见钙沉淀(图版Ⅰ,2)。一般情况下,细胞内的钙沉淀明显较质外体中的小而稀疏。退化珠心细胞电子密度很高,其中有一些细小的钙沉淀。退化珠心细胞与胚囊间有时形成一个电子透明的间隙,其中夹杂一些絮状物质,该部位钙沉淀很少(图版Ⅰ,7)。授粉后,内外珠孔及其周围的珠被细胞中钙沉淀的分布未见明显变化。

表1 胚囊成员细胞中钙沉淀的体密度和等效圆直径($X±SE$)

Table 1 Volume density (Vv) and equivalent circle diameter (D) of calcium deposits in cells of embryo sac ($X±SE$)

Parameter	Before pollination				After pollination
	Synergid 1	Synergid 2	Egg cell	Central cell	Degenerative synergid
$Vv(×100)$	1.64±0.18	1.59±0.16	0.64±0.08**	0.84±0.11**	3.93±0.46**
D(nm)	59.39±6.24	58.78±6.07	87.97±11.56**	88.95±9.97**	16.88±4.87**

Synergid 1 is located closer to the funicle and synergid 2 far from the funicle. ** Significantly different from the control, $P<0.01$.

第四章 有性生殖的超微结构与细胞化学研究

表 2 授粉前助细胞不同部位钙沉淀的体密度和等效圆直径（$X±SE$）

Table 2 Volume density (Vv) and equivalent circle diameter (D) of calcium deposits in different regions of synergid before pollination ($X±SE$)

Parameter	Filiform apparatus	Nucleus	Cytoplasm
$Vv(×100)$	2.26±0.51	2.11±0.22	1.42±0.27**
$D(nm)$	56.34±4.35	58.72±4.46	59.97±6.83

** Significantly different from the control, $P<0.01$.

助细胞 甘蓝型油菜的助细胞和一般植物的助细胞不同,没有明显极性。核位于细胞中部,无一般助细胞特有的合点端大液泡,只有众多分散的小液泡[15]。授粉前1对姊妹助细胞内钙沉淀分布没有明显差异(图版Ⅰ,4),丝状器和细胞核沉淀较多(图版Ⅰ,3、6),高于助细胞其它部位,也高于中央细胞和卵细胞(表1,表2)。需要注意的是助细胞核膜不可分辨(图版Ⅰ,6),已显露出退化的迹象。内质网和线粒体中也有钙沉淀分布,液泡中则可见絮状的钙沉淀,两个助细胞的公共壁(图版Ⅰ,4、5)以及助细胞与卵细胞相邻处也可以观察到有钙沉淀分布。授粉后22h,两个助细胞均已开始退化,细胞质电子密度明显加大。多数胚珠中花粉管尚未到达胚囊,少数胚珠中花粉管已进入胚囊的1个退化助细胞释放其内容物。花粉管尚未进入的胚囊中,两个助细胞的丝状器均含有颗粒状钙沉淀,但两个助细胞退化程度不同。其中1个退化程度较深一些,质膜破坏,核与细胞器几乎全部解体,少有典型的颗粒状或絮状钙沉淀。仔细观察可以发现,其中分布有许多极细密的粉末状钙沉淀,似乎与解体的细胞器的膜片段相联系。退化程度较浅的助细胞中线粒体、小液泡尚可分辨,钙沉淀主要是以颗粒形式分布,间或也有少量的粉末状沉淀(图版Ⅱ,9)。花粉管进入胚囊后,助细胞退化程度加剧,细胞器几乎全部解体。此时接受花粉管的助细胞电子密度较深,退化程度也较另一助细胞为甚。但2个助细胞中钙沉淀的分布形式无明显差异,皆无典型的颗粒状钙沉淀,主要是大量极细密的粉末状钙沉淀(图版Ⅱ,10)。进入助细胞的花粉管尖端壁也分布有钙沉淀,有趣的是这些钙沉淀并非随机分布,其中一侧管壁上钙沉淀颗粒明显为多(图版Ⅱ,11)。连续切片的观察表明,花粉管正是从钙沉淀多的一侧破裂,释放出内容物(图版Ⅱ,12)。

  卵细胞 卵细胞中钙沉淀的分布与其细胞的极性一致,授粉前后未见显著变化。卵细胞珠孔端大液泡中钙沉淀稀少,主要分布于液泡膜上。细胞质和核主要位于合点端,此处钙沉淀明显较珠孔端密集,多以颗粒形式分布于核中,核质中的钙沉淀明显较核仁中者为大。细胞质中也有一些大小与核仁中的颗粒相若的钙沉淀。卵细胞合点端仅以质膜与中央细胞分隔,其间有一宽窄不均匀的间隙层。以往观察曾发现,这一间隙层中有性质不明的电子致密物质[15]。本文证明,其中含有大量细密的钙沉淀,值得注意的是它们多分布在靠近卵细胞的一侧,授粉后可以看到这些沉淀与退化助细胞中的沉淀形成一个连续的系统(图版Ⅱ,14)。

  中央细胞 中央细胞中钙沉淀的分布极性与卵细胞中的正好相反。其珠孔端的沉淀物密集,2个极核中有明显的颗粒状沉淀。线粒体、内质网中也有,质体中较少(图版Ⅱ,13)。中央细胞合点端的沉淀分布稀疏,多以絮状形式出现在液泡中,传粉对这一极性分布没有明显影响。在与卵细胞相邻的部位,中央细胞的质膜呈波状,较卵细胞质膜曲折,其上钙沉淀物较相邻的卵细胞质膜上为少(图版Ⅱ,14)。胚囊壁近珠孔端部位有明显的内突,往合点端方向逐渐减少乃至消失,而角质层却显得异常明显。钙沉淀主要分布于壁内突外侧的胚囊壁上,而壁内突中钙沉淀较稀少。珠孔端胚囊壁近旁有时可以看到小泡存在,小泡的膜上分布有密集的钙颗粒,可能和钙的转移有关(图版Ⅰ,7)。合点端胚囊壁角质层的内外两侧都有钙沉淀,但内侧颗粒明显为大(图版Ⅰ,8)。

讨 论

  前人对向日葵和陆地棉雌蕊组织中花粉管整个生长途径中钙分布的超微细胞化学定位表明,凡在花粉萌发与花粉管生长所经的雌蕊组

第四章 有性生殖的超微结构与细胞化学研究

织中均有高含量的钙,并且钙主要分布在上述组织的质外体系统中[4,5]。毛节锜等对甘蓝型油菜柱头、花柱引导组织和子房假隔膜等处的钙离子定位研究也得到类似结果,但未报道花粉管生长途径的末段珠孔处是否有钙[3]。本文进一步证明,在甘蓝型油菜的内外珠孔处钙的密集程度较一般的珠被细胞显著为高,而且主要也是分布在胞间基质和细胞壁等质外体中。上述部位的钙无疑对花粉管的生长有重要作用。

助细胞是花粉管生长的终点。根据已有的研究,表明助细胞退化与引导花粉管进入胚囊、促使花粉管破裂、释放精子和帮助精子转移到受精靶区有关[13,16]。钙是引导花粉管定向生长的重要向化性因素[2,17]。甘蓝型油菜助细胞珠孔端丝状器是含钙量最高的部位,此处局部的钙梯度可能起吸引花粉管进入退化助细胞的作用。体外实验表明,花粉管在高钙环境中尖端壁增厚而终止生长,亚尖端易穿孔而破裂[18]。花粉管进入胚囊后,甘蓝型油菜助细胞中钙含量激增至授粉前的 2.4 倍。退化助细胞中超常含量的钙无疑对花粉管在其中终止生长、破裂与释放内容物有促进作用。本文发现,花粉管尖端壁一侧富含钙沉淀的部位,正是花粉管破裂的部位,看来不是一种巧合。

不同植物中助细胞退化具有各自的时间和空间特点[16]。向日葵的 1 个助细胞在传粉后受精前开始退化[19],与此相关,该助细胞也是在这一时期钙含量剧增[12]。禾本科植物小麦和珍珠谷的两个助细胞在受精前钙含量没有明显差别[8~10]。本文的观察表明,甘蓝型油菜传粉前两个助细胞钙含量和分布方式相似,显露出退化的迹象,传粉作用有助于加快助细胞的退化。花粉管进入胚囊前,两个助细胞在退化程度和钙沉淀大小上已出现差异,其中 1 个助细胞退化程度较大,而钙沉淀颗粒非常细小,正是该助细胞接受花粉管进入。这与 Tian 和 Russell 在烟草中的观察结果一致[20]。由此可见,除了钙浓度的增加,钙沉淀的大小似乎也与助细胞的退化程度相关。退化程度愈大,钙沉淀愈小。这种钙沉淀大小的变化可能涉及钙在细胞内的存在形式及其功能的关系,值得进一步探讨。

我们注意到在卵细胞和中央细胞两层质膜间的间隙层中有许多

钙沉淀，它们与退化助细胞中的钙连成一个体系。在烟草和百花丹中发现这种间隙区带中含肌动蛋白成分，在整体上呈"冠状"结构，被认为有引导精细胞到达受精靶区的作用[13]。关于钙与肌动蛋白微丝骨架的关系，前人曾有详细论述，认为它们的相互作用导致花粉管细胞驱动系统的形成[21]。此外，体外试验表明，钙是精卵融合的必要条件[22]。最近 Zhang 和 Liu 等发现钙可以诱导玉米精细胞体外融合[23]，他们认为较高浓度的钙可能诱导细胞膜发生有利于融合的变化。因此可以推测这一间隙层中的钙还可能对促使精卵细胞的融合起重要作用。

本文用图像处理结合细胞立体形态计量的方法对胚囊成员细胞中钙沉淀的体密度和截面圆的平均等效圆直径等参数进行了定量分析，较准确地反映了钙在不同细胞以及不同时期中的相对含量和钙沉淀的大小，这是在胚囊超微细胞化学研究中由定性、定位向定量方向发展的一次新的努力。

(作者：余凡立、梁世平、杨弘远、汪艳。原载：植物学报，1998，40(7)：591～597。图版2幅，表2幅，参考文献23篇，仅选录表1、2)。

5 水稻雌蕊与胚囊中钙的超微细胞化学定位

Ultracytochemical localization of calcium in the gynoecium and embryo sac of rice

Abstract

Potassium pyroantimonate precipitation method was used for ultracytochemical localization of calcium in rice (*Oryza sativa* L.) stigma, style, ovary and embryo sac before and after fertilization. To identify the nature of the pyroantimonate deposits, the energy-dispersive X-ray microanalysis (EDXA) was employed and the deposits observed were proved as calcium pyroantimonate. Transmission electron microscopic observations revealed that abundant deposits were present at the surface of stigma papillae and parenchymatous cells of style. One day before anthesis, when the embryo sac was immature and contained a few deposits, calcium distribution showed no difference between two synergids. By the time just prior to anthesis, one of the synergids that had degenerated accumulated more calcium than before, and than the other synergid. Six hours after anthesis at time when double fertilization had finished, the calcium deposits in the degenerated synergid increased. Before fertilization, the deposits in the egg cell were mainly localized in the vacuoles, whereas few deposits could be observed in the nucleus and cytoplasm. After fertilization, the amount of calcium increased dramatically in the zygote, especially in its nucleus. The results are discussed in relation to the role of calcium in mediating synergid degeneration and egg activation.

钙是植物体内重要的第二信使,其在植物体内的分布及浓度变化,调节着许多代谢和发育过程[1]。钙在植物有性生殖过程中的作用已成为植物生殖生物学研究的一个重要内容。从花粉在柱头上萌发到受精,是一个复杂的过程,涉及多种细胞组织间的相互作用。前人研究表明,钙在上述过程的各个环节中均起重要的作用[2~5]。但以往的研究多侧重于钙对离体花粉萌发和花粉管生长的作用及钙在助细胞中的分布与变化,而作为雌性生殖单位中最重要成员的卵细胞,钙在其中的分布及授粉、受精前后变化的研究却十分欠缺。另一方面,就研究对象而言,作为重要粮食作物的水稻,雌蕊及胚囊中钙的研究全属空白。本文用超微细胞化学方法比较系统地研究了水稻雌蕊组织尤其是胚囊中钙的分布特点。

材料与方法

供试材料为水稻(*Oryza sativa* L.)品种"春江05"。取即将开花和开花后3h的柱头和花柱,以及开花前1d、即将开花、开花后6h的子房,按焦锑酸盐沉淀法制样[6]。材料先后用含2%焦锑酸钾的戊二醛和锇酸固定,乙醇系列脱水,环氧丙烷过渡,Epon 812树脂包埋。在Sorvall MT-6000型超薄切片机上用玻璃刀切片,不染色或经醋酸双氧铀染色30min后,置于JEM-100 CX/Ⅱ型电子显微镜下观察与摄影。X射线能谱微区定性分析依张劲松等[4]的方法进行,结果证明所分析的沉淀确为焦锑酸钙沉淀(以下简称钙沉淀)。

结 果

一、柱头和花柱

即将开花时,柱头乳突细胞外表的细胞壁中含有许多钙沉淀,且主要分布在其外方靠角质层一侧。乳突细胞相连处的胞间隙中钙沉淀尤为密集,细胞内沿质膜也分布许多钙沉淀(图版Ⅰ,1)。柱头乳突细胞下的薄壁细胞内外也富含钙沉淀。花柱表皮细胞壁中则没有钙沉淀。花柱为不具引导组织的闭合型,除维管束外,均为结构相似的薄壁细胞组成(图版Ⅰ,2)。这些细胞纵向延长,其壁在局部地方向内形成突起,具传递细胞特征。从图版Ⅰ,3上可看出,细胞壁上有许多细小的钙沉淀,而壁内突上钙沉淀颗粒明显较大。细胞内钙沉淀主要分布在核中,胞质中相对较少,并主要贴近质膜分布。值得注意的是,在紧邻质膜处常分布有富含膜钙的小泡,显示细胞内外可能存在旺盛的钙交换。维管束中也含钙,主要分布在导管增厚壁的内侧(图版Ⅰ,4)。不同切段的花柱,钙的分布规律一致。开花后3h,柱头和花柱中的钙分布没有明显变化。

二、珠被和珠心

观察到的水稻珠被和珠心结构与前人报道的一致[7]。即将开花时,内、外珠被细胞中均有钙沉淀,主要集中于液泡中贴近液泡膜分布,而胞基质和细胞壁中皆无钙沉淀。珠孔端珠心细胞的细胞质浓厚,其中钙沉淀也很少。开花后3h,在子房壁和外珠被之间以及珠孔、珠心处观察到花粉管。图版Ⅰ,5 示在子房壁和外珠被间生长的一根花粉管顶端小泡密集区段的纵切面,可以看出与花粉管相邻的子房内表皮细胞含钙丰富;外珠被细胞钙沉淀仍然不多,但紧邻花粉管的壁上可见钙沉淀分布;花粉管本身的钙沉淀也不多,多糖小泡中几乎没有钙沉淀。图版Ⅰ,6 示另一花粉管富含细胞器的较后区段的纵切面,可见到除花粉管周围有密集的钙沉淀外,花粉管的管壁、线粒体中亦富含钙沉淀,而小泡中含钙较少。

三、助细胞

助细胞极性明显,细胞质主要分布在珠孔端,众多小液泡主要集中在合点端,细胞核居中部靠近侧壁[8]。开花前1d,2个助细胞在超微结构和钙沉淀分布上没有明显差别,均只含少量的钙沉淀(图版Ⅱ,7)。它们主要分布于细胞质中,线粒体、质体等细胞器中均有(图版Ⅱ,8)。合点端除液泡中含许多絮状物外,钙沉淀仍主要分布于胞基质中(图版Ⅱ,9)。细胞核、丝状器以及2个助细胞的公共壁上均少见钙沉淀分布。即将开花时,2个助细胞差别显著。其中一个助细胞呈退化状态,电子密度和钙沉淀明显增加,合点端延伸回抱卵细胞,原有的细胞极性丧失,小液泡均匀分布。核崩溃不可分辨,线粒体和质体等细胞器中钙沉淀明显增多,胞基质和小液泡中则无钙沉淀分布,并且液泡中的絮状物已消失(图版Ⅱ,10)。另一个助细胞(即宿存助细胞)保持原有的极性状态,钙分布也无明显变化。开花后6h,退化助细胞进一步加剧退化,细胞器已全部解体。其珠孔端除原有的电子致密物质外,还有许多不含钙沉淀的类圆形小泡,它们与图版Ⅰ,4中观察到的花粉管尖端中的多糖小泡的结构类似,推测是花粉管的释放物。珠孔端细胞质中钙沉淀也明显增多,这些钙沉淀颗粒一般很细小,聚集成簇,似乎与解体的线粒体等细胞器相联系(图版Ⅱ,11)。合点端主要是高电子密度的无定形物质,其中也有许多钙沉淀(图版Ⅱ,12)。宿存助细胞珠孔端中各种细胞器的状态看不出明显变化,钙沉淀未见明显增加,但其合点端液泡区的钙沉淀则较前两个时期明显增多。

四、卵细胞

水稻卵细胞呈梨形,无明显极性,核位于细胞中部,细胞质大多集中在核周围,而液泡则分散在其外围[8]。开花前1d,卵细胞中含有少量钙沉淀,多分布在液泡中,并且与液泡膜相联系。而细胞核、胞基质以及线粒体、质体等细胞器中几乎没有钙沉淀(图版Ⅲ,13)。临开花前,钙沉淀的分布变化不大,液泡仍是含钙最多的部位(图版Ⅲ,14)。此时,部分质体中出现淀粉粒,但仍无钙沉淀。开花后6h,卵细胞已受精形成合子[7,8],此时钙沉淀分布与前述两个时期相比有明显不同。最显著的变化就是细胞核中出现大量的含钙颗粒,其核质中的钙沉淀明显较核仁中大;质体和线粒体中也有钙沉淀;液泡中钙沉淀仍是贴近液泡膜分布(图版Ⅲ,15、16)。另外,即将开花时,卵细胞合点端和中央细胞珠孔端间有一个间隙层,其中电子密度较高的部分也含有钙沉淀。开花后6h,该间隙层中的钙沉淀似乎略有增加(图版Ⅲ,16)。

五、中央细胞

开花前1d,中央细胞内和胚囊壁上含钙均很少。即将开花和开花后6h,胚囊壁及相邻的珠心细胞中钙沉淀增多,中央细胞的液泡中也可观察到少量钙沉淀(图版Ⅲ,17)。

六、反足细胞

水稻反足细胞多数形成多细胞群体,其外有细胞壁包围,与珠心相邻的细胞壁与此处胚囊壁合为一体,形成显著加厚的多层结构,并有许多壁内突[8]。反足细胞胞质内的钙沉淀几个发育时期没有明显差异,均仅有少量零星的钙沉淀。开花后6h,反足细胞群与珠心相邻处的厚壁上有许多钙沉淀,与中央细胞相邻的细胞壁上也有少量,而群体内部各个细胞之间的界壁上则没有(图版Ⅲ,18)。

讨 论

前人研究表明,外源钙对离体条件下花粉萌发和花粉管生长起重要作用[9,10]。对甘蓝型油菜[11]、向日葵[12]、陆地棉[4]等几种植物雌蕊组织钙分布的超微细胞化学定位显示,凡是在花粉萌发与花粉管生长所必经的部位,均有高含量的钙,并且钙多分布于质外体系统中。但迄今关于不具引导组织的花柱中钙的分布则未见报道。本文观察表明,水稻柱头表面和不具花柱引导组织的花柱的薄壁细胞中均存在钙。花柱薄壁细胞普遍有壁内突现象,表明细胞内外存在旺盛的物质交流。而紧邻质膜处存在的富含钙沉淀的小泡,暗示其可能向细胞壁和胞间基质分泌含钙物质。花粉管进入子房后在子房壁与外珠被之间生长,研究结果似乎表明子房壁的内表皮较外珠被更多地向花粉管提供钙。

开花前1d,水稻的胚囊发育尚未成熟,此时整个胚囊中钙沉淀都很少,2个助细胞在钙分布上也未见明显差异,这与Tian和Russell在烟草中的观察结果一致[13]。Chaubal和Reger[14~16]发现珍珠谷授粉前2个助细胞均已开始退化,且均含有超量的钙;未授粉子房在植株上保留2~3d后,2个助细胞均进一步退化,并且钙的增加与助细胞的退化程度相关。由此认为珍珠谷助细胞退化是一种与钙相关的自发的程序化死亡过程。水稻即将开花时1个助细胞已退化,其钙含量也明显增加,高于宿存助细胞。这与同属禾本科的珍珠谷及小麦[17]2个助细胞同时退化的情况有所不同。但此时尚未授粉,助细胞的退化显然不是由花粉管所诱导的,这又和烟草[13]及向日葵[18]的情形不同。这是否暗示水稻助细胞的退化也是一种胚囊内钙调节的自然程序化死亡过程? 开花后6h,花粉管已进入退化助细胞并在其中释放内容物。此时该助细胞退化程度更深,钙含量亦进一步增加。这与Chaubal和Reger[17]在小麦中的观察结果不一致。他们发现小麦的助细胞在花粉管进入并释放内容物后,其钙含量大大降低,并认为这有助于阻止额外的花粉管进入胚囊。但是，在向日葵[18]、烟草[13]和油菜[8]等植物中，花粉管在退化助细胞中释放内容物后，该助细胞钙含量皆有所增加，水稻亦属同样情况。

卵细胞受精前处于生理上不活跃的状态，通常需要通过受精激活方能启动合子分裂和胚胎发生[2]。就被子植物而言，已有的研究工作尚未揭示钙在配子融合这一关键环节中的作用机制。前人对向日葵[18]和烟草[13]胚囊受精前后钙分布的超微细胞化学定位表明，卵细胞和合子中均有钙。但不同植物卵细胞中钙的分布方式有各自的时空特点。烟草受精前后，卵细胞中含钙量和分布方式变化不大，作者认为烟草胚囊中的钙可能主要与花粉管到达、释放内容物及配子转移相关，而与早期的合子发育关系不大[13]。向日葵授粉前，钙多分布在细胞核中，珠孔端液泡中则含有絮状钙沉淀。传粉和受精后，细胞核中钙含量略有增加，且合子核中有大的圆球体钙沉淀分布，而液泡中不再有絮状沉淀物[18]。本文观察表明，水稻卵细胞受精后其细胞核和细胞质中钙沉淀明显增加。这印证了 Dignnets 等最近在玉米中的实验结果。他们利用玉米离体受精系统，将荧光染料 Fluo-3AM 导入卵中观察到和动物中类似的现象，即卵细胞受精后其钙浓度也瞬间激增，并认为精卵融合是导致卵细胞钙激增的必要条件[19]。本文虽然没有捕捉到水稻体内精卵融合瞬间卵细胞中钙的变化状况，但是从合子中钙的明显增加，也可以从另一侧面推测受精是导致水稻卵细胞中钙增加的原因。动物方面的研究表明：受精卵内钙的瞬间激增是卵细胞激活的关键因素之一[20]，那么植物受精卵中钙的增加与卵细胞的激活是否存在类似关系？另外，动物中的研究还表明，精子产生某种耐热因子，诱导卵细胞释放贮存的结合钙，或开启钙通道促使外源钙进入卵内。植物受精卵中钙的增加是否也存在类似机制？这些都有待体内的原位研究和离体实验研究相结合深入探讨。

（作者：余凡立、赵洁、梁世平、杨弘远。原载：植物学报，1991，41(2)：125~129。图版3幅，参考文献20篇，均删去。）

6 钙调素 mRNA 和蛋白在水稻花药和雌蕊发育过程中的原位定位

In situ localization of calmodulin mRNA and protein in the developing anthers and pistils in rice

摘 要

利用 RNA 原位杂交和免疫组织化学定位技术分别检测了钙调素 mRNA 和钙调素蛋白在水稻(*Oryza sativa* L.)花药和雌蕊发育过程中的时空分布特征。钙调素基因在绒毡层、柱头、花粉管生长途径、退化助细胞以及维管薄壁细胞中大量表达,也可在小孢子母细胞、小孢子、花粉、反足细胞、卵细胞以及中央细胞中检测到。钙调素基因的表达强度随不同的发育阶段而变化:花药发育早期表达强,以后逐渐减弱并向特定部位集中,如绒毡层和花粉萌发孔等。胚胎发育早期,钙调素基因在胚乳细胞中的表达比原胚中强,而后期则在分化胚中比胚乳细胞中强。推测在有性生殖过程中,钙调素可能通过 Ca^{2+}-CaM 信号途径调节小孢子发育、花粉萌发、花粉管生长、受精以及物质运输等生理过程。

CaM, a well-characterized Ca^{2+}-binding protein presents in all plants and animals, is believed to play a key role in Ca^{2+}-signaling pathways in plants. CaM regulates many cellular and physiological functions, including ionic balance, microtubule behavior, gene expression, carbohydrate metabolism, protein storage, cell division and cell differentiation. Regarding sexual plant reproduction, CaM has been reported to have intimate relation to such processes as pollen germination and pollen tube growth[1,2], prefertilization events in sperms[3] and embryogenesis[4].

Although CaM is ubiquitous among eukaryotic cells, the cellular concentrations of CaM vary from tissue to tissue and can fluctuate depending on the physiological state of the cell[5] as well as with environmental and hormonal stimuli[6,7]. Localization of CaM mRNA and protein in plant tissues is one way to investigate the physiological functions of CaM during plant growth and development. Willemse[8] found that calcium and CaM were present in the tapetal cells during anther development in *Gasteria*, and postulated that the tapetum stored and supplied calcium during anther development. Tirlapur et al[9] visualized the presence of CaM in the egg, synergid, central and antipodal cells in *Petunia hybrida* and *Nicotiana tabacum* using chlorotetracycline (CTC) and fluphenazine (FPZ) as fluorescent probes. Breton et al[10] confirmed the CaM expression throughout maize embryogenesis at mRNA and protein levels using Southern and Western blots. However, they did not report the spatial changes of CaM distribution. *In situ* RNA hybridization is the most direct way of examining the modulation of gene expression during the development at cellular level and has been widely applied to localize RNAs in plant tissue. In the present report we describe the temporal and spatial expressions of CaM at the mRNA and protein levels in the developing anthers and pistils in rice. The relationship between CaM distribution and its possible physiological functions during sexual reproduction is discussed.

Materials and methods

Preparation of tissue sections

Field-grown rice (*Oryza sativa* L. cv. Chunjiang) anthers and pistils were collected before and after anthesis at intervals, fixed in FAA (3.7% formaldehyde, 50% ethanol, 3% acetic acid), dehydrated in a seven-step tertiary butyl alcohol (TBA) series[11] and embedded in paraffin (Paraplast

Plus, Sigma). Embedded tissues were sliced into serial 8 μm sections with an AO 820 microtome. Sections were attached to slides coated with poly-lysine hydrobromide (Sigma) and dried at 45℃ overnight.

RNA *in situ* hybridization

The CaM cDNA, inserting in *Sal* I /*Not* I sites of pBluescript II SK⁺, with T7/T3 RNA polymerase promoters, kindly provided by Dr. Nagamura, was isolated from rice callus by Sasaki et al[12]. Transformation, amplification and isolation of the plasmid were done as described by Sambrook et al[13]. The receptor strain was E. *coli* TG1.

The tissue sections were *in situ* RNA hybridized as previously described[14]. Plasmid was linearized with *Sal* I or *Not* I and transcripted *in vitro* into antisense-RNA or sense-RNA (control) with degoxigenin (DIG) labeling, using T3 or T7 polymerase, respectively. After prehybridization, the sections were hybridized with antisense RNA or sense RNA. Immunological reaction was carried out with anti-DIG-antibody conjugate followed with several rinses. Color reaction was done with nitro blue tetrazolium salt (NBT) and 5-bromo- 4-chloro-3-indolyl-phosphate (BCIP). The sections were examined and photographed under a microscope with blue-purple color as positive reaction.

Immunohistochemical localization

Paraffin removal and section hydration were done as described above. Sections were washed in phosphate buffered saline (PBS, 135 mmol/L NaCl, 2.7 mmol/L KCl, 1.5 mmol/L KH_2PO_4, 8mmol/L Na_2HPO_4, pH7.4) for 10 min, and then immersed in 4% H_2O_2 in PBS for 20 min to abolish endogenous peroxidase activity. After washing three times in PBS for 10 min each, the sections were incubated overnight at 4℃ with rabbit antiserum against bovine brain CaM[15] (a gift from Professor SUN Da-Ye) diluted (1:100) in PBS containing 5% BSA. Sections were washed in

PBS three times for 10 min each, and then incubated with peroxidase-labeled second antibody at a 1 : 100 dilution for 1 h at room temperature. After subsequent rinses, the sections were incubated with 0.05% 3,3'-diaminozidine-4HCl, 0.01% H_2O_2 in 0.05 mol/L Tris-HCl buffer (pH 7.2) for 5 to 15 min. The color reaction was stopped by immersing the sections in PBS for 20 min, and washed twice with H_2O for 10 min each. Then the sections were dehydrated, sealed and observed as described above. For controls, sections were incubated either with only the second antibody or with primary antibody treated with an excess of CaM.

Results

For *in situ* RNA hybridization, the sections using antisense mRNA probes were stained bluepurple in color, but the sections using sense mRNA probes (control) only showed a faint background (Pl. III:23). For *in situ* immunohistochemical localization, the positive sections showed brown but the controls colorless (Pl. III:24).

Anther and pollen

During early developmental stages of the anther, CaM mRNA and protein were detected ubiquitously in the anther wall, anther sac and anther septum, especially in the tapetum and the vascular bundle of the anther septum (Pl. I:3). As development proceeded, the expression level decreased gradually. By the mature pollen stage, as for anther wall and septum, CaM mRNA and protein were distributed only at the peripheral region of endothecium cells (Pl. I:4,6) and at the vascular bundle.

The difference of distribution patterns between CaM mRNA and protein was that CaM mRNA was localized mainly in the nuclei, especially in the nucleolei, but CaM protein in the cytoplasm.

From meiocyte, microspore tetrad, microspore to pollen, CaM mRNA was detected in both nuclei and cytoplasm but colored darker in the nuclei

(Pl.Ⅰ:1,2). In the nearly mature pollen, CaM mRNA was localized in the vegetative and generative nuclei, whereas in the mature pollen it was in the vegetative and two sperm nuclei (Pl.Ⅰ:4). In contrast, CaM protein was localized in the cytoplasm, and showed polar distribution. In the nearly mature pollen, CaM protein converged gradually toward the germination apertures (Pl.Ⅰ:5), and in the mature pollen, it was centred in the germination aperture zone (Pl.Ⅰ:6).

Ovary and embryo sac

In the ovary, CaM mRNA was distributed in its outer and inner epidermis, transfusion parenchyma cells of the vascular tissue, integuments and nucellus (Pl.Ⅲ:22). It was localized mostly in the nuclei of the nucellar cells (Pl.Ⅱ:10) but in the cytoplasm of other cells. The distribution of CaM protein was similar to that of CaM mRNA and showed more intense coloration in the cytoplasm.

As for the embryo sac, the antipodal cells always expressed CaM mRNA (Pl.Ⅱ:10) and protein (Pl.Ⅱ:13). CaM mRNA was localized in the nuclei of immature antipodal cells but showed similar level in both nuclei and cytoplasm at their later stages. All central cell, egg cell and synergids showed high level of CaM mRNA (Pl.Ⅱ:11,12) and protein (Pl.Ⅱ:14, 15). Among them, the highest level was observed in the degenerated synergid (Pl.Ⅱ:12,15). CaM mRNA was localized abundantly in the cytoplasm of synergid, but in the nucleoli of the others. In contrast, CaM protein always showed higher level in cytoplasm than in nuclei of the cells mentioned above (Pl.Ⅱ:14,15).

Pollination and fertilization

In young stigma cells, CaM mRNA was localized in the nucleoli (Pl. Ⅰ:1), while in the mature stigma cells, both CaM mRNA and protein were distributed abundantly in the cytoplasm (Pl.Ⅰ:7). During pollen germination on stigma, both pollen tube and stigmatic hair showed high levels of CaM mRNA and protein (Pl. I:8). From style to ovary, CaM mRNA was observed along the pollen tube track. Interestingly, CaM mRNA was expressed where pollen tube tip traversed (Pl. I:9).

During fertilization, when each of the two sperm nuclei adhered to the egg nucleus and polar nuclei respectively, CaM mRNA was substantially expressed in the nucleoli of the sperm cells, egg cell and polar nuclei (Pl. III:16), but CaM protein in the cytoplasm.

Embryogenesis

After fertilization, the zygote divided and developed into proembryo and then underwent embryo differentiation. CaM was expressed at a lower mRNA level and somewhat higher protein level (Pl.III:18). Fertilized polar nuclei divided and developed into endosperm (Pl.III:17). The expression level of CaM was higher in the endosperm than in the embryo during early development, but this situation was reversed during embryo differentiation (Pl.III:19). The outer and inner epidermis of the immature caryopsis as well as the transfusion parenchyma cells of the vascular tissue still expressed high levels of CaM (Pl.III:21,21).

Discussion

Although CaM distribution was studied in several cereal crops such as maize[10], corn[15] and barley[16], there were few reports on rice which is one of the most important cereal crops. In this study, we systematically localized the expression products of CaM gene at transcription and translation levels during sexual plant reproduction and embryogenesis in rice. The results revealed several temporal and spatial features of CaM gene expression.

Most of the tapetal functions, e. g. , nutrient supply, secretion, production of sporopollenin precursors, are related to pollen formation and its dispersal. Therefore, CaM in the tapetum might not only be related to the tapetum itself but also to the microspore and pollen formation, most likely involving in its secretory processes through Ca^{2+}-CaM pathway, as well as to the differentiation of endothecium wall thickenings (trabeculae), which are considered to be involved in dehiscence of the anther[8].

There was a maximum concentration of membrane-associated Ca^{2+} and also CaM in the vicinity of the germination apertures of the hydrated tobacco pollen cultured *in vitro*, and with the onset of germination, relatively higher amounts of Ca^{2+} and CaM were found to be regionalized towards the aperture through which the pollen tube would emerge[1]. Our results reveal that CaM protein converges toward the germination apertures in the nearly mature pollen and mainly distributes in the germination aperture zone in the mature pollen. These observations suggest that the polar distribution in the maturing pollen and the presence of CaM near the aperture are related to pollen germination through the Ca^{2+}-CaM pathway.

After pollen germinating on stigma hair and pollen tubes penetrating into stigma, the pollen tube and stigma have high expression level of CaM, indicating that it probably correlates with the pollen-pistil interaction. When the pollen tube enters the style, it continue to grow along the pollen tube track. It has been proposed that the extracellular matrix of the style assists pollen tube migration[17] and provides nutritive support for the growing pollen tubes[18]. It has been also proposed that extracellular CaM accelerates during the growth of pollen tubes[19]. Our finding of CaM expression along the pollen tube track suggests that CaM might also be involved in pollen tube migration. Since our results show that CaM mRNA is localized where pollen tube tip just traverses, probably the CaM expression is related to the stimulus from pollen tube penetration.

The degenerated synergid is the one with a significantly high level of calcium in sunflower[20], *Nicotiana*[21] and *Petunia*[9]. This degenerated synergid may regulate the attraction, arrest and discharge of the pollen tube, and the release of sperms into the receptive embryo sac[21]. This calcium regulation of synergid degeneration might be mediated by CaM[22] or ethylene[23]. Our finding also reflects that calcium regulates pollen tube behavior through the Ca^{2+}-CaM pathway in rice.

CaM is increased in proliferating cell population. It is suggested that the increase be required for cell cycle progression[5]. In this study the expression level of CaM is also high in the developing cells. Meanwhile, CaM mRNA is substantially distributed in the nuclei, especially the nucleoli of the developing cells, but expressed similar levels in both nuclei and cytoplasm of the mature cells. In contrast, CaM protein is always expressed a little higher in the cytoplasm than in the nuclei. It is postulated that a great quantity of transcripts of CaM gene is prepared for cell division and protein translation in the developing cells and it has reached a dynamic equilibrium between nucleus and cytoplasm in the mature cells.

Explanation of plates

Ac. Antipodal cell   Ec. Egg cell   En. Endosperm nucleus   Ga. Germination aperture   ms. microspore   Nc. nucellar cell   PN. Polar nuclei   Pt. Pollen tube   SN. Sperm nucleus   Sy. Synergid   Ta. Tapetum   Te. Tracheary element   tpc. Transfusion parenchyma cell   Vn. Vegetative nucleus

Plate I

Figs. 1 to 9. Localization of CaM mRNA and protein in sections of rice anther, stigma and style. All sections except the section shown in Fig. 3 (transection) are longisections. **Fig. 1.** Anther and stigma at meiocyte stage with antisense mRNA probe (ARP). ×350 **Fig. 2.** Anther at tetrad stage with ARP. ×640 **Fig. 3.** Anther at microspore stage with ARP. ×350 **Fig. 4.** Mature tricellular pollen with ARP. ×640 **Fig. 5.** Anther at nearly mature pollen stage with anti-CaM antibody (ACA) treatment. ×350 **Fig. 6.** Anther at mature pollen stage with ACA treatment. ×350 **Fig. 7.** Stigma before anthesis with ACA treatment. ×175 **Fig. 8.** Pollen germinating on stigma with ARP. ×350 **Fig. 9.** Pollen tube growing in style with ARP. ×350

第四章 有性生殖的超微结构与细胞化学研究

图版 I

plate I

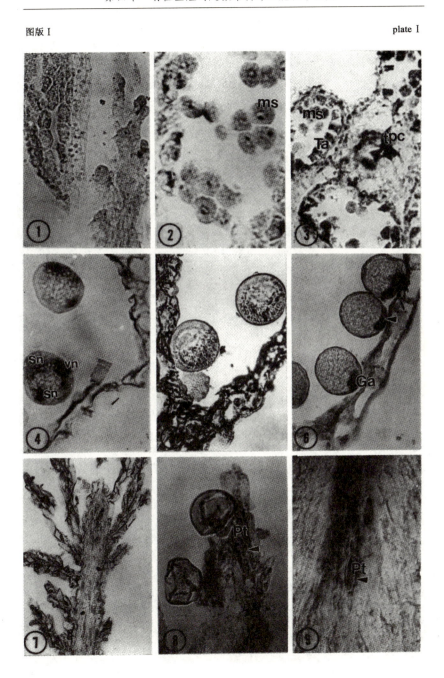

图版 II   plate II

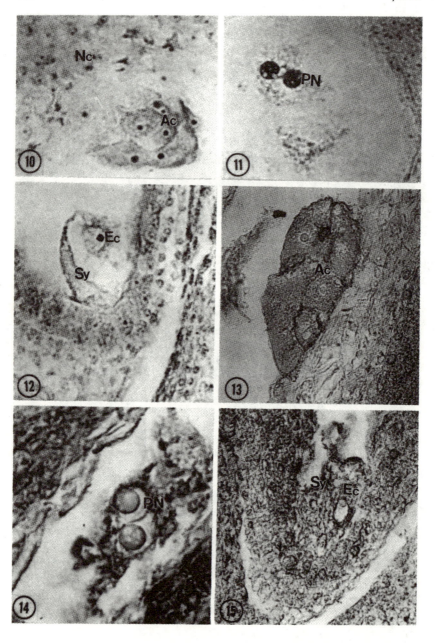

第四章 有性生殖的超微结构与细胞化学研究

图版III　　　　　　　　　　　　　　　　　　　　　　　plate III

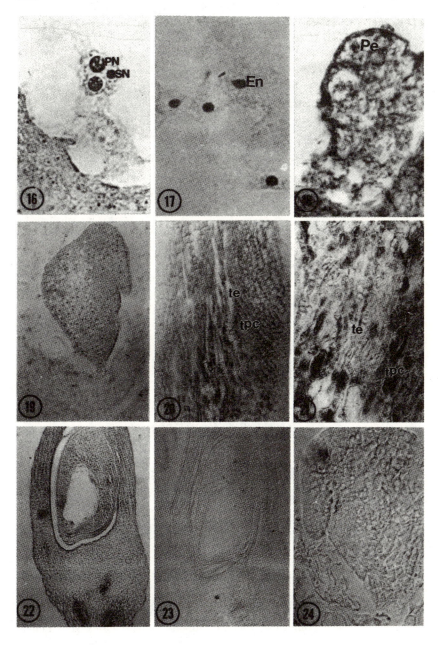

Plate II

Figs. 10 to 15. Localization of CaM mRNA and protein in sections of rice ovary. All are longisections. **Fig. 10.** Nucellus and antipodal cells with ARP. ×640 **Fig. 11.** Central cell with ARP. ×640 **Fig. 12.** Egg apparatus with ARP. ×640 **Fig. 13.** Antipodal cells with ACA treatment. ×350 **Fig. 14.** Central cell with ACA treatment. ×840 **Fig. 15.** Egg apparatus with ACA treatment. ×450

Plate III

Figs. 16 to 24. Fig. 16. One sperm nucleus adhered to two polar nuclei and another with egg nucleus on adjacent section, with ARP. ×640 **Fig. 17.** Endosperm cells with ARP. ×700 **Fig. 18.** Proembryo with ACA treatment. ×840 **Fig. 19.** Differentiating embryo with ARP. ×350 **Fig. 20.** Transfusion parenchyma cells with ARP. ×175 **Fig. 21.** Transfusion parenchyma cells with ACA treatment. ×350 **Fig. 22.** Section of ovary with ARP. ×100 **Fig. 23.** Section of ovary with sense mRNA (control). ×100 **Fig. 24.** Differentiating embryo without anti-CaM (control). ×350

(Authors: Chen SR, Han HM, Lu YT, Yang HY. Published in: Acta Botanica Sinica, 1998, 40(12):1087-1092, with 3 plates and 23 references. All the plates are retained here.)

7 烟草受精前后胚囊中钙调素的免疫细胞化学定位

Immuno-localization of calmodulin in unfertilized and fertilized embryo sacs in *Nicotiana tabacum* var. macrophylla

Abstract

Light microscopic immunohistochemical techniques with horse radish peroxidase (HRP) conjugated second antibody and protein A-gold immunoelectron microscopic techniques were used to study the distribution of calmodulin (CaM) in unfertilized and fertilized embryo sacs in *Nicotiana tabacum* var. *macrophylla*. Before fertilization, CaM was richer in the egg apparatus cells and antipodal cells than in the central cell. During the course from pollination to fertilization, the persistent synergid contained more CaM than the degenerated synergid. Meanwhile, two distinct bands rich in CaM were observed between the egg apparatus and the central cell, and gradually fused with each other appearing arc shape. When the two polar nuclei had fused, this CaM-rich band began to disappear. After fertilization, CaM level was still high in the zygote and the persistent synergid but low in the endosperm cells. Although there was no evidence about the polar distribution of CaM in the zygote, distinguishable difference, however, existed between the apical cell and the basal cell of a proembryo, being higher in the former than in the latter. The function of CaM during double fertilization and early embryogenesis as well as the temopral relationship between the CaM-rich band and the actin corona reported by other investigators are discussed.

　　钙调素(CaM)是一分子量小、耐热、耐酸、具高度保守性的钙调节蛋白,因其在细胞信号转导系统中所处的重要地位而有广泛深入的研究。前人对其在多种植物中细胞内定位已有详尽描述[1],细胞外钙调素的研究也引起了越来越多的重视,但对于植物胚囊在受精与胚胎发生这一重要过程中钙调素的定位迄今仅见于 Timmers 与 Schel[1]对萝卜体细胞胚与合子胚所作的荧光标记及免疫细胞化学定位的比较研究,以及 Tirlapur 等[2,3]在矮牵牛、烟草和玉米的雌性细胞中作的荧光染色观察。最近,赵洁等[4]应用氟备乃静(FPZ)荧光探针研究了小麦分离合子与幼胚中 CaM 的分布特点[4]。本文首次利用分离的烟草胚囊、合子与幼胚进行了 CaM 光镜水平的免疫细胞化学定位,并与电镜水平的工作相印证,为研究其对受精过程的调控提供线索。

材料与方法

一、光镜免疫细胞化学定位

取大叶烟草(Nicotiana tabacum var. macrophylla)开花前 1d 和授粉后 1~8d 的胚珠,在 0.1 mol/L 磷酸盐缓冲液(PBS,pH7.4)配制的 4% 多聚甲醛中固定过夜,PBS 换洗 3 次,转入含 1.5% cellulase R-10、1% macerozyme R-10、0.15 mol/L NaCl 的 PBS 中振荡酶解 3h,用不含酶的缓冲液换洗 3 次,以吸管吸打或以玻璃匀浆器研磨后,挑取受精前后的胚囊、合子或幼胚粘在预先涂有多聚赖氨酸(polylysine,Sigma)的玻片上。参照 Lin 等[6]的方法略加修改,进行 CaM 的免疫细胞化学反应;用 4% H_2O_2 抑制内源性过氧化物酶,样品经 3 mol/L 脲消化 30min,清洗,以卵清蛋白于室温下温育 20min 后,加兔抗 CaM 血清室温下 2h,PBS 洗 3 次,再加辣根过氧化物酶(HRP)标记的兔抗血清(goat anti rabbit IgG-HRP,SABC),室温下 90min,PBS 洗 3 次,然后用 0.2 mol/L Tris-HCI 缓冲液(pH 7.2)配制的 0.03% 3,3′-二胺基联苯胺(DAB)附加 0.01% H_2O_2 显色,PBS 洗后用 50% 甘油封片。Olympus BH2-RFCA 显微镜下观察摄影。

二、电镜细胞化学定位方法

按照李家旭等①的方法并加以修改:取开花当天、4d、8d 的胚珠,迅速浸入固定液(4% 多聚甲醛、1% 戊二醛溶入 0.1 mol/L 二甲胂酸钠缓冲液,pH 7.2),抽气后转入 4℃冰箱过夜,PBS 洗后,以 0.5% OsO_4 固定 30min,PBS 洗,按常规方法用各级乙醇脱水,环氧丙烷置换,Epon 812 渗透、包埋,在 Sorvall MT-6000 型超薄切片机上切片,捞在覆有 Formvar 膜的 50~100 目镍网上。将镍网载片面向下悬在 10% H_2O_2 小液滴上,室温下蚀刻 10min,吸干镍网上 H_2O_2 液,悬在阻断液滴(0.2% mol/L 甘氨酸、2% 卵清蛋白溶于含 0.02% Tween-20、0.02% NaN_3 的 PBS 中)上,室温下 20min,用上述 PBS 洗 30s,悬在兔抗 CaM 血清的液滴上,30℃温育4h,PBS洗,晾干后置金标 protein A 液滴上(胶体金直径10nm),30℃温育1h,PBS洗,双蒸水洗,经醋酸铀染15min与柠檬酸铅染3min后,在JEM-100CX/Ⅱ型透射电子显微镜下观察摄影。

三、对照实验

作3组对照:(1)用纯化的花椰菜(CaM预饱和抗CaM抗体);(2)用正常兔血清替代抗CaM抗体;(3)用PBS替代抗CaM抗体。

结　果

光镜免疫细胞化学观察显示,未受精胚囊中,位于珠孔端的卵细胞与助细胞和位于合点端的反足细胞染色较深,其CaM含量明显高于中央细胞(图版Ⅰ,1),电镜观察结果进一步显示,卵细胞的细胞壁、细胞质及其中的线粒体、质体、内质网等细胞器和占据细胞很大部分的细胞核中均分布有CaM免疫金颗粒,液泡中有时能看到少量金颗粒(图版Ⅱ,13)。助细胞中CaM主要集中于近珠孔端富含细胞质的部分,在线粒体、内质网等细胞器以及细胞核上均有分布(图版Ⅱ,14),其细胞壁上的金颗粒也多位于近珠孔部分,核仁中则呈负反应。中央细胞的珠孔端富含细胞质区域中能看到少许金颗粒(图版Ⅱ,13)。授粉至受精前的时期里,卵器细胞在整个胚囊中所占的比例有所增加,其在整个胚囊长度中的比例由原来约占1/4增加到约占1/2。酶标免疫反应显示,在卵器细胞与中央细胞之间出现了两条组成一夹角的染色极深的区带,标志着CaM集中分布于此处(图版Ⅰ,2、3),当两极核紧贴在一起将要融合时,上述CaM区带连结成一较宽的弧形,染色更深(图版Ⅰ,4)。当极核融合成一个次生核后,核周围有较多的CaM分布,而卵器细胞与中央细胞间的富含CaM区带开始消失(图版Ⅰ,5)。与此同时一个助细胞退化,其CaM含量明显少于宿存细胞(图版Ⅰ,6)。

受精后,初生胚乳细胞先于合子分裂,在2-细胞胚乳中可看见CaM围绕核且呈辐射状伸向细胞质边缘的细胞质索中(图版Ⅰ,7),此时上述富含CaM区带已不复存在。合子分裂前,胚乳细胞已分裂多次,并占据了胚囊中的绝大部分体积,但钙调素含量仍低于合子与宿存细胞(图版Ⅰ,8)。合子中 CaM 含量高但看不出极性分布(图版Ⅰ,9),细胞器丰富,偶有小的液泡,合子与胚乳间的细胞壁较受精前卵细胞与中央细胞的壁完整、清晰,并有所加厚(图版Ⅱ,15)。合子分裂后,2-细胞原胚的细胞间的 CaM 含量存在明显差异,一般顶细胞多于基细胞及其分裂产物(图版Ⅰ,10、11)。电镜下 3-细胞原胚亦呈现 CaM 含量由近珠孔端细胞向近合点端细胞递增的趋势,顶细胞中细胞壁、细胞基质、细胞器中均有丰富的 CaM(图版Ⅱ,16)。胚柄细胞的细胞质中有许多大小不一样的液泡,其中常有些絮状沉淀,质体略多于顶细胞,但 CaM 则分布较少,尤以最近珠孔端细胞中的含量最低。3 组对照实验均呈阴性,表明上述酶标与胶体金标记的免疫实验结果均为 CaM 的特异性染色。图版Ⅰ,12 示光镜细胞化学实验中的对照(2)。

讨 论

关于 CaM 在植物组织中的免疫细胞化学定位,前人的研究均在切片基础上进行。光镜水平的工作证明了 CaM 在植物体内分布具组织特异性[5],电镜观察则进一步揭示了 CaM 在细胞壁、细胞核、多聚核糖体、线粒体、内质网、质膜和质体上的定位。本文借鉴 Huang 等[6,7]以分离的胚囊为材料进行微管免疫荧光定位的经验,利用本实验室建立的烟草胚囊分离技术,分离出完整的胚囊、合子及幼胚,粘片后进行 CaM 的免疫细胞化学反应。和以往程序复杂、视野有限的切片工作相比,此法具有整体性强、操作简便的优点,其观察结果通过胶体金免疫电镜定位得到证实与补充。不过,此方法仅适用于受精前后的胚囊和几个细胞的原胚;对于细胞层数较多的小球形胚及更大的胚而言,仍以切片方法为宜。

CaM 在植物体内的分布具有时、空特异性已在多种植物中得到证实。有人[1,8,9]提出:CaM 的区域性分布和分布数量的多少是 CaM 起调节作用的限制因素之一。本文的研究结果亦支持这一观点。烟草受精前同样作为精子靶细胞的卵细胞与中央细胞之间、受精后宿存助细胞与退化助细胞间、合子与胚乳细胞间、2-细胞原胚的顶细胞与基细胞间，均存在 CaM 含量的差异，表明 CaM 与不同功能和发育前途的细胞有密切的联系。Dreier 等[10]认为，CaM 对植物生长、分化时淀粉的生物合成和降解起作用，质体上存在 CaM 即暗示了这一点。烟草原胚时期的胚乳细胞有很大的中央大液泡，细胞质中质体并不丰富，CaM 的含量也低于合子或原胚，我们的观察表明，此时期的胚乳细胞亦不积累淀粉（未发表资料）。烟草合子分裂前，细胞核、细胞质、细胞器具明显极性分布，而 CaM 的分布看不出明显极性，但当合子分裂成 2-细胞原胚后，顶细胞与基细胞间却存在 CaM 含量的差异。这与顶细胞将发育成一个胚，面基细胞仅分裂成 4 个细胞的胚柄有关。有趣的是，虽然胚柄细胞中质体丰富，但 CaM 含量仍大大低于顶细胞，说明后者中受 CaM 调控的生理生化动态较前者更为丰富，合子中细胞器的重新排列可能即是为此差异作准备。不同类型细胞中发生的生理生化反应不尽相同，可能也是合子与胚乳细胞间、卵细胞与中央细胞间 CaM 含量有不同的原因之一。

值得注意的是，在授粉后受精前，卵器细胞与中央细胞之间出现富含 CaM 的暂时性区带。Huang 与 Russell[7]曾观察到了相似部位肌动蛋白冠（actin corona）的存在，并认为这一结构对于配子转移和融合具有重要作用。已知钙通过调节肌动蛋白抑制蛋白（profilin）和凝溶胶蛋白（gelsolin）的活性影响肌动蛋白微丝的聚合与解聚[12]。由此可推测作为重要钙调节蛋白的 CaM 参与了此过程。Tirlapur 等[12,13]用 CaM 拮抗剂光氧化标记方法和免疫细胞化学定位方法，观察到内源 CaM 在花粉水合、萌发和顶端生长过程中呈极性分布。因此，本文所观察到的 CaM 区带有可能来源于花粉管释放的 CaM，或是由相关的雌性细胞（例如退化助细胞）内的 CaM 向该区带汇集而成，也有可能首先由胚囊内的 CaM 集中于该区域，调节肌动蛋白冠的形成，随后花粉管 CaM 亦释放至此区域导致肌动蛋白冠的解体。鉴于同样作为细胞骨架的微管与微丝的分布有重叠现象[14]，以及钙与 CaM 系统对微管行为的调控作用，有可能受到调节的微管以一定的方式排列，行使其蛋白质转运功能将 CaM 集中或疏散。关于在授粉后至受精前 CaM 区带的起源及动向,它同肌动蛋白冠的形成与消失和受精过程中其它生理活动的相互关系,是一个值得深入研究的问题。

图 版 说 明

A. 反足细胞 C. 中央细胞 CW. 细胞壁 DS. 退化助细胞 E. 卵细胞 EA. 卵器 EN. 胚乳 ER. 内质网 M. 线粒体 N. 细胞核 P. 质体 PN. 极核 PS. 宿存助细胞 SN. 次生核 V. 液泡 Z. 合子

图 版 I

1. 未受精胚囊,示卵器与反足细胞较中央细胞 CaM 丰富。×280 **2、3.** 授粉后胚囊,示卵器与中央细胞间出现两条夹角的富含 CaM 区带。×220 **4.** 授粉后胚囊,示卵器与中央细胞间的 CaM 区带变成弧形,染色加深,合点端的深色区域是胚囊外的附加物(图 5~7 同此)。×280 **5.** 授粉后胚囊,两极核已融合成一个次生核,富含 CaM 区带开始消失。×280 **6.** 宿存助细胞中 CaM 含量高于退化助细胞。×420 **7.** 2-细胞胚乳,示围绕核呈辐射状延伸的细胞质索中的 CaM。×200 **8.** 受精后胚囊,示合子与宿存助细胞中 CaM 含量高于胚乳细胞。×280 **9.** 伸长的合子,CaM 无极性分布。×450 **10.** 2-细胞原胚,示顶细胞中 CaM 高于基细胞。×450 **11.** 3-细胞原胚,示顶细胞中 CaM 高于两个胚柄细胞。×420 **12.** 2-细胞胚乳,示以正常兔血清代替抗 CaM 抗体的对照结果为阴性。×230

Explanation of plates

A. Antipodal cell C. Central cell CW. Cell wall DS. Degenerated synergid E. Egg cell EA. Egg apparatus EN. Endosperm ER. Endoplasmic reticulum M. Mitochondrium N. Nucleus P. Plastid PN. Polar nucleus PS. Persistent synergid SN. Secondary nucleus V. Vacuole Z. Zygote

Plate I

Fig. 1. An unfertilized embryo sac, showing its egg apparatus and antipodal cells with CaM richer than that in the cen tral cell. ×280 **Figs. 2,3.** Embryo sacs after pollination, showing two CaM-rich bands between egg apparatus and central cell. ×220
Fig. 4. Two CaM-rich bands fused into one arc-shape band wore intensively stained. The deeply stained area at the chalazal end is beyond the embryo sac. Similar cases are also

第四章 有性生殖的超微结构与细胞化学研究

图版 I　　　　　　　　　　　　　　　　　　　　　　　　plate I

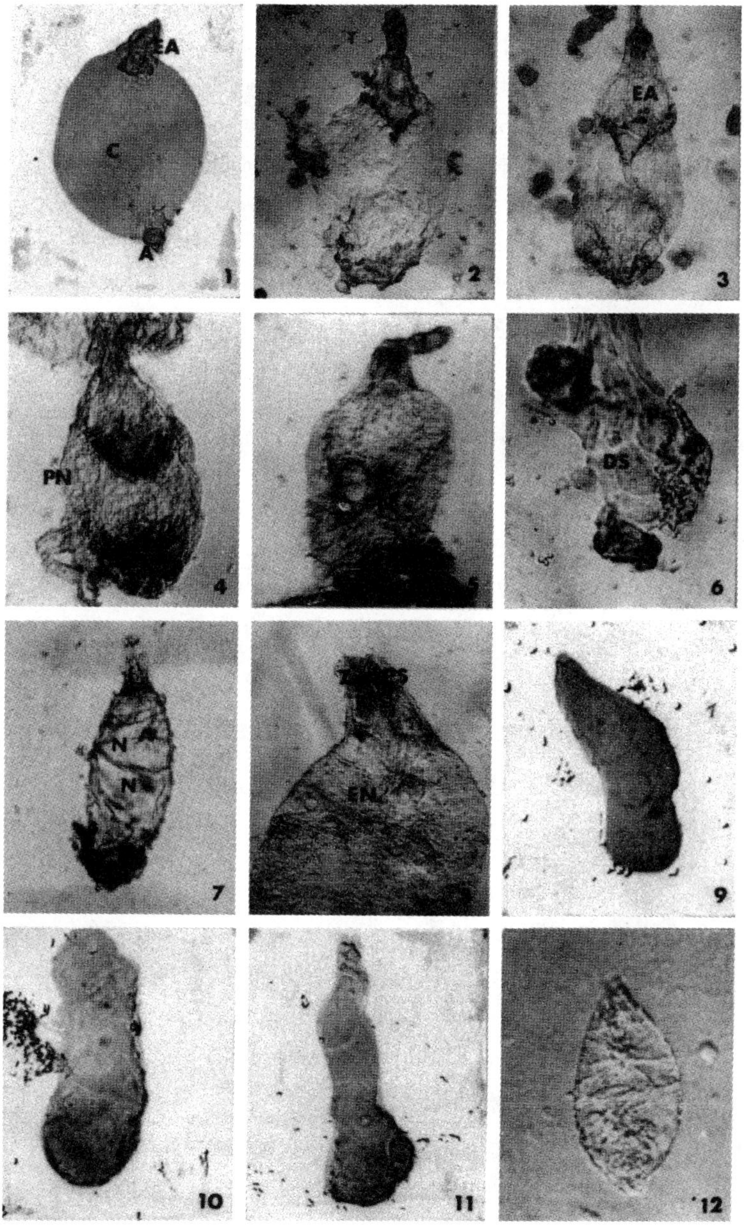

seen in Figs. 5 to 7. ×280 **Fig. 5.** The CaM-rich band begin to disappear while two polar nuclei has fused into a secondary nucleus. ×280 **Fig. 6.** A pair of synergids, showing in the persistent synergid than in the degenerated synergid. ×420 **Fig. 7.** Two-celled endosperm, showing CaM in cytoplasmic strains arround the nuclei. ×200 **Fig. 8.** A fertilized embryo sac, showing more CaM contained in its zygote and persistent synergid than in the endosperm cells. ×280 **Fig. 9.** An elongated zygote, showing its evenly distributed CaM ×480 **Fig. 10.** A two-celled proembryo, showing more CaM in its apical cell than in the basal cell. ×450 **Fig. 11.** A three-cell proembryo, showing a higher level of CaM in the apical cell. ×420 **Fig. 12.** A two-celled endosperm assayed with normal rabbit serum instead of anti-CaM serum as a negative control. ×230

(作者:傅缨、陈以峰、梁世平、杨弘远、周嫦。原载:植物学报,1998,40(8):683~687。图版2幅,参考文献15篇,仅选录图版Ⅰ。)

8 钙调素 mRNA 在受精前后分离的烟草胚囊中的定位

Using isolated embryo sacs and early proembryos for localization of calmodulin mRNA before and after fertilization in *Nicotiana*

摘 要

建立了一种新的 mRNA 原位杂交方法,适用于微量材料的整体观察。应用这一方法定位烟草(*Nicotiana tabacum* L. cv. W38)受精前后胚囊成员细胞中的钙调素 mRNA(CaM mRNA)。结果显示成熟胚囊中的 CaM mRNA 主要分布于珠孔极的卵器和合点极的反足细胞;中央细胞中较少。受精前后胚囊中 CaM mRNA 的分布发生显著变化,特别是授粉后到受精前极核与卵器之间出现一条暂时的钙调素 mRNA 条带。

受精前不久该带消失,CaM mRNA 扩展为占据胚囊珠孔端的扇形区域。受精后胚囊中钙调素 mRNA 主要集中于伸长的合子和原胚的合点端。讨论了钙调素 mRNA 表达与受精的关系。

Embryo sac (ES) is the female gametophyte of angiosperms that is deeply embedded in the ovule tissue. The success of ES isolation has provided a new route for observation on the ESs as a whole in contrast to previous sectioning methods[1]. Calmodulin (CaM), a well-characterized Ca^{2+} receptor, is believed to be involved in plant reproductive processes. Using isolated ESs as materials, CaM was localized by the fluorescent probe fluphenzine (FPZ) in the egg, synergid, central cell and antipodal cells in *Petunia hybrida* and *Nicotiana tabacum*[2]. More recently, a series of investigations have been carried out in our laboratory to localize CaM gene expression products in the ESs during sexual processes. Isolated ESs, zygotes and proembryos were used to detect the distribution of CaM protein in wheat[3] and tobacco[4] by fluorescent probe or immunohistochemical technique respectively. As for the localization of CaM mRNA, *in situ* RNA hybridization is the most direct way for examining the modulation of gene expression at cellular level. In this respect, Chen et al[5] observed the distribution of CaM mRNA in the ES component cells in rice by *in situ* hybridization on conventional paraffin sections. The whole-mount *in situ* hybridization technique has been developed and used for zygotes and embryos of animals[6], *Fucus*[7], as well as for higher plant pollen[8]. Due to the fact that only limited ESs can be yielded from artificial isolation, so far the whole-mount technique has not been applied to ESs yet. In the present report, by adopting a modified *in situ* RNA hybridization technique for the *in vitro* isolated tobacco ESs, we describe the preliminary results on temporal and spatial changes of CaM mRNA during fertilization process.

Materials and methods

1. Plants of *Nicotiana tabacum* L. cv. W38 were grown in a green-

house of Wuhan University. Flowers were harvested at intervals before and after pollination.

2. The ES isolation protocol was based on the technique established[1] and modified[9] in our laboratory. Ovules were fixed in 4% paraformaldehyde in phosphate-buffered saline (PBS), then dehydrated in a seven-step tertiary butyl alcohol (TBA) series[10], washed in PBS and diethylpyrocarbonate (DEPC)-H_2O for 10 min each, then transferred to 1.5 mL maceration mixture and incubated at 30 ℃ in a microshaker for 1 h. The maceration mixture contained 1.5% Cellulase R-10, 1.0% Macerozyme R-10 in PBS, pH 5.7. The ovules were then placed into a drop of DEPC-H_2O and pressed gently with a small glass pestle. ESs liberated from the ovules were collected with a micropipette under an inverted microscope. After three 5-min rinses with DEPC-H_2O to remove nucellus debris, the ESs were transferred directly to slides coated with poly-lysine hydrobromide (Sigma) and dried at 45℃ overnight.

3. The c419 clone provided by Dr Nagamura was isolated from the rice callus cDNA library and contains a 0.8-kb *Sal* I/*Not* I fragment encoding CaM in pBluescript II SK$^+$. Digoxigenin-labeled RNA transcripts were prepared from the linearized vector, according to the manufacturer's protocol (Boehringer Mannheim).

The *in situ* hybridization procedures were adopted according to the method developed by Dow and Mascarenhas[11]. The slides with attached ESs were hydrated in DEPC-H_2O for 2 min, then treated with 0.2 mol/L HCl for 20 min. After washing twice with 2×SSPE and DEPC-H_2O each for 5 min, the slides were incubated with 1% BSA in 10 mmol/L Tris-HCl, pH 8.0 for 10 min. Then BSA was removed by washing twice with DEPC-H_2O for 10 min, and the slides were dehydrated in 30%, 70%, 90% and 100% ethanol for 2 min each and dried at 45℃.

Hybridization was carried out at 60℃ in 10× salt (3 mol/L NaCl, 0.1 mol/L Tris-HCl (pH 6.8), 0.1mol/L Na-phosphate (pH 6.8), 50 mmol/L EDTA), 50% formamide, 12.5% dextran sulfate, 10 mmol/L dithiothreitol (DTT), 500 μg/ml tRNA, 625 μg/ml poly (A) RNA, 200 ng/ml heated denaturalized probe. Excess probe was removed by washing twice in 2×SSC (1×SSC: 150 mmol/L NaCl, 15 mmol/L sodium citrate, pH 7.0) for 1 h each, once in 1×SSC for 1 h, and once in 0.5×SSC for 30 min at room temperature. All the manipulations must be careful to prevent detaching the ESs from the slides.

4. Immunodetection and staining protocol was performed according to Tanimoto and Rost[10]. In short, the slides were washed in buffer I (100 mmol/L Tris-HCl, 150 mmol/L NaCl, pH7.5) for 5 min, incubated in 2% fetal serum (FS), 0.3% Triton X-100 in buffer I for 30 min. The anti-DIG-antibody conjugate was diluted (1 : 500) with buffer I containing 1% FS and 0.3% Triton X-100. Then they were incubated for 2 h with 200 μl of diluted antibody conjugate in a humidified chamber, followed by washing in buffer II (100 mmol/L Tris-HCl, 100 mmol/L NaCl, 50 mmol/L $MgCl_2$, pH 9.5) for 2 min and incubating in the same solution containing 0.5 mL of 0.34 mg/ml nitro blue tetrazolium salt (NBT), 0.175 mg/ml 5-bromo-4-chloro-3-indolyl-phosphate (BCIP) and 100 mg/ml polyvinyl alcohol (PVA) of 70-100 kD (Sigma). The color reaction was allowed to develop for 4-6h and stop in buffer III (10 mmol/L Tris-HCl, 1 mmol/L EDTA, pH 8.0) for 5 min. The slides were dehydrated in ethanol series and finally mounted with Permount (Fisher).

Results and discussion

1. The technique for non-radioactive *in situ* hybridization on isolated ESs reported here is as sensitive as the conventional *in situ* hybridization technique on tissue sections. It allows an easy, direct analysis of the spatial and temporal mRNA expression patterns in the whole ESs. Unlike other whole-mount techniques, it does not need many materials and can apply to such materials as ESs that are difficult to collect.

2. The isolated ESs of tobacco contains an egg cell, two synergids, a central cell and usually three antipodal cells. All these component cells display the distribution of CaM mRNA (Pl. I :1-5). In contrast, the control samples with CaM sense probe does not exhibit blue reaction (Pl. I : 9).

Before pollination, CaM mRNA is mainly distributed in the two poles of the ES, i. e. the egg apparatus (EA), including an egg cell and two synergids at the micropylar pole, and the antipodal cells at the chalazal pole of the ES exhibited more condensed CaM mRNA as compared to the central cell (Pl. I :1). In the egg cell and synergids, the large vacuoles give no staining. CaM mRNA accumulates mainly in the cytoplasm of EA and the position of filiform apparatus (FA) in the synergids (Pl. I :1). After pollination, the ESs show different pattern of CaM mRNA distribution and the polar nuclei exhibit CaM mRNA expression. A most striking change is that a prominent, temporary band of CaM mRNA appears at the position between EA and polar nuclei (Pl. I :2). Just prior to fertilizetion the band disappears and expands to a fan-shaped region that occupies the micropylar portion of the ES. At the same time, the polar nuclei move closer to the EA and position of the fan-shaped region (Pl. I :3). When a pollen tube enters the degenerated synergid, the micropylar portion of ES exhibits stronger color reaction and CaM mRNA accumulates in all ES component cells in which the large vacuoles have already dispersed. Unlike before fertilization, the polar nuclei show the highest level of CaM mRNA when fertilization occurs (Pl. I :4).

After fertilization, the central cell divides in advance of the zygote division, and the persistent synergid begins to degenerate. The zygote and persistent synergid show much higher level of CaM mRNA than the endosperm do (Pl. I :5). At first, CaM mRNA is uniformly distributed in the zygote. As the zygote elongates, it shows uneven distribution of CaM mRNA with higher concentration in its chalazal portion (Pl. I :6). After first division of the zygote, CaM mRNA is more abundant in the apical cell than in the basal cell of the two-celled proembryo derived from this division (Pl. Ⅰ:7). Similar pattern of distribution seems to maintain in the successive multicellular proembryos (Pl. Ⅰ:8).

3. Calcium is believed to be involved in a cascade of signaling events during fertilization. Extremely high level of total Ca^{2+} has been reported in the synergids in which pollen tube discharges its male gametes[12~15]. Ca^{2+} is also indispensable for the *in vitro* intergametic fusion in higher plants[16,17]. Digonnet et al[18] recently reported the evidence of a transient elevation of free cytosolic Ca^{2+} during *in vitro* fertilization. CaM is a ubiquitous Ca^{2+}-binding protein that plays a key role in Ca^{2+}-signaling pathway. Data presented here show that high CaM mRNA level maintains in the EA of ESs that might trigger a series of signaling process related to fertilization. From pollination to fertilization, the distribution of CaM mRNA changes dramatically in the ES, especially forming a prominent band between EA and polar nuclei. Fu et al[4] observed similar arch bands of CaM protein, but with some different shape, in the ES of *Nicotiana tabacum* var. *macrophyllar* by immunohistochemical technique, and postulated that it might have a relationship to the actin corona at the same position which was speculated to play a role in transport of the male gametes toward the target zone of double fertilization[19]. The spatial and temporal relationship of CaM mRNA, CaM protein and cytoskeleton during plant fertilization is still mysterious and requires further investigation.

Upon division of a polarized cell containing localized determinants such as mRNA or proteins, the resulting daughter cells possess different cytoplasmic contents that can specify the fate of each cell and hence the ensuing distinct cell lineages[20]. In the *Fucus* zygotes 8 to 12 h after fertilization, poly (A)$^+$ RNA is asymmetrically redistributed in a gradient which is maintained in two-celled embryos[7]. Recently Love et al[21] reported that photopolarization of *Fucus* zygotes was enhanced following microinjection of recombinant calmodulin. In *Nicotiana*, CaM mRNA accumulation in the chalazal portion of the zygote may possibly guide the preferential distribution of a number of molecules and organelles which later are segregated into the apical cell of a two-celled proembryo. Actin mRNA shows asymmetrical distribution in *Fucus* zygote and plays a key role in its polarization[22]. The role of CaM gene expression in zygote asymmetrical division and early embryogenesis in higher plants needs further study.

(Authors: Li ST, Chen SR, Lu YT, Yang HY, Zhou C. Published in: Acta Botanica Sinica, 1999, 41(7): 686-689. with 1 plate and 22 references, all omitted here.)

第三节 雌蕊中 ATP 酶与植物激素的细胞化学定位

提要

ATP 参与生命活动中的物质代谢、运输、能量转换和信息传递。ATP 酶作为水解 ATP 的酶类在上述各种生命活动中有重要的作用。应用超微细胞化学技术定位植物细胞内 ATP 酶的分布已有广泛报道，其中也包括少数在生殖器官中的定位研究。我们以金鱼草与向日葵为材料，研究了 ATP 酶在胚珠各部组织中的分布状况，重点研究了珠被绒毡层、承珠盘、胚囊壁与壁外间隙层等与胚囊营养有关的部位。结果发现，在金鱼草中，胚囊壁和珠被绒毡层之间的"间隙层"是一个包围胚囊的质外体系统，它和承珠盘中的 ATP 酶构成一个连续的通道，可能起向胚囊输送养料的作用（可参看第二章第二节中"金鱼草珠被绒毡层壁囊的分离与鉴定"一文）。在向日葵中同样存在胚囊壁外周的质外体系统,但其中的 ATP 酶与胚囊壁角质层的间断区域中存在的 ATP 酶相连续,暗示胚囊可以通过这些间断区由外方吸收营养。应该指出,阐明胚囊的营养途径无疑主要依靠生理学方法,但生理学方法难以细致到阐明各种细胞组织在其中所担当的作用,在这一点上,超微细胞化学可以从结构与功能关系的角度予以弥补。

植物激素在受精前后雌蕊与胚囊中的免疫细胞化学定位,在文献中尚未见到。由于植物激素分子量较小,在制样过程中容易流失,最好应用冷冻制样技术。我们应用常规的电镜制样方法进行免疫金定位,在烟草上作了尝试。即使考虑到存在一定的流失,仍然可以对激素的分布有一个大致的认识。所涉及的激素包括生长素、细胞分裂素、赤霉素、脱落酸四大类。研究范围包括从柱头到胚珠的花粉管轨道(配合前期)、受精前后的卵细胞(配合期)与原胚(胚胎发生早期)。这样一次"扫描性"的研究为了解各种植物激素在受精与胚胎发生早期的分布及其消长提供了初步的资料。研究结果是不够成熟的,有待改进定位技术、增加实验处理以及进行定量测定。沿着这一方向继续深入下去无疑对探索受精与胚胎发生的机理有重要意义。

1 金鱼草胚珠中 ATP 酶活性的超微细胞化学定位

The ultracytochemical localization of ATPase activity in the ovules of *Antirrhinum majus* L.

Abstract

The ultracytochemical localization of ATPase activity was carried out by the method of lead precipitation in the ovules of *Antirrhinum majus* L.

No ATPase activity is observed in the egg apparatus, but some in the polar nuclei, cytoplasm and plasma membrane of the central cell. Between the embryo sac wall and the cuticle surrounding it, there is a gap where some filament and vesicle-like structures were demonstrated by conventional staining method, and much of ATPase activity is found there. At the chalaza of the ovule, a lot of ATPase particles are found in the nuclei, plasma membranes and the thick and loose wall of the hypostase cells. The particles of ATPase in the hypostase and those in the gap surrounding embryo sac are continuously distributed through the intervals of the cuticle at the chalazal end of the embryo sac. Some of ATPase particles are found on the plasma membranes and plasmadesmata of integument cells, noticeably much more in the nucleoplasm of the integumentary tapetum. According to the ATPase distribution pattern in the ovules, we suggest that the function of the integumentary tapetum and hypostase is secretion, and that the gap surrounding the embryo sac may be an apoplastic channal for nutrient flow into the embryo sac.

胚囊的发育需要大量的营养。营养物质进入胚囊的途径,以前主要根据胚珠的形态学推测:维管束将可溶性化合物送到胚囊的合点端、助细胞、反足细胞、承珠盘以及胚囊的整个表面都能吸收营养[6]。Mogensen(1981,1985)通过对非洲紫罗兰(*Saintpaulia ionantha*)和烟草胚珠 ATP 酶活性的细胞化学定位后认为:助细胞和反足细胞在吸收营养方面的作用不大,而珠被绒毡层具有关键作用[17,18]。但绝大多数植物的胚囊外有一角质层,它显然是珠被绒毡层与胚囊之间物质交流的障碍。因此,对胚囊的营养途径尚有深入研究的必要。本文以金鱼草为材料,通过对胚珠中 ATP 酶活性的细胞化学定位研究,就此提出自己的看法。

第四章 有性生殖的超微结构与细胞化学研究

材料与方法

样品制备与超薄切片工作在武汉大学生物系植物细胞胚胎学研究室进行;电镜观察在武汉大学分析测试中心完成。供试材料为金鱼草(Antirrhinum majus L.),取开花前2d和开花当天2个时期的完整胚珠,按简令成(1980)的方法[8]制样(略有改动),即:用50mmol/L,sodium-cocadylate 缓冲液(pH7.2)配制的2.5%的戊二醛和4%的多聚甲醛混合液,在22℃下固定1.25h,用50mmol/L sodium-cocadylate 缓冲液洗涤2次,再用50mmol/L tris-maleate 缓冲液洗涤3次,共3h,然后在22℃的酶反应液中培育3h。酶反应液的组成是:50mmol/L tris-maleate 缓冲液(pH7.2)中包含 5′-ATP 2mmol/L,$Pb(NO_3)_2$ 3mmol/L,$MgSO_4$ 5mmol/L。以反应液中不加底物 ATP 和反应液中加 NaF 抑制剂作为对照。酶活反应后,用50mmol/L sodium-cocadylate 缓冲液洗涤3次,共2h,1%锇酸4℃过夜,双蒸水洗涤3~4次,历时3h。然后按常规方法用各级乙醇脱水,环氧丙烷过渡,Araldite(Durcupan ACM,Fluka)渗透和包埋,在 Sorvall MT-6000 型超薄切片机上切片。用光镜检查经甲苯胺蓝染色的半薄切片,选取有用的材料作厚度约750Å的超薄切片,不经染色,直接在 JEM-100cx/Ⅱ型透射电子显微镜下观察和摄影。部分胚珠采用常规的电镜制样方法,经醋酸双氧铀和柠檬酸铅先后染色,作超微细胞形态学观察。

结 果

金鱼草具倒生胚珠、蓼型胚囊,但胚囊成熟时反足器已退化[1],珠心组织也已经完全解体。本文按细胞中 ATP 酶反应产物磷酸铅沉淀的多少和部位来判断 ATP 酶的活性和部位。这些沉淀在细胞中界限分明,表明基本上没有反应产物的扩散现象。在酶活反应液中不加反应底物 ATP 和加入抑制剂 NaF 的两个对照中均没有发现明显的酶活反应(图版Ⅲ,5、6),进一步表明实验结果是可靠的。

一、胚囊细胞

卵细胞和助细胞(包括助细胞丝状器)内均没有观察到 ATP 酶活反应(图版Ⅰ,1)。我们同意 Mogensen(1981,1985)[17,18]的观点:助细胞丝状器的主要功能是分泌向化性物质吸引花粉管,卵器在营养物质的吸收方面处于相对被动的状态。但在胚囊中部和亚合点区,中央细胞的质膜、细胞质基质和两个极核内均有明显的酶活反应(图版Ⅰ,3),这与 Mogensen 在上述两种材料中的观察结果不同。

二、胚囊壁和壁外间隙层

常规染色制片发现:胚囊壁大体上与一般的纤维壁相似,但它与其外围的角质层之间有一间隙层,这在亚合点区尤为明显,表现为质地疏松,其中含有一些丝状和小囊泡状的物质(图版Ⅰ,4)。这些结构特征在亚珠孔区有各自不同的变化:胚囊壁加厚并形成明显的壁内突,壁外间隙层变得狭窄以至模糊,角质层在接近卵器的部位变得不连续(图版Ⅰ,5)。引人注目的是壁外间隙层中表现出强烈的 ATP 酶活反应(图版Ⅱ,1、2),在亚珠孔区的角质层及其内外也发现酶活反应,这可能是处于该部位的角质层稀薄和不连续所致(图版Ⅰ,1)。从图版Ⅱ,3 中还可看出间隙层中 ATP 酶反应产物呈小团块状分布,这似乎与其中的絮状物质的状态有关。有时可观察到胚囊中部和亚合点区的胚囊壁向胚囊腔内陷而形成囊状结构,其中包裹着絮状物质(图版Ⅱ,4),并有强烈的酶活反应(图版Ⅱ,5、6)。

三、承珠盘

图版Ⅲ图 1 是承珠盘的常规染色图像。它们看来是生活细胞,疏松的厚壁与一般的传递细胞的细胞壁类似。细胞化学定位发现,承珠盘呈现强烈的酶活反应,其细胞核、质膜和厚壁内均有表现(图版Ⅲ,2、3)。分隔承珠盘与胚囊合点端的角质层则出现不连续区段(图版Ⅲ,2),并可观察到承珠盘厚壁内的酶反应产物通过角质层的不连续部位与胚囊壁外间隙层中的酶反应产物连成一片。这与胚囊中部及亚合点区的连续角质层区段明显不同(图版Ⅱ,1、2)。

四、珠被绒毡层及其外围的珠被细胞

珠被绒毡层细胞具有酶活反应,且表现出珠孔-合点的极性:在亚珠孔区角质层不连续区段外围的珠被绒毡层细胞内酶反应产物很少(图版Ⅰ,1);在亚合点区酶活反应则较强,主要存在于质膜、内质网、细胞质基质和核质中(图版Ⅱ,1、5、6),其中以核质中的活性最强。细胞壁内基本上没有反应。Mogensen在非洲紫罗兰和烟草胚珠中的珠被绒毡层质膜上曾发现ATP酶活性反应[17,18],他认为这是珠被绒毡层向胚囊主动输送养料的佐证,但他无法对珠被绒毡层与胚囊之间存在致密而疏水的角质层的作用作出圆满的解释。我们认为,金鱼草这一发育时期的珠被绒毡层的主要功能是分泌各种酶来分解外围的珠被细胞,而角质层与后来形成的抗乙酰解的珠被绒毡层内壁则可防止酶侵蚀胚囊[2]。ATP酶活反应是这一功能的表现。珠被绒毡层以外的一般珠被细胞仅在质膜和与质膜相连的液泡膜及胞间连丝上发现酶活反应(图版Ⅲ,4),但靠近珠孔道的珠被细胞内的反应产物则较少,这与珠孔道内出现大量反应产物的情况形成鲜明对照(图版Ⅰ,2)。

讨 论

一、ATP酶在植物细胞中的分布特点及其功能

几乎所有的生命体系都以ATP作为能量载体,它参与生命物质的代谢、运输和信息传递等有关的能量转换过程。水解ATP是每个细胞的基本功能[19]。因而研究ATP酶的种类、特性和它们在细胞内的分布是了解细胞功能的重要手段。

运用磷酸铅沉淀技术在超微水平上研究高等植物细胞内ATP酶的分布状况已有不少报道。在不同的材料或/和不同的细胞类型中相继发现:细胞间隙、局部细胞壁、胞间连丝、质膜、线粒体、叶绿体、内质网、高尔基体及其分泌小泡、液泡膜及其内含物、细胞质基质、核膜、核质、核仁都有 ATP 酶的分布[7~10,14,16,18]。植物细胞壁所含的 ATP 水解酶的类型比动物和微生物所含的都要多[16]。对它们的功能尚不完全了解。大体上认为:细胞间隙和细胞壁上的 ATP 酶主要参与物质的质外体运输过程[7,9,10];胞间连丝、质膜、内质网、高尔基体、液泡、小泡、核膜等膜系上的 ATP 酶主要参与物质的分泌和共质体运输过程[8,14,19,22];线粒体、叶绿体片层内的 ATP 酶则主要参与物质的分解和合成[19]。最近对于核质内的 ATP 酶的功能研究有了新进展:Almouzni 等(1988)[11]证实它们在染色质组装时核小体与核小体之间生理空间的形成与稳定方面起重要作用;代谢活跃的细胞具有较高的与染色质相联系的 ATP 酶活性。以前发现的那些核质内有较高 ATP 酶活性的细胞也确实有代谢活跃的特点:如正在分化的山毛榉次生维管束细胞[20]、燕麦[16]、玉米[13]、大豆、蚕豆[12]、豌豆[21]的根尖细胞、小麦分蘖节细胞[8]、番茄子叶细胞[9]和小麦胚珠中正在进行核穿壁的细胞[5]等,其细胞核内均有较高的 ATP 酶活性。我们观察到金鱼草胚珠亚合点区珠被绒毡层细胞核质内有较强的酶活反应,显然这与此时这些细胞活跃的分化和分泌作用[15]是相关的。

二、从胚珠中 ATP 酶的分布看胚囊的营养

一般认为胚囊吸收营养有两条途径:其一是由维管束将营养物质送入合点区,经反足细胞加工改组后向珠孔端运输;其二是胚囊的整个表面吸收周围珠心细胞的解体产物,某些薄珠心的植物则由珠被绒毡层吸收营养供给胚囊[3,6,15]。但金鱼草胚囊成熟时反足器已经退化,胚囊外致密而疏水的角质层又构成了珠被绒毡层向胚囊输送养料的障碍[仅在靠近卵器部位的一段很短的角质层出现不连续的现象(图版Ⅰ,5),而此部位的珠被绒毡层细胞内基本上没有酶活反应,看不出它们通过这一不连续部位向胚囊大量提供养料的迹象(图版Ⅰ,1)]。看来在这一材料中,上述胚囊吸收营养的两条途径都缺乏令人信服的根据。

在小麦胚囊周围曾发现珠心解体产物呈丝状结构,这被认为是物质流动的迹象[3];在蒜苔衰退时,其质外体系统中也发现了类似的物质,据此推测:在某些生理状态下质外体可成为大分子物质远距离运输

的通道[4];大量报道证实胞间隙中存在 ATP 酶活性[7,9,10],有时可见到其沉淀产物附着在囊泡式包裹的物质上[10],并认为与物质的运输有关。本文发现金鱼草承珠盘细胞壁结构疏松,胚囊壁外间隙层中存在着絮状和囊泡状物质,并有强烈的 ATP 酶活反应。因此可以认为:在金鱼草开花前胚囊成熟过程中,承珠盘细胞壁、胚囊壁外间隙层和珠孔道相互连通,形成了一个复杂的质外体运输系统,承担向卵器运输养料的任务。

(作者:何才平、杨弘远。原载:植物学报,1991,33(2):85~90。图版3幅,参考文献22篇,均删去。)

2 向日葵胚珠中 ATP 酶活性的超微细胞化学定位

The ultracytochemical localization of ATPase activity in the ovules of sunflower

Abstract

The ultracytochemical localization of ATPase activity was determined employing the method of lead precipitation in the ovules of sunflower (*Helianthus annuus* L.). No ATPase activity is observed in the egg and synergids except some at the filiform apparatus. Much ATPase activity is localized on the plasma membrane and wall of the central cell. In the antipodal cells, ATPase activity is also found on the plasma membranes, but only a little in their walls. In the integumentary tapetum, besides the plasma membranes, most of the nuclei are rich in ATPase. Between the integumentary tapetum and uncontinuous cuticle surrounding the embryo sac, there is a gap where a lot of ATPase are found. These ATPases are continuously linked with those in the central cell wall through the intervals of the cuticle. At the sites of the wall ingrowths of the central cell, abundant vesicles and other structures with high ATPase activity aggregate noticeably in the gap region. According to the ATPase distribution in the ovules, we propose that the whole surface of embryo sac functions in absorbing nutrients directly from the apoplast outside the cuticle, especially via the wall-membrane apparatus of the central cell.

关于胚囊的营养途径,迄今在很大程度上依然依靠从超微结构方面所获得的信息来推测。尽管 ATP 酶活性的定位被认为是研究细胞参与物质运输方式的重要手段[5,7],且在植物各种器官中得到了广泛运用,但在胚珠和胚囊方面的研究还不多见。迄今仅有在非洲紫罗兰(*Saintpaulia ionanatha*)[15]、烟草[16]和金鱼草上[2]有这方面的报道。各种植物胚囊的超微结构特征不尽相同,其胚囊的营养途径也会有所差异。本文通过对向日葵胚珠中 ATP 酶的超微细胞化学定位,发现了其胚囊营养途径的一些特点。

材 料 与 方 法

供试材料为向日葵(*Helianthus annuus* L.)早熟品种"B-11"。取开花前一天和开花当天人工授粉 3h 左右的子房,将其中的胚珠迅速剖出,投入含 0.5% 戊二醛和 4% 的多聚甲醛混合液的 50mmol/L 二甲砷酸钠缓冲液(临用前配制,pH7.2)中,用小刀切除合点端大部,将珠孔端含胚囊的一小部分在上述固定液中继续固定 75min(22℃),其中包括抽气约 15min。嗣后,处理和对照的酶反应、锇酸后固定、脱水、渗透、包埋、切片等步骤均与我们以前研究金鱼草胚珠的方法相同[2]。切片不经染色,直接在 JEM-100cx/Ⅱ型透射电子显微镜下观察和摄影。

第四章 有性生殖的超微结构与细胞化学研究

结 果

向日葵具单珠被、薄珠心,倒生胚珠,蓼型胚囊。胚囊成熟时,两个助细胞和一个卵细胞构成卵器,助细胞具楔形丝状器;中央细胞的两个极核融合形成次生核,并向卵器靠近;中央细胞壁约在与卵器平行的部位形成发达的壁内突;2个多核的反足细胞呈纵向线形排列,在其合点端端壁上也有发达的内突,偶尔还能在反足细胞与中央细胞间壁靠反足细胞一侧观察到内突。以上描述基本上与 Newcomb(1973)[17]和阎华(1990)[6]的超微结构观察结果一致。胚囊成熟时,珠心组织已完全退化,数十层珠被细胞包围着胚囊。珠被的最内层细胞分化为珠被绒毡层,直接与胚囊相邻。但胚囊的珠孔端和合点端则伸出珠被绒毡层进入外层珠被组织中。由珠柄进入胚珠的维管束位于胚囊的一侧,终止于远离胚囊的合点区。本实验根据细胞中 ATP 酶反应产物的部位和多少来判断 ATP 酶的部位和活性。在酶反应液中不加反应底物 ATP 和加入酶抑制剂 NaF 的两个对照组中,基本上没有观察到酶活反应(图版Ⅲ,14、15)。前者表明处理组中磷酸铅的沉淀来源于 ATP 的水解,后者排除了 ATP 非酶水解的可能性,因而证明沉淀确系来源于 ATP 酶的特异性反应。

一、卵器

在本实验中虽然没有进行常规的醋酸铀、柠檬酸铅双染色,但还是能分辨细胞的大体结构,这可能是酶反应液中含有的 Pb^{2+} 在一定程度上增强了细胞基础物质的电子密度。图版Ⅰ,1 示传粉后、受精前的两个助细胞,可看出它们的深浅不同。常规的超微结构观察表明,助细胞之一在传粉后受精前即开始退化,且退化时电子密度显著增加[6,13],因而推测本照片中电子密度深的是正在退化的助细胞。两个助细胞内部均没有观察到 ATP 酶反应,质膜上似乎有一定的酶活性,但大量的酶存在于珠孔端的丝状器上。在已做过 ATP 酶超微结构定位的非洲紫罗兰、烟草和金鱼草三种材料中,助细胞丝状器及其质膜上均未见到

ATP酶活性。Mogensen(1981,1985)曾据此推测助细胞的主要功能是分泌向化性物质引导花粉管,而在胚囊吸收营养方面的作用不大[15,16],但是,没有足够的证据表明分泌向化性物质不需要ATP酶的参与。同时,本文证明向日葵的丝状器是有ATP酶活性的,因此上述推论还需要商榷。向日葵与上述三种材料在这一点上的差异是否归因于材料本身的差异还有待进一步研究。我们还在一个胚珠中发现,宿存助细胞在珠孔端形成一个类似吸器的延伸物,伸到丝状器外侧与退化的助细胞相对,这个延伸物的外侧有更强烈的ATP酶反应(图版Ⅰ,2)。花粉管内的物质和它释放到助细胞内的物质相似,并有少量ATP酶活性。图版Ⅰ,3是花粉管进入退化助细胞的局部的图像。其中的ATP酶可能与花粉管内含物的运动及其释放机制有关。在动物中已经证实细胞质流的形成、速度和方向均受ATP酶活性的控制[20,21]。至于卵细胞中则没有观察到ATP酶活性(图版Ⅰ,4),这与在其它三种材料中的观察结果一致[2,15,16]。说明卵细胞处于代谢不活跃的状态。

二、中央细胞

次生核的核质和核膜上有零星的ATP酶分布,但核仁及其液泡中没有酶活反应(图版Ⅰ,5)。次生核周围的细胞质中有大量油滴分布,只观察到极少量的ATP酶活性,有时可见一些具有酶活反应的小泡,它们大多靠近具有强烈酶活反应的质膜(图版Ⅰ,5;图版Ⅱ,8)。引人注目的是在亚珠孔区约与卵细胞平行部位的中央细胞壁,其内突反复分支形成迷宫似的结构,并将细胞质包裹在其中,沿壁内突边缘分布的质膜上显示强烈的ATP酶活性。在壁内突的外侧,即中央细胞和珠被绒毡层之间的胞间隙中,可见到具有酶活反应的小泡和其它结构物质集聚(图版Ⅱ,6、7;图版Ⅲ,12),暗示此处是养料进入胚囊的重要渠道。在亚合点区中央细胞壁不形成内突的部位,其壁中也观察到大量的ATP酶(图版Ⅱ,8)。我们也曾在金鱼草胚珠承珠盘的细胞壁中观察到大量的ATP酶[2]。关于细胞壁中含有ATP酶这一现象的可能的功能,将在后文详细讨论。

三、反足细胞

反足细胞核质中有一定的酶活反应,细胞质基质中基本上是负反应,但质膜上显示强烈的 ATP 酶活性(图版 Ⅱ,9、10)。反足细胞与中央细胞之间的隔壁是由正常的胞质分裂过程中通过成膜体形成的,而反足细胞相互之间的隔壁则多来源于壁的自由生长。这与 New-comb 的观察结果基本一致[17],只是偶尔也能看到反足细胞与中央细胞的隔壁来源于自由生长的情况(图版 Ⅱ,9)。自由生长形成的壁不整齐,呈结节状,有大量看来不太正常的胞间连丝,其上附有大量的 ATP 酶。与中央细胞侧壁相比,反足细胞侧壁内的 ATP 酶很少,但其合点端端壁与亚珠孔区的中央细胞壁相似,形成多重分支的壁内突,沿壁的质膜上也有大量的 ATP 酶(图版 Ⅲ,11)。看来这里也是胚囊吸收养料的重要部位。

四、珠被绒毡层及其外围的珠被细胞

珠被绒毡层细胞壁中很少有 ATP 酶活性,质膜和绝大多数细胞核内均有 ATP 酶活性(图版 Ⅱ,6、8;图版 Ⅲ,13)。在珠被绒毡层外围的珠被细胞中,质膜和胞间连丝上有 ATP 酶活性,偶尔也能看到个别细胞的核内有酶活反应,某些部位的细胞间隙内也有大量 ATP 酶分布。如前所述在珠被绒毡层与胚囊壁之间的间隙层中有大量的 ATP 酶(图版 Ⅱ,6;图版 Ⅲ,12)。推测它们可能由珠被绒毡层细胞所分泌,也可能由少数外围的珠被细胞分泌后,经胞间隙向中央细胞壁内突集聚。间隙层在胚囊细胞不形成壁内突的部位也含有大量的 ATP 酶,它们一侧与几乎没有酶活反应的珠被绒毡层细胞壁形成鲜明对比,另一侧则通过角质层的间断处与胚囊壁内的 ATP 酶连成一片(图版 Ⅱ,8)。我们也正是通过这个电子密度较大的角质层才能辨别间隙层与胚囊壁的界限。在金鱼草中,这一角质层是比较完整的[2],而向日葵的角质层很不完整,有些部位特别是助细胞丝状器和中央细胞壁内突的外侧已经基本解体(图版 Ⅰ,1;图版 Ⅱ,6;图版 Ⅲ,12、13)。还观察到珠被绒毡层细胞形成延伸物穿越角质层进入胚囊壁的现象(图版 Ⅲ,13)。这

些都可能与向日葵胚囊吸收养料的特点有关。

讨 论

一、胚囊壁的结构特点

Carpita 等(1979,1982)[9,10]用渗透技术测得一般生活细胞壁的孔径约为 3.8~4.2nm,能通过壁孔的球形蛋白质分子的分子量约为 17kU。许多实验证明,分子量大于此数的大分子的确不能通过正常的细胞壁。事实上,酸性磷酸酶[8,19]、过氧化物酶[8]、非特异性酯酶[8]和 α-淀粉酶[12]都只能通过细胞壁局部水解的区域向外释放。如大麦糊粉层细胞中的 α-淀粉酶的分子量为 45kU,它们只有在适当生理条件下,诱导细胞壁局部消化后,才能经过这些被消化的部位向外运输[12]。可见即使不存在胞间连丝,细胞壁也不单纯是一个机械的外壳;对于某些大分子来说,它们可能是受生理控制的门户。

就目前所知,在 ATP 酶家族中,分子量最小的成员也不会小于 70kU[18]。因此,在一般情况下 ATP 酶应该不能穿越细胞壁。在本文中,向日葵的珠被细胞和珠被绒毡层细胞壁内也很少观察到 ATP 酶活性。与此形成鲜明对比的是胚囊壁(包括助细胞丝状器)内有大量 ATP 酶存在。在金鱼草的承珠盘细胞壁中也曾观察到类似现象[2]。张伟成等(1990)[4]在研究小麦胚乳细胞形成时发现:胚囊壁内部包含大量纤丝、电子致密物和各种泡状物,时而可见到某些结构物质在其间迁移。作者推测大量尚未彻底降解的珠心物质直接参与胚乳自由壁的生长。这说明胚囊壁具有特殊的结构,推测其壁孔比一般的细胞壁大得多,其中所含 ATP 酶可能与物质的运输有关。应该指出:除胚囊壁和承珠盘细胞壁较为特殊外,类似的还有柱头乳突细胞和花柱通道细胞的壁。如唐菖蒲的乳突细胞壁孔径是一般细胞壁孔的 10 多倍[11]。又如麝香百合的乳突细胞和通道细胞都能分泌多种蛋白质,绝大多数的分子量大于 17kU,其中最大的超过 170kU。这些蛋白质分泌到胞外之前还可在壁内贮存一段时间[14]。此外,朱彤等(1990)还发现大叶杨柱头表面细胞壁存在分泌 ATP 酶的通道[1]。看来这些特殊细胞壁的结构和功能是值得深入研究的课题。

二、胚囊的营养途径

Mogensen（1981,1985）[15,16]发现在非洲紫罗兰和烟草中，珠被绒毡层的细胞质膜和多数细胞核中 ATP 酶活性较强。他据此推测由维管束输送到珠被的养料首先进入珠被绒毡层，经加工转化后再向胚囊分泌，但并不清楚养料从珠被绒毡层向胚囊转移的细节。因此这一推测还有待进一步用实验证实。我们在金鱼草与向日葵上进行了比较细微的研究，发现它们的胚囊营养途径可能分别代表两种不同的类型（图1），在金鱼草中，胚囊与珠被绒毡层之间有一层完整的角质层。在这种情况下，角质层成为珠被绒毡层与胚囊之间直接进行物质交换的障碍[2]。但角质层与胚囊壁之间的间隙层中含有具强烈 ATP 酶活性的丝状和泡状物质，它们通过胚囊合点端角质层的间断区与承珠盘细胞壁中的 ATP 酶活性物质连成一片。因而我们推测由承珠盘细胞壁到上述间隙层可能构成一个连续的质外体系统向胚囊输送养料。开花后，珠被绒毡层的内壁逐渐变成抗乙酰解的囊状结构包围胚囊，进一步加强了这种障碍。仅合点和珠孔两端留有开口可以进行内外物质交流[3]。在向日葵中，环绕胚囊的角质层是不完整的。与此相关，在亚珠孔区环绕卵器的中央细胞侧壁上和反足细胞合点端的端壁上都有发达的内突，构成具有强烈 ATP 酶活性的壁膜器，无疑是胚囊吸收养料的主要部位。其它部位的胚囊壁内部也有大量的 ATP 酶分布，并通过角质层的间断区域与外围胞间隙中的 ATP 酶连成一片。这说明整个胚囊表面可以不同程度地直接从角质层外围的胞间隙中吸收养料。看来，即使像金鱼草和向日葵这样同属薄珠心、具珠被绒毡层的胚珠中，由于许多细微结构上的差异，它们在胚囊的营养途径上也有显著的区别。在这方面，应用 ATP 酶定位及其它方法进一步研究各种不同类型胚珠养料运输途径的特点，并加以归纳总结是有意义的。

图 版 说 明

A. 反足细胞 C. 中央细胞 DS. 退化助细胞 E. 卵细胞 PS. 宿存助细胞
PT. 花粉管 S. 助细胞 SN. 次生核 T. 珠被绒毡层 c. 角质层 f. 丝状器
g. 间隙层 w. 胚囊壁 wi. 壁内突 wt. 珠被绒毡层细胞壁

图 版 Ⅰ

1. 胚囊珠孔端,禾助细胞丝状器、中央细胞壁内突以及珠被绒毡层细胞的质膜和核内有 ATP 酶活性。×2300 **2.** 在一个胚珠中发现宿存助细胞形成一个延伸物伸到丝状器外方,这一延伸物的外侧有强烈的 ATP 酶活性。×2100 **3.** 花粉管进入退化助细胞,花粉管及其释放在助细胞中的内容物中有一些 ATP 酶活性。×4000 **4.** 卵细胞内无 ATP 酶活性,但中央细胞壁内突上酶活性较强。×2100 **5.** 中央细胞的细胞质和次生核中 ATP 酶活性极弱,但质膜上活性较强(箭头)。次生核周围的细胞质中有大量油滴分布。×2100

图 版 Ⅱ

6. 中央细胞壁内突及其外侧胞间层中的 ATP 酶活性。×5400 **7.** 上图局部放大,示胞间层中具酶活反应的小泡和其它结构物质集聚在中央细胞壁内突的外侧。×15000 **8.** 中央细胞的亚合点区,胚囊壁及其外侧间隙层中的 ATP 酶活性。注意珠被绒毡层细胞的质膜及其分泌小泡上有酶活性,但其壁内无酶活性;角质层是胚囊壁与壁外间隙层的分界线。箭头示角质层的间断处。×5400 **9.** 中央细胞与反足细胞的间壁。×5400 **10.** 反足细胞间壁自由生长。×5400

图 版 Ⅲ

11. 反足细胞合点端壁内突及其外侧间隙层中具有较强的 ATP 酶活性。×7200 **12.** 珠被绒毡层质膜上和间隙层中具 ATP 酶活性的小泡可能正向具酶活反应的中央细胞壁内突外侧移动。×8700 **13.** 珠被绒毡层形成具酶活反应的延伸物经角质层的间断处进入胚囊壁内;珠被绒毡层细胞核内的酶活反应较强。×5400 **14.** 反应液中不加反应底物 ATP 的对照,示助细胞丝状器部位。×3200 **15.** 反应液中加入酶抑制剂 NaF 的对照,示中央细胞壁内突及其外侧间隙层与珠被绒毡层部位。两种对照中均没有观察到酶活反应。

Explanation of plates

A. Antipodal cell   C. Central cell   DS. Degenerated synergid   E. Egg cell

第四章 有性生殖的超微结构与细胞化学研究

PS. Persistent synergid PT. Pollen tube S. Synergid SN. Secondary nucleus T. Integumentary tapetum c. Cuticle f. Filiform apparatus g. Gap between the cuticle and the integumentary tapetum w. Embryo sac wall wi. Wall ingrowths wt. Integumentary tapetum cell wall.

Plate I

Fig. 1. The micropylar portion of an embryo sac. ATPase are localized in the filiform apparatus of synergids, wall ingrowths of central cell and the nuclei and plasma membranes of integumentary tapetum. ×2300 **Fig. 2.** In an ovule, an extension from the persistent synergid protrudes outside, where a lot of ATPase are found. ×2100 **Fig. 3.** A growing pollen tube in the degenerated synergid, showing some ATPase activity in the tube and its discharged contents. ×4000 **Fig. 4.** No ATPase activity is observed in the egg cell, whereas much of that is found in the wall ingrowths of central cell. ×2100 **Fig. 5.** Little enzyme reaction is observed in the secondary nucleus of central cell. Note a lot of oil drops distribute in the cytoplasm around the secondary nucleus. Arrowheads show plasma membrane with ATPase activity. ×2100

Plate II

Fig. 6. A lot of ATPase are localized in the wall ingrowths of central cell and in the gap between embryo sac wall and integumentary tapetum cell walls. ×5400 **Fig. 7.** The magnification of a part of the gap in Fig. 6, where many vesicles and other structured substances with ATPase activity aggregate noticeably to the outside of the wall ingrowths of the central cell. ×15000 **Fig. 8.** The subchalazal part of an embryo sac, showing ATPases on the plasma membrane, the embryo sac wall and in the gap between embryo sac wall and integumentary tapetum cell walls. The cuticle discriminates the gap from the embryo sac wall. Arrowheads show the intervals on the cuticle. In integumentary tapetum cells, no ATPase in the walls are seen but some vesicles with ATPases activity on the plasma membranes are found. ×5400 **Fig. 9.** Some ATPases are found in the wall boundering central cell and antipodal cells and on the plasma membranes of the these cells. ×5400 **Fig. 10.** ATPases located in the nucleus and the free growing wall of an antipodal cell. ×5400.

Plate III

Fig. 11. The chalazal end of an antipodal cell, showing a lot of ATPase in its wall

图版 I plate I

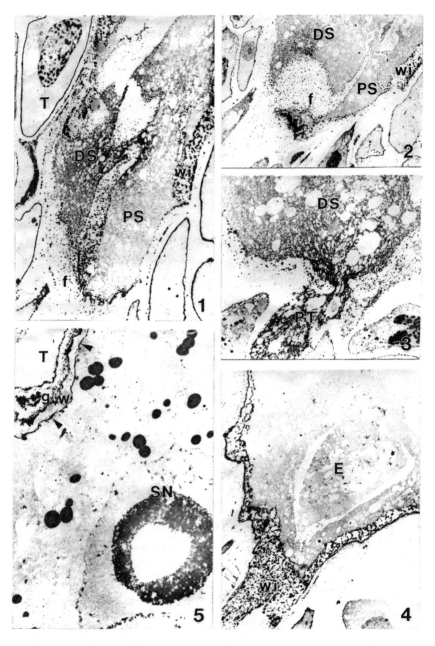

第四章 有性生殖的超微结构与细胞化学研究

图版 II   plate II

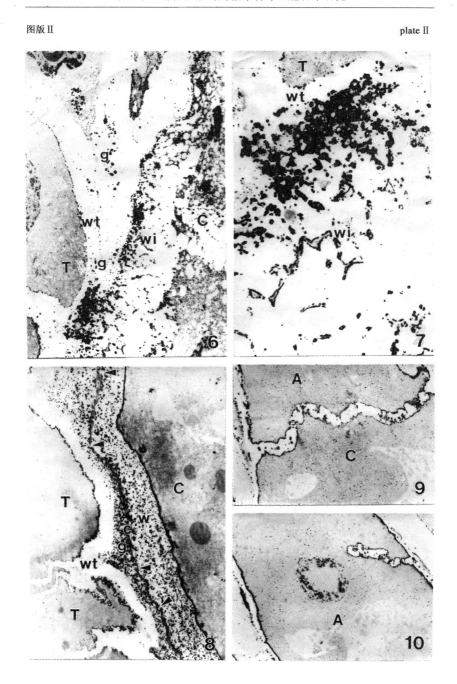

图版Ⅲ plate Ⅲ

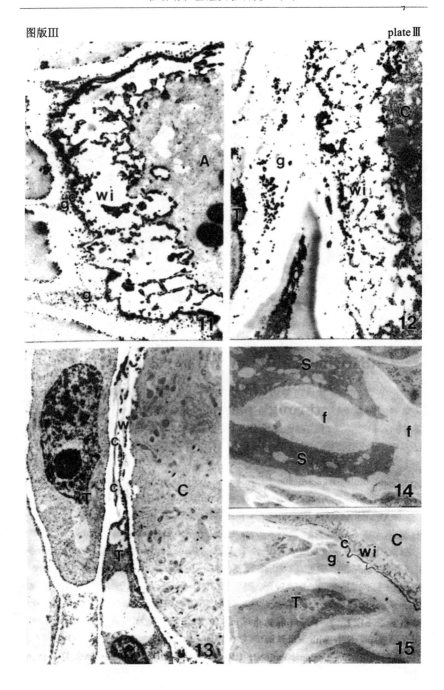

ingrowths and in the gap outside of it. ×7200 **Fig. 12.** A lot of vesicles showing ATPase activity on the plasma membranes of integumentary tapetum cells and in the gap outside the wall ingrowths of the central cell. The vesicles seem to move toward the wall ingrowths where much of ATPase activity is also found. ×8700 **Fig. 13.** An extension from a integumentary cell with ATPase activity going throuth the embryo sac wall via the uncontinuous cuticle. ×5400 **Fig. 14.** Control Ⅰ: Incubation in a medium without substrate ATP. ×3200 **Fig. 15.** Control Ⅱ: Incubation medium containing NaF. ×3200. In both cases, no ATPase was observed.

(作者:何才平、杨弘远。原载:植物学报,1991,33(6):574~580。图版3幅,图1幅,参考文献21篇。选录全部图版。)

3 烟草雌蕊的花粉管生长途径中几种植物激素的免疫金电镜观察

Immunogold electron microscopic observations on distribution of several phytohormones in pollen tube growth track through pistil in tobacco

Abstract

Up to now, little is known about distribution and function of phytohormones in the pollen tube growth track through pistils. Distribution of trans-zeatin (t-Z), indele-3-acetic acid (IAA), gibberellins A7 and A4 ($GA_{7/4}$), and (+) abscisic acid [(+)ABA] in the style, micropyle and synergids of *Nicotiana tabacum* var. *macrophylla* were observed with immunogold electron microscopy. Segments from middle part of the styles and ovules at various stages at and after pollination were fixed with 2% EDC [1-ethyl-3-(3-dimethyl aminopropyl) carbodiimide] and mixed paraformaldehyde and glutaraldehyde for acidic phytohormones (or only with the latter for t-Z), then stightly postfixed in 0.5% OsO_4 solution for 30 min. After dehydration, samples were embedded in Epon 812 resin. Following etched in 10% H_2O_2 for 10 min, ultrathin sections were immunostained respectively with rabbit anti-t-Z polyclonal antibody (PAb), anti-IAA methyl ester PAb, mouse anti-GA monoclonal antibody (MAb, specifically recognizes methyl esters of GA_7 and GA_4), or anti-(+) ABA methyl ester MAb. Protein A colloidal gold (Φ10 nm) was used as a marker for rabbit PAbs while sheep anti-mouse IgG-colloidal gold (Φ10 nm) for mouse MAb. At pollination plenty of t-Z but a few IAA and $GA_{7/4}$ were observed in cells and intercellular matrix of the transmitting tissue. One day after pollination, when pollen tubes had penetrated the transmitting tissue, the gold granules of t-Z remarkably decreased, but those of IAA slightly increased; and those of $GA_{7/4}$ obviously increased in the same region. IAA in cells and intercellular matrix of stylar parenchyma greatly increased from just at pollination to one day after pollination. Increased IAA and $GA_{7/4}$ were mainly localized in amyloplasts, vacuoles and intercellular matrix. IAA appeared in the walls of pollen tubes passing the transmitting tissue. High level of t-Z in themicropyle and the synergids (particularly abundant t-Z in the filifrom apparatus) greatly decreased from just at pollination to 4 days after pollination. $GA_{7/4}$ were less in the synergids at pollination but increased in both the degenerate and persistent synergids 4 days after pollination. $GA_{7/4}$ and IAA in the micropyle or IAA in the synergids maintained at lower levels 4 days after pollination. (+) ABA in the pollen tube growth track was rarely observed during the whole course. According to the results, the relationship of the distribution mode of these four phytohormones in tobacco pistil with pollen tube guidance is discussed.

近年的研究揭示出花粉管在雌蕊中定向生长的可能机理[1]。与花粉粒直径相当的乳胶珠粒可在花柱引导组织内向子房移动,其移动速率与花粉管生长速率相当[2]。花柱组织内一种类玻璃粘连蛋白(vitronectin-like protein)可能是花粉管生长所需的一种驱动分子[3]。烟草引导组织中特异的糖蛋白分子呈梯度分布,可引导花粉管生长[4]。拟南芥胚珠突变体实验显示胚囊内存在吸引花粉管的特异因子[5]。这些结果启发人们多方面探索雌蕊内特异的调节因子。

早期的证据已初步显示雌蕊内的生长素可能与花粉管定向生长有关[6]。但由于激素定位技术上的限制,迄今尚未获得雌蕊内花粉管生长途径中生长素与其它植物激素分布的直接证据。我们用免疫金电镜方法对烟草花柱、珠孔及助细胞内的玉米素(t-Z)、吲哚乙酸(IAA)、赤霉素(GA_7、GA_4)及脱落酸[(+)ABA]分布状况进行了探查,并讨论它们与花粉管生长的相关性。

材 料 与 方 法

材料为大叶烟(*Nicotiana tabacum* var. *macrophylla*)刚授粉及授粉后24h的花柱中段(3mm)、刚授粉及授粉后96h的胚珠。花柱切成1mm小段,参照文献[7,8]的方法制备透射电镜样品和进行免疫细胞化学反应。一部分材料用2%EDC[1-乙基3-(3-二甲氨丙基)碳二亚胺]室温下抽气固定2h后,转入4%多聚甲醛与1%戊二醛,室温抽气2h并于4℃过夜;再以0.5%锇酸处理30min,常规乙醇脱水,环氧丙烷过渡,Epon812包埋,此部分样品用于酸性激素定位。另一部分材料除省去EDC外,其余制样步骤不变,用于玉米素定位。前人研究[8~10]表明醛试剂主要固定细胞分裂素中的游离碱基形式。在MT6000型超薄切片机上切片,捞在覆有Formvar膜的镍网上,载片镍网经10%H_2O_2蚀刻10min,阻断液(0.2mol/L甘氨酸、2%鸡卵清蛋白溶于PBST液,pH7.2)阻断20min(PBST液组成:10mmol/L磷酸钠缓冲液,0.15mol/L NaCl,0.02%Tween-20,0.02%NaN_3,pH7.2);在兔抗t-Z多克隆抗体液滴(1:1500,PBST稀释,并含1%BSA)上30℃温育3h,或在兔抗IAA

多克隆抗体、鼠抗 $GA_{7/4}$,(+)ABA 单克隆抗体液滴(分别为 1∶400,1∶25 000,1∶12 000,PBST 稀释,并含 1% BSA)上 30℃温育 4h,PBST 洗涤后在金标 A 蛋白(识别多克隆抗体)或金标羊抗鼠抗体(识别单克隆抗体)液滴(1∶40,PBST 稀释,并含 1% BSA)上 30℃温育 60min 或 70min,充分洗涤后醋酸铀与柠檬酸铅双染,在 JEM100/Ⅱ型透射电镜上观察与摄影。设置 3 个阴性对照:①省去 4 种特异识别激素的抗体;②用正常兔血清代替兔抗 t-Z 及 IAA 多克隆抗体;③预先用 2mmol/L t-Z、1mmol/L IAA 甲酯、1mmol/L GA_4 甲酯、1.5mmol/L(+)ABA 甲酯与相应的抗体 37℃温育 2h,其余步骤不变。

10nm 胶体金及金标 A 蛋白与金标羊抗鼠抗体的制备按文献[11]的方法进行。

特异识别 t-Z、IAA 甲酯、$GA_{7/4}$ 甲酯、(+)ABA 甲酯的抗体制备方法及抗体质量参数见文献[12~15]。

结　果

一、花柱

发育完全的大叶烟花柱全长约 3.8cm。中段花柱引导组织及薄壁组织细胞的超微结构与前人的报道[16]一致。两种组织的细胞存在明显差异,即横切面上引导组织细胞较小,胞间基质层较厚,细胞质较浓,液泡多位于细胞中央,富含粗糙内质网、线粒体、淀粉体或质体;而薄壁细胞较大,胞间基质层较薄,细胞质较稀,胞内大部分空间被液泡占据,细胞器较少。

刚授粉时,引导组织胞间基质、细胞壁、质膜、细胞质中有较多 t-Z(图版Ⅰ,1);薄壁组织细胞壁上有一些 t-Z,细胞质中则较多,与引导组织细胞质中无明显差异。授粉 24h 后,引导组织与薄壁组织的胞间基质、细胞壁及细胞内 t-Z 均明显减少(图版Ⅰ,2)。此时已见花粉管穿过花柱中段引导组织(图版Ⅰ,5;图版Ⅱ,11)。

刚授粉时,引导组织与薄壁组织的胞间基质、细胞壁与细胞内 IAA

均较少(图版Ⅰ,3、4)。授粉 24h 后,引导组织胞间基质与细胞质基质内 IAA 略增加(图版Ⅰ,5),淀粉体中有增多的 IAA(图版Ⅰ,6);薄壁组织胞间基质、细胞壁、液泡及淀粉体上 IAA 明显增加(图版Ⅰ,7、8)。花粉管壁有 IAA 分布(图版Ⅰ,5)。

刚授粉时,引导组织胞间基质、细胞壁、细胞核、胞质基质及液泡中均有一些 GA_7 与 GA_4 分布,但不多(图版Ⅰ,9);授粉 24h 后,引导组织胞间基质、细胞壁及淀粉中 GA_7 与 GA_4 明显增多(图版Ⅱ,10),淀粉体上这种增加尤为明显,其它部位(液泡与细胞核)内 GA_7 与 GA_4 的分布密度与刚授粉时相近。而授粉前后薄壁组织内 GA_7 与 GA_4 均较少。

授粉前后引导组织与薄壁组织的各个部位(+)ABA 均较少或极少,花粉管厚壁内有少量(+)ABA(图版Ⅱ,11)。

二、珠孔与助细胞

刚授粉时,珠孔、胚囊壁、助细胞内有较多的 t-Z,丝状器中则存在大量 t-Z(图版Ⅱ,12、14;图版Ⅲ,16),并且总体上 2 个助细胞之间 t-Z 分布水平存在差异(图版Ⅱ,14、15)。授粉 96h 后,助细胞、胚囊壁及珠孔内 t-Z 均较少(图版Ⅱ,13)。

刚授粉时,珠孔、胚囊壁、助细胞内 GA_7 与 GA_4、(+)ABA 及 IAA 均不多或很少(图版Ⅱ,18~21)。授粉 96h 后,退化助细胞与宿存助细胞内 GA_7 与 GA_4 较多,IAA、(+)ABA 仍然不多,其它部位这 3 类激素仍然较少。

三、阴性对照

针对多克隆抗体(识别 t-Z 或 IAA)设定了 3 个阴性对照,针对单克隆抗体〔识别(+)ABA 或 GA_7 与 GA_4〕设定 2 个阴性对照。在所有激素的阴性对照中,省去抗体或用正常兔血清替代多克隆抗体后,花柱、珠孔、助细胞等处几无金标记,用过量的 t-Z、GA_4 甲酯、IAA 甲酯及(+)ABA 甲酯预先封闭各自特异抗体的抗原结合部位后,切片上金标记极少(图版Ⅲ,17),表明上述定位检测结果的真实性。

讨 论

在实验中,用免疫电镜法对雌蕊内花粉管生长途径及其周围细胞内 4 类植物激素的分布情况进行观察与比较。根据初步实验结果,提示花粉管生长途径及其周围细胞内的植物激素,可能与花粉管定向生长存在一定的相关性。具体可能包括以下几种情况:

(1)由于位置关系,花粉管生长途径中的 t-Z 与 $GA_{7/4}$ 可能与花粉管生长有着一定联系。这 2 类激素的分布水平随着授粉进程而变化,在变化方式上 2 类激素之间存在差异。花粉管生长途径的各个部分(特别是丝状器)中的 t-Z 在花粉管到达之前保持较高水平,穿过之后显著减少。丝状器内异常高的 t-Z 分布及其与授粉进程相协调的变化是否提示 t-Z 对花粉管生长起着引导作用,值得今后进一步研究。花柱内的花粉管生长途径中,GA_7 与 GA_4 在花粉管穿过之后增加。花粉管可能通过分泌 GA 并运输到助细胞,诱导后者退化[6],从而吸引花粉管向胚囊生长。花柱引导组织胞间基质内存在各种囊泡,其内的植物激素(图版Ⅰ,4、9;图版Ⅱ,11)可能也有某种生理功能。

(2)花柱引导组织细胞与薄壁细胞内的 IAA、$GA_{7/4}$ 可能与花粉管生长的营养供应有着一定联系。曾经推测,GA_4 活化淀粉酶为花粉管生长提供碳源[6],本文揭示的淀粉体上存在 $GA_{7/4}$ 与 IAA 为此提供了一定证据。液泡内的 IAA 可能还调节其它水解酶的活性。

(3)番茄根冠细胞质外体(包括细胞壁、粘液层)中分布较多的 $ABA^{[17]}$,而烟草花粉管生长途径中 ABA 分布水平较低,说明质外体内 ABA 水平存在组织差异性。这可能反映了 ABA 功能上的差异,前者可能与根对环境刺激的反应有关[17],而后者可能与花粉管生长相适应。离体萌发试验表明,外源 ABA 抑制烟草花粉萌发[18]。已知 ABA 能拮抗 GA_4 对淀粉酶的活化作用,因而,淀粉体上 ABA 较少从另一个侧面反映淀粉体的水解作用加强。

(作者:陈以峰、梁世平、杨弘远。原载:武汉植物学研究,1999,
    17(4):297~302。图版 3 幅,参考文献 7 篇,均删去。)

4 受精前后烟草卵细胞内玉米素和三类酸性植物激素分布的免疫金电镜观察

Distribution of trans-zeatin, $GA_{7/4}$, (+)ABA and IAA in tobacco egg cells before and after fertilization: immunogold electron microscopic observations

Abstract

Changes in distribution of trans-zeatin (t-Z), gibberellin A_7 and A_4 ($GA_{7/4}$), (+) abscisic acid [(+)ABA] and indoleacetic acid (IAA) in the egg cells of *Nicotiana tabacum* var. *macrophylla* before and after fertilization were studied with immunoelectron microscopy. The ovules just at pollination or 96 h after pollination were fixed with 2% EDC [1-ethyl-3-(3-dimethylaminopropyl) carbodiimide] and then with the mixed paraformaldehyde and glutaraldehyde for acidic phytohormones (or only with the aldehydes for t-Z), then slightly postfixed in 0.5% OsO_4 solution for 30 min. After etched in 1% H_2O_2 for 10 min, the ultrathin sections embedded in Epon 812 resin were immunostained with rabbit anti-t-Z (and t-ZR) polyclonal antibody (PAb), anti-IAA methyl ester PAb, mouse anti-GA_7 and GA_4 methyl esters monoclonal antibody (MAb), or anti-(+) ABA methyl ester MAb, respectively. Protein A-or sheep anti-mouse IgG-colloidal gold (Φ10 nm) were used to indicate rabbit PAbs or mouse MAbs respectively. In the model system of nitrocellulose membrane via immunogold-silver enhancement, the authors ascertained that immunostaining results at the basis of 1 ng per phytohormone (t-Z, IAA, GA$_4$, or (+) ABA) were comparable among the four-kind phytohormones and that t-Z riboside was far less fixed than t-Z with aldehydes. So the anti-t-Z PAb mainly recognized t-Z in aldehyde-fixed tissues. Immunogold electron microscopic observations showed that t-Z was rich in the egg cells before fertilization. In contrast the amounts of GA$_{7/4}$ and (+) ABA were lower in egg cells before fertilization but slightly increased after fertilization. Less IAA in egg cells was found either before or after fertilization. t-Z in unfertilized egg cells appeared to concentrate on the nucleus, endoplasmic reticulum and mitochondria. t-Z is rarely observed in the nuclei of synergids before fertilization but is abundant in the chalazal end of synergids and micropylar end of the central cell adjacent to the unfertilized egg cell. After fertilization, t-Z decreased obviously in the zygotes and the persistent synergids, but appeared in the thickened walls of the zygotes.

受精是植物有性生殖过程的中心环节。学者们曾提出植物激素参与植物受精过程的调控[1]。但由于研究手段的限制,迄今仍然缺乏植物激素在受精前后卵细胞中定位的直接证据[2]。自 Zavala 与 Brandon[3] 和 Sotta 等[4] 分别创立细胞分裂素与脱落酸免疫细胞化学定位方法以来,经过十多年的摸索,激素免疫定位方法已扩展到生长素[5]、赤霉素[6]与油菜素内酯[7]上,并在多种营养器官[3~5,8,9]、花药[6,7]和体细胞胚[10]的研究中得到应用,标志着这项技术开始趋于成熟。本文在此基础上,根据抗体的特异性,分别用醛试剂固定玉米素(t-Z),用 EDC [1-乙基-3(3-二甲氨丙基)碳二亚胺]固定 GA$_7$ 与 GA$_4$、(+) ABA 与 IAA,通过免疫金电镜技术,对烟草受精前后卵细胞内这4类植物激素的分布进行初步的比较研究,旨在探讨其中何种激素可能与受精作用密切相关。

材 料 与 方 法

实验材料为大叶烟(*Nicotiana tabacum* var. *macrophylla*)刚授粉及

第四章 有性生殖的超微结构与细胞化学研究

授粉后96h的胚珠。

10nm胶体金及金标A蛋白与金标羊抗鼠抗体的制备按文献[11]的方法进行。

特异识别t-Z、IAA甲酯、GA$_7$甲酯与GA$_4$甲酯、(+)ABA甲酯的抗体制备方法及抗体质量参数见文献[12~15]。

用Zhang等[1]提出的硝酸纤维素膜模式系统优化本实验中免疫细胞化学反应条件。主要流程：在6mm×6mm硝酸纤维素膜片上点样(20μgBSA与1ng激素溶于2μl缓冲液)——置入盛有固定剂的干燥器中，膜片与固定剂不接触，抽真空使固定剂挥发，气相固定（酸性激素先用EDC固定2h，再用多聚甲醛与2%戊二醛固定2h，t-Z与t-ZR仅用醛固定2h）——阻断——洗涤——加一抗（含5%BSA）——洗涤——加金标二抗（含5%BSA）——洗涤——加物理显影液（含AgNO$_3$），黑暗下显色，呈棕色圆斑——蒸馏水冲洗、晾干。设置2个阴性对照：(1)膜片上仅吸附BSA；(2)用过量激素（均为1mmol/L）预先温育一抗。

根据上述模式系统预备实验的结果，并参照文献[8,16]的方法，制备透射电镜样品和进行免疫细胞化学反应。一部分材料用2%EDC室温下抽气固定2h后，转入4%多聚甲醛与1%戊二醛，室温抽气2h并于4℃过夜，再以0.5%锇酸处理30min，常规乙醇脱水，环氧丙烷过渡，Epon 812包埋，用于酸性激素定位。另一部分材料除省去EDC外，其余制样步骤不变，用于玉米素定位。在MT 6000型超薄切片机上切片，捞在覆有Formvar膜的镍网上，10% H$_2$O$_2$蚀刻10min，阻断液(0.2mol/L甘氨酸和2%鸡卵清蛋白溶于PBST液，pH7.2)作用20min(PBST液组成：10mmol/L磷酸钠缓冲液、0.15mmol/L NaCl、0.02% Tween-20、0.02% NaN$_3$，pH7.2)；在兔抗t-Z多克隆抗体液滴(1:1500，PBST稀释，并含1%BSA)上30℃温育3h，或在兔抗IAA多克隆抗体、鼠抗GA$_{7/4}$、(+)ABA单克隆抗体液滴(分别为1:400；1:25 000，1:12 000，PBST稀释，并含1%BSA)上30℃温育4h，PBST洗涤后在金标A蛋白(识别多克隆抗体)或金标羊抗鼠抗体(识别单克隆抗体)液滴(1:40，PBST稀释，并含1%BSA)上30℃温育60min或

70min,充分洗涤后醋酸铀与柠檬酸铅双染,在JEM100/Ⅱ型透射电镜上观察与摄影。设置3个阴性对照:(1)省去4种特异识别激素的抗体;(2)用正常兔血清代替兔抗t-Z及IAA多克隆抗体;(3)预先用2mmol/L t-Z、1mmol/L IAA 甲酯、1mmol/L GA_4 甲酯、1.5mmol/L(+) ABA甲酯与相应的抗体37℃温育2h,其余步骤不变。

结 果

一、模式系统中4类植物激素免疫定位的可比性

运用免疫金增强原理,在硝酸纤维素膜上激素点样量均为1ng,反应条件保持一致时,t-Z、IAA、(+)ABA、GA_4 的点样圆斑显色程度比较接近,提示这4种激素的免疫定位结果具有可比性(图版Ⅰ,1C)。省去激素或先用1mmol/L过量激素(t-Z、IAA甲酯、(+)ABA甲酯或 GA_4 甲酯)预先封闭相应抗体的抗原结合位点,结果膜上激素点样圆斑几乎没有显色(图版Ⅰ,1A、1B),表明上述免疫染色反应是特异的。在相同条件下,醛试剂对t-Z的固定能力明显大于t-ZR(图版Ⅰ,1C、1D)。

二、受精前后卵细胞的超微结构特征

授粉后96h,合子体积已经增大,比刚授粉时的未受精卵细胞约增大50%。未受精卵与合子的超微结构与文献报道[17]基本一致。二者的细胞质基质均较浓,内质网、线粒体均较丰富(图版Ⅰ,3、4;图版Ⅱ,8),合子内细胞器分布极性也与未受精卵大体相同,细胞核位于合点端,液泡位于珠孔端。所不同的是,受精前卵细胞合点端没有细胞壁、珠孔端细胞壁也较薄(图版Ⅰ,3、4),合子合点端细胞壁已形成,并且似正加厚(图版Ⅰ,6);合子内一部分线粒体的结构似乎正处在发育之中(图版Ⅲ,14、15);此外,合子内出现脂质体和淀粉体(图版Ⅲ,14、18)。

三、受精前后卵细胞内t-Z的分布

受精前卵细胞内有大量的t-Z,主要分布于细胞核、内质网、线粒体

与细胞质基质上(图版Ⅰ,3、4),细胞核内 t-Z 以较密集形式分布于核被膜、核基质与染色质上(图版Ⅰ,3)。与未受精卵相接的中央细胞珠孔端,细胞质基质、液泡、淀粉体及线粒体内有较多 t-Z(图版Ⅰ,2),与卵细胞相接的助细胞合点端液泡内也有较多 t-Z(图版Ⅰ,4),卵细胞和中央细胞与助细胞之间似有 t-Z 的跨壁转运(图版Ⅰ,2、4)。助细胞除珠孔端 t-Z 较多外,近中央的细胞核与细胞质内 t-Z 很少(图版Ⅰ,5)。与胚囊壁相接的中央细胞胞质基质内有一些 t-Z。与反足细胞相接的中央细胞部分内 t-Z 较少。看来,成熟胚囊中,在卵细胞及其周围形成一个富含 t-Z 区域。

与未受精卵细胞相比,合子内 t-Z 显著减少。细胞核内各处 t-Z 均较少,偶见一处核被膜的内、外侧密布 t-Z,似乎存在跨核膜转运(图版Ⅱ,7)。粗糙内质网上 t-Z 较少,其它细胞器(线粒体、液泡等)中 t-Z 也较少,细胞质基质中稍多一些(图版Ⅱ,8)。合子珠孔端与合点端正加厚的细胞壁中有一些 t-Z(图版Ⅰ,6)。退化助细胞与宿存助细胞充满大量的高尔基小泡及其它无定形囊状物,t-Z 很少或较少。在宿存助细胞与合子交界处 t-Z 似有跨壁转运(图版Ⅱ,9)。

四、受精前后卵细胞内其它几种激素的分布

受精前卵细胞合点端 GA_7 或 GA_4 较少(图版Ⅱ,12),珠孔端比合点端略多(图版Ⅲ,13)。与卵细胞相接的助细胞合点端及细胞壁上 GA_7 与 GA_4 很少(图版Ⅲ,13)。合子内 GA_7 与 GA_4 比受精前有所增加(图版Ⅲ,15),增加的 GA 在合子内有着广泛分布,如细胞核、细胞质基质、线粒体等上(图版Ⅲ,15)。宿存助细胞内 GA_7 与 GA_4 较之助细胞内也有增加。

受精前卵细胞内各处(+)ABA 总体上较少(图版Ⅱ,11;图版Ⅲ,16)。合子内(+)ABA 略多于未受精卵(图版Ⅲ,14)。合子内增加的(+)ABA 广泛分布于细胞核、脂质体、淀粉体、线粒体等上(图版Ⅲ,14)。受精前助细胞内(+)ABA 略多于未受精卵,但受精后宿存助细胞内(+)ABA 少于助细胞。

受精前后卵细胞内各处 IAA 均很少(图版Ⅱ,17、18),少量的 IAA

分布在线粒体与细胞核等处,但内质网、淀粉体与脂质体上罕见 IAA(图版Ⅲ,18)。

五、阴性对照

在正常免疫染色程序中省去 4 种特异一抗(即抗 t-Z 抗体、抗 IAA 抗体、抗 $GA_{7/4}$ 抗体或抗(+)ABA 抗体)或用正常兔血清代替特异多克隆抗体后,切片上几乎不出现非特异吸附的金颗粒,用过量的 t-Z、GA_4 甲酯、(+)ABA 甲酯及 IAA 甲酯预先封闭各自特异抗体的抗原结合部位后,切片上金标记极少(图版Ⅱ,10)。表明上述卵细胞内的激素定位结果可信。

讨 论

Eberle 等[8]利用过量产生异戊烯基腺嘌呤的苔藓突变体,结合 HPLC 分析及番茄冠瘿瘤组织阳性对照,证实戊二醛能有效固定细胞分裂素碱基。Atzorn 等[18]对苔藓原丝体先用醛试剂固定,然后用或者不用碘酸固定,结果也表明醛试剂主要固定细胞分裂素碱基,而细胞分裂素核苷可被过碘酸盐固定,并估算出细胞分裂素的固定率达 70% 左右。我们在硝酸纤维素膜片上得到的结果与此一致。这样,尽管本文所用的兔抗 t-Z 多克隆抗体也能识别 t[9R]Z[12],但通过醛试剂的选择固定,同样能保证固定的 t-Z 被该抗体特异识别。

本文所用的其它 3 种抗体具有较高或很高的专一性[12,14,15],IAA 多克隆抗体可识别 IAA 甲酯,ABA 单克隆抗体只识别(+)ABA 甲酯,GA 单克隆抗体主要识别 GA_7 甲酯与 GA_4 甲酯。由于酰胺与甲酯存在电子等排关系能被同一抗体识别[15],因而,这些抗体能够特异识别用 EDC 通过形成酰胺键固定的 IAA、(+)ABA、GA_7 或 GA_4。

随着工作的深入,需要对激素之间或激素与蛋白质之间进行定位比较。Atzorn 等[18]率先比较了苔藓原丝体内 IAA 与细胞分裂素分布水平,发现前者高于后者;Ivanova 等[10]则用双荧光标记发现细胞分裂素与细胞增殖相关蛋白共同定位在分生细胞的核内。前人[4~6,9]以及我们在硝酸纤维素膜模式系统上的工作均一致地表明,EDC能有效固定IAA、ABA与GA,由此本研究中实现了对同一个固定细胞相近部位IAA、GA、ABA进行初步比较,同时也能对不同的固定细胞相近部位内t-Z及3类酸性激素进行初步比较。

受精前卵细胞内原来存在大量的t-Z,受精后显著减少。这启示t-Z可能与受精作用的早期事件相关。集中分布在细胞核、粗糙内质网及线粒体上的t-Z可能促进转录、翻译及能量代谢活动,为受精准备物质和能量基础。合子内GA_7与GA_4增加可能与合子体积扩大有一定联系,因为GA可以促进分生细胞和年幼细胞生长[19]。体细胞培养实验表明,ABA对胚性细胞诱导或胚性状态维持起着重要作用[20]。这是否也启示合子内(+)ABA增加与其分化状态有某种联系?总之,本文结果可为深入研究激素对受精作用的调控机制提供有益的线索。在这一研究方向上还需要作更深入的探索。

图 版 说 明

a. 淀粉体　CC. 中央细胞　d. 高尔基体　E. 未受精卵细胞　er. 内质网　En. 胚乳细胞　l. 脂质体　m. 线粒体　n. 细胞核　ne. 核被膜　no. 核仁　p. 质体　pm. 质膜　PS. 宿存助细胞　Sy. 助细胞　v. 液泡　ve. 小泡　w. 细胞壁　Z. 合子

刚授粉时的未受精卵与授粉后96h的合子的纵切面。箭头指示金颗粒。

图 版 I

1. 硝酸纤维素膜上玉米素(t-Z,i)、吲哚乙酸(IAA,ii)、脱落酸[(+)ABA,iii]及赤霉素A_4(GA_4,iv)的免疫金增强法的建立。用多聚甲醛与戊二醛气相固定t-Z(B,C)与t-ZR(D),3种酸性激素先启用EDC与醛试剂气相固定。BSA作为载体蛋白,每一小膜片上加样2μl,内含20μgBSA和1ng植物激素。A,对照,仅BSA点样于膜片;B与C,一种激素与BSA组成的混合液点样于膜片,其中B为对照,将特异性抗体用相应的1mmol/L激素预饱和。注意到醛固定t-ZR的能力较弱(D),提示4种植物激素显色结果具有可比性(C)。**2.** 中央细胞与未受精卵相邻部分,示中央细胞内淀粉体、液泡、线粒体上较多的t-Z,细胞壁中t-Z似被转运。×32 000

3. 未受精卵合点端,示核被膜、染色质、核基质及内质网上密集分布t-Z。

×32 0004. 未受精卵珠孔端,示胞质内较多的 t-Z。卵细胞与助细胞间似有 t-Z 跨壁转运。×32 000 **5.** 助细胞近中央部分,示胞核与细胞质内 t-Z 较少。×28 000 **6.** 合子合点端正在加厚的壁中存在 t-Z。×32 000

图 版 Ⅱ

7. 合子合点端,示细胞核各个部位 t-Z 较少,偶见一簇 t-Z 分布于核被膜,似在转运。×32 000 **8.** 合子中央部分,示粗糙内质网、线粒体等细胞器上 t-Z 很少,胞质基质内有一些 t-Z。×32 000 **9.** 合子与宿存助细胞相邻部分,示较多的 t-Z 似在宿存助细胞与合子间跨壁转运。×38 000 **10.** 阴性对照。正常兔血清代替兔抗 t-Z 多克隆抗体后,未受精卵核内几乎未见金颗粒。×32 000 **11.** 未受精卵珠孔端部分,示液泡、线粒体及胞质基质内(+)ABA 较少。×32 000 **12.** 未受精卵合点端,示胞质基质与细胞核内 GA_7 与 GA_4 较少。×32 000

图 版 Ⅲ

13. 未受精卵珠孔端部分,示胞质基质及线粒体中有少量 GA_7 与 GA_4,与助细胞相接的壁中 GA_7 与 GA_4 极少。×32 000 **14.** 合子合点端部分,示细胞质与细胞核中(+)ABA 略增加,脂质体、淀粉体、线粒体中有(+)ABA。×32 000 **15.** 合子珠孔端部分中略增加的 GA_7 与 GA_4,脂质体中也有。×32 000 **16.** 未受精卵合点端,示细胞质与细胞核中(+)ABA 很少。×32 000 **17.** 未受精卵珠孔端部分,示极少的 IAA 分布在液泡、线粒体、质体及细胞质基质中。×32 000 **18.** 合子近中央部分,示细胞核及细胞质内 IAA 极少。×32 000

Explanation of plates

a. Amyloplast   CC. Central cell   d. Dictyosome   E. Unfertilized egg cell   er. Endoplasmic reticulum   En. Endosperm cell   l. Lipid body   m. Mitochondrium   N. Nucleus   ne. Nuclear envelope   no. Nucleolus   p. Plastid   pm. Plasma membrane   PS. Persistent synergid   Sy. Synergid   v. Vacuole   ve. Vesicle   w. Cell wall   Z. Zygote

Longitudinal sections of unfertilized egg cells were prepared from materials sampled at pollination and of zygotes sampled 96 h after pollination in tobacco. Some of the gold granules are indicated with arrowheads in all longitudinal section micrographs.

Plate Ⅰ

Fig. 1. Establishment of immunogold-silver enhancement methods for trans-zeatin

第四章 有性生殖的超微结构与细胞化学研究

(t-Z, i), indoleacetic acid (IAA, ii), (+) abscisic acid [(+) ABA, iii] and gibberellin A_4(GA_4, iv) on nitrocellulose film (NCF) model system. t-Z (B, C) and t-Z riboside (D) were treated with gas-phase fixation in paraformaldehyde and glutaraldehyde. Three acidic hormones were treated with gas-phase fixation successively with EDC [1-ethyl-3-(3-dimethyl-aminopropyl) carbodiimide] and aldehydes. BSA was selected as carrier protein with 20 μg and 1 ng phytohormone in 2 μl buffer per small piece of NCF in all tests. A, control, only BSA added to NCF; B and C, mixture of one phytohormone and BSA added to NCF (B as the negative control, i.e., presaturation of specific antibody with 1 mmol/L phytohormone). Note t-Z riboside is faintly fixed with aldehydes (D) and the comparable results among the four phytohormones (C). **Fig. 2.** Adjacent portion between a central cell and an unfertilized egg cell, showing the plentiful t-Z in the amyloplasts, vacuole and mitochondria of CC; and some seeming being translocated t-Z in the cell wall between CC and E. ×32000 **Fig. 3.** The chalazal end of an unfertilized egg cell, showing the quite abundant t-Z in the nucleus; note the concentrated t-Z localizations in the nuclear envelope, chromatin, nucleomatrix, and endoplasmic reticulum. ×32000 **Fig. 4.** Micropylar end of an unfertilized egg cell, showing many t-Z in the cytoplasmic matrix. There may be translocation of t-Z across the wall between E and Sy. ×32000 **Fig. 5.** The nearly central portion of a synergid, showing a few t-Z in the nucleus and cytoplasm. ×28000 **Fig. 6.** Chalazal portion of a zygote, showing some t-Z in the thickening wall of the zygote. ×32000

Plate II

Fig. 7. Chalazal end of a zygote, showing a few t-Z in various parts of the nucleus; occasionally a cluster of t-Z at the nuclear envelope as being translocated. ×32000 **Fig. 8.** Central portion of a zygote, showing some t-Z in the cytoplasmic matrix but few in the rough endoplasmic reticulum, mitochondria and other organelles. ×32000 **Fig. 9.** The neighboring portion of a zygote and a persistent synergid, showing concentrated t-Z apparently translocated between PS and Z. ×38000 **Fig. 10.** Negative control. Almost no gold granules is visualized in the nucleus of an unfertilized egg cell when anti-t-Z polyclonal antibody was replaced by normal rabbit serum. ×32000 **Fig. 11.** Micropylar portion of an unfertilized egg, showing a few (+)ABA in the vacuole, mitochondria and cytoplasmic matrix. ×32000 **Fig. 12.** The chalazal end of an unfertilized egg, showing a few GA_7 and GA_4 in the cytoplasm and nucleus. ×32000

图版 I
plate I

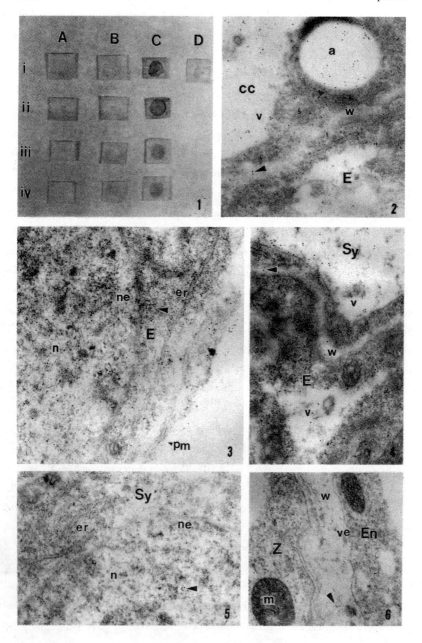

第四章 有性生殖的超微结构与细胞化学研究

图版 II plate II

图版Ⅲ   plate Ⅲ

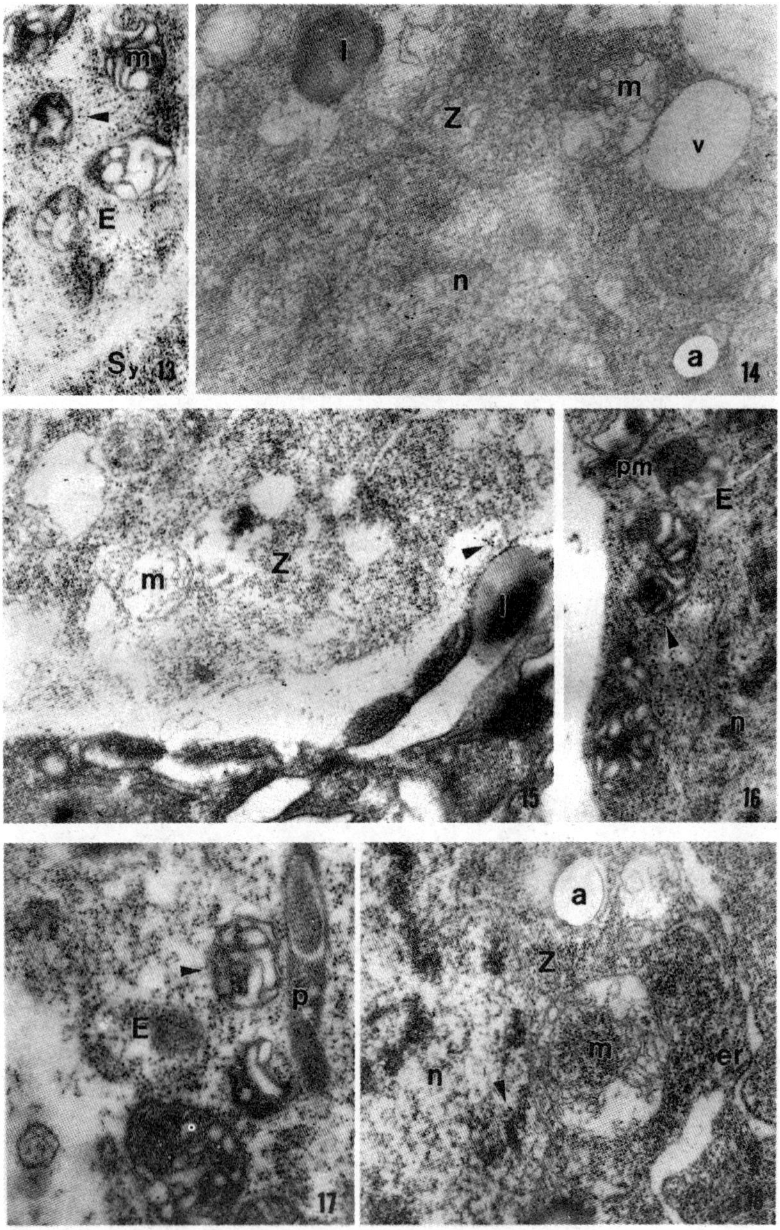

第四章 有性生殖的超微结构与细胞化学研究

Plate III

Fig. 13. The micropylar portion of an unfertilized egg cell, showing some GA_7 and GA_4 in the cytoplasmic matrix and mitochondria; note very few GA_7 and GA_4 in the cell wall adjacent to the synergid. ×32000 **Fig. 14.** The chalazal portion of a zygote, showing the slight increase of (+) ABA in cytoplasm and nucleus in comparison with those in the unfertilized egg cell (Fig. 17); note the (+) ABA localized in the lipid bodies, amyloplasts and mitochondria. ×32000 **Fig. 15.** The micropylar portion of a zygote, showing the slight increase of GA_7 and GA_4 in comparison with those in unfertilized egg cell (Fig. 14); note the GA_7 and GA_4 localizations in mitochondria and lipid bodies. ×32000 **Fig. 16.** The chalazal end of an unfertilized egg cell, showing few (+) ABA in the nucleus and cytoplasm. ×32000 **Fig. 17.** The micropylar portion of the unfertilized egg cell, showing very few IAA in the vacuole, mitochondria, plastids and cytoplasmic matrix. ×32000 **Fig. 18.** The nearly central portion of a zygote, showing very few IAA in the nucleus and various parts of cytoplasm. ×32000

(作者:陈以峰、梁世平、杨弘远。原载:植物学报,1999,41(11):1145~1149。图版3幅,参考文献20篇。选录全部图版。)

5 烟草原胚中赤霉素 GA_7 与 GA_4 分布的免疫电镜观察

Distribution of gibberellins A_7 and A_4 in tobacco proembryos using immunoelectron microscopy

Abstract

The ovules of *Nicotiana tabacum* var. *macrophylla* 8 days after pollination were fixed successively with 2% EDC [1-ethyl-3-(3-dimethyl-aminopropyl) carbodiimide] and a mixture solution of paraformaldehyde and glutaraldehyde, then slightly postfixed in 0.5% osmium solution. Ultrathin sections of 6-, 9- or 12-celled proembryos embedded in Epon 812 resin were stained in an anti-GA MAb and sheep anti-mouse IgG-colloidal gold (10nm). This MAb specifically recognizes methyl esters of GA_7 and GA_4. Therefore, it could be used as a probe to localize GA_7 and GA_4 in cells after EDC fixation. The 12-celled proembryo is composed of a 9-celled embryo and a 3-celled suspensor. Wide distribution of GA_7 and GA_4 was observed in all proembryo cells and most organelles at subcellular level, including walls, plasmodesmata, plasma membrane, cytoplasmic matrix, mitochondria, plastids, vacuoles, endoplasmic reticulum and nuclei. Clusters of gold granules were found in nuclear envelope, nucleomatrix, nucleolus, chromosome, cytoplamic matrix. In a region composed of cytoplasmic matrix, a vacuole and a mitochondrium, such concentrated gold granules were particularly obviously observed. There appeared a gradient distribution of GA_7 and GA_4 from embryo cells decreasingly to suspensor cells. GA_7 and GA_4 could be translocated via intercellular walls, plasmodesmata and vesicles between embryo and suspensor. The authors suspect the direction of GA translocation in proembryo may be from suspensor to embryo. To the author's knowledge, this is the first report to indicate subcellular and gradient distributions of bioactive gibberellins in plant proembryos.

植物胚胎发生可大致分为3个时期:(1)合子-原胚时期;(2)球形-心形时期;(3)鱼雷形-成熟时期[1]。在这3个时期中先后发生的最显著变化,分别是顶-基轴形成、辐射对称轴向两侧对称轴转变以及器官扩大、贮藏物积累、脱水与休眠[1,2]。脱落酸能够促进成熟胚的贮藏蛋白积累、脱水、维持其休眠状态[3],球形胚中生长素的极性运输为形成心形胚的两侧对称轴所必需[4,5]。但迄今尚未见到报道植物激素在胚胎发生中的免疫细胞化学定位研究。本文用免疫电镜技术对赤霉素在原胚中的分布作了探索。

专一识别 GA_7 甲酯与 GA_4 甲酯的单克隆抗体制备方法见文献[6]。按照文献[7]的方法制备胶体金及金标羊抗鼠抗体,胶体金直径为 10nm。参照文献[8]的方法制备透射电镜样品和进行免疫细胞化学反应。取授粉 8d 的大叶烟(*Nicotiana tabacum* var. *macrophylla*)胚珠用 2%EDC[1-乙基-3-(3-二甲氨丙基)碳二亚胺]室温下抽气固定 2h 后,转入 4% 多聚甲醛与 1% 戊二醛,室温抽气 2h 并于 4℃ 过夜,再以 0.5% 锇酸处理 30min,常规乙醇脱水,环氧丙烷过渡,Epon812 包埋。在 MT-6000 型超薄切片上切片,捞在覆有 Formvar 膜的镍网上,载片镍网经 10% H_2O_2 蚀刻 10min、阻断液(0.2mol/L 甘氨酸、2% 鸡卵清蛋白,溶于 PBST 液)阻断 20min(PBST 液组成:10mmol/L 磷酸钠缓冲液、0.15mol/L NaCl、0.02% Tween-20、0.02% NaN_3、pH7.2);在鼠抗 GA 单克隆抗体液滴(1∶25 000,PBST 稀释,并含 1% BSA)上 30℃ 温育 4h,PBST 洗涤后再在金标羊抗体液滴(1∶40,PBST 稀释,并含 1% BSA)上 30℃ 温育 70min,充分洗涤后醋酸铀与柠檬酸铅双染,在 JEM-100/Ⅱ 型透射电镜上观察、照相。为确保结果准确,设定 2 个阴性对照:(1)鼠抗 GA 抗体预先用 1mmol/L GA_4 甲酯 37℃ 温育 2h;(2)省去鼠抗 GA 抗体,其余步骤不变。

观察了 6-细胞、9-细胞和 12-细胞原胚。重点观察的 12-细胞原胚中,胚有 9 个细胞、胚柄有 3 个细胞。所有原胚的胚与胚柄细胞的胞质均较浓,核均较大(图1)。在这些胚细胞中线粒体与质体似乎少一些,胚柄细胞尤其是基部细胞中线粒体与质体丰富;胚细胞内液泡较小而多,胚柄细胞内液泡较大而少(图1)。在所有原胚细胞的亚细胞水平上观察到了 $GA_{7/4}$ 的广泛分布。细胞壁、胞间连丝、质膜、细胞质、线粒体、质体、液泡、内质网(图版Ⅰ,1、2)、细胞核中均存在或有大量的 $GA_{7/4}$。质体中的 $GA_{7/4}$ 位于膜与基质中(图版Ⅰ,1),线粒体中的 $GA_{7/4}$ 位于外膜、内膜及基质(图版Ⅰ,3、7)上。细胞核中的 $GA_{7/4}$ 分布于核基质、染色体、核仁(图版Ⅰ,4、5)以及核被膜的外侧、内侧及双层膜之间(图版Ⅰ,6)。暗示 $GA_{7/4}$ 可能参与原胚细胞中广泛的细胞生物学过程。胚柄基部与中部细胞中 $GA_{7/4}$ 不多(图版Ⅰ,8);顶部细胞中 $GA_{7/4}$

较多(图版Ⅰ,1),而胚细胞中则有大量的 $GA_{7/4}$(图版Ⅰ,1、3~7)。总体上 $GA_{7/4}$ 水平似存在从胚到胚柄由高到低的梯度分布,类似的梯度分布也在6-细胞与9-细胞原胚中观察到。胚细胞中的 GA,除了呈分散状态分布外,还以成簇形式分布在细胞质基质(图版Ⅰ,1)、染色体、核仁(图版Ⅰ,4)、核基质(图版Ⅰ,5)、核膜(图版Ⅰ,6)等处。在一处由相互接近的液泡-线粒体-细胞质基质组成的区域,成簇分布的金粒尤为密集(图版Ⅰ,7)。胚与胚柄细胞的间壁存在 $GA_{7/4}$,胞间连丝的各个部位(包括两端及内部)均有 $GA_{7/4}$ 分布,且在间壁的胚柄细胞一侧,小泡明显多于胚细胞一侧,小泡中含有金粒(图版Ⅰ,1),提示胚与胚柄细胞之间存在赤霉素分子的跨壁转运,转运方向很可能是从胚柄到胚。省去单克隆抗体后原胚中没有出现金标记,用 1mmol/L GA_4 甲酯封闭单克隆抗体的抗原结合部位,原胚细胞中金标记显著减少(图版Ⅰ,9),表明上述定位结果确是以抗 GA_4 单克隆抗体的特异性结合为基础的。

Hasegawa 等用抗 GA_1 甲酯多克隆抗体作探针定位水稻花药中的 GA,但只识别出结合态 GA_4 与 GA_7(17-葡萄苷)组织水平分布[9],未能定位游离 GA_4 与 GA_7。迄今尚未对植物细胞内活性赤霉素(如 $GA_{7/4}$)进行准确定位,重要原因之一在于特异识别一种或两种活性 GA 的抗体极难获得。本研究采用的单克隆抗体是以琥珀酸酯-13-GA_3-7-CONH-HSA(HSA:人血清白蛋白)为免疫原,通过特定的杂交瘤技术筛选而获得,基本上只与 GA_7 甲酯与 GA_4 甲酯发生交叉反应[6]。该抗体也同样能够识别经 EDC 固定的细胞内 GA_4 与 GA_7(以 $GA_{7/4}$-7-CONH-proteins 形式存在),为本研究建立游离 $GA_{7/4}$ 免疫细胞化学定位方法奠定了基础。运用这一新方法,在烟草原胚的亚细胞水平上揭示了赤霉素的广泛分布。这些结果将有助于深入探讨亚细胞水平上赤霉素的调节功能。

(作者:陈以峰、梁世平、杨弘远、郑志富、周燮。原载:植物学报,1998,40(5):478~480。图版1幅,图1幅,参考文献9篇,均删去。)

其它论文目录

1. Yan H. Yang HY. Jensen WA. 1991. Ultrastructure of the developing embryo sac of sunflower (*Helianthus annuus*) before and after fertilization. Canadian Journal of Botany, 69:191-202.

2. 陈绍荣、毕学知、吕应堂、杨弘远. 一种优化的植物组织 RNA 原位杂交技术. 遗传,1998.20(3):27~30.

3. 陈绍荣、吕应堂、杨弘远. 钙调素 mRNA 在烟草花粉发育过程中的原位定位. 科学通报,1998.43(20):2202~2205.

4. 陈以峰、梁世平、杨弘远. 烟草胚囊胞外钙调素的免疫电镜定位. 电子显微学报,1998.17(4):383~384.

5. 陈绍荣、吕应堂、杨弘远. 磷酸化酶 mRNA 在水稻雌蕊中表达的时空动态. 遗传,1999.21(3):31~33.

第五章　植物实验生殖生物学专题论述

提要

在本书"前言"中已经概括性地描绘了由植物胚胎学到植物生殖生物学、特别是由实验胚胎学向实验生殖生物学演变的趋势。在第二、第三、第四章中,以我们自己的研究具体展示了这一趋势的几个方面。但是,一个单位的工作毕竟是有限的,只是科学海洋中的一叶小舟。为了了解国际上这一研究方向的全貌,还需要概括性的综合评述。本章从我们历年发表的综述论文中选编了若干有代表性的文章供读者参考。这些综述论文撰写的年代有早有迟,只能反映当时的研究进展,但可为读者提供一些阶段性的历史总结及作者对这些问题的观点。这里还要重复本书"序"中的声明,即原综述论文中所附参考文献一律删去,以节省篇幅。

1989年我们撰写了"植物实验生殖生物学与生殖细胞工程:现在与未来"一文,就本分支学科的特点和发展趋势提出了总的看法。该文在本书中被采用作为"前言",仅在个别词句上作了修改。1990年,作者应邀在美国一次学术会议上以该文主旨作了报告。以后修改成文,补充了我们自己的具体研究内容,撰写成英文稿,以特邀论文形式发表于1992年的 *American Journal of Botany*,该期还以我们提供的花粉原生质体显微摄影作为封面照片。(如书末彩色图版Ⅲ中的幼嫩花粉原生质体群体)。本章即以该文作为第一篇文章。

第二篇文章是1982年发表于 *Theoretical and Applied Genetics* 中的未传粉子房与胚珠培养综述。当时这项研究在国际上起步不久,有关原始论文散见于法文与中文期刊,鲜为人知。我们这篇综述首次在国际刊物中以英文介绍了这个研究方向的历史背景、现状与前景,在国际上起了较大的推动作用。后来于1990年,我们应邀为 *Plant Tissue Culture: Applications and Limitations* 一书撰写了"离体雌核发育"专章,系统地总结了10多年来国际上这方面的研究进展。本书以此作为第三篇文章。以上两文一先一后,大体上反映了这项研究高潮时期的来龙去脉。

第五章 植物实验生殖生物学专题论述

从20世纪80年代中开始，国际上掀起植物性细胞操作的研究热潮。我们及时地总结了这方面的研究进展，以"花粉原生质体、精子与生殖细胞的实验研究"为题，发表于1989年的《植物学报》。此后，又陆续在其他专家主编的几部专著中分别撰写了有关专题论文。其中，1993年发表的"植物性细胞操作的研究进展"，内容最为详细，但因篇幅超过两万字，故未收入本书（见本章末的其它论文目录）。另一篇1997年发表的"植物性细胞原生质体的分离、培养和融合"，反映了较近的研究成果，现收入本章。上述第四、第五两篇文章，大体概括了20世纪80年代中至90年代中这段时期国际上有关植物性细胞操作的研究进展。这一研究被纳入本书的最后一篇文章，是2001年应邀为波兰科学院主办的 Acta Biologica Cracoviensia Series Botanica 年刊撰写的英文综述，其中反映了我们直至20世纪末在植物性细胞操作方面的主要工作；对于国际上有关研究动态，限于篇幅，仅择其要者加以引证。

关于受精生物学的研究，本章收入了三篇论文。其中"受精过程中助细胞退化机理的研究进展"发表较早，内容较专，但总结出了一个关于助细胞退化机理的假设。1998年发表的"被子植物离体受精与合子培养研究进展"概括了这一重要实验研究的历史渊源和90年代以来的研究进展，从操作、细胞生物学、分子生物学三个方面加以阐述。1999年发表的"钙在有花植物受精过程中的作用"则从钙信号系统与受精的关系角度进行了专题论述。这一专题的有关文献甚多，尤其是关于钙在花粉萌发与花粉管生长的关系，研究报道层出不穷。而从钙在受精过程各个环节中的作用的角度进行综合性评述则尚属首次尝试。

本章最后三篇文章是关于实验技术的综合评述。其中关于荧光显微术的两篇发表较早，初学者可从中获取必要的研究背景和基本知识。整体透明作为一种简捷技术在植物胚胎学中有广泛用途，我们对此法作了全面而扼要的介绍。

除了本章所选用的上述文章以外，我们还先后撰写了20多篇综述论文，内容涉及面较广而杂，篇幅过大，无法编入本书，仅以论文目录方式附于章尾，供读者查考。

1 Experimental plant reproductive biology and reproductive cell manipulation in higher plants: now and the future

During the last two decades, plant embryology has undergone rapid and profound changes that have promoted its integration with other related sciences and increased its experimental nature. The initial product of this is the multidisciplinary field known as plant reproductive biology which incorporates the techniques of plant tissue culture with floral organogenesis. During the course of this development at the tissue level, a parallel trend has occurred at the cellular level that integrates techniques of plant cell culture with the isolation of the male and female gametes, their precursor cells, and gametophytes. We tentatively call this new branch of experimental embryology "experimental plant reproductive biology."

Two major features characterize experimental plant reproductive biology. The first is the modification of culture techniques from the level of organ and tissue culture to that of cell and protoplast manipulation. The techniques previously adopted for culturing the flower, anther, ovary, ovule, nucellus, endosperm, and embryo, as summarized in the book *Experimental Embryology in Vascular Plants* (Johri, 1982), are designed primarily for organ and tissue manipulation. Although these techniques are still valuable, they are being refined toward a more advanced technology. For example, from anther culture has developed free pollen cell culture capable of controlling the development of pollen along either gametophytic or sporophytic pathways (Heberle-Bors, 1989). More recently, manipulation of pollen protoplasts, generative cells, and sperm has been put on the agenda.

第五章　植物实验生殖生物学专题论述

A series of recent reports in experimental plant reproductive biology has been made that would permit these accomplishments, e. g. , generation of triploid hybrid plants via the fusion of tetrad protoplasts with somatic protoplasts (Lee and Power, 1986; Pirrie and Power, 1986; Pental et al. , 1988), isolation of large populations of pollen protoplasts in the living state (Baldi et al. 1987; Tanaka et al. 1987; Zhou, 1989a), induction of cell division in pollen protoplast culture (Zhou, 1989b), isolation of large populations of sperm cells (Russell, 1986; Dupuis et al. , 1987; Shivanna et al. , 1987; Cass and Fabi, 1988; Nielsen and Olesen, 1988; Theunis et al. 1988; Southworth and Knox, 1989; Yang and Zhou, 1989), isolation of large populations of generative cells (Tanaka, 1988; Zhou, 1988), microinjection of sperm cells into the embryo sac (Keijzer et al. 1988), and fusion of isolated sperm and egg (Kranz, et al. 1990), to name some of the major contributions to date. With respect to the female counterparts, the earlier techniques of ovary and ovule culture have expanded in recent years to include the isolation of viable embryo sacs and their component cells or protoplasts, either by enzymatic maceration (Zhou and Yang, 1985; Hu, Li, and Zhu, 1985; Mól, 1986; Huang and Russell, 1989; Wagner et al. 1989a; Wagner et al. , 1989) or by means of a micromanipulator (Allington, 1985), and their subsequent culture (Mól, 1986; Keijzer et al. 1988).

The second feature of experimental plant reproductive biology is the increased reliance on and necessity of a multidisciplinary approach in order to resolve plant reproductive problems at the cellular level. As is evident in the title of the review "Embryology in relation to physiology and genetics," by Maheshwari and Rangaswamy (1965), the previous technology of experimental embryology was a consequence of the integration of embryology with plant physiology and genetic breeding. Contemporary experimental plant reproductive biology, however, will have to represent an integration of more modern plant sciences, among which cell biology and molecular biology play the most crucial roles. For instance, in manipulating sperm cells, detailed knowledge is required regarding the isolation and purification of cells and organelles, direct observation of living cells, maintenance of cell viability, ultrastructural observation, immunological and biochemical assays, and protoplast culture and fusion.

The manipulation of reproductive cells shows great potential in plant biotechnology, and as Linskens (1988) pointed out: "Application of tissue and cell culture techniques marked the entrance of plant reproduction into the area of biotechnology." So far, the manipulation of individual cells (i. e., "cell engineering") in higher plants relies mainly on somatic protoplast systems. However, a new trend of manipulating reproductive protoplasts is emphasized in the symposium volume entitled *Plant Sperm Cells as Tools for Biotechnology* (Wilms and Keijzer, 1988), leading to an attractive new field of research in the manipulation of haploid and gametophytic cells (i. e., "plant reproductive cell engineering"). Its value lies in the nature of the cells involved in reproductive cell lineages. First, the sexual reproductive system that has arisen over the course of evolution is a natural source of offspring that cannot be entirely substituted by vegetative reproduction. Second, the reproductive cells in broad sense (spores, gametophytes, and gametes) are natural haploid cells and thus are of great value in genetic manipulation. Haploid protoplasts obtained directly from pollen, for example, offer a simpler method of obtaining haploid cells than regenerating plants indirectly via anther culture. Furthermore, natural haploid protoplasts, the "gametoplasts" as Russell (1986) called generative and sperm cells, already exist within the male gametophyte. Third, structural and functional differences in reproductive cells, especially in cytoplasmic organelles, as is shown in organelle dimorphism in sperm cell pairs (Russell, 1984), may provide useful materials for various goals of cell manipulation.

A new stage of experimental plant reproductive biology is possible in the area of reproductive plant cell manipulation based on recently published work. The urgent task is to establish various *in vitro* models that will be indispensable for further academic as well as applied research. In this paper, we present some recent advances in our laboratory along this direction and our ideas for future work.

Manipulation of pollen protoplasts

***Isolation*-**The pollen protoplast, *in situ*, is enclosed by a double-layered wall consisting of the exine and the intine. The exine consists of enzyme-resistant sporopollenin and is considered to be the main barrier against isolation of pollen protoplasts. However, recent experiments indicate that by hydrating pollen in a medium containing cellulase and pectinase, the exine of some pollen grains can be caused to dehisce at the germinal aperture, and as the intine is exposed to enzymatic action, the protoplast is released (Zhou, 1989a). Using this method, large populations of viable pollen protoplasts have been isolated in our laboratory from young pollen grains (late uninucleate to early binucleate stage) in *Hemerocallis fulva*, *H. minor*, *Gladiolus gandavensis*, and *Iris tectorium*, and from mature pollen grains in six monocots, including the above-mentioned four species, *Zephyranthes grandiflora* and *Hippeastrum vitatum*, and in two dicots, *Brassica campestris* var. *purpurea* and *B. napus*, Figure 1 shows the released pollen protoplasts of *Zephyranthes grandiflora*. Similar methods and results have been reported by Tanaka et al. (1987) in *Lilium longiflorum*.

The success of this method depends on suitable hydration and enzymatic maceration protocols, as well as the nutrients, osmoticum, and protectants used. A critical factor seems to be the characteristics of the pollen wall itself. Pollen grains in all above-mentioned taxa are characterized by long furrows on the exine, which upon hydration expand to a considerable extent, making the exine easily detached from the intine-coated protoplast. Whether this method can be adapted to the pore-bearing pollen grains such as those in graminaceous species still must be resolved.

Culture-The successful isolation of pollen protoplasts enabled us to carry out culture experiments on a large scale. These experiments were aimed in two directions: maintaining gametophytic growth and switching the pattern of growth to the sporophytic developmental program. Significant progress has been made in both directions (Zhou and Wu, 1990). For maintaining gametophytic development, mature pollen protoplasts of *Gladiolus gandavensis* were cultured in K_3 medium (Nagy and Maliga, 1976) supplemented with 32% sucrose, 0.1 mg/L 2,4-dichlorophenoxyacetic acid (2,4-D), 1mg/L naphthalene-acetic acid (NAA), and 0.2mg/L benzyladenine (BA). After 18 hr of culture, cell walls were reestablished (as viewed by electron microscopy), and a large number of protoplast-derived pollen tubes were produced (47.7%-59.3% calculated under an inverted microscope; Wu and Zhou, 1990). The tubes appear normal in both morphology and growth behavior (Fig. 2). Localization of actin filaments with FITC-conjugated phalloidin demonstrates that the newly released pollen protoplasts contain an irregular network of actin filaments. Upon germination, however, they became polarized, converging into the tube and becoming distributed longitudinally along the tube (Fig. 3), resembling the case of normal pollen grain germination (Zhou, Yang, and Zee, 1990).

Pollen protoplasts that develop small embryogenetic masses of cells are examples of entry into the sporophytic pathway. Protoplasts released from late uninucleate microspores of cold-pretreated flower buds in *Hemerocallis fulva* are examples of such embryogenetic potential. These protoplasts were cultured in K_3 medium supplemented with 1 mg/L 2,4-D, 2 mg/L NAA, 1% calf serum, 1% Polybuffer 74, 9% mannitol and 17% sucrose. A frequency of up to 7.2% protoplasts was induced to divide (Zhou, 1989b), giving rise to proembryo-like structures and cell clusters (Figs. 4-7). Ultrastructural observations on pollen protoplast division are currently being conducted. Figures 8-11 show the ultrastructural organization of young pollen protoplasts of *H. fulva* and their mitotic derivatives. This provides experimental evidence that pollen protoplasts can be induced to follow the sporophytic developmental program and undergo divisions that may lead to plant regeneration. Similar results recently have been obtained in another species, *H. minor*.

Manipulation of generative and sperm cells

Isolation of generative cells- In about 70% of angiosperm species, the mature pollen is in a bicellular state, consisting of a vegetative cell and a generative cell—the latter being the precursor of the two sperm cells (Maheshwari, 1950). Isolation of generative cells is of importance for studies on both sperm formation and for cell manipulation. Zhou et al. (1986) released generative cells in *Haemanthus katherinae* by simply squashing the pollen under a coverslip and conducted observations on the condition of the cells (Zhou et al., 1988). For the goal of cell manipulation, however, large quantities of concentrated generative cells are needed. Zhou (1988), therefore, developed a two-step osmotic method in *Vicia faba* extending the original one-step osmotic shock method of Russell (1986) for isolation of *Plumbago* sperm cells. The pollen grains were incubated in a 20% sucrose solution for a period and then "shocked" by the sudden addition of an equal volume of water, causing the pollen to burst and release generative cells. Subsequent large-scale investigations have established several modified methods adaptable to various groups of plant materials (Wu and Zhou, 1991): 1) For pollen grains that readily burst, e. g., *Gossypium hirsutum*, a one-step osmotic shock is adequate. 2) For pollen grains in which bursting is difficult but germination is easy, the two-step method of osmotic shock is effectively applied just after germination, e. g., *Zephyranthes candida*, *Z. grandiflora*, *Licoris rediata*, *Impatiens balsamina*, and *Nicotiana tabacum*. 3) In the case of pollen grains that are either difficult to burst or to germinate in a simple sucrose solution, a weak enzymatic treatment is used to promote rapid bursting. *Iris tectorium*, *Gladiolus gandavensis*, *Hemerocallis minor*, *Allium tuberosum*, and *Hippeastrum vitatum* belong to this last group. Figures 12 and 13 show representative results of generative cell isolation.

Within the pollen grain, the generative cells are mostly spindle-shaped *in situ*. After liberation, though, the cytoplasm appears to condense gradually from the two poles of the cell toward the center, causing the whole cell to round and contract significantly. As a result, the cells become ellipsoidal and, ultimately, spherical (Zhou et al., 1988). We have studied microtubule organization in the generative cells of *Zephyranthes grandiflora*, both *in situ* and in released cells, by immunofluorescence microscopy using a monoclonal antitubulin (Zhou et al. 1990; Zhou, et al. 1990). *In situ*, the microtubule bundles are oriented mainly longitudinally, in parallel with the long axis of the spindle-shaped cell (Figs. 14, 15). After isolation, spindle-shaped and ellipsoid cells display transitional forms consisting of a mixture of longitudinally oriented microtubule bundles and meshes (Figs. 16-18), whereas spherical cells display predominantly meshlike distributions of microtubules (Fig. 19). This indicates that the microtubule cytoskeleton of generative cells can change easily from one structural form to another in accordance with environmental conditions and that microtubules may play an important role in cell shaping as Heslop-Harrison et al. (1988) have suggested.

Culture and fusion of generative cells-Isolated generative cells of *Hemerocallis minor* were cultured using an agarose medium nursed with anthers of the same species (Wu and Zhou, 1990). Microscopic observations on a total of 5961 cells revealed that a number of the cells underwent nuclear divisions resulting in two (rarely three or four) nuclei (Figs. 20-25). The division frequency is 3.27% on average, and 11.46% in the best case. These results indicate that free generative cells are capable of surviving and undergoing limited development under *in vitro* conditions. Sustained division, however, is by no means easy to obtain and will undoubtedly need further refinement of culture conditions.

Fusion experiments are also being carried on with isolated generative cells. Using the polyethylene glycol (PEG) method, generative cells of *Zephyranthes candida* were fused with each other and gave rise to viable homokaryons (Figs. 26-29). Fusions between the generative cells of *Hippeastrum vitatum* and the somatic protoplasts of *Nicotiana tabacum* also resulted in heterokaryons (Wu and Zhou, unpublished data). These preliminary experiments provide optimism with regard to the prospect of generative cell manipulation.

Isolation of sperm cells-Isolation of sperm cells has become a routine practice in many laboratories (see Introduction). Nevertheless, in only some cases has the viability of isolated sperm cells been confirmed by the fluorochromatic reaction (FCR) (Dupuis et al., 1987; Matthys-Rochon et al., 1987; Theunis, van Went, and Wilms, 1988; Yang and Zhou, 1989). With a view of obtaining sperm cells available for cell manipulation, maintaining the viability of sperm cells is of particular importance. In our experience, the isolated sperm cells assayed by fluorescein diacetate may be stored in a refrigerator and display FCR positivity (FCR+) for 3 d in *Brassica napus*, 1 d in *Zea mays*, but not more than 30 min in *Secale cereale* (Yang and Zhou, 1989). Figure 30 shows an isolated maize sperm cell population stained by the nuclear fluorochromes Hoechst 33258. Figures 31 and 32 demonstrate pairs of sperm cells exhibiting FCR+. More recently, this technique has been further improved by supplementing the medium with such protectants as polyvinylpyrrolidone and bovine serum albumin, as well as potassium dextran sulfate. Southworth and Knox (1989) have recently recommended the use of such protectants in sperm cell isolation. We have obtained high frequency of isolated viable sperm cells in *Brassica campestris* var. *purpurea* (up to 86%), and they could survive for up to 6-7 d in a refrigerator. Figures 33 and 34 demonstrate isolated sperm cells of *B. campestris* var. *purpurea* showing strong FCR+.

Manipulation of embryo sacs

Culture of unpollinated ovaries and ovules-Beginning 10 yr ago, we initiated research on culture of unpollinated ovaries and ovules aimed at inducing development of haploid plants; this subsequently led to the manipulation of the embryo sac itself. Haploid plants have been produced by this method in *Oryza sativa*, *Hordeum vulgare*, *Helianthus annuus*, and *Allium tuberosum*. The origin and development of these haploid products, as revealed by embryological observations, appear to be parthenogenetically derived *in vitro* from the development of the unfertilized egg cell in barley, sunflower, and *A. tuberosum*, synergid apogamy in rice, and antipodal apogamy in *A. tuberosum*, respectively. The advances in this field in general and our laboratory in particular have been reviewed elsewhere (Yang and Zhou, 1982; Hu and Yang, 1986; Yang, Yan, and Zhou, 1990; Yang and Zhou, 1990). An ultrastructural study on *in vitro* parthenogenesis in sunflower was recently carried out confirming and extending these observations (Yan, et al. 1989). All these investigations, taken together, are not only centered on manipulation of ovaries and ovules, but also provide a logical technical basis for the further manipulation of embryo sacs.

Isolation of embryo sacs-The enzymatic method for isolating embryo sacs has been worked out in more than ten species for fixed materials and three species (*Antirrhinum majus*, *Nicotiana tobacum*, and *Helianthus annuus*) for fresh materials (Zhou and Yang, 1985). In sunflower, some events of fertilization were observed in freshly isolated and Hoechst 33258-stained embryo sacs (Zhou, 1987). These results together with those obtained by other investigators (see Introduction) point to the possibility of embryo sac culture in the future.

Prospects

Early in 1950, when experimental embryology first originated, Maheshwari (1950) envisioned that the future of plant embryology would be related to the experimental direction. Forty years have now elapsed, and experimental embryological research is still one of the most active frontiers in plant reproductive biology. Keijzer, Wilms and Mogensen (1988) recently outlined some of the prospects in sperm cell manipulation: sperm cell culture as a possible source for haploid plants; fusion between isolated sperm cells and egg cells; fusion between sperm cells and somatic protoplasts; sperm cells used for DNA uptake and transformation; and long-term storage of sperm cells as a gene bank, among others. These goals are undoubtedly among the most fascinating in this area of endeavor at the present time.

Our ideas are focused on creating various models for reproductive cell manipulation, as well as for basic biological research. Reproductive cell manipulation, as currently envisioned, would include the following:

Pollen protoplast culture: The success of controlling development in young pollen protoplast cultures, namely the induction of cell divisions along a sporophytic pathway, is a hopeful step toward reaching the eventual goal of plant regeneration. This system, if fully realized, may be useful for fusion, mutation, or transformation, and will be superior to a number of alternatives since it is a haploid protoplast system.

Pollen protoplast germination: *In vitro* pollination and fertilization may be further utilized to generate protoplast-derived pollen tubes that develop along the gametophytic pathway. Since the pollen tubes derived in this manner apparently function successfully in both transportation of the male gametes and uptake of materials from the medium, it seems that foreign genes introduced into the medium could be transported readily to the fertilized egg (for reference see Chapman, Mantell, and Daniels, 1985). Considering that uptake of DNA by pollen protoplasts as wall-less cells should be much easier than through thick-walled pollen grains, the rate of foreign DNA incorporation could be much higher in the case of pollen protoplast pollination.

Generative cell culture: Although nuclear divisions have now been observed in cultured generative cells, these cells are small in size and poor in cytoplasm. To induce cultured generative cells to regenerate plants seems to be a complicated task. More difficult still may be sperm cell culture, since the cytoplasm of these cells, being smaller and more specialized, is further impoverished.

Fusion of generative and sperm cells with somatic or pollen protoplasts: Such fusions are likely feasible, because the somatic protoplasts and even pollen protoplasts are totipotent in development. These fusions, as such, may be considered a mimic of *in vitro* fertilization. Again, in this case, the generative or sperm cells can be utilized as vectors for foreign genes that one desires to introduce into the recipient cells.

Female cell culture and intergametic fusion: To realize intergametic fusion, i. e., true *in vitro* fertilization in higher plants, is the most attractive goal in plant reproductive cell manipulation, but as far yet is a formidable task. The difficulty lies in handling female rather than male cells. Culture of delicate and recalcitrant embryo sacs and egg cells is a significant challenge. Egg cell culture was, in fact, tried many years ago without success (Walker and Dietrich, 1963). Recently, Kranz, Bautor, and Lörz (1990) successfully fused an isolated sperm cell with an isolated egg cell using electroporation; by using an elaborate single cell culture technique, they achieved, for the first time, "artificially produced zygotes" that divided to form small multicellular structures (microcalli). This is a significant accomplishment that provides a stimulus for future research in this direction.

第五章 植物实验生殖生物学专题论述

Storage of reproductive cells: In order to enhance flexibility in experimentation, the cryopreservation of gametoplasts will be needed. Preliminary results indicate that *Brassica* sperm cells can survive up to 1 wk in a 4 C refrigerator (unpublished data), and that corn sperm cells can withstand freezing at −75℃ for up to 2 wk without special cryoprotection (Cass and Fabi, 1988). These successes seem to be an encouraging sign in regard to developing generative or sperm cell banks in the near future, just as pollen banks developed in the past.

The prospect of experimental research deepening our understanding of plant reproductive biology should also be noted. For example, the pollen protoplast systems may be helpful in throwing some light on such fundamental problems as pollen wall formation and pollen protoplast wall regeneration, germ aperture determination, pollen membrane structure and physiology, cytoskeletal organization, dedifferentiation and redifferentiation, and numerous others. The gametoplast systems are valuable for studies of the male and female germ units, gene expression in gametogenesis, gametic recognition, preferential fertilization, and responses of isolated gametes to environmental factors. These cells may also represent ideal cell sources for the application of molecular biology to the problems of the control of gene expression during reproduction, in the phase transition between the gametophyte and sporophyte, and during embryogenesis in general. It is certain that during the next 10 years, reproductive biology in higher plants will be greatly assisted by these and other new experimental systems.

(Authors: Yang HY, Zhou C. Published in: American Journal of Botany, 1992, 79: 554-363, with 34 figures and 53 references, all omitted here.)

2 *In vitro* induction of haploid plants from unpollinated ovaries and ovules

Historical Description and Present Status

Since the discovery of the first haploid in *Datura stramonium* in 1921, many attempts have been made to induce the parthenogenetic development of the unfertilized egg or other embryo sac cells. Many methods have been tried and these can be classified roughly into two types. One is *in vivo* induction of haploidy by various physical, chemical or biological stimulants. Much experience has been gained and important achievements have been made in some cases in these (Lacadena 1974). However, none of them can be regarded as a widely applicable means. Beginning in the late 1950's, the Maheshwarian school of plant embryology pioneered another method of haploid induction via *in vitro* culture. Culture of unfertilized ovules or ovaries has been tried in many materials, e. g. *Cooperia* (Sachar and Kapoor 1958), *Zephyranthes* (Sachar and Kapoor 1959), *Althaea rosea* (Chopra 1958, 1962; see Maheshwari and Rangaswamy 1965), *Papaver somniferum* (Maheshwari et al. 1961), *Allium cepa* (Guha and Johri 1966), *Nicotiana tabacum* (Siddiqui 1964) etc. All experiments failed except in one particular case in which an obligate apomict, *Aerva tomentosa*, was dealt with (Murgai 1959; Puri 1963; see Maheshwari and Rangaswamy 1965). Summarizing the work done during this period, Maheshwari and Rangaswamy (1965) concluded: "Ovule culture may also prove useful in the artificial induction of parthenogenesis. A direct handling of the eggs of angiosperms is by no means easy because of the problems involved in removing them without injury. However, there is no such bar to the use of unfertilized ovules. While ovules excised at the zygote or 2-to 4-

celled proembryo stage can be reared *in vitro*, unpollinated and unfertilized ovules have not proved amenable."

While ovule and ovary culture suffered setbacks, Guha and Maheshwari (1964) made a breakthrough in anther culture. This attracted the attention of researchers and as a result the culture of female tissues was neglected for about a decade. Yet from scattered reports during that period (e. g. Nishi and Mitsuoka 1969; Uchimiya et al. 1971; Mullins and Srinivasan 1976; Jensen et al. 1977; Pareek et al. 1980) some interesting results can be found. Uchimiya et al. (1971) cultured unpollinated ovaries of *Zea mays* and ovules of *Solanum melongena* and observed the division of haploid cells in their callus tissues. They believed that *in vitro* induction of haploid plants from angiosperm megagametophytes was possible. Jensen et al. (1977), dealing with cultured ovules of *Gossypium hirsutum*, found that the unfertilized polar nuclei could undergo free nuclear division and subsequent precocious cell formation; in addition, degeneration of one of the synergids took place just as it occurred after pollination.

So far as we know, the culture of unpollinated ovaries and ovules met with no success in haploid plant production until 1976 when San Noeum reported her first result in the ovary culture of *Hordeum vulgare*. Subsequently, Zhu and Wu (1979) obtained haploid plants from cultured ovaries of *Triticum aestivum* and *Nicotiana tabacum*. Yan et al. (1979) also raised a haploid albino plantlet in wheat ovary culture. Asselin de Beauville (1980) and Zhou and Yang (1980), using different methods, both succeeded in obtaining haploid plants from the ovary culture in *Oryza sativa*. In the meantime, ovule culture giving haploid plants was achieved by Cagnet-Sitbon (1980) in *Gerbera jamesonii* and by Ran (1980) in *Nicotiana tabacum*. Recently, haploid plants from ovaries have been also produced in barley (Wang and Kuang 1981), rice (Kuo 1982), tobacco (Wu and Chen 1982), *Lilium daviddii* (Gu and Chen, unpublished) and *Zea mays* (Ao et al., unpublished). Successes within such a short period indicate that the induction of haploids from female gametophytes via *in vitro* culture is not so inaccessible as had been thought before. Some data in this new field are shown in Table 1 for the convenience of reference.

Factors Affecting Culture Results

Genotype of donor plants

As in anther culture, a difference in response also exists among donor cultivars in ovary and ovule culture. For example, among 12 rice cultivars tested, 9 *japonica* and 2 of 3 *indica* cultivars could produce gynogenic calli with a percentage of ovaries ranging from 1.1% to 12%; 'Nang Ken No. 4', a *japonica* cultivar, showed the highest response in all experiments (Zhou and Yang 1982). The percentage of ovaries producing gynogenic calli in 4 wheat cultivars varied from 1.3% to 10.9% (Zhu et al. 1981). In *Nicotiana tabacum*, two cultivars had an induction frequency as high as 75% and 80%, but in another species, *N. rustica*, it was only 8% (Wu and Chen 1982). In ovule culture of 4 cultivars of *Gerbera jamesonii*, the percentage of ovules producing gynogenic calli ranged from 8% to 17% and the percentage of ovules regenerating plantlets was 0%-5% (Cagnet-Sitbon 1980). These facts indicate that genotype plays an important role in culture.

Embryo sac stage

It is not easy to observe directly the embryo sacs at the time of inoculation; an indirect judgement by pollen stage is more feasible. To do this, it is better to determine the corresponding embryo sac stages later by paraffin sections. In barley and rice some authors had good results only with late-staged ovaries, e.g. with nearly mature embryo sacs (San Noeum 1976, 1979; Asselin de Beauville 1980; Wang and Kuang 1981); others reported success with ovaries ranging from uninucleate to mature embryo sac stages (Zhou and Yang 1981b, 1982; Kuo 1982; Huang et al.

1982). *Gerbera* ovules responded only when their size was larger than half that of the ovary cavity (Cagnet-Sitbon 1980). Tobacco ovaries inoculated at the uninucleate (Zhu et al. 1981; Wu and Chen 1982) or binucleate (Wu and Chen 1982) pollen stage resulted in haploid production. It seems that quite a wide range of embryo sac stages are responsive to gynogenic development, but in most cases the later stages give better results. This is unlike anther culture in which mature pollen usually can not be induced to androgenesis.

Cold-treatment

Little data are available on this problem. Rice ovaries have been induced to gynogenesis when treated with low temperature (12-13℃) for 6 days after inoculation (Zhou and Yang 1980); however, repeated experiments indicate that there is no merit in cold treatment, either on panicles before inoculation or on cultured ovaries after inoculation (unpublished). Tobacco flower buds could be pretreated at 0-4℃ for 12h, but no comparative data were presented (Wu and Chen 1982). In *Gerbera*, cold pretreatment at 4℃ for 48h did not increase the number of gynogenic calli (Cagnet-Sitbon 1980). It appears that low temperature is not likely to be as effective a treatment in ovary and ovule culture as it is in anther culture.

Culture media

Basic media Most early work in the 1950s used Nitsch medium for ovule and ovary culture. However, since the 70s. Miller, MS or N_6 media have been used in successful experiments. Few experiments have been carried out to compare the effects of various basic media. In *Gerbera*, MS seems better than Knop and Heller (Cagnet-Sitbon 1980). An increase in the content of B group vitamins and glycine in H medium has been reported to have promoted induction-frequency in tobacco ovary culture (Wu and Chen 1982).

Exogenous hormones IAA (0.5-1mg/L) and KT (2-4mg/L) were supplemented in ovary culture of tobacco (Zhu and Wu 1979; Wu and Chen 1982). In *Gerbera* ovule culture, an auxin (IAA 0.5mg/L) plus two cytokinins (BA and KT 2mg/L each) proved most effective (Cagnet-Sitbon 1980). In graminaceous species, stronger auxins were adopted, e.g. 2,4-D 2mg/L (San Noeum 1976) or 2,4-D 0.5mg/L plus NAA 1mg/L (Wang and Kuang 1981) for barley; NAA 3mg/L (Asselin de Beauville 1980), MCPA 0.125mg/L (Zhou and Yang 1980, 1981a) or 2,4-D 2mg/L (Kuo 1982) for rice; 2,4-D 2mg/L for wheat (Zhu and Wu 1979). Comparative experiments in rice showed that when young flowers were cultured on liquid medium in the absence of exogenous hormone, there was a failure in the enlargement of the ovaries as well as in the production of gynogenic calli; an increase of MCPA concentration from 0.125 to 8mg/L did favor ovary swelling, but high auxin levels (2mg/L) stimulated callus formation from the ovary wall rather than from the embryo sac. Therefore, regulation of hormone level to enhance gynogenesis and inhibit proliferation of somatic tissue was considered to be a critical point in ovary culture (Zhou and Yang 1981b, 1982).

Sucrose concentration Sucrose concentration used in ovary and ovule culture was 3%-10% in barley, 8%-14% in wheat. 3%-6% in rice, 2% in tobacco and 3%-6% in *Gerbera*. No exactly comparative experiments were reported except in rice float culture, in which 1% and 9% sucrose proved unsuitable and 3%-6% were recommended (Zhou and Yang 1981b, 1982).

Solid and liquid media Almost all experiments succeeded on solid media, the exception being rice, which has been induced to gynogenesis either on solid (Asselin de Beauville 1980) or liquid (Zhou and Yang 1980, 1981a, b) media. A comparison showed the advantages of the latter over the former (Zhou and Yang 1982).

Modes of inoculation

Usually ovaries were randomly orientated on the solid medium. However, San Noeum noticed that better results could be obtained when ovaries were inoculated so that their placentae side was facing downwards to the medium (personal communication). Wang and Kuang (1981) placed barley ovaries in a vertical situation with the cut surface in contact with the medium. Another question is which parts are involved in the inoculum. Generally saying, the term ovary culture is in its exact sense a pistil culture. Some researchers have used whole flower buds or even pieces of inflorescence (Murgai 1959; Puri 1963; see Maheshwari and Rangaswamy 1965). Evidence has been given that in rice the best results of gynogenesis were obtained when an unhusked flower with pistil and stamens attached to the receptable was inoculated as a unit on liquid medium, the culture was less effective when the stamens were removed and was the worst when single pistils were used (Zhou and Yang 1981b, 1982). In barley, whole flowers (with or without stamens) inoculated vertically on solid medium proved to be a better culture source than randomly placed single pistils (Huang et al. 1982). It is of special interest that in *Gerbera* ovary culture failed whereas ovule culture succeeded (Cagnet-Sitbon 1980). It is unknown in this case whether the ovary wall has a harmful effect or whether it mechanically hinders the growth of gynogenic callus inside.

Culture conditions

Ovaries or ovules have been incubated usually at 25-28℃, either in darkness (Zhou and Yang 1980, 1981; Cagnet-Sitbon 1980) or with a 10-16h photoperiod of 500 (Cagnet-Sitbon 1980), 1,000 (Wang and Kuang 1981), 1500 (Wu and Chen 1982), 2000 (Kuo 1982) or 3000 (San Noeum 1976; Asselin de Beauville 1980) lux. No precise comparison has been reported.

Embryology of gynogenesis

The induction of haploid plants from cultured ovaries or ovules indicates that sporophytic development of megaspores or megagametophytes occurs in the absence of amphimixis. However, direct embryological observation is necessary in order to know the origins of the embryos and the developmental process.

Starting points of gynogenesis

San Noeum (1979), trying to analyze the origin of gynogenic proembryos in barley ovaries, classified them into eight groups, i. e. from (a) egg, (b) egg and antipodals, (c) antipodals, (d) egg and one synergid, (e) egg and two synergids, (f) one or two synergids, (g) synergids and antipodals and (h) synergids, egg and antipodals. She also mentioned that the best results were obtained with proembryos from the egg or antipodal cells; synergids gave only a proliferation of the callus type. These conclusions were drawn from observations on dissected ovaries under a binocular microscope and thus are puzzling. Zhou and Yang (1981 a, c) carried out observations on paraffin sections of cultured rice ovaries. They saw that while inoculation occurred at the 1-to 4-nucleate embryo sac stages, gynogenesis was not initiated until the gametophytes matured during culture; the proembryos were mainly located at the micropylar end and were derived from the egg apparatus. These observations were basically confirmed by Kuo (1982) in rice and Huang et al. (1982) in barley. Recently, based on large scale observations on rice ovary culture, Tian and Yang al. (unpublished) were able to trace the detailed apogametic development from the very beginning of the synergids into proembryos and calli; they also observed the proliferation of the egg cell and antipodals. On the other hand, in tobacco, ovary culture gynogenesis was reported to be initiated either directly from megaspores (Zhu et al. 1981; Wu and Chen 1982) or mature egg cells (Wu and Chen 1982). It is interesting to note that unfertilized polar nuclei could divide into endosperm-like structures in cotton ovule culture (Jensen et al. 1977), and in rice (Zhou and Yang 1981 a, c) and barley (Huang et al. 1982) ovary culture. Nevertheless, these structures did not seem to serve as a nurse tissue for the gynogenic embryoids nor were they themselves regenerated into plantlets.

Usually there was only one gynogenic unit inside one embryo sac but several units inside one embryo sac have been found in barley (San Noeum 1979; Huang et al. 1982) as well as in rice (Zhou and Yang 1981 a, c). Two possibilities can be supposed for the origin of such multiple units: polygenetic origin from different cells of the embryo sac, and monogenetic origin from a single cell, cleaving into several parts during subsequent development.

Characteristics of gynogenic development

As in pollen androgenesis, gynogenic plantlets may develop via embryoid or callus formation. To date in barley (San Noeum 1976, 1979; Huang et al. 1982) and tobacco (Zhu et al. 1981) only embryoids have been observed, but in rice (Asselin de Beauville 1980; Zhou and Yang 1981) and wheat (Zhu et al. 1981) both cases were identified. In barley ovary culture, the unfertilized eggs divided transversely into 2-celled proembryos, grew further into multicellular proembryos and then differentiated into embryoids more or less similar to *in vivo* zygotic embryos (Huang et al. 1982). In rice, under float culture conditions, proembryos were characterized by various morphological abnormalities, often leading to callus formation (Zhou and Yang 1981); detailed morphogenetic processes were further studied (Tian and Yang unpublished). Nagato (1979) has also observed various morphogenetic anomalies of rice embryos in caryopsis culture. It appears that *in vitro* culture may cause either zygotic or gynogenic embryos to behave quite differently from the *in vivo* developmental pattern.

Thus, the problem of how to control gynogenic development, shifting it to the pathway of normal embryogenesis, still remains to be solved.

Somatic calli or embryoids

A troublesome problem in ovary and ovule culture is the proliferation of somatic tissues which results in callus or embryoid formation. This not only makes it difficult to identify the gametophytic products from the sporophytic ones, but also may place inhibitory effects on the normal growth of the former. In fact, much previous work on ovary and ovule culture have merely induced somatic tissue proliferation (e.g. Maheshwari 1961; Mullins and Srinivasan 1976; Hsu and Steward 1976; Beasley 1977; Jensen et al. 1977; Pareek et al. 1980). Even in the cases of successful haploid induction, calli or embryoids might be simultaneously produced from such somatic tissues as ovary wall (Zhou and Yang 1982), funicle (Cagnet-Sitbon 1980), integument (Kuo 1982) or nucellus (Huang et al. 1982). Therefore, one should pay attention to the challenge of how to induce gynogenesis without the overabundant proliferation of somatic tissues. Hormonal regulation appears to play an important role in maintaining an optimal physiological balance (Zhou and Yang 1981b, 1982).

Cytological and genetic characteristics of regenerated plants

Ploidy Level

Under conditions when ovary or ovule culture produced haploid plants, haploids were sometimes obtained exclusively (San Noeum 1976, 1979; Zhu and Wu 1979; Yan et al. 1979; Asselin de Beauville 1980; Wang and Kuang 1981; Zhu et al. 1981), but in other cases both haploids and non-haploids were observed. For instance, plantlets regenerated from gynogenic calli of rice embryo sacs were observed to have haploid as well as diploid and polyploid roots (Zhou and Yang 1980, 1981a). A further large scale investigation of plantlets regenerated from 111 ovaries confirmed that 73.9% of them were haploid, 15.3% were diploid and 10.8% were mixoploid (Liu et al., unpublished). These non-haploids seemed to have originated more from a doubling during callus proliferation than directly from somatic tissues. Ovule culture of *Gerbera* induced 16 clones from gynogenic calli, of which 14 were haploid and 2 were diploid (Cagnet-Sitbon 1980). In tobacco, plants emerging directly from cultured ovaries were haploid, but those redifferentiating from calli were mostly diploid or mixoploid and a few were haploid (Wu and Chen 1982).

Albinism

Among species in which regeneration took place from cultured ovaries or ovules, *Hordeum vulgare*, *Nicotiana tabacum*, *N. rustica* and *Gerbera jamesonii* exclusively yielded green plants; *Oryza sativa* and *Triticum aestivum* produced both green plants and albinos. In wheat, plants regenerated from six ovaries were all green (Zhu et al. 1981). However, in another author's experiment, only one albino plantlet was reared from the same species (Yan et al. 1979). According to Asselin de Beauville (1980), rice plants regenerated from nine ovaries were all green. Dealing with rice ovary culture in three successive years, Zhou et al. found that green plants were dominant, although albinos did exist. In the first year, 5 ovaries were induced, among which 1 gave green plantlets and 4 gave albinos (Zhou and Yang 1980, 1981a); in the second year, 24 of 30 ovaries gave green plantlets and the other 6 gave albinos (Zhou and Yang 1981c); in the third year, from a total of 178 ovaries induced, 87.1% gave rise to green plantlets, 9.0% gave only albinos and the remaining 3.9% produced both green and albino plantlets (Liu et al., unpublished). Kuo (1982) also raised 12 green and 3 albino plants from rice ovary culture. It is evident that green plants are found in higher numbers in cereal ovary culture as compared with anther culture, in which albinism is a well known serious problem. For instance, San Noeum and Ahmadi (1980) have shown in barley cultivar 'Bérénicé' that 100% of the gynogenic plants were green; in contrast, 99% of the androgenic plants were albinos. Similarly, in rice cultivar 'Zao Geng No. 19', the percentage of green plantlets in ovary culture was 89.3%, but that in anther culture was only 36.4% (Liu et al., unpublished).

Other Characters

San Noeum and Ahmadi (1980) have compared two gynogenic and three androgenic doubled haploid lines with their original donor plants of the barley cultivar 'Bérénicé'. Block-trials in the field were made and the main agronomic characters were measured. Multivariate analysis showed that the groups of gynogenic doubled haploids were located close to the donor plants, while the androgenic ones were significantly distant from them. Analysis of progenies from selfing and reciprocal crosses confirmed the variability and showed important maternal and reciprocal effects, which led to the conclusion of new nucleocytoplasmic interactions.

Perspectives

The breakthrough of *in vitro* culture of unpollinated ovaries and ovules indicates that it may become an alternative way to haploid breeding. Its special contribution may be found in the following cases: (a) in certain species where anther culture has not succeeded or has a too low response to be actually applied to breeding, ovary or ovule culture may offer a useful approach. (b) in male-sterile plants, the value of haploid induction from female parts is obvious and this possibility has really been proved by the experiments in tobacco (Zhu et al. 1980). (c) while albinism is a limiting factor to anther culture of some cereal crops, ovary culture can provide a relatively higher proportion of green plants, as mentioned above. (d) In such cases when pollen plants show some unwelcome variation in ploidy or other characters, perhaps the gynogenic offsprings may behave otherwise. In fact, it is yet too early to talk about the validity of these merits. A great deal of work has to be done before it will become a practical and fruitful technique in plant breeding.

Other prospects are in the theoretical stages. Almost all stages in the reproductive cycle of angiosperms can now be studied by *in vitro* culture. Culture experiments of microspore mother cell, anther and pollen, embryo, endosperm and pollinated ovary or ovule have significantly pushed and are pushing forward our knowledge about microsporogenesis, pollen androgenesis, fertilization, embryogenesis, endosperm development and fruit formation. The understanding of such processes has a relation to the genetics and breeding of plants. However, *in vitro* study of megasporogenesis and the female gametophyte appears to be a unique gap among them. Taking this into account, Heslop-Harrison (1980) has recently made an appeal for enhancing the genetical and physiological researches on the "forgotten generation" – the angiosperm gametophytes. It is understandable that the culture of unpollinated ovaries and ovules may play an important role in filling this blank.

As reviewed by Raghavan (1976), the spontaneous origin of embryos from haploid components of the embryo sac has been recorded in about 100 species, but in the majority of cases it was merely a sporadic event and lacked cytologically substantiated accounts. As for induced parthenogenesis or apogamy, the existing methods are too laborious for routine work and the extremely low frequency of production of haploids has prevented more critical examination of their potential use. "Owing to such limitations, haploids originating from the embryo sac have had relatively little impact on contemporary studies and were regarded more as abnormalities than as basic tools for research. Technical considerations of isolating and culturing the delicate female gametophytes of angiosperms have thus far precluded attempts to study *in vitro* embryoid induction on them." Taking account of this, the significance of the culture of unpollinated ovaries or ovules is selfevident, not only due to its reproducible efficiency, but also because it can provide a basis for further approach to such important but yet unknown problems as why and how female gametophytic cells are directed to sporophytic development in the absence of fertilization.

(Authors: Yang HY, Zhou C. Published in: Theoretical and Applied Genetics, 1982, 63:97-104 with 1 table and 38 references, all omitted here.)

3 *In vitro* gynogenesis

"*In vitro* gynogenesis", a parallel term to "androgenesis", means the process of plant regeneration from the unfertilized egg cells (in broad sense also including the other haploid cells of the female gametophyte) in the cultures of unpollinated ovaries or ovules. The researches on *in vitro* gynogenesis have opened a new field in plant tissue culture and provide an alternative approach to haploid breeding. In a previous review we have described the historical background and early research status on this subject (Yang and Zhou, 1982). Since that time a considerable progress has been made, which can be summarized as follows: (a) regeneration of haploid plants via *in vitro* gynogenesis has been recorded in additional plant species, (b) culture techniques have been improved to raise the induction frequencies to a considerable degree in some species, and (c) the embryological aspects of this process have been studied in detail. Based on these advances the present article is aimed to make a more comprehensive assessment on this topic with emphasis placed on the recent results.

Records of haploid induction via *in vitro* gynogenesis

To our knowledge, so far gynogenic haploids through ovary/ovule culture have been obtained in sixteen species belonging to seven families (Table 1). However, the progress of researches is not so balanced: in some materials positive results have been obtained repeatedly and the works have continued, whereas in others only scattered reports are seen. Here we would rather like to describe briefly the situation family-wise so as to give a concrete impression to the readers.

Gramineae

Gramineae includes quite a number of species responsive to the *in vitro* gynogenesis. Barley (*Hordeum vulgare*) was the first species in which unpollinated ovary culture gave rise to haploid plants (San Noeum, 1976). Four cultivars were used and 0.2%-1.1% of the cultured ovaries produced one or two, rarely up to eight, plantlets (San Noeum, 1979). With one cultivar, the agronomic characters of the doubled haploid lines originated from gynogenesis and androgenesis were compared (San Noeum and Ahmadi, 1980). Other research groups were also successful with ovary culture of various cultivars of barley, using somewhat different methods (Wang and Kuang, 1981; Huang et al., 1982; Gu and Zheng, 1984).

The production of gynogenic haploid plantlets in wheat (*Triticum aestivum*) was first reported by Zhu and Wu (1979). In their second report, one cultivar and one hybrid among five materials used regenerated plantlets with the frequencies of 1.95% and 6.67%, respectively (Zhu et al., 1981).

Table 1 Records of induction of haploid plants via *in vitro* gynogenesis

Family	Species	Author(s) *
Gramineae	*Hordeum vulgare*	San Noeum, 1976 Wang and Kuang, 1981 Huang et al., 1982 Gu and Zheng, 1984
	Triticum aestivum	Zhu and Wu, 1979
	Oryza sativa	Asselin de Beauville, 1980 Zhou and Yang, 1980 Kuo, 1982
	Zea mays	Ao et al., 1982 Truong-Andre and Demarly, 1984
	Coix lacryma-jobi	Li and Zhang, 1984
Solanaceae	*Nicotiana tabacum*	Zhu and Wu, 1979 Wu and Cheng, 1982
	Nicotiana rustica	Wu and Cheng, 1982
	Petunia axillaris	De Verma and Collins, 1984
	Solanum tuberosum	Tao et al., 1985
Compositae	*Gerbera jamesonii*	Sitbon, 1981 Meynet and Sibi, 1984 Ahmim and Vieth, 1986 Cappadocia et al., 1988
	Helianthus annuus	Cai and Zhou, 1984 Gelebart and San, 1987
Chenopodiaceae	*Beta vulgaris*	Hosemans and Bossoutrot, 1983 Bornman, 1985 D'Halluin and Keimer, 1986 Van Geyt et al., 1987 Doctrinal et al., 1989
Liliaceae	*Lilium davidii*	Gu and Cheng, 1983
	Allium tuberosum	Tian and Yang, 1989
Salicaceae	*Populus X simonigra*	Wu and Xu, 1984
Euphorbiaceae	*Hevea brasiliensis*	Chen et al., 1985

* only the first report from a research group is listed in the table.

Several investigators succeeded independently in inducing *in vitro* gynogenesis in rice (*Oryza sativa*). In contrast to the standard ovary culture method adopted by Asselin de Beauville (1980) and Kuo (1982), Zhou and Yang (1980) developed a 'young floret float culture technique' for this species (see Zhou et al., 1983, 1986 for details). Among 19 rice cultivars tested with this method, all the 15 *japonica* and 2 of 4 *indica* cultivars responded positively. Using the same method, embryological investigations have been repeatedly carried out, and hundreds of the "embryo sac plants" were compared with the corresponding "pollen plants" with regard to ploidy level as well as green-albino ratio (Liu and Zhou, 1984). The method was adopted by Cai et al. (1988) for inducing haploid plantlets from a special photosensitive male-sterile rice line.

Two research groups have worked on corn (*Zea mays*) ovary culture (Ao et al., 1982; Troung-Andre and Demarly, 1984). Haploid plantlets were obtained and, in the latter case, the progeny of a gynogenic plant was studied.

Ovary culture also yielded haploid plantlets in another graminacious species, *Coix lacryma-jobi* (Li and Zhang, 1984).

Solanaceae

Zhu and Wu (1979) have induced haploid plantlets from cultured ovaries of *Nicotiana tabacum*. Further efforts of this group were pointed at the application of this method to various tobacco materials such as male-sterile lines (Zhu et al., 1980), and the gynogenetically derived haploid plants as the new donor plants for ovary culture (Zhu and Wu, 1981) as well as the propagation of large number of haploid plants (Zhu et al., 1984). In another lab, Wu and Cheng (1982) also obtained haploid plantlets of *N. tabacum* and *N. rustica* by ovary culture. Liu et al. (1986) reported their cytological observations on pollen abortion in the gynogenetically derived haploid plants in *N. rustica*.

There is only one report on *in vitro* gynogenesis in *Petunia axillaris* (De Verna and Collins, 1984). These authors cultured ovules attached to the placental tissue and their results showed that the ovules before anthesis yielded more haploids than those after anthesis.

Tao et al. (1985, 1988) have induced *in vitro* gynogenesis in *Solanum tuberosum*. Two of the five cultivars regenerated a few haploid plantlets. By cutting their stem segments, large number of test-tube plantlets and small potato tubers could be produced.

Compositae

In this family, *in vitro* gynogenesis has been repeatedly studied in two species, *Gerbera jamsonii* and *Helianthus annuus*.

Sitbon (1981) succeeded in the production of haploid plants in *G. jamesonii* by ovule culture, but failed in ovary culture. Meynet and Sibi (1984) obtained 200 gynogenic plants from ovule culture in 19 cultivars with an induction rate ranging from 4% to 7%. Ahmin and Vieth (1986) reported that *Gerbera* ovule culture yielded haploids with a frequency of 5%. More recently, Cappadocia et al. (1988) also obtained positive results in *Gerbera* ovule culture.

Cai and Zhou (1984) reported the results of young floret and ovule culture in sunflower. Eventually, four embryoids grew into haploid as well as diploid plantlets. Since the induction efficiency was low, Yan et al. (1987, 1988) continued this work with a large-scale culture experiments leading to significant increase in the induction frequency of gynogenic embryoids. The gynogenic embryoids, usually underdifferentiated, regenerated plants after several transfers to fresh medium (Hua and Yang, unpublished data). Gelebart and San (1987) also worked on *in vitro* gynogenesis in sunflower. They used eight genotypes including male-sterile and restoring lines and their F_1 hybrids. The method comprised ovary culture followed by culture of the ovules excised from the ovaries. On an average, five plants per thousand ovaries cultured were produced.

Chenopodiaceae

In this family only one species, *Beta vulgaris*, has been used for inducing gynogenic plants. Nevertheless, it has been studied by a number of research groups and has become one of the hopeful examples in respect to the application of this technique.

Hosemans and Bossoutrot (1983) cultured unfertilized ovules from male-sterile sugarbeet plants and obtained 0.3% yield of haploid plantlets. The same method was also applied to male-fertile plants (Bossoutrot and Hosemans, 1985). Bornman (1985) in his study on sugarbeet ovule culture induced gynogenic embryos with a rate of 0.3%-3%, and a part of the embryos regenerated haploid plants. D'Halluin and Keimer (1986) also carried out successful work on sugarbeet ovule culture and used it in practical breeding programs (cited in Olesen et al., 1988). Van Geyt et al. (1987) reported that either ovule or ovary culture could be used to induce haploid plantlets, but in the latter case the somatic cells derived from the ovary tissue must be removed or, alternatively, charcoal should be added to the medium to inhibit the callusing of somatic tissues. More recently, Doctrinal et al. (1989) used five male-sterile lines to study the factors affecting *in vitro* gynogenesis in sugarbeet ovule culture. A yield of 6%-10% gynogenic plants was obtained and among them 81% were haploid.

Other families

There are still several scattered reports on *in vitro* gynogenesis in some species belonging to other families.

Gu and Cheng (1983) induced 35 plantlets from cultured ovaries of *Lilium davidii*. Among them approximately 66% plants were haploid and the other 34% were diploid.

Allium tuberosum is a special species in which polyembryony derived from zygote and antipodal cells occurs naturally, but no haploid plants are produced *in vivo*. Using the method of unpollinated ovary culture, Tian and Yang (1989) were able to induce large number of haploid plants in this species.

Two woody plant species have been reported to exhibit *in vitro* gynogenesis. Wu and Xu (1984) cultured unpollinated ovaries of *Populus* X *simonigra* and obtained 26 viable plants, among which 12 plants were haploid. Chen et al. (1985) carried out ovule culture in *Hevea brasiliensis*, the rubber tree, and obtained 4 haploid plants.

Factors affecting induction of gynogenesis

Data accumulated to date are not rich enough to make a generalization about the techniques for the induction of *in vitro* gynogenesis. However, briefly summarizing some key factors is possible and useful for future works. Factors involved in unpollinated ovary/ovule culture are very complicated and can be broadly classified into two categories: the preculture factors and the factors during culture period.

Preculture factors

The efficiency of ovary/ovule culture appears to be largely dependent on preculture factors, such as genotype and physiological status of the donor plants, stage of embryo sac development, kind of explants, etc.

Genotype is most decisive, since various varieties of the same species usually responded differently to the same culture conditions. For instance, in rice the *japonica* cultivars were easier to induce gynogenesis as compared to the *indica* cultivars (Zhou et al., 1983). In sunflower, 12 cultivars tested could be classified into three groups: responsive to gynogenesis, recalcitrant to gynogenesis but responsive to somatic callusing, and recalcitrant to both (Yan et al., 1988).

Selecting ovaries or ovules at a suitable embryo sac stage for inoculation is also important. In most cases so far investigated a nearly mature embryo sac stage gave better results than either partially or fully mature stages. Rice is an exception where inoculation of ovaries at free nuclear embryo sac stage proved to be more suitable (Zhou et al., 1983).

Young flower, ovary, placenta-attached or isolated ovule can be used as explants in different cases. Generally, culture of flowers or ovules attached to placenta responded better than single ovule culture. However, in *Gerbera jamesonii* so far only ovule culture has been successful. In sunflower, ovule culture was superior to ovary or floret culture, although the latter two methods were also effective (Yan et al., 1988). In sugarbeet, ovule culture has been mostly adopted but ovary culture worked even better under certain conditions as reported by van Geyt et al. (1987).

Several reports have emphasized the significance of physiological state of the donor plants on the induction results. Meynet and Sibi (1984) noticed a significant variation in callus formation frequency in *Gerbera* ovule cultures raised between August and September. Cappadocia et al. (1988) reported the influence of fall versus spring sampling on callus formation and shoot regeneration, respectively, in *Gerbera*. Doctrinal et al. (1989) divided their four months' experiments on sugarbeet ovule culture into four periods, among which June gave the best results.

A few reports have pointed out the beneficial role of cold-pretreatment. For *in vitro* gynogenesis in sunflower pretreating the capitula at 4℃ for 24-48 hours significantly increased the induction frequency (Yan et al., 1987). A recent investigation on rice ovary culture also indicated promotive effect of cold pretreatment of the young panicles at 7℃ for one day (Cai et al., 1988).

Factors during culture

Factors influencing gynogenesis during culture period include: kind of basic medium, kind of supplements; kind, concentration and combination of exogenous hormones; sucrose concentration; solid or liquid medium;

orientation of explants; density of explants; temperature and light conditions; etc. For a given species, to make a search of the optimal combination of all the factors is quite complicated. It would be better to start with a series of comparative experiments to make sure of the effects of individual factors, followed by a synthetic experimentation involving the key factors so as to find out the optimal conditions. Here we will not comprehensively discuss all the factors concerned, but only mention the key points.

Exogenous hormone plays a crucial role in ovary/ovule culture not only because of its direct effect in inducing gynogenesis, but also due to the fact that excessive application of it usually causes vigrous callusing of the somatic tissues, which makes it difficult to distinguish the gynogenic derivatives from the somatic ones and also exerts harmful effect on the growth of the gynogenic units. For instance, in rice ovary culture 0.125-0.5mg/L MCPA (2-methyl-4-chlorophenoxyacetic acid) gave the highest induction frequency of gynogenesis and excluded somatic callusing; at a concentration of 2 mg/L the auxin stimulated callus formation from the ovary wall (Zhou et al., 1983). In sunflower, ovules cultured in MCPA-supplemented medium often yields, besides gynogenic embryoids, endothelial and integumentary embryoids or calli. On the other hand, a hormone-free medium gave rise to higher induction frequency of gynogenesis and at the same time eliminated the somatic derivatives (Yan et al., 1987).

Sucrose concentration is also important in regulating the balance between gynogenesis and somatic proliferation. In sunflower, different sucrose levels proved optimum for different responses: 1% sucrose for integumentary callus, 3%-9% for endothelial embryoid, and 12% for gynogenic embryoid formation (Yan et al., 1987).

Some comments should also be made on the role of light. In general, the ovaries or ovules were cultured in light. However, several works on rice (Zhou et al., 1986) and sunflower (Gelebart and San, 1987) kept the cultures in darkness during the early induction period and then transferred them to light for regeneration. Yan et al. (1988) observed that culturing sunflower ovules in dark favoured gynogenesis and inhibited somatic callusing. He and Yang (1988) confirmed the advantage of maintaining the cultures in dark over light in rice. Embryological observations revealed that 12 hours photoperiod (800 lux illumination) caused serious degeneration of the gynogenic proembryos, leading to their final abortion within one month of culture.

It should be emphasized that *in vitro* gynogenesis is a multistage process. The special requirements of each stage ought to be studied carefully. As a simple example, rice ovary culture comprises at least two stages: the first stage of induction is characterized by float culture on liquid medium, supplemented with exogenous auxin and culture in dark; while the second stage of regeneration requires transfer to solid medium, decrease in auxin concentration and keeping the cultures under illumination (Zhou et al., 1986).

Embryological aspects of *in vitro* gynogenesis

Embryological study is indispensable for elucidating the origin and morphogenetic features of *in vitro* gynogenesis. Nevertheless, due to the laborious task of processing large quantities of cultured ovaries or ovules, so far only a few embryological researches have been done in precise and systematic manner. In order to make the task easier, in recent years an *en bloc* staining followed by either paraffin sectioning (Zhou and Yang, 1981; Tian and Yang, 1983, 1984; Li and Yang, 1986; Yan et al., 1985, 1987, 1988) or methyl salicylate clearing (Yang, 1986; Yan et al., 1988) procedures have been adopted, which facilitated preparation of tens of thousands of embryo sac sections.

Origin of haploids

Each component cell of the embryo sac (egg, synergids, central cell and antipodal cells) might, theoretically, be the source of the haploid plants induced *in vitro*. Strictly speaking, there are two kinds of origin: *in vitro* parthenogenesis, where the egg cell is triggered to develop into a sporophyte, and *in vitro* apogamy, where the other cells of the embryo sac are induced to form embryo. For most of the cases investigated so far, as in barley (Huang et al., 1982), sunflower (Yan et al., 1985, Gelebart and san, 1987) and sugarbeet (olesen et al., 1988), parthenogenesis is the source of gynogenic plants. Yang et al. (1986) and Yan et al. (1989) have described the process of *in vitro* parthenogenesis in sunflower in detail. However, in ovary culture of rice an apogamic development from the synergid has proved to be the main, or even unique, source of the haploid plants (Tian and Yang, 1983; Li and Yang, 1986; He and Yang, 1988). This process is described in detail by Zhou et al. (1986). He and Yang (1988) confirmed the consistency of synergid apogamy in rice ovary culture irrespective of many kinds of alteration in culture conditions (basic media, exogenous hormones, culture modes, inoculation in light or dark etc). The central cell of the embryo sac could undergo limited divisions resulting in early endosperm-like structures (Jensen et al., 1977; Zhou and Yang, 1981; Huang et al., 1982; Yan et al., 1985). As for the antipodal cells, occasional formation of multicellular masses was reported (San Noeum, 1979; Huang et al., 1982; Tian and Yang, 1983), which could seldom serve as a main source of haploid plants. The unique example of antipodal apogamy was seen in a polyembryonic species *Allium tuberosum*, in which the antipodal cells as well as the unfertilized egg were induced to regenerate haploid plants by ovary culture. It is interesting at embryological level that the three antipodal cells *in vivo* bear close resemblance to an egg apparatus in that one of them usually is similar to the egg and the other two are similar to the synergids morphologically. It was the egg-like antipodal cell that gave rise to haploid plant (Tian and Yang, 1989).

Morphogenetic features in gynogenetic process

The gynogenic (parthenogenic or apogamic) proembryos develop via either direct embryogenesis or callus formation followed by regeneration of plantlets. The pathway involving completely normal embryo development can occur only rarely; a certain degree of callusing of the proembryo is often found. In rice, the proembryos usually grew into big protocorm-like structures which are intermediate between embryo and callus and, subsequently, regenerated shoots and roots via organogenesis (Tian and Yang, 1984). Exogenous auxin played important role in switching the pathway towards embryo differentiation or callusing. When picloram (4-amino-3,5,6-trichloropicolinic acid) was used as a substitute for MCPA, that was conventionally used in rice ovary culture, the callusing tendency decreased and the embryo differentiation improved (He and Yang, 1987). In sugarbeet, deviations of gynogenic embryo development occurred, which included broadening and callusing of the suspensor, variation of embryo morphology and underdevelopment of the cotyledons. Such embryos could develop through precocious germination, but a subculture was usually required for plantlet regeneration (Olesen et al., 1988). In sunflower, poor differentiation of the gynogenic embryos made it necessary to transfer them for several times through callus formation toward redifferentiation (Hua and Yang, unpublished data). In short, at present in many cases a pathway via callus formation has been an easier way leading to regeneration until more suitable culture conditions are worked out for direct differentiation into plantlets.

Ultrastructural changes in gynogenic development

To date ultrastructural studies of *in vitro* gynogenesis are only restricted to sunflower (Yan and Yang, 1989; Yan et al., 1989). Observations revealed that in some cases after several days of ovule culture the unfertilized egg cell, showed certain features common to the fertilized egg, such as migration of the nucleus from the chalazal end to the centre of the cell, increase in the number and activity of organelles, wall formation at the formerly naked chalazal surface of the cell, etc., indicating that an activation mechanism really occurred in the eggs preparing for parthenogenesis. The parthenogenic proembryos formed subsequently were also similar to the zygote proembryos in many cases, but they often exhibited some unusual features such as partitioning of the proembryos into two parts differing sharply in electron density, inversion of the polarity of the proembryos, occurrence of abundant autophagic vacuoles, thickening of cell walls, free growing mode of wall formation often leading to incomplete walls, occurrence of free nuclear divisions resulting in coenocytic structures, amitotic division of nuclei and so on. These peculiarities are believed to be the effects of *in vitro* culture rather than of gynogenesis itself.

Current limitations and future trends of research

Limitations

The works on *in vitro* gynogenesis have gone through a dozen years of its juvenile life. Although a lot of progress has been made, it still lies far from its fruitful application in plant breeding. Limitations seem to come from two sides:

First, the number of species where success has been achieved is too small to support the idea that *in vitro* gynogenesis has become a widely accepted technique applicable to a wide range of plant taxa. May be this situation reflects the real nature of the recalcitrance of female haploid cells to the induction of sporophyte development! That would be passimistic. We would rather believe that it is only a consequence of insufficient efforts made by so few investigators working in this field. Moreover, many of the studies did not involve sufficient manipulations. If we leave the ovaries or ovules in culture and wait for the emergence of plantlets, very often gynogenesis is induced but the haploid embryo degenerates prematurely and, consequently, the experiment is considered unsuccessful. A better way for exploring the potential of gynogenesis appears to be fixing, sectioning and making microscopic examination of the materials during early stages of culture to make sure whether proembryos are present inside the embryo sac. If there are proembryos, further efforts must be focused on how to make them grow and develop. To combine tissue culture with embryological observations is our recommendation.

Second, even in those species where gynogenesis has been reported, not every case can be considered to be a fully developed system. The induction frequencies are still low and the results are not reliable. Such a situation apparently limits the application of this method.

When taking account of the current limitations of *in vitro* gynogenesis itself, we should also recognize another situation. Induction of haploid plants has become easy via androgenesis, so that people tend to follow the line of least resistance, and gynogenesis seems to be superfluous; only in species such as sugarbeet or male-sterile lines where anther culture meets difficulty, the significance of gynogenesis becomes evident in respect of practical usage in plant breeding.

Future trends of research

First of all, sustained efforts to improve ovary/ovule culture techniques are likely to bring *in vitro* gynogenesis into practical use in such traditional areas of plant breeding as production of homozygotic lines for the usage of heterosis or acceleration of selection of recombinants, as anther culture has worked well in some cases.

Another direction may be using ovary/ovule culture system for gene engineering. Ultrastructural observations have revealed the absence of a complete chalazal wall around unfertilized egg (for reference see Kapil and Bhatnagar, 1981). Such a state of 'partial protoplast' makes it possible to introduce foreign DNA into the egg. In fact there have been some preliminary reports on microinjection of DNA into the ovaries or ovules to achieve transformation (Hepher et al., 1985; Steinbiss et al., 1985). *In vitro* gynogenesis relying on its haploid property might serve this purpose better.

The third trend is to develop the manipulation technique from ovary/ovule culture towards culture of embryo sacs or even egg cells. The successful isolation of viable embryo sacs and its component cells or protoplasts during recent years (Zhou and Yang, 1985; Hu et al., 1985; Allington, 1985; Mol, 1986; Wagner et al., 1989; Huang and Russell, 1989) has been carrying out a way towards this goal. The progress of cell and protoplast culture techniques may be of great help in realizing such sophisticated cultures.

Concluding remarks

Early, in 1950s and 1960s culture of unpollinated ovaries and ovules was tried by several experimental embryologists (for references *see* Yang and Zhou, 1982). However, success of this technique for induction of haploid plants came in the middle of 1970s, approximately ten years later than that of anther culture. During recent years advances in this field give us confidence that the unfertilized egg and other ephemeral megagametophytic cells can be triggered *in vitro* from their dormant state to sustained divisions without the involvement of fertilization. Tissue culture techniques once again exhibited its power in interfering in the natural program of plant reproduction. In order to push the works forward, an integration of embryology, physiology, genetics, cell and molecular biology is necessary. Only after a deeper fundamental research, we will be able to understand the mechanism of induction of gynogenesis, and make it a viable means for plant biotechnology.

(Authors: Yang HY, Zhou C. Published in: Plant Tissue Culture: Applications and Limitations. Bhojwani SS ed. Elsevier, Amsterdam, 1990, pp. 242-258, with 1 table and 66 references. Table 1 is retained here.)

4 花粉原生质体、精子与生殖细胞的实验操作

Experimental manipulation of pollen protoplasts, sperms and generative cells

当前,一个可以称为"植物实验生殖生物学"的分支学科正在形成之中。其主要特点是:由以往实验胚胎学的器官、组织水平的操作发展为细胞、原生质体水平的操作;由传统的比较单一的研究发展为多学科的综合性研究;因而无论在研究的手段与内容方面均进入更高的层次。例如,就雄性系统而言,已由花药培养发展到花粉粒培养,最近又向花粉原生质体、精子和生殖细胞的操作深化。就雌性系统而言,亦由子房、胚珠培养向胚囊及其成员细胞的操作迈进。以上两方面的汇合,迟早将会实现雌、雄配子的体外融合(即高等植物的确切含义的体外受精)以及有关的遗传操作;并随着相应的生物学研究的深入,使人类对植物有性生殖过程的认识与控制达到新的境地。

关于花粉原生质体、精子与生殖细胞的实验生殖生物学研究近年进展尤为迅速。这是因为:它们均属单倍性的原生质体或细胞,而在结构与功能上又各具特色;能提供较大而同步的群体;操作技术较雌性细胞相对容易。因而无论作为细胞工程或生物学研究的实验体系均有很大的优越性。不过,虽然早在十多年前研究者们已认识到这类工作的意义并开始进行探索,但直到最近才相继出现操作技术的重大突破。本文拟分如下三个方面系统地回顾有关研究历史,着重反映最新研究动态,并对发展前景提出一些看法。

花粉原生质体的操作

20世纪70年代初期,当花药培养和体细胞原生质体培养成功不

久,花粉原生质体的分离与培养即已受到重视,因为它兼具前二者的优点,是一种理想的单倍性原生质体实验系统。曾一度掀起热潮,但几经尝试技术不过关,此后研究进入低谷。最近几年,出现了几项重要突破,标志着第二次研究热潮的到来。鉴于 Bajaj(1983)[9]和胡适宜(1986)[6]已介绍过本领域过去的研究概况,本文将侧重近年的研究进展。

广义的花粉原生质体包括由四分体、花粉粒和花粉管分离的原生质体。它们的细胞壁成分不同,分离的方法与难易程度各异。其中,由小孢子到成熟花粉粒时期所分离的原生质体重要性与难度均最大。

一、分离

四分体原生质体 从 Bhojwari 与 Cocking(1972)开始,有较多的工作涉及四分体原生质体的分离[7,8,10,12,15,21,25,26,27,33,35,36]。四分体的壁由胼胝质组成,用蜗牛酶[8,15,36]、纤维素酶[21]、崩溃酶[26,27,33,35]等均可使之降解。只要分离介质的渗透压及其它条件适宜,在多种植物中只需 1~2 小时的短时间酶解即可获得大量的原生质体,分离率最高可达 100%。

花粉粒原生质体 早期分离花粉粒原生质体的试探大多是不成功的[15,21,34,36]。Bajaj 与 Davey(1974)曾用多种酶结合机械压力的方法由矮牵牛等植物中分离出有限的花粉粒原生质体[11]。以后在被子植物方面的研究停顿了相当长时间;而在裸子植物方面也仅见 Duhoux(1980)分离亚利桑那扁柏(*Cupressus arizonica*)花粉粒原生质体取得部分成功的一篇报道[22]。

由花粉粒分离原生质体的困难在于花粉外壁的基本成分孢粉素迄今没有酶可使之降解,而强烈的化学试剂在溶解外壁的同时也破坏了原生质体本身。直到近年,才在化学分离与酶法分离两方面找到一些突破口。Loewus 等(1985)发现一种多糖溶剂 MMNO(4-methylmorpholine N-oxide)在 75℃下处理麝香百合(*Lilium longiflorum*)花粉,可使外壁与内壁溶解,快速释放出大量原生质体,但药品毒性和高温使原生质体失活[28]。以后他们改在常温条件下用 MMNO 与酶解结合试验,使原生质体有60%成活[13]。不过在这种条件下MMNO的作用并非溶解外壁,而只是削弱内、外壁之间的联系,从而有利于内壁的酶解。另一方面,其他作者单用酶解方法也取得更好的效果:Tanaka 等(1987)用果胶酶和纤维素酶由同一种植物麝香百合的花粉粒中分离出大量原生质体,分离率高达70%～80%[42]。周嫱(1988)也是用果胶酶和纤维素酶分离出鸢尾(*Iris tectorum*)、萱草(*Hemerocallis fulva*)、风雨花(*Zephyranthes grandiflora*)三种植物的大量成熟花粉原生质体和萱草不同发育时期的幼嫩花粉原生质体[4]。以上两项研究所获得的原生质体均具高度的生活力。酶法分离技术成功的原理,是利用花粉粒水合后的吸胀作用撑破外壁,使内壁大面积地处于酶的作用之下而得以顺利降解。迄今取得成功的物种属于百合科、石蒜科、鸢尾科,其花粉外壁较薄、萌发沟较宽大等生物学特点也是成功的重要因素。

花粉管原生质体　花粉管壁由果胶质、纤维素、胼胝质三种成分构成,用酶法可以降解。Power(1973)最早由花粉管分离原生质体成功[34]。以后有些研究者以此作为细胞生物学研究的实验系统[19,31]。朱澂等(1984,1985)研究了金鱼草与烟草花粉管原生质体的分离,由一根花粉管可以分离出二三个大小不等的亚原生质体[1,2]。Power 与 Chapman(1985)描述了欧洲白桦花粉管原生质体分离的操作流程[35]。

二、纯化

刚分离的原生质体混有花粉壁等残渣,需要加以纯化。四分体原生质体和花粉管亚原生质体可根据体积的差异用一定孔径的筛网过滤以与残壁分开。花粉粒原生质体与花粉壁大小相近,用过滤法难以分开,可根据比重的差异,用蔗糖[4,14]或 Percoll[42]间断密度梯度离心法或漂浮法[5]加以纯化。

三、培养

四分体原生质体　早期的四分体原生质体培养工作报道了细胞壁再生、出芽、念珠状或管状结构的形成以及细胞核有限的分裂等现象[8,10,15]。Babbar 与 Gupta(1979)[7]曾对 Rajasekha(1973)[36]观察的曼陀罗四分体原生质体培养中的核分裂现象提出怀疑,因为在体内小孢子发生过程中就可能产生异常的多核细胞。Imamura 与 Potrykus (1983)指出:只有在逐日更换新鲜无糖培养基的条件下才能诱导烟草四分体原生质体形成大量的二核细胞;如转入含糖培养基则产生管状结构[25]。

虽然四分体原生质体单独培养未有显著进展,但用它开展融合实验却有很大成功。Deka 等(1977)曾试验木豆(*Cajanus cajan*)等的四分体原生质体之间的融合,融合率高达 70%~80%;培养后有 5%的融合细胞发生了核的融合[21]。最近,Pirrie 与 Power(1986)[33]、Lee 与 Power(1986,1988)[26,27]利用四分体原生质体和体细胞原生质体融合,开展"配子-体细胞杂交"(gameto-somatic hybridization),培养出三倍体杂种植株。这在烟草属的种间杂交和矮牵牛属的种间与种内杂交组合中已经成功。

花粉粒原生质体 Bajaj 与 Davey(1974)早就探索花粉粒原生质体的培养[11]。但由于在分离和培养两方面均有很大困难,一直没有突破。Tanaka 等(1987)培养麝香百合近成熟花粉的原生质体,1 天内再生细胞壁,10~12 天内形成类似花粉管的结构,并观察到生殖细胞分裂为一对精子[42]。周嫦(1988)培养萱草成熟花粉的原生质体再生了细胞壁,形成各种管状、结节状、念珠状、哑铃状以及出芽等结构,生殖核可分裂 1~2 次形成 2~4 个细胞核[4],结果和 Tanaka 等相似。这些工作表明:利用成熟花粉原有的配子体发育倾向,使其萌发成具有受精能力的花粉管,以便进一步开展将外源 DNA 导入卵细胞的实验,是可能的;但企图诱导其脱分化走上孢子体发育途径则不现实。为了达到后一目标,应该开展幼嫩花粉原生质体的培养实验,并结合采用花药、花粉培养和体细胞原生质体培养两方面的成功经验。周嫦(1989)将萱草含单核花粉的花蕾经低温预处理后分离出原生质体,在培养中启动了细胞分裂,形成了多细胞团及类似原胚的结构,突破了花粉原生质体胚胎发生(孢子体发育)的第一关[5]。在此基础上,预计再生植株是可望成功的。

花粉管原生质体 朱澂等(1984,1985)开展了金鱼草和烟草花粉管亚原生质体的培养实验。亚原生质体不论有核或无核,均能再生壁和形成管状结构。作者指出,在细胞工程中有核亚原生质体可作为类似核质体、无核亚原生质体可作为胞质体加以利用[1,2]。

四、细胞生物学研究

关于花粉原生质体的细胞生物学研究,迄今多限于与脱壁和壁再生有关的事态。关于脱壁,不少研究者结合自己所采用的分离方法进行了光镜或电镜观察[1,2,4,11,12,14,28]。关于壁再生,这里仅引述两项有趣的观察:Bajaj等(1975)在烟草四分体原生质体培养中发现可以再生两种类型的壁,一种是和一般体细胞原生质体培养中相似的纤维素性质的简单细胞壁;另一种是电子不透明的、可能由孢粉素构成的复杂细胞壁。作者认为后一事实暗示构成外壁的物质除了起源于绒毡层这一公认的看法外,还可能包括小孢子原生质本身的活动[12]。Miki-Hirosige等(1988)用超微组化技术研究了麝香百合花粉粒原生质体再生壁的过程。在培养初期形成无定形的多糖壁。以后形成含纤丝的壁,但其成分与结构和花粉内壁并不相同。没有看到类似外壁的形成。培养后期由原生质体萌发的花粉管,其管壁的成分与结构均与由花粉粒萌发的正常花粉管相似。作者还研究了高尔基体、光滑小泡、衣被小泡、小窝等在壁再生过程中的作用[30]。

<h1 style="text-align:center">精子的操作</h1>

种子植物的精子包藏于花粉粒或花粉管中,过去对精子的研究都是在原位进行的。Cass(1973)最早用分离精子的方法研究其在离体条件下的形态变化与运动性能[16]。最近几年,分离精子成为国际上实验生殖生物学中的一个热点,是有其时代背景的:一是关于精子结构与功能及双受精的研究有了重大突破,提出了雄性生殖单位、精子二型性、倾向受精等新概念;二是关于雌、雄识别的研究由花粉与雌蕊的识别向配子间识别的层次深入;三是在体细胞操作的基础上,生殖系统细胞的操作日益受到重视。建立分离精子的实验体系,对上述几方面的研究

都是十分需要的。

一、分离

三细胞花粉的精子分离 三细胞花粉成熟时已经形成精子,可以由花粉粒中分离出来。迄今大多数实验以此为材料。有两种分离方法:一为研磨法,即将悬浮于一定介质中的花粉用玻璃匀浆器轻轻研磨,使花粉壁破裂而释放精子。玉米[18,29]、油菜[18,44]、甘蓝[29]、非洲菊(*Gerbera jamesonii*)[40]曾用此法成功地提取了精子。其优点是操作较简便,不太依赖于花粉的成熟度与生理状况。但研磨时需要手工技巧,研磨后的花粉壁碎片较难除去。另一种方法是渗透压冲击(osmotic shock),即将花粉置于适宜渗透压条件下,任其吸水后自行破裂。白花丹(*Plumbago zeylanica*)[37]、玉米[17,23,44]、小麦[29]用此法获得成功。其优点是较少损伤精子,花粉壁较易过滤除去。但花粉的种类、发育时期、生理状况等因素对渗透压冲击的效果影响很大。

二细胞花粉的精子分离 二细胞花粉的精子是在花粉管中形成的,因此必须由花粉管分离精子。Shivanna 等(1988)采用"活体-离体技术"(*in vivo/in vitro* technique)成功地分离了一种杜鹃花(*Rhododendron macgregoriae*)与唐菖蒲(*Gladiolus gandavensis*)的精子[39]:先将花粉授于柱头上,隔一定时间当花粉管中已形成精子后,切下花柱,将切口插入培养基,待花粉管由切口长出后,用渗透压冲击法或酶解法促使管尖破裂而释放精子。此法除了可以收集大量而纯净的精子外,还有一个优点,即便于研究接近受精之前的精子的状况。因此,不仅二细胞花粉,而且三细胞花粉亦可考虑用此法研究不同阶段的精子的特点。

二、纯化

纯化精子的方法一般是先用一定孔径的筛网过滤以除去花粉壁,然后用间断密度梯度离心法除去花粉内含物中的淀粉粒与其它细胞质颗粒。间断梯度离心所用介质有蔗糖[37]、含蔗糖的培养基[17]、Percoll[23]、Nycodenz[18]等,方法亦各不相同。纯化的效果迄今仍是相对的。Russell(1986)的白花丹精子分离率为60%,纯化后花粉壁完全被清除,但尚有少量细胞质颗粒与营养核等杂质,精子产量为 8.8×10^6 个/ml[37]。Dupuis 等(1987)的玉米精子分离率为 20%,纯化后花粉壁和淀粉粒被除尽,仅余少量细胞核与细胞器杂质,精子产量为 3×10^6 个/ml[23]。Cass 与 Fabi(1988)的玉米实验中花粉破裂率(代表精子分离率)为 30%,精子产量为 1.5×10^6 个/ml[17]。

三、生活力

分离的精子若以遗传操作为目标,则要求具有足够的生活力。精子生活力的测定方法主要是:(1) Evans blue 染色反应(生活细胞排斥染料,死细胞吸收染料);(2) FDA(荧光素二醋酸酯)荧光染色反应(生活细胞有荧光,死细胞无荧光)。Russell(1986)用 Evans blue 鉴定分离的白花丹精子,保存 20h 后多数仍不染色;但用 FDA 鉴定同样的材料,数分钟后荧光反应即消失。两种方法的结果差异极大[37]。Matthys-Rochon 等(1987)报道用 FDA 鉴定甘蓝、玉米、小麦的精子,在保存 48h 后仍有荧光反应。但作者没有明确交代这是哪种植物的实验结果[29]。Dupuis 等(1987)分离的玉米精子经 FDA 鉴定保存 20h 后仍有 50% 存活[23]。Cass 与 Fabi(1988)用 Evans blue 鉴定玉米精子,在 3h 内约 50% 精子不染色,有时可存活 24h。如将精子在-75℃条件下冷冻保存,两周后仍有 10% 存活[17]。杨弘远与周嫦(1989)用 FDA 鉴定贮存于冰箱中的精子,油菜精子可存活 3d;玉米精子 1d;黑麦精子仅半小时[44]。作者认为:进一步研究分离精子的保存技术,将会建立高等植物的"精子库"。

四、细胞生物学研究

Cass(1973)最早研究大麦精子在体外环境中的形态变化和运动问题。精子刚释放时呈原有的长形且保持成对状态,但过一定时间均变成球形。根据变圆时体积并不增大的特点,作者认为变形并不简单地起因于低渗效应,而可能是由于微管的变化。没有观察到离体精子的运动[16]。Russell 与 Cass(1981)在白花丹上继续了这方面的研究,发现刚释放的二个精子与营养核是相连的,其中,与后者直接相连的一个精子体形较大且具长形延伸部分[38]。这是第一次在离体条件下发现了以后称为雄性生殖单位的现象。以后，Matthys-Rochon 等（1987）也在甘蓝、小麦、玉米上观察到分离的雄性生殖单位[29]。Shivanna 等（1988）将杜鹃花与唐菖蒲的刚释放的精子立即固定，然后进行扫描电镜观察，看到成对精子间有指状联系，单独的精子上也有这种联系的痕迹。作者还测量了一对精子的大小，证明它们相差很大[39]。

关于分离精子的超微结构，Dupuis 等（1987）和 Cass 与 Fabi（1988）报道：玉米的分离精子被一层质膜包围，没有营养细胞的内质膜，更没有壁，胞质中有各种细胞器，但未见到质体和微管，总之，呈原生质体状态[17,23]。

为了研究精子的识别机制，开展了初步的免疫学研究。Hough 等（1986）描述了将油菜分离精子与杂交瘤上清液或单克隆抗体作免疫学鉴定的方法[24]。Pennell 等（1987）将白花丹精子注入家鼠以诱发并制备单克隆抗体，再与抗原精子作免疫学鉴定，证明有部分的专一性反应[32]。这类工作目前还处于技术探索阶段，但显然是很有前途的。

生殖细胞的操作

生殖细胞是精子的前身。在占被子植物多数的二细胞花粉类型中，花粉成熟时只含生殖细胞和营养细胞。早在20世纪70年代初期就有人分离生殖核与营养核，但当时的目的只是获取供生化分析的细胞核。真正分离生活的生殖细胞是最近几年才开始的。

一、分离

有以下三种分离生殖细胞的方法：

压片法　周嫦等（1986）用压片法首先分离出一种网球花（*Haemanthus katherinae*）生活的生殖细胞[46]，以后在三个科的七种植物中试验均获成功[3]。此法用于细胞生物学观察是简便而有效的，但不适于收集供实验用的大量生殖细胞。因而作者以后又提出了渗透压冲击法。

渗透压冲击法 采用分离精子的渗透压冲击法对多数植物效果不够好。周嫦(1988)对此作了改进,先将花粉在一定浓度的蔗糖液中水合一段时间,再加入等量蒸馏水造成渗透压骤降,促使花粉大量而较为同步地破裂。这种改进的方法称为"二步渗透压冲击法",以示与原来只经一步的方法相区别[45]。二步渗透压冲击法在多种双子叶和单子叶植物中的试验结果是良好的,但适用于每种植物的具体方法各有不同。

花粉原生质体释放法 Tanaka(1988)先用酶法由麝香百合近成熟花粉中分离原生质体,再用机械搅拌法使原生质体破裂,释放生殖细胞[41]。这种方法虽然也是可行的,但需以分离花粉原生质体为前提,技术上受到很大限制,操作程序也较复杂。

前述几种方法分离的生殖细胞,经 FDA 鉴定均有很高的生活力[3,41,45,46]。

二、纯化

生殖细胞的纯化方法与精子相似,也是用筛网过滤除去花粉壁,用间断密度梯度离心法除去细胞质杂质。Tanaka(1988)用 Percoll 作离心介质,由一朵麝香百合花得到 1×10^5 个生殖细胞,分离率为 33.3%[41]。

三、融合

吴新莉与周嫦用 PEG 诱导玉帘生殖细胞互相融合,得到了含二至多个细胞核的融合体。FDA 鉴定融合体是生活的。这一试验为今后开展生殖细胞之间及与体细胞原生质体之间的融合与培养提供了可能性。

四、细胞生物学研究

分离的生殖细胞不受花粉粒的遮掩,便于直接进行活体观察。周嫦(1987)对此作了多方面的细胞生物学研究[3]。其中一个有意义的问题是关于细胞的变形。生殖细胞与精子的变形涉及其在花粉管中和释放到胚囊后的运动机制，而分离方法可为此提供一种实验系统。在几种植物中的观察均表明刚分离的生殖细胞呈原有纺锤形，以后逐渐变成球形，和分离精子的表现是一致的。介质的渗透压对变形有重要影响，但仅仅低渗的作用不足以解释即使在高渗下细胞终究变圆的事实。扫描电镜下的测量也显示不同形状的细胞尽管长度相差很大，而宽度却很接近，证明变圆并非由于低渗导致的细胞吸胀。用视频增差显微术对变形过程作了录像，发现这是细胞（及核）弯曲收缩而由长变圆的过程。看来和细胞内微管排列的变化有关。免疫荧光鉴定证明分离的生殖细胞中有纵向平行排列的微管。在分离介质中加入微管稳定剂 EGTA 和 $MgSO_4$ 确可延缓细胞的变圆。这方面的研究今后还值得继续深入。

生殖细胞是否有壁，过去的超微结构观察结果是有分歧的[43]。最近采用快速冷冻固定和冷冻替代法表明烟草和凤仙花花粉管中的生殖细胞被壁复合体包围[20]。分离的生殖细胞究竟是具壁的细胞，还是无壁的原生质体，也没有一致的结论。周嫦等（1986）用偏光显微术、荧光增白剂染色、PAS 反应和钌红染色等方法观察绣球百合的分离生殖细胞，均未显示壁物质的存在[46]。前述生殖细胞的融合实验，同样证明其原生质体的性质。另一方面，Tanaka（1988）用荧光增白剂和脱色苯胺蓝染色法鉴定麝香百合的分离生殖细胞，肯定其具备纤维素与胼胝质壁，并且还用酶法由其分离出生殖细胞的原生质体[41]。看来可能在不同的植物间存在着差异，不能一概而论。

展　　望

花粉原生质体、精子与生殖细胞操作的研究近年进展很快,是和植物生殖生物学研究的深化和细胞工程技术的进步分不开的。无疑,它们反过来必将推动有关基础研究与高技术的发展。

从植物生殖生物学学科发展的角度来看,一系列生殖系统体外实验体系的建立将有助于深入探讨下列各项理论问题,如:花粉两条发育途径(或分化与脱分化)的机理;花粉发育过程中细胞骨架的演变;花粉壁的形成与再生；花粉壁和膜的特征；花粉萌发的机理；生殖细胞与精子的变形及细胞骨架的变化；生殖细胞分裂成精子的机理；雄性生殖单位与精子二型性问题；精子的识别物质等。这些问题的阐明，仅靠体内的研究是不够的；体外实验会有很大帮助，我们从本文的引述中已可略见端倪。

从推进遗传工程技术的角度来看：当前的细胞工程主要依赖体细胞原生质体培养，而生殖系统中各种原生质体的开发将会建立起生殖工程新技术领域。从本文所反映的进展趋势可以设想今后进一步研究的课题，如：由花粉原生质体再生完整植株；花粉原生质体用于融合与转化；用携带外源基因的花粉原生质体授粉与受精；生殖细胞的培养与融合；生殖细胞或精子与体细胞原生质体的融合；精子库的建立；雌、雄配子的体外融合等。这些诱人的前景并非可望而不可即的空想。

(作者：周嫦、杨弘远。原载：植物学报，1989，31(9)：726~734。参考文献46篇，删去。)

5 植物性细胞原生质体的分离、培养和融合

Isolation, culture and fusion of sexual plant protoplasts

植物细胞包括体细胞和性细胞两大类。当前，体细胞原生质体研究已趋成熟，并在植物科学的基础研究与生物高技术领域中发挥着越来越明显的作用。性细胞原生质体则因其技术上的困难，20世纪80年代以来才逐渐提上日程。

本文所述性细胞是广义的概念，包括减数分裂之后形成的、参与有性生殖过程的一系列细胞类型。与体细胞相比，性细胞具有其独自的特色：第一，它们具单倍性基因组，遗传组成单一。第二，它们经历两性

分化与一系列的发育程序,由大、小孢子发生,雌、雄配子体发育直至精、卵细胞形成,产生形态与生理各异的多种细胞类型。第三,它们适应于有性生殖,能通过传粉受精,繁殖种子与后代。因此,利用性细胞的特点与有性生殖方式,开展性细胞原生质体的操作与遗传工程,将有可能开拓植物性细胞工程新技术,与体细胞工程相辅相成,使植物生物技术能更加充分地发挥其潜力[14]。近年,性细胞原生质体的研究进展令人瞩目,已有许多文献从不同角度作出综述[19,22~24,54,81,92],可供参考。本文将概括介绍花粉原生质体、生殖细胞、精细胞、卵细胞等原生质体的实验操作及其研究进展。迄今研究表明,除少数例外,分离的生殖细胞、精细胞与卵细胞一般均呈原生质体状态,它们也可概括地称为配子原生质体。本文为了叙述方便起见,常在其后省略"原生质体"四字。

原生质体分离

一、花粉原生质体的分离

花粉具有结构简单、单倍体遗传组成、发育同步、数量众多、参与传粉受精作用等特点。因此,花粉原生质体被视为植物性细胞工程中的重要实验体系。20世纪70年代初期,花粉原生质体的研究即开始受到重视。先后成功地分离出四分体原生质体和花粉管原生质体[4,5,32,34,36,41]。但是,游离小孢子至成熟阶段的花粉原生质体的分离却面临困难[33],直至20世纪80年代中后期才有所突破。目前花粉原生质体分离大致有三条技术路线:一是化学药品降解花粉壁[35,51],此法往往导致花粉原生质体本身严重失活;二是非酶脱壁,迄今仅有一篇报道[81],并且缺乏脱壁的证据,看来其分离产物可能并非真正的原生质体;三为酶解法[17,79],是目前唯一有效的分离方法。下面着重介绍酶解法。

分离方法 花粉壁包括外壁与内壁。外壁的基本成分为孢粉素,迄今尚无一种酶能加以降解。内壁的基本成分为纤维素与果胶质,可被常规的细胞壁降解酶所分解。显然,外壁是分离花粉原生质体的主要障碍。因此,花粉原生质体的酶解分离与体细胞原生质体分离时直接降解细胞壁是不同的。它不能直接降解外壁,而需要设法避开外壁以便降解内壁。针对不同植物花粉壁的化学成分、结构(包括萌发孔、沟的结构)以及花粉发育时期的特点,需要采取相应的措施,才能达到分离花粉原生质体的目的。

(1)一步酶解法 它适用于具宽大萌发沟的花粉。花粉置于酶液中,通过水合膨胀,使外壁沿萌发沟裂开,内壁大面积地暴露在酶的作用之下而分解,释放出原生质体。采用这一方法,Tanaka 等[79]分离出麝香百合近成熟花粉原生质体。周嫦[17]分离出萱草不同发育时期的花粉原生质体以及鸢尾、风雨花成熟花粉原生质体(图5-1)。以后单子叶植物百合科、鸢尾科、石蒜科中更多的植物以及双子叶植物中芸苔属一些植物均用此法取得成功[9,10,29,30,95]。

(2)水合-酶解二步法 在上述工作的基础上,将水合与酶解分为两个步骤。第一步,花粉在蔗糖液内充分水合,使外壁裂开,有的材料外壁完全脱落,成为仅具内壁的脱外壁花粉。第二步,转入酶液,降解内壁,获得原生质体。甘蓝型油菜与紫菜苔成熟花粉曾用此法分离出原生质体[7]。凌霄花用此法分离花粉原生质体的效果较一步法为佳(周嫦)。

(3)萌发-酶解二步法 第一步是将成熟花粉置于花粉萌发培养基中,待花粉大量萌发,长出短花粉管,及时转入第二步,在酶液中酶解。由于花粉管尖端的壁为果胶质与纤维素所组成,酶液由此处开始降解花粉管壁,再扩展至降解花粉内壁,从而释放出原生质体。烟草成熟花粉原生质体用此法分离成功[6]。此法可能对其它分离花粉原生质体尚未成功的植物有参考价值。

影响分离效果的因素

(1)植物种类 如上所述,植物种类不同,花粉壁的结构与成分各异,需要采取不同的分离方法。迄今获得成功的植物尚不多,特别是具萌发孔花粉的植物(如禾本科),还不能分离大量花粉原生质体,需要摸索新的途径。

(2) 花粉发育时期 花粉发育时期不同,其生理状态以及花粉壁结构与成分会有所变化,势必影响分离效果。Tanaka 等[79]比较了麝香百合不同发育时期花粉原生质体的分离率,以近成熟花粉的分离率最高。在紫菜苔与甘蓝型油菜中,成熟花粉比幼嫩花粉的原生质体分离率明显为高[29,30]。

(3) 酶的种类 在一步酶解法中,纤维素酶与果胶酶足以分离大量百合科与鸢尾科一些植物的花粉原生质体,但不能分离芸苔属植物的花粉原生质体[7,17]。后者需用长时期的水合-酶解二步法方能奏效。如果加入 Pectolyase Y-23,则也能用一步酶解法分离成功[29,30]。此外,酶的浓度、酶解时间也需按照实际情况予以调整。

(4) 酶液渗透压 花粉原生质体对于溶液渗透压很敏感。在一定范围内,随酶液渗透压提高,花粉原生质体分离速度减慢,但原生质体破裂频率则降低。

(5) 新鲜与保存花粉 一般常用新鲜花粉分离花粉原生质体。经超低温保存的花粉,在适宜保存条件后也可分离出花粉原生质体,分离频率接近新鲜材料。其中幼嫩花粉比成熟花粉敏感,对保存条件要求严格,保存后需经适当温育处理,再行分离花粉原生质体[29,30]。

纯化 花粉酶解液中除花粉原生质体外,尚残存未能酶解的花粉粒、空外壁、亚原生质体以及细胞质内含物等杂质。需要通过纯化步骤获得纯净的花粉原生质体群体。花粉粒与花粉原生质体的大小与重量相近,是纯化的难点。目前多借鉴体细胞原生质体的纯化方法,如蔗糖密度梯度离心、Percoll 密度梯度离心、离心漂浮、筛网过滤等方法加以解决[17,18,57,79]。

二、生殖细胞与精细胞的分离

两个精细胞分别与卵细胞、中央细胞融合,形成胚与胚乳,此即为被子植物的双受精。精细胞及其前身——生殖细胞的分离是开展离体受精的必要先决条件之一,具有重要意义。不仅如此,它们本身在生物学上也颇有特色,如细胞核大,细胞质与细胞器稀少,核质比很高。类似天然的单倍性的核质体,在植物性细胞工程中有特殊作用。

生殖细胞与精细胞位于花粉粒或花粉管内,沉浸于营养细胞的细胞质之中,被称为"细胞中的细胞"。因此,必须加以分离,否则无法对其进行直接的操作与研究。Cass 首先观察了分离的大麦精细胞[38]。Russell 首次建立白花丹精细胞的大量分离技术[72]。周嫦等首次分离出绣球百合的生殖细胞[96]。此后,在多种植物中精细胞与生殖细胞的分离陆续有新的报道[23,73,81,97]。其分离方法大致可分为两大类:一是机械分离法,利用机械力量使花粉壁破损,以获得生殖细胞[16]或精细胞;二是渗透压冲击法,利用低渗溶液使花粉或花粉管尖端破裂,释放出生殖细胞或精细胞。下面介绍具体方法。

生殖细胞的分离 简要介绍吴新莉与周嫦研制的几种渗透压冲击法[12]。

(1) 一步渗透压冲击法 花粉置于低渗蔗糖溶液中保温一段时间后,花粉自行破裂,生殖细胞随营养细胞质流出。用该法分离出棉花生殖细胞。

(2) 二步渗透压冲击法 花粉置于等渗蔗糖溶液中培养,当大部分花粉开始萌发时,加入蒸馏水使渗透压骤降,导致花粉破裂释放生殖细胞。石蒜、风雨花、玉帘、烟草、凤仙花适用此法(图5-2)。蚕豆用此法亦获成功[94]。

(3) 低酶法 花粉在低浓度酶(一般为0.1%纤维素酶与0.1%果胶酶)与低渗蔗糖溶液中保温,促使花粉破裂释放生殖细胞。黄花菜、韭菜、朱顶红、唐菖蒲、鸢尾适用此法。

(4) 花粉原生质体释放法 首先分离出花粉原生质体,然后用渗透压冲击法或机械法,使花粉原生质体破裂,获得生殖细胞。鸢尾、唐菖蒲、黄花菜、韭菜、朱顶红、风雨花也可用此法。

不同植物的花粉需用不同分离方法。第一类植物(如棉花)的花粉在蔗糖溶液中容易破裂,适用一步渗透压冲击法。第二类植物(如石蒜、烟草)的花粉在蔗糖溶液中不易破裂而易萌发,可用二步渗透压冲击法。第三类植物(如黄花菜)在蔗糖溶液中既难破裂又不易萌发,可用低浓度酶法。

在分离生殖细胞时,花粉发育时期、溶液成分、蔗糖浓度、保温时间、加水量、酶的浓度等因素均对分离效果有影响,分离方法也可适当调整。

精细胞的分离 精细胞的分离方法与生殖细胞类似。但被子植物成熟花粉有两种类型:三细胞花粉与二细胞花粉。三细胞花粉的营养细胞中含一对精细胞,而二细胞花粉的营养细胞中仅含一个生殖细胞。精细胞是在花粉管内形成。因此,精细胞的分离因花粉类型而异。

(1)三细胞花粉的精细胞分离方法

研磨法:用玻璃匀浆器轻轻研磨花粉悬浮液,使花粉壁破碎,释放出精细胞。甘蓝型油菜[91]、甘蓝[60,71]、紫菜苔[26]、非洲菊[75~77]、菠菜[80,89]均用此法分离出精细胞。

渗透压冲击法:将花粉置于低渗溶液中,使花粉自行破裂。白花丹[72]、玉米[40,43,71,81,91]、小麦[60]、甜菜[64]等植物适用此法。

(2)二细胞花粉的精细胞分离 在这类植物中,精细胞的分离必须待花粉管的生长至适当阶段,生殖细胞已分裂为一对精细胞以后,才能从花粉管中分离。

活体-离体法:首先将花粉授于柱头上,切下一定长度的花柱设法插入液体培养基中。当花粉管由花柱切口中长出后,用渗透压冲击或酶液处理,促使花粉管尖端破裂释放精细胞。杜鹃花、唐菖蒲[74]、鸢尾、棉花、朱顶红、烟草、黄花菜、萱草、玉帘[27]等用此法分离精细胞可取得良好效果。

离体萌发法:花粉置于培养基中人工萌发,待生殖细胞在花粉管中已分裂形成精细胞之后,用渗透压冲击等方法使花粉管破裂,释放出精细胞[48,76]。应用此法时,需注意诱导花粉高频率的萌发与同步生长,否则,分离的产物中容易有生殖细胞混杂。

三、卵细胞及其它胚囊成员细胞的分离

胚囊含卵细胞、中央细胞、助细胞与反足细胞。卵细胞与中央细胞直接参与双受精,和精细胞分别融合后发育成胚与胚乳。卵细胞具单

倍性细胞核与丰富的细胞质;中央细胞较大,内含一对极核或一个次生核,富含营养物质;助细胞具单倍性细胞核和高度活性的细胞质,有吸引花粉管的功能。

胚囊位于胚珠内方,被珠被与珠心组织重重包围,胚珠外方又有子房壁包裹。因此,卵细胞及其它胚囊成员细胞的分离,一般先用人工解剖法除去子房壁,然后以胚珠为单位分离出胚囊,再由胚囊分离出其成员细胞。

酶法 胚珠置于酶液中酶解,使珠被与珠心细胞离散,裸露出胚囊。一般需在酶解的基础上,适当辅加机械处理,如吸打、压片或解剖,以提高分离胚囊的效果。

Тырнов 等首先用酶法分离出烟草生活胚囊[98],进而分离出列当生活胚囊[99]。周嫦与杨弘远[15,21]先在金鱼草中,后又在向日葵和烟草中用酶解振荡法分离出生活胚囊。胡适宜、李乐工等[8,25]用酶解压片法分离出烟草和颠茄生活胚囊及其成员细胞原生质体。以后其他作者在蓝猪耳[62]、玉米[84,87]、麝香百合[86]、白花丹[47]、矮牵牛[83]、烟草[90]等植物中均用酶法分离出胚囊及其成员细胞的原生质体。

最近,孙蒙祥等[1]建立了一种新的酶解-渗透压冲击法,显著地提高了分离烟草生活胚囊的效果,进而获得卵细胞等胚囊成员细胞的原生质体。其方法之一是将胚珠置于含等渗甘露醇的酶液中保温,再转入低渗甘露醇溶液中冲击,辅以轻缓吸打。方法之二是将胚珠直接置于含低渗甘露醇的酶液中处理。在酶与低渗双重作用下,胚囊由胚珠的珠孔或珠柄断面逸出。进而可在收集胚囊时,或者连同残余酶液继续酶解一段时间,即可离散出胚囊各成员细胞,或者从胚囊中直接解剖也可(图 5-3)。

解剖法 应用显微操作器直接解剖胚囊及其成员细胞,技术上较难,效率较低[31]。但最近 Holm 等[46]用此法分离大麦卵细胞取得良好结果。特点是先从胚珠顶端将珠被解剖除去,通过解剖镜的透射光可以看到卵与中央细胞位置。然后刺穿中央细胞的大液泡,导致中央细胞破损,使卵细胞与助细胞从珠孔端逸出。

原生质体培养

一、花粉原生质体的培养

花药培养与花粉粒培养的大量研究表明,花粉在离体条件下有两条发育途径:配子体发育途径(继续体内原有的发育途径,直至花粉成熟,萌发出花粉管)和孢子体发育途径(改变体内原有发育途径,转向脱分化,启动细胞分裂,直至再生植株)。从目前初步资料分析,花粉原生质体也有配子体发育与孢子体发育两条途径[95]。

成熟花粉原生质体培养与离体配子体发育 目前已有一些资料报道,成熟花粉原生质体培养能够萌发出花粉管,其中生殖细胞分裂形成精细胞,表现出配子体发育的趋势。Tanaka 等[79]在麝香百合近成熟花粉原生质体培养中发现,10~12d 后少数花粉原生质体长出花粉管结构,有的生殖细胞分裂成一对精细胞。周嫦[17]培养萱草成熟花粉原生质体,也有少数萌发出少数花粉管状结构,生殖细胞分裂1~2次,形成2~4个精细胞。吴燕与周嫦[10]将唐菖蒲成熟花粉培养在含32%蔗糖、0.1mg/L 2,4-D、1mg/L NAA、0.2mg/L BA 的 K_3 培养基上,47%~59%花粉原生质体萌发出花粉管。花粉管形态正常,内有活跃的细胞质流动。在花粉管内生殖细胞分裂形成一对精细胞。

幼嫩花粉原生质体培养与离体孢子体发育 幼嫩花粉多指单胞中后期至二胞早期的花粉,这是许多植物诱导脱分化的适宜时期。周嫦于1989 年[18]首次报道了萱草幼嫩花粉原生质体培养启动细胞分裂,形成多细胞团,突破了其离体孢子体发育的第一关。其方法是:将含单胞后期花粉的花蕾经低温预处理后,用一步酶解法分离出花粉原生质体。在含有1%小牛血清、1%Polybuffer 74、1mg/L 2,4-D、2mg/L NAA、9%甘露醇、17%蔗糖的 K_3 培养基上培养。4~5d 后,有的原生质体发生第一次细胞分裂,逐渐形成多细胞团(图5-4)。以后,与萱草同属的黄花菜幼嫩花粉原生质体培养也诱导了细胞分裂(程红与周嫦)。

二、生殖细胞培养

生殖细胞在体内条件下被包围在营养细胞内,且其本身细胞质稀少,故带有半寄生性质。分离的生殖细胞能否在体外存活与发育,是生殖细胞操作需要解决的重要问题。吴新莉与周嫦[11]将分离的黄花菜生殖细胞在含多种附加物以及花药作为饲养物的 K_3 或 MS 琼脂培养基中离体培养,部分细胞发生 1~2 次核分裂,形成 2~4 核的细胞。在总数为 5961 个生殖细胞中,平均分裂率为 3.27%,最高分裂率达 11.46%。个别生殖细胞中似乎在核分裂的同时发生胞质分裂,形成二细胞结构。

三、胚囊与卵细胞培养

未受精的胚囊与卵细胞培养虽经多方努力,迄今未成功。近年,带有珠心组织的受精后胚囊培养有初步成功的报道。Campeиoт 等[37]将授粉后 1d 含受精卵与少数胚乳核的玉米胚囊,从珠心组织的珠孔端切下。将带有相当数量的珠心组织(胚囊在内)作为外植体进行离体培养。在此情况下受精卵能发育成胚,并在培养基上萌发,最终再生植株至性成熟。Mol 等[63]培养带数层珠心细胞的玉米受精后胚囊(内含受精卵或二胞原胚),约两周后,部分外植体内形成不同发育状态的胚。将它们从外植体中解剖出来,继续进行胚胎培养,有部分胚能进一步器官分化,少数能再生植株。

最近,Holm 等[46]培养大麦未受精与受精的卵细胞,前者不能分裂,后者发生了细胞分裂。单独的受精卵培养能够发育到小愈伤组织;用大麦小孢子作饲养物,受精卵能发育成胚状体,最后达到成熟植株,开花结实。

原生质体融合

性细胞融合有多种组合,大致可分为性细胞与性细胞融合(简称"性性融合")和性细胞与体细胞融合(简称性体融合)。性性融合包括雌、雄性细胞之间的融合,雄性细胞之间的融合以及雌性细胞之间的融合。性体融合包括以雄性细胞或雌性细胞为一方与体细胞之间的融合(表5-1)。

性性融合是从四分体融合的试验开端的。Deka 等[42]诱导了木豆四分体原生质体融合,其中少数融合体发生了核的融合。性体融合在20世纪80年代中期首先实现了四分体与体细胞融合[68]。最近,又实现了花粉与体细胞融合[9]。20世纪90年代以来,性性融合有重要进展,精卵体外融合获得突破,并再生植株[49,51,52]。至今,性性融合试验已扩展到多种组合(表5-1),发展形势令人瞩目。

一、性细胞与体细胞之间的融合

四分体与体细胞融合　Pirrie 与 Power 于 1986 年[68]首次报道粘毛烟草(*Nicotiana glutinosa*)四分体原生质体和普通烟草(*N. tobacum*)硝酸还原酶缺陷型的叶肉原生质体的融合及其融合体再生植株。根据形态性状、染色体数目、同工酶谱和蛋白质分析等方面的鉴定,表明再生植株为三倍体种间杂种。他们称这一方法为"配子-体细胞杂交"(gameto-somatic hybridization),以与体细胞杂交相区别。以后类似工作在矮牵牛属的种内与种间组合也获成功[55,56]。Petal 等[66]进一步研究了普通烟草四分体与黄花烟草(*N. rustica*)体细胞融合时,四分体的线粒体与质体基因组能否传递给杂种的问题。

花粉与体细胞融合　最近,在幼嫩花粉原生质体分离成功的基础上,游离时期的花粉原生质体与体细胞原生质体的融合获得成功。李昌功等[9]用聚乙二醇(PEG)诱导了青菜(*Brassica chinensis*)处于单胞后期至二胞初期的幼嫩花粉原生质体与甘蓝型油菜(*B. napus*)下胚轴原生质体融合,再生了小植株。经根尖染色体与同工酶谱测定,初步表明它们具杂种性质(图5-5)。

二、性细胞与性细胞之间的融合

性性融合有多种组合(表5-1),其中以精卵融合工作进展最快,是当前研究的热点。

精卵体外融合　Kranz 等[51,52]曾首次报道用微电融合(microelectrofusion)技术诱导了玉米单个精细胞与单个卵细胞成对融合,融合体在离体条件下发生细胞分裂,形成多细胞结构。他们进一步报道了玉米精卵融合体的离体胚胎发生过程和植株再生[49]。由 28 个融合体培养再生了 11 个植株,根尖染色体计数显示它们是二倍体。它们兼有双亲的形态特征,在 8 个再生植株中有 7 株所结的籽粒颜色呈 3∶1 的分离比例。这些试验结果确证了精卵体外融合的成功。Faure 等[45]用光镜、电镜以及三维重构技术研究了玉米精卵体外融合时细胞核的融合过程。并且初步制出精卵融合的时间表:融合后 20min 之内,精核处于卵细胞质内;20~45min 之间,精核靠近卵核;45~60min 发生两核的融合。

一个新的发展趋势是采用非电融合诱导精卵融合。Faure 等[44]指出,电融合是利用外力诱导融合,如能采用非电融合方式,比较接近体内受精条件,有利于研究受精理论。他们采用低浓度 $CaCl_2$(1mmol/L、5mmol/L 与 10mmol/L)溶液进行玉米精卵融合。其中以 5mmol/L $CaCl_2$ 条件的效果最好,融合率达 79.7%。他们用玉米卵细胞、精细胞与叶肉原生质体为材料,做了各种组合的融合试验。结果,精卵融合率最高即 79.7%,精细胞之间融合率为 16.7%,精细胞与叶肉原生质体之间融合率仅 1.8%,其它组合为零。这一结果表明,精卵细胞之间的融合是有特异性的。

Kranz 与 Lörz[50]用高钙高 pH 法(即 50mmol/L $CaCl_2$,pH11)也实现了玉米的精卵融合:精卵细胞粘连率约 25%,粘连后的融合率约 90%,融合体培养发育至 30~50 个细胞的结构。低钙(即 10mmol/L $CaCl_2$)时无论 pH 值高低如何,均不能诱导融合。这一结果与上述 Faure 等的试验结果是有一定矛盾的。

最近,孙蒙祥等[2,3]用 PEG 法诱导烟草单个雌性细胞与单个雄性细胞或体细胞原生质体成对融合获得成功。图 5(6~9)显示烟草性细胞融合的不同组合[3]。

显微注射　用显微操作器将分离的精细胞或精核注入胚囊,是体外受精的另一种尝试。Keijzer 等[48]用蓝猪耳作试验材料,利用其胚囊裸露在珠孔外的特点,将精细胞注入,但未有受精迹象。最近,Mathys-Rochon等[61]将玉米分离的精细胞核注入分离胚囊的卵细胞与中央细胞内。细胞学观察显示14%卵细胞和14%中央细胞内含精细胞核。其中有一个卵细胞,雌雄核接触,似乎正在进行着核的融合。

三、融合技术

从现有资料分析,性细胞融合与体细胞融合遵循共同规律。前者的工作正是在借鉴后者的经验基础上才能迅速发展。但是,性细胞融合具有自身的特点,尚需建立与之相适应的特殊技术。

融合方法 性细胞融合一般可用电融合或化学诱导融合方法。以雌性细胞为一方开展融合实验时,由于分离的雌性细胞数量有限,不可能采用常规的大群体的融合方法,而必须采用单对原生质体的融合技术。前文所述的电融合[52]、高 Ca^{2+} 高 pH[50]、Ca^{2+}[44]和 PEG[3]等诱导雌雄性细胞融合的方法,均是单对原生质体之间的融合。

当精细胞或生殖细胞和其它细胞进行融合时,双方细胞往往大小悬殊,不易融合。吴新莉与周嫦[13]、莫永胜与杨弘远[27]在诱导生殖细胞或精细胞与花瓣原生质体融合时,采取加大性细胞一方数量的措施,明显地提高了融合率。这一措施也被应用于青菜幼嫩花粉原生质体与甘蓝型油菜下胚轴原生质体的融合[9]。

融合体鉴定 雄性细胞与其它材料融合时,由于雄性细胞体积很小,很难鉴别融合体。吴新莉与周嫦[14]利用石蒜生殖细胞质呈黄褐色的形态特征,与花瓣原生质体融合时,能够较好地分辨融合体,但绝大多数生殖细胞与精细胞并无明显形态特征,在此情况下,采用荧光染料预先标记雄性细胞的方法,效果很好[93]。吴新莉与周嫦[13]、莫永胜与杨弘远[27]用荧光染料DAPI预先标记雄性细胞核,再与未染色的体细胞原生质体融合,融合体内可见性细胞核的明亮荧光。李昌功等[9]用荧光染料 FITC 标记花粉原生质体后,再和体细胞原生质体融合,也可达到鉴别融合体的目的。

杂种筛选 以四分体与体细胞融合开展"配子-体细胞杂交"时,由于四分体不能单独发育,因此,筛选工作中只需阻止体细胞单独的发育即可,比一般体细胞杂交时杂种筛选简便。例如,Pirrie 与 Power[68]用硝酸还原酶缺陷型的体细胞与四分体融合,在仅含硝酸盐的培养基上即可筛选杂种。Petal 等[67]利用抗卡那霉素显性基因的烟草四分体与黄花烟草的体细胞融合,在含卡那霉素培养基上,黄花烟草的体细胞不能发育,只有杂种才能发育,取得了良好的筛选效果。

在精卵成对融合试验中,可以直接挑选融合体进行培养,免去杂种筛选步骤,准确地获得杂种细胞。Kranz 等[49]用此法成功地得到了玉米品种间杂种植株。

融合体培养与植株再生 迄今已实现了多种组合的融合(表 5-1),但融合体培养达到再生植株者为数不多。性体融合方面,仅在烟草属、矮牵牛属的四分体与体细胞融合以及芸苔属的幼嫩花粉与体细胞融合中能够再生植株。精卵融合体的培养方面,由于融合体数量甚少,只能采用微量培养技术。Kranz 等[49]采用 Millicell 微室培养玉米人工合子,并用玉米悬浮细胞饲养物,细胞分裂率高达 83%[52],且顺利地分化植株。孙蒙祥等[3]用烟草叶肉原生质体作材料,试验了 Millicell 微室和微滴培养(以大量烟草叶肉原生质体作为饲养物)两种方法。结果表明两者均可诱导原生质体分裂,形成多细胞团,为性细胞融合体的培养作了技术准备。

综上所述,植物性细胞原生质体已取得长足进步。特别是各类性细胞原生质体的分离,花粉原生质体离体条件下的两条发育途径,四分体与体细胞以及花粉与体细胞融合再生植株,雌雄配子体外融合再生植株等方面的成就,令人鼓舞。当前的研究除继续改进分离技术外,重点已转向性细胞原生质体培养、融合与遗传转化,以期植物性细胞工程作为新的植物生物技术在作物的遗传操作中早日发挥作用。

(作者:周嫦。原载:植物原生质体培养和遗传操作(许智宏、卫志明主编),上海:上海科学技术出版社,1997,42~54。

图版 5 幅,参考文献 99 篇,均删去。)

6 花粉原生质体与配子原生质体操作的研究进展与前景

Advances and perspectives of the manipulation of pollen protoplasts and gametoplasts

有性生殖是植物个体发育中结构与功能变化最为复杂与曲折的阶段。有性生殖的研究和农业生产密切相关,是遗传育种的理论基础之一和高新技术的重要源泉。性细胞具有自然的生殖能力和单倍性等特点,在遗传工程中有特殊的价值。植物性细胞的操作,是当前国际上植物生殖生物学与发育生物学中的前沿研究方向。迄今在植物生殖过程的所有环节上,均已开展了实验操作的研究,其结果不仅加深了人类对植物传宗接代的规律性认识,还在单倍体育种、远缘杂交、基因工程等方面提供了不少新的技术手段。

近10年来,国际上的有关研究趋势是由性器官与组织的操作向性细胞及其原生质体的操作发展,如:由花药、花粉培养到花粉原生质体、生殖细胞、精细胞的操作;由子房、胚珠培养到胚囊、卵细胞的操作;由胚珠离体授粉到精卵体外融合;由胚胎培养到合子培养等,操作技术日益精密化,与此相关的研究手段也日益多学科化。我们曾先后著文概述了本领域的研究动态,指出植物实验胚胎学已上升为"植物实验生殖生物学"的更高阶段,并正在开拓植物生殖工程(或性细胞工程)新技术领域。

我们实验室在长期从事子房培养等性器官操作的基础上,近10年发展到性细胞原生质体操作。在此过程中,曾先后数次获得国家自然科学基金的支持,特别是在"八五"重大项目"植物性细胞的发育生物学研究与操作系统的创建"(1992—1996)中,承担了花粉原生质体和配子原生质体的操作与研究两项课题,在以往的基础上又取得了新的进展,共发表论文近60篇。本文旨在回顾我们迄今的研究成果,并联系国际动态,提出今后研究的战略设想。

花粉原生质体的操作与研究

花粉原生质体是一种理想的单倍性原生质体,但因包围它的花粉外壁由不受酶解的孢粉素组成,因此用一般分离原生质体的方法难以分离花粉原生质体。直至20世纪80年代后期才由日本和我们分别在几种单子叶花卉植物中取得分离大量花粉原生质体的成功。此后,我们开展了一系列关于花粉原生质体的分离、培养、融合、外源基因导入以及细胞生物学研究,同时还首次创建了脱外壁花粉实验系统。

一、花粉原生质体的分离

迄今,我们共在9种植物中分离出花粉原生质体,是国际上分离成功的植物种类最多的。尤其是国家自然科学基金重大项目启动以来,突破了双子叶植物中芸苔属和烟草两类经济植物与模式植物花粉原生质体的分离。此外,国外只能由成熟或近成熟的花粉分离原生质体,我们则除此以外,还由幼嫩花粉(单核至二核初期)分离出原生质体,后者在培养与融合中有特殊的价值。《美国植物学报》1992年第3期选用了我们分离的幼嫩花粉原生质体显微图像作为该期的封面照片。在各类花粉原生质体的分离中,我们针对不同植物与不同发育时期的花粉的生物学特点,创建了5种不同的分离方法,研究出由超低温保存的花粉中分离原生质体的方法,还研究出挑选单个花粉原生质体制备透射电镜样品的两种方法,使有关细胞学研究建立在精密操作的基础上。

二、花粉原生质体的培养

成熟花粉原生质体培养只能产生花粉管。幼嫩花粉原生质体则可在离体条件下脱分化启动细胞分裂。我们首次培养萱草幼嫩花粉原生质体形成多细胞结构,并提出花粉原生质体离体发育两条途径的设想,即成熟花粉原生质体继续体内的配子体发育途径;幼嫩花粉原生质体被诱导转向孢子体发育途径,这两条途径均可望在遗传操作中加以利用。近年,在青菜与烟草的重复试验中证明,幼嫩花粉原生质体确有孢子体发育的潜能。围绕花粉原生质体在培养中的细胞学行为,进行了超微结构与细胞骨架的研究,揭示了一些关于细胞壁再生、细胞分裂、微丝骨架动态方面的现象。

三、花粉原生质体与体细胞原生质体的融合

由于花粉原生质体迄今尚不能单独发育再生植株,因而开展了其与体细胞原生质体融合的试验,以图借助后者的发育能力达到再生三倍体杂种的目的。20世纪80年代国际上曾报道小孢子四分体与体细胞原生质体融合产生"配子-体细胞杂种",而本实验室则首次以游离花粉时期的原生质体进行试验获得成功:青菜幼嫩花粉原生质体与甘蓝型油菜下胚轴原生质体融合,再生了3个小植株,根据染色体、同工酶分析,判断其中1株为异源三倍体,2株为异源四倍体;后又根据RAPD分析及成熟植株的形态与育性鉴定,证明了上述判断的正确性。烟草抗卡那霉素品系的幼嫩花粉原生质体与黄花烟草叶肉原生质体融合,也再生了异源三倍体杂种植株,而用成熟花粉原生质体作融合实验则仅产生管状结构。迄今,国际上仅在烟草属中有一篇报道[1]。

四、脱外壁花粉的分离、人工萌发与离体授粉

用人工方法脱去外壁后仅为内壁覆盖的花粉,我们称为脱外壁花粉,是介于完整花粉粒和花粉原生质体之间的结构。我们在芸苔属中首次分离出批量的脱外壁花粉,以后认识到它的价值而开展了系统的研究。最近,专文报道了烟草脱外壁花粉的制备方法,开展了其人工萌发和离体授粉实验。结果表明,脱外壁花粉对萌发条件有较严格的要求,但可顺利产生花粉管;授粉与子房培养实验证明,其花粉管可在花柱中生长,并使胚珠受精结籽,最终萌发幼苗,从而创建了脱外壁花粉离体授粉受精的完整实验系统。

五、花粉原生质体与脱外壁花粉的电激基因转移

花粉在遗传转化中的潜在重要价值受花粉壁对外源DNA的屏障所限制,而花粉原生质体和脱外壁花粉由于完全或部分地脱去花粉壁,应是较理想的受体系统。最近,国际上报道用电激法将外源GUS基因导入百合花粉原生质体,其瞬间表达水平远较对照花粉为高[2]。同年,本室以电激法将外源GUS基因导入紫菜苔花粉原生质体,在有花粉特异启动子Zm13—260的条件下,花粉原生质体的GUS活性为对照花粉的100倍以上。在烟草方面,首先研究了花粉发育过程中内源GUS活性变化的背景,在此基础上开展了其脱外壁花粉的电激实验。结果表明,脱外壁花粉的外源GUS表达活性约为对照萌发花粉的5倍和未萌发花粉的30倍。由此可见,无论花粉原生质体或脱外壁花粉,确有利于外源基因的导入。这一优点与前述脱外壁花粉离体授粉实验系统结合,将有可能开拓一种通过有性过程获得转基因植株的新途径。

配子原生质体的操作与研究

人工分离的雄配子(精细胞)及其前身生殖细胞通称配子原生质体,扩而言之,分离的雌配子(卵细胞)亦应属此范畴。国际上于1986年首次分离出大量的白花丹精细胞。同年,我们报道了绣球百合生殖细胞的压片分离,继之以蚕豆生殖细胞的渗击大量分离。至于雌配子的分离,首先要经过胚囊分离的阶段。20世纪80年代以来,我们先后在金鱼草、向日葵与烟草中分离出生活胚囊。胡适宜等在烟草和颠茄中分离出胚囊细胞原生质体,带动了国际上此后在更多的植物中胚囊与卵细胞的分离研究,并成为后来精、卵体外融合的技术基础之一。以下仅介绍本实验室近年在雌、雄配子原生质体操作方面的研究进展。

一、生殖细胞与精细胞的分离

我们先后在7科12种植物中分离出大量生殖细胞,并针对不同植物花粉的生物学特点研究出4种分离技术;首次试探了分离的生殖细

胞的培养并实现了其在离体条件下的核分裂;应用扫描电镜、视频增差显微活体观察、免疫荧光显微术等研究了生殖细胞离体后的形态结构的变化。其中,关于生殖细胞在体内发育与离体后的微管骨架的变化规律的研究,是起步较早与较为系统的。

国际上多数由三细胞型花粉分离精细胞,仅在少数植物中是由二细胞型花粉分离。我们由玉米、油菜、黑麦、紫菜苔等三细胞型花粉分离出精细胞,以后又在5科8种具二细胞型花粉的植物中开展了精细胞分离的研究,其中多数植物是首次记录。

二、卵细胞与合子的分离

我们关于胚囊分离的研究,曾与离体雌核发育研究一起,获得1991年度国家自然科学奖。近年,在烟草上又进一步研究出分离胚囊的新技术,可以一次分离和收集数十个胚囊,进而从中分离与收集一定数目的卵细胞与其它胚囊细胞原生质体。分离烟草受精后的胚囊及其中的合子的技术也已研究成功。

三、配子原生质体的融合

20世纪90年代国际上植物有性生殖研究中的一项突破是玉米精、卵体外融合及培养的成功。为开展这一高难度的课题,我们先试验了生殖细胞和精细胞与其它原生质体的融合,研究了荧光预标记雄性细胞核以鉴定异核体的方法。为了适应精、卵单对融合的精密要求,创建了在PEG微滴中以显微操作法融合单对原生质体的技术,并用于烟草雌性细胞与各类原生质体(包括雄配子原生质体)融合。

四、合子的培养

无论是精、卵体外融合产生的"人工合子"或由体内直接分离的自然合子,数目均很有限,必须采用微培养技术。我们以烟草叶肉原生质体为先导,试验了微滴培养和微室饲养两种方法,在一个微滴或一个微室中培养5～10个原生质体,可分裂为多细胞团与愈伤组织。应用微室饲养法培养烟草合子,已分裂为二细胞与多细胞结构。

五、钙与花粉管生长和受精

钙对花粉管生长有重要作用。但国际上多以离体萌发的花粉管为实验系统。我们从另一角度研究雌蕊中钙的分布规律,以探讨其与花粉管生长的关系。超微细胞化学与 X 射线微区分析表明:向日葵与棉花雌蕊的整个花粉管生长途径(如柱头乳突、花柱引导组织、珠孔、珠心退化细胞柱、退化助细胞)中,均含有较相邻组织丰富的钙;并且钙多定位于花粉管所生长的质外体系统(胞间基质、细胞壁)中。助细胞是胚囊中含钙最多的细胞,这和前人在小麦[3]和珍珠谷[4]中的观察一致。不同的是,我们发现在向日葵与棉花中,当花粉管到达胚囊前即诱导两个助细胞之一钙量剧增,继之以该助细胞退化并吸引花粉管进入其中释放精子,而前述两种禾本科植物的两个助细胞没有如上差别。

关于钙对生殖核分裂的影响,前人尚未着重研究。我们应用钙离子通道阻滞剂 Nifedipine 对烟草人工萌发花粉管进行处理,表明钙离子通道的畅通不仅对花粉管生长,而且对生殖核的分裂有重要影响。钙离子载体 A23187 的实验也表明,过高或过低的钙离子浓度对生殖核分裂不利。

国际研究动态与我们的对策

20 世纪 90 年代以来,国际上植物性细胞操作与研究达到了新的水平,取得了不少突破性成就。了解国际上有关研究的热点与趋势,对于我们今后的战略部署是必要的。

一、精、卵体外融合

20 世纪 90 年代首次实现玉米精、卵体外融合并培养成多细胞结构[5];接着,用来自不同玉米品种的精、卵融合,再生了杂种植株[6]。这一突破带动了几方面的研究:在融合技术上,除最初的微电融合外,在高钙、高 pH[7],甚至一般有钙条件下[8],精、卵亦易融合。在实验材料上,除玉米外,小麦精、卵融合[9],玉米卵细胞与其它禾本科植物精细胞融合[10]亦培养到多细胞结构。在细胞学研究方面,对玉米精、卵融合及人工合子早期发育事态进行了观察,揭示了合子壁再生、极性形成、合子不对称分裂等特点[10~12]。精、卵体外融合的成功标志着高等植物受精过程的研究从此可以置于人工控制的离体的条件下进行。

二、合子与卵细胞培养

最近的另一项进展是合子培养。除前述"人工合子"培养外,大麦与小麦的自然合子培养,亦再生了植株[13]。这就将植物胚胎培养的起点由本来就很困难的原胚阶段进一步提早到合子阶段。未受精卵细胞的培养也有初步进展:分离的玉米卵细胞经外源生长素 2,4-D 短时处理后再培养,已发育至多细胞构造[10]。下一步将是朝实现诱导未受精卵细胞离体雌核发育为单倍体植株的目标前进。

三、受精与胚胎发生初期的基因表达

植物生殖过程中的基因表达研究,在花粉发育、胚胎发育方面已取得许多成就,然而涉及有性生殖的中心环节,即受精和胚胎发生的启动,由于研究难度很大,目前仅仅开始。最近,通过突变体分析,已分离出决定合子第一次分裂不对称性的基因[14]。应用 RT-PCR 技术,分别由 100 多个分离的玉米卵细胞和"人工合子"构建 cDNA 文库[15],通过差异筛选分离出卵细胞受精前、后的特异基因。可以预期,这将形成植物发育生物学研究中的一个新热点。

四、雌、雄配子识别

关于植物传粉受精中的雌、雄识别,过去集中在配子体与孢子体组织之间的识别研究上,特别是对自交不亲和的识别机理有很深入的研究。近年开始探讨配子水平上的识别,以求最终解开双受精之谜。20世纪 80 年代曾企图由分离的精细胞制备单克隆抗体以寻找与识别有关的物质,但未能解决好细胞纯化问题。20 世纪 90 年代以来取得若干新进展:发现单克隆抗体 JIM8 与油菜精细胞结合而不与中央细胞结合,并特异地标记细胞膜[16]。还发现单克隆抗体 LIG62 专一地与百合生殖细胞及精细胞结合[17]。在我国，曹宗巽等近年也着手从大批量分离的玉米与百合精细胞中提取其质膜并对质膜蛋白进行生化分析，初步发现了若干可能的精细胞特异蛋白。

五、生殖工程

生殖工程是植物有性生殖研究在高新技术中的应用。如前所述，植物生殖器官与组织的操作，如花药培养、胚胎培养等已在育种中发挥作用，而性细胞及其原生质体的操作则为今后的生殖工程提供新的技术源泉。利用有性途径进行植物遗传转化的研究，前人曾有专文评述，其中着重介绍了小孢子和花粉的转化[18]。但以性细胞原生质体与合子为对象的研究可以说尚未开始。我们曾撰专文，从宏观的角度论述了植物改良方法，由常规有性杂交到体细胞工程再到性细胞工程的螺旋式历史发展趋势，预期一二十年内将会出现植物生殖工程研究的高潮。纵观近两年召开的国际植物有性生殖会议(1994,1996)，这一趋势已经日渐明朗。

六、对策

植物有性生殖研究是植物发育生物学中的核心内容之一，而性细胞操作则是性细胞发育、受精与胚胎发生研究的重要手段。通过近十多年的努力，我国在这方面已经打下了良好的基础，取得了富有特色的研究成果，其中有些方面具有一定的优势，但迄今取得的成绩仍是阶段性的。为了迎接21世纪的挑战，今后的战略应是：瞄准国际研究前沿，立足国内优势，依仗有自身特色的实验系统，选准新的突破口，从分子、细胞到个体各个层次上深入开展研究。其中，精卵体外融合和合子培养、受精和胚胎发生早期的基因表达、花粉原生质体和配子原生质体的遗传转化等，应是既顺应国际趋势又符合我国国情的主攻方向。

(作者：杨弘远、周嫦。原载：中国科学基金，1997，(2):85~90。参考文献76篇，删去。)

7 被子植物离体受精与合子培养研究进展

Recent advances in *in vitro* fertilization and zygote culture of angiosperms

历 史 背 景

自从19世纪末发现被子植物双受精现象以来,关于植物受精的研究大致沿三个方向发展:(1)细胞胚胎学方向,由早期的石蜡切片、光镜观察发展到20世纪60年代以后的以超微结构研究为主。80年代以来进一步采用三维重构、图像分析等精密方法,揭示了雄性与雌性生殖单位、精细胞二型性、偏向受精等新现象[1~3]。(2)生理学方向,早期主要是研究传粉、受精前后花器生理变化,以后着重花粉与雌蕊组织间的识别反应,80年代以后开始探索配子间的识别等生理、生化机理[4]。(3)实验胚胎学方向,应用离体培养方法在人工控制的条件下研究受精和受精后的发育,从60年代开始,至90年代实现了重大的飞跃,并由此推动了整个受精与胚胎发生的发育生物学研究。本文重点综述这后一研究方向的新进展。

20世纪60年代开始成功的离体受精,当时称为试管受精,实际上是在胚珠(或柱头)上授粉,然后通过胚珠或子房培养完成受精和胚胎发育,形成有萌发力的种子。这样的受精仍然处于胚珠以至子房的孕育下,只能认为是离体受精研究的初级阶段,近来多数文献改称为离体授粉。90年代成功的精、卵离体融合,才是严格意义上的离体受精。由体内受精到离体授粉再到真正的离体受精,标志着人类对植物受精的实验控制能力不断提高。现在不仅可以像动物体外受精那样,在离体环境中完成被子植物精、卵融合,而且还可以将受精所产生的"人工合子"培养再生植株,做到整个受精与胚胎发育过程完全在离体条件下完成。这样一种实验系统的建立无论在基础研究与高新技术研究上均有重大意义。因此,把离体受精的成功看做植物受精研究史中的重要里程碑是毫不夸张的。

离体受精的突破有其学术思想背景与实验技术前提。近10多年,随着雌、雄配子分离技术的成熟,研究者们预见到配子离体融合指日可待,提出了开展这一研究的设想[5,6]。到20世纪80年代后期,精、卵离体融合的思路在国际上已经很明朗了,同时实现这一工作的技术前提也已具备了。Kranz等[7,8]把握了这一机遇,利用前人多方面的研究成果加以综合,在玉米上作出了首次突破。他们的成功是依靠三项技术前提:(1)雌、雄配子分离技术。不少实验室在80年代均建立了此项技术[6,9,10],其中包括玉米精细胞[11~13]和卵细胞[14]的分离。(2)单对原生质体融合技术。由于分离的卵细胞数目有限,很难将一般原生质体群体融合的方法用于精、卵离体融合,而微机操纵的单对原生质体微电融合技术[15]则可有目的地挑选一对原生质体进行融合。(3)微培养技术。由于同样的理由,精、卵融合产物的数目有限,不可能采用常规的原生质体群体培养方法,而必须依靠单个或微量原生质体培养技术,其中微滴培养法是比较有效的[16],以后又进一步改进,采用了商品生产的微室(millicell),将被培养的细胞置于微室中,通过室底的微孔滤膜吸取周围饲养细胞的活性成分,这种微室饲养法更有利于培养对象的持续发育而被用于玉米"人工合子"的培养[8]。总之,被子植物离体受精之所以直到90年代方才成功,是建立在近10多年来几方面的研究基础之上的应运而生的结果。

操作技术的研究进展

如前所述,植物离体受精包括三项技术基础,即配子分离、单对配子融合与人工合子的培养,这也是整个实验操作系统中的先后三个主要环节。

一、卵细胞分离技术

关于精细胞分离技术可参阅前节所引证的文献,此处不赘述。卵细胞分离相对困难。玉米离体受精程序中所成功使用的方法是:切取包含胚囊的胚珠部分,经酶处理约30min使之软化,然后用细玻璃针从其中解剖出卵细胞。在2h内可由100枚胚珠中获得20~25个卵细胞,供1d融合实验所需[17]。大麦与小麦自然合子培养中成功使用的卵和合子分离方法,则是不经酶处理的解剖技术,其特点是以镊尖由胚珠的准确部位刺入中央细胞,引起后者膨压剧变致使卵细胞逸出。在90min内,由20个子房可解剖出15个卵或合子,频率高达75%[18]。以上两种方法的共同点是其最终步骤均依靠手工解剖,但在是否经过酶解预处理上有差别。最近,在玉米受精后胚囊培养中,就酶解步骤的影响作了专门实验。结果表明,即使微量和极短暂(2min)的酶处理,均有害于胚囊的存活并导致以后发育异常[19]。关于酶处理有害的论断似乎还需要更充分的实验证明,但尽量在卵细胞分离过程中减少酶的影响是值得注意的。

二、精、卵融合技术

在精、卵融合实验中迄今共试验了四种融合技术:(1)微电融合。这是Kranz等[7,8]最早进行玉米精、卵融合取得成功的方法,融合率高达85%,以后一直有效地加以沿用。迄今仍然只有用这种方法融合的产物再生了植株。但此法依赖特制的自动化操作系统,尚未在国际上推广。另一个问题是:强制性的电融合作为细胞工程手段是适用的,但不适用于研究受精的机理。(2)高钙、高pH介导融合。Kranz与Lörz[20]进一步尝试了在高钙(0.05mol/L $CaCl_2$)与高pH(11)的条件下介导玉米精、卵融合,融合产物培养成含30~50个细胞的微愈伤组织。作者的本意是探索一种能借以研究受精机理的方法。然而,高钙、高pH这种化学诱导方法未必能模拟受精的自然条件,仍然带有人为强制的特点。(3)一般钙条件下的融合。Faure等[21]在一般含钙介质中观察玉米配子间与配子、体细胞间的融合情况,发现在5mmol/L $CaCl_2$ 条件下,精、卵在数分钟内粘贴,然后在 10s 的瞬间融合,融合率高达近80%,而其它组合的融合率则很低甚至完全不融合,认为这表明雌、雄配子间确有互相识别的能力。可惜这种融合产物的发育前途却不清楚。(4)PEG(聚乙二醇)诱导的融合。PEG 是原生质体融合最常用的诱导剂,但一般用于原生质体群体内的融合,具有随机、不定向的特点,融合后的异核体混杂于大量未融合的原生质体和同核体群体内,筛选也很麻烦。孙蒙祥等[22,23]对这一常规方法加以改进,创建了 PEG 诱导单对原生质体融合技术,用微吸管挑选一对原生质体置于 PEG 微滴内使之融合,以后又应用这一方法实现了包括雌、雄性细胞在内的多种组合成对融合。此法无需采用价格高昂的显微操作与微电融合设备即可有目的地选择单对性细胞进行融合,是一种便于推广的简易方法。但精、卵融合产物的培养前途仍不清楚。此外,和电融合一样,PEG 法在细胞工程上是有价值的,但也不适宜于受精机理的研究。

三、合子培养与植物再生

精、卵离体融合产生的"人工合子"和由体内直接分离的自然合子,现均已培养成植株,这是植物胚胎培养技术由成熟胚培养到幼胚培养再到原胚培养不断深化的必然趋势。成功的关键在于采用了有效的饲养系统。首先成功的是玉米人工合子培养,将微量人工合子置于微室中,以玉米悬浮细胞作为饲养物,合子分裂率高达 79%,并发育为多细胞结构[8]。以后进一步报道玉米品种间精、卵融合产物经胚胎发育途径(偶尔经多胚发育或器官发生途径)再生植株,自融合开始 100d 内达到开花。11 株的根尖染色体均为 2n;植株正常结实,穗部具双亲中间颜色性状;种子胚乳颜色表现 3:1 分离,与有性杂交的规律一致[24]。除玉米外,小麦精、卵融合[25],玉米卵细胞分别与高粱、薏苡、小麦、大麦等禾本科植物的精细胞融合,以及玉米卵细胞与玉米悬浮培养细胞原生质体等多种融合组合的产物[26],亦均培养到多细胞阶段。但玉米卵细胞分别与油菜精细胞、玉米卵细胞、玉米珠心细胞等组合的融合产物则不能分裂[26]。以上表明,禾本科的属间离体受精似无困难,而科间融合或非雌、雄配子间的融合产物则难以发育。由于缺乏遗传学鉴定,目前还不能对属间离体受精的结果作出评价。

关于自然合子培养,在其成功前不久曾报道含合子的玉米胚囊培养再生植株[27,28]。然而所培养的还不是完全分离的胚囊,而是带有或多或少的珠心组织,实际上介于胚珠培养与胚囊培养之间,尽管较近的一篇报道中,包围胚囊的珠心组织已减少到可以依稀透视合子的程度[29]。真正的合子培养是由 Holm 等[18]实现的:用解剖法由大麦和小麦胚珠中直接分离出卵细胞和合子原生质体。卵细胞培养未能启动分裂,而合子在大麦小孢子饲养下发育成胚状结构,频率高达 75%,其中约 50% 胚状体再生为可育植株。实验材料中包括在花药培养中反应迥异的品种,但其在合子培养中再生植株频率相近,似乎表明基因型对培养结果影响不大。小麦的合子在大麦小孢子饲养下也再生了一个可育植株。

迄今无论人工合子或自然合子培养的实验均局限于禾本科植物。在双子叶植物中,仅在烟草中开展了合子培养试验:用酶解-研磨法从烟草受精后胚珠中分离出胚囊,再由胚囊分离合子,以叶肉原生质体作饲养物,在微室中培养合子启动了第一次分裂[30]。随后在 64 个微室中培养了 242 个合子,进一步研究了合子分离方法、合子发育时期、饲养细胞发育状态以及培养方式等对合子培养的影响,经改进技术后,约 60% 合子完成第一次分裂,培养 12d 形成少数原胚或细胞团(李师翁等)。

四、未受精卵细胞与中央细胞的培养

在未传粉子房与胚珠培养诱导单倍体植株成功以后,研究者曾试图直接培养分离的胚囊而未获进展。但最近未受精卵细胞的培养却取得了可喜的初步结果。玉米品系 Garant 的卵细胞经高浓度(25mg/L、30mg/L、40mg/L)2,4-D 处理 1d,再在 2mg/L 2,4-D 条件下微室饲养,6% 分裂成多细胞团,品系 A188 则不分裂[26]。由于有的多细胞团已达到含 100 个以上细胞,因而预期应有继续发育与分化的可能,但是否最终能诱导单倍体植株有待研究。

中央细胞与精细胞融合发育为胚乳,是被子植物特有的现象。胚乳培养已有较多研究。玉米中央细胞与精细胞离体融合产物的培养最初未能分裂[31],但最近在一篇综述中提及,玉米中央细胞离体受精后能够生长,不过中央细胞不易分离,难以获得其融合产物[17]。至于未受精中央细胞的培养则更无报道。最近,傅缨等[32]借鉴玉米未受精卵细胞培养的经验,在烟草胚珠酶解过程中伴以高浓度(20mg/L)2,4-D处理,由此分离的中央细胞在微室饲养条件下可启动分裂。共培养了184个中央细胞,最高分裂频率达14%,有的已形成小细胞团;以同样方法分离和培养的卵细胞与助细胞原位融合体及反足细胞亦诱导了一次分裂。

细胞生物学与分子生物学研究

一、细胞生物学研究

离体受精提供了直接观察受精和受精后发育动态的实验系统,初步揭示出若干新的现象。在玉米离体受精成功后,研究者先后以3种方法进行了有关细胞生物学研究:一是将精、卵融合材料固定与包埋后制作半薄与超薄切片,进行光镜、电镜及三维重构观察[33];二是应用荧光显微术对生活材料进行观察[26];三是应用图像分析和共焦激光显微术进行研究[34]。研究结果首先是得出了离体受精过程的时间表:精、卵细胞融合十分迅速,在1s内即已完成;细胞融合后30s开始合成细胞壁,20min内完成壁的再生;35~60min完成核融合;18~46h出现两个核仁;39~50h发生核分裂;40~60h完成合子分裂。这样一个时间表对于离体受精而言是比较精确的,便于在此基础上开展进一步的细胞生物学、分子生物学和遗传工程研究。实验表明,在离体受精后10~45min再以附加精细胞与人工合子融合是不成功的,这是由于人工合子再生细胞壁之后,构成了防止多精入卵的障碍[21]。在这里可以提出一个问题:如果同时以两个以上的精细胞对一个卵细胞进行离体受精,结局会如何呢?是否可以获得多精入卵的结果呢?可惜迄今还没有这方面的实验。离体受精系统还可以用于研究卵细胞通过受精而激活的机制。初步观察到卵细胞和 4h 的人工合子中有钙调素和膜结合钙的分布,这暗示植物受精可能如动物受精那样,钙在卵的激活中起重要作用[34]。离体受精系统还可以用于研究合子的极性与分裂模式。据观察,玉米的分离卵细胞即使变成圆球形依然保持极性,其表现之一是:未受精的卵细胞或两个卵细胞的融合产物,仅仅在一极形成壁物质,与体内状态相似[26]。受精后的人工合子的第一次分裂也和体内合子一样是不对称的,这不仅见于玉米离体受精产物,而且亦见于玉米与其它禾本科植物远缘离体受精产物[26],由此导致此后主要按胚胎发育模式发育。最后值得提及离体受精和体内受精一个显著的差异是,精细胞质恒定进入卵细胞[33],而体内受精时精细胞质的行为则因物种而异。研究雄性细胞质在离体受精植株的遗传规律将是一个有意义的方面。

二、分子生物学研究

利用离体受精系统研究受精与早期胚胎发生过程中的基因表达,是本研究由细胞水平进入分子水平的新趋势,也是植物发育生物学中的一个新热点。获得小批量卵细胞与人工合子的方法以及由微量细胞构建 cDNA 文库的方法已经建立,为开展这一研究奠定了技术基础。实验材料迄今仍然是玉米。在 Kranz 实验室,应用逆转录酶-聚合酶链式反应(RT-PCR)技术,由 128 个分离的卵细胞与 104 个离体受精后 18h 的人工合子构建了 cDNA 文库[35],并通过差异筛选从后者分离出若干在卵细胞中特异表达或由受精诱导表达的基因[17]。由人工合子文库中分离出编码钙结合蛋白 calreticulin 的克隆(该蛋白主要存在于细胞内质网及某些细胞的核或质中,与细胞分裂有密切关系),测序分析表明它和常规的 calreticulin 基因 DNA 序列一致。从而证明,所构建的 cDNA 文库是符合质量要求的[36]。此外,根据 1996 年德国汉堡大学"Plant Embryogenesis Workshop"论文摘要介绍,由以上文库中还分离出如下一些克隆:与核糖体蛋白 L39、S21 及 Po 族同源的克隆;与哺乳动物翻译起始因子 EIF4D 类似的克隆;编码 MCM/3/4/5 基因族转录因子的克隆;类似哺乳动物 DNA 聚合酶一个亚单位的克隆;哺乳动物 flap-endonuclease 的克隆等。在 Dumas 实验室中,由 100 个玉米过渡期幼胚构建了 cDNA 文库,从中克隆出在胚胎发生过程中表达的钙调素基因[37]。迄今取得的成果仅仅在分离受精前后与早期胚胎发生过程中特异或优势表达基因上迈出了第一步,距离揭示这些基因在这一发育过程中的功能还远。除了通过构建 cDNA 文库研究基因表达外,还尝试了将外源基因导入合子的实验:应用显微注射将两种报告基因(GUS 基因、花青素调节基因)分别导入玉米的分离合子,经短时间培养后,证明被导入的基因在合子中瞬间表达[38]。显微注射技术在动物受精机理和受精工程研究中是一项有力的手段,在植物中的这一尝试是一个有意义的开端。

战略思考

受精是植物亲代与子代之间的联系桥梁,在发育与遗传中具有多重的作用:激活卵细胞,启动胚胎发生;实现由配子体向孢子体,由单倍体向二倍体的世代转换;实现双亲遗传物质的结合。因此,在植物有性生殖研究中,受精始终占有中心的位置。受精又是一个十分复杂的自然过程,包含一系列精巧有序的发育事态;被子植物的受精过程是在深藏的雌配子体中进行并包含独有的双受精特点,这更增加了研究的难度。在 19 世纪植物胚胎学的奠基阶段,双受精的发现是在大、小孢子发生与雌、雄配子体发育和胚与胚乳发育这些生殖过程的前后两段已经基本弄清之后方才突破的。历史似乎在新的水平上重演:在 20 世纪 90 年代,离体受精也是在生殖过程前后两段的实验研究(花药与花粉培养、未传粉子房与胚珠培养、胚囊分离、雌雄配子分离、离体授粉、胚胎培养、胚乳培养)已经达到相当深度之后方才突破的。在生殖过程的基因表达研究中,似乎也有这种由前后两段向受精这个中心发展的趋势。这样,可以预见 21 世纪将迎来受精机理和受精工程研究的辉煌时代。

在近年发表的一些综述中,从不同角度对离体受精研究进展与前景进行了探讨[17,39~41]。本文在总结前述各方面研究成就的基础上对今后发展趋势提出一些战略思考:在玉米离体受精这一模式系统之后,创建更多离体操作模式系统仍将是未来发展的动力。将有更多的植物离体受精与合子培养成功。首先是其它禾本科植物,包括已经取得阶段性成果的小麦、大麦。作为重要粮食作物的水稻已有较好的胚胎培养基础,合子培养指日可待,而离体受精则由于其花粉生活力极弱可能遇到困难。双子叶植物中,烟草在现有基础上应是最有希望成功的;油菜的精细胞分离与原胚培养均较成熟,只要卵细胞与合子分离成功,也很有前途。除了种内的离体受精外,远缘离体受精、多精受精、未受精卵细胞培养、未受精与受精中央细胞培养等各种实验研究,因为在理论研究与遗传育种上各有独特的价值,将呈现广阔的研究天地。利用卵细胞、合子等作为基因工程的受体有旺盛的生命力。前述研究结果表明,无论人工合子或自然合子,在离体条件下均有很高的分裂频率并基本遵循胚胎发育途径,这是一般体细胞和其它性细胞系统所不可比拟的优点。考虑到合子数目有限,外源基因导入将主要采取显微注射的手段,这样一种在动物生殖工程中证明有效的方法,在植物中也必然开拓出强有力的新转化体系。

在各种实验系统建立的基础上,有关受精和早期胚胎发生的理论研究将深入到新的层次。体内的受精与早期胚胎发生包含的主要发育事态有:花粉管进入退化助细胞释放精子;精子由退化助细胞转移至受精靶区并分别趋向卵与中央细胞;雌、雄性细胞融合与核融合,卵细胞(及中央细胞)通过受精激活;合子的极性分裂与胚和胚柄的分化;原胚分化;胚乳发育等。对于这些发育事态,体内研究只能解决部分问题,离体研究也只能解决部分问题,而两方面研究互相参照则将获得较全面的认识,这由前述离体受精的细胞生物学研究已略显端倪。由于离体研究具有可控条件和生活状态两大优点,特别适于应用细胞生理学、生物物理学的方法,以研究受精和胚胎发生过程中的信号传递等体内研究难以奏效的问题,正如花粉离体萌发系统中有关研究所显示的那样。至于应用分子生物学方法研究这一过程中的基因表达更是方兴未艾。有朝一日将能解析决定上述各个发育环节的功能基因,然后应用转基因技术改变受精与胚胎发育过程。例如,如果找到决定卵细胞激活的有关基因,就有希望通过基因操作创造出无需受精而启动胚胎发生的全新植物类型。卵的激活与胚胎发生是植物个体发育的起始程序,如能掌握决定这一起始程序的基因,便找到了最终人工控制植物个体发育起点的入门钥匙,其重大意义是不言而喻的。

(作者:杨弘远、周嫦。原载:植物学报,1998,40(2):
95~101。参考文献41篇,删去。)

8 Some approaches to the experimental manipulation of reproductive cells and protoplasts in flowering plants

During the last two decades of the twentieth century, one of the most important trends in experimental plant embryology is that the techniques of *in vitro* manipulation have become more and more precise and refined, shifting from organ/tissue manipulation to ward the level of cell/protoplast manipulation, from large-scale culture toward microculture, with an increasing reliance on a multidisciplinary approach in order to resolve plant reproductive problems at the cellular level. A series of novel experimental systems have been developed to explore the underlying mechanisms during sexual plant reproduction as well as to open new ways of reproductive cell engineering in higher plants (Yang and Zhou, 1992).

Along with these worldwide trends of research, our group also has made considerable advances in the manipulation of plant reproductive cells and protoplasts, which include pollen protoplasts, de-exined pollen, male gametoplasts and female gametoplasts. To manipulate them, the inevitable first prerequisite is to isolate them from the tissues or cell walls that surround these cells or protoplasts. These isolated cells or protoplasts must be viable enough to undergo further manipulation with culture, fusion or transgenic treatments. The various kinds of reproductive cells and also the same kinds of cells in different species require special methods and conditions for their isolation and further manipulation. A number of experimental results of our researches have been published elsewhere in past years, but most were published in Chinese journals which might not be so available worldwide.

In this article we summarize these results briefly, and we can only cite the most relevant reports by other investigators in the same field.

Pollen protoplasts

Isolation of pollen protoplasts

Pollen protoplast is a useful system for pollen biology study and may be an ideal haploid protoplast of natural source. However, the difficulty of deleting the recalcitrant exine was not overcome until the late 1980s (Tanaka et al., 1987; Zhou, 1988a). Success relies chiefly on suitable methods coupled with the characteristics of the pollen. Up to now, isolation of a large quantity of viable pollen protoplasts has been restricted mainly to pollen grains bearing a relatively broad germ furrow, as in Liliaceae, Iridaceae, Amaryllidaceae, etc. The exine of such kinds of pollen can dehisce along the furrow when submitted to hydration. As a result, the intine is exposed to direct enzymatic treatment and the protoplast is released. By this hydration-enzymatic maceration method we have isolated protoplasts from mature pollen in *Iris*, *Zephyranthes*, *Hemerocallis* (Zhou, 1988a), *Gladiolus* (Wu Y and Zhou, 1990), *Brassica* (Li et al., 1992) and *Nicotiana* (Wang and Zhou, 1995), and from immature, unicellular or early bicellular pollen in *Hemerocallis* (Zhou, 1988a), *Brassica* (Liang et al., 1993) and *Nicotiana* (Xia et al., 1996a). Fresh-collected and also cryopreserved pollen of *Brassica* could yield viable protoplasts in large quantity

(Liang et al., 1993). Each kind of pollen requires its special modified methods for protoplast isolation. For instance, in tobacco, whose pollen lacks broad germ furrows, its pollen protoplasts were isolated enzymatically from just-germinated short pollen tubes, with a yield up to 50%-70% (Wang and Zhou, 1995).

In vitro germination of mature pollen protoplasts

Mature or nearly mature pollen protoplasts incubated in an appropriate medium (usually sophisticated protoplast culture medium rather than simple pollen germination medium) could germinate into nearly normal pollen tubes in which the generative cell could divide into sperm cells, as reported in *Lilium* (Tanaka et al., 1987), *Hemerocallis* (Zhou, 1988a) and *Gladiolus* (Wu Y and Zhou, 1990). Various abnormal structures such as budding, beads, dumbbells or coenocytes were observed too. The pollen protoplasts regenerated cell walls which were not so typical of pollen wall. In *Gladiolus* the wall regenerated from pollen protoplasts was multilayered and cellulosic, with germ-furrow-like thickenings where the pollen tube grew out (Wu Y and Zhou, 1990). The distribution of actin was investigated in *Gladiolus* pollen protoplasts, using FITC-phalloidin as the fluorescent probe. A relationship between wall regeneration and actin distribution was found. When the pollen protoplasts germinated, actin filaments became polarized and converged into the emerging pollen tubes, forming parallelly arranged bundles along the tubes (Zhou et al., 1990).

Culture of immature pollen protoplasts

Can the protoplasts isolated from microspores or early bicellular pollen be switched from gameto-phytic development to the sporophytic pathway, as occurs in anther/pollen culture? Our experiments on immature pollen protoplast culture of monocots or dicots gave positive answers. Protoplasts isolated from cold-pretreated microspores of *Hemerocallis* could divide into multicellular or proembryo-like structures *in vitro* (Zhou, 1989). A series of profound ultrastructural changes happened during this course of signatured cell dedifferentiation: disappearance of the original large vacuole characteristic of a microspore, locomotion of the nucleus from the periphery to the central position, increase and activation of organelles, diminution of starch grains in plastids, etc. Mitotic or amitotic figures were observed. Regeneration of cell wall might be incomplete, creating a major obstacle that restricted sustained cell divisions (Wu Y and Zhou, 1992). In tobacco, protoplasts isolated from middle-stage bicellular pollen could be induced *in vitro* to undergo cell division or to continue the original pathway leading to the germinating pollen tube. In rare case a pollen protoplast divided at first into two daughter cells, then one of them germinated a pollen tube, indicating a reversal from temporal sporophytic development to gametophytic development (Xia et al., 1996a). Recently, in cooperative research between our group and a Dutch group, microspore protoplasts of rapeseed cultured in millicells and co-cultured with embryogenic microspores were induced to divide into microcalli (Sun et al., 1999).

Fusion of pollen protoplasts with somatic protoplasts

Gameto-somatic hybrids have been obtained from fusion between microspore tetrad protoplasts and somatic protoplasts in *Nicotiana* (Pirrie and Power, 1986) and subsequently in other genera. The callose wall of tetrads can be degraded easily by snailase for release of the protoplasts. However, small size and short duration limit the use of tetrads. Pollen protoplasts may be more convenient in this respect. We used immature pollen protoplasts of *Brassica chinensis* to fuse with hypocotyl protoplasts of *B. napus*. Plants were regenerated, one of which was allotriploid, and the other two were allotetraploids, as identified by chromosome counting (Figs. 1-10), esterase isozyme analysis and RAPD analysis (Li et al., 1994, 1996). Similar experiments were carried out in *Nicotiana*. Pollen protoplasts of *N. tabacum* fused with mesophyll protoplasts of *N. plumbaginifolia* produced 17 hybrids and three cybrids (Desprez et al., 1995). Fusion between pollen protoplasts of *N. tabacum* and *N. rustica* yielded four triploid hybrid plants (Lu et al., 1996). These results confirmed the feasibility of "pollen-somatic hybridization."

Gene transfer into pollen protoplasts

The potential of pollen in genetic transformation has been limited by the thick wall hindering gene transfer into it. Pollen protoplasts deprived of the wall were supposed to be a better system for gene transfer. This was proved by electroporation experiments in *Lilium* (Miyoshi et al., 1995) and *Brassica* (Shi et al., 1995). For example, pollen protoplasts isolated from mature, tricellular pollen of *Brassica* were electroporated with β-glucuronidase (GUS) gene in the presence of the pollen-specific promoter Zm 13-260. The results showed that the frequency of transient expression was nearly one hundred times higher than that of pollen having a wall. In contrast, pollen protoplasts isolated from younger, uni-or bi-cellular pollen did not show such a capability; this was attributed to the promoter used, which is unadaptable to early-stage pollen.

The use of pollen protoplast for ion channel analysis

Pollen germination and tube growth is known to be regulated by ion transport across the plasma membrane of the pollen and pollen tube. For analysis of whole-cell inward currents, which may give more comprehensive information than that obtained by single-channel recording, the isolated pollen protoplast seems better than the entire pollen grain. Recently the patch clamp technique was used to identify and characterize whole-cell currents in *Brassica* protoplasts. The results demonstrated that it is the inward K^+ currents that mainly account for the recorded whole-cell inward currents. The possible regulation of the inward K^+ channels by Ca^{2+} was also investigated (Fan et al., 1999).

De-exined pollen

Isolation of de-exined pollen

So-called de-exined pollen is a man-made intermediate form between the entire pollen grain and the pollen protoplast. Its exine is detached and its protoplast is coated only with intine. We first isolated quantities of de-exined pollen in *Brassica* when our interest was focused on isolation of pollen protoplasts (Li et al., 1992). Later we paid attention to the value of this special form. For isolation of de-exined pollen, different methods were developed in different cases. In *Brassica* a procedure involving three steps (hydration, heat shock and osmotic shock) was developed, which yielded viable de-exined pollen with a frequency of over 60% (Xu et al., 1996). In *Nicotiana* we used another three-step procedure (cold-pretreatment and float culture of the anthers followed by enzymatic treatment of the pollen) (Xia et al., 1996b). Then the isolated de-exined pollen were used in the following manipulation and investigation.

In vitro germination of de-exined pollen

Isolated de-exined pollen of *Brassica* was cultured in an alkalized medium containing polyethylene glycol (PEG), sucrose and lactalbumin hydrolysate (LH). Although the germination rate (30%-40%) was lower than that of intact pollen (-80%), the pollen tubes produced were quite normal and comparable to those germinated from intact pollen. Profound ultrastructural changes were observed during the de-exine procedure. The newly isolated de-exined pollen already had one or rarely two vesicle-rich sites at the sites of the previous germ furrows. Chlortetracycline (CTC) probe showed strong fluorescence of membrane-bound calcium at these sites. New wall materials deposited outside the sites led to germination of the pollen tube (Xu et al., 1997). In tobacco, de-exined pollen cultured in a medium containing PEG and LH but without sucrose could germinate normally at a rate of up to 48%. After 24h of culture, in more than half of the tubes the generative cell already divided into two sperms. This meant that the de-exined pollen might function in fertilization.

In vitro pollination with de-exined pollen

In an attempt to carry out *in vitro* pollination with de-exined pollen, tobacco flowers were excised one day before anthesis, emasculated and then inserted into culture medium, and pollinated the next day. We used small suspension droplets each containing 30-40 purified de-exined pollen grains. One droplet was placed onto one stigma with the aid of a small piece of filter paper. The de-exined pollen germinated; nearly half of the pollen tubes grew along the style. The ovaries were excised three days later and cultured to maturation. Approximately four de-exined pollen used in pollination could yield one seed on average. The seeds germinated into seedlings. Thus, in tobacco a complete procedure of *in vitro* pollination with de-exined pollen was set up (Wang et al., 1997).

Gene transfer into de-exined pollen

Further experiments on genetic transformation based on the above-mentioned *in vitro* pollination system were conducted in tobacco. Electroporation experiments had showed that GUS gene transfer into de-exined pollen was much easier than into intact pollen either before or a while after germination (Shi et al., 1996). Subsequent microprojectile bombardment experiments confirmed a better response to gene transfer when the pollen was free of the exine barrier. The frequency of de-exined pollen showing transient expression of green fluorescence protein (GFP) gene or GUS gene was approximately three and six times higher, respectively, than that of intact pollen. More interestingly, the introduced gold particles could be seen in the generative nucleus of the pollen. Pollen tubes derived from pollinated de-exined pollen could retain GFP fluorescence when they were growing in the style. Seedlings were obtained from *in vitro* pollination with particle-bombarded pollen, but so far no transgenic plants have been screened out, probably due to the limited scale of the experiments (Wang et al., 1998).

Male gametoplasts

Isolation of male gametoplasts

The artificially isolated sperm cell and its predecessor generative cell are called male gametoplasts because they usually turn into protoplasts after release from the pollen. Quantities of sperm cells were first isolated in *Plumbago* (Russell, 1986), whereas generative cells were isolated in *Haemanthus* (Zhou et al., 1986) and *Vicia* (Zhou, 1988b). There are too many works on male gametoplasts in various plant species to be cited here; they have been reviewed by other authors (e. g., Chauboud and Perez, 1992). In our lab, generative cells in twelve species belonging to seven families have been isolated by different methods. We have developed three methods in accordance with the nature of the pollen dealt with: A "one-step osmotic shock" method suited pollen that are easy to burst in sucrose solution, as in cotton. A "two-step osmotic shock" method suited the pollen that are not easy to burst but are easy to germinate in sucrose solution, such as in *Vicia*, *Zephyranthes*, *Lycoris*, *Impatiens* and tobacco. In such a case we incubated the pollen to just germinate, then gave it a sudden osmotic shock to release the generative cells. Pollen that are difficult to burst or germinate in sucrose solution, as in *Iris*, *Gladiolus*, *Allium* and *Hemerocallis*, needed a method of weak enzymatic treatment (Zhou, 1988b; Zhou and Wu, 1990; Wu and Zhou, 1991). Generative cells could also be isolated indirectly from isolated pollen protoplasts in *Lilium* (Tanaka, 1988). This method was also suitable for other plants like *Iris*,

Gladiolus, *Hemerocallis*, *Allium* and *Zephyranthes* (Wu and Zhou, 1991).

As for isolation of the sperm cells, we had success in tricellular pollen in rapeseed, maize and rye by the method of osmotic shock or gentle grinding (Yang and Zhou, 1989). The methods were refined in the experiments on *Brassica* by adding protectants like potassium dextran sulphate, bovine serum albumin and polyvinylpyrrolidone to the medium. The yield of viable sperms could be increased up to 86% and the sperms could survive up to one week at 4℃ (Mo and Yang, 1991). For isolating sperm cells from bicellular pollen, an "*in vivo-in vitro* technique" was adopted successfully in *Hippeastrum*, *Iris*, *Gladiolus*, *Hemerocallis*, tobacco and cotton. Pollen tubes grown out of the cut end of the style were subjected to osmotic shock combined with weak enzymatic treatment, which could release sperms at a frequency of up to 80% (Mo and Yang, 1992).

Morphological and structural changes of generative cells before and after isolation

The dynamic changes occurring in generative cells after liberation from mature pollen in *Narcissus* were recorded by video-enhanced microscopy. *In situ* the generative cells were spindle-shaped. After isolation the cytoplasm gradually condensed from two poles toward the center, causing the whole cell to contract significantly and finally become spherical. Parallel strands of microtubules could be visualized at the periphery of the cytoplasm; this was assumed to account for the shape change of the generative cells (Zhou et al., 1988). Thus we paid attention to the changes in the microtubule skeleton in isolated generative cells. In *Zephyranthes* a similar morphological change from the original spindle shape to ward a spherical shape occurred. In spindle-shaped cells just after isolation the microtubule bundles were mainly longitudinally oriented. Soon after, a transitional form consisting of a mixture of microtubule bundles and meshes was observed in the cells that became ellipsoidal. At last the mesh structure became predominant in the spherical cells. This observation indicates that the microtubule organization can change easily from one form to another in accordance with environmental conditions, and that it plays an important role in determining the shape of generative cells (Zhou et al., 1990). Isolated generative cells were also used to investigate the relationship of cell shape with microtubule organization during *in situ* development. A simple and effective method for isolating generative cells at various stages, still preserving their original microtubule organization, enabled us to carry out a large-scale investigation on the full course of generative cell development. In general the microtubule organization changed in correspondence with the cell shape, from a network pattern in young, spherical cells to an axial bundle pattern in mature, spindle-shaped cells. Various intermediate patterns were seen during the transitional stages. It is interesting that the changes in microtubule organization in development *in situ*, on one hand, and *in vitro* after isolation, on the other, seem to resemble each other if viewed in reverse order (Zhou and Yang, 1991).

In vitro culture of isolated generative cells

To determine whether and how isolated generative cells survive independently *in vitro*, we tried to culture generative cells of *Hemerocallis* in a thin layer of agarose medium with various additives and nursed with young anthers. In about 6000 cells observed during 4 to 20 days of culture, nuclear division occurred in about 200 cells, with an average frequency of 3.27% and the highest 11.46%. Usually two and rarely four nuclei were observed. The results show the ability of isolated generative cells to survive and undergo limited divisions even after liberation from the pollen (Wu XL and Zhou, 1990).

Female gametoplasts and their products after fertilization

Isolation of female gametoplasts

On the basis of our previous work on isolation of embryo sacs (Zhou and Yang, 1985), our research has pushed forward to the isolation and manipulation of female gametoplasts in the broad sense, including the egg cell, central cell and other component cells in the embryo sac and their products after double fertilization, that is, the zygote and the primary endosperm cell. In tobacco a new method combining enzymatic maceration with osmotic shock was developed to release the embryo sac. About 50-70 embryo sacs could be collected in one hour at the final step. Then protoplasts of the egg cell and other embryo sac cells could be isolated further from the embryo sacs, either by manual dissection or by weak enzymatic treatment (Sun et al., 1993). For isolation of the fertilized embryo sac and its component cells in tobacco, the method should be modified because the ovular tissue becomes more rigid than before. A brief enzymatic treatment followed by gently grinding the softened ovules on a slide could release the embryo sacs. Further, zygotes and primary endosperm cells could be isolated by a second enzymatic treatment or manual dissection (Fu et al., 1996).

In rice, a method of non-enzymatic manual dissection was established for isolation of the egg cell and zygote. Usually 5-8 of these cells could be isolated from 20 ovaries in two hours. The cells usually became spherical after release, but the original pear shape was retained if the zygotes were late-staged in development (Han et al., 1998). The method was further improved to isolate all component cells in the rice embryo sac either before or after fertilization by mere manual dissection or else dissection combined with enzymatic treatment. Enzymatic isolation was more rapid but had some negative effect on cell viability. The central cell, however, was too delicate due to its large vacuole to be isolated by manual dissection alone, and therefore needed enzymatic treatment (Zhao et al., 2000). Zygotes were also isolated in wheat by a similar method. Chlortetracycline and fluphenazine fluorescence showed membrane-bound calcium and calmodulin distribution, respectively, in the isolated zygotes in both wheat and rice (Zhao et al., 1980; Han et al., 1998).

In vitro fusion of female gametoplasts with other protoplasts

In vitro fertilization in flowering plants was first realized in maize by electromicrofusion between a pair of isolated egg and sperm cells (Kranz et al., 1991). This achievement opened wide prospects for research on the mechanism and control of fertilization in plants (Kranz and Dresselhaus, 1996; Dumas et al., 1998). Up to now the material used in most of the studies has been maize (Dumas and Morgensen, 1993). We have initiated similar studies in tobacco, a dicot. A microfusion technique mediated by PEG was developed to fuse a single female gametoplast with a male gametoplast or somatic protoplast (Sun et al., 1995; Figs. 11-22). This method was recently adopted in cooperative research between our group and an Italian group to trace the dynamic changes of *in vitro* nuclear fusion between a central cell and sperm cell in tobacco by video-enhanced microscopy. Observations revealed that the whole process went through a series of touch, adherence, membrane fusion and content mixing. The whole process could be completed within two seconds (Sun et al., 1999; 2000).

Culture of fertilized and unfertilized female gametoplasts

The zygotes produced from *in vitro* fertilization in maize underwent sustained development when cultured in millicells and fed with embryonic suspension cells (Kranz et al., 1991). Further experiments using different cultivars as the parents for *in vitro* fertilization gave rise to fertile hybrid plants (Kranz and Lörz, 1993). Plant regeneration was also reported from *in vivo* fertilized, isolated zygotes in barley (Holm et al., 1994) and wheat

(Kumlehn et al., 1998). The achievements of zygote culture in these cereals encouraged us to deal with zygote culture in tobacco and rice. In tobacco the zygotes isolated from embryo sacs were embedded in agarose and cultured in millicells with dividing mesophyll protoplasts as the feeder system. About 3-5 zygotes were cultured in one millicell. The frequency of first zygotic division was high enough, up to about 60%. A few proembryos and multicellular clusters were acquired (Li et al., 1998). In rice, zygotes cultured in millicells and fed with embryonic suspension cells were also developed *in vitro* (Figs. 23-28). The frequency of first division was up to 64%, and the frequency of multicellular structures produced was up to 31%. Compared to the enzymatic method, the non-enzymatic method of zygote isolation was superior in zygote development (Zhao et al., 2000). Recently, plants were successfully regenerated from isolated zygotes in rice (Zhang et al., 1999).

Besides zygotes, culture of isolated central cells, either unfertilized or fertilized, was tried as well. In tobacco, unfertilized central cells (Fu et al., 1997) and the fertilized central cells (primary endosperm cells; Li et al., 2000) were induced to divide up to multicellular structures. The frequency of first division was 4.5% in the former and 7.5% in the latter, much lower than in zygote culture. In rice, both unfertilized and fertilized central cells were also induced to divide, giving rise to free nuclei (Zhao et al., 2000). Real endosperm-like structures were obtained from *in vitro* fertilized central cells in maize (Kranz et al., 1998).

The zygote culture system makes it possible to try gene transfer at the initial stage of plant on-togenesis. *In vitro* produced zygotes of maize were micro-injected with two reporter genes (the GUS gene and the anthocyanin gene) and showed transient expression of them after brief culture (Leduc et al., 1996). Since the micro-injection technique is rather difficult, we tried to use electroporation technique for this purpose. The isolated zygotes are small in size and limited in number. This makes conventional electroporation unfeasible in such a case. We designed a special simple device in which several zygotes could be placed, submitted to electroporation and transferred for culture. The electroporated zygotes underwent first division at a frequency of 54.6% and developed to multicellular clusters. Introduced GFP gene and GUS gene demonstrated transient expression in 2.6% of the zygotes (Li and Yang, 2000).

Concluding remarks

The progress so far achieved in the space of a few years opens up wide prospects for utilization of plant reproductive cells/protoplasts for both basic research and applications. For example, pollen protoplasts and de-exined pollen can be used to study the role of the pollen wall or its exine alone in pollen germination and pollination. At least in tobacco it is now known that the exine appears unnecessary in these processes. The capabilities of pollen protoplasts and de-exined pollen that are much more accessible to gene transfer than their original intact form, the success of *in vitro* pollination with de-exined tobacco pollen, and the potential of *in vitro* sporophytic development as shown in immature rapeseed pollen protoplast culture, all together herald a new route for genetic transformation using these reproductive cells/protoplasts as starting materials. The successes in regenerating triploid hybrid plants from pollen protoplast-somatic protoplast fusion in *Brassica* and *Nicotiana* broaden the scope of previous gameto-somatic hybridization.

The use of male gametoplasts in a series of basic studies on the male germ unit, sperm cell dimorphism, sperm cell mobility and intergametic recognition is evident. In this review we have demonstrated the relationship between morphological changes in the generative cell and its microtubule organization pattern during its development and in the isolated state. The realization of *in vitro* intergametic fusion is of particular significance in deepening studies on fertilization and providing new means for plant breeding in the future. One century ago the discovery of double fertilization in *Lilium* and *Fritillaria* led to a wide search and final confirmation of this process in al all angiosperms. Today the breakthrough of *in vitro* fertilization, first in maize, will lead to further experimental research on fertilization in other species. Isolated egg cells and *in vitro* produced zygotes of maize have been used to construct cDNA libraries and to screen dominantly expressed genes before and after fertilization (Dresselhaus et al., 1996; Sauter et al., 1998). The use of an *in vitro* fertilization system combined with fluorescent Ca^{2+} indicator loading revealed a transient elevation of free cytosolic Ca^{2+} following sperm-egg fusion in maize. This is the first evidence of a calcium wave induced by fertilization in a flowering plant (Digonnet et al., 1997). Success in zygote culture which places the whole course of embryogenesis under controlled conditions will help in the study of gene expression and transformation from the earliest stage of embryo development. We are confident that experimental manipulation of reproductive cells and protoplasts will lead to more fruitful achievements in the 21st century.

(Authors: Zhou C, Yang HY. Published in: Acta Biologica Cracoviensia Series Botanica, 2000, 42:9-20, with 28 figures and 66 references, all omitted here.)

9 受精过程中助细胞退化机理的研究进展

Recent advances in research on the mechanism of synergid degeneration during fertilization process

被子植物的受精过程,从花粉与柱头相互作用开始,到雌、雄配子融合为止,经历一系列前后关联的事态。其中,助细胞退化是一个不可缺少的环节。根据现有的知识,助细胞退化至少和受精过程中以下三方面的作用有关:(1)引导花粉管进入胚囊。超微结构观察表明,在已研究过的具助细胞的植物中,花粉管进入胚囊时均是进入退化的助细胞;不仅珠孔受精时如此,即使在合点受精与中部受精的场合同样如此。(2)促使花粉管释放精子。花粉管进入退化助细胞以后不久即停止生长,管尖破裂,将精子和其它内含物释放到退化助细胞中。(3)帮助实现精子的转移。精子被释放以后,随着花粉管与助细胞的细胞质转移到退化助细胞的合点端。由于助细胞退化后质膜破坏,精子可以无阻碍地进入受精的"靶区",即卵细胞与中央细胞两层质膜之间的位置,然后分别和二者融合。由此可见,助细胞退化实际上是受精过程接近最后关头的一个十分精巧的、关键性的程序。

但是关于助细胞退化的机理研究还很薄弱,还有许多不明了的问题。例如:什么原因引起助细胞退化?为什么两个助细胞中一般只有一个退化而另一个宿存?两个助细胞中哪一个退化,仅仅是随机的,还是有倾向性的?助细胞退化为什么能起吸引花粉管和促进释放精子的作用?虽然目前无法圆满回答上述各个问题,但近年的研究进展已经得出若干有意义的启示。本文试图对此作出初步总结和讨论,作为进一步研究的基础。

助细胞退化大多由花粉管所诱导

根据近20多年对20多种植物所作的超微结构研究,助细胞退化的时间因植物而异,大致包括以下三种情况(表1):第一种情况是在授粉前助细胞即已开始退化。迄今仅在小麦[34]和水稻[2]中观察到这种现象。第二种情况是助细胞退化发生在授粉后至花粉管到达胚囊前这一段时间。多数已研究的材料属于这一类型。第三种情况,花粉管进入助细胞后导致其退化。有少数几种植物属于这种类型。但这些均系20世纪60年代至70年代初的早期报道,有的需要重新审订。例如就向日葵而言,早期报道助细胞退化是在花粉管进入以后[24],而近年更仔细的研究则认为退化是发生在花粉管到达以前[1,32]。造成判断失误的原因之一,可能是由于有些植物由授粉到受精的时间间隔很短,取样不易准确。

表1　　　　　　　　助细胞退化的时间

退化时间	植物	研究者
授粉以前	*Oryza sativa*	董健与杨弘远,1989[2]
	Triticum aestivum	You and Jensen, 1985[34]
授粉后至花粉管到达以前	*Crepis capillaris*	Kuroiwa, 1989[17]
	Epidendrum sculella	Cocucci and Jensen, 1969[7]
	Glycine max	Dute et al., 1989[9]
	Gossypium hirsutum	Jensen and Fishex, 1968[14]
	Helianthus annuus	阎华等,1990[1]
	Hordeum vulgare	Cass and Jensen, 1970[3]
		Mogensen, 1984[21]
	Linum usitatissimum	Vazart, 1969[30]
	Lycopersicum spp.	Nettancourt et al., 1973[23]
	Nicotiana spp.	Mogensen and Suthar, 1979[22]
		Huang and Russell, 1992[12]
	Paspalum spp.	Chao, 1971[4]
	Populus deltoides	Russell et al., 1990[25]
	Quercus gambelli	Mogensen, 1972[20]
	Spinacia oleracea	Wilm, 1981[31]
	Stipa elmer	Maz and Lin, 1975[19]
	Triticale	Hause and Schröder, 1987[10]
	Zea mays	Diboll, 1968[8]
花粉管进入以后	*Capsella bursa-pastoris*	Schulz and Jensen, 1968[27]
	Helianthus annuus	Newcomb, 1973[24]
	Petunia hybrida	Van Went, 1970[29]
	Torenia fournicri	Van dex Plyum, 1964[28]

为了使研究更为精确化,Huang 与 Russell(1992)应用定量统计分析方法研究烟草助细胞退化的时间[12]。他们采用酶法分离胚囊技术获得大量的生活胚囊,用荧光素二醋酸酯(FDA)鉴定助细胞生活力,退化助细胞为负反应,宿存助细胞为正反应。根据授粉后定期取样的观察结果,助细胞退化大多发生于授粉后 42~48h,即花粉管进入子房而尚未到达胚囊之前。如果不授粉或授粉后 36h 以前切断花柱阻止花粉管进入子房,则助细胞不退化。此外,绝大多数胚珠中只有一个助细胞退化,而两个助细胞均退化者仅占 2%~3%。以上烟草中的研究结果表明,助细胞退化确系由花粉管所诱导,并且是在花粉管已进入子房以后的短距离内发生作用,而以往对棉花的观察结果则是:当花粉管尚在花柱中生长时期已引起助细胞开始退化[14]。

花粉管可能通过传递激素诱导助细胞退化

花粉管在尚未到达助细胞之前即已诱导助细胞发生退化,必然通过某种信号的传递。对此,Jensen 等通过棉花的胚珠离体培养实验试图作出解释[15,16]。他们将未授粉的胚珠培养在含不同植物激素的条件下,结果发现,在含赤霉素的条件下培养 2~3d 后,助细胞之一发生退化,其超微结构上的变化特征和在体内授粉条件下所发生的情况十分相似。而培养在含生长素吲哚乙酸的条件下则不发生上述变化。由此推测,在体内的助细胞退化也可能是起因于赤霉素的诱导。花粉管在生长途中可以刺激雌蕊增加赤霉素,赤霉素扩散到胚囊诱导助细胞退化。这一离体实验系统及其推论是很有启发性的,可惜由于未能提供较翔实的实验数据和缺乏相应的体内分析资料,迄今只能作为一种假设。同时,除棉花以外的植物中是否存在类似的规律,也还有待证实。

在有些植物中助细胞退化具有位置效应

为什么两个助细胞中通常只有一个退化呢?是否助细胞在发育过程中发生某种分化,从而预定其中一个较易退化呢?如果是这样,则应在结构或生理上反映出助细胞的异型性。然而迄今的研究资料并未发现这种异型性。在亚麻中,通过图像分析术查明,两个助细胞在其丝状器的表面积和体积上有所差异,其中行将退化的助细胞具有较小的丝状器(Russell and Mao, 1990[26])。然而目前尚难以解释丝状器的大小和助细胞退化之间是否有因果联系,也不知道在其它植物中是否存在类似的现象。

尽管目前尚未肯定助细胞本身存在着决定其是否退化的异型性,但在另一方面,至少在一部分植物中,由于两个助细胞在胚珠中所处的位置不同,它们在退化的倾向性方面有所差异。可以将这称为助细胞退化的"位置效应"。Mogensen(1984)首先注意到,在大麦中,当只有一个助细胞退化的情况下,通常是近胎座的助细胞倾向退化,而远胎座的助细胞倾向宿存[21]。阎华等(1990、1991)在向日葵方面的研究表明这种助细胞退化的位置效应更为明显,在所观察的64个胚珠中,有52个胚珠的退化助细胞是位于近珠柄侧的一方,占总数的81%[1,32]。阎华等(1991)进一步发现,向日葵的助细胞退化和珠孔的结构有一定的关系。向日葵的珠孔是闭合型,为不对称结构,其近珠柄侧具发达的引导组织,而远珠柄侧则否。花粉管沿着近珠柄侧的引导组织生长,直达位于同侧的退化助细胞。由此推测,花粉管所产生的信号主要沿近珠柄侧扩散到胚囊,从而诱导该侧的助细胞退化[32]。

在另一些植物中没有看到助细胞退化的位置效应。如烟草的两个助细胞退化频率大致相等,表明其中哪一个退化是随机的[12]。在亚麻中,虽然助细胞退化表现一定的倾向性,但不明显[26]。因此还需要更多的研究才能完满回答为什么只有一个助细胞退化的问题。

助细胞退化与钙含量的超常增高有关

在花粉管的向化性生长中,钙被认为是重要因素之一。那么,助细胞之所以吸引花粉管是否由于它含有丰富的钙呢? Jensen(1965)首先用实验方法探索这一问题。应用显微灰化法(microincineration)发现,棉花助细胞富含灰分,推测其中有大量的钙[13]。进一步推论:当助细胞退化时,其液泡膜破坏,储存在液泡中的钙盐大量释出,在珠孔区形成钙的梯度,诱导花粉管进入。花粉管在退化助细胞的高钙环境中停止生长,破裂而释放精子[16]。但是,显微灰化法不足以确证助细胞所含灰分主要是钙。还需要应用相应的细胞化学方法证实钙的存在。

李天庆与曹慧娟(1987)在研究胡桃的合点受精时应用X射线微区分析法观察到花粉管生长途径的雌蕊组织含有较多的钙,其中退化助细胞的钙含量最高[18]。Chaubal与Reger(1990)应用冰冻替代制样和X射线微区分析,证明小麦两个助细胞均有高含量的钙,而卵细胞中则不多[5]。最近他们在珍珠谷中又应用X射线微区分析和焦锑酸盐沉淀(Pyroantimonate Precipitation)两种方法得出结论:在所测定的胚珠组织中,只有助细胞的钙含量最高。助细胞的细胞质中,退化的细胞器(线粒体、质体)有特多的钙盐沉淀。这种超常的高钙含量可能是导致助细胞退化的原因[6]。何才平与杨弘远(1992)应用焦锑酸盐沉淀法研究了向日葵授粉前后助细胞中钙含量的变化。传粉前1d,两个助细胞中钙的分布相近,且均呈现由珠孔端向合点端递增的梯度。传粉后至花粉管到达以前,上述梯度依然保持,但两个助细胞的钙含量出现差异,其中近珠柄侧助细胞(即退化助细胞)中钙含量激增。受精后,退化助细胞中钙沉淀密集不可分辨,宿存助细胞则无明显变化[11]。Huang与Russell(1992)应用金霉素(CTC)荧光染色法同样显示,烟草的退化助细胞比宿存助细胞中的膜钙远为丰富[12]。

根据以上研究进展,可以对至少在一部分植物中助细胞退化的机理作出如下假设(图1):在胚囊发育过程中,助细胞的含钙量原本高于其它细胞。传粉后,在花粉管所产生的信号(可能是激素)的诱导下,其中一个助细胞由于位置效应或其它未知原因发生反应,其结果钙含量超常增加。过高的钙含量导致细胞器的破坏和整个细胞的退化。退化助细胞产生向化性物质(其中也可能包含钙)吸引花粉管进入。花粉管进入退化助细胞后,在超高钙环境中破裂释放精子。精子沿退化助细胞转移到受精靶区实现双受精。

(作者:杨弘远。原载:植物学通报,1994,11(1):1~5。图1幅,
　　表1幅,参考文献34篇。选录图1与表1。)

10 钙在有花植物受精过程中的作用

The role of calcium in the fertilization process in flowering plants

钙作为第二信使在植物信号转导中的作用一直是植物生理学、细胞生物学和发育生物学研究的热点。近年已有不少综述和专著从不同角度对此作了详细评论[1~5]。虽然这些文章中只有部分内容涉及本文的主题——钙在植物受精中的作用,但是它们所论述的关于钙信使系统在植物生活中的作用的一般研究概况,为我们进入这个主题提供了必要的背景和向导。迄今研究的结果大致可以概括为以下几个要点:(1)钙在植物生理活动中起广泛的作用。由各种外界与内在的信号因子(如光、温度、盐、触动、重力、辐射、激素等)所导致的植物反应,大多和钙信号传导有关。钙在细胞分裂、极性形成、生长、分化、凋亡等生命过程中均起重要的调节功能。(2)胞质游离钙([Ca^{2+}]c)的瞬间变化,是细胞响应各种刺激信号的初始事态,由此而诱发以后一系列信号传导的下游事态。[Ca^{2+}]c 的瞬间增加有两种可能的来源:一是胞外 Ca^{2+} 通过 Ca^{2+} 通道的开启进入胞内;二是胞内钙库(如内质网、液泡等)向胞质释放出 Ca^{2+}。(3)钙信号通过钙靶蛋白进行信号传导。钙调素(CaM)是迄今已证明的分布最广、功能最多、分子结构高度保守的钙靶蛋白。钙靶蛋白又通过和其它靶蛋白分子(如各种蛋白激酶)结合而启动基因表达与细胞生命活动,从而构成钙信号传导的复杂体系。有人将所有与信号传导有关的蛋白质整体称为 transducon,包括各种受体、通道、CaM、蛋白激酶、磷酸酶、Ca^{2+}-ATP 酶等等[3]。(4)钙信号系统中的各种分子在细胞中具有严格的空间发布,它们在细胞壁、质膜、胞质溶胶、细胞器、细胞骨架、细胞核等部位定位,是复杂的细胞网络(cytonet)中的重要组成成分,以保证各项细胞生命活动的平衡,实现细胞结构与功能的统一[4]。(5)在植物中,用于研究钙信号的材料主要有:禾谷类幼苗的根、胚芽鞘、下胚轴、胚乳糊粉层、紫露草雄蕊毛、百合与百子莲等植物的花粉管,以及低等植物墨角藻(Fucus)的合子和假根等。在植物有性生殖系统中,研究最多和最深入的是离体萌发的花粉管,其它则较为薄弱。

本文试图概括有关钙在植物受精过程中的作用的研究进展。这里所说受精过程是广义的,即从花粉与柱头接触开始到双受精结束为止的全部过程。我们从后文可以看出,在这个过程的所有环节中,钙信号都是参与调节的。

钙与花粉管的离体生长

离体萌发的花粉管之所以成为研究钙信号的最佳材料是因为它有多方面的优点:萌发与生长速度较快,可以在短时间内完成观测;可以在显微镜下直接观测其变化;可以严格控制培养基与培养条件,形成较为稳定的实验系统;可以方便应用各种实验方法获取钙信号变化与花粉管生长之间的关系的资料。因此,研究者用离体花粉管研究钙信号,主要是看中了这一优良的实验系统。随着研究的深入,日益揭示出钙与受精的关系。

关于钙在花粉管离体生长中的作用,可以追溯到20世纪60年代的研究。当时有两项重要的发现:一是钙可以补偿花粉离体萌发时"花粉群体效应"的不足,被认为是"花粉生长因素"的主要成分,并由此推出含钙的"万能培养基",即至今仍然广泛采用的BK培养基[6]。另一项发现是在特殊设计的花粉培养实验中证明花粉管朝钙源定向生长,由此推测钙是花粉管生长的广泛向化因素[7]。这两项发现,加之当时在个别植物中观察到雌蕊中存在钙浓度的梯度以及推测助细胞富含钙(见后文),从而掀起了研究钙在植物受精中作用的高潮。近20年来,钙信号概念的提出与广泛研究、动物细胞中钙信号研究的深入以及各种先进实验方法的开发,对于钙信号与花粉管生长的关系有了日益深化的认识。近年有不少综述对此作过介绍[8~13],最近又有不少新的进展。现从以下 3 方面对主要研究结果加以简略概括。

钙在花粉管尖端的分布

许多证据表明,花粉管尖端是钙集中分布的区域。放射自显影显示,百合花粉管尖端约 20μm 处是 ^{45}Ca 密集的区域;同时这也证明 Ca^{2+} 向花粉管尖端的内流十分活跃[14]。在烟草花粉管尖端附近安置振动电极也测得 Ca^{2+} 的向内流动[15]。采用金霉素(CTC)作为膜结合钙的荧光探针,发现百合花粉管尖端存在由前向后递减的膜钙梯度[16],这可能是最早揭示花粉管尖端钙梯度的报道。X 射线微区分析同样证明花粉管尖端的总钙梯度[17]。但是作为钙信号最主要的指标还不是膜钙和总钙,而是胞质游离钙,这就需要应用测定 $[Ca^{2+}]c$ 的荧光指示剂进行研究。在这方面已有大量的工作[18~20]。所使用的荧光指示剂包括 Quin、Indo、Fura 等系列,并且逐渐由原来的温育法发展为显微注射法,以便取得更为精确的定位与定量效果。

除了 Ca^{2+} 外,也有愈来愈多的实验证明钙信号系统中的下游靶分子同样分布于花粉管尖端区域。最早是用荧光探针氟奋乃静(fluphenazine)显示花粉管尖端有 CaM 分布[21]。其后,这一结果亦被免疫细胞化学和共焦激光显微术所证实[22]。最近,应用 FITC 标记的 CaM 蛋白与 CaM mRNA 显微注入法和共焦激光显微镜观察,发现 CaM 蛋白与 RNA 在花粉管尖端 50μm 范围内均匀分布,只是在近尖端处有 CaM 的 V 形集中分布,从而对这种 CaM 的特殊分布状态及其与花粉管生长的关系提出了新的观点,认为 CaM 对花粉管尖端生长起重要作用,但与花粉管生长的导向没有关系[23]。在这里还要提及:Ca^{2+} 和 CaM 不仅与花粉管生长有关,而且与花粉萌发有关。应用视频图像分析方法发现,烟草花粉在水合过程中,Ca^{2+} 与 CaM 聚集于萌发沟附近并于花粉萌发时向萌发沟处汇集[24]。在水稻花粉发育过程中,甚至在接近成熟的花粉中也观察到 CaM 向萌发孔集中的现象[25]。

除 CaM 外,也有证据表明 Ca^{2+}-ATPase 与依赖钙的蛋白激酶(CDPK)在花粉管尖端的分布。Ca^{2+}-ATPase 主要分布于管尖,可能对维持 Ca^{2+} 梯度稳定起作用[10]。最近,以荧光标记的蛋白激酶抑制剂

BODIPY EL bisindolylmaleimide 作为探针,在花粉管尖端定位到较高的 CDPK 活性,并且 CDPK 活性与[Ca^{2+}]c 的变化相关。联系到同样在花粉管尖端分布的 RhoGTPase 和 H^+-ATPase,推测 CDPK 在管尖的分布有促进此处钙调节的胞吐及 Ca^{2+} 通道的作用[26]。

钙对花粉管生长的调节

花粉管的"生长点"在其尖端,而尖端生长要求精巧的[Ca^{2+}]c 动态平衡。打破这种平衡导致花粉管生长的抑制。实验表明,在花粉培养基中加入各种影响[Ca^{2+}]c 升高或降低的试剂,均使花粉管生长发生可逆甚至不可逆的停顿。这些试剂包括:Ca^{2+} 载体(如 A23187)、Ca^{2+} 螯合剂(如 EGTA、BAPTA)、Ca^{2+} 通道阻断剂(如 nifedipine、verapamil、TMB-8、La^{3+}、Gd^{3+})等。施加影响钙信号系统的 wbn 试剂亦获类似的效果,如:CaM 拮抗剂(TFP、CPZ、W7)、Ca^{2+} 泵抑制剂(vanadate)等。在以前的综述中对此已有较全面的引证[13],本文仅择举其中一些主要研究进展。

在实验方法上,已由在培养基中添加各种试剂发展到应用更精密的技术对单个花粉管进行试验,例如,显微注射 Ca^{2+} 敏感染料(Indo-1、CG-1)、电场刺激、注射束缚 Ca^{2+}(Caged Ca^{2+})等。束缚 Ca^{2+} 在紫外线激发下活化,释放 Ca^{2+}。以上 3 种方法均诱使花粉管中的[Ca^{2+}]c 瞬时升高,从而导致花粉管生长受抑,并可诱导花粉管朝刺激信号的方向生长[27]。进一步,采用微电极在花粉管外侧施放束缚 Ca^{2+} 或 Ca^{2+} 载体,使管尖一侧的[Ca^{2+}]c 提高。这种[Ca^{2+}]c 浓度的不均衡分布导致花粉管生长轴的重新定向,即朝施加刺激的一侧弯曲[28]。最近,沿着类似的思路将束缚的三磷酸肌醇[Ins(1,4,5)P_3]显微注入花粉管的不同部位,通过其光激活造成局部 Ins(1,4,5)P_3 的增加,结果发现,在花粉管的核区或亚顶端区的实验导致该区[Ca^{2+}]c 的局部瞬间增加和花粉管生长轴的重新定向;而在花粉管顶端的实验则常导致花粉管生长的停滞或破裂。以 Ins(1,4,5)P_3 受体阻断剂 heparin 处理可抑制上述效应。因此认为 Ins(1,4,5)P_3 是调节花粉管生长的钙信号系统中的重要参与者[29]。

CaM 在调节花粉管生长方面的作用已有不少实验证据,其中一个比较新颖的发现是外源 CaM 的作用。用抗 CaM 血清、CaM 拮抗剂 W7-agarose、Ca^{2+} 螯合剂 EGTA、外源 CaM 分别试验,结果前三者均抑制花粉萌发与花粉管生长,而外源 CaM 促进萌发与生长,并可清除前三者的抑制效应[30]。在花粉管生长的适当时期施加适当浓度的花椰菜 CaM 或牛脑 CaM-agarose,不仅促进花粉管生长,还促进其中生殖核的分裂。由于琼脂糖颗粒大于花粉粒,只能通过胞外将信号传导到胞内,从而对胞外 CaM 的作用提供了进一步佐证[31]。此外,采用类似的在花粉管生长不同时段施加影响的方法,也分别研究了外源 Ca^{2+} 浓度、Ca^{2+} 载体、Ca^{2+} 通道阻断剂等因素对花粉管生长特别是生殖核分裂的影响[32,33]。钙信号对生殖核分裂的影响是以往尚少注意的方面。

钙调节花粉管生长的机理是很复杂而尚未研究透彻的。公认的看法是:Ca^{2+} 与花粉管中的肌动蛋白微丝密切相关。花粉管顶尖缺乏微丝可能由于此处高浓度的 Ca^{2+} 破坏了微丝骨架,从而阻止后者导致的胞吐而保证尖端生长。花粉管亚尖端区段梯度减弱,保证微丝将细胞运送到管尖,而此处的高浓度 Ca^{2+} 促使细胞器分泌的小泡融合到质膜形成新壁[10]。有人提出 Ca^{2+} 与花粉管生长关系的模型,对此作了进一步补充:外源 Ca^{2+} 通过 Ca^{2+} 通道流入花粉管尖端,促进管尖的高尔基小泡融合。这一过程可能被蛋白激酶、RhoGTPase、CaM 等调节;激酶又被磷酸肌醇途径调节。当 Ca^{2+} 流入不均衡时,管尖花粉管中 $[Ca^{2+}]c$ 增高的一侧小泡融合加强,致使该侧生长受抑,结果花粉管朝该侧转向[3]。

花粉管中的钙振荡

花粉管尖端的 $[Ca^{2+}]c$ 浓度并非恒量,而是有节律地波动的。如百合花粉管尖的 $[Ca^{2+}]c$ 水平在 700~3 000nmol/L 范围内波动;与此相关,花粉管生长速率也在 $0.1~0.38\mu m \cdot s^{-1}$ 范围内波动[34]。虞美人花粉管生长亦被钙波所调节[35]。这种钙波或钙振荡的现象并非花粉管所特有,亦见于其它一些尖端生长的细胞(如某些真菌菌丝)中;在动物卵的受精过程中同样发生钙振荡(见后文)。花粉管是研究这一现象的很好材料。

为了取得更精确的结果,在测试方法上作了各种探索。有人应用水母发光蛋白(aequorin)测定百合花粉的$[Ca^{2+}]c$有40s的脉冲;而用calcium greendextran作指示剂,分辨率有所提高,可测得在5s内$[Ca^{2+}]c$的升高与花粉管生长率的提高呈正相关[36]。但另有作者报道,采用双波长的比率测量染料fura-2-dextran比用单波长指示剂calcium green更佳,既可获得高分辨率的空间分布,又可获得高分辨率的定量信息。用此法同样证明$[Ca^{2+}]c$与花粉管生长的相关,但发现钙振荡是位于花粉管的顶尖处[37]。应用fura-2-dextran测量花粉管内的钙离子($[Ca^{2+}]i$),同时用离子选择振动电极测量花粉管外的钙离子内流($[Ca^{2+}]o$),结果表明$[Ca^{2+}]i$振幅与花粉管生长同步,而$[Ca^{2+}]o$约迟11s。为解释这一延后现象,提出了两种模型:内存模型和外存模型。前者指$[Ca^{2+}]o$用于补充花粉管内的钙库;后者指$[Ca^{2+}]o$和花粉管壁结合[37]。

最近发现,花粉管尖端的pH值梯度也不是恒定的,同样显示脉冲变化。将H^+指示剂cSNARF-1 dextran显微注射到百合花粉管中,测知pH值在脉冲高峰时为7.05,脉冲停止时为6.0;pH脉冲高峰比花粉管生长脉冲延后7.5s以上,而与离子内流(指H^+与K^+)高峰同步。由此认为H^+对花粉管的脉冲生长很重要:它可能通过减弱Ca^{2+}结合蛋白和Ca^{2+}的结合,从而抑制依赖Ca^{2+}的小泡融合,也可能还影响其它生理生化过程[38]。

钙在雌蕊中的分布及其与花粉管体内生长的关系

花粉管的离体生长系统可以模拟但不能完全代替花粉管的体内生长状况。要了解受精的配合前期花粉管生长与钙的关系,还需要将雌蕊纳入研究的对象。早在20世纪60年代,已有研究者注意到这个方面。根据对金鱼草雌蕊各部组织中钙含量的测定,由花柱上部到胚珠与胎座,钙占干重的百分率由0.51递升为2.17,呈现明显的梯度,并且远高于在花瓣中的0.18和雄蕊中的0.36。这似乎支持了钙作为向化因素的推测[7]。但这种梯度未在其它植物的研究中被证实。同时,在切断的花柱中花粉管可以逆向生长的事实也不符合花柱中存在向化物质梯度的观点。看来,花粉管的体内生长远较离体生长的机制复杂。几十年来为了探讨花粉管在雌蕊中定向生长的机理,提出过向化性、向触性、向电性等理论,并且均有一定的离体实验为依据。近年更着重于应用多学科,特别是分子生物学手段,研究雌蕊与花粉管导向(guidance)的关系。下文从3个层次加以概括,在讨论钙的作用这一主题时,不免还涉及近年其它相关方向的重要进展。

钙与花粉-雌蕊间的识别

研究花粉-雌蕊间的识别主要依靠自交不亲和系统。根据近年有关综述[11],在分子水平上研究比较深入的有三类代表植物。一类是以芸苔属为代表的孢子体不亲和,已知其 S 基因产物是 S 糖蛋白。一类是以烟草属为代表的配子体不亲和,也证明其 S 基因产物为糖蛋白,并进一步查明其氨基酸序列与真菌 RNase 同源,表明 S-RNase 基因控制其不亲和性。蔷薇科和玄参科的某些植物也属于这类情况。还有一类以罂粟科植物虞美人为代表,情况较特殊,虽然它属于配子体不亲和类型,但花粉抑制区在柱头又与孢子体不亲和相似。它的一个优点是不亲和反应可以在离体实验中验证。在这种类型中,S-RNase 的探索未获预期结果;而另一方面,研究表明不亲和反应受[Ca^{2+}]c 调节。当在花粉培养基中加入不亲和柱头提纯物或重组的 S 蛋白质时,花粉管中出现瞬间[Ca^{2+}]c 激增[39,40]。由此推断不亲和反应是柱头 S 蛋白质诱导花粉管[Ca^{2+}]c 升高而致花粉管生长受抑。

花粉管在雌蕊中的生长

这里涉及两个问题:花粉管在雌蕊中的生长是否需要后者提供钙的营养?花粉管的导向是否与雌蕊中的钙有关?

首先需要弄清在雌蕊的"花粉管轨道"(pollen tube track)中钙的分布状况。CTC 荧光检测表明甘薯柱头表面有丰富的膜钙和 CaM[4]。假叶树柱头表面富含钙被 CTC 荧光检测与 X 射线微区分析所证明[42]。^{45}Ca 同位素示踪揭示柱头上的钙被花粉所吸收[43]。在国内,先后有不少研究者应用焦锑酸盐沉淀法查明,甘蓝型油菜[44]、向日葵[45]、陆地棉[46]、水稻[47]等的花粉管轨道(柱头表面、花柱引导组织、子房壁内表面、珠孔)中均有较相邻组织更为丰富的钙;而且在这些组织中的钙多分布于质外体系统(细胞壁、胞外基质),后者也正是花粉管在其中生长的部位。同时还观察到在雌蕊组织中生长的花粉管尖端的细胞器区段的胞质、细胞器及管尖外周也分布较多的钙[45~47]。尤其需要指出,在花粉管进入胚珠的入口——珠孔中,钙的密集程度特别高[45,46,48,49]。

定位研究可以说明钙与花粉管在雌蕊中的生长相关,尚不足以证明其与花粉管导向有关。花粉管导向是很复杂的问题,涉及更广泛的因素。近年有些综述对此作了精辟的讨论[50~52]。大体上提出了以下一些新颖的观点。其中一组实验得出意外的结果:用无生命的乳胶珠代替有生命的花粉对雌蕊进行授粉,发现乳胶珠可以模拟花粉管的行为在花柱中向下运动,由此解释花粉管在雌蕊中的生长是被动地由雌蕊组织中的胞外基质所驱动的[53]。进一步查明,在 4 种植物的雌蕊中均存在类似动物组织中的基质粘附分子 vitronectin[54]。另一组实验发现:烟草花柱中的"引导组织特异蛋白"(transmitting tissue specific protein, TTS 蛋白)起引导花粉管的作用。TTS 蛋白吸引花粉管生长,粘附于花粉管表面,参入后者并被后者去糖基化。TTS 蛋白中的阿拉伯半乳聚糖是花粉管的重要成分,而花柱中存在自上至下递增的糖基化梯度。以转基因方法减少花柱中的 TTS 蛋白导致花粉管在其中生长受抑[55,56]。在百合和一种苋属植物的花粉管与雌蕊组织中也定位了阿拉伯半乳聚糖蛋白质[57,58]。但是,有的研究者对上述观点提出置疑:因为有一种 TTS 同系物并不能吸引和促进花粉管生长,不被花粉管去糖基化,也不在花柱中呈现递升的糖基化梯度[59]。

通过突变体研究产生了另一种思路,即花粉管导向的控制机构主要是胚珠。在正常拟南芥中,花粉管进入子房后通常总是进入最先接近的胚珠;而在胚珠缺陷型突变体中,花粉管越过缺陷的胚珠在子房内漫游[60]。胚珠对花粉管的导向可能既包含向化性因素,又包含粘附因素。一种拟南芥突变体由于花粉管与胚珠的粘附发生缺陷导致不能进入胚珠,说明了后一种因素的重要性。综合已有拟南芥突变体的遗传分析结果,在胚珠导引花粉管的机理方面,有些突变体是雌配子体基因型起决定作用;另一些则是由雌、雄双方的孢子体基因共同决定的[60]。

钙与花粉管进入胚囊及精子的释放与转移

花粉管在雌蕊中生长的目的地是胚囊。花粉管通常进入胚囊珠孔端的一个助细胞并在其中破裂以释放精子。该助细胞或在花粉管到达前预先退化,或在花粉管进入后退化,因植物而异。精子释放后通过助细胞合点端转移到双受精的靶区,即卵细胞与中央细胞之间的位置。由此可见,助细胞在受精过程中至少担任3项功能:吸引花粉管进入胚囊;促成精子的释放;帮助精子转移到受精靶区。

助细胞在执行上述功能中均有钙参与。助细胞的一个显著特征是有高含量的钙。早在20世纪60年代,应用显微灰化方法即已发现陆地棉助细胞富含灰分,推测其主要成分是钙[61]。20世纪90年代以来,应用X射线微区分析、超微细胞化学、荧光检测等方法,在多种植物中证实了这一推测。然而授粉受精前后助细胞中钙的变化及其与助细胞退化的关系则有不同类型:一类是授粉前一对助细胞的钙量相近,授粉后花粉管到达前,其中一个助细胞中的钙量显著增多,该细胞随即退化,花粉管进入该退化助细胞。这种情况见于向日葵[49]、陆地棉、烟草[62,63]、甘蓝型油菜[48]。其中,在甘蓝型油菜中应用图像分析技术对焦锑酸钙沉淀进行了定量统计,结果表明授粉前助细胞中的钙沉淀的体密度显著高于胚囊中的其它细胞,为卵的2.5倍和中央细胞的1.9倍;钙沉淀颗粒的等效圆直径则较小,约为卵和中央细胞的2/3;授粉后2个助细胞均开始退化,其中一个退化程度较高,其钙沉淀的体密度增长为授粉前的2.4倍,沉淀颗粒的等效圆直径则减小至授粉前的1/3以下[48]。另一类情况见于小麦和珍珠谷。授粉前一对助细胞钙含量相近,均显著高于胚囊中其它细胞;授粉后2个助细胞的钙量仍然相仿。授粉前2个助细胞均有一定程度退化,细胞器的退化程度和钙含量呈正相关。看来这两种禾本科植物助细胞退化是一种钙调节的凋亡过程,主要由母体控制而与授粉刺激无关[64~68]。同属于禾本科的水稻情况又有所区别:其2个助细胞授粉前已出现差异,其中一个已呈退化状态,合点端环抱卵细胞呈牛角状,细胞中钙明显增多而胞基质和小液泡中则无钙沉淀分布,另一助细胞保持原状[47]。水稻助细胞退化不由授粉诱导,这和小麦、珍珠谷一致;但授粉前2个助细胞在退化与否上有差异又和小麦等明显不同。无论助细胞退化是由授粉所诱导还是与授粉无关,无论是1个助细胞退化还是2个助细胞同时退化,退化和钙的超常水平有关是肯定无疑的。已知钙的增多越过一定限度会导致内质网解体,释放的水解酶会破坏细胞骨架[5]。

花粉在退化助细胞中释放精子的机理,除了曾经推测的其它各种原因外,显然也与助细胞内的高钙环境有关。在这种条件下花粉管生长停止,尖端破裂而释放内含物,正如许多离体实验所证明的那样。不仅如此,游离精子由退化助细胞转移到受精靶区也许亦有钙的参与。被子植物的精子没有自主运动能力,其在花粉管中的运动主要依赖钙调节的肌动蛋白-肌球蛋白驱动系统[8]。游离的精子最初可能被花粉管释放内含物时的冲力带至助细胞合点端。由此处通向受精靶区,在卵细胞和中央细胞的两层质膜间存在弧形的间隙带。超微细胞化学在此间隙带中定位了钙[47~49]。鬼笔环肽荧光检测发现,受精前烟草胚囊中的相应部位出现暂时的冠状肌动蛋白带,推测它负责精子的转移[69]。最近,应用免疫细胞化学与原位杂交方法在烟草胚囊中分别定位到卵器与中央细胞之间出现暂时的 CaM[70] 和 CaM mRNA[71] 区带,似乎从钙信号系统角度为上述推测提供了相关的佐证。

钙与精卵融合及卵的激活

钙在动物受精中的重要作用已有许多深入的研究[72]。精子入卵后通常诱导卵内 $[Ca^{2+}]c$ 水平瞬间升高,形成各种形式的钙波:棘皮动物、鱼类、蛙类等形成单一钙波;哺乳类则形成重复钙波。将完整精子或由精子提取的一种不耐热因素注入卵中亦可诱导钙波。$[Ca^{2+}]c$ 急剧增加的来源,可能是外源 Ca^{2+} 经由 Ca^{2+} 通道的开启进入卵内,也可能是由于内源钙库的释放。钙波是受精卵激活的关键信号。

钙在低等植物墨角藻的受精中也起重要作用。外源 Ca^{2+} 的流入导致 Ca^{2+} 升高被认为是卵激活与细胞壁合成的必需前提[73,74]。Ca^{2+} 的升高恰好是在精子入卵处的质膜内方,由此导致以后合子极性(假根极与原叶体极)的形成[75]。

在被子植物中研究钙对受精的影响比较困难,因为受精发生在胚囊中,难以像动物和低等藻类那样直接操作。近年离体受精系统的建立为此开辟了道路。已知玉米精卵离体融合需在有钙的条件下进行[76,77]。最近,用钙指示剂 fluo-3AM 标记玉米卵细胞,在含 Ca^{2+} 溶液中进行精卵离体融合,根据图像分析追踪了融合过程中卵内 $[Ca^{2+}]c$ 的变化,发现融合开始后 4~8s,荧光由卵细胞向精细胞扩散;12s 后卵内 $[Ca^{2+}]c$ 迅速升高;至 85s 时达最高峰,继续保持 2min 复又降低;融合后 29min $[Ca^{2+}]c$ 降至融合前水平。由受精诱导的是单一的钙波,类似海胆受精时的单一钙波,而有别于哺乳动物中的多重钙振荡[78]。体内研究很难捕捉到钙的瞬间变化,然而在水稻中根据焦锑酸盐沉淀法在超微水平上观察到受精后的卵细胞中钙量显著增加[47]。这一观察结果是初步的,尚需进一步研究证明。

依赖离体受精实验系统研究受精前后的基因表达也有所突破。由玉米的"离体合子"构建 cDNA 文库,分离出编码钙网蛋白(calreticulin)的克隆。钙网蛋白是一种主要的储钙蛋白,通常位于内质网及某些细胞的核和/或质中,在 $[Ca^{2+}]c$ 的调节上起重要作用,并且与细胞分裂密切相关。研究表明,该基因在受精后的表达加强[79]。这是由高等植物的合子中所分离的第一个基因,标志植物离体受精的研究进入分子水平,也表明钙在植物受精与合子形成中的重要性。

结　　语

综上所述,关于钙信号在植物受精中的作用,可以得出以下几点结论:(1)钙在花粉萌发与花粉管生长中的重要作用已为花粉管离体实验所证明。花粉管尖端的 $[Ca^{2+}]c$ 梯度,通过影响微丝骨架、小泡运输与融合、新壁形成等细胞学事态,维持花粉管生长的动态平衡,若是各种因素的影响干扰了这种梯度,则导致花粉管生长的抑制。管类$[Ca^{2+}]c$浓度分布的对称性决定花粉管生长的方向;单侧刺激引起其不对称分布,导致花粉管生长轴的重新取向。花粉管尖端存在钙振荡现象,其节律与花粉管生长的脉冲节律相关;(2)钙与花粉管在雌蕊中的生长,即受精过程配合前期事态的关系,已有多方面的,然而大多是间接的证据。在花粉-雌蕊识别方面,迄今已能肯定罂粟科代表植物的自交不亲和反应是受柱头诱导花粉管$[Ca^{2+}]c$的变化所调节的。雌蕊花粉管轨道组织中存在相对丰富的钙,提示后者对花粉管生长至关重要。至于雌蕊中花粉管导向的机理,涉及许多因素,除离体实验表明钙的参与外,在体内尚缺乏直接证据。助细胞中尤其是在其退化过程中钙的超常含量,被认为在吸引花粉管、雄配子释放甚至雄配子转移等功能中有重要作用,这一论点与从突变体分子生物学研究所得出的胚珠/胚囊是花粉管导向的控制机构的论点不谋而合;(3)被子植物离体受精实验系统的建立为研究钙与精卵融合及卵的激活奠定了基础。离体受精的玉米卵细胞中存在瞬间激增的钙波,表明高等植物卵细胞的激活也和动物及藻类卵的激活一样有钙的参与。离体受精还为研究受精前后基因表达提供了构建卵细胞与合子 cDNA 文库的可能性,初步结果表明,一种钙结合蛋白在受精后出现优势表达。

  钙信号在受精过程中的作用还有不少需要进一步探索的方面:

  在实验方法上,$[Ca^{2+}]c$指示剂的导入、图像分析、微电极与精密传感器等手段在体内研究中很难应用,而离体实验结果又难以全面反映体内的自然状态,因此有待开拓新的思路。作为一条思路,水母发光蛋白转基因材料可能是研究授粉-受精过程中体内钙变化的有用方法。活体/离体授粉也许可以提供比离体实验更接近体内状态的系统。此外,将百合花柱的管道细胞组织分离出来,研究花粉管在其表面的生长行为,是否便于钙的检测等等,均有待探索。

  为了取得钙在助细胞退化与吸引花粉管、雄配子释放与转移等方面的更确切的实证,可以考虑采用蓝猪耳这类具有裸露胚囊珠孔端的材料,以研究$[Ca^{2+}]c$在其中的变化规律。研究受精过程的最终环节——双受精时期钙的作用,除了继续发挥离体受精系统的优越性外,也还可以尝试利用蓝猪耳或其它具有透明胚珠的材料(如某些兰科植物、*Jasione monatana*、*Galanthus nivalis* 等),以便进行体内钙的直接检测。中央细胞受精是双受精的一个组成部分,钙在这方面的作用尚无人涉足,应该予以重视。为了探讨钙在卵激活中的作用,还可以考虑利用孤雌生殖、雌核发育等实验材料,在排除受精的条件下观察$[Ca^{2+}]c$的变化。

最后,还应该加强对钙信号传导系统中各种下游分子在受精中的作用的研究,并且将钙信号系统与细胞网络中的其它方面的研究联系起来。此外,在研究受精时,了解与借鉴其它生命过程(如神经传导、光周期、气孔开闭等)中钙信号系统的研究进展,肯定有助于开阔思路。

(作者:杨弘远。原载:植物学报,1999,41(10):1027~1035。参考文献79篇,删去。)

11 荧光显微术在当代植物细胞生物学研究中的应用

Application of fluorescence microscopy in contemporary studies of plant cell biology

自从20世纪80年代第一次在显微镜下观察到生物组织经紫外线照射后发射荧光的现象以来,荧光显微术不断获得进步,现已发展成细胞生物学中一个重要的研究手段。高度的灵敏性和专一性、制样与观察程序的简便、尤其是适宜于活细胞研究等特点,是它所具有的独特长处。荧光显微术特别是免疫荧光技术在现代医学生物学研究中的应用是一个十分活跃的领域。近年来它也逐渐渗入植物学研究中。Kapil

等(1978)曾详细综述了植物胚胎学中引入荧光显微术后所取得的新进展[56]。但正如O'Brien和McCully(1981)在《植物构造的研究:原理与方法选集》一书中所指出的[82],这一技术在植物学中的作用迄今尚未引起足够的重视。本文的目的是就植物细胞生物学的广泛范围内荧光显微术近年的应用动态作一初步的总结。

染色体与DNA的研究

染色体荧光分带

Casperson(1969)发现植物染色体经芥子奎吖因(quinacrine mustard)染色后呈现荧光明暗不同的区段[12],由此开创了染色体荧光分带技术。不久,Vosa(1970)用价格低廉的奎吖因(quinacrine)同样获得成功[112]。他和其他研究者随后在一系列植物染色体材料上用此法开展了研究。1972年报道了另一种荧光染料Hoechst33258(H33258)对哺乳类染色体的显带效果,很快也被引入植物染色体分带[113]和姊妹染色单体交换[58]的研究。Comings等(1975)曾对奎吖因[18]和H33258[17]两类荧光分带的机制作过研究。近年关于植物染色体荧光分带的工作仍时有所见,但一般来说由于它不如Giemsa分带那样只需要简单的仪器和可以制作永久标本,因而退居次要地位。在某些场合荧光分带也表现特殊的价值,例如百合属的染色体用奎吖因显带可以发现一系列类似哺乳动物染色体的横带,而用Giemsa C-带技术则显示不出来[60,61]。近年还应用了其它一些新型的荧光染料进行染色体分带。例如在一种假葱属植物 Nothoscordum fragrans 的染色体上比较了C-带和quinacrine、H33258、DAPI、mithramycin、chromomycin A3等荧光显带的效果[99]。

DNA的细胞荧光测定

用细胞荧光测定术(Cytofluorimetry)测量DNA,是荧光显微术和细胞光度术(Cytophotometry)相结合的产物,标志着前者由定性向定量的重要进展。荧光测定具有较高的灵敏性和专一性,常能测出常规孚尔根光度术所难以反映的DNA含量。另一优点是它适用于生活细胞中DNA的测定,在细胞培养研究中特别有用。细胞荧光测定术的进步,固然和仪器的改进有关[96],而荧光染料的不断更新也是其重要原因。20世纪60年代,DNA的测定主要依靠吖啶橙(acridine orange)荧光染色。70年代以后,其它各种与DNA结合的高效荧光染料相继问世,使面目为之一新。

一类是"荧光锡夫型试剂"(fluorescent schiff-type reagent),如副品红(pararosaniline)、金胺O(auramine O)、吖黄素(acriflavine)、吖啶黄(acridine yellow)、BAO等的应用。它们的染色程序和常规孚尔根反应相似,也要经过盐酸水解;染液也要用二氧化硫脱色。但染液浓度很低,须在荧光显微镜下观察。例如副品红[63]和金胺O[66]曾分别被用于显示真菌的细胞核。BAO(bis-[4-aminophenyl]-1,3,4,-oxadiazole)是被认为较好的一种荧光锡夫型试剂。它对紫外线的照射比较稳定,检示DNA的灵敏度较高[98]。藻类细胞核DNA含量通常仅及高等植物的十分之一,用常规方法较难定量测定,而用BAO荧光测定则效果不错[93,49]。

Hoechst33258是一种双苯并咪唑(bisbenzimidazole)衍生化合物,近年常用于DNA的荧光测定。其特点是荧光明亮而不易衰退、毒性较低、染色程序简便,特别适于活细胞的鉴定。哺乳动物的活细胞经H33258或H33342染色后,可根据DNA含量作细胞分拣(cell sorting),然后重新培养,存活率高达90%以上[9]。被疟原虫感染的血细胞经H33258染色后亦可用流式荧光计分拣[47]。在植物材料上,Coleman(1978)用该染料和DAPI(见后文)分别显示绿藻细胞核DNA以及细胞质中的含DNA小体——叶绿体、多磷酸颗粒、共生细菌[14]。Laloue等(1980)研究了高等植物细胞核的H33258荧光染色技术[64];该法被Galbraith(1981)应用于烟草原生质体培养初期DNA变化的荧光测定[31]。但以上两项工作系用固定材料实验。另一方面,Meadows等(1981)则研究了用它活染原生质体的技术[76],并在此基础上作定量测定和细胞分拣[75]。但Puite等(1983)认为H33342染色仅适于固定的原生质体而尚不适于生活原生质体的流式荧光测定[92]。除植物体细胞外,我们最近在酶法分离胚囊的研究中用 H33258 显示固定和生活胚囊内的细胞核亦获得很好的效果[5,116]。

DAPI(4′,6-diamidino-2-phenylindole)是测定 DNA 的另一种高效荧光染料,其灵敏度在现有荧光染料中仅 H33258 可与之比拟,而其荧光衰退甚至比后者更慢[16],唯价格比较昂贵。Lin 等曾研究了它和 DNA 分子结合的机制[67]。除显示核 DNA 外,它尤其适于显示核外 DNA,如细胞质中的支原体污染、酵母菌的线粒体 DNA 和液泡中的多磷酸颗粒、藻类的叶绿体 DNA、细菌 DNA、以至噬菌体的极微量的 DNA(根据 Coleman 等引证[26])。白克智等用它测定了满江红鱼腥藻分化细胞的 DNA 含量[3]。在高等植物细胞学研究中,DAPI 亦可用于显示细胞器如叶绿体中的 DNA[51]。Scott 等(1984)用它观察马铃薯块茎细胞中的核和质体 DNA[100];Griesbach 等(1982)用它显示烟草原生质体摄取百合染色体的过程[35]。据最近的报道,DAPI 对花粉和花粉管中的细胞核也有很好的染色效果(Coleman 和 Goff 1984[15];朱澂等,)。

光辉霉素(mithramycin)亦可充当显示 DNA 的荧光染料。尽管它在灵敏度和荧光衰退两方面不如 DAPI 优越,但专一性很高,光度测定时读数稳定,同样适合于核 DNA 和核外 DNA 的研究[16]。贮存 48 年之久的水绵,经光辉霉素染色后所呈现的 DNA 荧光亮度和新鲜标本接近[49]。

此外,在动物细胞流式荧光术中应用较广的溴化乙锭(ethidium bromide,EB)和碘化丙锭(propidium iodide,PI)最近有时亦见于植物细胞学文献中。例如 McCully(1976)用 EB 染新鲜植物组织细胞(根据 O'Brien 与 McCully[82]引证);Hakman 等(1984)用 PI 染云杉组织培养中的不定芽和再生苗尖并作了定量测定[37]。

这里还要提到一种无毒的洗涤剂 Triton-x-100,它在低浓度时有增强活细胞荧光染色效果的作用,对 EB、PI、DAPI、H33342、光辉霉素等的荧光活染均为有效[110]。在植物生活原生质体的 H33258 荧光染色实验中,加入该洗涤剂被证明有良好作用[76];但另外也有实验表明它的作用不明显[92]。

细胞壁的研究

荧光显微术在植物细胞学中应用的一个突出方面,是用它鉴定细胞壁的成分、研究壁的形成与再生以及细胞壁在植物发育过程中和在环境影响下性质的变化。

用荧光增白剂显示细胞壁

荧光增白剂和细胞壁成分有强烈的亲和力。当前在植物细胞学中常用的荧光增白剂有:Calcofluor White ST(以下简称 ST)、Calcofluor White M_2R(以下简称 M_2R)、国产荧光增白剂 VBL。

ST 开始是被用来显示细菌、放线菌与真菌的细胞壁[22]。在粘菌材料上,它被证明与纤维素和几丁质有亲和力[38]。Nagata 等(1970)首先将它用到高等植物原生质体的研究中,观察细胞壁的降解与再生[81]。以后陆续有这方面的报道[31,80,84,101]。

M_2R 的化学性能是能与多种 β 构象的吡喃己多糖(包括 β-1,3-葡聚糖和 β-1,4-葡聚糖等)结合,而与 α 构象的多糖(淀粉、糖原等)不结合[71]。Hughes 等(1975)在植物组织的徒手切片、GMA 半薄切片和生活幼苗的实验中,对它的染色效果作了详细的研究[48]。特别有趣的是:Hahne 等(1983)将 M_2R 加入烟草原生质体培养基。使之对培养中的原生质体进行长时间的活体染色,观察了细胞壁再生和细胞分裂的过程[36]。

VBL 染细胞壁的效果也好。黄祥辉等(1980)用它观察了烟草原生质体的壁再生、出芽和细胞分裂,并比较了其与 ST 的异同[7]。王辅德等(1981)也作了类似的研究[1]。

胼胝质的鉴定

在植物的特殊组织、发育时期和生理条件下,胼胝质常以细胞壁的重要成分出现。例如它常见于筛板、纹孔区、孢子母细胞、四分体、花粉生殖细胞、花粉管等的细胞壁,以及不亲和授粉时的柱头表面、受伤或衰老的组织等处。在植物胚胎学中,用荧光显微术研究胼胝质在生殖过程中的变化及其作用占有突出的地位,有关资料详见 Kapil 等[56]和 Dumas 等[25]的综述。至于在各种异常生理条件(伤害、机械、超声波、冷冻、热、重力等刺激)下胼胝质的出现,则可参看 Jaffe 等的论文[50]。本文仅就技术的角度介绍有关研究动态:

当前流行的显示胼胝质的荧光染色技术是 Currier(1957)首创的水溶性苯胺蓝(aniline blue, w. s)脱色溶液染色法[21]。此法以后又经若干细节上的改良,以适应不同的情况。几种主要的配方可参阅 Dumas 等的文章[25]。一般荧光染色标本不能永久封藏,但 Ramanna (1973)提出用 Euparal 胶作为苯胺蓝染色标本的封藏剂,可保存荧光达半年之久[93]。Polito 等(1981)则提出了另一种半永久封藏的介质[89]。苯胺蓝染色还可以和其它染色方法合用,例如半薄切片先经甲苯胺蓝或 PAS 反应染色后,再用苯胺蓝荧光染色,可以消除木质或细胞壁其它成分的荧光,突出胼胝质荧光[89,104]。有些植物如李属(Prunus)单独用苯胺蓝染色不易显示花粉管,而将苯胺蓝和 Calcofluor White M_2R 混合染色则效果较好[53]。此外,它也可以和吖啶橙、溴化乙锭等复染[25]。目前商品生产的苯胺蓝染料中起荧光色素作用的成分为 $C_{25}H_{18}N_2Na_2$[103]。最近该荧光色素已被提纯并人工合成。可望投入商品生产(根据 Dumas 等引证[25])。

其它细胞壁成分的鉴定

细胞壁中的木质、酚化合物、角质、栓质、孢粉素等成分能呈现自发荧光。在木材解剖学和植物化石的研究中可以充分利用这一特性而获得很佳的荧光标本(参看 O'Brien 与 McCully[82])。

角质的荧光染色鉴定过去是用苯并芘(benzpyrene)。鉴于该化合物有致癌作用,Heslop-Harrison(1977)改用金胺 O 染色法鉴定柱头表面的角质层,研究花粉与雌蕊的相互作用[44]。我们最近应用这一方法鉴定酶法分离的金鱼草胚囊,证明胚囊壁中有角质成分[5,116]。

孢粉素是孢子与花粉外壁的主要成分。Waterkeyn 等(1971)用樱草素(primulin)荧光染色法鉴定花粉原外壁与乌氏体[114]。Heslop-

Harrison 等(1982)则用金胺 O 染色鉴定孢粉素[45]。最近,朱澂(1983)由派罗宁(pyronin B)中分离出一种对孢粉素有良好染色效果的荧光成分[4]。

细胞中其它成分与构造的鉴定

RNA

一般采用吖啶橙荧光染色鉴定 RNA。吖啶橙具有荧光异色(fluorescence metachromasia)的性能,即与 DNA 双链分子结合呈绿色荧光,而与 RNA 单链分子结合呈红色荧光。在动物材料上,用吖啶橙染色可进行 RNA 含量的细胞荧光测定[23]和流式细胞荧光测定[10,108]。但在植物方面尚未见到有关定量测定的报道。

蛋白质

1-ANS(1-anininonaphthylsulphonic acid)是与蛋白质结合的一种荧光染料。Heslop-Harrison 等(1974)曾用它证明花药绒毡层制造的蛋白质转移到花粉外壁构成在亲和反应中起识别作用的蛋白质[42];还用它鉴别了柱头表面起接受器作用的蛋白质表膜[74]。

异硫氰酸荧光素(fluorescein isothiocyanate, FITC)也是一种与蛋白质结合的荧光染料,在免疫荧光技术中常用作抗体的标记物(详见后文)。

调钙蛋白(calmodulin)是与钙离子有关的生命活动过程的调节因子。而吩噻嗪(phenothiazine)类化合物是调钙蛋白的抑制剂,它们可与后者专一结合并在光照下氧化成荧光色团。根据这一原理,Hauber 等最近用吩噻嗪类化合物对调钙蛋白进行荧光定位观察,在几种植物材料的尖端生长细胞(如花粉管的生长尖端等)上取得成功[40]。

多糖

按照常规 PAS 反应的程序,材料经高碘酸氧化后用荧光锡夫型试剂(副品红、吖黄素等)染色,可显示细胞壁多糖和淀粉粒(根据 O'Brien 与 McCully 引证[82])。另一种荧光锡夫型试剂 tripaflavin 曾被用于研究百合属胚囊发育过程中多糖的动态(根据 Kapil 与 Tiwari 引证[56])。

应用荧光标记的外源凝集素(lectin)鉴定多糖是一种很巧妙的方法。各种外源凝集素能有选择地与特异的糖结合,因此如用荧光染料加以标记,即可作为鉴定多糖残基的探针。Rougier 等(1979)曾以此法测定了玉米和水稻根冠粘液的多糖成分[13,97]。用类似的方法也研究了衣藻有性生殖过程中的碳水化合物变化[77]。

油脂

早期应用苯并芘(benzpyrene)或磷化氢 3R(phosphine 3R)荧光染色法鉴定油脂(参看 Jensen[54])。最近仍有人用苯并芘显示小麦原生质体的培养过程中出现的油滴(Sethi 等,1983[101])和薯蓣原生质体膜的磷脂(Onyia 等,1984[84])。但该药品有致癌作用,使用时须加注意。

叶绿体

叶绿素在紫外线或蓝光激发下发射红色荧光,因此新鲜(或戊二醛固定)的组织无需染色即可在荧光显微镜下清楚地观察到叶绿体的分布。Elkin 等(1975)发现,C_4 植物维管束鞘中的无基粒叶绿体与一般叶肉组织的有基粒叶绿体有不同的荧光特点:前者的荧光谱主要在红外区段,后者的荧光谱主要在远红区段;用红外线照相底片很容易区分这两种叶绿体的荧光[26]。

叶绿素的自发荧光很容易消退,观察时间较长即仅留下类胡萝卜素的黄色荧光。不同种的植物,叶绿素荧光衰退的速率差别很大。当叶绿素荧光太弱时,可在封藏剂中加入二氯苯二甲脲(3-(3-4 dichloropheny1)-1,1-dimethylurea,DCMU)以加强叶绿素的荧光[82]。

线粒体

用荧光显微术可观察活细胞中的线粒体。一种方法是 DASMPI(dimethyl amino styryl methyl pyridinium iodine)染色[11,78]。另一种方法

是罗丹明 123(rhodamine 123)染色,该法对线粒体有高度的专一性,而对细胞中的其它构造(如质膜、核膜、溶酶体、内质网、高尔基体等)则不起作用,具高度分辨率,无毒性[55],值得在植物细胞学研究中试用。

质膜

四氯荧光素(4,5,6,7-tetrachlorofluorescein)染生活的洋葱鳞片叶细胞,可显示原生质膜及胞间连丝,观察质壁分离现象[107]。

免疫荧光鉴定

免疫荧光技术是利用抗原抗体反应的高度专一性,用荧光染料(通常是 FITC)标记抗体,以便对特异的抗原进行精确的定位。在医学生物学中这是一项用途很广的技术;在植物细胞生物学方面则方兴未艾,于下述各例可见一斑。

种子中的蛋白质

各种植物的种子蛋白质各有其特异性,用免疫荧光技术可以鉴别。例如近年用这一技术研究了菜豆子叶中的豆球朊(legumin)与豌豆球朊(vicilin)[19]、豌豆种子中的贮藏蛋白质[20]、燕麦胚盾片中的球蛋白[6]等。此外,还对菜豆种子中的外源凝集素(lectin)作了免疫荧光定位研究,发现不同组织中的外源凝集素具有不同的功能[72]。

花粉与柱头中的识别蛋白质

Knox 与 Heslop-Harrison 等应用免疫荧光技术与其它方法,发现了花粉壁中存在着识别蛋白质;后来又发现柱头表面有起"接收器"功能的蛋白质表膜,从而深入揭示了受精不亲和性的机制(参看 Kapil 等的综述[56])。

酶

应用免疫荧光技术研究光合作用代谢中的酶作了一系列工作,如

Hattersley 等(1977)研究了 C_3 和 C_4 植物叶中的二磷酸核酮糖羧化酶(RuBP carboxylase)[39]；Madhavan 等(1982)进一步研究了该酶在分属 C_3、C_4、景天科植物代谢三种类型的 41 个物种的叶保卫细胞中的分布[70]；Perroy-Rechenmann 等(1984)则对豌豆属的四个种(分属 C_3、C_4 和 C_3—C_4 中间型)叶中的磷酸烯醇丙酮酸羧化酶(phosphenol pyruvate carboxylase)和二磷酸核酮糖羧化酶的分布作了定位观察[86]。

此外 Suatter 等(1982)研究了西瓜子叶中的苹果酸脱氢酶(malate dehydrogenase)，分别鉴定了存在于乙醛酸循环体、线粒体与细胞溶质中的该酶的三种同工酶[106]。Verncoy-Gerritsen 等(1983)研究了大豆种子萌发期间子叶中的脂氧合酶(lypoxygenase)的分布特点及其可能的功能[111]。

微管蛋白

微管蛋白的免疫荧光定位是研究细胞骨架的一项重要方法。Frank 等(1977)曾用于观察胚乳细胞纺锤体的微管结构[28]。Marchant (1978)观察了一种绿藻原生质体中与质膜相联系的微管[73]。Lloyd 与 Powell 等在这方面的研究比较系统，他们首先观察了胡萝卜细胞中微管蛋白的整体分布情况(Lloyd 等,1979[68])，然后研究了在制备胡萝卜原生质体过程中由原来的长形细胞变为球形原生质体时、以及在秋水仙素处理后的微管系统的变化(Lloyd 等,1980[69])。同时，他们研究了一种藓类细胞中的微管骨架(Powell 等,1980[90])，还发现在该植物原生质体中有一种与细胞核相联系的、抗洗涤剂作用的细胞骨架系统(Powell 等,1982[91])。最近，Dickinsen 等(1984)对百合小孢子进行免疫荧光研究，发现有一种微管系统由核膜辐射状延伸至质膜，推测它与花粉壁特殊成分的形成有关[24]。

细胞生活力测定

测定细胞生活力有各种方法，其中荧光显微术是比较简便而准确的一种。Rotman 等(1966)在研究哺乳动物细胞时发现：荧光素的酯类化合物如荧光素二醋酸酯(fluorescein diacetate, FDA)本身不发荧光,但进入活细胞后可在细胞内的酯酶作用下分解成发荧光的荧光素(fluorescein),这一现象被称为 fluorochromasia 或 fluorochromatic reaction (FCR)。当细胞具有酯酶活性和完整的细胞膜时,根据 FCR 可判断其为生活的。当细胞不具酯酶活性或当细胞膜破坏而荧光素流失时,细胞均不发荧光。因此根据 FDA 处理后细胞的 FCR 可以有效地鉴定细胞生活力[95]。这一方法被引入植物细胞学,在测定花粉、体细胞和原生质体等的生活力方面发挥了很大作用。

花粉生活力测定

Heslop-Harrison 等(1970)首先用 FDA 测定花粉生活力成功[41]。这一方法后来被其他研究者采用,认为优于另外的染色测定法[52,83]。近年,Heslop-Harrison 等继续研究 FCR 和花粉膜状态的关系[102],并通过几种测定花粉萌发力方法的比较,进一步肯定了 FCR 与萌发力有最高的相关性[43]。此外,FCR 还被用于鉴定植物四分体时期小孢子的育性[59]和柱头表面的脂酶活性[74]。最近,我们用 FDA 鉴定酶法分离的胚囊生活力亦获成功[116]。

培养细胞与原生质体的生活力测定

Widholm (1970)首先用 FDA 技术测定植物培养细胞的生活力[115]。Zilhak 等(1978)在研究有毒化合物引起悬浮培养细胞死亡时应用了此法[117]。Smith 等(1982)肯定了它对测定悬浮培养细胞衰老过程中的生活力变化是有效的[105]。在原生质体生活力测定方面,Larkin(1976)应用了 FDA 技术[65],随后也被其他研究者采用[2,101]。

分离液泡的鉴定

由植物细胞中大规模分离液泡的方法已告成功。要鉴定分离液泡的纯度,用 FDA 处理,非液泡(即带有原生质的)部分显荧光,而液泡本身则呈暗区[8,33]。这是从另一个角度应用此项技术。

活体染色与荧光示踪

许多荧光染料在低浓度时没有毒性,因而适于超活染色(supravital staining)和活体染色(vital staining)。前文列举的生活细胞中细胞核、细胞器、细胞膜、细胞壁的荧光染色以及细胞生活力测定等方法均属超活染色;如果荧光染色后的细胞或原生质体再进行培养则属于活体染色,在此不重复赘述。以下再谈谈荧光活体染色的其它几个重要方面。

原生质体融合式摄取外源遗传物质

用荧光染料标记不同来源的原生质体或外源遗传物质,可以把已融合或已摄取外源物质的原生质体和其它原生质体区分开来,作为筛选细胞杂种的一项方法。Galbraith 等在这方面作了有意义的探索。他们用 FITC 和罗丹明 B(rhodamine B)分别标记大豆的不同原生质体,利用该两种染料荧光的差异(前者呈黄色,后者呈红色)鉴别融合的原生质体(同一细胞呈现两种荧光)[29]。在另一个实验中,他们用 FITC 和异硫氰酸罗丹明(rhodamine isothiocyanate)分别标记异种烟草的原生质体[30]。后一方法被 Redenbaugh 等(1982)采用,对几种植物的原生质体及其融合产物作流式细胞测定和细胞分拣[94]。Galbraith 等还将 FITC 标记的烟草原生质体经过流式细胞分拣后培养成苗,证明荧光标记并不损害原生质体的生活力[32]。Patnaik 等(1982)用 FITC 标记矮牵牛悬浮培养细胞的原生质体,使之与烟草叶肉细胞的原生质体融合,杂种异核体兼有 FITC 荧光和叶绿体自发荧光;然后用显微操作器进行分离培养[87]。在动物细胞融合实验中也有类似的设计,如将罗丹明 123 标记一种细胞或其胞质体(cytoplast)中的线粒体,用 H33258 标记另一种细胞或其核体(karyoplast)的细胞核,用以鉴别所得到的杂种细胞、胞质杂种和核质杂种[46]。

在原生质体摄入外源物质的实验方面,Griesbach 等(1982)用 DAPI 标记百合染色体,观察了其被烟草原生质体摄取的过程[35]。

核质间物质转移的荧光示踪

应用荧光染料作为示踪物,可以研究物质在细胞质与细胞核之间的变换。迄今这类实验主要以动物细胞为材料。如 Pain 等用 FITC 标记各种蛋白质(血清蛋白、卵清蛋白、肌球蛋白、溶酶体、细胞色素)。将它们分别用显微注射器注入蟑螂卵母细胞的细胞质中,然后观察其进入细胞核的可能性与速度[86]。以后他们又将该技术应用于体细胞实验:将荧光示踪物注入摇蚊的唾腺细胞,观察其由核向质和由质向核的转移以及核膜的选择作用[85]。关于将荧光标记的蛋白质显微注入活细胞中,以便进行细胞骨架动态的研究,可参看 Kreis 等的综述[62]。在高等植物中,虽有类似的设想[56],但尚未见到实验报道。

细胞间物质转移的荧光示踪

植物细胞间的物质转移通过质外体与共质体两条途径,二者皆可应用荧光染料示踪。早期曾用樱草素或硫酸小檗碱作为植物蒸腾液流的示踪染料。后来发现荧光增白剂作为质外体的示踪剂更适合,因为它不会进入共质体(根据 O'Brien 与 McCully 引证[82])。

Goodwin 等用荧光示踪法研究了物质在共质体系统中的转移。将 FITC 标记的几种氨基酸以及羧基荧光素(6-carboxy fluorescein)作为示踪物注入伊乐藻属植物的叶细胞中,以上分子均不能通过原生质膜,但其中有些可通过胞间连丝在细胞间运动。还研究了阳离子、代谢抑制剂、细胞松弛素等对这种运动的影响[27,34]。Tyree 等(1975)用荧光素钠(fluorescein disodium, uranin 观察了物质通过紫露草雄蕊毛细胞共质体的运动[109]。Mogensen(1981)也用该染料研究了物质进入不同类型的胚珠的途径[79]。

结 束 语

O'Brien 和 McCully 曾强调指出:一切想炫耀植物材料的人都应该懂得荧光显微镜的价值[82]。纵观荧光显微术在植物学中应用的历史,早期主要是利用植物体内得天独厚的自发荧光。当20世纪50年代发明了胼胝质的荧光染色技术以后,相当长一个时期所应用的只集中在这一技术和其它少数几种有限的方法。直到20世纪70年代以来才出现各种荧光鉴定技术争妍竞艳的局面。从目前状况来看,荧光显微术在植物细胞生物学中可以发挥作用的主要方面有:(1)利用荧光技术的高度专一性与灵敏性,对细胞组织中的各种成分与构造进行精确的定位,以补充常规组织化学技术的不足。尤其是免疫荧光技术,它在植物研究中还有很大的潜力。(2)借助细胞荧光测定术和流式细胞荧光测定术对核DNA含量进行定量测定,并扩及核外DNA的测定。(3)利用荧光染色的无毒性这一独特的优点,开展活细胞的研究。由于植物细胞具有全能性,荧光显微术和细胞工程研究相结合大有用武之地。本文已经引述了许多饶有兴趣的实验;今后还可以按照研究的目的设计出其它各种巧妙的实验。总的看来,似乎可以认为荧光显微术的应用由描述阶段正在走向实验阶段。这种变化趋向是值得我们注意的。

(作者:杨弘远。原载:武汉植物学研究,1986,4(1):80~90,参考文献117篇,删去。)

12 荧 光 显 微 术

Fluorescence microscopy

荧光显微术(fluorescence microscopy)是利用短波光照射被测物质,以激发其发射荧光而在荧光显微镜下加以检视的方法。Köhler(1904)第一次用显微镜观察到生物组织经紫外线辐射所产生的荧光。Lehman(1911)观察了叶绿体和花粉的自发荧光。Haitinger(1938)发明了荧光染料,从而使荧光技术得到广泛应用。Coons等(1941)开创了荧光抗体技术,大大推动了免疫学和有关科学技术的发展。目前,荧光显微技术在生物学研究中日益显示其重要性,这是由于它具有如下几个优点:①高度的专一性和灵敏性,能以很低浓度的荧光染料检测出组织中存在的含量极微的物质。②制样与观察程序简便,适于快速鉴定。③随着新荧光染料的发明,不断扩大被测物质的范围。④低毒性的荧光染料适于作培养细胞的活体染色。在现代植物细胞学研究中,荧光显微技术的用途也日益广泛,概括说来包括如下许多方面:

组织化学与细胞化学鉴定

现在已有各种相应的荧光染料可以鉴定细胞组织中的许多成分,如:

DNA　　吖啶橙(acridine orange, AO)

　　　　吖黄素(acriflavine)

　　　　Hoechst 33258 (H33258)

　　　　碘化丙锭(propidium iodide, PI)

　　　　溴化乙锭(ethidium bromide, EB)

　　　　光辉霉素(mithramycin)

　　　　DAPI (4′6-diamidino-2-phenylindole)

　　　　BAO(2,5-bis[4′-aminophenyl(1′)]-1,3,4-oxadiazole)

　　　　morin dehydrate (3,5,7,2′,4-pentahydroxyflavanol)

RNA　　吖啶橙

　　　　溴化乙锭

　　　　morin dehydrate

蛋白质　异硫氰酸荧光素(fluorescein isothiocyanate, FITC)

　　　　1-ANS (1-anilinonaphthylsufonic acid)

油脂　　3,4-benzpyrene phosphine 3 R

多糖　　tripaflavin

β吡喃己多糖　荧光增白剂(Calcofluor white)

角质　　金胺 O（auramine O）
孢粉素　金胺 O
　　　　樱草黄（primulin）

特别值得指出的是，不少荧光染料已可用于 DNA 的显微荧光光度测定，从而使荧光显微技术进入定量的水平。

染色体分带

奎吖因（quinacrine）、芥子奎吖因（quinacrine mustard）、Hoechst 33258 等荧光染料可用于显示染色体的异染色质带，称为荧光显带技术。

细胞识别的研究

如用异硫氰酸荧光素（FITC）标记的荧光抗体技术研究花粉壁蛋白质与柱头表膜蛋白质的识别反应。

鉴定细胞生活力

用荧光素二醋酸酯（fluorescein diacetate, FDA）进行活染，以鉴别花粉或其它细胞的存活或死亡（详见后文）。

原生质体培养与融合中的应用

如用荧光增白剂显示原生质体再生细胞壁的过程；用 DAPI 标记染色体，研究其被原生质体吸收的过程；用不同荧光染料分别标记两种原生质体，用以鉴别杂种细胞。

物质运输途径的观察

用荧光素钠盐（uranin, disodium fluorescein）等荧光示踪剂，研究物质在器官与组织中的运输途径。

荧光显微技术在植物细胞学中的应用，几年前 O'Brien 与 McCully（1981）曾有介绍[24]，最近笔者就当前应用动态作了综述[2]，并推荐

Kapil 与 Tiwari(1978)[16]关于荧光显微术在植物胚胎学中的应用一文,供读者参考。

荧光显微术的原理与方法

关于荧光显微术的基本原理与方法,迄今国内尚未见系统的阐述。以下主要参考 Rost(1980)所著"荧光显微术"一文[27],并结合其它参考文献和作者的经验,从三方面作一简明的介绍。

荧光与荧光染料

某些物质经短波光照射后,分子呈激发状态,其所吸收的能量一部分转化为热或用于光化学反应,另一部分则以波长较长的光能形式重新发射出来。重新发射的光称为荧光。能发射荧光的物质称为荧光色团(fluorophore)。一般的规律是:荧光谱恒较激发光谱的波长为长。根据不同荧光色团的特点,可分别采用紫外、紫、蓝、绿等波段的激发光源,以获得紫、蓝、绿、黄、橙、红等波段的荧光。例如用波长为 450～500 纳米的蓝光激发 FITC,可发射波长为 500～550 纳米的黄绿荧光(图1)。

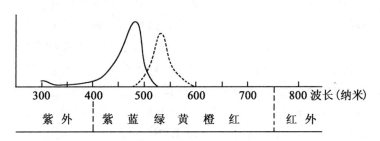

图 1　FITC 的激发光谱与荧光谱
（——激发光谱,……荧光谱）

在生物学中,有各种产生荧光的情况。

自发荧光　生物材料中自然存在的某些物质可在激发光的作用下

发射荧光,称为自发荧光(autofluorescence)或原发荧光(primary fluorescence)。例如叶绿体因其组成成分中有卟啉而呈红色荧光。木质、孢粉素、脂类中溶解的某些物质以及树脂等亦常呈现自发荧光。但是生物体中多数物质则需在某些外源物质的作用下或与外源荧光物质相结合才能发射荧光,这称为次级荧光(secondary fluorescence),如下所述。

诱发荧光　细胞组织中某些物质经一定的化学处理后可转化为荧光色团,由此产生诱发荧光(induced fluorescence)。最常见的是甲醛诱发蛋白质中所含的芳香乙胺基团转化为荧光色团,因此经甲醛固定的材料虽不染色亦显示一定程度的荧光。

荧光染料染色　应用含荧光色团的物质作为染料,使之与细胞中和该物质有特殊亲和力的物质相结合,即可被激发出荧光,这类染料称为荧光染料(fluorescent dye)或荧色素(fluorochrome)。许多常规染色剂(如刚果红、曙红、碱性品红等)兼有荧光染料的性能,也有许多专用的荧光染料(如 H33258、DAPI、FITC 等)。一般讲,用于荧光观察的染料,其使用浓度比常规染色时的浓度要低得多。

荧光染料中有一些希夫型的试剂,如碱性品红、吖黄素、BAO、tripaflavin 等,称为荧光希夫试剂。它们可代替常规的孚尔根反应或 PAS 反应而收到更灵敏的效果,因此被用于作精确的定量测定。

有的荧光染料能显现不止一种颜色的荧光,称为荧光异色现象(fluorescence metachromasia)。这是由于染料分子与被染物质分子的结合方式不同所致。例如吖啶橙,和 DNA 分子结合呈绿色荧光(原色形式),而和 RNA、变性 DNA 或酸性多糖分子结合则呈红色荧光(异色形式)。技术上的变化(如过度染色时)亦可导致荧光异色现象。

某些荧光染料毒性较小,加之使用浓度很低,因而特别适于用作超活染色(supravital staining)。例如活材料可用下列荧光染料标记:

Hoechst 33258——染色质

8-aniliro-1-naphthalene sulphonic acid(ANS)
——核仁与细胞质

dimethylaminostyrylmethylpyridiniumiodine

（DASMPI）——线粒体

rhodamine 123——线粒体

中性红（neutral red）——溶酶体

4,5,6,7-tetrachlorofluorescein——原生质膜

4-acetamido-4′-isothiocyanostilbenene-2,2′-disulphonic acid——质膜外表成分

Calcofluor white M_2R——初生壁

荧光染料品种繁多且与日俱增。研究者了解各种荧光染料的特点，对于选择和正确应用荧光染料染色很有必要。附录Ⅰ列举了几十种荧光染料的名称与性能，可供参考。

免疫荧光 以荧光染料标记特异的抗体，制成荧光抗体（fluorescent antibody），用于检验相应的抗原的存在，称为免疫荧光（immunofluorescence）技术或荧光抗体技术。这是荧光技术在生物学中应用最广、研究最活跃的领域。常用 FITC 作为标记抗体的荧光染料。被标记的抗体可以是直接由被测抗原注入家兔体内而产生，但用得更多的是间接的抗体，如在山羊体内产生的抗家兔球蛋白的抗体。后者可以大量商品生产，并事先用 FITC 加以标记，制成可以随时取用的荧光抗体。将这种商品生产的荧光抗体和被测抗原在家兔体内诱发的抗体相结合，即可对被测抗原物质进行定位。这一流行的免疫荧光技术被形象化地称为"三明治技术"（sandwich technique），以表示被测抗原—直接抗体—间接抗体三者之间的关系。其优点为：①每一直接抗体分子能与数个间接抗体分子结合，从而可以增加被标记的荧光染料分子，增强荧光反应。②荧光抗体是商品生产的，无需研究者自己制备。在植物细胞学中，应用免疫荧光抗体技术鉴定细胞中某些特异的蛋白质如微管蛋白、酶、外源凝集素、花粉壁蛋白等，取得了不少新的进展。

酶促荧光 在细胞内的酶的作用下，进入细胞的非荧光物质被转化为荧光色团而发射荧光的现象称为酶促荧光（enzymatically produced fluorescence）。利用这一现象可以进行酶的鉴定和细胞生活力的测定。后文中用荧光素二醋酸酯测定花粉生活力的方法即属此类。

荧光显微镜

一架普通光学显微镜配备一套特殊的光源和滤光装置,构成一个用于荧光观察的照明系统,就成为荧光显微镜。

光源 荧光显微镜要求提供波长较短的光源。一般采用高压汞灯,也有用氙弧灯或卤钨灯者。光源中包括紫外线与可见光线。实际上紫外线只在少数场合有用;一般荧光观察多系利用紫、蓝以至绿色的可见光作为激发光源。灯泡在工作过程中还散发大量热能,因此还需配备滤热装置。此外当然还包括调焦、集光装置。

滤光装置 上述光源所发射的光谱范围很广,而需要用于激发特定荧光色团的只是其中特殊的波段。同时,人们所观察的又应该只限于该荧光色团所产生的特有的荧光而不应该包括其它的光波段。因此需要有一套滤光装置来保证上述两个目的。滤光装置是荧光显微镜的关键部件,它由两种互补的滤光片组成(图2):①激发滤光片(exciter filter),其作用是除掉光源中无用的光谱成分,仅容许所需要的激发光波段通过而到达被检标本。②阻挡滤光片或称压制滤光片(barrier filter),其功能是除掉多余的激发光,仅容许特有的荧光波段通过而到达人眼,这样既避免杂光的干扰,又防止紫外线对人眼的伤害。

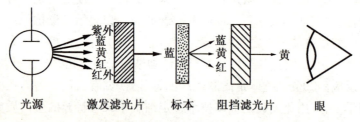

图2 滤光片在荧光显微镜中的作用(以 FITC 为例)

按照光线通过的波段范围不同,滤光片可大体分为四类(图3):

(1)广带滤光片(wide-band filter) 能容许较广的光波段通过。

(2)狭带滤光片(narrow-band filter)或带通滤光片(band pass filter) 仅容许狭窄的光波段通过。

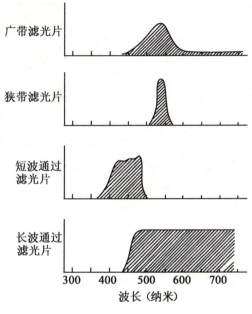

图3 几种滤光片的光波通过曲线

(3)短波通过滤光片(shortwave pass filter) 仅容许一定波长以下的短波通过,是一种截止滤光片(cut-off filter)。

(4)长波通过滤光片(longwave pass filter) 仅容许一定波长以上的长波通过,也是一种截止滤光片。

就激发滤光片而言,当荧光色团的激发光谱与荧光谱二者的峰值较远时,可应用广带滤光片;但当二者的峰值较近时,则宜应用狭带滤光片或短波通过滤光片。至于阻挡滤光片,通常是应用长波通过滤光片。图4表示用狭带滤光片作为激发滤光片和用长波通过滤光片作为阻挡滤光片时的相互关系。实际上,在现代荧光显微镜的设计中,各种滤光片已经配套,以适应不同荧光染料的需要。附录Ⅱ列举了几种荧光显微镜的滤光组合。

照明系统 荧光显微镜可采用三种不同的照明系统:

(1)透射照明(dia-illumination) 与普通光学显微镜所采用的明

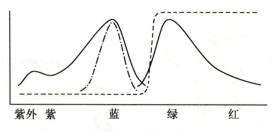

图4 按照荧光色团的特点选择滤光片
—— 左:激发光谱;右:荧光谱
—·—·— 左:激发滤光片(狭带滤光片)
-------- 右:阻挡滤光片(长波通过滤光片)

视野透射光照明系统一致,唯在光源与聚光镜之间设置激发滤光片,在物镜与目镜之间设置阻挡滤光片:

光源——→激发滤光片——→聚光镜——→标本——→物镜——→阻挡滤光片——→目镜

(2)暗视野照明(dark-ground illumination) 与上述光路相同,唯以暗视野聚光镜代替明视野聚光镜。

(3)落射照明(epi-illumination) 采用入射光照明方式。其主要特点是:①在激发滤光片与物镜之间设一个二向色镜(dichroic mirror)或称分色镜(chromatic beam splitter),将激发光中的有用部分自上而下地反射到物镜。②物镜兼有聚光镜的功能,激发光通过物镜聚焦到标本上。③标本发射荧光自下而上地返回到物镜,然后再通过二向色镜和阻挡滤光片剔除杂光抵达目镜。新式的落射荧光显微镜常将激发滤光片、二向色镜、阻挡滤光片三者组装成一套滤光片组合,使用者可以根据需要选择其中任何一套而无需自己分别选配三种滤光片(附录Ⅱ)。落射照明系统的光路可简示如下(参看图5):

光源——→激发滤光片——→二向色镜——→物镜(聚光镜)——→标本——→物镜——→二向色镜——→阻挡滤光片——→目镜

三种照明系统各有其优缺点。一般讲,透射照明是以往荧光显微镜的常规照明系统,迄今仍然应用。暗视野照明是为适应FITC荧光抗

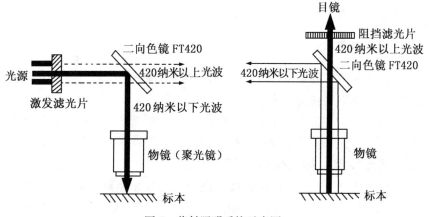

图 5　落射照明系统示意图
左:激发光路线　右:荧光路线

体技术的兴起而应用的,因为它可以将峰值接近的激发光与荧光二者分开;但它有光度很弱、操作不便等缺点。落射照明系统是 Ploem(1967)发明的装置,它和上述两类相比具有多方面的优点:

(1)激发光与荧光的发射方向相反,避免了多余的激发光到达人眼。

(2)二向色镜的运用可以将激发光与荧光准确地区分开来,适用于 FITC 那样的情况。

(3)照明和观察是由标本的同一方向进行,避免了透射照明系统中光线透过标本时的损失,尤其有利于厚标本的观察。

(4)物镜代替聚光镜,因而便于调中与调焦。

(5)物镜即是聚光镜,因而物镜放大倍数愈高,数值孔径愈大,荧光的可见亮度亦愈大,这一点与透射照明恰好相反。特别适于作弱荧光的高倍观察、摄影与光度测定。

(6)同一显微镜中镜台上方的落射照明系统和镜台下方的透射照明系统可以同时使用。例如可将荧光观察和相差观察结合以获得荧光相差图像(fluorescence-phase contrast image)。

落射照明也有缺点，即当使用低倍物镜（如×4）时，由于数值孔径降低致使荧光太弱，这一点不及透射照明好。

荧光显微术

荧光显微术的操作程序包括制样、观察、摄影三个环节。每个环节中都有一些需要注意的事项：

制样 （1）应当了解所使用的荧光染料的性质。一般荧光染液应置于低温无光条件下；有些荧光染料有毒，操作时须加小心。荧光染液通常浓度很低，要防止在制样过程中损失。

（2）用于观察的生物材料视不同情况可采用新鲜或固定的标本。如系固定标本，应考虑到固定剂中是否含有会诱导蛋白质荧光的甲醛等类物质。

（3）用荧光染料染色时，溶液的 pH 值常对荧光的强度有影响，宜应用缓冲液配制染液。

（4）染色的标本大多临时封藏，常用蒸馏水、缓冲液、甘油或其它低荧光介质作封藏剂。有的标本可用 D. P. X. 或 Euparal 作永久封藏，但加拿大树脂因会自发荧光不可采用。

观察 （1）荧光显微观察应在光线较暗的室内进行。

（2）高压汞灯所发射的光源中包含紫外线，切忌直视光源。

（3）不可使易燃物品接近正在工作的光源。

（4）高压汞灯启动后需稍等候片刻方能达到稳定的最大亮度。

（5）应避免反复启闭光源。一旦开启后，宜在每个单元工作完毕后关闭。如需中途关闭，则应静候 10 分钟以上再行开启。

（6）针对荧光色团的特点选用适宜的滤光片组合，是正确运用荧光镜检术的关键。

（7）许多荧光染料被照射后会因光化学反应而出现荧光消退现象（fluorescence fading）。故搜索与观察标本宜多用普通光，俟一切就绪后再换用荧光观察，力求在尽可能短的期间完成观察与摄影。标本停止照射后，在黑暗条件下经一定时间可有某种程度的荧光恢复。有的荧光染料（如 H33258）很少发生荧光消退，宜于光度测定。

(8)如用油镜观察,不能用会自发荧光的香柏油,只能用低荧光油。

摄影 (1)荧光显微摄影时,由于光度较弱和荧光消退现象,宜用快速感光的底片(如27DIN,400ASA)。

(2)为使荧光尽可能通过,应使用数值孔径大的物镜和放大倍数低的目镜;不宜用大底片摄影。

(3)在每次曝光后应隔断激发光源,减少标本被照射的时间。

(4)自动曝光装置是根据整个视野的平均亮度决定曝光时间,而荧光显微镜视野偏暗,当发射荧光的标本在视野中所占比例很小时,根据测光表确定的时间曝光往往不准确,除非使用定点测光法方能奏效。

植物细胞学中几种荧光染色技术

目前已经投入使用的荧光染料品种繁多且不断有新的增加。列举各种荧光染料的使用方法不是本文的任务,本文仅就植物细胞学中常用者,选择四种有代表性的染色剂,介绍其性能、用途与使用方法,俾收举一反三之效。

Hoechst 33258

Hoechst 33258 是较新发明的荧光染料,属于双苯并咪唑类的化合物。它和DNA的重复序列有特殊的亲和力。它的优点是:染色明亮而不易消退;宜于光度测定;毒性低,可用于细胞的活染,因此其应用日渐增多。在植物学方面,最早是用于染色体分带[8,28,32,33],其显带机理曾被研究阐明[4]。近年 Laloue 等(1980)用其进行植物培养细胞中细胞核与染色体的荧光观察[19]。Meadows 与 Potrykus(1981)用于原生质体活染[22]。Hightower 等(1981)用 H33258 标记一种哺乳动物细胞的染色质,用 rhodamine133 标记另一种细胞的线粒体,以此鉴别融合后的细胞杂种[12]。此外,最近已在动物材料上用 H33258 进行 DNA 的显微光度测定[5]和流式光度测定[13]获得成功。以下介绍 Laloue 等的技术。

材料和药品

(1) 植物的生活组织(根尖、表皮、柱头、胚囊、培养细胞等)。
(2) 卡诺(3∶1)固定液。
(3) 柠檬酸-磷酸盐缓冲液(pH5):
   A 液:0.1mol/L 柠檬酸溶液(21g 溶于 1000ml 蒸馏水);
   B 液:0.2mol/L $Na_2HPO_4 \cdot 12H_2O$ 溶液(72g 溶于 1000ml 蒸馏水);
   pH5 缓冲液:A 液 48.5ml 加 B 液 51.5ml 调整至 pH5。
(4) Hoechst 33258。
贮备液:H33258 1mg 溶于 1ml 蒸馏水,贮于 4℃黑暗中可达数月。
染液:临用前用上述缓冲液将贮备液稀释 1000 倍,成为 1μg/ml 溶液。稀释时防止缓冲液温度过低引起染料沉淀。
(5) 甘油。

操作程序

(1) 卡诺液固定,70%乙醇保存。切片、压片或整体材料均可。
(2) 自来水彻底洗去多余固定液。
(3) 柠檬酸-磷酸盐缓冲液(pH5)浸泡(时间因材料而异)。
(4) H33258(1μg/ml)染色。时间因材料而异:组织块染色过夜;悬浮细胞 30min 至 1h;切片 15～30min。
(5) 加入等量甘油透明,用甘油封片。
(6) 荧光显微镜观察。如 Opton STANDARD 型落射荧光显微镜可采用 G436(激发)、FT510(二向色镜)、LP515(阻挡)滤光片组合(附录Ⅱ)。细胞核染色质与染色体呈绿色荧光(图6)。

水溶性苯胺蓝

低浓度的水溶性苯胺蓝在碱性条件下可用于植物组织中胼胝质的鉴定。Currier(1957)首次提出这一技术[6]。Martin(1959)用于显示花柱中的花粉管[22]。其后许多研究者用它研究植物胚胎学的其它方面(如大、小孢子发生过程中的胼胝质变化)[17]。在技术上也有若干改进,如用 Euparal 对染色标本进行永久封藏[26];与 PAS 反应及甲苯胺蓝染色结合应用[25,31]等。以下介绍 Jensen(1962)[16]所描述的方法。

材料和药品

(1)花粉(或胚囊)母细胞、花粉(或胚囊)四分体、花粉管。

(2)FAA 固定液。

(3)磷酸盐缓冲液(0.15mol/L,pH8.2)。

    K_2HPO_4 26g

    蒸馏水   1000ml

(4)脱色苯胺蓝染液(0.005%) 用上述 pH8.2 的磷酸盐缓冲液配制。刚配制的染液有色,在碱性条件下经数小时变为脱色溶液即可应用。用偏碱(pH9~10)的缓冲液配制染液可加强荧光,但不能用于活染。染花柱中的花粉管宜用较浓的染液(如 0.1%)[21]。

操作程序

(1)FAA 固定,切片或整体材料。

(2)自来水彻底冲洗。

(3)0.005%苯胺蓝溶液(pH8.2)染色。切片或悬浮细胞染 20~30min;组织块染数小时。

(4)用染液封藏,临时观察。如需制作永久片,可按 Ramanna[26] 的下法进行。亦可用 Polito 与 Luza[25] 的介质作半永久封藏。

(5)经 50%乙醇、70%乙醇、95%乙醇、100%乙醇逐级脱水,转入无水乙醇与正丁醇等量混合液,再用纯正丁醇透明,Euparal 封藏。在低温黑暗条件下可保存半年仍有荧光。

(6)荧光显微镜观察。如 AO 落射荧光显微镜可用 436(激发)、450(二向色镜)、475(阻挡)滤光片组合(附录Ⅱ)。胼胝质呈黄绿色荧光(图7)。

荧光增白剂

荧光增白剂因其与植物纤维有亲和力,在细胞学中可用作植物细胞壁的荧光染料。有三种类型:

Calcofluor white ST——4,4′-bis〔4-anilino-6-di(hydroxyethyl)-amine-S-triazin-2-ylamino〕-2,2′-st-ilbene disulfonic acid

Calcofluor white M2R——4,4′-bis〔4-anilino-bis-diethyl amino-S-triazin-2-ylamino]-2,2′-stilbene disulfonic acid

VBL(国产)——4,4′-bis[4-anilino-6-hydroxet-hyl-amine-S-triazin-2-ylamino]-2,2′-stilbene disulfonic acid

Darken(1961)首次应用 Calcofluor white ST 显示细菌、放线菌和真菌的细胞壁[7]。以后,在粘菌上证明它和纤维素、几丁质有亲和力[10]。它在藻类方面也可作为细胞壁的荧光染料[29]。在高等植物中,Nagata 与 Takebe(1970)用它显示原生质体的再生壁获得成功[23]。Maeda 与 Ishida(1967)研究了另一种增白剂 Calcofluor white M2R 的特点,证明它能与各种 β 吡喃己糖的多糖结合,而不和 α 构象的多糖结合[20],因而适于作为细胞壁中纤维素与胼胝质等成分的荧光染料。M2R 可用于对植物组织的徒手切片、GMA 半薄切片和生活的幼苗进行染色[14];它还可以与前述苯胺蓝染色法相结合,用于显示李和梨花柱中的花粉管[15]。最近,Hahne 等(1983)将 M2R 加入培养基,对烟草原生质体进行活染获得成功[9]。以上两种增白剂均为国外产品。在国内,黄祥辉等(1980)[3]、王辅德等(1981)[1]应用上海试剂厂生产的荧光增白剂 VBL 观察原生质体的再生壁,亦获得良好的结果。

材料和药品

(1)植物的薄壁细胞(如叶肉)或原生质体。

(2)0.1% Calcofluor white ST 或 VBL 溶液(如作一般细胞壁检查,可用接近 pH7 的蒸馏水配制;如用于原生质体再生壁的检查,则用 0.7mol/L 甘露醇溶液配制)。VBL 在溶解后如有杂质,应取上清液使用。

操作程序

(1)制备新鲜的叶肉细胞或其原生质体悬浮液。

(2)离心去上清液后,在 0.1% 的 Calcofluor white ST 或 VBL 溶液中染色 5～10min。

(3)用蒸馏水(对叶肉细胞)或 0.7mol/L 甘露醇(对叶肉原生质体)换洗 3～4 次。

(4)用蒸馏水或甘露醇溶液重新悬浮。

(5)将悬浮液滴于载玻片上封片。

(6)置荧光显微镜下观察。如 Opton STANDARD 型落射荧光显微镜可用 G436（激发）、FT510（二向色镜）、LP520（阻挡）滤光片组合。细胞壁呈蓝绿色（ST）或绿色（VBL）荧光；叶肉细胞中的叶绿体呈红色自发荧光。

荧光素二醋酸酯

荧光素二醋酸酯（FDA）是目前鉴定细胞生活力的重要染料。Rotman 与 Papermaster（1966）在研究哺乳类细胞时发现：本身不发荧光的荧光素酯类能透过细胞膜进入细胞，被细胞内的酯酶分解而释放出能发荧光的荧光素。后者不能通过细胞膜，故当细胞膜完整时可保存于细胞内而被荧光显微镜检出。这一现象称为 fluorochromasia。当细胞失去酶活力或膜受损时，即不显荧光。Heslop-Harrison 等（1970）在植物细胞学中应用 FDA 测定花粉生活力[11]，并进而研究了有关机理[30]。FDA 亦被用于研究花粉不育的过程[18]和其它体细胞及原生质体的生活力。

材料和药品

（1）花粉（或悬浮细胞）。

（2）荧光素二醋酸酯（FDA）。

贮备液：FDA 的丙酮溶液（5mg/ml），贮于冰箱中。

染液：用 0.5mol/L 蔗糖溶液（或其它等渗溶液）将贮备液稀释 50～500 倍，成为 10～100μg/ml 的浓度。配后立即应用。

操作程序

（1）在载玻片上加一滴临时配制的 FDA 染液。

（2）将供试花粉撒在染液中使之悬浮。

（3）在带湿润滤纸的培养皿中静置 5min。

（4）加上盖玻片，置荧光显微镜下观察。采用 BG12 激发滤光片与 500 纳米的阻挡滤光片。活细胞呈绿色荧光，死细胞无荧光。在 40min 内，荧光持续加强（图 8）。

（作者：杨弘远。原载：《植物细胞学研究方法》（孙敬三、钱迎倩主编）。北京：科学出版社，1987，219～248 页。图 8 幅，参考文献 33 篇，附录 2 份。仅选录图 1～5。）

13　植物胚胎学中的整体透明技术

Whole clearing techniques in plant embryology

20世纪70年代以来,透明技术作为一种快速制样与观察的手段,在植物胚胎学中得到日益广泛的应用。它不仅大大减轻切片技术所需的人力和物力,而且便于在观察时获得三维图像,还可以在不破坏整体结构的条件下对所研究的目标作原位的考察。

Herr(1971)发明了"4½"复合透明剂。胚珠经它透明后,在相差(或干涉差)显微镜下可以观察其中的大孢子发生与胚囊不同时期的图像[5]。此法又经过若干细节上的改进而适用于不同的材料[6]。有关应用此法于胚胎学研究的文献资料可以参考杨弘远[1]的引述。Crane(1978)用另一种透明剂水杨酸甲酯(即冬青油),观察整体胚珠中的胚囊[4]。用此法观察某些禾本科植物子房中的无融合生殖,效果良好[12,13]。但单纯的透明,由于反差不够强,只能采用干涉差或相差观察。Stelly等(1984)提出了"染色-透明"技术,即先用梅氏苏木精明矾染色,再用冬青油透明,这样增强了反差,在普通明视野下即可观察[11]。杨弘远(1986)则改用爱氏苏木精短时间染色与冬青油透明相结合,不仅观察到胚囊的图像,而且还可以观察到受精后子房与胚珠中的幼胚与胚乳,以及离体培养子房中的胚状体[1]。

花粉也可以采用透明法观察。成熟的花粉,由于富含贮藏物质且有厚壁包围,用一般方法有时难以察见其中的细胞核。近来采用与DNA结合的荧光染料显示花粉细胞核,取得较好的结果[3,7]。但花粉壁中的孢粉素及色素能发射强烈自发荧光,遮掩细胞核的荧光。在许多厚壁的花粉,这种情况尤其严重。如将Hoechst 33258(H33258)荧光染色与冬青油透明结合起来,则可使花粉壁荧光大为减弱,花粉内含物透明度增加,从而改善细胞核荧光。用此法还可以观察花粉核由花粉粒向花粉管的转移及其在花粉管中的动态变化(杨弘远,1988)[2]。以下分别介绍胚珠和花粉的透明方法。

胚珠的"$4\frac{1}{2}$"透明技术

"$4\frac{1}{2}$"透明剂及其改良配方

1. "$4\frac{1}{2}$"透明剂：由乳酸(85%)、水合三氯乙醛、酚、丁香油、二甲苯五种透明剂按2∶2∶2∶2∶1的重量比配合而成,故简称"$4\frac{1}{2}$"。配制时须按上述顺序依次混合。混匀后在加塞玻瓶中可以久贮。

2. "IKI-$4\frac{1}{2}$"透明剂：将碘100mg、碘化钾500mg溶于"$4\frac{1}{2}$"9g中配合而成,用以增强反差,并显示组织中的淀粉粒。

3. "PP-$4\frac{1}{2}$"透明剂：将高锰酸钾3mg溶于"$4\frac{1}{2}$"1g中,用以增强反差。此剂需临用前配制。

4. "BB-$4\frac{1}{2}$"透明剂：将苯酸苄酯(benzyl benzoate)1g溶于"$4\frac{1}{2}$"9g中,用以增强反差。

5. "PPBB-$4\frac{1}{2}$"透明剂：将高锰酸钾3mg溶入"BB-$4\frac{1}{2}$"1g中。高锰酸钾须临用前加入。

制样与观察程序

胚珠固定于FPA中,保存于70%乙醇(低温)中——直接转入或经95%乙醇稍加脱水后转入"$4\frac{1}{2}$"或其改良配方中透明24h或更长时间——将胚珠连同透明剂转到特制的Rajslide[5]上,加盖玻片封藏；也可用普通载玻片,但最好在盖玻片四隅下加垫盖玻片碎片,以免将胚珠压散——在相差或干涉差显微镜下,通过调焦观察胚囊的不同光学切面。

预处理

有些材料直接透明效果不佳,可在透明前采取各种预处理方法加以改善：

1. 在85%乳酸中室温下浸泡4d,或50℃下24h,再转入透明剂(Smith,1973)[10]。

2. 在10% KOH中处理2min,水洗4次,脱水至95%乙醇后,再转入透明剂(Smith,1973)[10]。

3. 用4%淀粉转葡糖苷酶(amyloglucosidase溶于pH4.5的0.1mol/L、醋酸盐缓冲液中)在35℃下处理3h,以除去胚珠中的淀粉。水洗后,再用Stockwell液(水90ml、重铬酸钾1g、铬酸1g、冰醋酸10ml)在室温下处理20h,以除去组织中的鞣酸。水洗数次去掉黄色,再转入透明剂(Kenrick等,1986)[8]。

胚珠的冬青油透明技术

冬青油透明程序

胚珠或幼小子房固定于FPA中,保存于70%乙醇(低温)中──→95%、100%乙醇彻底脱水──→100%乙醇与冬青油2∶1与1∶2混合液依次过渡(亦可仅经过一次1∶1混合液)──→冬青油透明,换液2次,第2次最好透明24h──→将材料连同冬青油转移到Raj slide或浅凹玻片上,加盖玻片封藏──→在干涉差或相差显微镜下通过调焦观察胚囊的不同光学切面。

梅氏苏木精明矾染色-冬青油透明程序

将FPA固定的胚珠经逐级乙醇下行至蒸馏水,在水中浸泡2~24h──→梅氏苏木精明矾(Mayer's hemalum)染色1~2d──→自来水(或0.1碳酸氢钠溶液)换洗2~24h──→乙醇逐级脱水──→无水乙醇与冬青油混合液过渡──→冬青油透明──→封藏后,用明视野柯勒照明法观察。

爱氏苏木精染色-冬青油透明程序

将FPA固定的胚珠经逐级乙醇下行至蒸馏水──→用稀释的爱氏

苏木精(Ehrlich's hematoxylin)在20℃下染色5～120min。染色时间因植物材料和胚珠大小而异。爱氏苏木精原液的氧化成熟度对染色效果影响很大,应用充分成熟的原液,加等量或两倍的醋酸乙醇(45%酯酸与50%乙醇等量混合液)稀释成染液——→蒸馏水换洗1～2d——→自来水换洗1～2d——→乙醇逐级脱水——→无水乙醇与冬青油混合液过渡——→冬青油透明——→用明视野柯勒照明法观察。适当缩小孔径光栏,相应提高照明电压可以增强反差。

花粉的荧光染色、冬青油透明技术

H33258染液的配制

用蒸馏水溶液Hoechst 33258(Sigma)配成1mg/ml的贮备液,贮于冰箱中备用。数毫升的贮备液即可使用一年,出现沉淀、杂质则不可再用。用pH5的柠檬酸—磷酸氢二钠缓冲液将贮备液稀释为20μg/ml的染液,此染液在冰箱中亦可保存月余。注意不可将冷藏的贮备液和缓冲液立即混合,需用水浴略加温使其恢复为室温后方可混合,否则H33258会析出沉淀。贮备液与染液均不可长久见光。

花粉粒的染色透明程序

将含成熟或幼嫩花粉的花药固定于Carnoy(3∶1)液中,低温保存于70%乙醇中——→经逐级乙醇下行至蒸馏水,并用水换洗数次除去酸液——→捣碎花药,将花粉置离心管中,用pH5的柠檬酸—磷酸氢二钠缓冲液浸泡数小时[9]——→离心去缓冲液,用20μg/ml的H33258染液(溶于同样缓冲液)染色过夜至1d。染色应在25℃、黑暗条件下进行——→乙醇逐级脱水——→无水乙醇与冬青油混合液过渡——→冬青油透明。透明时间宜在一天以上。材料在冬青油中低温避光保存一个月,荧光染色依然保持——→将花粉连同冬青油滴于载玻片上,加盖玻片,置荧光显微镜下,用U或V激发滤光组合观察,花粉核呈紫或蓝色。

花粉管的染色透明程序

新鲜花粉接种在适宜的培养基中进行人工萌发。在培养基中事先加入 H33258 贮备液使成 $20\mu g/ml$ H33258 的浓度。在 25℃、黑暗条件下培养，一边萌发一边染色──→吸取花粉管置离心管中，用 Carnoy(3∶1)液固定 1h ──→离心去固定液，用 50% 乙醇换洗数次以除去酸液──→逐级乙醇脱水──→无水乙醇与冬青油混合液过渡──→冬青油透明──→荧光显微观察（同上方法）。

结　　语

整体透明技术可用于观察植物的大、小孢子发生与雌、雄配子体发育，早期的胚和胚乳发育，以及无融合生殖。在鉴定胚囊、花粉、胚或胚乳的败育方面，透明法很可能是十分简便有效的。在观察诸如各种吸器之类的整体构造时，它也可能是一种有用的手段。但透明技术更能发挥作用的领域应是实验胚胎学。在花药培养、子房与胚珠培养、离体受精等研究工作中，由于胚胎发生频率通常不高，而实验处理规模又大，依靠切片技术完成制样与观察是十分繁重的任务；而透明法则可大大节约人力物力，便于对培养结果进行快速的鉴定与统计。在观察了透明的材料之后，还可以挑出其有用的部分作石蜡包埋与切片，而大量无用的材料则予废弃。

(作者：杨弘远。原载：植物学通报，1988，5(2)：114～116。参考文献13篇，均删去。)

其它论文目录

1. 周嫦.培育花粉植株研究的某些进展.武汉大学学报(自然科学版),1973.(1):69~79.
2. 武汉大学生物系植物遗传育种专业.水稻的幼穗发育与开花结实.武汉:湖北人民出版社.1975.(杨弘远撰)
3. 杨弘远.花粉研究的新进展.武汉大学学报(自然科学版),1977.(2):78~87.
4. 杨弘远.植物染色体显带技术及其在遗传学与细胞学中的应用.武汉大学学报(自然科学版),1978.(3):110~118.
5. 周嫦.植物组织培养与遗传育种.湖北农业科学,1980.(10):36~40.
6. 杨弘远.花粉两条发育途径中的细胞生物学问题.生物科学动态,1980.(4):17~23.
7. 杨弘远.花粉雄核发育中的几个问题.武汉大学学报(自然科学版),1981.(2):90~96.
8. 周嫦、杨弘远.未传粉子房与胚珠的离体培养.武汉大学学报(自然科学版),1982.(3):61~72.
9. 杨弘远.植物的细胞周期及其控制.生物科学动态,1983.(1):11~18.
10. 杨弘远.植物生殖的细胞生物学:一个新的学科生长点.生物科学动态,1984.(4):1~5.
11. 杨弘远.荧光显微镜技术的基本原理与方法.植物学通报,1984.2(6):45~48.
12. 杨弘远.胚珠组织的实验操作.武汉植物学研究,1987.5(1):97~100.
13. 周嫦.胚囊分离技术.植物细胞学研究方法(孙敬三、钱迎倩主编).北京:科学出版社,1987.46~59.
14. 周嫦.花粉原生质体培养.植物原生质体培养(孙勇如主编).北京:

科学出版社,1991. 35~41.

15. 周嫦、杨弘远. 植物性细胞操作的研究进展.《细胞生物学进展》第三卷(郑国昌、翟中和主编). 北京:高等教育出版社,1993. 201~222.

16. 胡适宜、杨弘远. 植物胚胎学与生殖生物学——发展现状及近期战略的初步设想.《植物科学》. 北京:中国林业出版社,1994. 30~57.

17. 周嫦. 植物性细胞融合进展. 植物学通报,1994. 11(4):12~16.

18. 范六民、杨弘远、周嫦. 钙-钙调素信使系统在花粉萌发、花粉管生长及生殖细胞分裂中的作用.《细胞生物学动态》第二卷(翟中和主编). 北京:北京师范大学出版社,1998. 63~71.

19. 杨弘远. 植物受精生物学研究的世纪回顾与前瞻. 世界科技研究与发展,1998. 20(3):9~13.

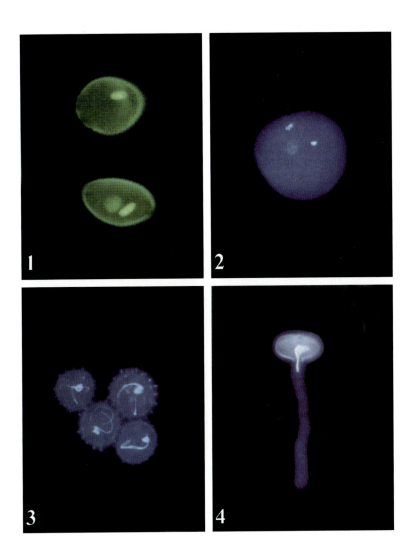

图版 I. 花粉的荧光染色与冬青油透明
Plate I. Fluorescent staining-clearing of pollen grains

　　1.石蒜的二细胞花粉，示明亮的生殖核与暗淡的营养核. 2.玉米的三细胞花粉，示一对明亮的精核与一个暗淡的营养核. 3.向日葵的三细胞花粉，示一对细长如丝的精核与一个营养核. 4.蚕豆的花粉管，示生殖核正由花粉粒移进花粉管.

　　1.Bicellular pollen of *Lycoris radiata*. showing a bright generative nucleus and a dim vegetative nucleus. 2.Tricellular pollen of *Zea mays*, showing a pair of bright sperm nuclei and a dim vegetative nucleus. 3.Tri-cellular pollen of *Helianthus annuus*, showing a pair of thread-shaped sperm nuclei and a vegetative nucleus. 4.Pollen tube of *Vicia faba*. The generative nucleus is moving from the pollen grain into the tube.

(参看第一章第二节：2. See Chapter 1, Section 2:2.)

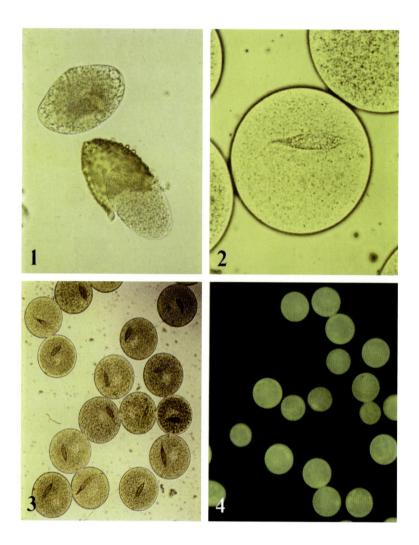

图版Ⅱ. 风雨花的成熟花粉原生质体
Plate Ⅱ. Mature pollen protoplasts of *Zephyranthes grandiflora*

    1.脱外壁花粉由外壁中逸出. 2.花粉原生质体，内含显著的生殖细胞. 3.纯化后的花粉原生质体群体. 4.FDA荧光染色，示花粉原生质体的生活力.

    1.release of de-exined pollen from the exine. 2.A pollen protoplast containing a prominent generative cell. 3.Purified population of pollen protoplasts. 4.Viability of pollen protoplasts visualized by fluorochromasia.

(参看第一章第三节 :1. See chapter 1, Section 3:1.)

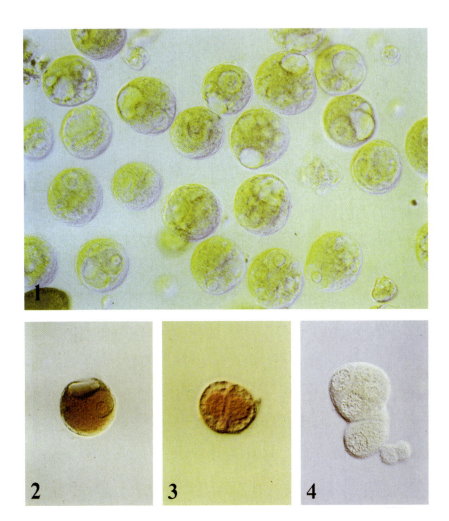

图版Ⅲ. 萱草的幼嫩花粉原生质体
PlateⅢ. Young pollen protoplasts of *Hemerocallis fulva*

 1.分离的幼嫩花粉（小孢子）原生质体群体. 2.一个小孢子原生质体，示细胞核与液泡. 3.小孢子原生质体在培养中分裂为两个细胞. 4.形成的多细胞团.

 1.Population of young pollen (microspore) protoplasts. 2.A microspore protoplast, showing its nucleus and vacuole. 3.Two cells divided from a microspore protoplast. 4.A multicellular structure derived from a microspore protoplast.
(参看第一章第三节：2. See Chapter 1, Section 3:2.)

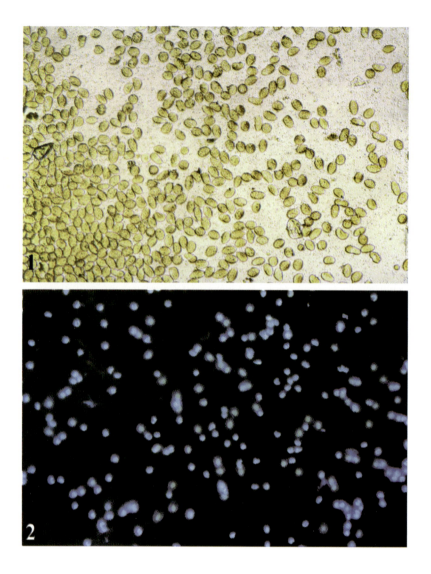

图版Ⅳ. 生殖细胞与精细胞
PlateⅣ. Generative cells and sperm cells

1.分离的玉帘生殖细胞群体. 2.分离的玉米精细胞群体,H33258荧光染色示精细胞核.

1.Population of isolated generative cells of *Zephyranthes candida*. 2.Population of isolated sperm cells of *Zea mays*, showing sperm nuclei stained by H33258.
(参看第一章第五节:3;第六节:1. See Chapter 1, Section 5:3; Section 6:1.)

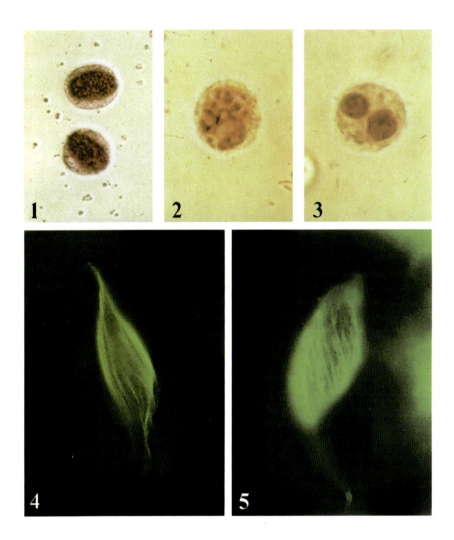

图版 V. 生殖细胞
Plate V. Generative Cells

1.分离的玉帘生殖细胞，注意其核质比. 2.分离的黄花菜生殖细胞在培养中启动核分裂. 3.黄花菜生殖细胞分裂而成的二核细胞. 4.朱顶红生殖细胞，示其中纵行排列的微管. 5.同上，示其中螺旋状排列的微管.

1.Isolated generative cells of *Zephyranthes candida*. Note its nucleus/cytoplasm ratio. 2.Nuclear division of an isolated generative cell of *Hemerocallis minor* during *in vitro* culure. 3.A binuclcate cell derived from a generative cell of *H. minor*. 4.A generative cell of *Hippeastrum vittatum*, showing longitudinal microtubules in it. 5.Ibid, showing spirally arranged microtubules.

(参看第一章第五节：4, 5, 10. See Chapter 1, Section 5:4, 5&10.)

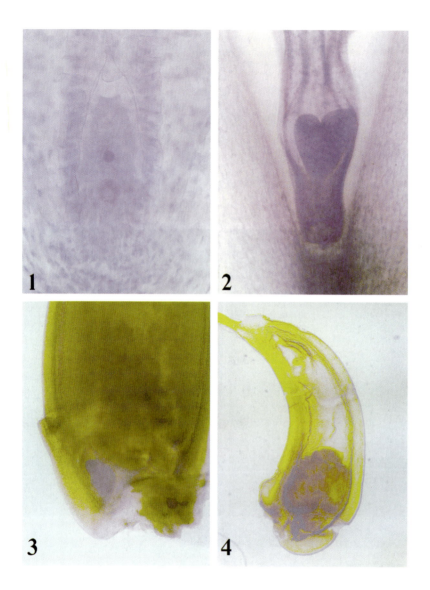

图版Ⅵ. 胚珠与子房的整体染色与冬青油透明
Plate Ⅵ. Whole staining-clearing of ovules and ovaries

　　1.向日葵胚珠整体染色与透明，示包围在珠被绒毡层中的胚囊结构. 2.向日葵胚珠整体染色与透明，示包围在胚乳中的心形胚. 3.水稻子房整体染色透明，示开始分化的幼胚. 4.水稻离体培养的未传粉子房，染色透明后示其中由无配子生殖发育而成的愈伤组织.
　　　1.A whole stained and cleared ovule of sunflower, showing its embryo sac surrounded by integumentary tapetum. 2.A whole stained and cleared ovule of sunflower, showing a heart-shaped embryo enclosed in en-dosperm. 3.A whole stained and cleared ovary of rice, showing a differentiating young embryo in it. 4.An unpollinated rice ovary during culture, showing an induced apogamic callus in it.
　　(参看第二章第一节：16. See Chapter 2, Section 1:16.)

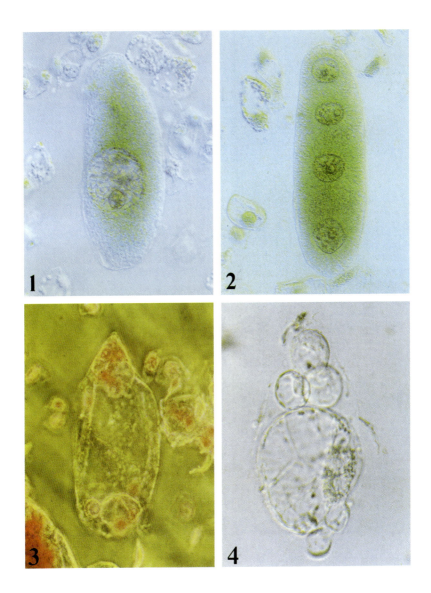

图版Ⅶ. 胚囊
PlateⅦ. Embryo sacs

1.分离的卷丹大孢子,示减数分裂前期Ⅰ. 2.分离的卷丹大孢子四分体,示四个大孢子核直线排列. 3.分离的烟草胚囊. 4.分离的烟草胚囊原生质体,示卵细胞、助细胞、中央细胞、反足细胞原生质体.

1.An isolated megaspore of *Lilium langifolium*, showing meiotic prophase Ⅰ. 2.An isolated megaspore tetrad, showing four longitudinally arranged megaspore nuclei in *L. langifolium*. 3.An isolated embryo sac of tobacco. 4.An isolated tobacco embryo sac, showing protoplasts of the egg cell, synergids, central cell and antipodal cells.

(参看第二章第二节:1, 8, 及其它论文目录. See Chapter 2, Section 2:1, 8 and list of other papers.)

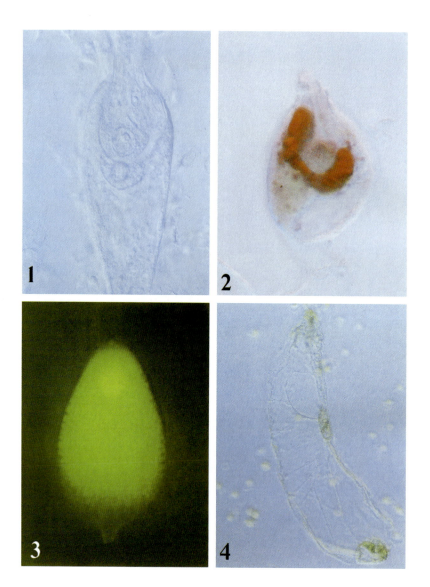

图版Ⅷ. 胚囊

PlateⅧ. Embryo sacs

  1.分离的向日葵胚囊，示卵细胞、助细胞、中央细胞（其中含一个次生核）. 2.分离的向日葵胚囊，苏丹Ⅲ染色示其中丰富的油滴. 3.分离的向日葵受精后胚囊，FDA染色示球形胚与胚乳生活力. 4.分离的蚕豆胚囊，示中央细胞中的细胞质索.

  1.An isolated embryo sac of sunflower, showing its egg cell, synergid and central cell which contains a prominent secondary nucleus. 2.An isolated sunflower embryo sac stained by SudanⅢ, showing abundant oil drops in it. 3.An isolated sunflower embryo sac after fertilization. Fluorochromasia shows the viability of the globular embryo enclosed in endosperm. 4.An isolated embryo sac of *Vicia faba*. Note the cytoplasmic strands in it.

（参看第二章第二节：1, 5. See Chapter 2, Section 2:1&5.）

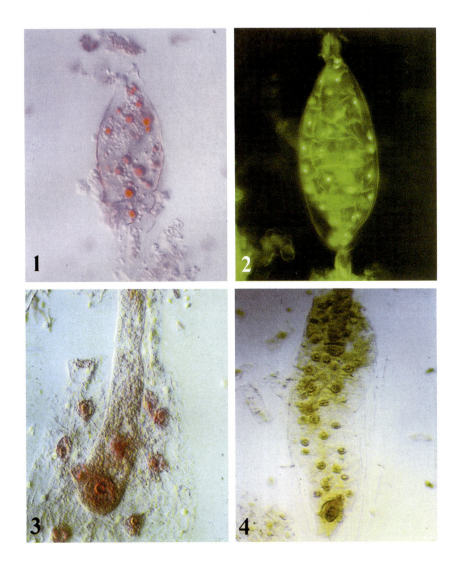

图版Ⅸ. 胚囊
PlateⅨ. Embryo sacs

1.分离的金鱼草胚囊，苏丹Ⅲ染色示其中的油滴. 2.分离的金鱼草受精后胚囊，示胚乳细胞核荧光染色. 3.分离的杏叶沙参的细长合子. 4.同上，示二细胞原胚与胚乳.

1.An isolated embryo sac of *Antirrhinum majus* stained bv SudanⅢ, showing oil drops. 2.An isolated embryo sac of *A.majus* after fertilization. Fluorescent staining shows endosperm nuclei. 3.An isolated elongated zygote of *Adenophora axillifolia*. 4.An isolated embryo sac of *Adenoplzora axillifolia* after fertilization, showing an elongated 2-celled proembryo enclosed in endosperm.

(参看第二章第二节：3，及其它论文目录. See Chapter 2, Section 2:3 and list of other papers.)

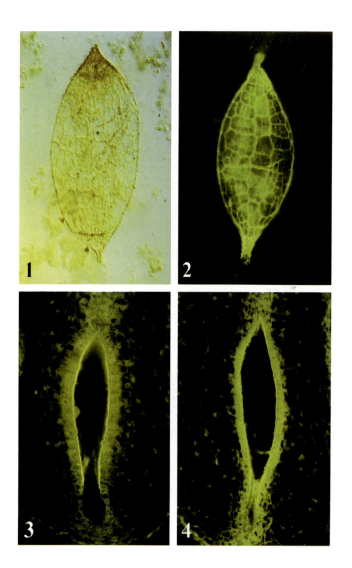

图版 X. 珠被绒毡层壁囊
Plate X. Wall sac of integumentary tapetum

1. 金鱼草胚珠经乙酰解分离的珠被绒毡层壁囊. 2. 金鱼草胚珠经酶解分离的珠被绒毡层壁囊, 金胺O荧光染色. 3、4. 金鱼草胚珠切片, 金胺O荧光染色示明亮的珠被绒毡层壁囊包围胚囊 (暗区) 及合点端的开口.

1. An integumentary tapetal wall sac of *Antirrhinum majus* isolated by acetic anhydride H_2SO_4. 2. A " wall sac" isolated by enzymatic treatment and stained by auramine O. 3 and 4. Sections of *Antirrhinum* ovule stained by auramine O, showing brightly fluoresced "wall sac" enclosing the embryo sac (dark area) and its chalazal aperture.
(参看第二章第二节: 7. See Chapter 2, Section 2:7.)